Quality and Safety
in Radiotherapy

IMAGING IN MEDICAL DIAGNOSIS AND THERAPY

William R. Hendee, Series Editor

Quality and Safety in Radiotherapy
Todd Pawlicki, Peter B. Dunscombe, Arno J. Mundt, and
Pierre Scalliet, Editors
ISBN: 978-1-4398-0436-0

Forthcoming titles in the series

Image-Guided Radiation Therapy
Daniel J. Bourland, Editor
ISBN: 978-1-4398-0273-1

Adaptive Radiation Therapy
Allen X. Li, Editor
ISBN: 978-1-4398-1634-9

Informatics in Radiation Oncology
Bruce H. Curran and George Starkschall, Editors
ISBN: 978-1-4398-2582-2

Adaptive Motion Compensation in Radiotherapy
Martin Murphy, Editor
ISBN: 978-1-4398-2193-0

Image Processing in Radiation Therapy
Kristy Kay Brock, Editor
ISBN: 978-1-4398-3017-8

Proton and Carbon Ion Therapy
Charlie C.-M. Ma and Tony Lomax, Editors
ISBN: 978-1-4398-1607-3

Monte Carlo Techniques in Radiation Therapy
Jeffrey V. Siebers, Iwan Kawrakow, and
David W. O. Rogers, Editors
ISBN: 978-1-4398-1875-6

Informatics in Medical Imaging
George C. Kagadis and Steve G. Langer, Editors
ISBN: 978-1-4398-3124-3

Stereotactic Radiosurgery and Radiotherapy
Stanley H. Benedict, Brian D. Kavanagh, and
David J. Schlesinger, Editors
ISBN: 978-1-4398-4197-6

Cone Beam Computed Tomography
Chris C. Shaw, Editor
ISBN: 978-1-4398-4626-1

Quantitative Magnetic Resonance Imaging in Cancer
Thomas Yankeelov, Ronald R. Price, and
David R. Pickens, Editors
ISBN: 978-1-4398-2057-5

Handbook of Brachytherapy
Jack Venselaar, Dimos Baltas, Peter J. Hoskin, and
Ali Soleimani-Meigooni, Editors
ISBN: 978-1-4398-4498-4

Targeted Molecular Imaging
Michael J. Welch and William C. Eckelman, Editors
ISBN: 978-1-4398-4195-0

IMAGING IN MEDICAL DIAGNOSIS AND THERAPY

William R. Hendee, Series Editor

Quality and Safety in Radiotherapy

Edited by

Todd Pawlicki
Peter B. Dunscombe
Arno J. Mundt
Pierre Scalliet

 CRC Press
Taylor & Francis Group
Boca Raton   London   New York

CRC Press is an imprint of the
Taylor & Francis Group, an **informa** business

A TAYLOR & FRANCIS BOOK

CRC Press
Taylor & Francis
6000 Broken Sound Parkway NW, Suite 300
Boca Raton, FL 33487-2742

First issued in paperback 2020

© 2011 by Taylor and Francis Group, LLC
CRC Press is an imprint of Taylor & Francis Group, an Informa business

No claim to original U.S. Government works

ISBN-13: 978-0-367-57703-2 (pbk)
ISBN-13: 978-1-4398-0436-0 (hbk)

Library of Congress Cataloging-in-Publication Data

Quality and safety in radiotherapy / editors, Todd Pawlicki ... [et al.].
    p. ; cm. -- (Imaging in medical diagnosis and therapy)
  Includes bibliographical references and index.
  ISBN 978-1-4398-0436-0 (hardcover : alk. paper)
  1. Radiotherapy--Safety measures. 2. Radiotherapy--Quality control. I. Pawlicki, Todd. II. Dunscombe, Peter. III. Mundt, Arno J.
IV. Scalliet, Pierre. V. Series: Imaging in medical diagnosis and therapy.
  [DNLM: 1. Radiotherapy. 2. Quality Assurance, Health Care. 3. Radiation Injuries--prevention & control. 4. Safety. WN 250]

RM854.Q83 2010
615.8'42--dc22   2010025581

Visit the Taylor & Francis Web site at
http://www.taylorandfrancis.com

and the CRC Press Web site at
http://www.crcpress.com

We dedicate this textbook to radiotherapy patients worldwide. Our hope is that it will result in further advances in the quality and safety of radiotherapy.

Contents

PART I Quality Management and Improvement

PART II Patient Safety and Managing Error

PART III Methods to Assure and Improve Quality

PART IV People and Quality

PART V Quality Assurance in Radiotherapy

PART VII Quality Control: Patient-Specific

Series Preface

Advances in the science and technology of medical imaging and radiation therapy are more profound and rapid than ever before, since their inception over a century ago. Further, the disciplines are increasingly cross-linked as imaging methods become more widely used to plan, guide, monitor, and assess treatments in radiation therapy. Today, the technologies of medical imaging and radiation therapy are so complex and computer-driven that it is difficult for the people (physicians and technologists) responsible for their clinical use to know exactly what is happening at the point of care, when a patient is being examined or treated. The people best equipped to understand the technologies and their applications are medical physicists, and these individuals are assuming greater responsibilities in the clinical arena to ensure that what is intended for the patient is actually delivered in a safe and effective manner.

The growing responsibilities of medical physicists in the clinical arenas of medical imaging and radiation therapy are not without their challenges, however. Most medical physicists are knowledgeable in either radiation therapy or medical imaging and expert in one or a small number of areas within their discipline. They sustain their expertise in these areas by reading scientific articles and attending scientific talks at meetings. In contrast, their responsibilities increasingly extend beyond their specific areas of expertise. To meet these responsibilities, medical physicists periodically must refresh their knowledge of advances in medical imaging or radiation therapy, and they must be prepared to function at the intersection of these two fields. How to accomplish these objectives is a challenge.

At the 2007 annual meeting of the American Association of Physicists in Medicine in Minneapolis, this challenge was the topic of conversation during a lunch hosted by Taylor & Francis Publishers and involving a group of senior medical physicists (Arthur L. Boyer, Joseph O. Deasy, C.-M. Charlie Ma, Todd A. Pawlicki, Ervin B. Podgorsak, Elke Reitzel, Anthony B. Wolbarst, and Ellen D. Yorke). The conclusion of this discussion was that a book series should be launched under the Taylor & Francis banner, with each volume in the series addressing a rapidly advancing area of medical imaging or radiation therapy of importance to medical physicists. The aim would be for each volume to provide medical physicists with the information needed to understand technologies driving a rapid advance and their applications to safe and effective delivery of patient care.

Each volume in the series is edited by one or more individuals with recognized expertise in the technological area encompassed by the book. The editors are responsible for selecting the authors of individual chapters and ensuring that the chapters are comprehensive and intelligible to someone without such expertise. The enthusiasm of volume editors and chapter authors has been gratifying and reinforces the conclusion of the Minneapolis luncheon that this series of books addresses a major need of medical physicists.

Imaging in Medical Diagnosis and Therapy would not have been possible without the encouragement and support of the series manager, Luna Han of Taylor & Francis Publishers. The editors, authors, and, most of all, I are indebted to her steady guidance of the entire project.

William R. Hendee
Series Editor

Preface

Quality and safety have always been important topics in the radiotherapy community. However, a single text that focuses on quality and safety does not exist. In attempting to provide such a reference, we decided to take the broadest possible view of quality and safety. We strove to encompass not only traditional, more technically oriented, quality assurance activities, but also general approaches of quality and safety. In order to achieve these goals, we sought contributions from experts both inside and outside our field. Through the choice of authors, we have also attempted to present a global view of the topics covered.

We have been in the midst of a transformation of quality and safety in radiotherapy. Among the key drivers of this transformation have been new industrial and systems engineering approaches. These approaches have come to the forefront in recent years following revelations of system failures (in both the scientific and public press) that have harmed radiotherapy patients and compromised quality. The task of assuring quality is no longer viewed solely as a technical, equipment-dependent endeavor. Instead, it is now recognized for depending as much on processes and the people delivering the service.

This text is divided into seven broad categories. Part I, "Quality Management and Improvement," includes essays on the management and improvement of quality with topics such as lean thinking, process control, and access to services. Topics such as reactive and prospective error management techniques are covered in Part II, "Patient Safety and Managing Error." In Part III, "Methods to Assure and Improve Quality," we have grouped chapters that deal broadly with techniques to monitor, assure, and improve quality. The chapters in Part IV, "People and Quality," focus on human factors, changing roles, staffing, and training. The final three sections of the book deal with what one may traditionally view as quality assurance and quality control in radiotherapy. Part V, "Quality Assurance in Radiotherapy," addresses the general issues of quality assurance with descriptions of the key systems used to plan and treat patients. Several chapters provide specific recommendations on the types and frequencies of certain tests where these are not covered elsewhere. In Part VI, "Quality Control: Equipment," and Part VII, "Quality Control: Patient-Specific," we have collected chapters that provide explicit details of quality control relating to equipment and patient-specific issues.

The contributions dealing with topics that are still in development provide our best current knowledge of, and guidance for, quality assurance and control in these areas. Recommendations may change in the future but we expect that many components presented here will remain relevant. While the technical details of this book may be superseded by future advances, we hope that the core contents presented here will be useful to the field for many years to come.

Acknowledgments

We would like to thank all the authors who gave so much of their time to contribute to this book. During the final editing process, we realized a few important topics were still omitted. We would like to extend our sincerest gratitude to the authors who stepped in to fill those gaps on short notice—notably, Parminder Basran, Laura Cerviño, Lijun Ma, and William Song. We would also like to give a special appreciation to Sasa Mutic who produced an additional chapter beyond what was originally agreed upon. Phuong La undertook the huge task of reformatting 100 chapters, diligently checking over 1500 references, and keeping the editors on track. She managed to accomplish this while remaining cheerful throughout. Finally, we acknowledge Luna Han, our publishers, and Bill Hendee for their guidance and support from the inception of this project to its successful completion.

Contributors

Edwin Aird
Medical Physics Department
Mount Vernon Cancer Centre
Mount Vernon Hospital
Northwood, Middlesex, United Kingdom

Christos Antypas
Department of Radiology, Aretaieion
    Hospital
University of Athens
and
CyberKnife Center, Iatropolis–Magnitiki
    Tomografia
Athens, Greece

Bulent Aydogan
Radiation and Cellular Oncology
The University of Chicago
Chicago, Illinois

Parminder S. Basran
British Columbia Cancer Agency
Victoria, British Columbia, Canada

A. Sam Beddar
Department of Radiation Physics
The University of Texas MD Anderson
    Cancer Center
Houston, Texas

Stanley H. Benedict
Department of Radiation Oncology
University of Virginia Health System
Charlottesville, Virginia

Jeffrey Bews
Division of Medical Physics
CancerCare Manitoba
Winnipeg, Manitoba, Canada

Jean-Pierre Bissonette
Department of Radiation Physics
Princess Margaret Hospital
Toronto, Ontario, Canada

Cari Borrás
Dosimetry and Nuclear Instrumentation
    Group
Department of Nuclear Energy
Federal University of Pernambuco
Recife, Brazil

Terence Bostic
CMA
St. Louis, Missouri

Scott Brame
Department of Radiation Oncology
Washington University School of
    Medicine
St. Louis, Missouri

Stephen L. Breen
Radiation Medicine Program
Princess Margaret Hospital
Toronto, Ontario, Canada

Megan Bright
Department of Radiation Oncology
Wake Forest University Baptist
    Medical Center
Winston-Salem, North Carolina

Kristy K. Brock
Department of Radiation Physics
Princess Margaret Hospital
Toronto, Ontario, Canada

Derek Brown
Department of Oncology
University of Calgary
Calgary, Alberta, Canada

Antonella Bufacchi
UOC Medical Physics
S. Giovanni Calibita
    Fatebenefratelli Hospital
Rome, Italy

Jing Cai
Department of Radiation Oncology
Duke University Medical Center
Durham, North Carolina

Martin Carrigan
College of Business
The University of Findlay
Findlay, Ohio

Ken Cashon
.decimal, Incorporated
Sanford, Florida

Laura Cerviño
Department of Radiation Oncology
University of California, San Diego
La Jolla, California

Liam Chadwick
Centre for Occupational Health and
    Safety Engineering and Ergonomics
National University of Ireland Galway
Galway, Ireland

Sha Chang
Department of Radiation Oncology
University of North Carolina Medical
    School
Chapel Hill, North Carolina

Lili Chen
Department of Radiation Oncology
Fox Chase Cancer Center
Philadelphia, Pennsylvania

Indrin J. Chetty
Department of Radiation Oncology
Henry Ford Health System
Detroit, Michigan

Kin Yin Cheung
Department of Clinical Oncology
Prince of Wales Hospital
Sha Tin, Hong Kong, China

Charles W. Coffey
Department of Radiation Oncology
Vanderbilt University
Nashville, Tennessee

Mary Coffey
Discipline of Radiation Therapy
School of Medicine, Trinity College Dublin
Dublin, Ireland

David L. Cooke
Haskayne School of Business
University of Calgary
Calgary, Alberta, Canada

Gregory W. Cotter
Division of Radiation Oncology
The University of South Alabama College
    of Medicine
Mobile, Alabama

Sonja Dieterich
Department of Radiation Oncology
Stanford University Hospital
Stanford, California

Laura Drever
Department of Medical Physics
Kingston General Hospital
Kingston, Ontario, Canada

Peter B. Dunscombe
Department of Oncology
University of Calgary
Calgary, Alberta, Canada

Sara C. Erridge
Edinburgh Cancer Centre
University of Edinburgh
South Edinburgh, Scotland

Carlos Esquivel
Department of Radiation Oncology
University of Texas Health Sciences
    Center San Antonio
San Antonio, Texas

Enda F. Fallon
Centre for Occupational Health and Safety
    Engineering and Ergonomics
National University of Ireland Galway
Galway, Ireland

Jiajin Fan
Department of Radiation Oncology
Fox Chase Cancer Center
Philadelphia, Pennsylvania

Stella Flampouri
University of Florida Proton Therapy
    Institute
University of Florida
Jacksonville, Florida

Gerald B. Fogarty
Radiation Oncology Associates
Mater Hospital
Sydney, Australia

Benedick A. Fraass
Department of Radiation Oncology
University of Michigan
Ann Arbor, Michigan

Vincenzo Frascino
Department of Radiation Oncology
Università Cattolica S. Cuore
Rome, Italy

Ashley A. Gale
Department of Radiation Oncology
Mayo Clinic
Jacksonville, Florida

Karine Gérard
Department of Medical Physics
Alexis Vautrin Cancer Center
Vandoeuvre les Nancy, France

Michael Gossman
Medical Physics Section
Tri-State Regional Cancer Center
Ashland, Kentucky

Alonso N. Gutierrez
Department of Radiation Oncology
University of Texas Health Sciences Center
San Antonio, Texas

Per Halvorsen
Alliance Oncology LLC
Newport Beach, California

William R. Hendee
Whitefish Bay, Wisconsin

Jessica Hiatt
Department of Radiation Oncology
Rhode Island Hospital
Providence, Rhode Island

Michelle Hilts
Department of Medical Physics
British Columbia Cancer Agency
Victoria, British Columbia, Canada

Ola Holmberg
Radiation Protection of Patients Unit
International Atomic Energy Agency
Vienna, Austria

Annie Hubert
Centre National de la Recherche
    Scientifique
Université de la Méditerranée
    Aix-Marseille-II
Marseille, France

Geoffrey D. Hugo
Department of Radiation Oncology
Virginia Commonwealth University
Richmond, Virginia

Sandeep Hunjan
Texas Oncology
Sugar Land, Texas

Mohammed Saiful Huq
Department of Radiation Oncology
University of Pittsburgh Cancer Institute
Pittsburgh, Pennsylvania

Amjad Hussain
Department of Medical Physics
Tom Baker Cancer Centre
Calgary, Alberta, Canada

Andrew Hwang
Department of Radiation Oncology
University of California San Francisco
San Francisco, California

Geoffrey S. Ibbott
Department of Radiation Physics
MD Anderson Cancer Center
Houston, Texas

Edmond W. Israelski
Abbott Laboratories
Abbott Park, Illinois

Joanna Izewska
Dosimetry and Medical Radiation
    Physics Section
International Atomic Energy Agency
Vienna, Austria

Swamidas V. Jamema
Department of Medical Physics
Tata Memorial Hospital
Mumbai, India

Darren Kahler
Department of Radiation Oncology
University of Florida
Gainesville, Florida

Guy Kantor
Department of Radiation Oncology
Institut Bergonié
Bordeaux, France

Ben-Tzion Karsh
Department of Industrial and Systems
    Engineering
University of Wisconsin–Madison
Madison, Wisconsin

Christopher S. Kim
Department of Internal Medicine and
    Pediatrics
University of Michigan
Ann Arbor, Michigan

Siyong Kim
Department of Radiation Oncology
Mayo Clinic
Jacksonville, Florida

Christian Kirisits
Department of Radiotherapy
Medical University of Vienna
Vienna, Austria

Eric E. Klein
Department of Radiation Oncology
Washington University School of
    Medicine
St. Louis, Missouri

Tommy Knöös
Radiation Physics
Skåne University Hospital and Lund
    University
Lund, Sweden

Anantha Kollengode
Quality Management Services
Mayo Clinic
Rochester, Minnesota

Stine Sofia Korreman
Department of Radiation Oncology
Rigshospitalet, University of Copenhagen
Copenhagen, Denmark

Nina Kowalczyk
OSU-Radiologic Sciences and Therapy
    Division
The Ohio State University
Columbus, Ohio

Tomas Kron
Department of Physical Sciences
Peter MacCallum Cancer Centre
East Melbourne, Australia

Santanam Lakshmi
Department of Radiation Oncology
Washington University School of Medicine
St. Louis, Missouri

Kathy Lash
Department of Radiation Oncology
University of Michigan
Ann Arbor, Michigan

Theodore S. Lawrence
Department of Radiation Oncology
University of Michigan
Ann Arbor, Michigan

Robert C. Lee
Neptune and Company, Inc.
Albuquerque, New Mexico

Jan Willem H. Leer
Department of Radiation Oncology
Radboud University Nijmegen Medical
    Centre
Nijmegen, the Netherlands

Zuofeng Li
University of Florida Proton Therapy
    Institute
University of Florida
Jacksonville, Florida

Bruce Libby
Department of Radiation Oncology
University of Virginia Health System
Charlottesville, Virginia

Yolande Lievens
Department of Radiation Oncology
University Hospital Gasthuisberg
Leuven, Belgium

Alexander Lin
Department of Radiation Oncology
University of Pennsylvania
Philadelphia, Pennsylvania

Chihray Liu
Department of Radiation Oncology
University of Florida
Gainesville, Florida

Thomas J. LoSasso
Department of Medical Physics
Memorial Sloan-Kettering Cancer Center
New York, New York

Michael Lovelock
Department of Medical Physics
Memorial Sloan-Kettering Cancer Center
New York, New York

Daniel A. Low
UCLA School of Medicine
Los Angeles, California

Wei Lu
Department of Radiation Oncology
University of Maryland
Baltimore, Maryland

Lijun Ma
Department of Radiation Oncology
University of California, San Francisco
San Francisco, California

Daniel J. Macey
Certified Medical Physics Services
Birmingham, Alabama

Kathy Mah
Department of Medical Physics
Odette Cancer Centre and Sunnybrook
    Health Sciences Centre
Toronto, Ontario, Canada

Julian Malicki
Department of Electroradiology
University of Medical Sciences
Poznan, Poland

Marco Marchetti
Health Technology Assessment Unit
Università Cattolica S. Cuore
Rome, Italy

Peter Martelli
School of Public Health
University of California, Berkeley
Berkeley, California

Richard L. Maughan
Department of Radiation Oncology
University of Pennsylvania
Philadelphia, Pennsylvania

Osama R Mawlawi
Department of Imaging Physics
MD Anderson Cancer Center
Houston, Texas

Ann Scheck McAlearney
Division of Health Services Management
   and Policy
College of Public Health
The Ohio State University
Columbus, Ohio

Brendan McClean
Department of Physics
St. Luke's Hospital
Dublin, Ireland

Paul M. Medin
Department of Radiation Oncology
University of Texas Southwestern
   Medical Center at Dallas
Dallas, Texas

Robert J. Meiler
Department of Radiation Oncology
University of Rochester Medical Center
Rochester, New York

Ben Mijnheer
Department of Radiation Oncology
The Netherlands Cancer Institute
Antoni van Leeuwenhoek Hospital
Amsterdam, the Netherlands

Michael D. Mills
Department of Radiation Oncology
University of Louisville
Louisville, Kentucky

Andrea Molineu
Radiological Physics Center
MD Anderson Cancer Center
Houston, Texas

Manuel A. Morales-Paliza
Department of Radiation Oncology
Vanderbilt University
Nashville, Tennessee

Sasa Mutic
Department of Radiation Oncology
Washington University School of Medicine
St. Louis, Missouri

William H. Muto
Abbott Laboratories
Irving, Texas

Wayne Newhauser
Department of Radiation Physics
MD Anderson Cancer Center
Houston, Texas

Kwan Hoong Ng
Department of Biomedical Imaging
University of Malaya
Kuala Lumpur, Malaysia

Camille Noel
Department of Radiation Oncology
Washington University School of Medicine
St. Louis, Missouri

Pedro Ortiz López
Safety and Security Coordination Section
International Atomic Energy Agency
Vienna, Austria

John Øvretveit
Medical Management Centre
Karolinska Institutet
Stockholm, Sweden

Jatinder R. Palta
Department of Radiation Oncology
Davis Cancer Center
University of Florida
Gainesville, Florida

Evaggelos Pantelis
Medical Physics Laboratory
University of Athens Medical School
and
CyberKnife Center, Iatropolis–Magnitiki
   Tomografia
Athens, Greece

Niko Papanikolaou
Department of Radiation Oncology
University of Texas Health Sciences
   Center San Antonio
San Antonio, Texas

Parag J. Parikh
Department of Radiation Oncology
Washington University School of
   Medicine
St. Louis, Missouri

Todd Pawlicki
Department of Radiation Oncology
University of California, San Diego
La Jolla, California

Paula L. Petti
Taylor McAdam Bell Neuroscience
   Institute
Washington Hospital Healthcare System
Fremont, California

Tarun K. Podder
Department of Radiation Oncology
Thomas Jefferson University
Philadelphia, Pennsylvania

Philip Poortmans
Department of Radiation Oncology
Institute Verbeeten
Tilburg, the Netherlands

Robert A. Price Jr.
Department of Radiation Oncology
Fox Chase Cancer Center
Philadelphia, Pennsylvania

James A. Purdy
Department of Radiation Oncology
University of California, Davis Cancer
   Center
Sacramento, California

Chester Ramsey
Department of Radiation Oncology
Thompson Cancer Survival Center
Knoxville, Tennessee

Frank Rath
Department of Engineering Professional
   Development
University of Wisconsin–Madison
Madison, Wisconsin

A. Joy Rivera-Rodriguez
Department of Industrial and Systems
   Engineering
University of Wisconsin–Madison
Madison, Wisconsin

Karlene H. Roberts
Haas School of Business
University of California, Berkeley
Berkeley, California

Jean-Claude Rosenwald
Medical Physics Department
Institut Curie
Paris, France

Eeva Salminen
Finnish Radiation and Nuclear Safety
    Authority
and
Department of Radiotherapy and
    Oncology
Turku University Hospital
Turku, Finland

Lakshmi Santanam
Department of Radiation Oncology
Washington University School of
    Medicine
St. Louis, Missouri

Cheng B. Saw
Division of Radiation Oncology
Hershey Cancer Institute
Hershey, Pennsylvania

Pierre Scalliet
Department of Radiation Oncology
Université Catholique de Louvain
University Hospital St Luc
Brussels, Belgium

Michael C. Schell
Department of Radiation Oncology
University of Rochester Medical Center
Rochester, New York

David Schlesinger
Department of Radiation Oncology
University of Virginia Health System
Charlottesville, Virginia

Christopher F. Serago
Department of Radiation Oncology
Mayo Clinic
Jacksonville, Florida

Amish Shah
Department of Radiation Physics
MD Anderson Cancer Center Orlando
Orlando, Florida

Michael B. Sharpe
Radiation Medicine Program
Princess Margaret Hospital
Toronto, Ontario, Canada

Ke Sheng
Department of Radiation Oncology
University of Virginia Health System
Charlottesville, Virginia

Chengyu Shi
Department of Radiation Oncology
University of Texas Health Sciences
    Center San Antonio
San Antonio, Texas

Claudio H. Sibata
Radiation Oncology Department
East Carolina University School
    of Medicine
Greenville, North Carolina

Thomas Simon
Sun Nuclear Corporation
Melbourne, Florida

Roelf Slopsema
University of Florida Proton Therapy
    Institute
University of Florida
Jacksonville, Florida

Timothy D. Solberg
Department of Radiation Oncology
University of Texas Southwestern
    Medical Center at Dallas
Dallas, Texas

William Y. Song
Department of Radiation Oncology
University of California, San Diego
La Jolla, California

Fanny Soum-Pouyalet
Institut Bergonié
Comprehensive Cancer Center
Bordeaux, France

Benjamin A. Spencer
Departments of Urology and
    Epidemiology
Columbia University
New York, New York

Sotirius Stathakis
Department of Radiation Oncology
University of Texas Health Sciences
    Center San Antonio
San Antonio, Texas

Robin L. Stern
Department of Radiation Oncology
University of California, Davis Medical
    Center
Sacramento, California

Richard Sweat
.decimal, Incorporated
Sanford, Florida

Bruce Thomadsen
Departments of Medical Physics, Human
    Oncology, Engineering Physics, and
    Biomedical Engineering
University of Wisconsin–Madison
Madison, Wisconsin

Michael D. Thomas
Department of Radiation Oncology
Wake Forest University Baptist Medical
    Center
Winston-Salem, North Carolina

Alexander Usynin
Department of Radiation Oncology
Thompson Cancer Survival Center
Knoxville, Tennessee

Vincenzo Valentini
Department of Radiation Oncology
Università Cattolica S.Cuore
Rome, Italy

Laura A. Vallow
Department of Radiation Oncology
Mayo Clinic
Jacksonville, Florida

Uulke A. van der Heide
Department of Radiotherapy
University Medical Center Utrecht
Utrecht, the Netherlands

Wil J. van der Putten
Department of Medical Physics and
    Bioengineering
Galway University Hospitals
and
National University of Ireland Galway
Galway, Ireland

Leo van der Reis
Department of Health Management
    and Informatics
School of Medicine
University of Missouri
Columbia, Missouri

Prathibha Varkey
Department of Medicine
Mayo Clinic
Rochester, Minnesota

Jack L. M. Venselaar
Department of Medical Physics
    and Engineering
Instituut Verbeeten
Tilburg, the Netherlands

Dirk Verellen
Medical Physics
Radiotherapy, AZ-VUB Radiotherapy
Brussels, Belgium

Jose Eduardo Villarreal-Barajas
Department of Oncology
University of Calgary
Calgary, Alberta, Canada

Vincenza Viti
Department of Technology and Health
Istituto Superiore di Sanità
Rome, Italy

Tony Wang
Department of Radiation Oncology
Columbia University
New York, New York

Chris Warner
.decimal, Incorporated
Sanford, Florida

Twyla Willoughby
Department of Radiation Physics
MD Anderson Cancer Center Orlando
Orlando, Florida

Wensha Yang
Department of Radiation Oncology
University of Virginia Health System
Charlottesville, Virginia

Daniel Yeung
University of Florida Proton Therapy
    Institute
University of Florida
Jacksonville, Florida

Cheng Yu
Department of Neurological Surgery
University of Southern California
Los Angeles, California

Yan Yu
Department of Radiation Oncology
Thomas Jefferson University
Philadelphia, Pennsylvania

Jian-Ming Zhu
Department of Radiation Oncology
University of North Carolina
Chapel Hill, North Carolina

I

Quality Management
and Improvement

1

Perspective on Quality and Safety in Radiotherapy

Peter B. Dunscombe
University of Calgary

David L. Cooke
University of Calgary

Introduction

Maintaining and improving the quality and safety of radio-therapy involves many interwoven activities. Here we discuss a structure for illuminating the relationships between these activities and illustrate the connection between quality and error management, mainly by using the technical component as the context.

Definitions

In exploring the general structure of a comprehensive quality management program, the first difficulty we encounter is that of definitions. The first definition that we should address is that of "quality." Arriving at a concise, all-encompassing definition of quality for radiotherapy is difficult—it depends on perspective. For a radiation oncologist, quality might include the ability to offer a full range of available technologies; for a physicist, quality would probably include being confident in the output of a linear accelerator to within 2%; and for a patient, it could include being able to park immediately outside the treatment facility. However, one dimension of quality that all stakeholders would agree upon is the avoidance of adverse events resulting from treatment.

A challenge of functioning effectively in a multidisciplinary environment is arriving at consensus decisions that satisfy all participants' (including patients') frequently unspoken specific requirements for quality. Some would argue that it is only the patients' perception of quality that is important, as they are the consumers of the service. In many healthcare jurisdictions, such as Canada or the United Kingdom, the patient cannot choose between service providers, so the question of quality-based decisions on the part of patients is a moot point. It could also be argued that the majority of the patient population does not have the knowledge to assess technical quality and must therefore rely

on the systems of the particular healthcare organization, possibly endorsed by an accrediting body. Whether the nature of competition for patients in the U.S. healthcare system leads to a higher quality of service delivery overall is still an open question. A general discussion of quality, its dimensions, management, and improvement is presented in Part I of this book (Quality Management and Improvement).

However, even given this initial difficulty of defining "quality," we suggest that it is possible to arrange the terms quality management, quality assurance, and quality control (QC) into a structure that serves to illuminate their meanings and, hence, provides a framework for improving the organization of quality management in the radiation therapy department. The relationship between quality management, quality assurance, and quality control is shown in Figure 1.1.

A commonly used basic definition of quality control is "the operational techniques that are used to fulfill requirements for quality." When we think of quality control in radiation therapy, we tend to think of the measurements and adjustments made on the linear accelerators and other hardware (the infrastructure) before the start of the treatment day or during more extensive scheduled downtime. However, with the increasing integration of the infrastructure used for the treatment of a patient, there is greater emphasis on the quality control of processes. An example of this is the quality control of intensity modulated radiation therapy (IMRT) beams. Agreement between measurement and prediction in this case depends not only on whether the accelerator is performing adequately but also on its interpretation of the instructions it receives from the treatment planning system and the integrity of data transfer between the planning system, the information system, and the treatment machine, as well as other factors. Whether quality control is applied to infrastructure or processes, performance has to be measured against some standard. A standard for linear accelerator output promulgated

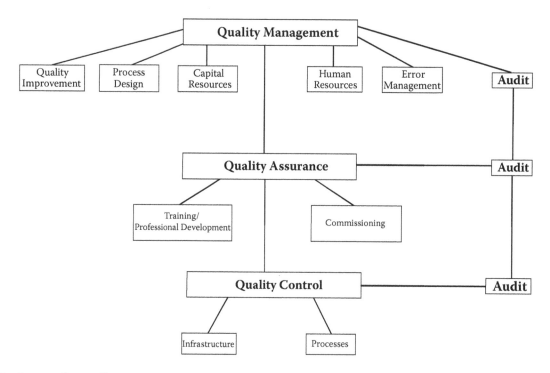

FIGURE 1.1 Structure for a quality management program.

by many organizations is, for example, 2% maximum deviation in output. Processes or infrastructures that do not meet stated performance standards should be removed from service or restricted to those applications in which the performance deficiency is of known and acceptable consequences. Quality control related to clinical infrastructures and processes is discussed in Parts VI and VII, respectively, of this book.

It is worth noting that the need for quality management extends well beyond the confines of the radiotherapy clinic. Although outside the scope of this book, a focus on quality in the diagnosis, staging, and follow-up of cancer patients is vital if the modern techniques described here are not to be misapplied to the detriment of the patient.

An interesting point to ponder is whether quality control is applicable to human resources. If the defining characteristics of quality control are the measurement of performance against stated standards and the implementation of corrective actions to address deficiencies, then quality control can be, and is, applied to human resources to some degree. Academics and others are subject to regular performance appraisals. A well-designed performance development program will include the setting of objectives (performance expectations) at the beginning of the cycle and evaluation of performance against these expectations at the end. Fortunately, and for good reason, suboptimal performance does not generally mean removal from service, as it would for equipment although corrective actions might be implemented. The use of the term "quality control" applied to human resources would be offensive to many, so it's probably better to stick with performance appraisal.

Quality control is one of the activities that falls under the umbrella of quality assurance (QA). Commonly used definitions

of QA encompass the following principle: all those planned and systematic actions necessary to provide adequate confidence that a product or service meets the requirements for quality. Clearly, quality control, being a regular check of performance, is a planned and systematic action designed to establish that the equipment or process is capable of meeting the stated quality requirements. Another component of QA is equipment and process commissioning (see Figure 1.1). Including acceptance testing as part of the overall commissioning process, this stage can be viewed as establishing that the equipment/process is capable of delivering the expected quality and that its behavior and characteristics are sufficiently understood so as to form the input to other equipment and processes (Part V: Quality Assurance in Radiotherapy). If commissioning and acceptance testing establish that the equipment/process is capable of delivering the required quality, and quality control confirms through regular checks that it is doing so, then all that is left to be required to provide adequate confidence that requirements for quality will be met are the human inputs. In this interpretation of quality assurance, the term encompasses the ongoing training and professional development of human resources. Aspects of this vital input to a quality program are discussed in Part IV (People and Quality). Only with well-tuned equipment, processes, and staff can adequate confidence in the quality of the service be provided.

Management is yet another term that, from its use, seems to be subject to a variety of interpretations. A simple definition that encompasses the functions generally associated with the word is that it describes the activities involved in the acquisition and allocation of resources. Management involves decision making and is distinguished from administration, which, in the context of quality, is the monitoring of inputs and outputs. The administrative function is necessary to provide the information upon which management decisions are made. The resources in a radiation treatment program are mainly allocated to human (70%) and capital (30%) categories. Quality management decisions encompass establishing the equipment inventory and the number and disciplines of the staff within the program. This is where most of the resources go. Managing for quality also involves resource commitments to several other important activities as indicated in Figure 1.1. Designing processes is clearly a quality-driven (and efficiency-driven) activity. Process design involves decision-making based on an understanding of the system and the requirements for quality. The documentation generated at the conclusion of process design is essential as it forms the basis for the QA that will follow.

Quality Improvement

Technical quality improvement and error reduction can be seen in the context of quality management in Figure 1.2. Taking a restricted view of quality as the degree to which the oncologist's prescribed dose is actually delivered to the target; Figure 1.2 shows the distribution of delivered doses for a population of patients. There is some evidence to confirm that a distribution of this form reflects clinical reality.

Technical quality improvement encompasses those activities directed toward narrowing the width of the distribution in Figure 1.2. Recent developments in radiation therapy suggest that the current focus of quality improvement is on meeting the volume component, as opposed to the dose component, of the oncologist's prescription. Image-guided radiation therapy (IGRT) and proton therapy are directed toward a higher degree of conformality to the target with better normal tissue sparing. IGRT, in its most basic manifestation, is a quality control activity that addresses the variability in routinely meeting the

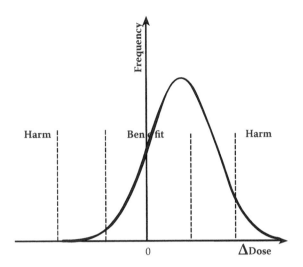

FIGURE 1.3 Schematic for the frequency distribution of dose deviations in the presence of both random and systematic errors.

oncologist's prescription as it relates to the treatment volume. The discussion in this chapter is largely focused on technical quality. A broader discussion on assuring and improving quality can be found in Part III of this book (Methods to Assure and Improve Quality).

An error or mistake could result in a departure from a quality treatment sufficient to cause harm to a patient. The tails of the distribution represent error states and error reduction strategies are directed toward minimizing both the severity and probability of significant departures from a quality treatment. Figure 1.3 illustrates the situation where a systematic dose error is present in addition to normal system variability. Under such circumstances, not only is the frequency of occurrence higher, that is, more patients will be affected, but the dosimetric severity is as well. An effective error reduction program will have input into both quality assurance and quality control activities. Commissioning of infrastructure and processes will need to include safeguards against the inadvertent incorporation of systematic errors. Patient safety and managing errors are discussed in Part II of this book.

Standards and Judgments

Our final comment relates to standards and judgment. In an ideal world, quality control would be based on written standards of performance with little room for judgment. Technical standards are specified in terms of tolerance, action levels, etc. Such terms are also being interpreted differently by different individuals. Clinical standards should be based on statistically reliable evidence and not anecdotal experience. In the real world, although there should be little room for judgment once the quality control program has been established, there will always be situations where minor deviations in equipment performance have to be balanced against continuity of a highly fractionated course of radiotherapy or situations where patient-specific factors require departures from evidence-based treatment protocols.

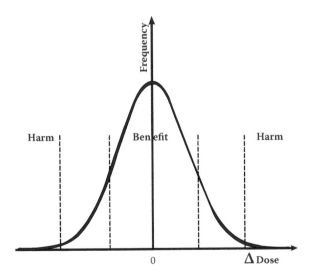

FIGURE 1.2 Schematic for the frequency distribution of dose deviations in the presence of random errors only.

Focusing on the technical aspects of radiotherapy quality management standards is of decreasing importance and focusing on judgment is of increasing importance, moving up from quality control through quality assurance to quality management, as shown in Figure 1.1. With rapid technological development and process change, quality assurance programs will be based on the knowledge and expertise of users in individual treatment facilities. Consensus opinions on appropriate commissioning and training activities, for example, cannot be formed quickly enough in a rapidly changing environment. The clinical parallel to this observation is that reliance on standards, derived from an evidence basis, for quality control of patient treatment prescriptions, is no longer possible in an increasing number of cases as the availability of advanced techniques presents, in the judgment of practitioners, opportunities for quality improvement that should not be denied to patients.

At the quality management level, judgment becomes paramount. While there may be guidelines on staffing levels and equipment utilization, operational decisions at the clinic level currently reflect local circumstances and the local understanding of quality. Which is more important: a short waiting time for treatment or daily in-room imaging? And is the decision based on the treatment site? When it comes to the allocation of resources for error reduction and quality improvement, there seem to be no guidelines in healthcare. We know little about how much time and resources institutions currently devote to these related activities and we have no data on how one clinic performs compared to another clinic. Judgments in this regard will likely reflect, among other things, recent institutional history, such as a serious recent adverse outcome as a result of an error.

Summary

The radiation therapy community could improve its service to patients by engaging in discussions on these issues. The contexts provided in this chapter's figures are just some ways of viewing quality- and safety-related issues and their relationships to one another. In a medical area where accuracy is crucial, speaking a common language cannot be anything but beneficial for patients.

2

Quality as Viewed and Lived by the Patient

Fanny Soum-Pouyalet
Institut Bergonié
Comprehensive Cancer Center

Annie Hubert*
Université de la Méditerranée
Aix-Marseille-II

Guy Kantor
Institut Bergonié
Comprehensive Cancer Center

Introduction

Analyzing quality, as viewed and lived by a patient undergoing radiotherapy, leads us to consider the main points of sociopsychological quality theories. The concept of quality of life (QOL) is currently used in oncology. However, no systematic overviews or guidelines issued for QOL assessment have been produced yet, even on the international level (Shimozuma et al. 2002). A consensus on QOL has been settled, and this relates QOL to symptoms, functioning, psychosocial and social well-being, and even to fulfillment. Thus, the concept has undertaken a multidimensional meaning and has been renamed health-related quality of life (HRQOL) (Kaasa and Loge 2003). These new dimensions of the concept of quality lead to linking the very intimate experience of the patient undergoing radiotherapy to the quality of cure and care. It also underlines the close link existing between the following three concepts: quality, satisfaction, and well-being. These are the three keywords around which the question of quality as viewed and lived by patients must be considered. Nevertheless, these criteria have not been taken into account enough in the field of radiotherapy. Many attempts have already been made toward the systematization of toxicity; however, they were mainly focused on physical and functional criteria rather than on multidimensional ones (Ciabattoni et al. 1997). Perception of quality and patients' satisfaction are closely connected.

Patients' Perceptions of the Environment of Radiotherapy

Ergonomic and environmental factors of radiotherapy treatments obviously affect patients' perception of quality. It has been underlined that times, spaces, and care coordination have great impact on patients undergoing radiotherapy (Hoarau, Kantor, and Dilhuydy 2000). Some papers point out the importance of waiting times and the variability of local demand of patients, referring to the schedule of treatments (Calman et al. 2008). Other papers stress the impact of the physical environment, which can significantly influence one's sense of well-being (Jarvis 2003). Radiotherapy treatments can cause individual discomfort due to a lack of confidentiality or a lack of privacy, for example, in the waiting areas. Difficulty can occur in sharing private information with the medical team. Several studies underline the importance of the entire hospital organization, and especially the treatment session itself, as the waiting times and the interaction with fellow patients are often considered potentially stressful (Dilhuydy et al. 2002).

As a consequence, considering radiotherapy as an ambulatory treatment neglects the impact of its context on patients' experience, although quality of cure and care are strongly linked to QOL. In fact, several studies demonstrate that the patients' perception of the burden of treatment contributes to coping less well with the radiotherapy treatment itself and could explain the long duration of side effects, such as fatigue, even after the end of treatments (Smets et al. 1998). The subjective dimension of coping with radiotherapy needs to be seriously considered.

Actually, initiatives that stress the improvement of patients' well-being during radiotherapy have demonstrable effects on

---

* In memoriam: With regard and respect to Annie Hubert's constant and courageous fight against breast cancer as well as her personal connection with all the patients and human beings involved.

their experience and evaluation of quality as in, for example, the role of supportive care. In radiotherapy, supportive care can integrate the key aspects of diagnosis and treatment and alleviate physical and psychosocial comorbidities inherent to the disease as well as to the treatment (Perez Romasanta and Calvo Manuela 2005).

Even minor improvements of patients' experience during radiotherapy can have a great effect on their perception of quality. For example, socio-aesthetic care, cosmetology, and relaxation therapy have demonstrated their genuine impact on patients' mood and states of well-being (Jereczek-Fossa, Marsiglia, and Orecchia 2002; Titeca et al. 2007), and especially on patients' self-perception of the disease and treatments.

Patients Coping with the Complexity of Radiotherapy

Radiotherapy has undergone major evolutions during the past three decades. Advanced technology and the necessary multidisciplinary composition of its medical teams set it apart. Modes of treatment are complex and require various medical skills (Hogle 2006; Perez Romasanta and Calvo Manuela 2005). This situation makes the patient feel lost and powerless while undergoing treatment. Studies of patients' perceptions of radiotherapy reported that many patients feel alienated by the techniques (Hoarau and Kantor 2000) and could not handle radiotherapy without being stressed or depressed and needed to be helped by one or various members of the clinical team.

Studies of nursing in radiotherapy also pointed out the necessity of guiding patients during the entire process (Carper and Haas 2006). As a matter of fact, recognition of each patient's individual needs is necessary to deal with the treatment and to cope with the disease. Obviously, patients would rather feel like the most important actor of the therapeutic relation than like a medical object (Soum-Pouyalet et al. 2005).

Importance of the Medical Team Role and the Impact of Communication

Several studies emphasize the importance of the entire medical team's (physicians, therapists, and other professionals) understanding in helping patients cope with radiotherapy treatments (Gamble 1998). Moreover, the lack of care delivered in some radiotherapy departments has been pointed out (Long 2001). Follow-up during radiotherapy and after completion of treatment appears essential. These deficits are essentially due to the lack of communication and information related to the objectives of the radiotherapy treatment or to the inappropriate attitude of the medical team in meeting the needs of patients. Failure of coordination within the radiotherapy department can also explain why patients may receive confused or contradictory messages, especially about side effects or the care received (Dilhuydy et al. 2002). It appears that, according to the medical referents,

improvements are needed to provide the right information to patients (Sandoval et al. 2006). For example, a study about radiotherapy-induced nausea concludes on the high percentage of patients who would have liked to receive more information about this specific side effect (Enblom et al. 2009). This example also stresses the importance of identifying and adequately treating side effects related to radiotherapy. Also, the patients' perception of quality is strongly related to the information given and the means of communication (Hogle 2006).

Quality Related to Information in Radiotherapy

According to the actions of collective patient associations and the evolution of the national laws concerning the rights of cancer patients, the public demand for and interest in information about radiotherapy has increased. Therefore, the necessity of providing patients with adequate information has become more and more difficult for the medical teams (Schäfer et al. 2005). Many studies are devoted to the issue of informing patients about their illness and treatment. For example, a pilot study examined patients' understanding of their illnesses and their expectations from palliative radiotherapy. It has been shown that a significant proportion of the patients have misconceptions regarding their illness and unrealistic expectations for their treatment (Chow et al. 2001). Inadequate information can also cause anxiety for the patient and might lead to legal action against the physician (Schäfer et al. 2005). The link between anxiety, side effects, and information resources is obvious.

Patient information about radiotherapy has many ethical implications that must also be considered. The most important ethical principles of patient information are truth, autonomy, informed consent, and hope (Schafer and Herbst 2003). For each of these, a detailed discussion of various typical situations while undergoing radiotherapy (such as adjuvant therapy or palliative treatment) is desirable. Besides patient information, expectations seem to be the most important in the biophysiological, functional, and social fields (Siekkinen et al. 2008). The necessity of informing patients of the possible side effects of their radiotherapy treatments has already been underlined (Chow et al. 2001).

Many studies mention the specific question of delivering information to patients. Some regional variations on the topic have also been studied. The question of information requires an entire development in itself. Information expectations may vary from time to time during the duration of the treatments (Siekkinen et al. 2008). Therefore, it is very difficult to point out the true information needed for particular and singular maturation of patient status (Hoarau and Kantor 2000).

Although the types of information provided to patients appeared to fit their needs, health professionals and patients placed different levels of importance on information. The priority given to specific information may not be optimal from the perspectives of patients, as can also be the case for a wide range of information deliveries (Halkett, Short, and Kristjanson 2009).

Impact of Information Materials

Educational booklets (Dilhuydy et al. 2003; SFRO 2000) and educational strategy concerning information appear to be a strong necessity. Improvements are needed in counseling and education of patients and their relatives, especially concerning the different sequences of radiotherapy and the follow-up care required after completing the treatment (Sandoval et al. 2006).

In fact, educational bases are highly regarded by a large majority of patients, especially when they are based on patients' experiences (Bonnet et al. 2000; Dilhuydy et al. 2002; Hoarau and Kantor 2000). The practical and technical knowledge provided by booklets and other patient information materials give reassurance about treatments and the medical teams (Fervers et al. 2003). Satisfaction with information materials in general leads patients to consider them as real necessities (Bonnet et al. 2000; Dunn et al. 2004). At this point, forums and medical information on the Internet do not seem to substitute for the traditional information modes, even if they have great impacts on the therapeutic relationship between patients and physicians (Siekkinen et al. 2008).

However, educational materials that have excellent face validity and that are well received by patients may fail to fit the information expectations of the patient regarding his or her specific needs (Dunn et al. 2004). Information is best accompanied by professional caregivers (especially physicians) (Hubert et al. 1997). As a matter of fact, oral and direct communication remains the preferred mode of information delivery (Bonnet et al. 2000).

Summary

The whole context of radiotherapy has a deep impact on the quality perceived and experienced by patients. The high technical specificity and complexity of radiotherapy contribute to make the patient feel powerless during treatments. On top of that, the coordination between the different members of the medical team may not be well identified by the patient and that could contribute to making that person feel lost. In spite of the progress made in the field of patient education and information, the singular relationship between patient and physician for the quality of communication in the different therapeutic steps is still considered a conclusive factor in the satisfaction and well-being of the patient. The very specific needs and expectations of each patient during the different periods of treatment and the great variability of each radiotherapy department should lead to develop a sharper view of each and every context to improve the quality perceived and experienced by a patient undergoing radiotherapy. From this perspective, the particular approaches of the social sciences could provide a useful contribution (Soum-Pouyalet, Hubert, and Dilhuydy 2008).

References

Bonnet, V., C. Couvreur, P. Demachy, F. Kimmel, H. Milan, D. Noel, M. Pace et al. 2000. Evaluating radiotherapy patients' need for information: A study using a patient information booklet. *Cancer Radiother.* 4 (4): 294–307.

Calman, F., L. White, E. Beckingham, and C. Deehan. 2008. When would you like to be treated? A short survey of radiotherapy outpatients. *Clin. Oncol. (R. Coll. Radiol.)* 20 (2): 184–190.

Carper, E., and M. Haas. 2006. Advanced practice nursing in radiation oncology. *Semin. Oncol. Nurs.* 22 (4): 203–211.

Chow, E., L. Andersson, R. Wong, M. Vachon, G. Hruby, E. Franssen, K. W. Fung et al. 2001. Patients with advanced cancer: A survey of the understanding of their illness and expectations from palliative radiotherapy for symptomatic metastases. *Clin. Oncol. (R. Coll. Radiol.)* 13 (3): 204–208.

Ciabattoni, A., A. Scopa, M. R. Spedicato, and A. Turriziani. 1997. Organ preservation and quality of life: The quality of life of patients undergoing radiation therapy. *Rays* 22 (3): 490–498.

Dilhuydy, J. M., H. Hoarau, B. F. Delices, N. Bonichon, C. Laporte, M. Minsat, N. Pontet, and L. Votron. 2002. The patient care experience in radiotherapy: perspectives for better patient support. *Cancer Radiother.* 6 Suppl 1: 196s–206s.

Dilhuydy, J. M., E. Luporsi, L. Leichtnam-Dugarin, P. Vennin, and H. Hoarau. 2003. Radiotherapy of breast cancer. *Cancer Radiother.* 7 (3): 213–221.

Dunn, J., S. K. Steginga, P. Rose, J. Scott, and R. Alison. 2004. Evaluating patient education materials about radiation therapy. *Patient Educ. Couns.* 52 (3): 325–332.

Enblom, A., A. B. Bergius, G. Steineck, M. Hammar, and S. Börjeson. 2009. One third of patients with radiotherapy-induced nausea consider their antiemetic treatment insufficient. *Support Care Cancer* 17 (1): 23–32.

Fervers, B., L. Leichtnam-Dugarin, J. Carretier, V. Delavigne, H. Hoarau, S. Brusco, and T. Philip. 2003. The SOR SAVOIR PATIENT project—An evidence-based patient information and education project. *Br. J. Cancer* 89 Suppl 1: S111–S116.

Gamble, K. 1998. Communication and information: The experience of radiotherapy patients. *Eur. J. Cancer Care (Engl.)* 7 (3): 153–161.

Halkett, G. K., M. Short, and L. J. Kristjanson. 2009. How do radiation oncology health professionals inform breast cancer patients about the medical and technical aspects of their treatment? *Radiother. Oncol.* 90 (1): 153–159.

Hoarau, H., and G. Kantor. 2000. Understanding the information booklet "For a Better Understanding of Radiotherapy." *Cancer Radiother.* 4 (4): 308–316.

Hoarau, H., G. Kantor, and J. M. Dilhuydy. 2000. An anthropological study of radiotherapy care experience. *Cancer Radiother.* 4 (1): 54–59.

Hogle, W. P. 2006. The state of the art in radiation therapy. *Semin. Oncol. Nurs.* 22 (4): 212–220.

Hubert, A., G. Kantor, J. M. Dilhuydy, C. Toulouse, C. Germain, G. Le Pollès, R. Salamon, and P. Scalliet. 1997. Patient information about radiation therapy: A survey in Europe. *Radiother. Oncol.* 43 (1): 103–107.

Jarvis, J. A. 2003. Transforming the patient experience in radiation therapy. *Radiol. Manage.* 25 (6): 34–36.

Jereczek-Fossa, B. A., H. R. Marsiglia, and R. Orecchia. 2002. Radiotherapy-related fatigue. *Crit. Rev. Oncol. Hematol.* 41 (3): 317–325.

Kaasa, S., and J. H. Loge. 2003. Quality of life in palliative care: principles and practice. *Palliat. Med.* 17 (1): 11–20.

Long, L. E. 2001. Being informed: undergoing radiation therapy. *Cancer Nurs.* 24 (6): 463–468.

Perez Romasanta, L. A., and F. Calvo Manuela. 2005. Supportive care in radiation oncology. *Clin. Transl. Oncol.* 7 (7): 302–305.

Sandoval, G. A., A. D. Brown, T. Sullivan, and E. Green. 2006. Factors that influence cancer patients' overall perceptions of the quality of care. *Int. J. Qual. Health Care* 18 (4): 266–274.

Schäfer, C., B. Dietl, K. Putnik, D. Altmann, J. Marienhagen, and M. Herbst. 2005. Patient information in radiooncology results of a patient survey. *J. Psychosoc. Oncol.* 23 (4): 61–79.

Schafer, C., and M. Herbst. 2003. Ethical aspects of patient information in radiation oncology. An introduction and a review of the literature. *Strahlenther. Onkol.* 179 (7): 431–440.

SFRO, The French Society for Radiotherapy and Oncology and The National Syndicate of Radiotherapy Oncologists. 2000. To better understand radiotherapy: An information booklet. *Cancer Radiother.* 4 (4): 285–293.

Shimozuma, K., T. Okamoto, N. Katsumata, M. Koike, K. Tanaka, S. Osumi, M. Saito et al.. 2002. Systematic overview of quality of life studies for breast cancer. *Breast Cancer* 9 (3): 196–202.

Siekkinen, M., S. Salantera, S. Rankinen, S. Pyrhonen, and H. Leino-Kilpi. 2008. Internet knowledge expectations by radiotherapy patients. *Cancer Nurs.* 31 (6): 491–498.

Smets, E. M., M. R. Visser, B. Garssen, N. H. Frijda, P. Oosterveld, and J. C. de Haes. 1998. Understanding the level of fatigue in cancer patients undergoing radiotherapy. *J. Psychosom. Res.* 45 (3): 277–293.

Soum-Pouyalet, F., J. M. Dilhuydy, A. Hubert, and G. Kantor. 2005. Patients and physicians: Cross representations anthropological study of the construction of individuality in radiation therapy. *Bull. Cancer* 92 (7): 741–745.

Soum-Pouyalet, F., A. Hubert, and J. M. Dilhuydy. 2008. Interests of applied anthropology to oncology. *Bull. Cancer* 95 (7): 673–677.

Titeca, G., F. Poot, D. Cassart, B. Defays, D. Pirard, M. Comas, P. Vereecken, V. Verschaevec, P. Simon, and M. Heenen. 2007. Impact of cosmetic care on quality of life in breast cancer patients during chemotherapy and radiotherapy: An initial randomized controlled study. *J. Eur. Acad. Dermatol. Venereol.* 21 (6): 771–776.

3

Quality Management: An Overview

Martin Carrigan
The University of Findlay

Introduction

Different approaches to assuring quality in organizations have evolved over the years dependent on the needs of the industry, the definitions of quality, and the level of management of the quality process. In high-performing organizations, assuring that quality exists in products and services is an integral part of the organizational and department-wide strategy. A quality product or service is essential to an organization's reputation, liability, and potential for growth or dominance in a market. Quality and approaches to quality, however, are not without significant cost. The cost of quality includes prevention costs, appraisal costs, internal failure, and external costs. All have to be weighed to formulate an effective quality assurance program.

Evolution of the Concept of Quality

Business and manufacturing quality assurance evolved from the early industrial revolutions in England and America. Quality in the United States started as assuring uniform manufacturing processes, as advocated by the scientific management movement of Frederick Taylor and adopted by Henry Ford's assembly line. It moved to statistical process control for production during wartime and has evolved into quality management systems like those advocated by W. Edwards Deming and the Six Sigma program popularized by General Electric. Quality has also become an accepted management protocol that has led to quality awards such as the Malcolm Baldrige award.

Quality assurance has migrated to healthcare services, but with different emphases than for manufacturing because of the different populations and customer expectations. Quality in healthcare can include judgments about having the latest technology, the most accurate procedures, the least invasive options, the appropriate surgical techniques, the lowest cost, correct differentiation, and the best outcomes. For example, a patient may regard a quality service as one that "works" or has an effective outcome. A provider may regard a quality service as one that is technically proficient and that utilizes the latest technology, without regard to patient comfort or convenience. Management may regard quality as a service that results in the greatest revenue at the least cost and liability, despite recent innovations or legacy systems. As a result, there are many different approaches to quality assurance in healthcare and many different management perspectives of the process.

Historical Development of Quality

Quality assurance is the process of being aware of quality at every level of delivery in an organization. How that process is developed and managed has taken different iterations throughout different industries and time periods.

Early in history, skilled craftsmen and artisans gained their knowledge through an apprenticeship program where they were supervised by masters of their trade until they themselves became masters. Here, quality was assured through the apprentice system and the measurement was the individual judgment of the master.

In 1798, Eli Whitney designed and manufactured firearms with interchangeable parts. The parts had to be precisely crafted to assure that they would fit the same model of firearms. Thus, unlike skilled craftsmen of the time who created unique items, Whitney's production had to be uniform. A measure of quality control was created.

Quality Comes of Age in the Twentieth Century

In 1911, Frederick Winslow Taylor published his *Principles of Scientific Management* and defined the role of a quality inspector from the perspective of an industrial engineer. Henry Ford created the assembly line for his automobiles and integrated Taylor's ideas on a large scale where quality became associated with the idea of inspection.

Throughout the early twentieth century, a number of individuals contributed to the quality assurance process. Joseph M. Juran added the human dimension to quality management. He pushed for the education and training of managers at the middle management and top management levels. For Juran, human relations problems were the ones to isolate. Resistance to change—or, in his terms, cultural resistance—was the root cause of quality issues.

Juran used Pareto's law, (the 80%–20% rule) to identify the quality problems. During quality planning, one engages in a process of understanding what the customer needs and then designs all aspects of a system to meet those needs reliably. Quality control is used to constantly monitor performance for compliance with the original design standards. If performance falls short of the standard, plans are put into action to deal quickly with the problem. Quality improvement occurs when previously unattained levels of performance (breakthrough performance) are achieved.

W. Edwards Deming had great success improving production in the United States during World War II, although he is perhaps best known for his work in Japan, where he taught how to improve design, product quality, and testing through various methods, including the application of statistical methods.

Although both Juran and Deming were Americans who taught in Japan, their messages on quality assurance and management went largely unnoticed until Japan started making dramatic increases in the quality of products that it exported, particularly in automobiles and electronics. As a result, in the late twentieth century a number of quality assurance and management emphases arose in the United States. Certainly many others have had great influence on quality, including Philip Crosby with *Quality is Free* (Crosby 1996), Dr. Kaoru Ishikawa with *Quality Circles*, Noriaki Kano with his two-dimensional model of quality: "must-be quality" and "attractive quality," Robert Pirsig, Genichi Taguchi ("uniformity around a target value" and "the loss a product imposes on society after it is shipped"), and Peter Drucker ("Quality in a product or service is not what the sup-plier puts in. It is what the customer gets out and is willing to pay for" among others.).

Quality Systems and Awards

New concepts also arose during the twentieth century, among them total quality management (Beer 2003). Six Sigma (number of defects per million opportunities) and its variations including lean manufacturing (elimination of waste), the focus on the customer, value chain management, strategic planning, and leadership have all recently led to a renewed emphasis on quality assurance and management (Evans and Lindsay 2005).

Quality in Healthcare

In healthcare, one of the quality emphases involving technology focuses on statistical process control (SPC), which monitors standards, makes measurements, and takes corrective action during the delivery of healthcare services (Carey 2003; Kelley 1999). SPC first achieved prominent use as a result of manufacturing production methods during World War II. SPC uses control charts, which are graphical presentations of data over time that show the upper and lower limits of the process sought to be controlled (Ryan 2000). SPC charts were originally called Shewhart charts after Walter A. Shewhart, the father of statistical quality control.

Regular plotting of data on an SPC chart will tell if the process is out of control, that is, subject to special causes (Woodall 2000). Some of the tools of quality assurance, as articulated by Dr. Kaoru Ishikawa, include: flow charts; process maps; check sheets; histograms; scatter plots; control charts; cause and effect diagrams; and the Pareto analysis.

One of the criticisms of quality assurance in radiotherapy is that radiotherapy is not uniform and end results differ widely. As a result, there is a greater call for quality assurance and management (de Andrade et al. 2008). Some of the recommended assurance procedures are site visits, in vivo dosimetry, dummy-run procedures, and individual case reviews (Ikeda 2002).

Organizations

Many state-wide, national and international organizations exist today that require adherence to their own quality assurance standards. Some of the better known organizations are the International Organization for Standardization's ISO 9000:2000 series, which describes standards for a quality management system (QMS) to address the principles and processes surrounding the design, development, and delivery of a general product or service. Organizations can participate in a continuing certification process to ISO 9001:2000 to demonstrate their compliance with the standard, which includes a requirement for continual planned improvement of the QMS.

The European Foundation for Quality Management's (EFQM) Excellence Model supports an award scheme similar to the Malcolm Baldrige Award for European companies. In Canada, the National Quality Institute presents the Canada Awards for Excellence on an annual basis to organizations that have displayed outstanding performance in the areas of quality and workplace wellness and that have met the Institute's criteria with documented overall achievements and results.

In radiotherapy, the International Atomic Energy Agency and the World Health Organization perform audits on radiotherapy, as does the European Organization for Research and Treatment of Cancer. In the United States, the Radiological Physics Centers and the Radiation Therapy Oncology Group's protocols form a standard for assuring quality procedures.

Quality Awards

The Malcolm Baldrige National Quality Award (MBNQA) is a competition to identify and recognize top-quality U.S. companies (Brown 2004). This model addresses a broadly based range of quality criteria, including commercial success and corporate leadership. Once an organization has won the award it has to wait several years before being eligible to apply again. In 2002, through hard work, focus, and perseverance, SSM Healthcare was the first hospital to receive the Malcolm Baldrige National Quality Award (Ryan 2007). Additional details on the Baldrige Award are presented in Chapter 6.

The Alliance for Performance Excellence is a network of state, local, and international organizations that uses the Malcolm Baldrige National Quality Award criteria and model at the grassroots level to improve the performance of local organizations and economies. NetworkforExcellence.org is the Alliance Web site; browsers can find Alliance members in their state and get the latest news and events from the Baldrige community.

Companies invest considerable time, resources, and money in testing and quality assurance. Costs can reach up to 300% of the total product development budget. In order to reduce test development time and improve test coverage and efficiency, many companies have created automated testing systems using skilled, in-house human resources, or have invested in third-party test automation solutions. Although automated tests developed in-house are tailored to an organization's specific requirements, this solution suffers from a number of disadvantages.

Lack of standardization significantly impacts data management efficiency, test development time, and quality management process as a whole on an enterprise-wide basis. One way of addressing these issues is to view testing as an enterprise-wide total quality management system similar in concept to enterprise resource planning (ERP). This combines a top-down system that considers quality control across the entire organization with a bottom-up approach. This approach commences with discrete tests and uses them as building blocks to create automated processes for full test coverage. The result is a total automated quality management system.

Objectives of Any Quality System

The common element of the business definitions is that the quality of a product or service refers to the perception of the degree to which the product or service meets the customer's expectations (Prahalad and Krishnan 1999). Quality has no specific meaning unless related to a specific function and/or object. Quality is a perceptual, conditional, and somewhat subjective attribute.

The one common element of all of these approaches is that quality cannot be an adjunct to the process. Quality must be an integral part of the development. For this to happen, a philosophy of quality improvement must be ingrained into the corporate culture (Aikens 2006). Quality starts in the design phase and continues throughout.

Traditionally, quality was an adjunct effort. Quality inspections were made as the product went out of the door. The philosophy expressed by Deming, Juran, and others is that strategy is integrally attached to the entire process. Strategic planning implies long-term planning. The Malcolm Baldrige National Quality Award stresses that continuous quality control is part of the entire process. Changing the name from "Strategic Quality Planning" to "Strategic Planning" emphasizes this point.

Summary

Generally, many approaches to quality assurance and management exist in healthcare and in business. Clinical aspects of quality can differ and organizations should both define and integrate quality assurance into all aspects of management and operational efficiency.

References

Aikens, C. H. 2006. *Quality: A Corporate Force: Managing for Excellence*. Upper Saddle River, NJ: Prentice Hall.
Beer, M. 2003. Why total quality management programs do not persist. *Decis. Sci.* 34 (4): 623–642.
Brown, M. 2004. *Baldrige Award Winning Quality*. University Park, IL: Productivity Press.
Carey, R. G. 2003. *Improving Healthcare with Control Charts: Basic and Advanced SPC Methods and Case Studies*. Milwaukee, WI: ASQ Quality Press.
Crosby, P. 1996. *Quality Is Still Free*. New York, NY: McGraw-Hill.
de Andrade, R. S., J. W. Proctor, S. M. Rakfal, E. D. Werts, L. L. Schenken, C. B. Saw, M. Dougherty, and D. Stefanik. 2008. Assessment of "best practice" treatment patterns for a "radiation oncology community outreach group" engaged in cancer disparities outcomes. *J. Am. Coll. Radiol.* 5 (4): 571–578.
Evans, J., and W. Lindsay. 2005. *An Introduction to Six Sigma and Process Improvement*. Mason, OH: Thompson-Southwestern.

Ikeda, H. 2002. Quality assurance activities in radiotherapy. *Jpn. J. Clin. Oncol.* 3 (12): 493–496.

Kelley, D. L. 1999. *How to Use Control Charts for Healthcare.* Milwaukee, WI: ASQ Quality Press.

Prahalad, C. K., and M. S. Krishnan. 1999. The new meaning of quality in the information age. *Harv. Bus. Rev.* 77 (5): 109–118.

Ryan, Sister M. J. 2007. *On Becoming Exceptional: SSM Health Care's Journey to the Baldrige and Beyond.* Milwaukee, WI: ASQ Quality Press.

Ryan, T. 2000. *Statistical Methods for Quality Improvement.* New York, NY: John Wiley.

Woodall, W. 2000. Controversies and contradictions in statistical process control. *J. Qual. Technol.* 32 (4): 341–350.

4

Quality Management: Radiotherapy

John Øvretveit
Karolinska Institutet

Introduction

This chapter gives a nontechnical introduction and an overview of the different approaches to quality assurance and management in radiotherapy. Providing safe, high-quality radiotherapy services requires the management of a complex combination of sophisticated equipment and personnel, in interaction with vulnerable and often anxious patients. It is this combination that poses the challenge for quality management. Approaches for equipment quality are different from those for quality in the interaction between personnel and patients, which are different from those for managing the combination of elements in a radiotherapy service. Another challenge is to know which of the many regulations apply to the service and to prove the service has met them—doing this may not leave much time and resources for other quality improvement.

Quality management (QM) is widely used as a generic term to describe any activity that aims to ensure and improve quality

---

WHAT IS QUALITY IN RADIOTHERAPY SERVICES?

Providing services that:

- Meet and exceed patient expectations (patient quality)
- Follow best professional practice and achieve the highest clinical outcomes (professional quality)
- Meet higher level regulations at the lowest cost without waste (management quality)

(Pohly 2008). It has been defined as "that aspect of the overall management function that determines and implements the quality policy" (ISO 2000) and as including "strategic planning, allocation of resources, and other systematic activities for quality, such as quality planning, operations, and evaluations."

Approaches to Ensuring Quality

The language and literature of quality can be daunting for some and is confusing to many. This is because the terms are used in many different and overlapping ways and because advocates over-emphasize the benefits of one method or another: Quality is a big business. To give a simple orientation, the different approaches used internationally can be divided into the four categories of individual-, standards-, process-, and risk-based.

Individual-Based: Individual Regulation and Development

One approach to ensuring quality is to set standards for individuals to practice and to ensure continued professional development. This is based on an assumption that the source of quality is the individual practitioner. As services have become more team-based and complex, this approach has proved insufficient, but is still necessary. Perhaps the weakest part of this approach is in detecting and taking action against practitioners who provide poor quality services or act in unsafe ways. Services need effective systems for performance appraisal, personnel development, and for raising and resolving issues concerning an individual's performance.

Standards-Based: Quality Control and Quality Assurance (QA)

This approach is to set standards for different aspects of the service and how it is provided, assess compliance, and take corrective action where necessary. The older industry approach to quality control and quality assurance was to use standards and measurement to detect and reject products with faults. More modern approaches aim to be more proactive and preventative: Standards are for what is thought to be necessary to avoid poor quality in the practice, operations, and physical aspects of the service. Assessment is by internal documentation, regular audit, and external inspection. This approach is useful in radiotherapy services for QA of equipment and physical infrastructure, and is a required part of running a service.

Process-Based: Quality Improvement

Both the individual- and standards-based approaches involve improvement, but the term is now more often used to describe a combination of methods and a philosophy (Adams et al. 2009; Berwick 1989). The central idea is that quality problems are caused by how care is organized. The solutions are to enlist personnel to work in teams using quality methods to improve processes and systems of care. This approach tends to focus on a few problems and on testing changes and making wider changes if the tests are successful. It is useful for understanding and changing how the different elements of a radiotherapy service come together to produce the results. Lower costs and improved quality are possible with the right selection of problems and effective management of improvement teams to streamline processes (Øvretveit 2005).

Risk-Based: Risk Management (RM)

There are people- and finance-based approaches to risk management, but these overlap and some methods are common to both. Risk is a function of the expected harm or losses that may be caused by an event and the probability of this event. The people approach aims to assess and reduce the risks of avoidable adverse events occurring to patients or personnel. The finance approach focuses on which risks could cause the greatest short- or long-term financial losses for the organization, and on protecting the organization from this financial exposure. Some strengths are the recognition that risk is inherent, that reducing risk below a certain point is prohibitively expensive, and in methods for calculating potential people and financial dangers (Carroll 2006; Rozovsky and Woods 2005; Vincent 2001).

Challenges for Quality Management in Radiotherapy Services

There is an increasing focus on quality and safety in healthcare accompanied by new regulations, new technologies, and shortages in skilled personnel. These have not been matched with extra resources and support for radiotherapy services to respond (Hendee 2008). Similar challenges in managing quality are faced by radiotherapy services in most countries:

- Responding to increasing public concerns and patient anxiety, in part due to wider media coverage of adverse events
- Keeping up to date with regulations, standards, best practice, and recent research on effectiveness, and ensuring these are known and incorporated into the service
- Deciding which data and indicators to collect and track routinely and ensuring systems to do this
- Prioritizing which specific problems to focus improvement work on, and which activities are most necessary to respond to regulatory requirements
- For managers and senior clinicians, how best to raise awareness of quality and safety and to equip personnel with the skills and support to ensure and improve quality
- Ensuring problem diagnosis is followed through with testing of a change and then developing systems to ensure successful changes are maintained and work for all patients at all times

Given these and other challenges for leaders, perhaps the most important task is to decide how the limited resources available for QM can best be allocated. The chapters in this book provide guidance and help in addressing these challenges, but each service will need to decide priorities and adapt approaches to their setting.

Change Challenge

A more specific challenge is how to ensure successful change and implementation. This is difficult enough for relatively simple changes, such as ensuring personnel know and follow new procedures for high-dose-rate brachytherapy, or meet other new standards. Some changes are more complex and sizable, such as installing new radiotherapy equipment with new patient receiving and processing arrangements and documentation. The quality literature has been slower to draw on change management methods and research and this field of social science and management practice can be useful for ensuring change is fully carried through (Iles and Sutherland 2001; Silvey and Warrick 2008).

Simple Strategy for Quality Management in Radiotherapy Services

A synthesis of the research suggests that the following principles guide all successful improvement leaders:

- Believing in the need for improvement and demonstrating this belief more in their behavior than in their words
- Aligning incentives and systems to support improvement and the improvement to organizational goals
- Inspiring and motivating all personnel to take responsibility and action for improvement and influencing those who are slowing improvement

- Defining what "responsibility and action for improvement" means for each role, developing competencies, developing themselves, and putting in the time to make change real
- Defining the constraints within which personnel must work for improvement and involving them in setting priorities and targets so as to maximize their ownership of process and changes
- Providing resources, especially for data collection, analysis, and expertise
- Paying persistent attention to implementation and follow-through, especially by ensuring activities and projects use effective methods and are accountable, and that data is collected and used for measuring progress

More details on how to lead quality management are provided in a research-based guide (Øvretveit 2009).

Summary

Quality management and assurance can be confusing with different uses of terms and approaches and is certainly challenging for radiotherapy services, with the complex combination of technology, professionals, and patients. Given the many demands for safety and quality assurance and the limited resources and expertise, radiotherapy leaders and quality specialists will need to prioritize action. Choosing the right approach, methods, and combination of approaches is necessary for making sure the efforts make a difference for patients. Approaches and systems will always need to be tailored to the setting of the service. The chapters in this book provide useful detailed guidance for what needs to be done and how best to do it.

References

Adams, R. D., S. Chang, K. Deschesne, S. Freeman, K. Karbowski, D. LaChappelle, D. E. Morris, B. F. Qaqish, J. L. Hubbs, and L. B. Marks. 2009. Quality assurance in clinical radiation therapy: A quantitative assessment of the utility of peer review in a multi-physician academic practice. *Int. J. Radiat. Oncol. Biol. Phys.* 75 (S133).

Berwick, D. M. 1989. Continuous improvement as an ideal in health care. *New Engl. J. Med.* 320 (1): 53–56.

Carroll, R. 2006. *Risk Management Handbook for Health Care Organizations.* 5th ed. San Francisco, CA: Jossey-Bass.

Hendee, W. R. 2008. Safety and accountability in healthcare from past to present. *Int. J. Radiat. Oncol. Biol. Phys.* 71 (1 Suppl.): S157–S761.

Iles, V., and K. Sutherland. 2009. Organisational change: A review of for health care managers, professionals and researchers. NCCSDO 2001 [cited March 12 2009]. Available from www.sdo.lshtm.ac.uk/publications.htm.

Øvretveit, J. 2005. What are the advantages and limitations of different quality and safety tools for health care? WHO Regional Office for Europe 2005 [cited March 12 2009]. Available from Health Evidence Network report http://www.euro.who.int/document/e87577.pdf.

Øvretveit, J. 2009. Leading improvement effectively. Part 2: Research informed guidance for leaders. Health Foundation/Karolinska Institutet, MMC, Stockholm 2009 [cited March 12 2009]. Available from http://homepage.mac.com/johnovr/FileSharing2.html.

Pohly, P. 2009. Glossary of Managed Care Terms 2008 [cited March 12 2009]. Available from http://www.pohly.com/terms.html.

Rozovsky, J., and J. Woods. 2005. *The Handbook of Patient Safety Compliance: A Practical Guide for Health Care Organizations.* San Francisco, CA: Jossey-Bass.

Silvey, A. B., and L. H. Warrick. 2008. Linking quality assurance to performance improvement to produce a high reliability organization. *Int. J. Radiat. Oncol. Biol. Phys.* 71 (1 Suppl.): S195–S199.

Vincent, C., ed. 2001. *Clinical Risk Management: Enhancing Patient Safety.* London: BMJ Books.

5

Development and Operation of a Quality Management Program: A Case Study

Jan Willem H. Leer
Radboud University
Nijmegen Medical Centre

Introduction

In the past decades, safety and quality assurance have become more and more important issues on the agenda of hospital and departmental management for the following reasons: the patient wants more transparency about the performance of health organizations and even of individual doctors; trust in doctors and medical care is no longer self evident; healthcare has become part of a market-driven economy and, consequently, insurance companies want to know whether they get value for money; and, finally, politicians feel accountable to their voters and the population for optimal and best available care for those who need it. All of this has increased the responsibility of department heads in an area that they are not trained for. Heads of departments are chosen because they are good or excellent physicians or other healthcare professionals, not because they are excellent safety and quality managers.

In this chapter, I describe some personal experiences in setting up a quality system in a radiation oncology department and making it operational. The aspects described are based on Dutch regulations and systems and also on the Radboud University Medical Centre internal quality and safety system, which was recently developed and is still a work in progress.

Safety

In 2007, the Dutch society of hospitals, the Dutch society of medical specialists, and the Dutch society of nurses cosigned a charter on a safety management system for hospitals and institutions that deliver medical care. The aim of this system is to reduce the risks and the potential damage to patients. The main issue of this system is to perform a regular risk inventory, to create a system for blame-free reporting, to develop a system for data analysis, and to put a system in place for implementation of improvements.

What was the consequence of this for the radiotherapy department? The board of the Radboud University Medical Centre has created an institute for quality and safety. Through this institute, the department heads who are given the integral responsibility for the departments, including quality and safety, now regularly receive directives on these issues.

An example of this is the directive that coworkers who had direct patient contact were no longer allowed to wear wrist watches and hand jewelry to reduce the risk of infections. Implementing such a directive turned out not to be that simple in the radiation oncology department. Coworkers did not understand the directive. They said, "We have never seen a patient infected by us in our department." This is very likely to be true, but cannot be proven and, of course, the board of directors cannot make exceptions on the basis of unknown criteria. So, the department head had to attempt to convince people to follow the rule, set the right example himself, and correct coworkers that were not following the directive by walking around the department.

A second, more important, consequence was the introduction of blame-free reporting. This is now a Web-based system in which every incident can be reported anonymously. We extended this system by not only reporting incidents in which patients could be harmed, but also every incident that was a deviation from an agreed-upon procedure or an intended situation. Examples of the latter could be a patient record that could not be found in time, a too long waiting time in the outpatients' clinic, or a dirty toilet.

When we started the system, the number of reported incidents almost tripled. The challenge was to keep the system going. This could only be achieved if the management of the department was visibly taking action on the reported incidents. We created a multidisciplinary committee with a doctor, technologists, and physicists to collect the data and analyze them on the basis of severity and frequency. Based on this analysis, which is regularly reported to the management, we decide whether action should be taken and what that action should be. Of course, the hospital has a system in place through which fatal and severe incidents are directly reported to the board of directors.

Quality Management

In a previous European Society for Therapeutic Radiology and Oncology (ESTRO) publication (Leer et al. 1998), we have described a way to set up a quality system in a radiation oncology department. The system we have described was based on the so-called NIAZ system (Nederlands Instituut voor Accreditatie van Ziekenhuizen, the Dutch institute for accreditation of hospitals).

In 2008, NIAZ published a new quality standard (NIAZ 2009), which I will describe here with its consequences for departments. The aim of a quality system is that the set standards create a justified confidence in the quality of the delivered healthcare. Professionals should be constantly aware that quality is important and be stimulated in a continuing process of improvement. I briefly summarize the standards as they were laid down in this document and give some examples.

Leadership

The first standard is related to leadership. Translated to the departmental level, it means that the management of the department should have a long-term vision on the issues of quality and safety for the department. The role of leadership in ensuring quality is further discussed in Chapter 36.

An excellent leadership tool is the vision document, which the department writes every year and in which targets are set that should be reached. A long-term vision document can also be very helpful. It is important that the vision document also pays attention to safety and quality, apart from medical, scientific, economic, and other professional targets. It is also important that all disciplines in the department are involved in creating this document, so the document and the targets will become a creation of everyone. Regular oral presentations on the document by the head of the department are an important tool to reach this point. In order to come to a vision, the NIAZ standard offers many evaluation points.

The vision document describes how the department sees its involvement in the process of quality improvement and how it guarantees the continuity of care. It also describes the choices that should be made when resources are limited. The department takes care that it knows what external partners expect and the department compares its own performance with others (benchmarking).

The department should have a quality manual (nowadays mostly Web-based) in which all documents, procedures, protocols, and work instructions are put into an easily accessible form for all coworkers in the department. A description of how such a quality manual could be set up is described in the ESTRO booklet (Leer et al. 1998). The major challenge is not only the creation of such a quality manual, but also to keep it up to date, to keep it orderly, and to stimulate its use. To achieve this, it should not be too extensive or too detailed and complicated.

Management is also responsible for ensuring that the structure of the organization is clear, that mandates and responsibilities are clear and well-described in the quality manual (level 1 documents). Also, the aim, participation, and frequency of meetings should be described.

Strategy

The second standard is related to strategy with respect to a constant improvement of quality and risk management. An example of an evaluation point is that the hospital and, consequently, the department are aware of changing laws and directives. As a consequence of this, recently all specialists involved in clinical research, including those who have done this for decades, had to go to a course in which they were taught about all the new regulations on privacy protection, trial conduct, etc.

Management of Coworkers

There are many evaluation points for this standard. The most important one is the guarantee that the coworkers have the competencies for their function and responsibility. The department management should put a system in place in which an annual, documented meeting with each coworker is held where the personal performance of the coworker is discussed and agreements are made on continuing medical or professional education.

Supportive Processes

For radiation oncology, it is important that this standard describes the way radiation protection is organized. The department should have a qualified responsible staff member for radiation protection and a multidisciplinary committee to discuss issues related to radiation and should take care that rules and regulations are followed. All this should be part of the quality manual.

Infrastructure

As a consequence of this standard, all departments in the Radboud University Medical Centre recently had to register every piece of equipment to check its lifespan, state of maintenance, and safety. For large equipment in radiation oncology departments, the condition and performance was mostly satisfactory. However, we were surprised by the sometimes bad state of smaller equipment. It is important to realize that a few accidents in hospitals, like fires, were not caused by a defect in large equipment but by a coffee machine.

Control

The standard of control of care processes might be considered as one of the most important of all standards, because it relates directly to patient care. At a time when patient care is increasingly fragmented due to subspecialization and part-time workers, control of the care process has become the Achilles heel of the medical system. This is also very evident in oncology where many specialists are involved in the diagnosis and treatment of cancer patients.

A few points for evaluation of this quality standard are the following. It should be clear who is responsible for a multidisciplinary chain. Recently, we have started appointing a coordinator mandated by all heads of departments involved in a certain oncological chain, for example, breast cancer. This coordinator has to start with a description of the care process of his chain and develop protocols and standard operating procedures. The participants of such an oncological chain then have to define so-called performance indicators.

Performance indicators have become an obligatory part of hospital quality systems. They are also used by the Dutch Health Inspectorate to check the quality of delivered care in a hospital. By law, hospitals have to report on the performance indicators on a yearly basis. A national performance indicator in the Netherlands for radiation oncology is, for example, the use of a complication registration. As a consequence of this, the department had to set up a Web-based system to regularly, for example, yearly, register medical complications. At a yearly follow-up for each patient, the late effects with a score of 3 or higher are registered. In quarterly meetings, they are discussed and incidences are compared with the available literature. An important performance indicator for an oncological chain is the existence of a multidisciplinary meeting in which all cases are discussed. Also, time intervals between diagnosis and start of treatment could be a performance indicator for an oncological chain.

We are presently setting up systems to register several of these performance indicators. For each performance indicator, one has to define a standard. When the standard is not met, the chain coordinator has to discuss this with the participants in the chain and start a process for improvement. Auditing is also part of this aspect of the quality system. We can distinguish internal and external auditing systems. A hospital should have an operational internal auditing system. At our institution, all departments are audited every 4 years. The audit is based on the NIAZ standards discussed above. External audits are performed by the Dutch Society of Radiation Oncology on a 5-year basis. Audits should lead to improvements.

The implementation of this way of thinking is not simple. It is important to take care that there remains a good balance between a rigid system in which everything must be part of a protocol, and freedom and flexibility. Finally, quality should be based on the performance of individuals and not on the existence of protocols.

Patient Satisfaction

The standard of patient satisfaction is very useful. By regular questionnaires, we measure the satisfaction of patients. We have to be aware that patients might not always give you a true answer, but these questionnaires are helpful in detecting areas for improvement of the service.

Employee Satisfaction

As part of the quality system, regular employee satisfaction measurement is also important. A periodic risk inventory is part of it. This standard also helps to inform the department management on whether coworkers feel that management is transparent, communication is open and sufficient, and what the general atmosphere in the department is.

Summary

Quality systems have evolved over the past years. The structure of the standards for a quality system has been changed, but all the original elements are still there, as is the basic philosophy. The standards still do not prescribe how you should organize your department, only what should be organized and what should be laid down in protocols and work instructions. The standards consist of extensive checklists with points of evaluation, which are helpful for department management to see whether they meet the necessary criteria, and are also useful for the obligatory internal and external audits. It is a challenge for department management to keep the system vivid, practical, and useful. It is important that every coworker in the department is in some way involved in creating and using the quality system. A simple Web-based document system—with a responsible document manager who takes care that the documents are regularly updated, old documents are removed, and new documents are added—is very useful for this aim.

References

Leer, J. W. H., A. L. McKenzie, P. Scalliet, and D. I. Thwaites. 1998. Practical guidelines for the implementation of a quality system in radiotherapy. In *ESTRO Quality Assurance Committee Sponsored by "Europe Against Cancer," Physics for Clinical Radiotherapy, Booklet no. 4.*

NIAZ. 2009. Dutch Institute for Accreditation of Hospitals (NIAZ).

6

Methodologies for Quality Improvement

Prathibha Varkey
Mayo Clinic

Anantha Kollengode
Mayo Clinic

Introduction

While the first steps for quality in healthcare are traceable to Codeman in the early 1900s, much of the twentieth century focused on identifying the "odd practitioner" and weeding them out. In the past several decades, other industries such as manufacturing, automotive, aviation, and nuclear have moved to encompass a broader approach to quality including systems thinking, involvement of employees at all levels of the organization, and creating a culture of continuous quality improvement. Healthcare has been late to adopt these approaches. Some of the philosophies commonly used in healthcare quality are summarized below.

Quality Assurance

Historically, healthcare has focused on quality assurance, which emphasizes the evaluation of the delivery of services or the quality of products, and quality control, which is essentially a system for verifying and maintaining a desired level of quality. Unfortunately, these methods used alone are not adequate to enhance outcomes and can often be looked upon negatively by frontline providers, who may feel that they are being "monitored" by leadership. Checking for defects and recommending changes without recognizing the effects of these changes on other parts of the organization may improve one process, but harm others. Finally, providing data to healthcare providers about what is not working or how poorly they are performing, without providing strategies for improvement, is usually not effective in enhancing outcomes.

Total Quality Management

Total quality management (TQM) is an integrated quality approach that is customer-centric and emphasizes data-driven decision processes managed by empowered staff to continually improve processes. TQM encourages experimentation, teamwork, shared leadership, continuous quality improvement, quick responses to customer needs, flexibility, and a long-term strategic perspective into improvement.

Continuous Quality Improvement

Continuous quality improvement (CQI) is a key component of TQM and is based on the principle that an opportunity for improvement exists in every process, every time. Within an organization, it becomes the commitment to constantly improve operations, processes, and activities in order to meet patient needs in an efficient, consistent, and cost-effective manner. The CQI model emphasizes the view of healthcare as a process and focuses on the system, rather than the individual, when considering improvement opportunities (Varkey, Cunningham, and Bisping 2007; Varkey 2010; Varkey, Reller, and Resar 2007).

Quality Improvement Methodologies

While there are several methodologies to achieve quality improvement (QI), we will focus on five commonly used methodologies in healthcare (Adams et al. 2009). The choice of methodology should be based on the nature of the project, among other considerations.

Plan-Do-Study-Act Methodology

Plan-Do-Study-Act (PDSA) is one of the most widely used QI methodologies in healthcare. The methodology promotes small, rapid tests of change to improve a process or product incrementally. In the plan phase, the team agrees on the area and the test for change being implemented, and tasks are delegated to the team members (Figure 6.1a). In the do phase, the small tests of change are carried out and the desired, as well as undesired, outcomes are documented. In the study phase, the data are analyzed to verify if the plan worked as intended and what actions, if any, need to be taken in the next cycle. In the act phase, necessary changes are made to the interventions for the next cycle. If desired outcomes are obtained, the interventions are disseminated. Nolan's Three Question Model is often used along with the PDSA cycle when starting a project (Figure 6.1b).

Varkey et al. used the PDSA methodology to devise a medication reconciliation process in their institution (Varkey, Cunningham, and Bisping 2007). Each PDSA cycle lasted about 24 hours. Each cycle of improvement lasted 24 hours, with changes made to the medication reconciliation process based on lessons learned from each previous cycle. The first iteration entailed the creation of a sheet for collection of medication information from patients. Based on provider and patient feedback, the second iteration modified the form to prompt patient response to a medication list from the most recent visit to the clinic, which significantly enhanced patient participation and efficiency of collection. Other iterations included further modification of the form to make it user- and provider-friendly, academic detailing of providers, and audit and feedback of provider reconciliation performance. At the end of a month, the new medication reconciliation process was standardized and implemented in the clinic. The number of discrepancies decreased from 5.24 to 2.46 per patient (a 50% reduction). The number of medications included in the electronic medical record increased from 47.3% to 92.6% in that period.

Six Sigma

Six Sigma is a data-driven, statistical process to drive process improvements. Sigma refers to the number of standard deviations

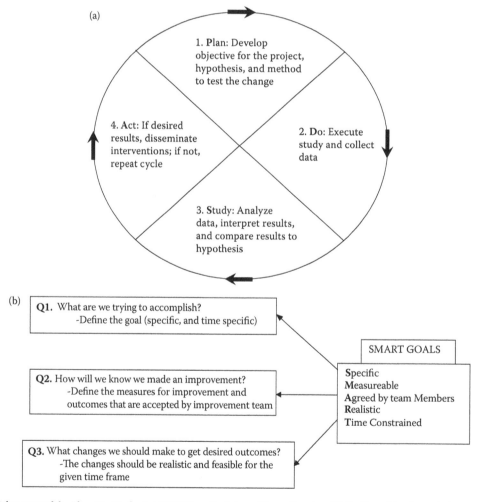

FIGURE 6.1 (a) Schematic of the Plan-Do-Study-Act (PDSA) methodology. (From Deming, W. E., *Out of the Crisis*, MIT Press, Cambridge, MA, 2000.) (b) Nolan's Three Question Model (often used along with the PDSA cycle from Gerald J. Langley, *The Improvement Guide*, Jossey-Bass, San Francisco, CA, pp. 9–10), which may be used when starting a project.

a process is from average performance. A process is said to be at Six Sigma if it is within six standard deviations of average performance, resulting in about 3.4 defects (or errors) per million opportunities (DPMO). At this stage, the process is virtually error-free (99.9996%). Six Sigma is customer-centric and is ideally used to improve existing processes to improve products or services by minimizing or eliminating variation in the process.

Six Sigma utilizes the five interconnected steps of the DMAIC framework, which includes the phases of define, measure, analyze, improve, and control. In the define phase, the strategic objectives are developed and include defining the scope of the project, customer needs, success metrics, timelines, team members, constraints, and assumptions. These are typically captured in the project charter document. In the measure phase, metrics to measure defects are developed to understand the current state of the process and to get an idea of the quality and extent of defects (DPMO) and/or opportunities for improvement (OFI). Inefficiencies, variations, errors, and key cost drivers are identified in this phase. In the analyze phase, the DPMO is measured, the variance of the process from the ideal state is computed, and an in-depth analysis of the sources of variation is performed to identify the underlying problems that adversely affect the outcomes of the project. A variety of descriptive data analysis and statistical tools are used in this phase. In the improve phase, unique solutions that help to improve the process to be efficient, economical, and safe are developed using creative ideas and are tested using small tests of change, simulation, or other means to validate the improvements. In the control phase, systems are developed to ensure that the improvements made are sustained and do not revert back to the old ways. Strategies such as mistake-proofing, standardization of processes and procedures, operating manuals, monitoring, education, and training tools are developed and implemented. The project is then transitioned to the new operational owner for dissemination or institutionalization of the improvements. Some of the common tools used in DMAIC are shown in Table 6.1.

TABLE 6.1 Some Common Six Sigma Tools Used in Different Phases of DMAIC Framework

Phase	Tool	Brief Description
Define	Charter document	Defines the goal, scope, assumptions, constraints, team members, timelines, etc. for the project
	SIPOC + R	Describes the suppliers, input, process, output, customers, and requirement for the process
	Voice of the customer	To understand the needs of the ultimate customer of the process
	Quality function deployment (QFD)	To elicit the voice of the customer
	Goals, roles, process, interpersonal (GRPI)	To articulate the goals of the project and the roles of the team members, as well as to define the processes for decision-making, communication, change management, and the relationships and individual styles of people
	Critical to quality (CTQ)	A tool used to define general customer requirements in specific measurable objectives
Measure	Data visualization tools	Tools such as pareto chart, histograms, run chart
	Process capability	Method of verifying if the process is capable of delivering customer requirements
	Measurement systems assurance	To verify measurement methodologies are valid and verify that they measure what is important for the project
	Control charts	To determine if the process is stable and identify when variation is due to special causes
	Pareto	A visual tool to identify the important data from the rest of the data using the 80/20 rule
	Failure mode effects analysis (FMEA)	Classifies failures by severity, detectability, and frequency of occurrence in order to proactively identify and rectify weak points
	Gauge R&R	Determines the consistency and stability of measurement
Analyze	Cause and effect/Ishikawa/fishbone diagram	Identifies underlying causes for an issue or output
	Data visualization tools	See Measure Phase
	Analysis of variance (ANOVA)	A statistical methodology used to determine if the differences in means of variables are statistically significant (usually based on a sample from the population)
Improve	Affinity diagram	Used to categorize ideas from brainstorming
	Brainstorming	Gathering ideas (emphasis on quantity) for a given problem
	Analysis of variance (ANOVA)	See details in Analyze Phase
	Affinity diagram	See details in Analyze Phase
	Design of experiment (Heydarian, Hoban, and Beddoe 1996)	A series of structured tests used to plan, conduct, analyze, and interpret controlled tests to determine effects of changes made to input variables on changes to predetermined output variables
	Data visualization tools	See details in Measure Phase
	Process capability	See details in Measure Phase
Control	Data visualization tools	See details in Measure Phase
	Control charts	See details in Measure Phase

Taking radiology as an example, sources of variation include image quality, radiation dosage, errors in judgment, anchoring bias, technique, and communication (Fitzgerald 2001). For example, the goal of a Six Sigma project may be to reduce variability and minimize exposure to radiation in a neonatal intensive care unit, while ensuring the quality of images or timely reporting of breast-imaging abnormalities to the referring provider (Cherry and Seshadri 2000). Commonwealth Health Corporation reported that within 18 months of implementing Six Sigma, there was a 27% reduction in radiology cost per procedure, resulting in savings of over $1.65 million, and a 90% reduction of errors in the radiology ordering process. Some other benefits reported include reduced patient wait times (arrival to time to exam), enhanced film jacket retrieval process, decreased MRI turnaround time, reducing patient identification errors through bar-coding, and reduction in staffing by 14 full time equivalents (FTEs) (Cherry and Seshadri 2000).

Lean

Lean, or agile, techniques involve a relentless pursuit of elimination of waste or inefficiencies by eliminating non-value-added activities from a given process. Womack and Jones (2003) outlines a five-step process for organizations interested in lean transformation.

1. Defining value from the end customer point of view.
2. Mapping all steps of the process as it stands at the start of the project (value-stream mapping).
3. Building processes and procedures that keep the customers' needs as the key consideration. For example, if the value-added time (from a patient perspective) at the admissions desk of the radiology department is only 5 minutes, then design the process so that the patient moves to the next step in 5 minutes from arrival at the desk. Womack refers to this as making value-creating step flow toward the customer.
4. Developing processes that take the product (or patient) from the previous step (pull), rather than processing the product and moving it to the next step (push). For example, in a push system, hospital patients are sent to the radiology department and wait to see the next available radiologist. In a pull system, a signal is sent to the nursing floor when the radiologist is available, and the patient will be roomed as soon as she arrives.
5. Pursuing perfection: When all four steps are implemented, profound transformation happens and the process becomes more predictable. For example, in a nonlean state, the wait-times are unpredictable. Following lean, one will be able to predict accurately the time for a patient to go through the entire imaging process. Similarly, with a lean process, errors and bottlenecks come to the surface easily.

TABLE 6.2 Some Commonly Used Lean Tools

Lean Tool	Description	Typical Use of Tool
5S	Sort, set in order, shine, standardize, sustain	To maximize the cleanliness, organization, and safety of all elements in a working environment
Standardized Work	Specific instructions that allow processes to be completed in a consistent, timely, and repeatable manner	To increase production, improve quality, safety, predictability, repeatability, ease of cross-training
Value Stream Mapping (VSM)	A value stream encompasses all the value-added and non-value-added actions required to bring a product or service to a customer	To get a visual representation of every step in the process that the product or service goes through; to develop a baseline for current processes and identify opportunities for improvement
Total Productive Maintenance (TPM)	Program to maximize overall equipment effectiveness and to reduce equipment downtime to zero, while improving quality and capacity	To maximize the productivity of machine life, to minimize unplanned downtime, and to improve productivity
Kaizen	Philosophy of highly focused improvement events designed to achieve orderly and continual improvements to resolve important business issues and/or constraints	To identify real constraints: A need for substantial improvements, lack of plans and responsibilities for longer range improvements, and expectations for bottom-line results
Error- or Mistake-Proofing (poka yoke)	Make it difficult (impossible) to do the wrong thing and easy to do the right thing	To eliminate production losses due to inconsistent processes, methods, materials, etc. (improved quality and cycle-times are nearly always achieved and waste is reduced or eliminated)
Single Minute Exchange of Die (Trump et al. 1961)	SMED processes are highly choreographed and rehearsed to minimize machine downtime	To reduce turnaround time, decrease downtime due to changeover, and increase throughput
Kan Ban	Customer order–driven production schedules based on actual demand and consumption rather than forecasting a signal to move or make an item: No item produced or moved unless there is a kan ban authorizing it	To reduce waste, inventory, work-in-progress, and storage and to reduce the lead time

Some of commonly used lean tools are summarized in Table 6.2.

Through the use of the lean methodology, the Cardiovascular Health Clinic (CVHC) at Mayo Clinic, Rochester, MN, reduced the number of process steps from 16 to 6 and made the process more customer focused by using "pull" scheduling for tests. Open slots were incorporated so that patients could have same-day echocardiograms and stress tests. As a result, no shows at the clinic decreased from 30% to 10%; wait time to get an appointment decreased by 91% (from 33 days to 3 days), while increasing the time patients spent with their physician by 45 minutes (Taninecz 2005).

Lean Six Sigma

More recently, many organizations use a lean Six Sigma blended approach to fast-track improvements and achieve superior results versus using either a lean or Six Sigma methodology alone. Lean tools are used to address issues with flow and waste. Lean Six Sigma is used to leverage the bias for action to discover hidden problems that cause variation. Lean Six Sigma also incorporates the tools from project management and business management principles to provide appropriate resources as the project develops. The lean methodology is presented in Chapter 7, using radiation oncology as an example.

Baldrige Award

Although not a QI methodology, many organizations use the Baldrige Award criteria to initiate organizational improvement. The Baldrige Award criteria were originally initiated in 1987 by Congress for the manufacturing industry to recognize U.S. companies that have implemented successful quality-management systems. Education and healthcare categories were added in 1999; government and nonprofit categories were added in 2007. Award recipients are selected by an independent board of examiners based on achievements and improvement demonstrated in seven criteria for performance excellence: leadership, strategic planning, customer and market focus, measurement, analysis and knowledge management, human resource focus, process management, and business/organizational performance results (NIST 2009).

Commonly Used QI Strategies in Healthcare

Commonly used QI strategies are summarized in Table 6.3. Typically, multipronged strategies work best for improvement as compared with single strategies, which may produce small to moderate improvements (Grimshaw et al. 2003). Re-engineering processes (based on the interventions used) typically have the most impact if they are done the right way.

Summary

The key elements to develop and implement a robust quality improvement program in an organization include patient-centricity, data-driven decision-making, an empowered staff, a pervasive culture of continuous improvement, systematic problem-solving, use of appropriate methodologies, obsession with failure, and a cultural willingness to embrace change and to be uncomfortable before process changes are initiated.

TABLE 6.3   Some Commonly Used Quality Improvement Strategies

Quality Improvement Strategy	Brief Description	Typical Uses
Academic detailing or educational outreach	Experts conducting face-to-face visits with the adopting provider to facilitate the adoption of desired process	To enhance practitioner's knowledge; quick dissemination of best practices (may also be conducted in small focus groups)
Expert and leaders	Tapping into key influencers to obtain desired changes	To harness the thought leaders' expertise, especially in breakthrough or innovation approaches (effective in conjunction with other quality improvement strategies)
Audit and feedback	Obtaining a snapshot of current process effectiveness and flagging opportunities for improvements	Useful for assessments and benchmarking within the organization or from other organizations
Reminders	A flag, either physical or electronic, to ensure key actions/steps are not overlooked	Effective for periodic occurrences (e.g., vaccinations, follow-up appointments, additional services) and best when reminders occur at the point of care
Patient education	To increase awareness and help increase compliance	Helpful in delivery of key information to all patients, consistently in a standardized format
Re-engineering	Process redesign to improve existing processes or develop innovative solutions to solve problems	Useful in creating novel solutions to reduce costs, improve productivity, and eliminate wastes from processes
Incentives	Positive or negative incentives to motivate desired outcomes or for achieving predetermined targets (e.g., pay for performance)	To increase target compliance for positive enforcement of key goals of the organization

References

Adams, R. D., S. Chang, K. Deschesne, S. Freeman, K. Karbowski, D. LaChappelle, D. E. Morris, B. F. Qaqish, J. L. Hubbs, and L. B. Marks. 2009. Quality assurance in clinical radiation therapy: A quantitative assessment of the utility of peer review in a multi-physician academic practice. *Int. J. Radiat Oncol. Biol. Phys.* 75 (S133).

Cherry, J., and S. Seshadri. 2000. Six Sigma: Using statistics to reduce process variability and costs in radiology. *Radiol. Manage.* Nov/Dec: 42–45.

Fitzgerald, R. 2001. Error in radiology. *Clin. Radiol.* 56 (12): 936–946.

Grimshaw, J., L. M. McAuley, L. A. Bero, R. Grilli, A. D. Oxman, C. Ramsay, L. Vale, and M. Zwarenstein. 2003. Systematic reviews of the effectiveness of quality improvement strategies and programmes. *Qual. Saf. Health Care* 12 (4): 298–303.

Heydarian, M., P. W. Hoban, and A. H. Beddoe. 1996. A comparison of dosimetry techniques in stereotactic radiosurgery. *Phys. Med. Biol.* 41 (1): 93–110.

National Institute of Standards and Technology (NIST). 2009. Boulder, CO: NIST.

Taninecz, G. 2009. Best in healthcare getting better with lean 2005. http://www.lean.org/Community/Registered/Article Documents/Mayo_Clinic_Lean_Final.pdf (accessed February 10, 2009).

Trump, J. G., K. A. Wright, M. I. Smedal, and F. A. Salzman. 1961. Synchronous field shaping and protection in 2-million-volt rotational therapy. *Radiology* 76: 275.

Varkey, P. 2010. Basics of quality improvement. In *Medical Quality Management: Theory and Practice*, ed. P. Varkey. Boston: Jones and Bartlett.

Varkey, P., J. Cunningham, and S. Bisping. 2007. Improving medication reconciliation in the outpatient setting. *Jt. Comm. J. Qual. Patient Saf.* 33 (5): 286–292.

Varkey, P., M. K. Reller, and R. Resar. 2007. The basics of quality improvement in healthcare. *Mayo Clin. Proc.* 86 (6): 735–739.

Womack, J. P., and D. T. Jones. 2003. *Lean Thinking. Banish Waste and Create Wealth in Your Corporation*. New York: Free Press.

7

Lean Thinking and Quality Improvement

Christopher S. Kim
University of Michigan

Kathy Lash
University of Michigan

Alexander Lin
University of Pennsylvania

Theodore S. Lawrence
University of Michigan

Introduction

Quality assurance and patient safety are of utmost importance, especially in an era where the care delivered is of increasing complexity and often requires coordination among multiple medical specialties. Despite the best intentions of all care providers, quality of care and adverse events leading to patient safety concerns still abound. The delivery of state-of-the-art radiation therapy is often a complex, multistep process. As a result of the increased complexity of therapy, additional processes have been added in the treatment of each patient. Efforts to establish a consistent approach to continuous quality and process improvement are therefore important. Industries outside of healthcare have pioneered different approaches to quality improvement. One such model, Lean Thinking, has been adopted by many healthcare organizations to improve the quality and efficiency of care. In this chapter, we will describe the basic tenets of Lean Thinking and how our experience in applying Lean Thinking has improved our overall ability to deliver better and more efficient radiation therapy for our patients.

Lean Thinking Management: Approach to Quality Assurance

Lean Thinking is a management model that started in the manufacturing industry. In their book, *The Machine That Changed the World*, Womack, Jones, and Roos (1990) chronicled how Japanese automobile companies, particularly the Toyota Motor Corporation, were able to build better cars with less time and resources. Building on the classic quality improvement method of the Plan-Do-Study-Act (PDSA) cycle, numerous manufacturing companies have incorporated the concepts of Lean Thinking to improve their quality and productivity. Those who have successfully implemented Lean Thinking not only as a means to improve efficiency, but as a corporate philosophy, have become industry leaders. The success of Lean Thinking in the manufacturing industry has led to its adoption by organizations in the service sector, including those in healthcare (IHI 2005).

The fundamental goal of Lean Thinking is to focus on transforming waste into value from the perspective of the customer (Womack and Jones 2003). Waste is defined as anything that does not add value to the final product or service, whereas value is the capability to provide the right product/service, at the right time, and at an appropriate price (Womack and Jones 2003). When patients undergo a course of radiation therapy, they enter into a system that generally operates in a series of processes. However, each successive step in a process may or may not provide the required efficiency and quality of care at all times. Those who are involved in the delivery of care are often specialists on an individual step, and can fail to appreciate the entire care delivery process. As a result, there can be discordance between a patient's expectations of care and the care that is actually provided, which can result in compromised efficiency and quality. We believe that by incorporating Lean principles into healthcare delivery processes, team-centered approaches could be used to improve the whole process and not just to optimize individual parts. This approach could significantly improve the quality, safety, and efficiency of healthcare.

Lean Thinking Approach

The initial step of a Lean Thinking project is to identify a target for improvement in the radiation oncology group. Workers from each step of the process are then asked to observe and analyze the entire series of process steps. This is done to emphasize the idea that meaningful quality improvement initiatives need to directly engage the frontline workers, and take place at the site of practice (Liker 2004). The summary of the current state of activities is then depicted in a standard flowchart called a value stream map (Rother and Shook 2003). The value stream map details

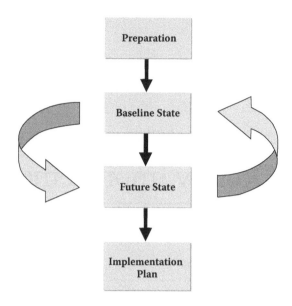

FIGURE 7.1 Once the initial set of improvement ideas has taken shape, further refinements are made to stabilize and further optimize the process through an ongoing continuous improvement cycle. (From Kim et al., *Journal of Hospital Medicine*, 1, 191–199, 2006. With permission.)

how work flows from the beginning to the end of a process, and how work at a particular step may affect work at other points in the process. Workers are encouraged to think about the entire process and how their area of focus fits within the context of the entire process. Modifications are then made from the initial value stream map to create the future state map, which is used as a tool to share and communicate ideas. When done appropriately, waste is minimized, whereas quality, safety, and efficiency are greatly improved.

The ideal outcome of implementing a Lean project is to have every worker who is involved in a process identify problems as they arise and work with others to solve them expeditiously. As individuals work and identify problems, rapid deployment of improvement ideas can lead to dissemination to other individuals doing the same type of work (Spear and Bowen 1999; Spear 2005). Once the initial set of improvement ideas has taken shape, further refinements are made to stabilize and further optimize the process through an ongoing continuous improvement cycle as shown in Figure 7.1 (Allway and Corbett 2002; Kim et al. 2006).

Example 7.1: Urgent Treatment of Palliative Care Cases

At the University of Michigan, Department of Radiation Oncology, we began to implement the principles of Lean Thinking to improve specific care delivery processes, in accordance with our quality mission. One of our initial projects was to improve the treatment for patients with bone and brain metastases. This patient group was chosen because they are typically the most symptomatic, they often have limited survival time (Gaspar et al. 1997), early palliation with radiation can provide

rapid relief of symptoms (Hoegler 1997; Lim et al. 2004; Serafini 2001), and the treatment planning and delivery procedures have a standard that is well accepted by the department faculty and staff. An evaluation of patients referred for bone and brain metastases treatments revealed 27 separate process steps by the radiation oncology staff. Patients could wait up to a week before starting therapy, and only 0.2% of the patients went through the entire process without at least one step requiring some rework (Kim et al. 2007). The department leadership convened a group of staff and faculty from different stages of the care delivery process and asked them to apply a Lean Thinking approach to improve treatment quality and efficiency. Using the value stream map tool, specific areas of waste and delay were identified. These were areas that compromised quality and often required rework. By implementing ideas that modified and, most importantly, standardized patient intake, scheduling, and treatment, the total number of steps required to complete the process was reduced. Having fewer steps allowed for increased focus on each step, ensuring high quality and minimizing rework. By applying these Lean-oriented modifications, patients were able to undergo evaluation, simulation, and treatment initiation within a day from the time of initial referral 95% of the time (compared to 43% before) (Kim et al. 2007). This project led to evaluation of other care delivery processes within our department where we felt quality and safety could be improved.

Example 7.2: Process of CT Simulation

The Lean Thinking approach was then used to examine the process of CT simulation for radiation treatment planning. Evaluation of baseline metrics found that the entire simulation process took 48 steps performed by different individuals who were not always cognizant of the other required steps of the process. Simulator staff was working overtime on 47% of the days and only 22% of the cases were ready to start simulation on time. Interviews with faculty and staff confirmed that hardworking clinical providers were working in a suboptimal arrangement that led to confusion, rework, treatment delays, and, in certain instances, cancelled cases.

As the team probed deeper to identify root causes of the quality and safety issues, they identified an inefficient and fragmented workflow and communication process that required rethinking and restructuring. Problems at the time of simulation included lack of a written informed consent, incomplete counseling or prep for contrast administration, lack of written directives for the simulation, and variability in the method of communication among staff. Standardized documentation and methods of communication were necessary to minimize interruptions and errors. A system of dynamic quality assurance was built into the process, which identified errors early and prevented their propagation downstream. This type of approach encouraged each team member to continuously monitor and develop ideas for quality improvement, and to eliminate waste (Womack and Jones 2003; Spear and Schmidhofer 2005). Forms were standardized with checklists

(a)

**University of Michigan
Health System**

PATIENT ACTIVITY DOCUMENT

Department of Radiation Oncology

If pediatric patient, call Child Life Service

MD initials _____ Date:_____/_____/_____

Patient diagnosis:_____

_____

❑   Interpreter (Language):_____

SIMULATION   FOLLOW-UP

Simulation: ❑ CT Sim ❑ Sim-on-set (60 min) (Ex-1, Ex-2, Ex-3, Ex-4)   FOLLOW-UP____Months____
Body Stereo - Fill out Sim directive only

Notes _____

IV Contrast	❑ Yes	❑ No
If yes, Creatinine within 2 months	❑ Yes	❑ No
Oral Contrast	❑ Yes	❑ No
Scan C	❑	
Volumen	❑	
Sim with full bladder	❑ Yes	❑ No
IV form (have pt complete back page)	❑ Yes	❑ No
IV team needed	❑ Yes	❑ No
Interpreter needed	❑ Yes	❑ No
Pacemaker patient *(If so call 7-7321 for urgent consult)*	❑ Yes	❑ No

Start: Schedule at Sim appt. or requested date: _____/_____/_____
Consent completed (if no, have patient come in 1   ❑ Yes   ❑ No
hour prior to Sim
Sim directive completed. If no, why
_____   ❑ Yes   ❑ No

TO BE FILLED OUT DURING SIMULATION BY RTT

Volumes completed and in Dosimetry by 9 am on:

Date ___/___/____   ❑ Done

Start scheduled:

❑ Next business or available day

❑ On this date _____

❑ Must treat on start date (60 min)

❑ BID: ___ 4hrs___ 6hrs___ 8hrs (RT appt)

❑ BID: ___ duration

❑ Films Thursday, Tx Friday acceptable

❑ Friday Tx acceptable

❑ Films only Friday

Chemo patient:

❑ Chemo must be prior to Tx

❑ Chemo may be pre- or post Tx

❑ Tx with full bladder (add to Tx notes)

Notes _____
_____
_____

Radonc Protocol   ❑ Yes ❑ No
If yes: Protocol No.   _____

Check Radonc Protocol Coordinator below:
❑ Kristin   phone # (pager #)
❑ Kate   phone # (pager #)
❑ Monika   phone # (pager #)
❑ Stacy   phone # (pager #)
❑ Jody   phone # (pager #)

❑ 30 min   ❑ 60 min   ❑ 90 min   ❑ 120 min

❑   IMRT (all machines)
*If IMRT schedule data date 5 business days from volume
date, schedule start 2 business days from data date*
❑   Anesthesia (Ex-3, Ex-4 or 600)
❑   Body stereo (Ex-3, Ex-4) # Fx's   _____
❑   Brain stereo (Ex-1)
❑   Calypso (Ex-1)
❑   Cone Beam (Ex-3, Ex-4)
❑   OBI (Ex-3, Ex-4)
❑   TBI (Ex-3)
❑   TLI (Ex-1, Ex-2, Ex-3, Ex-4)
❑   ABC (Ex-2, Ex-4)

FIGURE 7.2   (a) The Patient Activity Document that is used to collect key information about the patient. (b) Part of the Patient Activity Document also serves other purposes, such as helping to streamline scheduling and ensuring that patients are properly assessed for and counseled about contrast administration before simulation.

(b)

PATIENT QUESTIONNAIRE FOR CT

Department of Radiation Oncology

University of Michigan
Health System

Date: ____/____/_____

Completed by: ☐ Patient   ☐ Family (relationship) _____
☐ UMHS Staff   ☐ Other (specify) _____

DO YOU HAVE A HISTORY OF ANY OF THE FOLLOWING:

	Yes	No	Unknown
Any IV contrast (x-ray dye) allergy?	☐	☐	☐
A severe food or medication allergy?	☐	☐	☐
Asthma, wheezing or breathing problems?	☐	☐	☐
Are you pregnant or nursing (lactating)?	☐	☐	☐
Have you ever had chest pain, angina or heart failure?	☐	☐	☐

Have you had the following medical conditions or procedures:

	Yes	No	Unknown
Renal transplant?	☐	☐	☐
Kidney disease or failure?	☐	☐	☐
Myasthenia gravis?	☐	☐	☐
Thyroid cancer or overactive thyroid?	☐	☐	☐
Diabetes?	☐	☐	☐

Taking Metformin (i.e. Glucophage,

	Yes	No	Unknown
Glucovance, Avandemet, Metaglip)?	☐	☐	☐
Taken Interleukin-2 within the past 2 weeks?	☐	☐	☐
Is it difficult for caregiver to find your veins?	☐	☐	☐
Has an IV team previously been needed?	☐	☐	☐

_____
Signature of person completing this questionnaire

TO BE FILLED OUT BY HEALTHCARE PROVIDER I MEDICAL STAFF

Date of most recent Creatinine test _____/_____/_____

Creatinine level on that date_____

Healthcare Provider_____

Medical Staff_____

This portion will be reviewed and signed by medical staff (SP,PA, or Radiation Oncologist) if creatine level is >1.2 or if "yes" is checked in any of the above boxes.

FIGURE 7.2 (continued)

designed to ensure that information was collected and conveyed simply and clearly. This concept of standardizing a checklist to improve quality and safety was demonstrated in a recent study of a surgical safety checklist in a global population, where there was a reduction of 36% of postoperative complications (Haynes et al. 2009). In our case, the simulation process is now standardized and the need for overtime has been eliminated. The checklists we instituted to improve the simulation process include:

1. Patient Activity Document (Figure 7.2a and b): Used to collect key information about the patient. Simple check boxes are used to denote standard procedures, with space provided to note additional details or exceptions for the simulation. The document also serves other purposes such as helping to streamline scheduling, and ensuring that patients are properly assessed for and counseled about contrast administration before simulation.

2. Physician Simulation Order Form (Figure 7.3): Provides a detailed directive for the actual simulation. The order can be used to select standardized setups or to provide specific details on positioning, use of immobilization devices, and the area to be scanned. Before this standardization of work flow, there were occasions when a physician's specific intent for the simulation was not being clearly communicated to the simulation staff, which led to delay, suboptimal positioning or immobilization, and rework. From this Lean-based initiative, customized simulation order forms are made for each specific anatomic site or malignancy. These forms help improve the efficiency and overall quality of the simulation process.

3. Treatment Planning Directives (Figure 7.4): Used by physicians after simulation to define targets, organs at risk, and specific parameters such as dose per fraction, total dose, energy, and limits for normal structures. This form ensures that all the necessary information is available for

Patient Sticker

General Physician Simulation Orders

Pre-sim
☐ Consent obtained at consult or_____
☐ UM path available in Careweb or (circle one: Outside path report in RT chart/No slides available for review/Not needed per attending)
☐ No IV contrast or _____
☐ Do not page attending prior to immobilization or_____

Patient Set-Up: (check all that apply)
Position: ☐ Supine   ☐ Prone
Immobilization: ☐ Blue Pad   ☐ Thorax board   ☐ Breast board   ☐ 3pt mask
                    ☐ 5 pt mask   ☐ Cradle
Arms: ☐ Up   ☐ At side   ☐ Akimbo   ☐ On chest/abdomen
Legs: ☐ Kneefix   ☐ Toes banded
Head: ☐ Headrest   ☐ Pillow

Additional immobilization & simulation instructions:
(e.g. placement of markers, wires, bite block, pacemaker etc...)
_____
_____

Scan Parameters:
☐ Upper border: _____
☐ Lower border: _____
☐ Slice thickness 3mm or _____
☐ CT reference point: middle of site or _____

Field Parameters: (check all that apply & fill in appropriate information)
Field arrangement: ☐ AP/PA   ☐ AP   ☐ PA   ☐ Opposed Laterals
                      ☐ Obliques
Energy: ☐ 6X   ☐ 16X
Isocenter: ☐ MPD   ☐ 100 SSD (prescribed to desired depth)
Field borders: Superior @ _____, inferior @ _____
               lateral borders @ _____
Structures to be contoured by sim therapist: None or_____
Blocking: None or _____

☐ Page attending ☐ Resident to check scan
Simulation Directive Completed By: _____Date: _____
Attending Physician Signature: _____Date: _____

Simulation Note:
_____
_____

Signature or Attending: _____ Date: _____

Updated: 8/12/08 JSD

FIGURE 7.3 The Physician Simulation Order Form provides a detailed directive for the actual simulation.

Treatment Planning Directive: General Clinical Set Fields

Imaging: Tx planning CT

Field Borders: Set clinically by MD in Simulator

Target(s):	Drawn by	Dataset	Instructions
_____	MD	CT	_____
_____	MD	_____	_____

Normal structs:	Drawn by	Dataset	Param	Planning Limit
_____	_____	_____	_____	_____
_____	_____	_____	_____	_____
_____	_____	_____	_____	_____
_____	_____	_____	_____	_____
_____	_____	_____	_____	_____

Dose: Dose/Fx: _____Gy
       Tot Dose: _____ Gy

Plan: Energy   ☐ 16x   ☐ 6x   ☐ Electron_____
      Dose Calcs   ☐ MPD   ☐ Depth: _____   ☐Customized w/ Beam Wts, Wedges
      Tx Devices   ☐ MLC/block   ☐ Bolus   ☐ hand block   ☐ open fields

Considerations: ☐ Special Treatment Procedures
                  ☐ Previous tx ☐ Pacemaker ☐ Multiple tx sites ☐ Concurrent chemo
    Medical Necessity _____

*(Special Treatment procedures include: Hyperfractionation (BID treatment), planned combination with chemotherapy or other combined modality therapy, radiation response modifier, retreatment of same site, concurrent multiple site treatment, any other special time-consuming treatment plan)

Other Instructions:

_____

_____

_____ **See Script**

Staff Physician: _____   **Date:** _____

FIGURE 7.4 The Treatment Planning Directives used by physicians after simulation to define targets, organs at risk, and specific parameters such as dose per fraction, total dose, energy, and limits for normal structures.

our dosimetrists before initiation of treatment planning, minimizing the need to repeat the planning process due to missing information.

4. Patient Treatment Scheduling Form (Figure 7.5): Filled out by the radiation therapists and given to the patient and a centralized scheduler. The scheduler, working together with the patient, then schedules the remainder of the treatment sessions based on clinic availability and patient requests. This process is designed to be patient-centric, as

we realized that the required therapy for these patients is time-intensive and often could include therapy from other specialties such as surgery or medical oncology.

From the implementation of our initial Lean Thinking projects, we were able to improve the quality and efficiency of the delivery of palliative radiation therapy to patients with bone and brain metastases, and the process of simulation and treatment planning. More importantly, we realized that proper imple-

Patient Treatment Scheduling Form

Patient Sticker

_____ # of Total Treatments

- ☐ Next Business Day
- ☐ M T W T F (circle one)
- ☐ BID 4 6 8 Hrs (circle one)
- ☐ Chemo Pt (afternoon Tx time)
- ☐ Pacemaker Patient (9AM-5PM)
- ☐ Anesthesia
   (see attached schedule)
- ☐ Inpatient
- ☐ Other

Boost Date _____
Boost #2 _____
Boost #3 _____
Final Boost _____
Reassess ☐ ____ # of Treatments known
          ☐ ____ # of Treatments unknown

CHEMO
☐ Chemo prior to Tx
☐ Chemo prior or post Tx

Machine needed for Treatment
Team Maize

- ☐ EX1
- ☐ EX2

Team Blue

- ☐ EX3
- ☐ EX4
- ☐ Or 600CD

Activity	Corresponding Color
☐ ABC	White
☐ ABC IMRT	White
☐ Body Stereo	White
☐ Boost	Purple
☐ Daily Treatment	White
☐ IMRT	White
☐ OBI	Blue
☐ OBI ABC	Blue
☐ OBI Cone Beam	Blue
☐ OBI IMRT	Blue
☐ Other	White

Length of time needed for procedure

- ☐ 10 minutes
- ☐ 15 minutes
- ☐ 20 minutes
- ☐ 30 minutes
- ☐ 40 minutes
- ☐ 60 minutes
- ☐ ___minutes

Activity Category	Activity
TBI	TBI

Created by: _____
Scheduler's initials: _____

FIGURE 7.5 The Patient Treatment Scheduling Form is filled out by the radiation therapists and given to the patient and a centralized scheduler.

mentation of Lean principles could dramatically change the way we provide healthcare and that further initiatives were warranted for continued quality improvement.

Although we have presented some of the benefits of Lean Thinking, the difficulties of implementation should not be underestimated. Physicians are individualists who tend to resist attempts to standardize their work. Healthcare providers have become used to a system of inefficiencies with "work arounds" to bypass problems. They tend to be passive and feel that the system, although inefficient, cannot be changed. Lean Thinking challenges them to fix these issues and work as a team. Implementation can only succeed if the department's physician, physics, and administrative leadership are committed to it. It is necessary to constantly remind people that Lean Thinking does not increase the work; it decreases work by removing waste (especially rework). We have been actively applying Lean principles for nearly 4 years, and we are now only on the verge of having a "Lean culture" in which workers are constantly looking for ways to improve their work. Thus, we feel that we have only begun to scratch the surface of the benefits that Lean Thinking can provide.

Summary

Patients receiving radiation therapy can require high acuity care. This must often be coordinated with other medical subspecialists.

In complex organizations such as healthcare, pathways for care evolve over time with patchwork solutions applied to isolated problems at the expense of examining an entire process that may require major refinement. Efforts to establish a consistent approach to continuous quality and process improvement are therefore important.

Lean Thinking promotes quality improvement via a team-based approach. Lean Thinking concepts and tools such as standard work, visible work flow, value stream mapping, and identifying and eliminating waste will help the entire radiation oncology group strive toward assuring that quality care processes are implemented on an ongoing basis to their patients. The Lean Thinking approach requires all staff to be involved and empowered to identify problems and learn to become problem-solvers at the site of care. By using Lean Thinking as a systematic approach, busy healthcare providers may be able to improve the process of delivering a more optimally designed clinical care pathway (Kim et al. 2007). Individuals interested in learning more about Lean Thinking management and how this management model could help improve their healthcare practices are referred to the reading list in the references section.

References

Allway, M., and S. Corbett. 2002. Shifting to Lean Service: Stealing a page from manufacturers' playbooks. *Journal of Organizational Excellence* 21 (2): 45–54.

Gaspar, L., C. Scott, M. Rotman, S. Asbell, T. Phillips, T. Wasserman, W. G. McKenna, and R. Byhardt. 1997. Recursive partitioning analysis (RPA) of prognostic factors in three Radiation Therapy Oncology Group (RTOG) brain metastases trials. *International Journal of Radiation Oncology, Biology, Physics* 37 (4): 745–751.

Haynes, A. B., T. G. Weiser, W. R. Berry, S. R. Lipsitz, A. H. Breizat, E. P. Dellinger, T. Herbosa et al. 2009. A surgical safety checklist to reduce morbidity and mortality in a global population. *New England Journal of Medicine* 360 (5): 491–499.

Hoegler, D. 1997. Radiotherapy for palliation of symptoms in incurable cancer. *Current Problems in Cancer* 21: 129–183.

IHI. 2005. *GoingLean in Health Care*. Cambridge: Institute for Healthcare Improvement (IHI).

Kim, C. S., J. A. Hayman, J. E. Billi, K. Lash, and T. S. Lawrence. 2007. The application of Lean Thinking to the care of patients with bone and brain metastasis with radiation therapy. *Journal of Oncology Practice* 3: 189–193.

Kim, C. S., D. A. Spahlinger, J. M. Kin, and J. E. Billi. 2006. Lean healthcare: What can hospitals learn from a world-class automaker? *Journal of Hospital Medicine* 1: 191–199.

Liker, J. K. 2004. *The Toyota Way*. 1st ed. Madison, WI: McGraw-Hill.

Lim, L. C., M. A. Rosenthal, N. Maartens, and G. Ryan. 2004. Management of brain metastases. *Internal Medicine Journal* 34: 270–278.

Rother, M., and J. Shook. 2003. *Value-Stream Mapping to Create Value and Eliminate Muda*. 1.3 ed. Brookline, MA: The Lean Enterprise Institute, Inc.

Serafini, A. N. 2001. Therapy of metastatic bone pain. *The Journal of Nuclear Medicine* 42: 895–906.

Spear, S., and H. K. Bowen. 1999. Decoding the DNA of the Toyota Production System. *Harvard Business Review* 77: 97–106.

Spear, S. J. 2005. Fixing health care from the inside, today. *Harvard Business Review* 83: 78–91.

Spear, S. J., and M. Schmidhofer. 2005. Ambiguity and work-arounds as contributors to medical error. *Annals of Internal Medicine* 142: 627–630.

Womack, J. P., and D. T. Jones. 2003. *Lean Thinking. Banish Waste and Create Wealth in Your Corporation*. New York, NY: Free Press.

Womack, J. P., D. T. Jones, and D. Roos. 1990. *The Machine that Changed the World: The Story of Lean Production*. New York, NY: Rawson Associates.

8

Process Control and Quality Improvement

Stephen L. Breen
University of Toronto

Karine Gérard
Alexis Vautrin Cancer Center

Introduction

Statistical process control (SPC) is a method of controlling and improving the quality of a process through statistical analysis. SPC originates from Shewhart's work (Shewhart 1931) in the 1920s, when he suggested using graphical representations of statistical tools to detect the causes of quality drift of a product. SPC has been adopted since the 1960s by Japanese industry (Deming 1986) and, since the beginning of the 1980s, in Europe. In industry, SPC is integrated in a general methodology for dynamic improvement of quality control (Montgomery 1992; Wheeler and Chambers 1992) including management tools like the Six Sigma methodology (Pande 2003).

SPC has been applied to healthcare more recently (Boggs et al. 1998; Fasting and Gisvold 2003; Lighter and Fair 2000); the few published applications in radiotherapy are reviewed in the section "Measurement Accuracy."

Definition of a Process

A process is a collection of related, structured activities or tasks that produce a specific service or product. In radiotherapy quality control, examples include daily measurements of linear accelerator output, in vivo measurements in a series of patients, and intensity-modulated radiation therapy (IMRT) dose verification.

A process can be fully characterized with only three parameters: the shape of its data distribution, the position of its mean value (i.e., its centering), and its dispersion.

Need for Process Monitoring

A common method of quality control individually analyzes each measurement in relation to agreed tolerances. If the result is within the tolerances, the control is validated; if the result is outside the tolerances, the cause of the result is determined and corrective actions are undertaken to restore the result within the tolerances. This is a binary analysis (within/outside the tolerances) that categorizes the current measurement, but does not give information explaining why the result is within or outside the tolerances. The global performance of the process is not improved.

If we take an example of a result located within the tolerances (Figure 8.1), a measurement-by-measurement analysis will validate this result without noticing that it belongs to a process (represented by a normal distribution) whose mean value is not centered on the target and whose dispersion is much larger, implying a large proportion of data outside tolerance. This lack of information arises because each result is individually analyzed without taking into account historical data. Consequently, it is impossible to detect a trend in the evolution of the process, and to anticipate a drift; only an a posteriori analysis can be

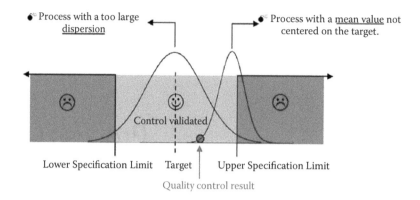

FIGURE 8.1 Scheme showing that an individual analysis of the quality control results does not allow detection of a drift of the mean or a widened dispersion.

performed. To move from an a posteriori analysis to an a priori analysis, it is necessary to use a global process approach that takes into account historical data.

Goal: To Reach the Target with Minimal Variation

There are two kinds of causes of variability that can act on a process: random causes and special causes. Random causes are inherent in the process and are responsible for small variations. Their behavior can be modeled with a normal distribution. Special causes are not inherent in the process and their effects significantly disturb the normal evolution of the process. A process that is subject only to random causes of variation is statistically predictable and statistically in control, that is, its mean and its dispersion are constant over time, within certain limits. A process that is subject to both random and special causes is statistically unpredictable and not statistically in control. The aim of SPC is to make processes in control (Figure 8.2).

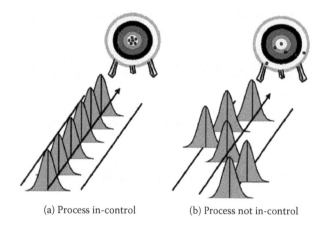

(a) Process in-control   (b) Process not in-control

FIGURE 8.2 (a) An in-control process (only subject to random causes), whose mean value and dispersion are operating at constant levels within the limits statistically defined. (b) A process that is not in control (subject to both random and special causes) leading to a drift of the mean, which is not centered on the target.

Control Charts and Capability Indices: Two Fundamental Tools of SPC

Control charts and capability indices are two tools to make a process in control. Control charts monitor the process over time using statistical control limits to distinguish random variations from significant changes so that special causes can be identified and reduced or eliminated from the process. Control charts allow a real-time analysis, so that drifts can be efficiently detected before the data are outside the specification limits. Capability indices quantify the ability of a process to produce data that are within tolerances at a precise moment.

Exploratory Data Analysis

Tukey (1977) developed exploratory data analysis to explore the nature of measurements by plotting the data and relying on simple statistical descriptors, without the requirement of a priori models or assumptions (Filliben 2006).

Figure 8.3 shows the data for 142 consecutive measurements of daily calibration of a 6-MV x-ray beam. The run-time plot shows that the data are stable over time, with no change in the mean or dispersion. The lag plot shows the data are uncorrelated (i.e., not falling on a line). The histogram shows the distribution while the normal probability plot assesses how the data compare to a particular distribution. In this case, higher values of calibration are measured more frequently than would be predicted by a normal distribution, but the Anderson-Darling test statistic (Anderson and Darling 1954) indicates that the data approximate a normal distribution. Normality is not a prerequisite for control charts, but is required for the calculation of some capability indices.

Measurement Accuracy

To detect changes in a process, the measurement accuracy must be much less than the tolerance and the variation in a process. Measurement devices are often very precise and accurate (e.g., a calibrated ionization chamber and electrometer pair), but precision is reduced when measurements are multiplied together;

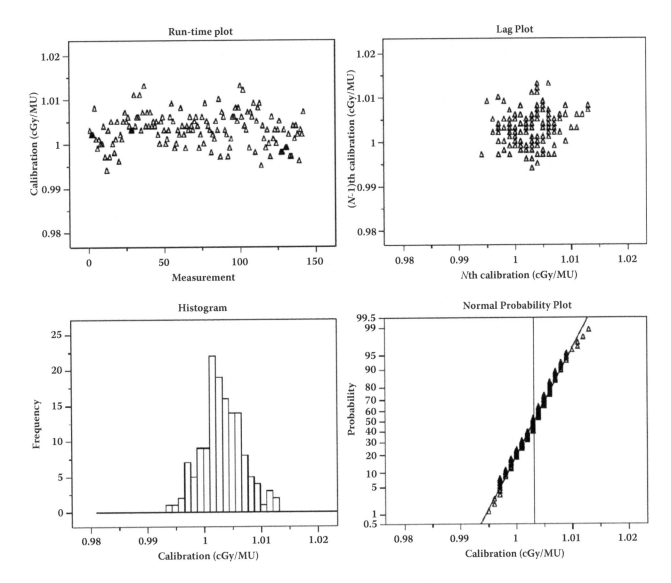

FIGURE 8.3 Clockwise from top left: run time plot, lag plot, normal probability plot, and histogram.

multiple observers may contribute inaccuracies and imprecision. To detect changes in performance, measurement accuracy should be about one-tenth of specification.

Control Charts

A control chart consists of a run-time plot, superimposed with a central line and upper and lower control limits. The center line is μ, the process mean, and the control limits are $\mu \pm k\sigma$, where σ is the standard deviation, and $k = 3$ is an accepted standard.

The collection of data for a control chart falls into two phases. In the initial phase, historical data are analyzed to assess the process's variation and stability and ensure that the process is in control. In the monitoring phase, data are continually added to the control chart to detect special-cause variation.

In the initial phase, subgroups of the data are sampled, and the means and dispersions (standard deviation or range) of the

subgroups are calculated. The value of σ is estimated as the average, s, of the subgroup standard deviations. The control limits for an $\bar{X}$ control chart are $\bar{\bar{X}} \pm \dfrac{3}{c} s$, where $\bar{\bar{X}}$ is the mean, s is the average of the subgroup standard deviations, and c is a constant whose value depends on the number of samples in each subgroup. Tables of c are available in textbooks on SPC (Lighter and Fair 2000; Wheeler and Chambers 1992).

During the monitoring phase, special-cause events must be investigated. Systematic patterns (e.g., data points outside the control limits, and runs of measurements that steadily increase or that repeatedly fall on one side of the central line) have a low probability of occurring when a process is in control; these are likely to have a special cause that should be addressed (Hoyer and Ellis 1996; Wheeler and Chambers 1992).

Analogous plots ($\bar{S}$-plots) can be made of the subgroup dispersion. This can be plotted as the standard deviation or range of the subgroups.

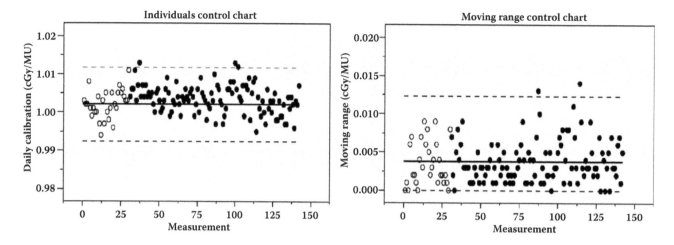

FIGURE 8.4 Individuals control chart (left) and moving range control chart (right) for linear accelerator output data. The moving range is $mR = |M_{i+1} - M_i|$.

Individual Control Charts

Most applications in radiotherapy use individual control charts because measurements such as daily linear accelerator calibration, flatness, or IMRT measurement discrepancies are not amenable to rational subgrouping.

In Figure 8.4, the first 30 data points correspond to the initial phase of the chart, and were used to determine the center line and control limits, based on Table 8.1. During the initial phase, the process is in control; no points fall outside the control limits, and any runs of consecutive measurements on one side of the control line are short. At the beginning of the monitoring phase, a series of 14 points is above the center line, indicating the process is out of control; the cause of this behavior should be investigated. If the output of the linear accelerator were adjusted, it would be reasonable to reinitiate the initial phase of the control chart and to calculate a new center line and control limits.

Moving Range Control Charts

The moving range control chart of Figure 8.4 shows the dispersion of the measurements over time, and is the partner of the individuals control chart. Moving ranges control charts make sudden changes in the process apparent.

TABLE 8.1 Calculations for Individuals and Moving Range Control Charts

	Individuals Control Chart	Moving Range Control Chart
Center line	$\bar{X}$	$\overline{mR}$
Lower control limit	$\bar{X} - \dfrac{3}{1.128}\,\overline{mR}$	0
Upper control limit	$\bar{X} + \dfrac{3}{1.128}\,\overline{mR}$	$3.268\overline{mR}$

Short- and Long-Term Capability Indices

The ability of a process to produce data that meet tolerances can be evaluated in the short term within a small group of data and in the longer term using all the available data. The short-term capability represents the *potential* of the process and is quantified with the capability indices, C_p, C_{pk}, and C_{pm}. The long-term capability represents the process's "real" performance and is quantified with the performance indices, P_p, P_{pk}, and P_{pm}. All these indices are designed to quantify the relation between the stated tolerances and the performance of the process (i.e., its real dispersion, 6σ).

Calculations of the short- and long-term capability indices differ only by the way σ is obtained. For short-term capability indices, σ is calculated for a very short period of time (if the process is stable) whereas for long-term capability indices, it is calculated for a longer period of time, and may include events such as shifts and drifts of the process.

Each index (Table 8.2) has a particular role to play. P_p and C_p compare only the process dispersion to the specifications. P_{pk}

TABLE 8.2 Equations for Capability Indices

Short-Term	Long-Term
$C_p = \dfrac{\text{USL} - \text{LSL}}{6\sigma_{\text{short term}}}$	$P_p = \dfrac{\text{USL} - \text{LSL}}{6\sigma_{\text{long term}}}$
$C_{pk} = \dfrac{\min\left\{(\text{USL} - \bar{X});\,(\bar{X} - \text{LSL})\right\}}{3\sigma_{\text{short term}}}$	$P_{pk} = \dfrac{\min\left\{(\text{USL} - \bar{X});\,(\bar{X} - \text{LSL})\right\}}{3\sigma_{\text{long term}}}$
$C_{pm} = \dfrac{\text{USL} - \text{LSL}}{6\sqrt{\sigma_{\text{short term}}^2 + (\bar{X} - T)^2}}$	$P_{pm} = \dfrac{\text{USL} - \text{LSL}}{6\sqrt{\sigma_{\text{long term}}^2 + (\bar{X} - T)^2}}$

USL and LSL are the upper and lower control limits, respectively. $\bar{X}$ and σ represent the mean and the standard deviation of individual data, respectively.

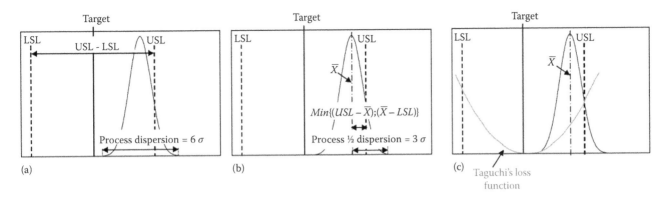

FIGURE 8.5 Example of a process whose dispersion fits within the specification limits but whose mean value is not centered on the target value, implying: (a) P_p (or C_p) > 1, (b) P_{pk} (or C_{pk}) < 1, and (c) P_{pm} (or C_{pm}) < 1.

and C_{pk} are considered more robust than P_p and C_p because of their ability to quantify process centering and dispersion (Kotz and Johnson 2002). P_{pm} and C_{pm} evaluate the position of the process compared to the target value, T, in addition to comparing the centering and the dispersion of the process. P_{pm} and C_{pm} treat discrepancies from the target as harmful for the product quality (Chan, Cheng, and Spiring 1988; Kane 1986; Pillet, Rochon, and Duclos 1997). This consideration is known as "Taguchi's loss function" (denominator of P_{pm} and C_{pm}), which models the loss due to poor process performance as a quadratic function having a minimum at the process target. Figure 8.5 illustrates the calculations of the capability indices.

A process is defined as statistically capable when it reaches a minimal threshold value of its capability index, set at particular values between 1 and 2, depending on the complexity of the problem and the statistical risks involved. A capability index of 1 implies that 0.27% of the data are statistically outside the specification limits and a value of 2 implies that 0.002 parts-per-million of the data are statistically outside the specification limits, for a normal distribution that is centered on the target.

Applications of SPC in Radiation Oncology

Pawlicki et al. (2005) have described the application of SPC to clinical and simulated data for linear accelerator QC. Breen et al. (2008) used SPC with a large series of head and neck IMRT dose measurements, demonstrating that it was possible to maintain IMRT dosimetric verification within tolerance and in control. Pawlicki et al. (2008) reported similar data in a multi-institutional setting and have also compared measurement and secondary calculations as a means of dosimetric verification for IMRT (Pawlicki et al. 2008). Gerard et al. (2009) demonstrated that the analysis of three selected control charts (individuals, moving-range, and exponentially weighted moving average control charts) allowed efficient drift detection of the dose delivery process for prostate and head and neck IMRT treatments,

before the quality control results were outside the clinical specification limits.

Summary

Control charts and capability indices utilize historical data to characterize processes. Performance of a process can be monitored with a control chart to detect and correct drift or shifts in process before they become out of tolerance. Capability indices compare performance of a process to specification. Clinical medical physicists have the task of ensuring quality in treatment delivery; SPC allows physicists to monitor quality to ensure that the performance of radiotherapy processes remains within specification and with minimal variation, without increasing the time devoted to the analysis.

References

Anderson, T. W., and D. A. Darling. 1954. A test of goodness of fit. *J. Am. Stat. Assoc.* 49 (268): 765–769.

Boggs, P. B., D. Wheeler, W. F. Washburne, and F. Hayati. 1998. Peak expiratory flow rate control chart in asthma care: chart construction and use in asthma care. *Ann. Allergy Asthma Immunol.* 81 (6): 552–562.

Breen, S. L., D. J. Moseley, B. Zhang, and M. B. Sharpe. 2008. Statistical process control for IMRT dosimetric verification. *Med. Phys.* 35 (10): 4417–4425.

Chan, L. K., S. W. Cheng, and F. A. Spiring. 1988. A new measure of process capability: Cpm. *J. Qual. Technol.* 20 (3): 162–175.

Deming, W. E. 1986. *Quality, Productivity and Competitive Position.* Cambridge, MA: Massachusetts Institute of Technology Centre for Advanced Study.

Fasting, S., and S. E. Gisvold. 2003. Statistical process control methods allow the analysis and improvement of anesthesia care. *Can. J. Anesth.* 50 (8): 767.

Filliben, J. J. 2009. Exploratory Data Analysis. NIST/SEMATECH, July 18, 2006 [cited February 12, 2009]. Available from http://www.itl.nist.gov/div898/handbook/eda/section1/eda1.html.

Gerard, K., J. P. Grandhaye, V. Marchesi, H. Kafrouni, F. Husson, and P. Aletti. 2009. A comprehensive analysis of the IMRT dose delivery process using statistical process control (SPC). *Med. Phys.* 36 (4): 1275–1285.

Hoyer, R. W., and W. C. Ellis. 1996. A graphical exploration of SPC Part 2: The probability structure of rules for interpreting control charts. *Qual. Prog.* 29 (6): 57–64.

Kane, V. E. 1986. Process capability indices. *J. Qual. Technol.* 18 (1): 41–52.

Kotz, S., and N. L. Johnson. 2002. Process capability indices - A review. *J. Qual. Technol.* 34 (1): 1992–2000.

Lighter, D. E., and D. C. Fair. 2000. *Principles and Methods of Quality Management in Health Care.* Gaithersburg, MD: Aspen Publishers.

Montgomery, D. C. 1992. *Introduction to Statistical Quality Control.* New York, NY: John Wiley & Sons.

Pande, P. S. 2003. *The Six Sigma Way Team Fieldbook: An Implementation Guide for Process Improvement Teams.* New Delhi: Tata-McGraw Hill.

Pawlicki, T., M. Whitaker, and A. L. Boyer. 2005. Statistical process control for radiotherapy quality assurance. *Med. Phys.* 32 (9): 2777–2786.

Pawlicki, T., S. Yoo, L. E. Court, S. K. McMillan, R. K. Rice, J. D. Russell, J. M. Pacyniak et al. 2008. Process control analysis of IMRT QA: Implications for clinical trials. *Phys. Med. Biol.* 53 (18): 5193–5205.

Pawlicki, T., S. Yoo, L. E. Court, S. K. McMillan, R. K. Rice, J. D. Russell, J. M. Pacyniak et al. 2008. Moving from IMRT QA measurements toward independent computer calculations using control charts. *Radiother. Oncol.* 89 (3): 330–337.

Pillet, M., S. Rochon, and E. Duclos. 1997. Generalization of capability index Cpm. Case of unilateral tolerances. *Qual. Eng.* 10 (1): 171–176.

Shewhart, W. A. 1931. *Economic Control of Quality of Manufactured Product.* New York, NY: Van Nostrand.

Tukey, J. W. 1977. *Exploratory Data Analysis. Addison-Wesley Series in Behavioural Science—Quantitative Methods.* Reading, MA: Addison-Wesley.

Wheeler, D. J., and D. S. Chambers. 1992. *Understanding Statistical Process Control.* 2nd ed. Knoxville, TN: SPC Press.

9

Access to Care: Perspectives from a Private Healthcare Environment

Leo van der Reis
University of Missouri

Introduction

The issue of access to medical care is a central concern for healthcare policy formulation. All people deserve access to similar levels and quality of healthcare services. Penchansky and Thomas (1981) identified five dimensions of access to medical services: affordability, acceptability, accommodation, availability, and accessibility. Andersen et al. (1983) claimed that a number of variables must be considered to assess access and that they generally can be divided into two groups. One group consists of those characteristics that influence the utilization of healthcare, such as individual resources like income or insurance that provides the means for using the system. The other group consists of realized indicators of access and includes those measures that reflect actual utilization, for example, physician visits, outcomes of care, patient satisfaction.

Access can be addressed from the customers' perspective or the providers' perspective. Customers' perspective emphasizes fiscal control, with a particular focus on the agencies that function as fiscal intermediaries and enablers such as insurance companies. Providers' perspective concentrates on providers of medical services, such as physicians, nurses, and technicians, as well as on various types of medical care facilities.

Limitations on Access

In practice, two determinants basically determines access: one is financial, the other is geographic. The latter can be subdivided on the basis of pure geographic factors or an admixture with demographic factors.

The financial determinant can be classified according to the conditions of "benefits." At the lowest level, charity is dependent on the goodwill of the community. The next level is the health maintenance organization (HMO), which entails limitations, then the preferred provider organization (PPO), which has limitations that are less strict than the HMOs and give a little more leeway in physician selection. Point of Service (POS) programs include aspects of both HMOs and PPOs, where the service system is not determined until the point of service has been reached. All three, HMOs, PPOs, and POSs are examples of managed care. Finally, there is the traditional "fee-for-service" system, providing choice of physician and hospital. The ultimate form of fee for service is the growing "boutique practice," in which a membership fee is paid to the physician insuring liberal access to that physician's services.

Geographic factors may play a significant role in access. Remoteness of medical facilities is a common form of impaired access due to geography. Demographic conditions such as poverty can, of course, contribute to the threshold of entry.

History of Health Insurance in the United States

In the United States, health insurance is complex, composed of subsystems interacting with each other. A general categorization of distinct insurance subsystems can be public, private, and a combination of the two. Extensive research demonstrates that people with insurance coverage, either private or public, have better access than the uninsured (Weissman and Epstein 1994). Health insurance in the United States began about 200 years ago with hospital care for seamen being paid through compulsory wage deductions (Scofea 1994). The continuous growth of health insurance is linked by many researchers to the growing industrialization of America (Scofea 1994). In 1970, more than % of Americans had some type of hospitalization coverage (Scofea 1994).

Blue Cross and Blue Shield

It was not until hospitals realized that they would benefit from prepaid health insurance plans that such plans began to be offered. The Blue Cross plans were the first actual health insurance plans to be offered on a widespread scale. Prepaid hospital care was mutually advantageous to both subscribers and hospitals, protecting consumers from unforeseen medical expenses while providing hospitals with a guaranteed stream of income.

In 1939, the American Hospital Association officially adopted a blue cross emblem. Blue Shield is the physicians' response to providing prepaid care. In the same year, California Physicians' Service (CPS) began to operate as the first prepayment plan designed to cover physicians' services. During the next decade, the Blue Shield was adopted by the Associated Medical Care Plans, eventually evolving into the National Association of Blue Shield Plans (Scofea 1994).

In 1982, the Blue Cross Association and the National Association of Blue Shield Plans merged, creating the Blue Cross and Blue Shield Association, and growing to coverage of more than 100 million Americans to date (BCBSA 2008). At present, some of the 39 Blue Cross Blue Shield agencies are still not for profit, while others, such as the 14 state WellPoint organizations, have become publicly traded companies. They continue to make available traditional fee-for-service programs as well as PPOs, HMOs, and POS products.

Kaiser Permanente

A major competitor of the Blue Cross Blue Shield group is the Kaiser Permanente organization. Started during WWII as a closed service plan to provide medical care for west coast shipyard workers by Henry Kaiser, it has since grown into a major player in the health insurance field. In California, it has one third of the health insurance market and has expanded into Hawaii and other states. Kaiser still adheres to the closed group program in contract with the Permanente medical group of physicians. In an effort to restrain costs and provide efficient care, Kaiser has taken advantage of implementing IT in its organization, leading to easing access to care for its members/patients.

Health Maintenance Organizations

The legislation establishing HMOs (Nixon 1973), as well as the introduction of managed care, had an impact on access by virtue

of their intent. Limitations on utilization of services (benefits) are an integral part of this mode of healthcare delivery.

The following three issues have emerged as important players influencing healthcare access. First, conducting medical care strictly as a business degrades the doctor–patient relationship by treating patients like customers (Salgo 2006). The unique relationship between patient and physician, fundamental for the practice of good medicine, cannot be equated with the purchase of material goods. The deterioration of the doctor–patient relationship, primarily the result of economic demands from fiscal intermediaries, has clearly had a negative impact on access.

Hospitals have commercialized the practice of medicine through contractual agreements with individual and groups of physicians. For all practical purposes, these agreements are commercial contracts subject to the conditions contained in the contract. When financial concern becomes a major object of hospitals, there is a shift in attitude within the medical profession (Salgo 2006).

Economics of Healthcare

The business plans of most hospitals and fiscal agencies have become geared to close scrutiny of the bottom line of efficiency. The increase in the overall cost of medical care encourages minimizing the use of facilities that do not make a profit or at least break even. This pressure has had the effect of narrowing the options that exist for a patient to be treated in a hospital. The economics of the situation favors increasing admission of new customers, but at the same time shortening lengths of stay. In the meanwhile, the greater use of emergency room facilities has made timely access to physicians a mirage. Publicly traded HMOs, for example, began restricting doctors to an average seven-minute "encounter" with each customer (Salgo 2006) and consequently reduced the doctor–patient relationship to a financial concept. Such performance measures that relate only to the cost concern and not to quality of care do not contribute to the practice of good medicine.

Improving Access to Care

Ease of access and free flow of patient movement are conducive to the satisfaction of both the patient and the provider of services. As soon as a patient passes the access threshold, all hands should be on deck to insure seamless travel through the intricacies of scheduling in today's complex scenario. Scheduling becomes a moot question if access is impeded. In order for access to be effective, a scheduling system needs to be integrated into patient care. All components of a care system should be included in the scheduling program: human resources, facility resources, provider data entry and retrieval, patient data entry and retrieval, and intrasystem as well as external communication.

It is essential that the scheduling/access model be geared to the needs of the care providers and the needs of the patients. This requires a great deal of insight into the practice of the potential user(s) of a system by the developer of the access/scheduling model, that is, recognition of the true needs of the caregivers and patients. Regrettably, it seems that all too often there is a disconnect between the system developer and the users of the system, a situation that has caused a great deal of misery and the waste of large amounts of time, energy, and money. Fortunately, there seems to be a trend toward greater emphasis on user–developer compatibility. After expenditures that run into many millions of dollars that did not provide the return on investment that was anticipated, both developers and users are belatedly coming to the conclusion that no one truly benefits from a situation in which the developer and the users are not fully attuned and cognizant of the real needs of the system. It is clear that implementation of IT hardware and software that can communicate unimpeded through the healthcare system is a *sine qua non* for modern and efficient medical care. Ready access, universal access, and error prevention are but a few outstanding benefits that are derived from a thoughtful informatics network.

Example of an Improved System for Access to Care

One model that does offer a system that fulfills the criteria for access/scheduling of a medical facility is DAVID (Visser and Rupert 2008), a database application for holding a patient's entire medical history, along with comments from various health care providers, which can be customized by each provider to display only the data of interest to them ("What You See Is What You Need").

In this model (Figure 9.1), the pathways to access and flow through the system are clearly delineated and can be followed seamlessly, insuring intercommunicability as well as provider and patient satisfaction (Visser 2008). Because a patient's entire medical history is available at the next point of service, DAVID facilitates differential diagnosis and speeds up selection of treatment. By virtue of this directional component, the patient is

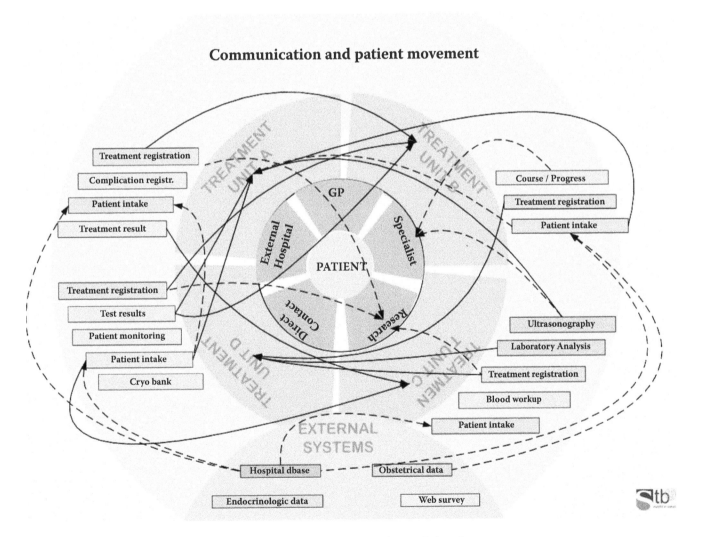

FIGURE 9.1 Communication and patient movement—DAVID concept applied to a radiology department.

more likely to reach the appropriate source for treatment or diagnosis. The model has been implemented at the University of Utrecht Medical Center's IVF department. A modification has been developed for use by a radiotherapy department (van der Reis et al. 2008).

Summary

Ease of access sets the stage for satisfaction of patient and provider alike. Success depends on the ability to navigate the system and reach providers and service facilities. This implies that the master plan for all medical care facilities, large and small, should include an up-to-date design for access and patient movement that relies heavily on informatics.

References

Andersen, R. M., A. McCutcheon, L. A. Aday, G. Y. Chiu, and R. Bell. 1983. Exploring dimensions of access to medical care. *Health Serv. Res.* 18: 50–74.

BCBSA. 2008. About Blue Cross Blue Shield [Web page]. Blue Cross Blue Shield Association [Accessed December 2008]. Available from http://www.bcbs.com/about/.

Nixon, R.M. 1973. Health Maintenance Organization Act of 1973. 42 U.S.C. § 300e.

Penchansky, R., and J. W. Thomas. 1981. The concept of access: Definition and relationship to consumer satisfaction. *Med. Care* 19: 127–140.

Salgo, P. 2006. The doctor will see you for exactly seven minutes. *New York Times*, March 22, A27.

Scofea, L. A. 1994. The development and growth of employer-provided health insurance. *Mon. Labor Rev.* 117: 3–4.

van der Reis, L., J. L. Rupert, and J. H. Visser. 2008. DAVID concept applied to a radiology department. Houten, The Netherlands: STB Automatisering and Advies BV.

Visser J. H., and J. L. Rupert. 2008. *DAVID: Database Application with a Variable Interface Design.* Houten, The Netherlands: STB Automatisering & Advies BV.

Weissman, J. S., and A. M. Epstein. 1994. *Falling through the Safety Net: Insurance Status and Access to Health Care.* Baltimore, MD: The Johns Hopkins University Press.

10

Access to Care: Perspectives from a Public Healthcare Environment

Sara C. Erridge
University of Edinburgh

Introduction

One of the principal requirements for a safe and effective radiotherapy service is sufficient capacity, in terms of both staff and equipment, to meet the clinical need. Understaffing and overworked machines have been cited as contributory factors in a number of incidents (Furlow 2009). In many countries, investment in radiotherapy decreased through the later years of the twentieth century, in part due to funding being diverted to systemic and novel therapies, resulting in under provision of radiotherapy capacity.

Consequences of under Provision

It is clear that in situations where there is a lack of radiotherapy capacity, fewer patients receive this treatment. Clinicians and patients are deterred by long waits for treatment or excessive travelling and, consequently, opt for alternative, potentially inferior, treatment options. Much work has been conducted trying to estimate the optimal proportion of cancer patients who should receive radiotherapy. Most authors conclude that around 50% of cancer patients need radiotherapy [Scotland 44–48% (Erridge et al. 2007), Australia 52–53% (Delaney et al. 2005)], but the exact requirements for any given population will depend on the case mix, stage at presentation, and patients' suitability for alternative treatments, such as surgery. For example, a population with a high rate of ovarian cancer but a low rate of head and neck cancer will need less radiotherapy capacity than one where the converse is true. Therefore, it is important that planning of

radiotherapy capacity should be conducted on a local level rather than simply adopting figures from other regions of the world.

Not only will the types of cancer in a population have an impact, but also the stage at presentation; generally patients with more advanced disease require radiotherapy rather than surgery, often to palliate symptoms. If the population is elderly, for example, the Scottish lung cancer population, fewer will be suitable for surgery and, hence, a greater proportion will be treated with radiotherapy (Erridge et al. 2008).

Long Waits

Waiting for treatment is a highly emotive issue, understandably causing significant anxiety to patients and their carers. Queen's Cancer Research Group, Ontario published a systematic review of the evidence in 2008. They identified only 44 studies of reasonable caliber, more than half of which were in breast cancer patients and the rest in head and neck or sarcoma patients. They concluded from these limited studies that delays to treatment had a detrimental effect, particularly on local control, and they felt this was likely to impact on overall survival, although the head and neck studies could only show a marginal impact (Chen et al. 2008).

Staff under Pressure

When there is lack of radiotherapy capacity, staff will inevitably feel pressured to plan and treat more patients in a working day. Stress, fatigue, and the requirement to perform routine tasks without sufficient breaks are all known contributory factors to human error (RCR 2008).

Slow Introduction of New Techniques

Radiotherapy has changed dramatically over the past 20 years but, when capacity is limited, it is difficult for staff to have time to develop new treatment protocols and have access to linear accelerators to work up new techniques. This has been cited as a cause for the slow introduction of IMRT in the UK (Jefferies, Taylor, and Reznek 2009).

Inability to Enter Clinical Trials

Clinical trials in radiotherapy are one of the most effective methods of ensuring a consistent quality of care (Moore, Warrington, and Aird 2006; Venables et al. 2006). The benefits of training and the development of a defined treatment protocol and quality assurance remain long after the trial has closed. However, in order to perform clinical trials to a good standard, clinical research organizations, such as the EORTC and RTOG, set minimum standards required for a center to be accredited (see, e.g., Table 10.1; Budiharto et al. 2008).

Methods of Estimating Optimal Capacity

So how can we estimate the radiotherapy capacity required now and in the future? There are three main approaches that have been used.

Linear Accelerators per Million Head of Population

This was one of the first measures used to compare infrastructure between countries (Lote et al. 1991; Barton, Frommer, and Shafiq 2006) and within countries (RCR 2003). Wide discrepancies have been observed in the European Union (EU) in 2004: Sweden had 6.8 linear accelerators per million, compared with

3.2 in England and 1.9 in Poland (Bentzen et al. 2005). Inevitably, those countries with fewer linear accelerators treat a lower proportion of patients. In order to calculate the optimal rate across the EU, the ESTRO QUARTS team combined the crude incidence of 23 cancers in each country, the optimal radiotherapy rate for each tumor site calculated by CCORE (see below), a 25% retreatment rate, and a complexity factor (e.g., 1.54 for head and neck, 0.7 for Hodgkin's, and 0.38 for colorectal) to reflect short course regimens. Then, using an average of 450 courses of radiotherapy per linear accelerator per year they estimated the requirement for each country. This ranged from 4.0 per million for Cyprus to 8.0 per million for Hungary, with figures of 7.0 for Sweden, 6.5 for the UK, and 5.0 for Poland.

Modeling

This is a bottom-up approach that was initially developed by Queen's Cancer Research Group, Ontario (Tyldesley et al. 2001). In 2005, the Australian Collaboration for Cancer Outcomes Research and Evaluation (CCORE) produced models for all cancer sites (Delaney et al. 2005), which have been subsequently adapted for a Scottish (Erridge et al. 2007) and an English population (Williams et al. 2007). In this technique, the cancer population is subdivided by factors that influence management decisions, for example, primary cancer site, stage, and performance status. These data are obtained, wherever possible, from the population for which the model is being developed, enabling the proportion of the patients who fit into each treatment group to be calculated. If the data are not available for the specific population, then data are obtained from other sources but must be, wherever possible, population-based data. A root and branch diagram is then built up (see Figure 10.1) and the percentage of patients requiring radiotherapy calculated. The optimal treatment rates for a Scottish population are shown in Table 10.2.

The main challenge of this technique is obtaining good population-based data on stage and performance status. Also, some management decisions can be very subtle and, hence, impossible to model. A prime example of this is the selection of chemotherapy or palliative radiotherapy for a patient with advanced lung cancer. The models suggest that all patients with

TABLE 10.1   Current Requirements for Accredited Radiotherapy Centers for EORTC Trials

FTE radiation oncologist per dept	Min 3
Number of patients treated per yr per FTE radiation oncologist	Max 250
Number of FTE-qualified radiation physicists	Min 2
Number of patients treated per year per FTE radiation physicist	Max 500
Number of radiation technologists per treatment unit	Min 2
Simulator (classical and/or CT scanner)	Min 1 (access to CT scanner)
Megavoltage units (preferably <10 years old)	Min 2
Patients per year per MV unit (in 8-hour day)	Max 600
Patients per year per conventional simulator	Max 1200
Patients per year per CT simulator	Max 2400

Source: http://groups.eortc.be/radio/MinimumRequirements.htm.

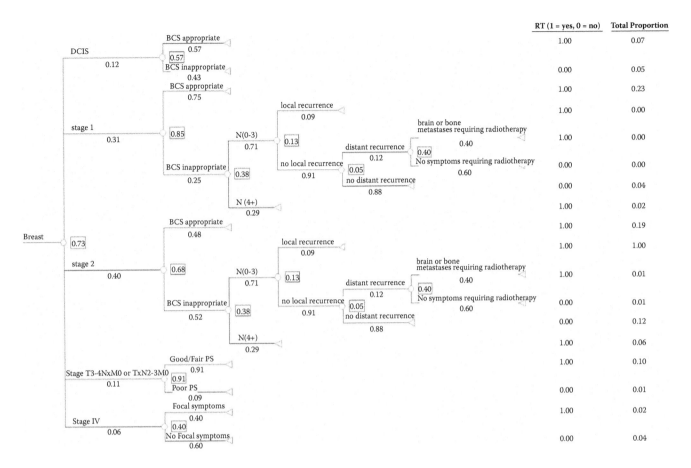

FIGURE 10.1   Scottish model for optimal radiotherapy for breast cancer.

chest symptoms, around 60% in one study (Gregor et al. 2001), should receive radiotherapy, but chemotherapy can also palliate chest symptoms and, unlike low dose palliative radiotherapy, may prolong survival. Which treatment is selected will be influenced by patient choice, general fitness, and other symptoms. This is the most likely reason that, though the models suggest an optimal radiotherapy rate of around 42% in Canada (Barbera et al. 2003), 64% in Scotland (Erridge et al. 2007), and 76% in Australia (Delaney et al. 2003), clinical audits usually suggest the actual treatment rates are considerably lower (Erridge et al. 2008; Jack et al. 2003).

Once the optimal treatment rate has been established for each cancer site, it is then possible to calculate the capacity required in terms of number of fractions per annum. In the Scottish model, all Scottish Radiation Oncologists supplied the fractionation schedule for each clinical scenario (Erridge et al. 2007), whereas in the English model a systematic review of fractionation was used (RCR 2006). This information is then applied to each scenario, multiplied by the number of cases in the population per annum, and then combined to give a grand total. For Scotland, which has a population of 5 million, a total of 195,000–256,000 fractions were required per annum (see Table 10.2), which equates to 39,000–51,000 fractions per million head of population. The English estimate was 54,000.

The advantage of quoting radiotherapy capacity as fractions per million head of population is that local working practices can be used to convert the figures into the number of linear accelerators required. The Scottish models examined the impact of different working patterns on the number of linear accelerators required, for example, length of the working day and number of public holidays/service days on the linear accelerator requirement. If the centers treat 5 patients per hour (what is actually currently delivered in Scotland, but higher than other studies, which have used 4–4.5 per hour) for 8 hours a day, 230 days per year, then each linear accelerator can deliver 9200 fractions per annum.

To ensure that waiting times are minimized, it is recommended that capacity of any service exceeds demand by 10% (Steyn, Silvester, and Lenden). So, based on standard working, this figure translates to a requirement of between 4.7 and 6.1 linear accelerators per million head of population for Scotland. If, practices however, a 10-hour working day was adopted (felt to be maximum acceptable to patients and staff) and nontreatment days minimized, then each machine can deliver 11,200 fractions per annum, reducing the requirement to 3.8–5.0 linear accelerators per million head of population. The disadvantage of this calculation is that it fails to recognize that there will need to be matching resources, especially staff, to run the machines for longer periods each day.

TABLE 10.2 Scottish Models Showing Optimal Treatment Rates, Fractionation (#) Schedules (Current and Future), and the Total Number of Fractions Required

	Optimal RT Rate	Current # for Radical Treatments	Current Total # Required[a]	Future # for Radical Treatments	Future Total # Required
Head and neck	78.6	34	26,724	35	34,370
Esophagus	53.9	20–25	7400–9505	20–25	10,670–13,705
Colon	1.0	20–25	35–4850	25	43–6245
Rectum	27.8–88.7[b]	5–25	6620–7720	25	8991–9931
Lung	62.8	20–36	27,247–39,261	24–39	34,698–41,382
Melanoma	15.7	20[c]	2035	25[c]	3850
Breast	70.0	20–27[c]	48,445–60,216	5–25[c]	40,290–70,590
Cervix	56.1	20–25	3810–4740	25	3185
Corpus uteri	46.3	20–25	3830–4740	25[c]	5810–5825
Prostate	61.4	20–41	14,516–29,195	32–41	30,930–39,426
Testis	46.0	15[c]	1350	0–15[c]	0–1830
Bladder	28.2	20–25	7183–9135	25–30	10,950–12,970
CNS	60.7–81.9[c]	28–32	5272–6818	28–32	6090–8010
Hodgkin's disease	71.4	15–20	1350–1800	15	1440
NHL	54.4	15–20	6780–9040	10–20	9090–12,555
Stomach	13.4	25[c]	195–4260	25	160–2805
Pancreas	41.9	25	6150	25	7390
Ovary	4.0	–	125	–	160
Leukemia	4.0	10	240	10	320
Kidney	24.0	–	212–1310	–	1060
Myeloma	33.1	–	500	–	595
Other and unspecified[d]	44.2–47.9	22	16,855–18,198	36	29,962–30,396
Deferred treatment[e]	44.2–47.9	–	7475	–	9522
Benign conditions[f]		15	934	15	1190
Total			195,283–256,321		248,766–318,752

[a] Includes radical, postoperative, and palliative fractions.

[b] Variation due to possible use of preoperative RT in all or only selected patients.

[c] Variation depends on whether all patients who have undergone an operation are fit for radiotherapy or only a proportion.

[d] Represent 10.5% of cancers, assuming 44.2–47.9% use chemotherapy with a distribution of 52% of courses having radical intent, 20% local palliative, 28% metastatic with current fractions of 22, 7, and 4, respectively. To take into account the general trend toward more prolonged fractionation, 30, 10, and 6 fractions are projected in 2011–2015.

[e] Taken as 4% of total fractions using data from NHS Radiotherapy Episodes Statistics.

[f] Taken as 0.5% of total fractions, as seen in the 2003 Scottish audit.

Benchmarking

This approach, adopted from industry, has been used recently by the Ontario team to look at the impact of capacity and travel on radiotherapy utilization. The premise is that if there is optimal access, multidisciplinary working, and no financial barriers or incentives, then all patients suitable for a particular treatment will receive it. In a study comparing lung cancer treatment rates across Ontario and the United States, those communities in Canada felt to meet the benchmark criteria had treatment rates of 41% compared with 23–43% in other regions of Ontario, where capacity was lower and travel times longer, and 43–61% in SEER counties in the United States, where financial incentives might have an impact.

Predicting the Future

Cancer epidemiology is not static and the types of tumors occurring in a community will change with differences in lifestyle (e.g., smoking rates, obesity, and diet), age, or racial profile. In Scotland, where there is an increasingly elderly population but a dropping smoking rate, the number of cases of cancer is predicted to rise by 18.9% over the period of 1996–2000 to 2011–2015 (Erridge et al. 2007). This is primarily due to a predicted increase in the number of patients with breast cancer (+23%), prostate cancer (+35%), endometrial cancer (+23%), non-Hodgkin's lymphoma (+50%), but also with a decline in the number of patients with lung cancer (−10%) and cervix cancer (−33%).

If the changes in the number of cases are entered into the models above, then the predicted requirement for radiotherapy in Scotland in 2011–2015 will be 248,000–319,000 (50,000–64,000 fractions per million head of population).

Summary

These methods concentrate particularly on the number of radiotherapy machines required, but it is important to appreciate that it is the complete radiotherapy service that must match the demand. The EORTC requirements (Table 10.1) clearly show that not only must there be sufficient machines, planning, equipment, etc., but also staff from across the multidisciplinary team. Often the purchasing of new equipment is relatively straightforward, but recruiting staff is the main hindrance to progress (Jefferies, Taylor, and Reznek 2009). The shortage of radiation physicists and radiation technologists is worldwide and, despite a number of initiatives, does not seem to be improving yet. It is important that professional bodies, training organizations, and governments collaborate to ensure that sufficient staff are trained now and in the future to meet the radiotherapy service capacity required.

References

Barbera, L., J. Zhang-Salomons, J. Huang, S. Tyldesley, and W. Mackillop. 2003. Defining the need for radiotherapy for lung cancer in the general population: A benchmarking approach. *Med. Care* 41 (9): 1074–1085.

Barton, M. B., M. Frommer, and J. Shafiq. 2006. Role of radiotherapy in cancer control in low-income and middle-income countries. *Lancet Oncol.* 7 (7): 584–595.

Bentzen, S. M., G. Heeren, B. Cottier, B. Slotman, B. Glimelius, Y. Lievens, and W. van den Bogaert. 2005. Towards evidence-based guidelines for radiotherapy infrastructure and staffing needs in Europe: The ESTRO QUARTS project. *Radiother. Oncol.* 75 (3): 355–365.

Budiharto, T., E. Musat, P. Poortmans, C. Hurkmans, A. Monti, R. Bar-Deroma, Z. Bernstein et al. 2008. Profile of European radiotherapy departments contributing to the EORTC Radiation Oncology Group (ROG) in the 21st century. *Radiother. Oncol.* 88 (3): 403–410.

Chen, Z., W. King, R. Pearcey, M. Kerba, and W. Mackillop. 2008. The relationship between waiting time for radiotherapy and clinical outcomes: A systematic review of the literature. *Radiother. Oncol.* 87 (1): 3–16.

Delaney, G., M. Barton, S. Jacob, and B. Jalaludin. 2003. A model for decision making for the use of radiotherapy in lung cancer. *Lancet Oncol.* 4 (2): 120–128.

Delaney, G. P., S. Jacob, C. Featherstone, and N. B. Barton. 2005. The role of radiotherapy in cancer treatment: Estimating optimal utilisation from a review of evidence-based clinical guidelines. *Cancer* 104: 1129–1137.

Erridge, S. C., B. Murray, A. Price, J. Ironside, F. Little, M. Mackean, W. Walker, D. H. Brewster, R. Black, and R. J. Fergusson. 2008. Improved treatment and survival for lung cancer patients in South-East Scotland. *J. Thorac. Oncol.* 3 (5): 491–498.

Erridge, S. C., C. Featherstone, R. Chalmers, J. Campbell, D. L. Stockton, and R. Black. 2007. What will be the radiotherapy machine capacity required for optimal delivery of radiotherapy in Scotland in 2015? *Eur. J. Cancer* 43 (12): 1802–1809.

Furlow, B. 2009. Radiotherapy errors spark investigations and reform. *Lancet Oncol.* 10 (1): 11–12.

Gregor, A., C. S. Thomson, D. H. Brewster, P. L. Stroner, J. Davidson, R. J. Fergusson, and R. Milroy. 2001. Management and survival of patients with lung cancer in Scotland diagnosed in 1995: Results of a national population based study. *Thorax* 56 (3): 212–217.

Jack, R. H., M. C. Gulliford, J. Ferguson, and H. Moller. 2003. Geographical inequalities in lung cancer management and survival in South East England: Evidence of variation in access to oncology services? *Br. J. Cancer* 88 (7): 1025–1031 [erratum appears in *Br. J. Cancer* 2004 91(10): 1852].

Jefferies, S., A. Taylor, and R. Reznek. 2009. Results of a national survey of radiotherapy planning and delivery in the UK in 2007. *Clin. Oncol.* 21 (3): 204–217.

Lote, K., T. Möller, E. Nordman, J. Overgaard, and T. Sveinsson. 1991. Resources and productivity in radiation oncology in Denmark, Finland, Iceland, Norway and Sweden during 1987. *Acta Oncol.* 30 (5): 555–561.

Moore, A. H., A. P. Warrington, and E. G. Aird. 2006. A versatile phantom for quality assurance in the UK Medical Research Council (MRC) RT01 trial in conformal radiotherapy for prostate cancer. *Radiother. Oncol.* 80: 82–85.

Royal College of Radiologists (RCR). 2003. *Equipment, Workload and Staffing in UK 1997–2000.* London: Royal College of Radiologists.

RCR. 2006. *Radiotherapy Dose-Fractionation.* London: Royal College of Radiologists.

RCR. 2008. *Towards Safer Radiotheraphy.* London: Royal College of Radiologists.

Steyn, R. S., K. Silvester, and R. Lenden. NHS Modernisation Agency. (www.steyn.org.uk). Accessed 3/26/2009.

Tyldesley, S., C. Boyd, K. Schulze, H. Walker, and W. J. Mackillop. 2001. Estimating the need for radiotherapy for lung cancer: An evidence-based, epidemiologic approach. *Int. J. Radiat. Oncol. Biol. Phys.* 49 (4): 973 Royal College of Radiologists 85.

Venables, K., E. A. Miles, E. G. Aird, and P. J. Hoskin. 2006. What is the optimum breast plan: A study based on the START trial plans. *Br. J. Radiol* 79: 734–739.

Williams, M. V., E. T. Summers, K. Drinkwater, and A. Barrett. 2007. Radiotherapy dose fractionation, access and waiting times in the countries of the UK in 2005. *Clin. Oncol.* 19 (5): 273–286.

11

Cost of Quality: Health Technology Assessment

Yolande Lievens
University Hospital
Leuven, Gasthuisberg

Peter Dunscombe
University of Calgary

Introduction

Over the past several decades, developed countries have been devoting an ever-growing proportion of their gross domestic product (GDP) to healthcare. According to the latest report of the Organization for Economic Cooperation and Development (OECD 2006), the annual growth rate in healthcare spending varied between 1.7% and 10.1% over the past 10 years. With a health expenditure of $6102 per year per person, the United States largely outspent the other countries surveyed (OECD 2006). Obviously, such cost increases are not sustainable in the long run and priorities will have to be set regarding both the share of society's resources to devote to healthcare and which interventions to implement and fund, taking cost and effectiveness into consideration.

Among the many reasons that have been put forward to explain the continuous rise in healthcare costs, the rapid diffusion of new technologies has been proposed as a major contributing factor (Bodenheimer 2005). Radiation oncology, a highly technological discipline, is obviously not immune to this general tendency of increasing costs. Alongside the expanding imaging and information–communication applications, novel treatment strategies such as intensity-modulated (IMRT), image-guided (IGRT) and stereotactic radiotherapy (SRT), to name only a few, have rapidly gained acceptance and established their place in daily practice. The cost consequences of this technological and clinical process evolution remain largely unexplored.

Even if radiotherapy's share of the total oncology budget is very modest (in the order of 5%) (Norlund 2003), particularly when considering the large proportion of cancer patients being treated with radiation (Delaney et al. 2005), resource limitations and tightening budgets urge the radiotherapy community to develop an interest in healthcare management, decision-making, and budgeting in close collaboration with government and insurance organizations.

In this chapter, we will briefly review the principles of both health technology assessment and economic evaluations, outline the current state of knowledge of costs in radiotherapy, and conclude with a discussion of activity-based costing as a tool to support health technology assessment in radiation therapy with a special emphasis on new emerging technologies and their associated quality assurance.

Health Technology Assessment

Health technology assessment (HTA) plays an essential role in modern healthcare by supporting evidence-based decision-making in policy and practice. HTA involves formulating answers to five different questions (Detsky and Naglie 1990):

1. Can a healthcare intervention achieve its expected goal when used in optimal circumstances? (Efficacy)
2. Does the intervention do more good than harm when used in routine practice? (Effectiveness)
3. What is the balance between the health outcome obtained and the resources required to deliver the intervention? (Efficiency or Cost-Effectiveness)
4. Is the supply of services matched to locations where they are accessible to persons that need them? (Availability)
5. Who gains and who loses by choosing to allocate resources to one healthcare program instead of another? (Distribution)

A statistically valid answer to the first question can be obtained through a clinical trial of appropriate level. The second question considers the specificities of daily clinical practice, such as the variability in patient characteristics and compliance, as well as potential differences in available resources and in treatment techniques. Fairly rigorous analytical methodologies can be brought to bear on the third question. Such methodologies are grouped under the heading of economic evaluations (EE) and are described briefly in this chapter. The last questions regarding where to introduce a new treatment and how to allocate funding among different treatment strategies by society entails ethical, jurisdictional, and social considerations besides purely economic and budgetary issues. Nevertheless, knowledge of the costs of healthcare interventions remains the cornerstone of both EE and HTA.

Economic Evaluations in Healthcare

Economics is the science of limited resources. It is implicit that an economic evaluation, to be of any value, will lead to decisions on the use of those resources. There are various levels of sophistication of economic evaluation as have been described and discussed in detail in the comprehensive text by Drummond et al. (2005). A full economic evaluation typically involves the quantitative examination of costs and outcomes or consequences, using an appropriate metric of competing interventions, possibly including no intervention. As can be imagined, full economic evaluations in radiotherapy and other branches of medicine are very difficult to perform. Fortunately, we can make some progress in understanding the implications of current and new technology and processes using more limited approaches.

The first two of these more limited approaches address costs and outcomes separately and are called cost and outcome descriptions, respectively (Drummond et al. 2005). Much of the clinical literature in radiotherapy deals with outcome descriptions. For example, in the early days of intensity modulated radiation therapy, there was a wealth of literature describing the improved dose distribution and organ sparing that was (in principle) achievable when the ability to modulate radiation beams in two directions relaxed the constraints on gantry angle without compromising dose uniformity in the target. Outcome descriptions generally were quite often related to dose volume histograms in terms of dosimetric quantities. More useful descriptions of outcome of a new treatment strategy, particularly from the patient's perspective, result from the conduct of Phase 1 and 2 clinical trials. A fully randomized and controlled Phase 3 trial essentially involves the comparison of outcome descriptions of two or more interventions. Such studies have been categorized as efficacy or effectiveness evaluations (Drummond et al. 2005). However, it can be argued that such comparisons may indirectly lead to the increased cost of healthcare, as funders come under pressure to broadly adopt new interventions with only marginal clinical benefit and no consideration of the opportunity cost. This observation emphasizes the importance of economic evaluation in healthcare resource allocation.

On the surface it would appear that cost descriptions, based as they are on the metric of monetary units, should be easier to develop than outcome descriptions. At its simplest level, a cost description could be developed by one analyst sitting at a desk compiling cost data. However, for a cost description to be useful to others it has to be clear on several issues.

The first of these issues is whose costs are included. Most cost analyses address the costs incurred by the provider of the service for a specific service. For example, publications in radiotherapy tend to be confined to analyzing implications for the radiotherapy budget. Yet, radiotherapeutic interventions increasingly occur in conjunction with surgery and/or systemic therapy. Some interventions such as low dose rate brachytherapy for gynecological malignancies require hospitalization. It is important to be clear on the boundaries of the cost analysis.

What about real and imputed patient costs? Are they and should they be included? Centralizing services, which can result in economies of scale for the service provider, can increase the costs borne by the patient particularly in areas of low population density. Some studies have adopted the societal perspective as an aid to decision-making on the distribution of services (Dunscombe and Roberts 2001). Again, it is important to be clear on the perspective of the analysis.

Looking just at the institutional costs incurred by the service provider, the next issues to be explicit about in a cost analysis are exactly which resources have been included and how. Lack of clarity here has made the interpretation and comparison of published work difficult. It is relatively straightforward to calculate the total salary costs within a radiation treatment program. However, many of us interested in such analyses work in academic centers and have responsibilities beyond purely clinical activities. Is this fact acknowledged and explicitly accounted for?

A challenge related to securing adequate capital budgets for radiotherapy is that the equipment and facility necessary to provide the service appear to be very expensive to funders and the public alike. A proper cost analysis deals with this issue through amortizing equipment over a set period of time—perhaps 10 years for a linear accelerator—and applying an annuity factor that is based on interest rates. A simple calculation yields the equivalent annual cost of the equipment, which can be interpreted as the (equal) amounts that would have to be paid yearly to an investor to pay off the initial cost over the amortization period, assuming no resale value. Such estimates can then be introduced in cost calculation algorithms, developed with the aim of allocating investment costs to the actual product or service.

However, in spite of this perception of high investment costs, it is the personnel cost that dominates the cost of the provision of the service with capital investment typically representing less than 30% of the total radiotherapy budget (Lievens, Kesteloot, and Van den Bogaert 2003; Norlund 2003; Perez et al. 1993). Although analysts broadly agree on the division of costs between infrastructure and process (Ploquin and Dunscombe 2008) many details are missing. For example, how do we account for the fact that specific equipment, such as electronic portal imaging, may be used more or less intensively depending of the complexity of the treatment? And how does the initial purchase cost of different pieces of equipment translate into the final treatment cost of each individual treatment? We will suggest below that these issues can be addressed through activity-based costing.

A functioning radiotherapy clinic requires more than just clinical staff and equipment. There are a variety of costs that are sometimes described as overhead. These include services that are frequently shared with other healthcare programs, such as housekeeping, human resources, and financial administration. Are these included, and if so how are they apportioned? Health economists have varying views on allocating overhead, so there is no agreed upon approach. Again, it is important to be clear if and how this was done in the cost analysis.

From the above discussion, it is apparent that there are different levels of completeness in a cost analysis. There are also different levels of uncertainty in the inputs used. A cost analysis that is going to be useful to others needs to acknowledge these input uncertainties through a sensitivity analysis. Inputs are varied individually or together over ranges that are considered reasonable and the impact on the cost estimate is presented. For example, if labor costs are a significant fraction of total costs, as they usually are in healthcare, then the final cost estimate will be very sensitive to this input. Cost estimates generated in one jurisdiction will not be transferable to another with a very different salary and staffing structure.

Full economic evaluations calculate the extra cost ensuing from a new treatment strategy compared to a standard alternative and divide this by the gain in outcome obtained with the new treatment. This results in the so-called incremental cost-effectiveness ratio (iCER). Such full analyses are relatively rare in radiotherapy. Apart from the aforementioned problems arising from the definition of the treatment cost, especially in the case of rapidly evolving technology and practice, it is even more difficult to estimate the denominator. This latter difficulty arises partly from the changing cancer treatment landscape with technology advances and increasing multimodality treatments, which can quickly make the strategy under study obsolete, and partly from the inherent success of radiotherapy in curing cancer and prolonging life, which makes it hard to factor out the impact of a better local treatment on improved survival that may only become apparent many years after treatment. This is, however, necessary for a cost-effectiveness analysis, which ideally employs life years gained (LYG) as the denominator.

Assuming a well-defined comparison of two radiotherapy approaches and a sufficiently long follow-up, prolongation of life can be measured but it is not necessarily the metric that has most relevance to patients. Two years spent in serious pain would not be regarded by most as a better outcome than 1 year in good health. Cost-utility analyses acknowledge the quality of life following an intervention by weighting the LYG with a utility factor ranging from 0 (death) to 1 (perfect health) and using quality adjusted life years (QALYs) as the denominator. Examples of cost-utility analyses in radiotherapy are comparative studies of two approaches to breast cancer treatment—whether to use postoperative radiotherapy; to treat the internal mammary chain (Dunscombe, Samant, and Roberts 2000; Lievens, Kesteloot, and Van den Bogaert 2005b)—and a study comparing the use of

a continuous hyperfractionated regime to conventionally fractionated radiotherapy in lung cancer (Lievens, Kesteloot, and Van den Bogaert 2005a). As can be imagined, assigning a quality factor to a health state is fraught with difficulty. However, it appears that using QALYs as opposed to LYGs has had little impact on the allocation of resources for cancer treatments (Tengs 2004).

The term cost benefit analysis is used in an all-embracing fashion to include much of what we have described here. In fact, it has a very specific meaning—that the outcome or consequence of the intervention is measured in monetary units. This approach has the theoretical advantage of allowing a direct comparison between input costs and savings. However, putting a dollar value on a quantity and quality of life is a highly specialized and quite controversial area. Regardless of our subjective assessments of what we would pay to prolong our lives, societies have implicitly made decisions on what they are prepared to pay. Here too, important variations exist among and within countries. By examining the costs and outcomes of those interventions that have been implemented and those that have not, it has been estimated that one QALY appears to be worth about US$50,000 in the United States (Weinstein 2008), £20,000 to £30,000 in the UK (NICE 2004), and Can$20,000 to Can$100,000 in Canada (Laupacis et al. 1992).

A cautionary note: while it would be advisable to consider economic evaluations whenever new radiotherapy equipment and strategies are developed, as they may support public policy, they are not necessarily the dominant factor in decision-making. For those of us who have radiation safety or licensing as part of our duties, the landmark cost-effectiveness study by Tengs (2004) is an eye-opener. The disparity between the amounts society spends to save 1 life year through toxin control (including radiation safety), medical interventions, and preventive measures is enormous.

Cost of Radiotherapy

As mentioned above, the cost of radiotherapy is often perceived as being high due to its technological environment, which requires significant financial investment in treatment machines and buildings. However, available data have shown that even in countries with optimal infrastructure, the budget consumed by radiotherapy comprises roughly only 5% of the total amount spent on cancer care (Norlund 2003). In the early nineties, the European Union estimated the average cost per course of radiotherapy at about €3000, compared to roughly €7000 and €17,000 for surgical and chemotherapy treatments, respectively. A comprehensive analysis of published data has arrived at a cost of radiotherapy within 10% of this European Union estimate (Ploquin and Dunscombe 2008). Although these calculations confirm that radiotherapy is not expensive in comparison with other treatment modalities, more accurate data on the actual resource costs of radiotherapy treatments are needed. Unfortunately, such data remain scarce, especially when it comes to the cost evolution that goes along with technological evolution.

Two reports give us some hint of radiotherapy cost evolution over the past decades. The Swedish Council on Technology Assessment in Healthcare (SBU) computed an inflation-corrected 16% increase in the total cost of external radiotherapy in Sweden between 1991 and 2000. Due to increased automation and computerization, however, the number of fractions increased even more, rendering external radiotherapy delivery overall more efficient (Norlund 2003). More recently—based on an analysis of the limited costing literature from high-income countries—the real annual increase in radiotherapy cost per patient was estimated at 5.5% over the past 15 years (Ploquin and Dunscombe 2008). The improved efficiency apparent from the Swedish study was not reflected in the literature analysis. Intuitively, one can easily understand that radiotherapy costs will evolve with time, in parallel with the technical evolution of our treatments. It is important to note, however, that the cost estimates presented in the two publications discussed above were made well before the recent and widespread introduction of more sophisticated treatment techniques such as IMRT, IGRT, and SRT. Although the cost consequences of these novel technologies can be expected to be significant, accurate cost estimates are, to date, virtually nonexistent.

Even after correcting for the time factor, the considerable variation between the computed costs in different studies is significant. This variation can, in part, be attributed to the different salary and cost structures of the countries of origin of the studies, making the interpretation of the computed treatment costs across country borders a delicate exercise. On top of that, the lack of a standardized costing methodology—institutional versus societal perspective, cost per patient versus cost per fraction, the inclusion of different cost components and radiotherapy activities—renders the comparison between the analyses extremely hard to do. Such difficulties have been discussed earlier in this chapter. In order to arrive at solid cost data for current and new radiotherapy techniques and strategies, the choice of an appropriate evaluation methodology is clearly crucial.

Activity-Based Costing

A technique that is very useful in evaluating the budgetary impact of new technologies and strategies in radiotherapy is activity-based costing (ABC). ABC was developed with the aim of capturing the economic consequences of product complexity (Baker 1998). In this approach, the course of treatment that a patient receives is made up of a basket of services (activities), each of which is costed, most frequently on the basis of time consumption. Thus, the cost of a course of treatment will depend on whether image guidance is included, for example.

Besides a more accurate calculation of the treatment costs, the use of ABC provides a better insight into the (cost-) structure of the treatment and of the RT department as a whole, which makes it suitable for evaluating the budgetary impact of new technologies and process changes. The major advantage of ABC is indeed found in the fact that it allows inclusion or exclusion of particular steps in the process, such as IMRT or portal verification, for

which the financial impact can then be studied. Different fractionation schedules can, for example, be modeled without relying on the very poor approximation that treatment cost scales linearly with the number of fractions. Also, an ABC model could be used to quantify the financial impact of the increased QA efforts resulting from the current emphasis on patient safety.

An example of such a modeling exercise was performed with an ABC cost-calculation model, developed in 2000, of the Leuven radiotherapy department (Lievens, Kesteloot, and van den Bogaert 2003). In the process of adapting the model to the evolving radiotherapy standards, a sensitivity analysis was performed evaluating the financial consequences of the extra time burden ensuing from more advanced treatment techniques such as IMRT and more frequent QA procedures using electronic portal imaging (Van de Werf et al. 2009). It was found that when using IMRT instead of 3D-CRT for prostate and head and neck irradiations, relative cost increases of roughly 30% can be expected. The impact of more rigorous QA would be even more considerable, with costs of IMRT, along with daily portal checks, being roughly 70% to 80% higher than the baseline 3D-CRT technique with field checks on the first treatment day. Furthermore, if higher doses and/or hyperfractionated regimes were adopted, the costs would almost double compared to the baseline costs. These examples underscore the dominant impact of total treatment time (i.e., daily treatment time × number of fractions) on the total cost picture, resulting from the use of expensive equipment and personnel involved with the day to day treatment.

Another example of ABC is the analysis of Ploquin and Dunscombe (2009), in which cost outcome estimates of image-guided patient repositioning for prostate cancer patients have been made. Different correction protocols have been examined for both incremental cost and incremental benefit, using a dosimetric measure of benefit. Within the constraints of the methodology used and assumptions made, this study should provide guidance to both purchasers and users of IGRT. Depending on the imaging protocol used, image-guided patient repositioning can be relatively expensive for a meager dosimetric gain. Similarly to the Leuven study, ABC allowed the inclusion or exclusion of certain activities, in this case related to image guidance.

Both examples demonstrate that ABC is particularly well-suited for the study of incremental costs and savings of process and technology changes.

Summary

To improve the service we provide to radiotherapy patients in a manner that is economically responsible to society as a whole, we need a far better understanding of the costs and benefits of new technologies and treatment strategies (Baumann, Holscher, and Zips 2008). Estimating costs, particularly with the aid of ABC, is not difficult, although, as stressed above, specification of the perspective of such estimates and the boundaries of the calculation are essential if the results are to be of value. Estimating cost effectiveness is appreciably more difficult, although possible within limits. The community could usefully engage in a discussion aimed at developing a consensus on the optimum approaches to health technology assessment and economic evaluation in radiotherapy.

References

Baker, J. J. 1998. *Activity-Based Costing and Activity-Based Management for Healllthcare.* Gaithersburg, MD: Aspen Publishers.

Baumann, M., T. Holscher, and D. Zips. 2008. The future of IGRT—cost benefit analysis. *Acta Oncol.* 47 (7): 1188–1192.

Bodenheimer, T. 2005. High and rising health care costs: Part 2. Technologic innovation. *Ann. Intern. Med.* 142 (11): 932–937.

Delaney, G., S. Jacob, C. Featherstone, and M. Barton. 2005. The role of radiotherapy in cancer treatment: Estimating optimal utilization from a review of evidence-based clinical guidelines. *Cancer* 104 (6): 1129–1137.

Detsky, A. S., and I. G. Naglie. 1990. A clinician's guide to cost-effectiveness analysis. *Ann. Intern. Med.* 113 (2): 147–154.

Drummond, M. F., M. J. Sculpher, G. W. Torrance, B. O'Brien, and G. L. Stoddart. 2005. *Methods for the Economic Evaluation of Health Care Programmes.* Oxford: Oxford University Press.

Dunscombe, P., and G. Roberts. 2001. Radiotherapy service delivery models for a dispersed patient population. *Clin. Oncol. (R. Coll. Radiol.)* 13 (1): 29–37.

Dunscombe, P., R. Samant, and G. Roberts. 2000. A cost-outcome analysis of adjuvant postmastectomy locoregional radiotherapy in premenopausal node-positive breast cancer patients. *Int. J. Radiat. Oncol. Biol. Phys.* 48 (4): 977–982.

Laupacis, A., D. Feeny, A. S. Detsky, and P. X. Tugwell. 1992. How attractive does a new technology have to be to warrant adoption and utilization? Tentative guidelines for using clinical and economic evaluations. *CMAJ* 146 (4): 473–481.

Lievens, Y., K. Kesteloot, and W. Van den Bogaert. 2005a. CHART in lung cancer: Economic evaluation and incentives for implementation. *Radiother. Oncol.* 75 (2): 171–178.

Lievens, Y., K. Kesteloot, and W. Van den Bogaert. 2005b. Economic consequence of local control with radiotherapy: Cost analysis of internal mammary and medial supraclavicular lymph node radiotherapy in breast cancer. *Int. J. Radiat. Oncol. Biol. Phys.* 63 (4): 1122–1131.

Lievens, Y., W. van den Bogaert, and K. Kesteloot. 2003. Activity-based costing: A practical model for cost calculation in radiotherapy. *Int. J. Radiat. Oncol. Biol. Phys.* 57 (2): 522–535.

Norlund, A. 2003. Costs of radiotherapy. *Acta Oncol.* 42 (5–6): 411–415.

OECD. 2009. OECD in Figure 2006–2009. Available from http://www.oecdobserver.org/news/fullstory.php/aid/1988/OECD_in_Figures_2006-2007.html.

Perez, C. A., B. Kobeissi, B. D. Smith, S. Fox, P. W. Grigsby, J. A. Purdy, H. D. Procter, and T. H. Wasserman. 1993. Cost accounting in radiation oncology: A computer-based model for reimbursement. *Int. J. Radiat. Oncol. Biol. Phys.* 25 (5): 895–906.

Ploquin, N., and P. Dunscombe. 2009. A cost-outcome analysis of image-guided patient repositioning in the radiation treatment of cancer of the prostate. *Radiother. Oncol.* 93(1): 25–31.

Ploquin, N. P., and P. B. Dunscombe. 2008. The cost of radiation therapy. *Radiother. Oncol.* 86 (2): 217–223.

Tengs, T. O. 2004. Cost-effectiveness versus cost-utility analysis of interventions for cancer: Does adjusting for health-related quality of life really matter? *Value Health* 7 (1): 70–78.

Van de Werf, E., Y. Lievens, J. Verstraete, K. Pauwels, and W. Van den Bogaert. 2009. Time and motion study of radiotherapy delivery: Economic burden of increased quality assurance and IMRT. *Radiother. Oncol.* 93 (1): 137–140.

Weinstein, M. C. 2008. How much are Americans willing to pay for a quality-adjusted life year? *Med. Care.* 46 (4): 343–345.

12

Past, Present, and Future of Quality in Radiotherapy

Pierre Scalliet
University Hospital St Luc

Introduction

Since consciousness emerged in humans many hundreds of thousands of years ago, the sense of danger was an important part of it. There was danger in the environment from wild neighbors (mammoths, tigers, etc.), danger from climatic convulsions, and danger from other humans. Managing this ubiquitous danger was a condition for survival and life propagation. Quite obviously, humans have been good (perhaps too good) at managing danger in their daily lives.

But danger management was not part of an elaborate behavior; it was merely a reaction to current conditions in a given place at a given time. Also, the essence of danger contained a mystical element. The gods or spirits were challenging humans by introducing complications into their daily lives. Danger was an impediment, an external factor, located outside the human mind, rather than a part of it.

A systematic approach to danger needed a leap of the mind toward understanding its many characteristics: level of danger, risk of danger (probability), categorization of these risks, etc. This deep evolution of the mind came in the seventeenth century with the description of the first laws of probabilities. This is how danger became risk. Risk is danger captured in numbers. It is a measurement of danger. And because it can be measured, it can be prospectively managed.

We speak today of technological risks because ways have been discovered to measure probabilities of undesirable outcomes in complex activities. An application of this is the management of quality in radiotherapy. Quality is the desired outcome of radiotherapy treatments, measured according to preestablished standards, and deviations from quality standards are treated as risks of nonquality or nonconformity to expected outcomes. Thus, quality can be captured in probabilities of nonquality, the goal of quality management being to minimize nonquality. It should be clear to all workers in the field that risk can be minimized but not eliminated.

As soon as ionizing radiation was discovered and applied to cancer treatment, measuring tools were developed to systematize the therapeutic advantage and minimize the risk of side effects. The following list gives examples of how the process of incorporating quality elements evolved over time.

Emergence of Systematic Quality Control Procedures

- The development of measurement equipment and the calibration of beams were the first steps toward mastering the medical use of ionizing radiation. This further developed in more elaborate QC of irradiation equipment (geometric stability, measurement of accessories) and, later on, on imaging equipment that is used for tumor imaging [computed tomography (CT), magnetic resonance imaging (MRI), positron emission tomography (PET)]. Clinical research in radiotherapy has experienced tremendous development with the advent of large, multicentric trials managed by the European Organization for Research and Treatment of Cancer, Radiation Therapy Oncology Group, and others. The publication of comparative beam calibrations between various research centers was a landmark. It revealed that large variations could exist between departments as to the dose actually delivered by the equipment (Horiot et al. 1986). This, later on, extended to the entire field of radiotherapy, outside the research set up, to routine clinical activity. Various organizations have taken over the regular auditing of equipment on a national or international basis (IAEA).

- Introducing simulators to relate the treatment area to internal bony landmarks refined the earlier technique of drawing fields on the patient's skin, a method relying on the anatomical knowledge of organ and tumor projection on the patient's surface. Simulators appeared concomitantly with megavoltage units (first telecobalt, then linear accelerators), and contributed enormously to the development of modern radiotherapy. With the dissemination of 3D-conformal irradiation techniques, there is a trend to replace simulators with CT scanners, and to move away from bony landmarks as a reference for volume definition. Indeed, CT (and MRI and PET) allows the definition of tumor regions (GTV, CTV) and organs at risk (OAR) directly in a volumetric manner, instead of indirectly from bony landmarks.
- Portal imaging, first with films (gammagraphy) and later with electronic portal imaging, permitted verification that the areas defined at simulation were effectively covered by the treatment beams. It also demonstrated that day-to-day variations exist in a fractionated treatment. This led to the concept and definition of the PTV.
- In vivo dosimetry has been developed and is promoted as the simplest and most efficient end-to-end test for individual QC of a treatment. In many countries, it is a legal requirement to proceed with in vivo dosimetry at the first treatment session. It does not apply, however, to recent developments in treatment techniques like intensity-modulated radiation therapy (IMRT) or rotational therapy (VMAT, RapidArc, tomotherapy) since there is no longer any use of static, uniform beams that can be simply measured at the entrance on the patient skin. More elaborate and indirect verification techniques are currently being developed for the verification of these particular techniques.
- Complex quality control (QC) of radiotherapy techniques are currently being developed (asymmetric phantoms, phantoms with density heterogeneities, etc.) to control for equipment performances and accuracy of dose calculations in complex treatments (nonuniform dose distributions, IMRT, rotational therapy), as well as for verification of the information quality of imaging that is used as a source for treatment preparation and dose calculation.
- Record and verify systems have been developed as a mean of securing data transfer from planning systems to treatment machines. It allowed the development, in a secure environment, of techniques for the transfer of many thousands of numerical parameters from the planning to the treatment unit.

Lessons from Accidents

It must be acknowledged that many of the QC elements have developed in reaction to accidents. In a sense, accidents are lessons on the insufficiencies of quality control, and should be considered as such from a quality management perspective. Accidents contributed both to the improvement of equipment

safety and to the indication of what kind of tests would give confidence that an expected outcome would be safely reached.

Emphasis should be put on the educational element. In reviewing a large number of accidents, the IAEA identified in virtually all of them a lack of proper training of the operating staff. This has been true for radiation oncologists, as well as medical physicists and radiation technologists (RTTs).

Emergence of Total Quality Management

All the elements that are listed above can be taken separately, addressing each element in the chain that links cancer diagnosis to cancer treatment. Early in the 90s, triggered by several reports of radiotherapy accidents, the need for a comprehensive approach to quality management emerged (Van der Schueren 1993).

The central idea is that every QC intervention on equipment or processes should be meshed in a comprehensive system that encompasses all steps, from the moment a patient enters the department until the moment that person leaves it (Thwaites et al. 1995).

The issue of quality in radiotherapy is dominated today by this concept of a comprehensive quality system covering all aspects, clinical (diagnosis, treatment indications, and procedures) and technical (obtaining and maintaining means and materials). It closely follows the philosophy of ISO 9001, with an exception regarding standards. In ISO 9001, the operator himself selects standards to which an activity is compared. On the contrary, standards in radiotherapy are, to a certain extent, internationally rather than locally defined and always aim at the best possible result.

Technology Helps

Tomotherapy, the Cyberknife, and hadron therapy permit very complex and elaborate treatment techniques. A lot of automation is included, by design, in these types of equipment/techniques. However, a common lesson is that automation does not provide complete safety of the radiotherapy process.

Equipment can fail in an infinite number of ways, and automation can never cover the entire spectrum of risk for malfunction. Therefore, although they have been designed to self-manage many of the risks, they eventually put more pressure on departments to elaborate more and more complex QA programs. The lesson of the past is that no completely safe system exists and that quality assurance programs have a complexity level that should mirror the complexity of the technique that needs to be monitored.

Future

Integration of radiotherapy in oncology, taking both surgery and medical oncology on board, is the next frontier. Today, many departments have developed, or are engaged in developing, a comprehensive quality management program. Tomorrow, these programs will need to be extended toward other cancer departments in the same hospital and across similar departments in different hospitals. The rational for a broader quality system comes from the developing treatment practices where patients receive different, concomitant, or sequential treatments for the same disease. Optimal outcomes in this context are conditioned by the existence of an approach to quality that crosses the radiotherapy department boundaries. This is the challenge facing radiotherapy.

References

Horiot, J. C., K. A. Johansson, D. G. Gonzalez, E. van der Schueren, W. van den Bogaert, and G. Notter. 1986. Quality assurance control in the EORTC cooperative group of radiotherapy. 1. Assessment of radiotherapy staff and equipment. European Organization for Research and Treatment of Cancer. *Radiother. Oncol.* 6 (4): 275–284.

Thwaites, D., P. Scalliet, J. W. Leer, and J. Overgaard. 1995. Quality assurance in radiotherapy. European Society for Therapeutic Radiology and Oncology Advisory Report to the Commission of the European Union for the 'Europe Against Cancer Programme.' *Radiother. Oncol.* 35 (1): 61–73.

Van der Schueren, E. et al. 1993. Quality assurance in cancer treatment. Report of a Working Party from the European School of Oncology. *Eur. J. Cancer* 29: 172–181.

13

Past, Present, and Future of Quality in Radiotherapy Physics

William R. Hendee
Medical College of Wisconsin

Introduction

Quality assurance has been an evolving concept in radiation therapy since the first time that radiation was used to treat cancer and other assorted maladies. Recognizing the evolution of quality assurance in radiation therapy is a prerequisite to understanding the present practice of radiation therapy and medical physics and to preparing for the future. In this brief chapter, we provide an overview of the concept of quality assurance in radiation therapy in the past, present, and future.

Past

Ionizing radiation was discovered and characterized in the laboratories of physicists, and some of the earliest experiments of applying radiation to humans were conducted by physicists. The use of radiation to diagnose and treat a spectrum of diseases and disabilities was explored by physicians and physicists over the first two decades of the twentieth century with little regard to quality assurance and radiation dose. It was not until short-term (skin burns and epilation) and longer-term (cancer induction) effects were noted that concern arose over the amount of radiation used and the medical justification for its use. In 1928, the roentgen was defined as a unit of exposure and rather simplistic measures began to be taken to limit the amount of radiation to which employees and, to a lesser extent, patients were exposed. Nevertheless, radiation therapy continued to be mainly an empirical practice for several decades.

All of this changed with the advent of 60Co teletherapy in the 1950s. When this technology with its skin-sparing advantage was introduced into radiation therapy, physicians lost the ability to judge the response to treatment by watching for reactions in the patient's skin. This loss required physicians to rely on responses predicted from computations of the internal distribution of radiation dose compiled from depth-dose data and isodose curves.

Physicists were the individuals who could acquire these data, and for the first time they became an essential part of the clinical practice of radiation therapy. The importance of physicists was accentuated in the 1970s when linear accelerators were introduced as replacements for 60Co machines. Linear accelerators are highly complex treatment machines that require careful calibration and continuous monitoring to ensure that they are functioning properly. The calibration process includes interpretation of radiation treatments delivered as a number of "monitor units" in terms of the actual radiation dose delivered to tumors and normal tissues. The monitoring process is termed quality assurance and with the introduction of linear accelerators, it became an essential ingredient in the use of radiation to treat cancer.

Many additional technologies have been introduced into radiation therapy since the 1970s, including computer-based treatment planning systems, image-guided therapy employing computed tomography, positron emission tomography, single-photon emission computed tomography, magnetic resonance imaging and other imaging technologies, conformal therapy and intensity-modulated radiation therapy—including arc therapy and radiosurgery, and seed implants and high-dose-rate brachytherapy applied to specific tumor sites. With adequate design, application, and follow-up, all of these technologies contribute to improvements in the likelihood of success in radiation therapy. They also increase the negative consequences of inadequate design, application, and follow-up. To gain the advantages of these technologies without their possible negative consequences, sophisticated quality assurance measures are required. Meeting this requirement is the responsibility of the medical physicist.

Present

Today, quality assurance (and, therefore, the medical physicist) is integral to the use of radiation to treat patients with cancer.

However, this role raises the question of just who is a medical physicist and what are the necessary credentials to ensure that a specific individual has the knowledge and skills necessary to ensure quality in the performance of radiation therapy. In the United States, the credentialing process in medical physics includes certification of individuals through required education and training and through examination by the American Board of Radiology (ABR). For physicists working in radiation therapy, the appropriate certification is in therapeutic radiological physics (or for older members, radiological physics). In Canada, a similar certification process is conducted by the Canadian College of Physicists in Medicine. To be eligible for ABR certification, an individual must have a graduate degree in medical physics or a related discipline, 3 years of experience working in clinically related therapeutic radiological physics, and passing scores on both written and oral examinations in medical physics. These requirements are described in detail at www.abr.org. In four states (Texas, New York, Florida, and Hawaii), medical physicists must also be licensed to practice in the discipline.

Beginning in 2012, eligibility for ABR certification in a field of medical physics will require that an individual has graduated from either an accredited graduate program in medical physics or has completed an accredited 2-year residency in the medical physics specialty (diagnostic radiologic physics, medical nuclear physics, or therapeutic radiologic physics) for which certification is sought. In 2014, the eligibility requirements will be tightened so that only individuals who have completed a 2-year accredited residency are admitted for ABR examination and certification. Accreditation of graduate programs and residencies is provided by the Commission on Accreditation of Medical Physics Educational Programs (CAMPEP) (www.campep.org).

Since 2002, physicists who are certified by the American Board of Radiology are granted a time-limited certificate that is valid for 10 years. Over this 10-year period (and in subsequent 10-year intervals), physicists must participate in a maintenance of certification program that consists of four categories: professional standing; lifelong learning and self-assessment; cognitive expertise; and assessment of performance in practice (practice quality improvement). In these categories, an individual must demonstrate six competencies: practice-based learning and improvement; patient care; professionalism; interpersonal and communication skills; medical knowledge; and systems-based practice. These competencies have specific meanings for medical physicists, and their interpretation should be guided by consultation with the ABR (www.theabr.org/moc/moc_rp_landing.html). Physicists are professionals whose work in a clinical environment, similar to that of physicians, has a direct impact on the care and welfare of patients. Consequently, every physicist working in such an environment should be certified and should maintain that certification as assurance that his or her knowledge and skills are up-to-date and sufficiently developed to provide quality care to patients.

In recent years, the healthcare professions, including medical physics, have been required to demonstrate a level of accountability to the public that is greater than in the past. There are many reasons for this requirement of a higher level of accountability, including the demands of patients and payers of healthcare services for greater disclosure of the quality and cost of the services that they receive and pay for. Some payers have withheld payment if disclosure is not provided, and others have offered a higher level of payment to those healthcare organizations that can quantitatively demonstrate that their services meet an agreed-upon level of quality. These actions are referred to colloquially as "pay for performance." Finally, press coverage of dramatic and preventable medical errors that have caused the death or disability of patients has led to a nationwide movement to improve the safety of patients in hospitals and other healthcare settings (www.npsf.org). Institutions today are expected to demonstrate their commitment to patient safety as a principal objective of the services they provide. Patient safety is simply one very important aspect of quality assurance.

In achieving a higher level of accountability and commitment to quality, several barriers must be overcome. The first barrier is the potential to compromise the quality of healthcare services as a consequence of efforts to constrain costs within an institution or organization. Some expenditures will have to be reduced as payments for healthcare services are brought under greater control of those who pay for the services. The quality of services must be protected from actions to control costs. Reduced costs and higher quality must be the goal, not reduced costs and lower quality, if an institution or organization intends to remain competitive in an increasingly cost- and quality-conscious environment.

A second barrier is the apathy of healthcare professionals in embracing change, and the feeling of comfort in keeping things as they are. These characteristics are exhibited in all disciplines, including medical physics. The forces for change are originating principally outside the professions and are largely coming from the patient, payer, and public sectors. Ignoring these forces is a recipe for irrelevance in the future, and professionals who do not embrace them will be increasingly isolated from the practice environment.

Finally, there is a level of activism in many areas of public life that is growing and demanding greater accountability in all types of professional services, including education, politics, and healthcare. This activism is in part a product of the "decline of deference," in which professionals such as teachers, political leaders, and physicians and other healthcare professionals are not held in the same level of esteem as they once were. It is interesting to note that nurses have not endured a lowered level of esteem and are still perceived as dedicated principally to the interests of patients and the public. Healthcare professionals, including medical physicists, should not ignore this enhanced level of activism because, in part, it is driving the demand for accountability in all areas of public service.

Future

The growth in complexity of radiation therapy is expected to continue with advances in many areas of the discipline. Some of these advances will facilitate the delivery of higher doses of radiation to tumors, with rapid falloff of dose beyond the tumor margins so that doses to normal tissues are reduced. Conformal radiation therapy (CRT), intensity-modulated radiation therapy (IMRT), tomotherapy, volume-modulated arc therapy (VMAT), robotic radiosurgery, and proton and heavy-ion therapy are examples of technologies that bring achievement of this objective closer and additional examples are expected to evolve in the future. Each of these technologies provides a sharp dose gradient between tumor and normal tissue and demands high accuracy and precision in treatment delivery. These demands require sophisticated quality assurance techniques to ensure that the dose is high and uniform in the treatment region and much less outside the margins of the region, and that the dose gradient is positioned precisely just outside the tumor margins.

The challenge of exact positioning of the high-dose region in radiation therapy will be enhanced by methods currently under development to adjust the treatment volume by tracking the tumor as it moves due to respiration or other causes. This ability will permit reduction in the size of the treatment volume currently necessary to ensure that the tumor does not move out of the high-dose region during treatment. As the superposition of the treatment volume over the tumor becomes increasingly exacting, the demands will increase on the medical physicist to provide quality assurance measures to verify the accuracy and precision of treatment.

Image-guided radiation therapy (IGRT) has been a major advance in the delivery of radiation treatments, as mentioned earlier. Some imaging modalities have been combined (e.g., PET/CT, SPECT/CT, and PET/MRI) so that the patient can be imaged with both modalities while in the position to be used for treatment. These hybrid imaging systems permit the delineation of the treatment volume with greater accuracy, facilitate the delivery of high radiation doses to the treatment volume, and allow rapid fall-off of the dose outside the volume. Imaging of the treatment volume during treatment is also possible with some modalities today (e.g., fluoroscopy with kilovoltage and megavoltage x-ray beams), and efforts are underway to add other imaging modalities (e.g., MRI and ultrasound). One might predict that, ultimately, imaging and treatment technologies will be offered in a single treatment platform with the patient in an unchanging geometry. Just as with other technological advances, IGRT demands a greater depth of quality assurance to ensure that the precision and accuracy promised by the technology are achieved in practice.

All of the treatment methods described above are modeled to provide a uniform dose throughout the treatment volume and a rapid dose fall-off outside the volume. However, it is widely acknowledged that the sensitivity of tumor cells to radiation is not uniform throughout the treatment volume. Reduced levels of oxygenation of tissues in different regions of the tumor (often in the central core of the tumor where blood flow is reduced) cause cells in these regions to be more resistant to radiation damage. Other, more subtle cellular characteristics (e.g., metabolic rate and apoptosis) in different regions also may affect cellular

radiosensitivity. These differences suggest that a uniform dose may not be the ideal distribution of radiation throughout the tumor volume. Instead, the dose should be delivered in a manner that furnishes greater doses to less radiosensitive cells to compensate for their reduced response to radiation. This approach is sometimes referred to as "dose painting."

The limitation in dose painting is the challenge in identifying and quantifying variations in radiosensitivity throughout the tumor. Progress in overcoming this challenge is occurring, in part through the use of molecules and nanoparticles that are tagged with radionuclides and that serve as targeted markers for specific characteristics of tumor cells. Examples of molecular markers include 18F [18-fluorodeoxyglucose (FDG) for glucose metabolism, and 18F-fluoromisonidazole (MISO) and 18F-fluoroazomycin arabinoside (FAZA) for hypoxia-specific imaging], 11C (11C methionine and 11C choline for prostate cancer activity), and 99mTc (99mTc Annexin V for apoptosis activity). Using molecular markers in combination with methods of radiation delivery that permit dose painting may ultimately furnish dose variations throughout the tumor volume that correspond to differences in cellular radiosensitivities. These technologies will require quality assurance measures in both marker chemistry and administration and in dose delivery that are well beyond the levels of quality assurance currently employed in radiation therapy.

Summary

In radiation therapy, quality assurance has always been the responsibility of the medical physicist. In the past, this responsibility has meant calibrating machines, accumulating dose information in phantoms, measuring doses in patients, designing treatment plans, checking patient treatment charts, and protecting both patients and personnel from excessive radiation exposure. Today it means all of these activities plus assurance that the steep gradient in dose between treated and untreated regions is positioned accurately and precisely for each treatment. This requirement will become even more exacting in the future as treatment guidance and delivery technologies become more sophisticated. When dose painting in response to varying cellular sensitivity over the treatment volume becomes clinically available, the demands for quality assurance will approach an entirely new level.

Future advances in radiation therapy will stretch the capacity of medical physicists to acquire the breadth and depth of knowledge and skills needed to enact quality assurance at the appropriate level. Because treatments will be more patient-specific and complex, medical physicists will likely be required to interact more directly with patients and oncologists. These requirements will place more demands on the time and expertise of medical physicists, and they will be held to a level of accountability that is greater than in the past. However, these responsibilities offer greater opportunities for medical physicists to improve the care and well-being of patients and to serve as a leader in the healthcare team providing patient care.

II

Patient Safety and Managing Error

14

Issues in Patient Safety

Mary Coffey
Trinity College Dublin

Introduction

Patients attending our radiotherapy departments have a basic right to expect high quality treatment delivered in a safe environment. This basic fundamental right to safe healthcare has been acknowledged internationally (Committee of Ministers 2006) and the failure to achieve this goal has been the focus of research for some time. In 1999, it was stated that 4% of patients in American hospitals incur adverse outcomes with between 44,000 and 98,000 Americans dying each year from preventable errors (Kohn, Corrigan, and Donaldson 1999). Recent surveys in the United Kingdom have highlighted the percentage of preventable errors still prevailing in the United Kingdom (Runciman et al. 2006; Raleigh et al. 2008).

Over the past two decades, there has been a significant increase in emphasis on safe delivery of healthcare with unfavorable comparisons being drawn between error management in healthcare and in other high-risk organizations (Sexton, Thomas, and Helmreich 2000). Recent high-profile errors in radiotherapy departments in several countries have further highlighted the necessity of addressing patient safety and ways in which it can be improved. Human factors are predominantly identified as the most frequent cause of error with the recent World Health Organization (WHO) publication on radiotherapy risk profile assessing 81 specific risks, of which 53 were associated with staff alone, and less than 10 with patients or the system (WHO 2008).

Safety Factors

When considering patient safety, a wide range of factors must be taken into account. Radiotherapy can be considered as a microcosm of the hospital as a whole. It reflects the characteristics of high technology, complexity, fragmentation, diversity of professionals and professional activities, levels of responsibility,

knowledge levels, and the individual clinical freedom of the wider healthcare setting. Complexity, of itself, increases the rate of error by increasing the number of steps in a process, the number of tasks that must be undertaken, the options available, the vulnerability of patients, and the higher number of human interactions necessitated (Nolan 2000; Reason 2004). Issues relating to patient safety must take these factors into account by identifying areas of highest risk from specific factors and devising appropriate means to manage them.

Patient safety is compromised, for example, by poor working conditions and time pressure to treat ever increasing patient numbers with the same or fewer resources. A focus on quantity rather than quality can result in stress, lack of concentration, and lack of attention to detail. In a comparison of physician and public perceptions on the possible causes of errors, understaffing, overwork, stress, or fatigue were identified by both groups, with the public also citing physicians not having sufficient time with patients (Blendon et al. 2002). Comparisons are often drawn between healthcare and the aviation industry and where this is, perhaps, not the most valid comparator given the complexities discussed previously, it is important to recognize that in the aviation industry, perceptions of fatigue and stress are acknowledged as risk factors and continue to be topics of training and targets for improvement (Sexton, Thomas, and Helmreich 2000). Healthcare can also learn from aviation by introducing more stringent systems, alerts, and constraints into routine practice (Kane and Mosser 2007).

Radiotherapy Safety Factors

Patient safety in radiotherapy encompasses delivery of the prescribed dose to the correct volume, the patient's physical safety within the department, and their physical and psychological care during, and subsequent to, a course of treatment. Achieving and maintaining patient safety is the responsibility of everybody in the organization and must be seen as a collaborative process.

Education

First and foremost is the accurate delivery of the prescribed course of treatment. Fundamental to this is an understanding of the principles and practice of radiotherapy that includes components of clinical decision-making, safety, and risk management and the importance of recognizing and minimizing the impact of errors (Patey et al. 2007). Education programs must be dynamic to meet the challenges of ever increasing levels of complexity of practice and evolving levels of responsibility, thus enabling all disciplines to take an active role in patient safety. Professional education at the undergraduate and postgraduate levels for all disciplines involved in the delivery of radiotherapy must be expanded to include a section on patient safety and the means by which it can be achieved. Close collaboration between the clinical and academic bodies is essential in defining how these goals are best met.

Communication

Patients attending for treatment will be seen by a number of different professionals in a wide range of settings, both within the department and the wider hospital setting; for each patient, this pathway is unique. Patient safety throughout the patient pathway is, to a large extent, dependent on the quality of the communication and associated documentation that accompanies them on this pathway, with poor communication considered to be a causative factor in more than 60% of errors (Yeung et al. 2005; Bourhis, Roth, and MacQueen 1989). Systems must be put in place to enable staff to raise issues of patient safety without fear of sanction or recrimination. Historically, healthcare was both hierarchical and bureaucratic with few channels for open communication. Public perception of one cause of medical error is the failure of health professionals to work together or communicate as a team (Blendon et al. 2002). As a specialty dependent on the multidisciplinary team, effective communication is fundamental. Clear definition of roles and responsibilities within the team, with time allocation for patient discussion, must be in place and where there is more than one shift per day, time must be allocated for information exchange between the teams.

Documentation

Documentation is an integral component of good communication. Practice must be evidence-based and all departments should have well-documented policies, procedures, and protocols. These documents should be regularly reviewed in the context of new findings, and deviations from the agreed protocols should be discussed and documented. All staff within the department should be familiar with the protocols and they should be readily available in written or electronic format. For all clinical trials and trial centers, Standard Operating Procedures (SOPs) are the norms and are an effective way to ensure consistency of practice. It has been clearly demonstrated that stringent quality assurance review can have a positive impact on everyday clinical practice (Haworth et al. 2009). However, there are also difficulties associated with introducing SOPs into routine practice, as discussed in an editorial in 2003 (Hopper 2003) that outlines the difficulties that exist in standardizing practice.

Accurate Practice

Care and attention to detail and working with awareness and alertness are essential components of patient safety. The individuality of patients and their treatment necessitates this approach. The treatment of patients should never be seen as routine, indeed "routine may be the enemy of good care" (Kane and Mosser 2007). Internationally, many departments will not have access to the most sophisticated technology, but a high level of accuracy can still be achieved.

Accuracy and, hence, safety in the patient pathway starts with identification: correct patient, correct site, and correct procedure (Australian Council for Safety and Quality in Healthcare 2004) and the accurate and clear documentation of this information. Errors in identification are compounded when patients have a number of procedures carried out in different venues by different staff members, a typical scenario for the radiotherapy patient. Departments should put in place standardized protocols for verification of identification, particularly where staff rotate frequently and these protocols should be adhered to. It has been recommended that three independent items are used as the most effective safety procedure for patient identification, with many departments now introducing safety measures such as bar coding or fingerprinting for patient identification (New York State Department of Health 2001). Photographs are now commonly used for both patient and treatment field identification. Consideration must also be given to managing the communication issues that may be encountered routinely in the patient population attending for radiotherapy. These include language barriers, cultural differences (Johnstone and Kanitsaki 2006), hearing problems, and psychiatric problems (Bartlett et al. 2008), and an appropriate range of measures must be introduced to address these.

Initial patient positioning for image acquisition and subsequent treatment is critical to the process. Table tops and immobilization devices should be indexed and the reference points clearly documented. Clear positioning instructions with details of the associated immobilization devices and any additional requirements must be produced either in written or electronic format and, if not, an integral part of the treatment sheet should be securely attached.

All patient specific accessory devices must be clearly labeled and stored with due care and attention. Immobilization masks, including the number and position of fixation points, should be selected to ensure immobilization of the treatment area. The information necessary to accurately position and treat each patient should be sufficiently clear to enable any member of the treatment team to carry out the procedure. Protocols and procedures should be in place for verification procedures before and during treatment to confirm accuracy of position, including clearly defined responsibility levels for field shift.

Ongoing vigilance throughout treatment is integral to patient safety. Effective patient monitoring will identify changes, such as weight loss or weight gain, increasing pain levels, etc., that can affect position.

Creating a Safety Culture

Management policy largely dictates the organizational culture which, in turn, exerts a strong influence on the safety culture and, therefore, on patient safety within the hospital and the radiotherapy department. "Safety should be an explicit organizational goal" (Kohn, Corrigan, and Donaldson 1999). Management is responsible for ensuring that patient safety risks are assessed and resources provided to implement safety improvements and to disseminate information to all staff (Hogan et al. 2008). A safety committee should be established with representatives from all disciplines. This committee should meet regularly, review patient safety infringements in the department and the current literature in the area, make and implement recommendations to improve patient safety, and disseminate information to all staff through regular feedback and safety reports.

As part of encouraging a safety culture, management should implement a system of incident and near-incident reporting with feedback and evidence-based revision of practice (Murphy et al. 2007). Staff must be educated and supported when such a system is implemented and the benefits emphasized. It should be seen as a learning initiative integral to the quality assurance program of the department. An incident reporting system can give confidence to patients that the department culture is one that puts safety as a priority. Patients can also be invited to have representation on the safety committee and to identify ways in which they can actively participate in safety awareness in departments (Hibbard et al. 2005).

Summary

The responsibility for patient safety rests with all staff of the organization at all levels. Management has a responsibility to ensure that adequate resources, within the confines of available finances, are provided and that all equipment is fit for purpose and maintained to a high standard. Cost savings are often made, for example, on accessory equipment that can result in poor positioning and immobilization and subsequent inaccurate treatment or unsafe practice. Within the radiotherapy department, working with vigilance, alertness, and awareness at all times will greatly enhance patient safety and increase the levels of satisfaction and motivation of staff.

References

Australian Council for Safety and Quality in Healthcare. 2004. *Ensuring Correct Patient, Correct Site, Correct Procedure.* Canberra: Australian Council for Safety and Quality in Healthcare.

Bartlett, G., R. Blais, R. Tamblyn, R. J. Clermont, and B. MacGibbon. 2008. Impact of patient communication problems on the risk of preventable adverse events in acute care settings. *CMAJ* 178 (12): 1555–1562.

Blendon, R. J., C. M. DesRoches, M. Brodie, J. M. Benson, A. B. Rosen, E. Schneider, D. E. Altman, K. Zapert, M. J. Herrmann, and A. E. Steffenson. 2002. Views of practicing physicians and the public on medical errors. *N. Engl. J. Med.* 347 (24): 1933–1940.

Bourhis, R. Y., S. Roth, and G. MacQueen. 1989. Communication in the hospital setting: A survey of medical and everyday language use amongst patients, nurses and doctors. *Soc. Sci. Med.* 28 (4): 339–346.

Committee of Ministers. 2006. Recommendation Rec (2006) Adopted by the Committee of Ministers on 24 May 2006. Council of Europe.

Haworth, A., R. Kearvell, P. B. Greer, B. Hooton, J. W. Denham, D. Lamb, G. Duchesne, J. Murray, and D. Joseph. 2009. Assuring high quality treatment delivery in clinical trials— Results from the Trans-Tasman Radiation Oncology Group (TROG) study 03.04 "RADAR" set-up accuracy study. *Radiother. Oncol.* 90 (3): 299–306.

Hibbard, J. H., E. Peters, P. Slovic, and M. Tusler. 2005. Can patients be part of the solution? Views on their role in preventing medical errors. *Med. Care Res. Rev.* 62 (5): 601–616.

Hogan, H., S. Olsen, S. Scobie, E. Chapman, R. Sachs, M. McKee, C. Vincent, and R. Thomson. 2008. What can we learn about patient safety from information sources within an acute hospital: A step on the ladder of integrated risk management? *Qual. Saf. Health Care* 17: 209–215.

Hopper, J. 2003. Left, right, left . . . 'forward march' towards standard operating procedures? *Knee* 10 (4): 309–310; discussion 310.

Johnstone, M. J., and O. Kanitsaki. 2006. Culture, language, and patient safety: Making the link. *Int. J. Qual. Health Care* 18 (5): 383–388.

Kane, R. L., and G. Mosser. 2007. The challenge of explaining why quality improvement has not done better. *Int. J. Qual. Health Care* 19 (1): 8–10.

Kohn, L. T., J. M. Corrigan, and M. S. Donaldson, eds. 1999. *To Err Is Human: Building a Safer Health System.* Washington, DC: The Institute of Medicine, National Academy Press.

Murphy, J. G., L. Stee, M. T. McEvoy, and J. Oshiro. 2007. Journal reporting of medical errors: The wisdom of Solomon, the bravery of Achilles, and the foolishness of Pan. *Chest* 131 (3): 890–896.

New York State Department of Health. 2001. *Pre-Operative Protocols to Enhance Safe Surgical Care.* Albany, NY: New York State Department of Health.

Nolan, T. W. 2000. System changes to improve patient safety. *BMJ* 320 (7237): 771–773.

Patey, R., R. Flin, B. H. Cuthbertson, L. MacDonald, K. Mearns, J. Cleland, and D. Williams. 2007. Patient safety: Helping medical students understand error in healthcare. *Qual. Saf. Health Care* 16 (4): 256–259.

Raleigh, V. S., J. Cooper, S. A. Bremner, and S. Scobie. 2008. Patient safety indicators for England from hospital administrative data: Case-control analysis and comparison with US data. *BMJ* 337: a1702.

Reason, J. 2004. Beyond the organisational accident: The need for "error wisdom" on the frontline. *Qual. Saf. Health Care* 13 Suppl 2: ii28–ii33.

Runciman, W. B., J. A. Williamson, A. Deakin, K. A. Benveniste, K. Bannon, and P. D. Hibbert. 2006. An integrated framework for safety, quality and risk management: An information and incident management system based on a universal patient safety classification. *Qual. Saf. Health Care* 15 Suppl 1: i82–i90.

Sexton, J. B., E. J. Thomas, and R. L. Helmreich. 2000. Error, stress, and teamwork in medicine and aviation: Cross sectional surveys. *BMJ* 320 (7237): 745–749.

WHO. 2008. *Radiotherapy Risk Profile—Technical Manual by the World Health Organization*. Geneva, Switzerland: World Health Organization.

Yeung, T. K., K. Bortolotto, S. Cosby, M. Hoar, and E. Lederer. 2005. Quality assurance in radiotherapy: Evaluation of errors and incidents recorded over a 10 year period. *Radiother. Oncol.* 74 (3): 283–291.

15

Overview of Risk Management

Robert C. Lee
Neptune and Company, Inc.

Introduction

Hippocrates admonished health care providers to "do no harm." Yet, patient harm does occur, despite the best intentions of providers and healthcare systems (Institute of Medicine 1999). Radiotherapy occurs in a complex environment that is subject to uncertainties and unintentional incidents, which in turn can compromise treatment outcomes and/or result in patient morbidity or mortality. The commissioning (testing in clinical settings) of technologies and treatment with ionizing radiation are tightly monitored and regulated by national, provincial/state, and local agencies, and a large amount of resources is expended at a typical cancer treatment center on quality control (QC) and quality assurance (QA). Yet despite great care to ensure accurate and precise administration of radiotherapy, there are notable reports of incidents in radiotherapy that have led to serious adverse events in patients, including death and disability (International Atomic Energy Agency 2000). Formal approaches to radiotherapy safety are likely to be useful in understanding the complex delivery systems and identifying influences that pose the greatest risk, which in turn inform risk management.

Literature Review

A brief literature review was performed to determine the state-of-the-art in patient safety risk analysis and management. PubMed and Internet searches using the terms "risk assessment," "risk analysis," "decision analysis," "patient safety," and "medical error" (no date limits) identified more than 200 documents, for which abstracts or Web summaries were reviewed.

Based upon this review, the state-of-the-art in patient safety risk assessment and risk management (aside from ad hoc approaches) appears to be represented by approaches such as failure mode and effects analysis (FMEA) (Stamatis 2003), which is a semiquantitative, proactive approach with engineering origins currently used by many health systems. FMEA qualitatively characterizes a system, identifies potential failure points, assigns probability and severity categories, and incorporates this information into a decision matrix/flowchart. The term "semiquantitative" is used here because FMEA does not actually model the relationships between system variables.

Other approaches used in patient safety include root cause analysis (Latino and Latino 2002), which is a qualitative way of tracing the causes of a particular incident, and microsystems analysis (Nelson et al. 2002), which is a qualitative way of characterizing clinical environments, their components, and their interactions. Neither technique estimates probability and severity, and both are typically used retrospectively (i.e., after an event has occurred). These approaches and FMEA are reasonable, important first steps in identifying causal factors and incidents that have led or can lead to adverse patient outcomes. However, characterization of the full range of human, organizational, technological, and information management sources of failure is the key to identifying risk management alternatives and strategies (Haimes 1998). One way to do this is by using fault trees to structure the sources of failure. Fault trees use Boolean logic to represent combinations of events that can lead to a failure, as well as potential mitigating actions or procedures. An example relevant to radiotherapy safety is given by Thomadsen et al. (2003).

Despite their utility, the methods outlined above do not completely inform a holistic risk assessment and management strategy in complex, technologically intensive healthcare systems such as radiotherapy, because they do not quantitatively model the relationships between events in the system, nor do they allow prospective prediction of the magnitude of risks and associated uncertainties. Despite much discussion regarding the utility of quantitative modeling approaches to patient safety (Battles et al. 2006; Marx and Slonim 2003), examples are limited in the literature. Notable published examples are the work by Pate-Cornell et al. (1997) who examined anesthesia related risks and Lee et al. who examined external-beam radiotherapy risks using probabilistic risk modeling (Ekaette et al. 2007; Lee et al. 2006). These modeling methods were developed in engineering and have been applied in many scenarios ranging from nuclear safety to environmental assessment (Cox 2002). Typically using simulation techniques, probabilistic risk modeling estimates the probability of adverse events by identifying potential incidents and propagating the uncertain probability of failure according to system characteristics. If severity of the events is incorporated, then this method is termed probabilistic risk analysis (PRA).

Ultimately, risk management involves tradeoffs between benefits, risks, and costs of particular risk management strategies. Decision analysis or economic evaluation models (Hunink et al. 2001) can directly inform risk management decisions by estimating comparative utility across different strategies. The utility of a strategy is an aggregated measure of how well that strategy attains decision objectives, i.e., quantitative representations of individual or organizational values (Keeney 1992). Examples of objectives in healthcare are maximizing patient survival and quality of life, minimizing patient harm, maximizing compliance, and staying within resource constraints. Tradeoffs across multiple objectives are almost inevitable in patient safety risk management. Based upon the author's literature search, decision analyses specifically addressing risk management in radiotherapy have not been published to date, although a number of economic models using decision analysis for evaluating radiotherapy technologies have been published (e.g., Sher et al. 2008).

A holistic quantitative risk analysis and risk management framework was developed by the author and others to address radiotherapy patient safety in the system represented in Figure 15.1. This work represents the first time, to the author's knowledge, that such a framework has been applied in a healthcare scenario. The process map that is the foundation of Figure 15.1 was developed in a collaborative and iterative fashion with clinical staff members at a large cancer treatment facility in Canada, who were all expert in some part of the process. Similar to other cancer treatment facilities, there was no person or persons who were expert in all parts of the process; this would not be humanly possible. The safety of the system is therefore in large part reliant upon a degree of "trust" that the information or decisions produced by earlier parts of the process are correct; this trust being reinforced by QA/QC procedures and safety programs. Detailed descriptions of analyses focused on different parts of the process map can be found in the references cited in Figure 15.1. Briefly, these examples include:

- An analysis of the impact of uncertainties associated with staging of breast cancer on subsequent radiotherapy decisions (Lee et al. 2006)
- Development and evaluation of an incident taxonomy and classification system for radiotherapy (Ekaette et al. 2006)
- A probabilistic fault tree analysis of a portion of the radiotherapy process (Ekaette et al. 2007)
- Introduction of an incident learning system (ILS) to the cancer treatment organization, and analysis of its impact (Cooke, Dunscombe, and Lee 2007)

Issues in Risk Management

There are a number of challenges that analysts face in the transfer of safety approaches and modeling methods to healthcare

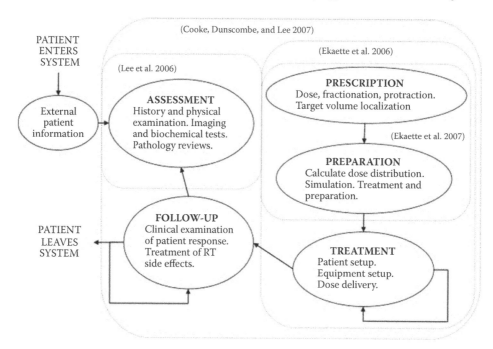

FIGURE 15.1 Process map and boundaries of cited analyses (dashed lines). Note that the outer boundary represents the usual sphere of responsibility of a cancer treatment center. Ovals represent broad areas of clinical activity/responsibility. Arrows represent patient and information flow and radiotherapy is denoted by "RT."

(Marx and Slonim 2003; Wreathall and Nemeth 2004). First, most models are data intensive and healthcare information systems are typically designed to meet clinical requirements, not risk management needs. As a result, most clinical data lack the emphasis or details that would be ideal for system modeling purposes. Second, each healthcare facility has a unique combination of service models, programs, and performance standards, which makes it difficult to develop generalizable models. Third, clinical procedures have to be flexible in order to accommodate different patients, which results in the interactions between the physicians, nurses, therapists, pharmacists, physicists, and other staff being quite complex. Finally, system changes such as technology upgrades or organizational changes can substantially modify the system to be modeled. However, despite these challenges, it is generally agreed that healthcare systems can benefit from rigorous approaches to risk analysis and management (Marx and Slonim 2003; Wreathall and Nemeth 2004).

A remaining question relates to how risky radiotherapy actually is; both intrinsically and compared to other forms of healthcare. Without long-term, detailed incident tracking and clear definitions of consequences (such as mortality due to incidents, reduction in patient quality of life, etc.), it is difficult to assign accurate and precise values to radiotherapy risk. However, general statements regarding comparative risk can be made. A recent paper on barriers to achieving "ultrasafe" healthcare (Amalberti et al. 2005) listed radiotherapy as one of the safest areas of medicine (at a risk of death per exposure between 1×10^{-5} and 1×10^{-6}), along with anesthesiology and blood transfusions. In contrast, the same article listed cardiac surgery as having a risk 100 to 1000 times higher.

One problem with these numbers, however, is that radiotherapy, or any form of cancer treatment, is not only susceptible to incidents that may harm patients in an overt fashion via overdosing or other means, but it is also susceptible to incidents that could lead to, for example, underdosing of a tumor. In such cases, it may be very difficult to tell whether an incident has occurred because cancer treatments are not always effective, even when administered properly. Furthermore, the outcomes of incidents are often idiosyncratic. For example, the same overdosing incident that may cause little change in treatment outcome in a healthy young person may kill a frail elderly person. Conversely, radiotherapy in a young person may indeed cause secondary cancers later in life (Raj, Marks, and Prosnitz 2005). This speaks to the necessity of further careful analysis in this field, especially in an environment of constantly changing technology and practice. Radiotherapy may be relatively safe compared to other areas of medicine, but that does not mean that improvements are not needed and should not be made a priority.

Additionally, radiotherapy is typically conducted within cancer treatment centers that offer a broad range of other treatments, including surgery and chemotherapy. In academic centers, a wide range of experimental treatments that may be associated with higher risks may be offered as well. Ideally, risk analysis and risk management should be conducted at a care pathway level, as patients often receive a broad range of treatments. A truly holistic approach would follow a patient cohort from the time that cancer is diagnosed to the time that they exit the system via cure, death, or voluntary withdrawal and include all of the tests and treatments in the care pathways that the patients receive. However, it would be quite complex to capture this. For example, the author led a micro-costing analysis that captured all of the financial costs associated with a cohort of lymphoma patients (Lee et al. 2008). The care pathways for this cohort were highly variable, even under a system of evidence-based care, and included a wide range of tests, chemotherapy, radiotherapy, stem cell transplantation, and palliative care. This analysis, although difficult and time-intensive, was a much simpler task than a full-scale risk analysis.

The issue of resource allocation is nonetheless of critical importance. There seems to be a pervasive notion that organizations can simply allocate resources to safety and system improvements will happen. This is unlikely to be the case, and in most organizations those safety resources are reallocated from existing budgets. In some cases, there may be a risk that a broad based safety program may actually reduce safety. For example, if frontline staff resource requirements are reduced to pay for or staff the safety initiative, then quality of care may be reduced and incidents in the process of care may increase. There is evidence from other industries that supports this phenomenon (Crites 1995). In most cases, well-planned safety programs will result in net benefits, but these programs must be integrated with other operational aspects of the system. There may not be easy technological "fixes" to improving safety. For example, implementing electronic health records or computerized physician order entry without appropriately engaging and incentivizing clinical staff can result in increased costs and reduced effectiveness (Sidorov 2006).

Broader and longer scale risk management issues (e.g., national policies on radiotherapy safety, allocating safety resources to radiotherapy compared to other areas of medicine, etc.) may be challenging to assess. In such cases, powerful dynamic methods such as system dynamics modeling (Cooke and Rohleder 2006; Lee, Cooke, and Richards 2009) may be appropriate to explore risk management options and inform decisions. System dynamics models can also be used to explore unintended effects such as risks associated with safety programs (as discussed above) and to analyze feedback effects. This is a fertile area for methodological and applied research.

Summary

A number of approaches currently exist to inform patient safety risk management in radiotherapy, ranging from very simple qualitative approaches to simulation modeling. Simple approaches may not be fully informative for this complex specialty. Those involved with the delivery, funding, and/or policy of radiotherapy should embrace all appropriate approaches while encouraging methodological and applied research. The field of patient safety in general will benefit as a result.

References

Amalberti, R., Y. Auroy, D. Berwick, and P. Barach. 2005. Five system barriers to achieving ultrasafe. *Ann. Intern. Med.* 142 (9): 756–764.

Battles, J. B., N. M. Dixon, R. J. Borotkanics, B. Rabin-Fastmen, and H. S. Kaplan. 2006. Sensemaking of patient safety risks and hazards. *Health Serv. Res.* 41 (4 Pt 2): 1555–1575.

Cooke, D. L., P. B. Dunscombe, and R. C. Lee. 2007. Using a survey of incident reporting and learning practices to improve organisational learning at a cancer care centre. *Qual. Saf. Health Care* 16 (5): 342–348.

Cooke, D. L., and T. R. Rohleder. 2006. Learning from incidents: From normal accidents to high reliability. *Syst. Dyn. Rev.* 22 (3): 213–219.

Cox, L. A. 2002. *Risk Analysis: Foundations, Models, and Methods.* Boston, MA: Kluwer Academic Publishers.

Crites, T. R. 1995. Reconsidering the costs and benefits of a formal safety program. *Prof. Saf.* 40 (12): 28–32.

Ekaette, E., R. C. Lee, D. L. Cooke, S. Iftody, and P. Craighead. 2007. Probabilistic fault tree analysis of a radiation treatment system. *Risk Anal.* 27 (6): 1395–1410.

Ekaette, E. U., R. C. Lee, D. L. Cooke, K. L. Kelly, and P. B. Dunscombe. 2006. Risk analysis in radiation treatment: Application of a new taxonomic structure. *Radiother. Oncol.* 80 (3): 282–287.

Haimes, Y. Y. 1998. *Risk Modeling, Assessment, and Management.* New York, NY: John Wiley & Sons.

Hunink, M., P. Glasziou, J. Siegel, J. Weeks, J. Pliskin, A. Elstein, and M. Weinstein. 2001. *Decision Making in Health and Medicine.* Cambridge, MA: Cambridge University Press.

International Atomic Energy Agency. 2000. Lessons learned from accidental exposures in radiotherapy. In *Safety Reports Series No. 17.* Vienna: International Atomic Energy Agency.

Institute of Medicine. 1999. *To Err Is Human: Building a Safer Health System.* Washington, DC: National Academy Press.

Keeney, R. L. 1992. *Value-Focused Thinking: A Path to Creative Decision Making.* Cambridge, MA: Harvard University Press.

Latino, R., and K. Latino. 2002. *Root Cause Analysis: Improving Performance for Bottom-Line Results.* Boca Raton, FL: CRC Press.

Lee, R. C., E. Ekaette, K. L. Kelly, P. Craighead, C. Newcomb, and P. Dunscombe. 2006. Implications of cancer staging uncertainties in radiation therapy decisions. *Med. Decis. Making* 26 (3): 226–238.

Lee, R. C., D. L. Cooke, and M. Richards. 2009. A system analysis of a suboptimal surgical experience. *Patient Saf. Surg.* 3 (1): 1.

Lee, R. C., D. Zou, D. Demetrick, L. DiFrancesco, K. Fassbender, and D. Stewart. 2008. Costs associated with diffuse large B-cell lymphoma patient treatment in a Canadian integrated cancer care centre. *Value Health* 11 (2): 221–230.

Marx, D. A., and A. D. Slonim. 2003. Assessing patient safety risk before the injury occurs: An introduction to sociotechnical probabilistic risk modelling in health care. *Qual. Saf. Health Care* 12 (Suppl 2): ii33–ii38.

Nelson, E. C., P. B. Batalden, T. P. Huber, J. J. Mohr, M. M. Godfrey, L. A. Headrick, and J. H. Wasson. 2002. Microsystems in health care: Part 1. Learning from high-performing front-line clinical unit. *Jt. Comm. J. Qual. Improv.* 28 (9): 472–493.

Pate-Cornell, M. E., L. M. Lakats, D. M. Murphy, and D. M. Gaba. 1997. Anesthesia patient risk: A quantitative approach to organizational factors and risk management options. *Risk Anal.* 17 (4): 511–523.

Raj, K. A., L. B. Marks, and R. G. Prosnitz. 2005. Late effects of breast radiotherapy in young women. *Breast Dis.* 23: 53–65.

Sher, D. J., E. Wittenberg, A. G. Taghian, J. R. Bellon, and R. S. Punglia. 2008. Partial breast irradiation versus whole breast radiotherapy for early-stage breast cancer: A decision analysis. *Int. J. Radiat. Oncol. Biol. Phys.* 70 (2): 469–476.

Sidorov, J. 2006. It ain't necessarily so: The electronic health record and the unlikely prospect of reducing health care costs. *Health Aff (Millwood)* 25 (4): 1079–1085.

Stamatis, D. H. 2003. *Failure Mode and Effect Analysis: FMEA from Theory to Execution.* Milwaukee, WI: ASQ Quality Press.

Thomadsen, B., S. W. Lin, P. Laemmrich, T. Waller, A. Cheng, B. Caldwell, R. Rankin, and J. Stitt. 2003. Analysis of treatment delivery errors in brachytherapy using formal risk analysis techniques. *Int. J. Radiat. Oncol. Biol. Phys.* 57 (5): 1492–1508.

Wreathall, J., and C. Nemeth. 2004. Assessing risk: The role of probabilistic risk assessment (PRA) in patient safety improvement. *Qual. Saf. Health Care* 13 (3): 206–212.

16

Tools for Risk Management

Pedro Ortiz López
*International Atomic
Energy Agency*

Introduction

The enormous benefits from radiotherapy are unquestionable. At the same time, from a safety perspective, radiotherapy has unique features: It is the only application of radiation in which humans are intentionally delivered very high radiation doses, on the order of 20 to 80 Gy. Often, normal tissue receives radiation doses that are on the upper edge of tolerable doses, as a result of which accidental overdosage has sometimes had devastating consequences. Underdosages, which may not always be detected in a timely fashion, can also have severe consequences. Improvements in radiotherapy are associated with ever increasing complexity in both equipment and techniques. Despite the fact that radiotherapy equipment is provided with standardized interlocks, safety in radiotherapy remains highly dependent on human actions, including a complex interaction among professionals of a multidisciplinary team involved in a large number of steps (ICRP 2000). The first step for hospital administrators and heads of radiotherapy departments is to stay aware of these facts when setting up a radiotherapy program and thereafter.

First Step: Key Issues Built on a Solid Base

The two key elements are: (1) the allocation of responsibilities to qualified professionals; and (2) an effective program of quality and safety. There is no question that without these key elements, no radiotherapy practice should be built-up and operated. In other words, key elements need to be mandatory by international standards and equivalent national regulations (IAEA 2006). The quality assurance program is much more than measurements on equipment. It should embrace the whole treatment process from prescription to delivery and patient follow-up, and include the organization, the clear definition of responsibilities, training, and professional development, as well as purchasing, acceptance testing, calibration, commissioning, and maintenance.

Accidental exposures can also occur in well-structured radiotherapy departments if some of the procedures or verifications are omitted. Therefore, a team should be charged with the responsibility of ensuring that the safety program works effectively and stays effective over time through evolutionary changes of the radiotherapy department. This includes changes in workload, and the incorporation of new technologies and techniques.

Second Step: Applying Lessons from Accidental Exposures

Hospital administrators and heads of radiotherapy departments need to gain reasonable confidence that the type of events that have caused major accidental exposure in the past are not likely to occur in their department. Lessons from these events are reported (IAEA 1998, 2000, 2001, 2003; ICRP 2000; Peiffert, Simon, and Eschwege 2007; ICRP 2010) and should be used to challenge the radiotherapy program with questions that would reveal vulnerable aspects. This method of using lessons from past events is a "retrospective approach." Examples of these lessons are the following:

- Mistakes in beam calibration and commissioning of radiotherapy equipment have led a number of radiotherapy departments to put in place a second independent determination of the absorbed dose to a reference point; some countries have even made it mandatory.
- Reports on improper use of the treatment planning system (TPS) or confusion related to its functionalities have

highlighted the need for thorough understanding of the TPS functionalities as well as for testing and ensuring that the TPS is used according to the instructions for use. For atypical use of the TPS, it is important to validate the use of the TPS for these situations.

- Reports on a repair error have provided the lesson that it is indispensable to notify the physicist about any repair and to report the nature of the repair so that the need for measurements can be identified before resuming patient treatment.
- Instructions and displays in a language that is not understood by the users, including equipment operators, have resulted in severe accidental exposures.
- New technologies, new techniques, and increased workloads makes revisiting staff needs, both in terms of number and qualifications, an essential task.
- Teaching the case histories of accidental exposures and their lessons to the radiotherapy staff as part of their training is an effective tool to raise awareness. Cost-free teaching material is available from the ICRP and the IAEA and training courses are regularly organized (http://rpop.iaea .org and http://www.icrp.org/educational_area.asp).

Third Step: Learning from Near Misses and Continuous Improvement

Learning the lessons from major past events is necessary and valuable but not sufficient because new errors occur that were not foreseen in the lessons from past major events (IAEA 2003). Initiatives are needed to collect and share information from other types of errors and near misses that otherwise may go unreported. Moreover, focusing only on major events with catastrophic consequences and very low probability of occurrence leaves the important gap of potentially overlooking other, more frequent errors with lower but still significant consequences. Sharing near misses helps to address these types of errors and forms the basis for keeping safety under review for continuous improvement. An example of this initiative is the radiation oncology safety information system, ROSIS (http://www.clin .radfys.lu.se/default.asp).

Next Steps: Anticipating, Quantifying, and Making Risk-Informed Decisions

Although retrospective approaches, expanded to include near misses, are an important step forward, they still have the limitation of being confined to reported experience. Other unreported, latent risks may remain unaddressed. There is a need to proactively ask the questions, "What else could go wrong?" or "What other potential hazards might be present?" in a systematic manner. This should be done by a multidisciplinary group (radiation oncologist, medical physicist, radiotherapy technologist) that asks and answers these questions for every step of the radiotherapy process.

Once the possible human errors and equipment faults have been identified and listed, an assessment of the probability of an accidental exposure and the severity of consequences is carried out. Additionally, with some of the approaches, there is an assessment of the likelihood that the event will not be detected by the procedures and checks that are part of the program. The three prospective approaches that are most commonly used and have recently been applied to radiotherapy are failure modes and effects analysis (FMEA) (Stamatis 1995), a probabilistic safety assessment (Huq et al. 2008; Ortiz-López et al. 2008), and a risk matrix (RM) (IAEA 1996). They are not totally independent, as FMEA is often used as the first step to probabilistic safety assessments, as described in the following paragraphs.

Failure Modes and Effects Analysis

An example of the application of a failure modes and effects analysis to radiation therapy is that performed by Task Group 100 of the American Association of Physicists in Medicine or the probabilistic safety study of electron linear accelerators (Vilaragut Llanes et al. 2008). Three numerical values were used to describe each failure mode. O describes the probability that a particular failure mode occurs. S is a measure of the severity of the consequences resulting from the failure mode if it is not detected and corrected. D describes the probability that the failure will be detected before the treatment commences or the failure is effective. In the TG100 implementation, O ranges from 1 (failure unlikely, <1 in 10^4) to 10 (highly likely, more than 5% of the time). S ranges from 1 (no danger, minimal disturbance of clinical routine) to 10 (catastrophic if persists through treatment). D ranges from 1 (very detectable, −0.01% or less of the events go undetected throughout treatment) to 10 (very hard to detect, >20% of the failures persist through the treatment course). An important point to note in the evaluation of detectability, D, is that the failure mode is assumed not to have been detected through quality control. Thus, in TG 100's implementation of FMEA, the likelihood of (lack of) detection downstream from failure is estimated.

Risk Matrix

The risk matrix approach is a systematic method for screening risks and focusing deeper analysis on the higher risk issues. A four-level scale (e.g., very low, low, high, and very high) is used to evaluate the likelihood and the severity of events and provide a risk level for each event. The risk level is compared against acceptability criteria. The global risk is also four-level scaled, and those events with high and very high risk will be addressed by a deeper analysis, while the others may be accepted as they are and not require further consideration. The deeper analysis of the higher risk events is done by systematically asking the questions:

- How robust are the existing safety provisions?
- Can the frequency of occurrence of the event sequence or its consequences be reduced?

- Is there a need to add one or more safety provisions to reduce the risk to an acceptable level (low or very low)?

The answers to these questions constitute the conclusions and recommendations from the study. An example of this type of study was that performed by a multidisciplinary group (radiation oncologists, medical physicists, radiotherapy technologists, regulators, and safety specialists) appointed by the Ibero American FORO of Nuclear and Radiation Safety Regulatory Agencies (Ortiz-López et al. 2008).

Probabilistic Safety Assessment

The most comprehensive approach is a probabilistic safety assessment, an approach well-recognized in the nuclear, aeronautical, and petrochemical industries, which was applied to radiotherapy within a program of the International Atomic Energy Agency (IAEA 2006) and more recently by a group also appointed by the Ibero American FORO (Vilaragut Llanes et al. 2008). The identification of initiating events was performed by means of FMEA. Once the list of initiating events had been obtained, the sequences of events that can lead to an accidental exposure were postulated and analyzed by means of fault and event trees. The frequency of occurrence of the accidental exposure from each sequence was quantified by using complex fault and event tree probability computation methods. The method provides an insight into the strengths and vulnerabilities of the radiotherapy process and reveals the dominant contributors to the overall risk as well as the options to reduce it. Probabilistic safety assessments also quantify the risk reduction brought about by the existence of a given safety measure or the increase in risk caused by its absence. With this information, decisions can be based on objective and precise analysis. Moreover, the combination of several tools to evaluate safety (qualitative, quantitative, and graphical tools) allows complementary inputs to cover the limitations of each tool if used separately.

Assessing Safety for New Technology

When a new technology or technique emerges, it is important that the first users share experiences in the first months and years and develop and disseminate, as soon as possible, concise advice from their experiences. This approach can minimize the probability of accidental exposures experienced by other users. It would be beneficial to perform at least one FMEA and probabilistic safety assessment for each new technology, which would provide a wealth of information on basic safety. Complementary to this, the use of more simple methods such as the risk matrix in each hospital would provide an opportunity for self-evaluation in individual hospitals and for managing the safety measures that are most suitable to the hospital's own conditions.

Summary

Basic key elements should be mandatory in the form of standards and regulations. Lessons from accidental exposure and near misses should be incorporated into the education and training curricula. Proactive safety assessments should be promoted for more comprehensive accident prevention and for new technologies. Pioneer users of a new technology should work together, collect experience, write advice, and disseminate it to other users. Professional bodies can be instrumental in gathering these efforts.

References

Huq, M. S., B. A. Fraass, P. B. Dunscombe, J. P. Gibbons Jr., G. S. Ibbott, P. M. Medin, A. Mundt et al. 2008. A method for evaluating quality assurance needs in radiation therapy. *Int. J. Radiat. Oncol. Biol. Phys.* 71 (1 Suppl): S170–S173.

IAEA. 1996. *International Basic Safety Standards for Protection against Ioniuzing Radiation and for the Safety of Radiation Sources.* Vienna: International Atomic Energy Agency (IAEA).

IAEA. 1998. *Accidental Overexposure of Radiotherapy Patients in San José, Costa Rica.* Vienna: International Atomic Energy Agency.

IAEA. 2000. *Lessons Learned from Accidental Exposures in Radiotherapy.* Vienna: International Atomic Energy Agency.

IAEA. 2001. *Investigation of an Accidental Exposure of Radiotherapy Patients in Panama.* Vienna: International Atomic Energy Agency.

IAEA. 2003. *Investigation of an Accidental Exposure of Radiotherapy Pin Bialystok.* Vienna: International Atomic Energy Agency.

IAEA. 2006. *Case Studies in the Application of Probabilistic Safety Assessment: Techniques to Radiation Sources.* Vienna: International Atomic Energy Agency.

ICRP. 2000. *Prevention of Accidental Exposure to Patients Undergoing Radiation Therapy.* Oxford: International Commission on Radiological Protection.

ICRP. 2010. Preventing accidental exposures from new external beam radiation therapy technologies. *Annals of the ICRP,* Publication 112. Amsterdam: Elsevier.

Ortiz-López, P., C. Duménigo, M. L. Ramírez, J. D. McDonnell, J. J. Vilaragut, S. Papadopulos, P. Pereira et al. 2008. Radiation safety assessment of cobalt 60 external beam radiation therapy using the risk-matrix method. Paper read at the 12th International Congress of the International Radiation Protection Association (IRPA), at Buenos Aires, Argentina.

Peiffert, D., J. M. Simon, and F. Eschwege. 2007. Epinal radiotherapy accident: Past, present, future. *Cancer Radiother.* 11 (6–7): 309–312.

Stamatis, D. H. 1995. *Failure Modes and Effects A.* Milwaukee: ASQ Quality Press.

Vilaragut Llanes, J. J., R. Ferro Fernández, M. Rodríguez Martí, P. Ortiz-López, M. L. Ramírez, A. Pérez Mulas, M. Barrientos et al. 2008. Probabilistic safety assessment of radiation therapy treatment process with an electron linear accelerator for medical uses. Paper read at 12th International Congress of the International Radiation Protection Association (IRPA), at Buenos Aires, Argentina. Oct 19-24, 2008.

Error and Near-Miss Reporting: View from Europe

Ola Holmberg
*International Atomic
Energy Agency*

Introduction

Errors and near-misses occur frequently in clinical practice. In a widely reported study (Brennan et al. 1991), it was estimated that injuries caused by human errors occurred in 2.5% of admissions to participating hospitals in an acute care setting. In another study (Wilson et al. 1995), it was judged that 8.3% of admissions were associated with highly preventable adverse events in the healthcare management of patients resulting in disability or a longer hospital stay. It should be noted that in both studies, the method to estimate these frequencies was to screen a large number of medical records and analyze those records meeting certain criteria more extensively to conclude that an iatrogenic injury had occurred. None of these studies based the estimated occurrence frequency on reports of errors and near-misses. Significant difference in frequency between specialties was reported in the second study, for example, three times as many adverse events in cardiac surgery as in medical oncology. While radiotherapy was not one of the specialties reported on, other studies have indicated that errors and near-misses also occur frequently here (Holmberg and McClean 2002; Huang et al. 2005; Yeung et al. 2005).

What Is and Isn't the Purpose of Reporting Errors and Near-Misses?

When considering the reporting of errors and near-misses in radiotherapy, it is important to first ask what is hoped to be achieved. If the frequency of these events occurring in radiotherapy is sought, counting submitted reports on errors and near-misses is not the method to use. Only a small fraction of all events are reported, even in situations involving mandatory reporting, and the full population from which the reports are drawn can be difficult to assess (Vincent 2007). An incident

reporting system on its own, no matter how good and user-friendly, does not increase the safety for radiotherapy patients much. The reporting system has to be a link in a longer chain of incident identification, reporting, investigation, analysis, management, and learning. Charles Billings, the designer of the Aviation Safety Reporting System in the United States, emphasizes that the narrative is what makes incident reports meaningful (Billings 1998). There must be expertise residing with the person entering the report, as well as with the person evaluating the report. Furthermore, there must be time (and thus resources available) to establish a system, gather and evaluate data, and ensure support of the system by all the relevant stakeholders. If this is the case, reporting of errors and near-misses can play an important role in highlighting critical problems, patterns of causes of these problems, as well as the safety critical steps in the radiotherapy pathway. Spreading knowledge on novel errors, involving new technology in a rapidly evolving discipline, and promoting a safety culture and awareness are further benefits.

Different Types of Error and Near-Miss Reporting Systems in Practice

Incident reporting in radiotherapy can be mandatory or voluntary, as is also the case in some other activities that have a major potential impact on human health (such as aviation). Mandatory reporting usually involves reporting incidents with actual (or potential) consequences above a specific minimum level to a national regulatory authority. Radiotherapy has a further level of added complexity due to the role of both health regulators and radiation protection regulators in this area. There can be regulations mandating reporting to health authorities and/or radiation protection authorities in a single country. Mandatory reporting should focus on incidents with serious actual or potential consequences only, such as injury or death, in order to prevent these

TABLE 17.1 Information Requested in 27 Local Incident Reporting Systems in European Clinics

Category	Description	Sub-Description	Frequency
Incident information	Description of incident		25 of 27
	Cause of incident		9 of 27
	Number of fractions affected		10 of 27
	When did it occur?	Date	18 of 27
		Time	12 of 27
		Weekday	1 of 27
	Detection of incident	How	4 of 27
		By whom	2 of 27
		Where in process	1 of 27
		Date	3 of 27
	Estimate of deviation	Absorbed dose	2 of 27
		Dose after correction	2 of 27
		Field location	1 of 27
		Correctable or not	3 of 27
	Clinical significance or risk to patient		12 of 27
	Contributing factors	General comment	4 of 27
		Treatment plan complexity	1 of 27
		Staffing levels	4 of 27
		Staffing composition	2 of 27
		Staff on leave	1 of 27
		Distractions	1 of 27
Action information	Corrective action	Action to be performed and/or already taken	22 of 27
		Responsible for this	3 of 27
		Date for completion	5 of 27
	Preventive action	Recommended action to prevent recurrence	10 of 27
		Procedural changes	2 of 27
		Confirmation of preventive action	3 of 27
	Communication	Patient informed	4 of 27
		Responsible physician informed	13 of 27
		Authority informed	9 of 27
		General	6 of 27

Source: Radiation Oncology Safety Information System, unpublished survey, 2002.

serious events from being obscured by a large number of reports on less serious events. Reasons for having mandatory reporting in a country or within a health system include ensuring that the providers of medical radiation exposures are held accountable to the public and that the protection of patients is held above a minimum level through follow-up investigations and actions relating to the most serious incidents.

Voluntary reporting of errors and near-misses constitutes the other type of reporting system. These systems are open to relevant parties, in this case to health professionals in the radiotherapy area, for the reporting of any safety-related events or conditions where there are lessons to be learned. Voluntary reporting systems are often confidential in relation to the reporter and focus on events where there have been less serious consequences to patients. It has been pointed out (Billings 1998) that it is very important that voluntary reporting systems are run by respected third-party bodies that are seen as independent from regulators

within the community reporting to the system, so that professionals will not be reluctant to report. A reason for having voluntary reporting is to encourage wider sharing of new knowledge on safety-related events and conditions and on lessons to be learned for the improvement of the safety environment.

While reporting systems in radiotherapy can be characterized as mandatory or voluntary, it is also possible to describe them as being either internal or external (Holmberg 2007). External reporting systems take the reported information outside the local organization (e.g., hospital, clinic, or department) for sharing with other relevant parties, such as a regulatory authority in the case of an external mandatory reporting system, or with other health professionals in the case of an external voluntary reporting system. External reporting has an advantage through the possibility of pooling information with other institutions and thereby providing an opportunity for learning from more events and from incidents that have not yet occurred in the local organization. Internal reporting systems, on the other hand, are those used within the local organization to learn from and act on events that happen locally. Advantages with internal reporting systems are that the lessons are more direct, due to the reports originating in the local environment with its specific equipment, procedures, and other characteristics. They also provide an opportunity to follow the specific management of the patients affected and to follow-up actions taken.

There are a number of examples of external systems for reporting errors and near-misses in radiotherapy. Many countries require mandatory reporting to health and/or radiation protection authorities. These reports can sometimes have a general format having been developed to cover any reportable events in the health sector or in any radiation application. Such a design, leading to general characteristics of events being captured, may not be specific enough to enable valuable lessons to be learned for radiotherapy specifically. Reporting to these systems should therefore trigger a more thorough investigation of the event, and not a general release of information. There are also examples of voluntary external reporting systems for radiotherapy. The Radiation Oncology Safety Information System (ROSIS 2009), is an external and voluntary reporting system that was developed by a small group of health professionals in Europe and made accessible on the Internet in 2003. Six years later, more than 1000 incidents have been reported to this system by clinics around the world and shared among the radiotherapy community. Presently, the International Atomic Energy Agency (IAEA) is developing a system intended to incorporate elements of reporting, analysis of reports, and prospective analysis, called Safety in Radiation Oncology (SAFRON).

As a step in the development of the ROSIS system, a number of clinics in Europe were queried on their local reporting systems and what information was asked for in the corresponding forms. With 27 responses from nine European countries, the results highlight which elements of reporting are seen as important in these clinics and can, therefore, be used when evaluating the report form characteristics locally. In addition to patient information of an administrative nature (e.g., name and identification number), the requested information has been categorized in Table 17.1 as belonging to incident information, relating to the characteristics of the reported incident, and action information, relating to the actions taken as a result of the event.

What Errors and Near-Misses Have Been Reported and How Are They Classified?

Some of the most comprehensive reports on errors in radiotherapy that have been made publicly available are the assessments of specific accidents published by the IAEA (1999, 2001, 2002). While these reports represent one end of the spectrum, detailing comprehensive investigations with extensive causal analyses, most errors and near-misses have been reported with far fewer details. As an overview of the type of radiotherapy errors and near-misses that have been reported to three systems (Ekaette et al. 2006), parts of the databases and collations held by ROSIS

(324 reports) (ROSIS 2009), IAEA (46 reports) (IAEA 2000), and the United States Nuclear Regulatory Commission (67 reports) (NRC 2009) were analyzed. IAEA and NRC collations were seen to contain a far higher percentage of reports related to infrastructure than ROSIS, with infrastructure here defined as the elements that are established for the treatment of multiple patients, for example, equipment, standard work procedures, and protocols. On the other hand, ROSIS contained more incidents related to process, here defined as the activity related to the definition and/or execution of a treatment plan for an identifiable patient. Another conclusion of this study was that there was insufficient contextual information in general in the reports in all collations to maximize the potential learning from the events. This highlights again the observation by Billings (1998) that the narrative is what makes incident reports meaningful.

To ensure a systematic approach to lessons learned from reports, and to find common patterns and similarities that enable the building of successful strategies of prevention, a consistent taxonomy is required for incident reporting. This includes a classification system that can link the reported incident to a well-defined step in the radiotherapy process, rate the incident or near-miss on an unambiguous severity scale, as well as group the underlying causes into distinct categories.

The World Health Organization's World Alliance for Patient Safety has recently published the International Classification for Patient Safety (WHO 2009), which is a taxonomy intended to aid consistent categorization of patient safety information with preferred terms and definitions. It is a conceptual framework and not a detailed taxonomy, but it could be used for the development of a more specific taxonomy for patient safety in radiotherapy.

An example of a classification system, tested in practice, for linking incidents to steps in the radiotherapy process is the radiotherapy pathway coding that is presented in the publication *Towards Safer Radiotherapy* (The Royal College of Radiologists et al. 2008). The intention here is to give a clearer picture of where errors occur through the provision of a coding system for the steps of the radiotherapy pathway. There are 21 activities or processes outlined in this system, each with an average of about 10 subactivities or processes. It can also be of interest to classify points where errors are detected, as done in the ROSIS system (ROSIS 2009), in order to find steps in the pathway where specific errors are likely to be detected, requiring heightened awareness.

Classification of severity is exemplified by the Autorité de Sûreté Nucléaire (ASN)-Société Française de Radiothérapie Oncologique (SFRO) scale for dealing with events affecting patients undergoing radiotherapy (ASN 2007), where events are rated between 0 and 7, ranging from events without dosimetric consequences for the patient (0), to events with associated fatal complications for more than 10 patients (7). In this scale, which is used in practice, it is not only confirmed consequences that are taken into account but also potential consequences. It is also recognized that provisional ratings could be modified with time, depending on clinical effects. Furthermore, the scale can accommodate and distinguish events with effects on more than one patient.

When it comes to the classification of causes, the International Commission on Radiological Protection identified seven different causes that were present when analyzing a number of major accidental exposures (ICRP 2000). These were: (1) deficiencies in education and training; (2) deficiencies in procedures and protocols; (3) equipment faults; (4) deficient communication and transfer of essential information; (5) lack of independent checks; (6) inattention and unawareness; and (7) unsecured long-term storage and abandonment of radiotherapy sources. It was recognized that accidents often had several underlying causes. There are also other models for classifying accidents according to why they happened, for example, in aviation. James Reason suggested the Swiss cheese model of system accidents (Reason 2000), where defensive layers have holes, sometimes lining up to give hazards opportunity to damage victims. These holes appear due to active failures (unsafe acts by people in direct contact with the patient or system) and to latent conditions (system flaws arising from design of the system).

Summary

Reporting of errors and near-misses is not a panacea for achieving safety for radiotherapy patients, but an instrument that can improve safety when used properly. It can be used to better identify system design flaws in the radiotherapy pathway and enable action to be taken to improve the system design. It can be used to gain new knowledge of problems appearing with the introduction of new technology or treatment processes, and to highlight this knowledge in the radiotherapy community. A reporting system can be used to identify safety critical steps in the whole radiotherapy pathway, where the system needs to be designed in a more robust way by all the stakeholders, including health professionals, regulatory bodies, educational facilities, hospital management, and others who have an influence over system-building. It can be used to heighten awareness and enhance the safety culture in an organization through involvement of and feedback to staff and managers. It should always be used together with investigation, analysis, management, learning, and improvement in an organization, so that the reporting on its own is not seen as an endpoint.

References

Autorité de Sûreté Nucléaire (ASN). 2007. *ASN-SFRO Experimental scale for dealing with radiation protection events affecting patients undergoing a medical radiotherapy procedure*. Paris: Autorité de Sûreté Nucléaire.

Billings, C. 1998. Incident reporting systems in medicine and experience with the aviation reporting system. In *A Tale of Two Stories: Contrasting Views of Patient Safety*, ed. R. I. Cook, D. D. Woods, and C. A. Miller. North Adams, MA: US National Patient Safety Foundation.

Brennan, T. A., L. L. Leape, N. M. Laird, L. Hebert, A. R. Localio, A. G. Lawthers, J. P. Newhouse, P. C. Weiler, and H. H. Hiatt.

1991. Incidence of adverse events and negligence in hospitalized patients. Results of the Harvard Medical Practice Study I. *N. Engl. J. Med.* 324 (6): 370–376.

Ekaette, E. U., R. C. Lee, D. L. Cooke, K. L. Kelly, and P. B. Dunscombe. 2006. Risk analysis in radiation treatment: Application of a new taxonomic structure. *Radiother. Oncol.* 80 (3): 282–287.

Holmberg, O. 2007. Accident prevention in radiotherapy. *Biomed. Imaging Interv.* J3: e27.

Holmberg, O., and B. McClean. 2002. Preventing treatment errors in radiotherapy by identifying and evaluating near misses and actual incidents. *J. Radiother. Pract.* 3: 13–25.

Huang, G., G. Medlam, J. Lee, S. Billingsley, J. P. Bissonnette, J. Ringash, G. Kane, and D. C. Hodgson. 2005. Error in the delivery of radiation therapy: Results of a quality assurance review. *Int. J. Radiat. Oncol. Biol. Phys.* 61 (5): 1590–1595.

IAEA. 1999. *The Overexposure of Radiotherapy Patients in San José, Costa Rica*. Vienna: International Atomic Energy Agency (IAEA).

IAEA. 2000. *Lessons Learned from Accidental Exposures in Radiotherapy*. Vienna: International Atomic Energy Agency (IAEA).

IAEA. 2001. *Investigation of an Accidental Exposure of Radiotherapy Patients in Panama*. Vienna: International Atomic Energy Agency (IAEA).

IAEA. 2002. *Accidental Overexposure of Radiotherapy Patients in Białystok*. Vienna: International Atomic Energy Agency (IAEA).

International Commission on Radiological Protection (ICRP). 2000. *Prevention of Accidental Exposure to Patients Undergoing Radiation Therapy*. Oxford: International Commission on Radiological Protection.

Reason, J. 2000. Human error: Models and management. *BMJ* 320: 768–770.

ROSIS. 2009. Radiation Oncology Safety Information System 2009. Available from http://www.rosis.info.

The Royal College of Radiologists, Society and College of Radiographers, Institute of Physics and Engineering in Medicine, National Patient Safety Agency, and British Institute of Radiology. 2008. *Towards Safer Radiotherapy*. London: The Royal College of Radiologists.

United States Nuclear Regulatory Commission (NRC) Event Notification Reports. 2009. Available from http://www.nrc.gov/reading-rm/doc-collections/event-status/event/.

Vincent, C. 2007. Incident reporting and patient safety. *BMJ* 334 (7584): 51.

World Health Organization (WHO). 2009. The conceptual framework for the international classification for patient safety. In *World Alliance for Patient Safety*. Geneva: World Health Organization.

Wilson, R. M., W. B. Runciman, R. W. Gibberd, B. T. Harrison, L. Newby, and J. D. Hamilton. 1995. The quality in Australian health care study. *Med. J. Aust.* 163 (9): 458–471.

Yeung, T. K., K. Bortolotto, S. Cosby, M. Hoar, and E. Lederer. 2005. Quality assurance in radiotherapy: Evaluation of errors and incidents recorded over a 10 year period. *Radiother. Oncol.* 74 (3): 283–291.

18

Error and Near-Miss Reporting: View from North America

Sasa Mutic
Washington University
School of Medicine

Scott Brame
Washington University
School of Medicine

Introduction

Due to the significant potential for serious and catastrophic errors, patient and employee safety is of critical concern in radiation oncology. One of the key components for developing a highly reliable organization (HRO) (Weick and Sutcliffe 2007) is the ability of an organization to assess its weaknesses and gaps in error prevention measures and identify processes for prevention of potential future errors. Clearly, this requires that an organization has a means of reporting and tracking errors. Due to this need, error and near-miss reporting have been the subject of a great deal of investigation and development for several decades and across a wide spectrum of industries, including some areas of healthcare. It is commonly accepted that the value of error and near-miss reporting for the purposes of enhanced system reliability, performance improvement, and error prevention cannot be overstated.

Implementing an effective reporting program is a challenging task requiring some or all of the following:

- An organizational culture in which employees are motivated to report errors and near-misses
- An understanding of what data should be collected and a standardized reporting terminology
- Efficient systems for data collection that are easy to use yet comprehensive enough to allow collection of all relevant data (Reason 1997)
- The competence to interpret reported data and identify important trends and weaknesses in the system

- A willingness to implement, when necessary, significant changes based on collected data and subsequent analyses
- The ability to share the collected data, not only within the individual organization, but also with a broader community so as to accelerate the development of a collective knowledge base

Error and near-miss reporting can be accomplished under statutory or voluntary reporting programs; both systems are used in radiation oncology. Numerous regulatory bodies legally obligate radiation oncology facilities to report errors through statutory reporting programs. These programs primarily require the reporting of actual errors, meaning near-misses are rarely addressed. As later described, near-miss reporting is extremely important and can result in process improvements before actual errors are made. Voluntary reporting programs are designed to collect both errors and near-misses. Since the programs are voluntary, they rely on the ability of an organization to create and sustain an atmosphere wherein employees value participation in the reporting process. Ultimately, voluntary programs have more potential, but are much more difficult to construct and maintain.

The existing error and near-miss reporting programs in radiation oncology are inconsistent and vary in sophistication and effectiveness. This is not surprising given the broad spectrum of local resources and expertise among radiation oncology facilities. However, this variety hinders community-wide communication and exchange of error and near-miss data, an exchange that holds great potential for expanded data collection, knowledge assimilation, and, ultimately, the provision of better, safer treatment to more patients. Several important healthcare bodies have discussed the need for this type of data exchange and collaboration. In a recent report entitled *Towards Safer Radiotherapy* authored by the Royal College of Radiology (2008), Sir Liam Donaldson, Chief Medical Officer of the British Department of Health stated:

> In my Annual Report on the State of Public Health in 2006 I drew attention to the problem of radiotherapy safety... in a number of unfortunate cases over the last few years, overdoses of radiation led to severe harm to patients. It is recognized that these are uncommon events, yet their impact on the patient, staff involved and the wider health service are devastating. Not only does it compromise the delivery of radiotherapy, it calls into question the integrity of hospital systems and their ability to pick up errors and the capability to make sustainable changes.... We still have not yet mastered the art of harnessing all available knowledge, both national and international, to reduce adverse events in healthcare. (Donaldson 2007)

The World Health Organization (2009) published a similar report outlining risks in radiation oncology and the need to collect error data and implement system improvements based on prior events.

A prerequisite for the development of a successful error and near-miss reporting program is the creation and cultivation of a *safety culture* within an organization. In addition to the right attitude and philosophy, an institution must also have mechanisms to efficiently process and analyze reported events and then make sustainable changes in response to what is learned. This chapter addresses the components of an effective reporting program and outlines how such programs can be created in radiation oncology departments. To simplify the text, errors and near-misses will be collectively referred to as events; distinction between the two will be made only when necessary.

Organizational Culture

The success of event reporting programs, especially voluntary ones, depends on the value and importance that an organization assigns to them. This value and importance is established through organizational culture. Every organization has a unique behavior and culture, which will ultimately determine the success of the reporting system. Historically, the term *culture* has been used to describe attributes of different nationalities. During the 1980s, culture became one of the concepts used to describe an organization (Reason 1997). In short, an organizational culture can be described as:

> Shared values (what is important) and beliefs (how things work) that interact with an organization's structures and control systems to produce behavioral norms (the way we do things around here). (Uttal 1983)

One of the terms often used to describe an organization's approach to safety is *safety culture*. As an in-depth description and analysis of safety culture is beyond the scope of this text, the reader is directed to other resources (Reason 1997). However, it is necessary to understand that, for an event reporting program to be effective, there has to be an atmosphere within the radiation oncology department that encourages and enables event reporting, especially in voluntary programs. It can be difficult to motivate individuals to report events that they have precipitated, or events resulting from the actions of colleagues and/or friends. Event reporting programs can create situations where individuals feel threatened to file a report. The entire organization (from top management down) is responsible for creating a *reporting culture* (one of the components of a safety culture). As described by Reason (1997), when a reporting culture is created, members of an organization are motivated to file reports by the existence of the following factors:

- Indemnity against disciplinarily proceedings and retribution
- Confidentiality
- To the extent practical, separation of those collecting the event data from those with the authority to impose disciplinary actions
- An efficient method for event submission
- A rapid, intelligent, and broadly available method for feedback to the reporting community.

A *just culture* is another component of a safety culture and is interwoven with a reporting culture. In a just culture, an organization strives to define acceptable and unacceptable actions. Since one of the necessary components of an effective reporting culture is indemnity against disciplinary proceedings, it is important to emphasize that the mere reporting of unacceptable behavior or action does not indemnify an individual against corresponding consequences. Establishing a well-defined distinction between acceptable and unacceptable events can be difficult. Many unacceptable situations are readily obvious; examples include negligence, deliberate dereliction of duties, and falsification of facts and records. These are most often discovered and reported through mechanisms outside an event reporting program. A separate group of events are those that require an in-depth analysis of circumstances and the consideration of multiple factors before a determination can be made. Unjust processing of such events can be detrimental to the reporting process and every precaution should be taken to ensure that disciplinary actions are taken only when all circumstances, and implications, are clearly understood. This means that in certain situations the conservative conclusion will be that an event was attributable to acceptable actions.

Culture holds the key to unlocking the potential of a reporting program. An environment must exist within which employees understand and appreciate the value of error reporting to the overall mission of the organization and the role they play in ensuring patient and staff safety. Management must understand as well that employee participation and reporting are the foundations of the program; reactions to reported events should be thoughtful, judicious, and always in the context of continual process improvement.

Errors and Near-Misses

Radiation oncology is a unique medical specialty with respect to the number, types, and spectrum of events that can occur. Potential events in radiation oncology include unintended exposures of staff and the public, tumor underdose, normal tissue overdose, target volume misses, and unintended exposure of critical structures. Dose-related events can range in magnitude from a few percent to lethal doses. Similarly, anatomical misses can range in magnitude from millimeters to completely inaccurate site treatments. Events can be random, but historically have been systematic, affecting dozens of patients (The Royal College of Radiologists et al. 2008; WHO 2009). Sometimes, events result in no harm to the patient but must still be reported pursuant to governing regulations. Staff, treatment planning and delivery software, and imaging and treatment hardware are all possible sources of events. With such an array of possible events—and with such variety in reporting requirements—it is helpful to understand what distinctions are typically made between errors and near-misses in radiation oncology.

Errors

Errors are commonly defined as events that result in harm or damage. Obviously, errors can be separated into minor deviations and actual errors of concern—that is, there is a spectrum of "harm." As noted earlier, errors in radiation oncology can range from a difference between delivered and intended dose of a few percent, all the way to catastrophic events that result in patient death. Often, minor deviations are considered acceptable and may be allowed to go uncorrected. For example, monitor units incorrectly calculated by 1% may not be corrected due to acceptance of minor variations between prescribed and delivered doses. As described in the next section, all errors should not be treated equally: it is necessary to define reportable errors, both in statutory and voluntary programs.

Near-Misses

Near-misses are preludes to errors and are generally defined as events that did not result in harm or damage, but had the potential to if they had gone undetected. Other terms for near-misses are near-hits, near-collisions, free lessons, and close calls. A near-miss can be a single event that is immediately discovered, or a chain of events that could have resulted in an error had the chain gone unbroken. Many near-misses pose as small, seemingly benign events. Reporting and analysis of such events may often appear to be of questionable value, especially given the percentage of small, seemingly benign events that are, in fact, benign. However, single event errors are rare: Most errors result from a chain of small events which, when combined, have disproportionately large effects. Organizations ignore near-misses at their own risk; some near-misses are valuable opportunities to learn what can go catastrophically wrong in a process—before the catastrophe occurs. In the words of Weick:

> We know that single events are rare, but we do not know how small events can become chained together so that they result in a disastrous outcome. In the absence of this understanding, people must wait until some crisis actually occurs before they can diagnose a problem, rather than be in a position to detect a potential problem before it emerges. To anticipate and forestall disasters is to understand regulations in the ways small events can combine to have disproportionately large effects. (Weick 1991)

Subsequent sections will describe systems and processes that can reduce the effort and expense associated with near-miss reporting, as well as techniques that assist in discovering the harmful potential of individual near-misses.

In a clinical setting, it may be difficult to establish where near-misses end and errors begin. There are events that cause minor harm to the patient but, due to timely discovery, did not result in a catastrophic error. For example, imagine a situation in which incorrectly calculated monitor units result in a patient overdose of 20% during the first treatment fraction in a course of therapy. A physician notices that the treatment monitor units appear unusual and further investigation reveals that the monitor units were indeed incorrectly calculated. The error is corrected and the patient is treated correctly for the remainder of the treatment fractions. While this event was an *error* which resulted in "tolerable harm" to the patient it was also a *near-miss* that could have resulted in a catastrophic error had the physician not discovered the incorrect calculation. It should be obvious that it is irrelevant whether this event is classified as an error or a near-miss. The important point here is that there is a weakness in the treatment planning process that allowed incorrect monitor unit calculation and delivery. This blending of an error and a near-miss further demonstrates the value of reporting of both types of events and, as will be discussed in the next section, highlights the difference in potential benefit between statutory and voluntary programs.

Reportable Events

Clearly communicated definitions and reporting expectations for errors and near-misses are needed for an effective event reporting program. Without this foundation, potentially valuable events will be considered insignificant, increasing the chances of a catastrophic event. Statutory programs communicate definitions and reporting requirements via regulations; voluntary reporting programs rely on the individual institution to define terms and set reporting policy. The following questions should be considered when implementing statutory and voluntary programs.

Statutory Reporting—Errors

What Agencies Subject an Institution to Statutory Reporting?

Regulatory agencies can include the hospital administration, state and federal agencies, and accreditation bodies.

What Are Reportable Errors?

Statutory reporting is typically based on a specific set of regulations that classify an error as reportable or not. Reporting requirements are based on error type and magnitude, some errors are automatically reportable regardless of the magnitude while others have to be of a certain magnitude before reporting is required. It is extremely important that the reporting program educates all staff on these regulations and develops or employs personnel with in-depth regulatory knowledge and expertise. Some errors are not easily classifiable; local expertise will be needed to make determinations in such situations. Regardless, the program should ensure that unnecessary error reporting is avoided and, more importantly, that all required errors are reported.

How Is an Error Reported?

In addition to identifying the relevant regulatory agencies and which errors are reportable, there should be a defined process for documenting and communicating errors to regulatory agencies. This process should ensure that all the necessary information is communicated in an adequate form, to the correct place, in timely manner. Regulatory agencies typically have specific reporting requirements regarding communication channels, reporting form, and timing. For example, the U.S. Nuclear Regulatory Commission in the U.S. Code of Federal Regulations Title 10 § 35.3045 Report and Notification of a Medical Event provides detailed requirements for reporting medical events that involve the use of byproduct material (NRC 2009).

Statutory Reporting—Near-Misses

Which Near-Misses Are Subject to Statutory Reporting?

Statutory reporting typically does not address near-misses and any such reporting is often due to concerns that practitioners may have with certain procedures or equipment. Requirements for such reporting are typically not well-defined and the decision to report such events must be based on the practitioner's professional judgment.

Voluntary Reporting

Effective voluntary reporting programs do not distinguish between errors and near-misses. With the exception of those errors subject to regulatory reporting requirements, these programs will use a unified reporting requirements and mechanisms. The questions that need to be answered are the same for all events (see below).

What Are Reportable Events?

In voluntary reporting systems it is difficult to clearly define reportable events. Reporting requirements could be based on statutory examples. However, these examples often have high thresholds that will result in few events being reported. Furthermore, events requiring statutory reporting typically deal with those that cause, or have the potential to cause, bodily harm. In voluntary systems, reporting of events that do not have potential to cause bodily harm but have potential to result in operational issues and affect staff and patient satisfaction may also be considered important. For example, failure to notify a patient of an appointment time change is a patient satisfaction event that many clinics would want reported so the event can be analyzed and the patient notification process modified. Examples of events that do not cause actual harm but have a deleterious effect on process performance are bountiful in most clinics. Defining which of these events is reportable is typically left to employee judgment. In the Washington University event reporting process, employees are instructed to report all events that they deem reportable, regardless of their magnitude and severity. As described later, this reporting program is quite manageable, thanks in part to the mechanisms used to determine if an event warrants further investigation or not. This low threshold for event reporting requirements ensures that employees always feel empowered to report events and that their input is extremely valuable. In the Washington University reporting culture, the potential benefits of discovering obscure error propagation mechanisms far outweigh the time and effort required to process insignificant events. In some situations there are *explicit events* that an institution would like to track and monitor. Explicit events have special significance for the overall operation of an organization. These could be events that occur routinely and, as a result, have been identified as opportunities for improvement (e.g., employees not washing their hands). These events could also be specific random problems for which investigation requires timely identification and communication (e.g., many software bugs need to be captured at the time of occurrence). Explicit event reporting identifies these events by name and these events are often specifically outlined in the reporting forms and explicitly described in the reporting procedures.

How Is an Event Reported?

In addition to a well-developed safety culture, the reporting mechanisms in voluntary programs are critically important to the program's success. The reporting process has to be efficient, requiring minimal time to complete a submission. Ideally, minor and explicit events should take no more than a few minutes to report. As described in the next section, this can be accomplished through electronic reporting.

What Should Be in the Event Report?

The Alberta Heritage Foundation for Medical Research published a report, *A Reference Guide for Learning from Incidents in Radiation Treatment* (Cooke et al. 2006), in which they discuss components of event reports in radiation oncology. Typically, the reporting process consists of two steps: an initial event report and, when necessary, an in-depth follow-up. The initial event report must be designed so that the report can be filed quickly, yet detailed enough to support initial classification of the event. The initial report should include who, when, what, how many, and where sections along with a free text option for description of the event as seen by the reporter. The report should also include items that will facilitate routing to correct supervisor(s) and processing priority.

Reporting Systems

While most radiation oncology event reporting programs currently rely on paper reports, the ongoing pursuit of paperless clinical environments will certainly lead to the proliferation of electronic reporting systems. While this trend is intact, it should be noted that most external statutory reporting programs rely on paper or verbal communication. Given that these agencies collect data from many different institutions, their transition to electronic communication represents both a critical need and a tremendous opportunity. One early example of community-based reporting is the Radiation Oncology Safety Information System (ROSIS 2009). ROSIS is a voluntary Web-based reporting system and safety information database that allows users to contribute events from their institution and review those submitted by others. This system and those that follow it hold the promise of accelerated learning based on data aggregation and sharing. As discussed later, electronic reporting systems facilitate rapid communication, processing, and event analysis. Electronic data can also be mined and used for trend detection, benchmarking, and standardization efforts.

Paper-Based Systems

Figure 18.1 shows an example of an initial incident reporting form (Cooke et al. 2006). The form contains all the essential components for further follow-up and analysis. The form makes clear this organization's definitions of errors (persons *actually affected*) and near-misses (persons *potentially affected*). The form also guides the event narrative by asking for an event description and subsequent actions. As always, there is a need to balance ease of use and degree of specificity. The form must contain enough detail to guide the user through the process and ensure that minimally sufficient detail is provided. However, too much detail can create reports that are burdensome or that, even worse, appear threatening. The form design should reflect employee understanding of the reporting process, contain visual cues and wording that can be easily understood, and encourage the reporter to provide sufficient detail for subsequent event analysis and corrective actions. Figure 18.2 shows a supervisor's report (Cooke et al. 2006) for the event presented in Figure 18.1. The supervisor's form and report build on the initial event report and provide additional information regarding the event. All subsequent forms and analysis reports should complement the

initial reporter and supervisor forms. All forms together should create a unified document that makes clear the event, analysis, and corrective actions. If the reporting process results in several disconnected reporting and analysis forms, it may be very difficult to analyze the event at a later time or to provide effective feedback to the reporter (see the "Feedback Mechanisms" section). Unified documents also facilitate easy organization, filing, and future access.

Paper-based systems require well-defined event distribution rules that ensure efficient information processing and timely delivery of documents to the appropriate places and personnel. Without well-defined distribution rules it is possible that some events will be overlooked or lost. Ideally, report forms will contain distribution rules and guidance for document processing.

Overall, paper-based systems can be very effective. The main shortcomings of these systems are the inability to distribute event reports rapidly and the labor required to copy, distribute, organize, and file reports. Given these issues, and the industry trends mentioned above, it seems inevitable that paper-based reporting systems will be replaced by electronic reporting systems. This process will likely be facilitated by wider availability of commercial event reporting systems that serve the individual institution while, at the same time, providing a means for community-wide data aggregation and sharing.

Electronic Systems

There are a number of electronic event reporting systems that have been designed for hospital-wide use. The use of these systems for voluntary reporting programs in radiation oncology is difficult due to the unique (as compared to other clinical disciplines) spectrum of possible events in radiation oncology and the necessary generality of any system designed for hospital-wide use. Personnel reporting radiation oncology events with these systems will likely find the workflow cumbersome and inefficient, making it unlikely that they will contribute either the volume or detail needed for voluntary reporting programs. Currently, custom-designed radiation oncology specific electronic reporting systems are not widely available and existing solutions are mainly in-house developed solutions.

A radiation oncology–specific electronic error reporting system has been in use at the Barnes-Jewish Hospital, Mallinckrodt Institute of Radiology, Washington University, Department of Radiation Oncology since July 2007. The Process Improvement in Radiation Oncology (PIRO) Web-accessible database tool was designed for the efficient collection, dissemination, storage, and analysis of error data that facilitate the application of systems engineering techniques in radiation oncology (Figure 18.3). The PIRO tool was written in the Asp.net environment using SQL Server 2005 as the database. A key design goal was to minimize event reporting time. This was accomplished through the use of context-sensitive menus and pull-down options. The tool design was informed by recommendations found in *A Reference Guide for Learning from Incidents in Radiation Treatment* (Cooke et al. 2006).

TOM BAKER CANCER CENTRE
RADIATION THERAPY INCIDENT REPORT—ORIGINATOR

Incident: an unwanted or unexpected change from a normal system behavior, which causes, or has the potential to cause, an adverse effect to persons or equipment.

WHO

Please indicate by checkmark *WHO* was potentially affected/actually affected by this incident.

Potentially affected	Actually affected	
☐	☒	*Patient
☐	☐	Public
☐	☐	Staff
☐	☐	Visiting Worker/Student
☐	☐	Not applicable

*Patient Number:_ Z.0000000

*Oncologist notified (for ACTUAL incidents only):
Name:_ Dr. R. Oncologist

Signature:_ R.Oncologist
Date: 2005/06/02   time: 11:30 am

WHAT

Please indicate *WHAT* system(s) were involved in this incident.
☒ Clinical
☐ Occupational
☐ Operational
☐ Environmental
☐ Security/Other:_____

WHERE

Please indicate *where* the indicent occurred

Room number:_____

Work process/area:_____ RT Unit #10

WHEN

Date incident *occurred*:
2005/05/06   time: 10:30 am

Date incident was *discovered*:
2005/06/02   time: 10:30 am

Description of incident

Please briefly summarize the incident:

First treatment: PA MLC visually checked against DDRs in room. PA field treated and EPI taken during treatment. EPI assessed immediately and it was recognized that a 1cm x 2cm area of MLC was missing.

What was your response?
PA MLC checked thoroughly - okay, thus PA field treated. Information taken to calc room upon completion of daily treatment to have MLC leaves adjusted

What other forms (if any) were filled out?

Signature

Name (print): J.Therapist

Signature:_ JTherapist

Date: 2005/06/02   time: 11:00 am

Please submit immediately to your supervisor:

FIGURE 18.1 Example of a paper-based radiation therapy incident form that is submitted by the initial reporter. (From Cooke et al., *Initative Series,* Alberta Heritage Foundation for Medical Resarch, Alberta, Canada, 2006. With permission.)

Figure 18.3a shows the page seen by the initial event reporter. The reporter needs only to enter the patient therapy number and the software populates the patient name and attending physician through an interface to the department's record and verify system. Optionally, all these fields can be manually entered. Event entry is facilitated through option buttons and pull down menus.

The option buttons and pull down menus allow the standardization of data entry and storage in a queryable database capable of providing statistical trends, summaries, and various reports to support the department's process improvement efforts. Explicit events can be reported without any text entry. Random events can be described in a narrative field. The majority of events can

TOM BAKER CANCER CENTRE

RADIATION THERAPY INCIDENT REPORT - SUPERVISOR

Incident Severity

Initial severity classification

Potential	Actual	Severity
	☐	Critical
☐	☐	Major
☐	☐	Serious
	☒	Minor

Additional information needed:

_____

_____

_____

_____

_____

_____

_____

_____

Details of initial response

Radiation Oncologist notified and viewed EPI identified area of MLC variation is small and thus no dose correction necessary.

_____

_____

_____

_____

_____

_____

_____

_____

_____

Individuals Notified

Name (Print): _Dr. R. Oncologist_
Date: 2005/08/02   time: 11:50

Name (Print): _P. Dosimetrist_
Date: 2005/08/02   time: 12:05

Name (Print): _____
Date: YYYY/MM/DD   time: HH:MM

Name (Print): _____
Date: YYYY/MM/DD   time: HH:MM

Name (Print): _____
Date: YYYY/MM/DD   time: HH:MM

Name (Print): _____
Date: YYYY/MM/DD   time: HH:MM

Signature

Name (print): _A. Manager_

Signature: _A. Manager_

Date: 2005/08/02   time: 12:05pm

Note: if you are not a member of the Quality Assurance Committee, please submit this form immediately to one of the following:

	Phone	Pager
RT Safety Officer	12345	6789
Head, Medical Physics	12345	6789
Electronics Dept	12345	6789
Supervisor, Dosimetry	12345	6789
Supervisor, RT	12345	6789
Supervisor, Nursing	12345	6789

FIGURE 18.2 Example of a form for a supervisor's analysis of the report presented in Figure 18.1 (From Cooke et al., *Initative Series*, Alberta Heritage Foundation for Medical Resarch, Alberta, Canada, 2006. With permission.)

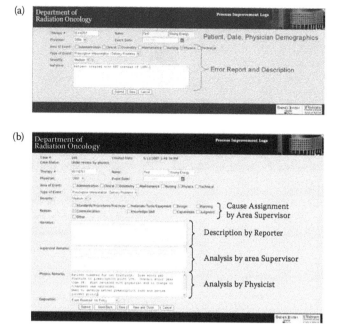

FIGURE 18.3 Example of Web-based event reporting and response database interface. (a) Initial screen as seen by the initial error reporter. (b) Example of the follow-up screen showing the cause assignment and descriptive analysis as defined by the area supervisor and the subsequent analysis by the physicist. The disposition of the event is shown below the analysis by the physicist.

be reported in less than three minutes. More complicated events that require extensive description will take longer to report, depending on the length of the written narrative. After the initial reporter submits an event, the software automatically e-mails the report to area supervisors. Depending on the severity of an event, as selected by the initial reporter, reports may also be automatically e-mailed or sent as a text message to the pagers and mobile telephones of departmental supervisors. This facilitates rapid communication of, and response to, critical events. Figure 18.3b shows the follow-up page with the cause assignment and descriptive analysis as defined by the area supervisor and the subsequent analysis by the physicist. The disposition of the event is shown below the analysis by the physicist. Most events can be processed and analyzed within one day of being reported.

The PIRO event reporting, processing, and disposition tools are designed to facilitate data collection, analysis, trending, and benchmarking. The departmental process improvement committee, consisting of representatives from all functional areas, analyzes reported events and statistics and uses this data to set departmental priorities and implement corrective actions. Subsequently, data is collected to evaluate the effectiveness of corrective actions, often resulting in further process modifications. Process improvement actions, event analysis, and trends are reported back to functional areas for further analysis and recommendations. The feedback mechanism is currently not electronic, but instead is performed by members of the process improvement committee. Faculty and staff have embraced the electronic reporting system; reporting compliance has been exceptional and unwavering, exceeding all expectations.

Feedback Mechanisms

The long-term success of an event reporting program is critically dependent on the feedback provided to event reporters and the broader organization. When institutional leadership implements significant operational and safety improvements in response to reported events, it is important that event reporters know that their efforts affected positive change. The more benefit they see from their reporting, the more involved and committed they become to the event reporting program. The relationship between reporting and corrective action is communicated via direct feedback mechanisms. The optimal feedback mechanism depends considerably on an individual institution. Factors such as size, physical layout, organization, and culture must be considered in the design of a feedback mechanism. Thus, while reporting mechanisms and data processing can be standardized amongst most institutions, feedback mechanisms will often demand customized solutions. Furthermore, the feedback mechanism should be perpetually evolving, adapting to institutional changes, always seeking to ensure that event reporting continues and that the organizational culture continues to evolve in a positive direction. At the minimum, each event report should be followed with a response to the initial event reporter that details the disposition of the event and, if appropriate, any resulting process improvement efforts.

Summary

Event reporting programs and related process improvement activities are critical for optimal and safe operation of radiation therapy facilities. Radiation therapy is a rapidly evolving specialty with continual advances in technology and procedures. Effective implementation of these technologies and procedures requires understanding of the issues and problems related to their use. An event reporting program can facilitate rapid feedback on the design of various processes and will enable significant changes and improvements. As described earlier, event reporting can be performed through statutory and voluntary mechanisms. Ideally, voluntary reporting mechanism should exist in each radiation oncology facility.

For the most part, paper-based reporting systems provide adequate support for event reporting, event analysis, and corrective actions. While they are the most common form of reporting today, eventually they will be replaced with electronic systems. Electronic event reporting methods support learning healthcare systems (Olsen, Aisner, and McGinnis 2007) and have the potential to affect the collective development of radiation oncology, standardization of practices, improvement in treatment safety, and more effective dissemination of best practices.

The success and acceptance of event reporting programs are critically dependent on institutional culture. Departmental management is responsible for developing a culture that supports event reporting and sustainable process and safety improvement efforts. The discussion presented here briefly outlines several components of safety culture. Other chapters in this book and work published elsewhere should also be used when designing and implementing event reporting systems. Ultimately, this is a learning process for each institution and event reporting processes should undergo continual modifications to ensure sustainable effectiveness.

References

Cooke, D. L., M. Dubetz, R. Heshmati, S. Iftody, E. McKimmon, J. Powers, R. C. Lee, and P. B. Dunscombe. 2006. A reference guide for learning from incidents in radiation treatment. In *Initative Series*. Alberta, Canada: Alberta Heritage Foundation for Medical Resarch.

Donaldson, Sir Liam. 2007. *2006 Annual Report of the Chief Medical Officer on the State of Public Health*. London: The Department of Health, UK. Available from http://www.dh.gov.uk/en/Publicationsandstatistics/Publications/AnnualReports/DH_076817.

NRC. 2009. Medical use of byproduct material. *Code of Fedral Regulations*, title 10, part 35.

Olsen, A., D. Aisner, and J. M. McGinnis, eds. 2007. *The Learning Healthcare System: Workshop Summary*. Washington, DC: Institute of Medicine.

Reason, J. 1997. *Managing the Risks of Organizational Accidents*. Hampshire, England: Ashgate Publishing Limited.

ROSIS. 2009. Radiation Oncology Safety Information System 2009. Available from http://www.rosis.info.

The Royal College of Radiologists, Society and College of Radiographers, Institute of Physics and Engineering in Medicine, National Patient Safety Agency, and British Institute of Radiology. 2008. *Towards Safer Radiotherapy*. London: The Royal College of Radiologists.

Uttal, B. 1983. The corporate culture vultures. *Fortune* 108 (8): 66–72.

Weick, K. E. 1991. The vulnerable system: An analysis of the Tenerife air disaster. In *Reframing Organizational Culture*, eds. P. J. Frost, L. F. Moore, M. Reis Louis, and C. C. Lundberg. London: Sage Publications.

Weick, K. E., and K. M. Sutcliffe. 2007. *Managing the Unexpected*. San Francisco, CA: Jossey-Bass.

WHO. 2009. *Radiotherapy Risk Profile*. Geneva: World Health Organization.

Yenice, K. M., D. M. Lovelock, M. A. Hunt, W. R. Lutz, N. Fournier-Bidoz, C. H. Hua, J. Yamada, M. Bilsky, H. Lee, K. Pfaff, S. V. Spirou, and H. I. Amols. 2003. CT image-guided intensity-modulated therapy for paraspinal tumors using stereotactic immobilization. *Int. J. Radiat. Oncol. Biol. Phys.* 55 (3): 583–593.

19

The Impact of Cultural Biases on Safety

Daniel A. Low
UCLA School of Medicine

Sasa Mutic
*Washington University
School of Medicine*

Terence Bostic
CMA St. Louis, Missouri

Introduction

The Institute of Medicine (IOM) published a report in 1999 entitled *To Err Is Human: Building a Safer Health System*, within which they examined the level of safety in medical institutions in the United States (Kohn, Corrigan, and Donaldson 1999; Baker et al. 2003). Using data gathered from the Harvard Medical Practice Study (HMPS) and the Utah–Colorado Medical Practice Study (UCMPS) (Rosenthal and Sutcliffe 2002), they concluded that medical errors cause between 44,000 and 98,000 deaths annually. This is more deaths than result from automobile accidents (43,458), breast cancer (42,297), or AIDS (16,516) (Kohn, Corrigan, and Donaldson 1999). The UCMPS and a study conducted in New York suggested that between 2.7% and 3.7% of all hospitalized patients experienced an adverse event, defined as an injury resulting from medical intervention (Hofmann and Mark 2006; Kohn, Corrigan, and Donaldson 1999).

Organizational factors as antecedents to major industrial accidents have gained significant attention (Perrow 1984; Hurst et al. 1991; Patecornell 1990; Wagenaar and Groeneweg 1987; Weick 1990; Wright 1986; Pidgeon 1991). Several consistent types of organizational factors have been identified as causal to these accidents. First, the accidents happened during the normal course of events and were considered a natural consequence of the hazardous nature of the industries. In spite of this, there were often no procedures to follow and so workers developed their own. The second factor was that there was pressure to get the work done as quickly as possible. Finally, there was a lack of effective communication regarding safety. Hofman and Stetzer (1996) pointed out that "workers who perceive a high degree of performance pressure will focus their attention on completing the work and less on the safety of the work procedures." They found that at the team level, there was a significant association between safety climate and unsafe behaviors and accidents.

High Reliability Organizations (HROs) are organizations that have potentially catastrophic consequences for failure, yet perform with very low levels of failure under demanding conditions (Gaba et al. 2003). Examples of HROs include commercial and military aviation, fire fighters, and the nuclear power industry. Medicine has many of the characteristics of these HROs in that it is often demanding (e.g., in the emergency room), complex, and failures can easily lead to permanent injury or death. Single critical errors in radiation therapy can lead to multiple deaths. What keeps medicine in general and radiation therapy in particular from being considered an HRO is its relatively high error rate.

Gaba et al. (2003) compared the naval aviation and medical industries by conducting a survey on the safety climate of both industries. Naval aviation was one of the first HROs studied in detail (Roberts, Rousseau, and La Porte 1994). Gaba et al. (2003) developed surveys that had parallel questions that queried similar subjects from each industry. In medicine, they selected employees from 15 hospitals; while in naval aviation, they selected aviators from 226 squadrons. The questions concerned the safety climate in each industry and naval aviation was used as the benchmark HRO to compare against medicine. The questions were all true/false and considered resources to adequately conduct the job, the attitude of leadership to changes in plans, the attitude concerning violations of standard operating procedures, safety in training, and the leadership's understanding of risks. The analysis included identifying the answers inconsistent with a safety climate (termed problematic responses). Gaba et al. (2003) found that there was a statistically significant difference in the problematic response rates between healthcare and naval aviators. More disturbingly, the discrepancy increased when only high-hazard healthcare domains were considered. They concluded that a climate in which safety and organizational processes aimed at safety had not been developed in the hospitals they tested.

Hofman and Mark (2006) investigated the relationship between safety climate and medication errors in hospitals. They examined nurse back injury rates and needle sticks as rule-based incidents, medication errors (Kohn, Corrigan, and Donaldson 1999; Bogner 1994) and urinary tract infections as more complex outcomes, and patient perceptions, including patient satisfaction and perceptions of nurse responsiveness. Regarding medication errors, they defined errors as those that resulted in some harm. They hypothesized that positive outcomes of these indicators would be correlated with a strong safety climate, which was measured using nine items from the Zohar (1980) measure of safety climate as revised by Mueller et al. (1999) coupled with 13 items from the Rybowiak et al. (1999) Error Orientation Scale. They found that the overall safety climate significantly predicted the rate of nurse back injuries, medication errors, and urinary tract infections, but not needle sticks. They also studied the role of patient treatment complexity using a 14-item scale assessing patient complexity. They found that the need for a positive safety climate increased as patient problems became more complex. The data suggested a two-factor solution. The first factor focused on the degree to which learning occurs from errors, while the second factor focused on the openness and communication regarding errors.

Cannon and Edmondson (2001) examined constructive responses to failure and how those responses could provide opportunities to learn from failure. One of the core themes of the Total Quality Management movement is turning analysis of error and failure into a positive act that is recognized and valued for its contributions to overall performance. They hypothesized that "The degree to which beliefs about failure are learning-oriented is likely to be shared within and to vary across work groups within the same organizational context," "Effective coaching, a clear direction, and context support are associated with learning-oriented beliefs about failure in working groups," "Shared beliefs about failure are associated with performance in work groups," and "Shared beliefs about failure mediate between antecedent conditions and group performance outcomes" (Cannon and Edmondson 2001).

It is clear from the previous work that considerable work has gone into identifying methods for improving safety records of HROs, and that medicine would like to be and should be an HRO. From Gaba et al. (2003), we learned that medicine has a long way to go before those working in medicine have the same safety-conscious attitude as personnel in naval aviation, itself an HRO. Finally, there was a strong correlation between medical errors and the lack of a safety climate as well as a correlation between medical errors and the complexity of a patient's case.

Parallels to Aviation

While many areas of medicine have taken advantage of the lessons learned from the success of HROs in general and aviation in particular, radiation therapy has not. A Pubmed search using the keywords "errors," "radiation therapy," and "aviation" found one relevant publication that suggested only that radiation therapy kept records of accidents in a similar fashion as aviation.

Radiation therapy safety beyond its historical approach is only beginning as an area of independent study. Traditionally, medical physicists have been responsible for setting up and monitoring the radiation therapy safety programs and have done an highly reliable job at assuring that the radiation therapy machines are producing the prescribed dose, that the dose calculations are accurate, that the patient is positioned relatively accurately, and that the radiation treatment plan is accurately delivered. However, the ever-increasing complexity of radiation therapy continues to challenge medical physicists to continually update their techniques and tools.

One of the more interesting findings in the study of aviation safety is the influence of culture on aviation safety. In the work of Helmreich and Merritt (1998), *Culture at Work in Aviation and Medicine*, they identify key cultural components that impact aviation safety. They divide culture into professional, national, and organizational, and identify key aviation accidents that were caused by cultural influences. They examined the process of standardizing operational procedures throughout the world and emphasized that procedures that worked for pilots for some countries failed utterly when applied to pilots of other countries due to cultural variations in human-to-human interactions and the underlying assumptions made by those that defined the aviation procedures.

While the standardization of procedures has had challenges, the result of the modern aviation safety programs has been nothing but amazing. There continue to be an increasing number of flights, yet the accident rate for commercial aircraft has plummeted, at least for the western air carriers. This is due in part to improved technology, but also due to the sensitivity of those that define training programs and operations procedures to human factors and the influence of culture in the cockpit. Pronovost et al. (2009) suggested that "Healthcare's slow and disappointing efforts to improve safety contrast with the remarkable success of aviation safety."

While medical physicists continue to do an excellent job at developing radiation safety practices and QA procedures, given what has been seen in aviation, one can ask, "Are there any significant cultural issues that could negatively impact radiation therapy safety?" It is our hypothesis that the answer is a definitive "Yes," but we are challenged by a fundamental issue: lack of data. In aviation, the operations of the key players are continuously recorded such that many, if not most, commercial aircraft accidents can be deconstructed into their component causes using the recorded voices and aircraft operations. These data are eventually made public, as are the conclusions of the investigation teams, so the entire aviation industry (and other industries) can learn from these tragic accidents (Pronovost et al. 2009). Medicine in general and radiation therapy in particular, have historically taken almost the opposite approach. The accidents and incidents are kept private, the underlying causes may not have been recorded, the records are often sparse, and there is no mechanism for collecting and analyzing accident data.

Culture

Culture in this context is defined as "The predominating attitudes and behavior that characterize the functioning of a group or organization" (Carroll 2006). The groups can be as small as a family, and as large as an entire country. For purposes of this study, we examine the issue of culture to identify if there are any cultural features relevant to radiation therapy safety.

In 1980, Geert Hofstede (Hofstede 1980) published a seminal book entitled *Culture's Consequences: International Differences in Work-Related Values*. In that book, he posed a theoretical reasoning and statistical approach to studying the influences of culture. Later, Hofstede and Hofstede (2005) were given the opportunity to study a large body of survey data about values of people in more than 50 countries. These people were employees of the multinational company IBM. This database enabled Hofstede and Hofstede (2005) to evaluate core human values across a broad spectrum of countries and is the compendium used by many psychologists and sociologists to define differences between cultures. This lead to defining cultural differences across four dimensions (quoted from Hofstede and Hofstede, 2005):

1. Social inequality, including relationship with authority
2. The relationship between the individual and the group
3. Concepts of masculinity and femininity: the social and emotional implications of having been born a boy or a girl
4. Ways of dealing with uncertainty and ambiguity and the expression of emotions

These are shortened to: power distance, collectivism versus individualism, femininity versus masculinity, and uncertainty avoidance.

While these dimensions were defined in a national context, each can be examined throughout the spectrum of cultural contexts, including families and organizations. In this work, we are interested in organizational aspects of culture, specifically through the eyes of the American culture.

Power Distance

We hypothesize that, of the four cultural dimensions in the previous section, the one most relevant to radiation therapy safety is power distance. In a corporate context, power distance describes the attitude of staff, both leaders and followers, to accede to the authority of the leaders and the subordination of the followers.

Studies of power distance are quantified by a power distance index (PDI), which is based on survey questions provided by the IBM database and subsequent updates to that database. The PDI has typically been normalized such that the countries with the lowest and highest power distances have been assigned values of 0 and 100, respectively. Rather than renormalize, when countries are added to the analysis through updates to the surveys, they may have PDIs outside the range. Higher scores indicate greater employee fear to express disagreement with his or her managers.

For reference, the United States has a PDI of 40 and ranks near 58 out of 74 countries (in decreasing PDI order). France has a PDI of 68 and ranks near 28. The countries with the largest PDIs are Malaysia and Slovakia and countries with the smallest PDIs are Israel and Austria.

Radiation Therapy Organization

Radiation therapy, like other areas of medicine, has a natural hierarchy, and is subdivided into five groups based on responsibility and job description. The five groups are: physicians, medical physicists, dosimetrists, radiation therapists, and nurses. One can argue that administrators also play an important role in the practice of radiation therapy, but they do not play as integral a role in patient treatments as do the other five groups.

Because this is medicine, the radiation oncologists have a well-defined leadership role. They are typically considered in charge in the department and in virtually every academic department, the chairman of a radiation therapy department is a radiation oncologist. The medical physicist plays a key role in the management of the patient's treatment and can be considered a partner with the radiation oncologist. The dosimetrist typically works under the direct supervision of the medical physicist to develop the treatment plans, preliminary paperwork, and data transfer. The radiation therapist (technologist) delivers the treatment and typically works somewhat independently from the other groups. The nurse works under the supervision of the radiation oncologist to coordinate the patient's treatment, medical care, and schedule.

Ideally, these groups work tightly together as a team to assure that the treatment plan is optimal, the treatment is delivered according to the plan, and that any deviation from the expected outcome is handled immediately and effectively. Barrett et al. (2001) proposed that formal teamwork training would produce an environment of safety and, consequently, reduce medical errors. The fact that a hierarchical relationship exists between the different members of the radiation oncology team, and that the hierarchical differences between team members is defined by the local organizational culture can lead to differences in the depth of teamwork between organizations. In high power distance organizations, a member in a position that is perceived as lower in the power structure would have difficulties asking questions or challenging those above them. Given the high level of complexity in modern radiation therapy, the hierarchical group structure, and the very many steps involved in the planning and delivery of treatments, there is a need to understand the role that culture has on radiation therapy safety. A large power distance could decrease the ability for the team to avert mistakes in treatments. Errors are likely to occur in high power distance departments during handoffs of the data used to develop the treatment. A high power distance can interfere with the ability of members in one group to ask questions or point out concerns.

Radiation Therapy and Power Distance

A large power distance, coupled with a strong desire to be "right" in one or more of the participants, can lead to both documented and undocumented errors. An error is defined as leading to either the overdosing of a normal organ, causing an unnecessary side effect, or the underdosing of the tumor, leading to an incomplete treatment and subsequent unnecessary cancer relapse, which is almost always fatal. Because there is little to no data to cite, anecdotes are required to examine this point.

Underdosing

It is surmised in the radiation therapy community that some radiation oncologists would prefer to underdose tumors rather than risk lawsuit-generating side effects that correlated with more aggressive therapy. The radiation therapy team members may know the modern dose protocols, yet be unwilling to step forward and point out the inappropriately low-dose prescription. A contributing factor is the lack of outcomes record keeping that would identify the poor outcomes record.

Uncorrected Errors

A relatively common error in radiation therapy is the mispositioning of a patient, due for example to an arithmetic error in shift definition or an error in the applied shift. This error is typically identified by a portal film or portal image that compares the patient's actual position to the planned position. Physicians may not spend sufficient time reviewing the films, and in a clinic with a high power distance and poor intrapersonal communications, another team member may be reluctant to point out the positioning error. Worse, the error may be pointed out but the physician may choose to ignore it and the rest of the team may not be willing to confront the physician or medical physicist to correct the error.

Suboptimal Treatment Plans

Medical physicists may not take the time to assure that a treatment plan has been fully optimized. Dosimetrists may be unwilling to get assistance from the medical physicist to improve the treatment plan. A physician may be impatient and approve a suboptimal treatment plan. Each of these will lead to a suboptimal treatment plan being delivered.

Even within an institution, standardization and formalization of radiation therapy protocols may not be widely practiced. Unfortunately, the history of medicine has led to a paucity of accident data as well as a lack of useful survey data that would provide critically important quality insights. The scenarios mentioned here are anecdotal, but they likely occur every day and

are due to a combination of large power distance and weak corporate leadership. It is our hypothesis that these two challenges combine to lead to significant numbers of suboptimal treatments and that successfully addressing these issues would substantially improve the quality of radiation therapy. More research regarding the influence on power distance as it relates to culture and safety in radiotherapy should be conducted.

Response

Medicine is only just starting to critically examine the roles of organization, training, and culture as causes of accidents. The history of medicine has led to the current lack of data on the number and causes of accidents and this includes radiation oncology. Even considering its excellent safety record, the commercial aviation industry in partnership with the United States government formed the Commercial Aviation Safety Team (CAST) with a goal of reducing the United States commercial aviation fatal accident record by 80% within 10 years (CAST). Pronovost et al. (2009) recently suggested that the lessons learned from CAST be implemented in medicine, including:

1. Standardize work processes
2. Use checklists to ensure patients consistently receive evidence-based interventions
3. Improve teamwork and communication to reduce errors
4. Use robust scientific methods in collaborative efforts to identify and mitigate risk

The standardization of work processes is challenged by attitudes in radiation oncology departments that each clinic should develop their own processes. Treatment guidelines provided by national associations such as the American Association of Physicists in Medicine (AAPM) and the American Society of Radiation Oncology (ASTRO) often include in their guidelines that clinics should customize procedures for their institutions. For example, in the recent AAPM report of the quality assurance of medical linear accelerators, one of the most basic functions of a medical physicist, the report included the following text:

> The recommendations of this task group are not intended to be used as regulations. These recommendations are guidelines for qualified medical physicists to use and appropriately interpret for their individual institution and clinical setting. Each institution may have site-specific or state mandated needs and requirements which may modify their usage of these recommendations. (Klein, Hanley, and Bayouth 2009)

During a recent symposium on quality assurance in radiation therapy sponsored by both AAPM and ASTRO, one of the conclusions of the conference was: "Industrial engineers and human factor experts can make significant contributions toward advancing a broader, more process-oriented, risk-based formulation of RT QA" (Williamson et al. 2008).

From the same symposium, Silvey, Warrick, and Chapin (2005) recommended that moving radiation therapy to becoming an HRO would require third-order change management approaches, triple-loop learning techniques, and committed executive leadership. Third-order change was defined by Bartunek and Moch (1987) as the development of the capacity of a group to change organizational frameworks for understanding events as those events required. Single-loop learning involves identifying errors so that the system can modify techniques to be more consistent, but not fundamentally modifying the procedures themselves. Double-loop learning involves changing the knowledge base and skills available to an institution, while triple-loop learning involves learning and developing structures for how to learn. Those institutions that actively conduct triple-loop learning incorporate process evaluation when developing new procedures and monitor the successes of past and present changes.

Silvey, Warrick, and Chapin (2005) stated that improving radiation therapy safety will require us to go beyond quality assurance and quality control to quality improvement by redesigning individual processes we use in treating cancer patients. We will need to implement programs such as failure modes and effects analysis and root-cause analysis to develop programs that question why specific processes are in place and whether new processes would be more effective. There are five QI dimensions that contain a number of change concepts necessary for creating successful process improvement systems (Silvey, Warrick, and Chapin 2005; Silvey and Warrick 2008).

1. Empowerment: Individuals are empowered by defining their QI responsibility and authority. They are equipped with the knowledge, skills, time, and resources required to perform their responsibilities. They are provided with feedback and evaluations regarding their performance and their efforts are acknowledged through recognition and reward when demonstrating desired QI behaviors.
2. Clinical management strategies: The use of tools, such as checklists, techniques, and technologies to facilitate patient care and optimize patient safety.
3. Quality management strategies: Apply principles such as data-driven decisions and evidence-based protocols. Examples include six-sigma and failure modes and effects analysis.
4. Monitoring: Monitoring the patient care experiences, processes, and outcomes through data collection, analysis, result documentation, and review.
5. Communication: Communicate QI activities, priorities, and results.

In order for radiation therapy to become an HRO, it will need to embrace the previous five QI behaviors as well as develop a committed executive leadership with the vision necessary to conduct the challenging but important evolution of the current radiation therapy culture. Part of that change will require training and enlightenment of the radiation therapy team to act as a team, not a set of independent silos. The complexity of current radiation therapy, the steep rise in automation, and the tightening of tumor margins and dose escalation will mandate that the radiation therapy team change its culture and embrace QI techniques.

Summary

Developing a safety culture will not be as simple as translating work from other industries such as steel factories, offshore environments, and the nuclear power industry (Katz-Navon, Naveh, and Stern 2005). The concept of a safety culture will require development because of the differences between medicine and the other industries for which safety cultures are relatively mature. For medicine in general, and radiation therapy in particular, many types of errors lead to harm to the patient, rather than the provider. In aviation and on the sea, the captain and crew may go down with the ship, so they have a vested interest in driving a safety culture. Second, radiation therapy is not only complex, but the goals and optimal approaches to achieving those goals are not universally agreed upon. There remains an art to medicine and radiation therapy is no exception, so human judgment will remain a key component of the radiation therapy process. Third, the radiation therapy process is not controlled entirely by the administration or regulatory bodies, but also by the radiation therapy team itself. These differences mean that implementing QI techniques and developing a culture of safety in radiation therapy will be a significant challenge, but one that we must engage in to significantly improve the quality and efficacy of this important medicine.

References

Baker, D. P., S. Gustafson, J. Beaubien, E. Salas, and P. Barach. 2003. *Medical Teamwork and Patient Safety: The Evidence-Based Relation*. Washington, DC: American Institutes for Research.

Barrett, J., C. Gifford, J. Morey, D. Risser, and M. Salisbury. 2001. Enhancing patient safety through teamwork training. *J. Healthc. Risk Manag.* 21 (4): 57–65.

Bartunek, J. M., and M. K. Moch. 1987. First-order, second-order, and third-order change and organizational development interventions: A cognitive approach. *J. Appl. Behav. Sci.* 23: 483–500.

Bogner, M. S. 1994. *Human Error in Medicine*. Hillsdale, NJ: L. Erlbaum Associates.

Cannon, M. D., and A. C. Edmondson. 2001. Confronting failure: Antecedents and consequences of shared beliefs about failure in organizational work groups. *J. Organ. Behav.* 22: 161–177.

Carroll, R. 2006. *Risk Management Handbook for Health Care Organizations*. 5th ed. San Francisco, CA: Jossey-Bass.

CAST. Available from http://www.cast-safety.org.

Gaba, D. M., S. J. Singer, A. D. Sinaiko, J. D. Bowen, and A. P. Ciavarelli. 2003. Differences in safety climate between hospital personnel and naval aviators. *Hum. Factors* 45 (2): 173–185.

Helmreich, R. L., and A. C. Merritt. 1998. *Culture at Work in Aviation and Medicine: National, Organizational, and Professional Influences*. Brookfield, VT: Ashgate.

Hofmann, D. A., and B. Mark. 2006. An investigation of the relationship between safety climate and medication errors as well as other nurse and patient outcomes. *Pers. Psychol.* 59 (4): 847–869.

Hofmann, D. A., and A. Stetzer. 1996. A cross-level investigation of factors influencing unsafe behaviors and accidents. *Pers. Psychol.* 49 (2): 307–339.

Hofstede, G. H. 1980. *Culture's consequences: International Differences in Work-Related Values. Cross Cultural Research and Methodology Series*. Beverly Hills, CA: Sage Publications.

Hofstede, G. H., and G. J. Hofstede. 2005. *Cultures and Organizations: Software of the Mind*, 2nd ed. New York, NY: McGraw-Hill.

Hurst, N. W., L. J. Bellamy, T. A. W. Geyer, and J. A. Astley. 1991. A classification scheme for pipework failures to include human and sociotechnical errors and their contribution to pipework failure frequencies. *J. Hazard. Mater.* 26 (2): 159–186.

Katz-Navon, T., E. Naveh, and Z. Stern. 2005. Safety climate in health care organizations: A multidimensional approach. *Acad. Manage. J.* 48: 1075–1089.

Klein, E. E., J. Hanley, and J. Bayouth, et al. 2009. Task Group 142 report: Quality assurance of medical accelerators. *Med. Phys.* 36 (9).

Kohn, L. T., J. M. Corrigan, and M. S. Donaldson, eds. 1999. *To Err is Human: Building a Safer Health System*. Report of the Committee on Quality of Health Care in America, Institute of Medicine. Washington, DC: National Academy Press.

Mueller, L., N. DaSilva, J. Townsend, and L. Tetrick. 1999. An empirical evaluation of competing safety climate measurement models. Paper read at 14th Annual Conference of the Society for Industrial and Organizational Psychology, Atlanta, GA, April 30–May 2.

Patecornell, M. E. 1990. Organizational aspects of engineering system safety—The case of offshore platforms. *Science* 250 (4985): 1210–1217.

Perrow, C. 1984. *Normal Accidents: Living with High-Risk Technologies*. New York, NY: Basic Books.

Pidgeon, N. F. 1991. Safety culture and risk management in organizations. *J. Cross Cult. Psychol.* 22 (1): 129–140.

Pronovost, P. J., C. A. Goeschel, K. L. Olsen, J. C. Pham, M. R. Miller, S. M. Berenholtz, J. B. Sexton et al. 2009. Reducing health care hazards: Lessons from the commercial aviation safety team. *Health Aff. (Millwood)* 28 (3): w479–w489.

Roberts, K. H., D. Rousseau, and T. La Porte. 1994. The culture of high reliability: Quantitative and qualitative assessment aboard nuclear powered aircraft carriers. *J. High Technol. Manage. Res.* 5: 141–161.

Rosenthal, M. M., and K. M. Sutcliffe. 2002. *Medical Error: What Do We Know? What Do We Do?* 1st ed. San Francisco, CA: Jossey-Bass.

Rybowiak, V., H. Garst, M. Frese, and B. Batinic. 1999. Error orientation questionnaire (EOQ): Reliability, validity, and different language equivalence. *J. Organ. Behav.* 20 (4): 527–547.

Silvey, A. B., L. W. Warrick, and C. Chapin. 2005. Identification and synthesis of component essential to achieving "high performer" status in various provider types: Report to CMS. HSAG 2005. Available from http://www.hsag.com/services/special/highperformers.aspx.

Silvey, A. B., and L. H. Warrick. 2008. Linking quality assurance to performance improvement to produce a high reliability organization. *Int. J. Radiat. Oncol. Biol. Phys.* 71 (1 Suppl): S195–S199.

Wagenaar, W. A., and J. Groeneweg. 1987. Accidents at sea—Multiple causes and impossible consequences. *Int. J. Man-Mach. Stud.* 27 (5–6): 587–598.

Weick, K. E. 1990. The vulnerable system—An analysis of the Tenerife air disaster. *J. Manage.* 16 (3): 571–593.

Williamson, J. F., P. B. Dunscombe, M. B. Sharpe, B. R. Thomadsen, J. A. Purdy, and J. A. Deye. 2008. Quality assurance needs for modern image-based radiotherapy: Recommendations from 2007 interorganizational symposium on "quality assurance of radiation therapy: Challenges of advanced technology." *Int. J. Radiat. Oncol. Biol. Phys.* 71 (1 Suppl): S2–S12.

Wright, C. 1986. Routine deaths—Fatal accidents in the oil industry. *Sociol. Rev.* 34 (2): 265–289.

Zohar, D. 1980. Safety climate in industrial organizations—Theoretical and applied implications. *J. Appl. Psychol.* 65 (1): 96–102.

Primer on High Reliability Organizing

Peter F. Martelli
University of California, Berkeley

Karlene H. Roberts
University of California, Berkeley

Introduction

By now, everyone in the health field should be familiar with the Institute of Medicine's 2000 report *To Err Is Human* and its astonishing finding that between 44,000 and 98,000 Americans die each year as a result of medical errors (Kohn, Corrigan, and Donaldson 1999). The report served as a wake-up call and particularly drew attention to safety as "primarily a systems problem" (Leape, Berwick, and Bates 2002). This short chapter will focus on the systems issue, and introduce the reader to high reliability organizing (HRO) theory, a systems approach developed out of decades of studying failure and resilience.

It should come as no surprise that safe and reliable operations in radiology are critical to ensuring patient safety. In diagnostic radiology, errors most commonly result from perceptual misses, errors in judgment, and communication deficiencies. Errors in diagnostic radiology are relatively common and can lead to "delayed diagnosis and treatment, the failure to recognize a complication of treatment, the performance of a study when not indicated or when contraindicated, or the failure to supervise or monitor a case" (Alpert and Hillman 2004). In radiotherapy, errors commonly result from alignment issues, calculation and input errors, and incorrect preparation and planning, as well as from poor documentation and communication. These errors may lead to wrong-side intervention, delays in delivering therapy, reduction in tumor control, increased normal tissue toxicity, or more serious injury (The Royal College of Radiologists 2008).

System View

From its beginnings, radiology and radiotherapy has been notable for its technological sophistication, high work specialization, professionalized and varied support staff, and interdependent relationship with other disciplines. When systems display high levels of technical complexity and interdependence, quality interventions that target components of the system independently are bound to be insufficient to prevent error. Aside from the most blatant violations, errors in both diagnostic radiology and radiotherapy are unlikely to occur in isolation; individual failures of observation (e.g., misreading) or lapses in procedure (e.g., inadequate preparation) compound when checkpoints in the system fail or when other members of the system are not cognizant of, allowed to, or willing to address them. Approaching safety in radiology practice requires addressing the situational and systemic, not just the individual, factors that facilitate errors.

A system is a set of interdependent parts that share a common aim. The Agency for Healthcare Research and Quality (AHRQ) defines the systems approach, noting that:

> most errors reflect predictable human failings in the context of poorly designed systems (e.g., expected lapses in human vigilance in the face of long work hours or predictable mistakes on the part of relatively inexperienced personnel faced with cognitively complex situations). Rather than focusing corrective efforts on reprimanding individuals or pursuing remedial education, the systems approach seeks to identify

situations or factors likely to give rise to human error and implement 'systems changes' that will reduce their occurrence or minimize their impact on patients (Agency for Healthcare Research and Quality (AHRO), *Patient Safety Network (PSNet) Glossary*).

A clinical practice is a system in which interdependence is notable through workflow and team work, and improving interactions of the components contributes to performance. Problems in a clinical system stem from issues in several domains, including cognitive capacity (e.g., required information not accessed), communication (e.g., errors in written or oral orders), therapy error (e.g., dosage), and patient involvement in the care plan (Kilo 2008). Creating systems of safe care will require overcoming strategic, structural, and technical barriers, as well as cultural dysfunctions related to professional protectionism, defensiveness, and deference to authority (Shortell and Singer 2008; Walshe 2004).

High Reliability Organizing Theory

High reliability organizing theory is a systems approach describing organizations that, despite high socio-technical complexity, operate nearly error-free over time (Roberts 1999). HRO theory began by asking what factors made some high-risk organizations fail at a far lower rate than would be expected considering their complexity. The lessons learned from operations on aircraft carriers, and in nuclear power plants, air traffic control, intensive care units, and many other organizations formed principles that have informed research in academia and application in industry. A few of the observations relevant to diagnostic radiology and radiotherapy are presented below.

First, HRO theory admonishes us to recognize that safety is not a badge, or even National Committee for Quality Assurance (NCQA) certification. Simply enacting regulations to mandate reliability, or achieving a low error rate for a short time, does not constitute being a high reliability organization. Extensive research demonstrates that high reliability *organizing* is paramount, using processes that focus the members of the organization not on success, but on consistently preventing errors.

That said, the most important aspect of any safety effort is to challenge assumptions about safe operations. Research shows that organizations exhibiting high reliability are generally more prepared than they report being, and those exhibiting low reliability are generally underprepared compared to their impression of themselves (Mitroff 1995). Perfect scores on personnel "slip-and-fall" safety metrics can often serve as false prophets, convincing an organization that it is doing well on system quality. Assumptions can be destructive to promoting safety. Consistent with this point is Groopman's (2007) statement that "the radiologists who performed poorly were not only inaccurate; they were very confident that they were right when they were in fact wrong."

In the following paragraphs, we briefly present four processes from HRO theory that might benefit radiology and radiotherapy

practice. These are: (1) how the work team or organization is structured; (2) teamwork; (3) open communication; and (4) fluid decision making and information migration.

Organization Structuring

Often, organizations fail to realize that the way they put themselves together can lead to calamitous outcomes. Healthcare organizations usually adopt stringent authority gradients and build organizations so that physicians exercise control with few or no inputs from subordinate resources—many wrong-site surgeries are attributable to this practice (Pennsylvania Patient Safety Authority 2007). High reliability theory suggests that healthcare settings, commercial nuclear power plants, financial institutions, and other complex organizations need to allow the possibility of moving from hierarchical structuring, which is beneficial during routine operations, to a more fluid and flexible structuring, beneficial under rapidly changing conditions and other uncertainties. These kinds of flexible systems cannot emerge in situations where interpersonal trust is weak, people jockey for position, or common goals are never adopted by organizational members.

Teamwork

Actively engaging in teamwork helps reduce the authority gradient, and allows team members to come to agreement on common goals. By encouraging teamwork, managers actively promote integration of both functions and perspectives. Too often in complex systems, organizations design specialized tasks and give little attention to integrating them and bringing them back together (Heath and Staudenmayer 2000). Lack of integrated teamwork was one factor that resulted in a U.S. submarine, the USS *Greeneville*, hitting a Japanese fishing boat, the *Ehime Maru*, resulting in the deaths of nine Japanese crewmen. The captain of the submarine micromanaged operations during the rapidly unfolding scenario, overasserting authority and undermining the effectiveness of teamwork (Roberts and Tadmor 2002).

Another benefit of teamwork is that it helps members develop a shared understanding of the problems and challenges they face. The different senses and meanings people draw from their experiences must be blended together and integrated across the organization through the process of "heedful interrelating" (Weick and Roberts 1993). Heedful behaviors include those done with care, consistency, vigilance, conscientiousness, and purpose. Activities of heedful interrelation include encouraging group members to comprehend the perspective of others as they approach common goals, and appropriately matching individual know-how (or expertise) to the demands of the situation at hand. We suspect that shared understanding processes were operating aboard US Air Flight 1549, which successfully crash-landed in the Hudson River in January 2009. That is, the crew's capacity to rapidly develop a common view of the unfolding event and integrate their behaviors accordingly allowed them to prepare, crash-land, and evacuate the plane safely.

Open Communication

Healthcare settings, which are characterized by numerous hand-offs across care providers and sites, multiple opinions, interactions across various specialties, and other complexities provide rich opportunities for communication breakdowns. Cultural, legal, and organizational barriers further exacerbate these problems in sharing information. Creating the conditions for open communication means both developing capacity with structuring and teamwork, as well as removing opportunities for confusions in work flow that might lead to poor monitoring, flawed intervention, or incorrect dosage.

As Alpert and Hillman (2004) report, "because radiology practice typically involves pivotal interactions with referring physicians regarding diagnoses and care decisions, deficiencies in communication can be devastating." Addressing communication deficiencies like those between radiologists and physicians involves developing a common language, both literally and figuratively. On the one hand, simple interventions, such as double-checking seemingly large or small doses, avoiding abbreviations or making explicit any look- or sound-alike treatments, help people literally speak the same language. For example, miscommunication about dosages contributed to the death of Betsy Lehman, the Boston Globe health columnist, at the Dana-Farber Cancer Institute in 1994.

On the other hand, developing a common language figuratively entails ensuring that one caregiver hears not only the words, but also the meaning and perspective that the other is conveying. For instance, what seems "within tolerance" to one person or profession might seem risky to another; or what counts as "common sense" in one setting might be ill-suited to another. Both heedful interactions and a climate of safety are key elements to overcome these kinds of communication deficiencies. Open communication of this sort means that caregivers take the perspective of each other, and consider why a given treatment or diagnosis might have been made. Moreover, it also means that a caregiver is able to, without embarrassment or sanction, question the treatment or diagnosis if he or she believes that it might have been in error.

Information Migration and Fluid Decision Making

Without fluid structuring, teamwork, and open communication, ensuring information migration and fluid decision making would be impossible. Management failures have taught us that the highest ranking person is not always the appropriate person to make the decision. In HROs, decisions flow freely to the part of the organization in which the expertise and information exists to make them. For example, when preparing a patient, radiologists should accept that a technician may be more familiar with the operation of a piece of machinery, or that a nurse might know more about the patient's willingness to endure a procedure than they do. Allowing these partners to make decisions requires

yielding authority and responsibility temporarily in favor of more fine-grained, harder-to-access knowledge. However, fluid decision making does not mean losing control. Accountability should migrate with the decision, such that the decision maker "owns the decision," and its consequences, until the primary decision maker regains command.

In addition to being a key part of HRO, fluid decision making is also appearing in healthcare credentialing: one criterion for being qualified as a magnet hospital is that lead nurses in these hospitals are supposed to make decisions with their subordinates (The Center for Nursing Advocacy 2008).

Summary

In this short chapter, we reviewed four systems approaches to patient safety in radiology suggested by HRO theory. Our discussion gave an overview of the processes of flexible structuration, teamwork, open communication, and fluid decision making, and hopefully provided the reader with the desire to learn more.

References

Agency for Healthcare Research and Quality (AHRO). Systems approach. In *Patient Safety Network (PSNet) Glossary.* May 18, 2009. Available from http://www.psnet.ahrq.gov/glossary.aspx.

Alpert, H. R., and B. J. Hillman. 2004. Quality and variability in diagnostic radiology. *J. Am. Coll. Radiol.* 1 (2): 127–132.

Groopman, J. 2007. *How Doctors Think.* Boston, MA: Houghton-Mifflin.

Heath, C., and N. Staudenmayer. 2000. Coordination neglect: How lay theories of organizing complicate coordination in organizations. *Res. Organ. Behav.* 22: 153–191.

Kilo, C. M. *Educating Physicians Systems-Based Practice.* Power-Point presentation for the American College of Physicians. 2008. Available from www.myacp.org/files/public/office/ppt (accessed May 18, 2009).

Kohn, L. T., J. M. Corrigan, and M. S. Donaldson, eds. 1999. *To Err Is Human: Building a Safer Health System.* Report of the Committee on Quality of Health Care in America, Institute of Medicine. Washington, DC: National Academy Press.

Leape, L. L., D. M. Berwick, and D. W. Bates. 2002. What practices will most improve safety? Evidence-based medicine meets patient safety. *JAMA* 288 (4): 501–507.

Mitroff, I. 1995. *Why Some Organizations Recover Faster and Better from a Crisis.* New York, NY: AMACOM.

Pennsylvania Patient Safety Authority. 2007. *Patient Safety Authority Releases Wrong-Site Surgery Data 2007.* Available from http://www.psa.state.pa.us/psa/lib/press_release/press_wrong_site_2.pdf (accessed May 18, 2009).

Roberts, K. H., and C. T. Tadmor. 2002. Lessons learned from non-medical industries: The tragedy of the USS *Greeneville.* *Qual. Saf. Health Care* 11 (4): 355–357.

Roberts, K. H. 1999. Some characteristics of one type of high reliability organization. *Organ. Sci.* 1: 160–176.

Shortell, S. M., and S. J. Singer. 2008. Improving patient safety by taking systems seriously. *JAMA* 299 (4): 445–447.

The Center for Nursing Advocacy. 2008. What Is Magnet Status and How's that Whole Thing Going? Available from www.nursingadvocacy.org/faq/magnet.html (accessed May 18, 2009).

The Royal College of Radiologists, Society and College of Radiographers, Institute of Physics and Engineering in Medicine, National Patient Safety Agency, British Institute of Radiology. 2008. *Towards Safer Radiotherapy.* London: The Royal College of Radiologists.

Walshe, K. 2004. When things go wrong: How health care organizations deal with major failures. *Health Aff.* 23 (3): 103–111.

Weick, K. E., and K. H. Roberts. 1993. Collective mind in organizations: Heedful interrelating on flight decks. *ASQ* 38 (3): 357–381.

Errors in Patient Information Flow

Eric E. Klein
Washington University

Introduction

Historically, for radiation oncology, one of the largest sources of errors has been the transfer of information regarding technical parameters to treat patients (Valli et al. 1994). Although streamlining of data transfer has increased dramatically over time, scenarios still exist for incorrect information to move from treatment planning to the treatment machine via conduits such as electronic medical record systems, record and verify systems, or other direct transfer methods. There have been numerous reports (Goldwein, Podmaniczky, and Macklis 2003; Klein et al. 2005; Macklis, Meier, and Weinhous 1998; Patton, Gaffney, and Moeller 2003) that analyze, retrospectively and prospectively, error occurrence. Although the methods for data transfer have improved by removing human intervention, there have been instances where users have taken for granted quality assurance reviews, which are still needed. As localization imaging within the treatment room is now becoming more of a standard method for image-guided radiation therapy, another source of potential error has arisen. If we map the evolution of data transfer through the end of the twentieth century, hard copy, paper, and film were the methods of data transfer. Coincidentally, the complexity of treatments in terms of number of fields and variations were fairly limited in 1999, compared to 2009. The advent of intensity modulated radiation therapy (IMRT) to many nonacademic facilities also occurred at the end of the twentieth century. There were many manual methods during this time period that customized treatment fields for radiotherapy such as: field-shaping apertures such as cerrobend blocks; compensation devices such as wedges and custom-made metal compensators; use of bolus, etc.; and the use of nonaxial fields requiring table rotations. Some of the errors most damaging to patients were related to wedge positioning and orientations.

At the beginning of the twenty-first century, the evolution of record and verify systems as part of the electronic medical record led to them becoming staple devices within radiotherapy departments. This was timely as IMRT evoked more pieces of information and more complexity in treatment fields. Leaf instructions for multileaf collimator (MLC)-based IMRT and an increase in the number of fields occurred at the same time. Table coordinates became vital parameters used to set up patients. The paradigm shift of data moving through a record and verify system as a conduit between treatment planning and the treatment machine ideally should have reduced errors. However, at the initiation of the use of record and verify systems (R&V), direct DICOM data transfer from treatment plans to the record and verify systems did not exist. Therefore, a great deal of manual entry was performed for the R&V systems. Facilities that systematically reviewed paper hardcopy entries of data, may have taken for granted that R&V systems could potentially propagate errors. Several reports (Klein et al. 1998; Macklis, Meier, and Weinhous 1998; Patton, Gaffney, and Moeller 2003) had alluded to this fact, although it was clear that the facilities had not developed a system of checks of data transfer. However, data transfer by different digital methods, such as DICOM (Germond and Haefliger 2001) and RTOG data transfer, evolved and became standards, requiring planning systems and R&V systems to adopt a common language and methodology for moving data from one system to another. This undoubtedly reduced error, but without checks in the system, there was the possibility of misinterpretation or even movement of corrupt or changed data without review. Once again, the community may not have evolved in time for this new methodology of moving patient information.

One of the largest areas of confusion and volatility is treatment coordinates. Despite efforts to maintain common language in terms of coordinate systems, this still does not exist. Planning systems may have unique coordinate systems from one to another, and these systems may be different from the R&V system, which may also be unique to the treatment machine system coordinates. On many occasions, the directions of *Y* or *Z* are flipped in terms of defining a longitudinal axis versus a transverse axis. In addition, whether the systems are pure Cartesian versus non-Cartesian coordinate systems, the actual sign of the positive versus negative may actually be reversed. Therefore, due diligence is required in checking transfer of coordinates. Also, the treatment isocenter may not necessarily correspond with the imaging scanning isocenter. Patients who are having data localization performed by cone beam CT may have to be positioned in yet another location for scanning due to collisions of the patient with the imaging system.

In the past 5 years, it has become more routine practice, certainly in the United States, for DICOM transfer of data from the treatment planning system to the electronic medical record (R&V), which is instrumental for machine setup, verification, and recording. However, there are still tasks that occur in the background that could lead to an incorrect patient position.

In this chapter we will examine different stages in recent years of both the introduction of record and verify systems and its impact on reducing or propagating errors. More important, in the era of DICOM data transfer, the possibility of errors remains and we will discuss how a particular analysis of severity, frequency, and detectability of errors, viz. failure modes and effective analysis, can be used.

Works Related to Data Transfer

Klein et al. (2005) published a paper entitled *Errors in Radiation Oncology: Study in Pathways and Dosimetric Impact*. In this work, the authors provided a history of errors that occurred over a 30-month period. This retrospective analysis allowed the group to capture errors happening most frequently. Simultaneously, they examined particular errors of high propensity in order to examine where most errors occurred, what were the detection methods, and, finally, the dosimetric impact of particular errors. They then created scenarios of how particular errors can happen, how they can be detected, and how they can be assessed for dosimetric impact. Many of the errors depicted were due to incorrect beam data transfer for a particular patient. We detail the findings for particular scenarios later in this section. It should be noted that this data was captured during a time when the R&V system was in place. However, manual entry of the data generated from treatment planning was the standard. As mentioned, one of the most common errors detected in this study was incorrect treatment coordinates due to incorrect shifting. From the findings, additional checks were put in place and, in some cases, existing check systems were reiterated, particularly for therapists to review all treatment data before the first fraction.

The facility in this study was treating approximately 2000 patients per year on seven linear accelerators, each equipped with an R&V system (Varis v1.4g, Varian Oncology Systems or Clinical Desktop v4.10, Elekta Oncology). Over a 30-month period, 3964 courses of therapy were initiated. It must be noted that neither of these systems, nor the various radiotherapy treatment planning systems at their stage of software version, were conducive to complete electronic transfer of treatment setup data. Therefore, there were processes in place for comprehensive reviews of manually entered setup parameters by an experienced medical physicist. Patients were treated on one of five Varian linear accelerators (three of which were equipped with MLCs), or two MLC-equipped Elekta linear accelerators. The Varian linear accelerators possess tertiary physical wedges and had the capability of delivering enhanced dynamic wedging. The Elekta machines use an internal universal wedge. The latter two wedge systems rely on external icons to assist with deciphering orientation on the treatment machine.

The clinic tracked errors by means of a notable event procedure implemented by our Continuing Quality Improvement (CQI) Committee. Any events discovered mandate a physics review to determine dosimetric impact, if any, and are reported to the CQI Committee. The system is strategically designed to be nonpunitive to encourage disclosure of all events with a potentially negative consequence. Neither the individual(s) responsible for the event nor the individual(s) who missed catching the error are reprimanded in order to encourage reporting. Rather, information gleaned from the incident is used to modify procedures or to increase in-services in the procedures where the error occurred. The committee examined events involving conventional photon beam treatment. Specifically, they investigated errors of incorrect source-to-surface distance (SSD), incorrect energy, omitted wedge (physical, enhanced dynamic, or universal) or compensating filter, incorrect wedge or compensating filter orientation, and geometrical misses due to incorrect gantry, collimator or table angle, reversed field size, and setup errors. Because of specific incompatibilities between the various systems, the R&V systems did not possess the capability of electronically verifying block or compensator trays or bolus. Many of these errors could be initiated in either simulation, treatment planning, on the treatment machine, or somewhere in between.

The committee found geometric misses to have the highest error probability for both photons and for electrons. They most often occurred due to improper setup due to coordinate shift errors, incorrect field-shaping, and reversed collimator jaws. The dosimetric impact is unique for each case and depends on the proportion of fields in error and the volume mistreated. These errors were short-lived due to rapid detection via port films. A summary of error types is found in Table 21.1. Examples of scenarios and pathways for some of the most common errors are presented here.

Scenario A: Shift Error

During CT simulation, an initial reference is scribed in image planes denoting a projected isocenter. If subsequent treatment planning requires that the isocenter be placed 4.0 cm to the patient's left, requiring a shift of the table's lateral coordinates by +4.0 cm, this will lead to the table coordinates to be used for treatments. However, due to a misinterpretation, a value of −4.0 cm could used to determine the lateral table coordinates.

There is no feasible electronic mechanism to catch this error before treatment, short of having the coordinates checked by the physicists or therapist staff pretreatment, or a more dramatic solution of placing the patient on a conventional simulator to confirm

TABLE 21.1 Summary of Error Types

Error Type: Typical No. of Fractions: Method of Detection	No. of Events (% of Starts)	No. of R&V	Notes
Incorrect treatment coordinate(s): most often 1 fraction: port film	19 (0.48%)	7	*One case R&V not used.* Led to change in process whereby R&V display turned off in treatment room
Wrong gantry angle: most often 1 fraction: port film	15 (1.38%)	8	*One case R&V not used.* Led to change in process whereby R&V display turned off in treatment room
Wrong or omitted cerrobend block: most often 1 fraction: port film	15 (1.38%)	N/A	Not part of R&V system
Incorrect calculation: 2 to 5 fractions: diode reading or physics chart review	11 (0.28%)	N/A	Not part of R&V system
Wrong field size: 1 fraction: port film	9 (0.23%)	8	*One case affected dose p.q. enhanced dynamic wedge factor.* Led to change in process whereby R&V display turned off in treatment room
Incorrect collimator angle: 1 fraction: port film	8 (0.20%)	3	Led to change in process whereby R&V display turned off in treatment room
Missing compensating filter: 2 to 5 fractions: diode of physics chart review	6 (0.15%)	N/A	
Incorrect MU: 2 to 5 fractions: diode check	5 (0.13%)	4	*One case without R&V.* Led to change in override rights in MUs for therapists
Wrong photon energy: 2 to 3 fractions: diode check	3 (<0.1%)	3	In-service provided to therapists in reading treatment plans
Missing or incorrect MLC Shape: 1 fraction: port film	3 (<0.1%)	3	
Incorrect wedge direction: 1, 5, or 16 fractions: physics check, later diode check	3 (<0.1%)	3	After the second event: (1) diode readings taken off-axis, (2) TP printout with rooms-eye-view placed in setup information
Incorrect number of fractions for given set of fields: 1 to 3 fractions: physics review	3 (<0.1%)	2	
Incorrect or rotated compensating filter: 2 or 11 fx fractions; therapist discovery for first case, diode check for second	2 (<0.1%)	N/A	Coinciding with wedge direction led to action, diodes taken off-axis
Patient treated head to gantry but scanned foot to gantry: 1 fraction: port film	1 (<0.1%)	N/A	Divergent wire built into immobilization system registration device

Note: R&V, record and verify; MU, monitor units; TP, treatment planning; MLC, multileaf collimator.

the isocenter location. This particular error was determined to happen rarely enough (<0.5%) that the clinic began to rely exclusively on the therapists visually checking the new isocenter against digitally reconstructed radiographs (DRRs) generated during virtual simulation or treatment planning. However, as discussed, incorrect coordinate use not necessarily due to shifts was the most common error in this clinic. Related to these errors, the R&V system sometimes contained the incorrect coordinate information (versus the paper chart) due to an incorrect shift instruction from the original reference point (projected isocenter) to the treatment isocenter. This led to a process change in regards to the R&V system, where the in-room R&V display monitors were turned off. Although this may seem draconian, it was thought vital to emphasize that the R&V system was not a treatment setup system and should only be used to verify and record, maintaining the paper record as the primary setup documentation. The majority of shift errors occurred somewhat independently of the R&V system, whereby the resultant table coordinates in both the paper and electronic record were derived incorrectly. Therefore, prudent checks of the DRRs from treatment planning are the primary method of preventing such an error prior to obtaining a port film.

The pathway for such an error is as follows:

1. Dosimetry staff incorrectly completes an information sheet concerning the shift. Incorrect table coordinates are calculated and entered. (Note: If the treatment plan was completed before treatment commenced and a direct DICOM transfer including table coordinates could be made, this error would be avoided.)
2. Physics staff member fails to check shift instructions filled out by dosimetrist (reasonable process/training change).
3. Therapist staff does not check DRR in room to ensure shift has been made correctly (definite process/training change).
4. Diode does not facilitate error discovery.
5. Port film taken and reviewed by physician who discovers error, which is corrected before subsequent fraction.

A graphic depicting this scenario is exhibited in Figure 21.1.

Scenario B: Incorrect Wedge Direction

A four-field technique is planned for a pelvis patient, including opposed 18 MV lateral beams requiring 60° wedges. The wedging is to be accomplished by the use of enhanced dynamic

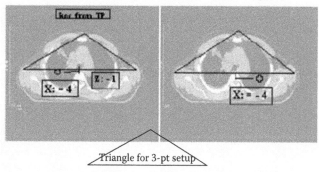

FIGURE 21.1   A graphic depicting the error of an incorrect shift applied in setting up a patient.

wedges, which create wedged dose distributions by motion of the collimating jaws while the beam is on. This technique is limited to only one jaw pair (the upper Y jaws on Varian linear accelerators), and therefore only one plane of wedging is possible unless a collimator rotation is applied. The collimator angle has to be chosen strategically. The chosen direction has a notation according to which jaw is moving (Y1in or Y2out) and is related to the "heel" end of the field. Because there is not an obvious confirmation method of the wedge orientation (heel) prior to treatment, as is the case with physical wedges, an incorrect collimator angle or choice of moving jaw (heel), could result in an incorrect wedge direction. There are external icons depicting the heel direction that are not as intuitive as for physical wedges.

An additional problem with this type of virtual wedging is that the nomenclature used by treatment planning systems is not intuitive and not necessarily correlated to the nomenclature used by the treatment machines and/or R&V systems. Therefore, an incorrect wedge direction could possibly go undetected by

physics checks and by therapists on the first day of treatment. Unfortunately, in vivo dosimetry performed with a diode on the central axis will not catch this particular error. Although the facility had the therapists check the final jaw position at the end of treatment (indicating the "toe" of the wedge), this policy was not carried out sufficiently to catch these potential errors. For the scenario described, the dosimetric error could be as high as 80% to a point 8 cm off-axis, for opposed 60° wedged fields. They found this type of error occurred when the wedge orientation was not obvious in a 2D representation of the treatment, as is the case for central nervous system treatments requiring nonaxial fields. An example of a hypothetical intracranial treatment with incorrect wedge directions is seen in Figure 21.2. In this case, there are a total of seven beams and the wedge was reversed from the intended orientations for four of the seven beams. This would have resulted in an increased high-dose region (7800 cGy) outside of the tumor volume, as compared with 7000 cGy in the correct plan. The pathway for such an error is as follows:

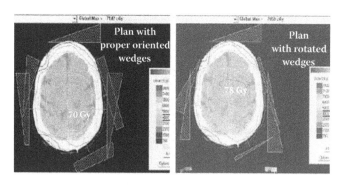

FIGURE 21.2   A hypothetical case showing the dose distribution due to incorrect wedge orientation.

1. Dosimetry staff prepares the intended plan for treatment based on MLC field-shaping. After the planning is complete and the paper and electronic charts prepared, there is a decision to switch to cerrobend field-shaping. To avoid having the weight of the block against the latch for some fields (a nonissue for MLC), the collimator is rotated 180° for these particular fields. Unfortunately, the original wedge direction is not corrected for the fields. (Mandated replanning would have avoided this error.)

2. A physicist checks individual R&V beam settings but does not detect the incorrect wedge direction.

3. Therapists do not closely compare wedge directions as indicated on the treatment plan and drawings with the icons on the treatment machine.

4. A diode check (Cozzi and Fogliata-Cozzi 1998) does not facilitate error discovery, as measurements are taken on the central axis. (Change in procedure implemented, whereby diodes are now placed off-axis.) Therapists are instructed to position the diode 2 to 3 cm toward the "heel" of the wedge as they perceive it, but then denote the direction of the diode placement according to anatomical direction (i.e., for a breast patient with treated wedges, the diode is placed anteriorly).

5. A subsequent weekly review of the chart by a physicist catches the error.

Starting in 2004, the AAPM assigned a task group, TG-100 (Huq et al. 2008), to design a method for evaluating quality assurance needs in radiation therapy. The emphasis of their paper was on process mapping of particular tasks from beginning to end and the application of failure modes and effects analysis (FMEA) to look for where emphasis should be placed on quality assurance reviews. In a similar vein, a group at Johns Hopkins lead by Eric Ford (Ford et al. 2009) published an evaluation of safety in the radiation oncology setting also using FMEA. They used FMEA as an assisting method for finding vulnerability in the process before an error occurs. In their manuscript they created a visual map of the processes mainly for data flow for external beam radiotherapy patients. From the mapping, they identified possible failure modes, assessed the risk probability, and incorporated scoring for severity, frequency of occurrence, and detectability. A composite scoring system was used in order to place importance on a given failure more for a particular task

or item in the mapping of the total process. As a result, their process map consisted of 269 nodes in which nearly half had failure mode possibility scoring. They focused on their 15 top failure modes and introduced process improvement steps so that errors would not occur (see Table 21.2). These changes to lower the probability scores of occurrence were incorporated into their process flow. It should be noted that such studies performed in recent years, as well as data flow, were accomplished by DICOM RT. However, there were still certain aspects that flowed by methods outside of direct DICOM transfer. The process map had multiple subsections. One particular failure mode was the generation of the digitally reconstructed radiograph, which are used for the initial setup and for verification of comparative films. Something as simple as identification on the film could be misinterpreted and therefore the patient could be sent up to the wrong location due to this misinterpretation. Therefore, the changes made in regards to movement of data for this particular antidote are simple things such as color coding and how the isocenter appears graphically on the digitally reconstructed radiograph. An explicit checklist for therapists was introduced, including interpretation of the treatment coordinates from the treatment plan that were marked on the patient from the scan. They also made a change in the system where previously there was a manual key entry of the DRR exported into the record and verify system, ironically by physics post-doctoral residents. Therefore, in cooperation with the R&V system vendor, they were able to come up with a method for directly importing the DRR from the treatment plan into the R&V system. This eliminated one person in the chain of data entry thereby decreasing the chances of miscommunication or error leading to this particular failure mode. This does not mean that this error will never happen again, but the likelihood decreases significantly.

Another example Ford used was actually pulling up the wrong patient for treatment from the R&V system. Again, the likelihood is very small, but the severity is enormous. Therefore, they came up with a few simple methods such as verifying a patient name prior to treatment, calling out the names—including the anatomy, displaying the patient's picture in the treatment room as well as on the console, and implementing barcode readers that automatically pull up the patient's data before treatment. Again, this does not completely eliminate the possibility of the wrong patient being treated, but certainly reduces the probability significantly. This example is of interest because no matter

TABLE 21.2 Examples of Failure Modes with Corresponding FMEA Risk Probability Numbers (RPN)

Failure Mode	Cause	RPN
Pt. Tx at incorrect location	DRR generated for wrong isocenter	160
Pt. Tx at incorrect location	Pt. aligned to wrong marks	128
Wrong plan used for Tx	Plan pulled up for wrong pt. in R&V system	96
Wrong plan used for Tx	Incorrect contours used in planning	96
Wrong plan used for Tx	Data entered for wrong pt at CT-SIM	96

how robust a record and verify system is employed or how robust the DICOM and data transfer are, calling up the wrong patient for treatment is something that still remains, though minute, a possibility.

Discussion

Although there has been significant progression in cleaner data transfer, elimination of hardcopy entries, and numerous people involved in the process, a new series of systematic data transfer errors has been introduced with image-guided radiation therapy. As previously discussed, there is a small bore condition that exists with cone beam CT (CBCT) as performed with kV imagers attached to the treatment machine. Therefore, the patient is often scanned, not at the treatment isocenter, but at the CBCT center. This imaging center is determined at the time of planning with a prior knowledge of the limitations and the shifts that would be needed. What is volatile is that the shifts that are found from the CBCT are not directly translatable to the actual treatment isocenter coordinates; instead, a translation is determined and defined anatomically. There are numerous image-guided radiation therapy devices being used for daily localization setup beyond CBCT, such as simple on-board imaging with x-rays, ultrasound devices, electromagnetic transponders implanted into a gland such as the prostate, radio-surface imaging, and other markers where images are used to not only look at bony anatomy but also implanted fiducial markers. All of these systems work in a very similar fashion, where one examines a nominal position of a patient as determined by the external contour, bony anatomy, or gland location. Depending on the image or scan of the day, translation instructions are given to translate the patient table coordinates. Therefore, it is imperative that all data transfer that relies on these imaging and localization devices has the proper reference image associated and denoted. This is yet another element that involves review, ideally by physics, before treatment, by therapists on the first day of treatment, and then on an ongoing basis by both groups.

Summary

Many data transfer problems have been resolved through the use of electronic transfer and electronic data entry. As a conduit from planning to the treatment machine, there still exist potential scenarios for incorrect data to be used for treatment and, more important, the use of image-guide devices adds another element of volatility due to the related data transfer.

References

Cozzi, L., and A. Fogliata-Cozzi. 1998. Quality assurance in radiation oncology. A study of feasibility and impact on action levels of an in vivo dosimetry program during breast cancer irradiation. *Radiother. Oncol.* 47 (1): 29–36.

Ford, E. C., R. Gaudette, L. Myers, B. Vanderver, L. Engineer, R. Zellars, D. Y. Song, J. Wong, and T. L. Deweese. 2009. Evaluation of safety in a radiation oncology setting using failure mode and effects analysis. *Int. J. Radiat. Oncol. Biol. Phys.* 74 (3): 852–858.

Germond, J. E., and J. M. Haefliger. 2001. Electronic dataflow management in radiotherapy: Routine use of the DICOM-RT protocol. *Cancer Radiother.* 5 (1): 172S–180S.

Goldwein, J. W., K. C. Podmaniczky, and R. M. Macklis. 2003. Radiotherapeutic errors and computerized record/verify systems. *Int. J. Radiat. Oncol. Biol. Phys.* 57 (5): 1509; author reply 1509–1510.

Huq, M. S., B. A. Fraass, P. B. Dunscombe, J. P. Gibbons Jr., G. S. Ibbott, P. M. Medin, A. Mundt et al. 2008. A method for evaluating quality assurance needs in radiation therapy. *Int. J. Radiat. Oncol. Biol. Phys.* 71 (1 Suppl): S170–S173.

Klein, E. E., R. E. Drzymala, J. A. Purdy, and J. Michalski. 2005. Errors in radiation oncology: A study in pathways and dosimetric impact. *J. Appl. Clin. Med. Phys.* 6 (3): 81–94.

Klein, E. E., R. E. Drzymala, R. Williams, L. A. Westfall, and J. A. Purdy. 1998. A change in treatment process with a modern record and verify system. *Int. J. Radiat. Oncol. Biol. Phys.* 42 (5): 1163–1168.

Macklis, R. M., T. Meier, and M. S. Weinhous. 1998. Error rates in clinical radiotherapy. *J. Clin. Oncol.* 16 (2): 551–556.

Patton, G. A., D. K. Gaffney, and J. H. Moeller. 2003. Facilitation of radiotherapeutic error by computerized record and verify systems. *Int. J. Radiat. Oncol. Biol. Phys.* 56 (1): 50–57.

Valli, M. C., M. Prina, A. Bossi, L. F. Cazzaniga, D. Cosentino, L. Scandolaro, A. Ostinelli, A. Monti, and P. Cappelletti. 1994. Evaluation of most frequent errors in daily compilation and use of a radiation treatment chart. *Radiother. Oncol.* 32 (1): 87–89.

22

Identifying and Reducing Risk

Frank Rath
University of Wisconsin

Introduction

The origins of quality management (QM) date back to the Middle Ages, when guilds created inspection committees to mark goods that met strict quality rules with special symbols. This QM model was widely followed until the middle of the twentieth century, when industrial engineers began developing a QM approach that was less dependent on the inspection of goods and services. This new QM model used tools to analyze processes and identify key or critical process parameters or steps. Quality controls were put in place to insure that these critical parameters or steps would not fail and result in unacceptable products or services. In this model, process maps and flow charts or trees are used to define and understand the process, and fault trees and failure modes and effects analysis (FMEA) (Stamatis 1995) are used to identify the critical process parameters or steps.

The QM model most commonly found in radiation therapy today is based mainly on inspection. Various guidelines published by the American Association of Physicists in Medicine (AAPM), the Nuclear Regulatory Commission (NRC 2009), the European Society for Therapeutic Radiation and Oncology (ESTRO), and the American College of Medical Physics (ACMP) all recommend that medical physicists measure, verify, and inspect a large number of radiation equipment and software parameters. This approach presents two problems: first, most radiation therapy facilities have limited resources and cannot possibly perform all of these recommended inspections and, second, it does not look at the overall process of prescribing and delivering radiation therapy. AAPM Task Group TG-53 states that the goal of a QM program in radiation therapy should be to "cure or locally control the disease (cancer) while minimizing complications in normal tissue" (Fraass et al. 1998). To meet this goal, a QM program must address the entire process, not just the equipment or software.

Steps to Identify and Reduce Risk

The first step in deploying the industrial-engineering-based QM model described above is to define the process. Flow charting,

process mapping, and process tress are powerful and often undervalued techniques. Too often, process improvement efforts focus on single steps in a process and not the overall process. Providing a picture or visual representation of the process using these techniques gives everyone involved a better understanding of the entire process and how their contributions affect it. Figure 22.1 shows a partial process tree for intensity-modulated radiotherapy (IMRT), the "treatment planning" branch.

FMEA and fault tree analysis (FTA) are the two techniques most often used to identify critical process parameters or steps. Of the two, FMEA is by far the more frequently used technique. A team of individuals familiar with the process being analyzed should be assembled to perform the FMEA. Each step from a process tree, flow chart, or process map is analyzed for potential failures. Table 22.1 shows a standard FMEA form. Steps in completing a FMEA include:

1. Identify all of the potential failure modes—ways in which a process could fail—for each process step. Each process step could, and usually does, have several failure modes.
2. Identify all of the potential causes for each failure mode. Each failure mode could, and usually does, have several causes.
3. Identify the end effect of each failure mode; that is, what will happen if a failure mode occurs.
4. List the current process quality controls. The team should list all of the process controls. Process controls typically fit into one of three categories: preventing causes of failure modes; detecting failure modes before they compromise the process outcomes; or steps that can be taken to moderate the severity of results following a failure mode. Examples of process controls include quality inspections, training, work instructions, standard procedures, and checklists.
5. Rank the current process controls' effectiveness in preventing the causes of a failure mode, detecting a failure mode if it does occur, and moderating the severity of the end effect if a failure mode occurs (the occurrence, detectability and severity columns from the FMEA form in Table 22.2). The ranking scale for each factor ranges from

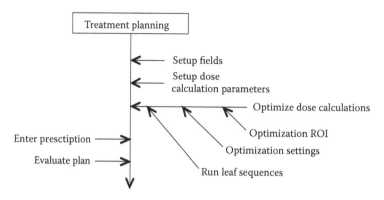

FIGURE 22.1 Partial IMRT process tree showing treatment planning branch.

1 (best case) to 10 (worst case). Table 22.2 shows a typical ranking scale for all three factors.

6. Calculate the risk priority number (RPN) (the product of the occurrence, detection and severity rankings) for each process step cause, failure mode, and end effect combination. An RPN indicates which process steps are critical (most likely to fail) or potentially hazardous. The higher the RPN, the more likely the failure mode is to occur, the less likely the failure mode will be detected, and the more severe the results of a failure mode will be.

7. The FMEA team should develop corrective actions for process steps with the highest RPN's or high severity rankings (regardless of the RPN value). Corrective actions should reduce the RPN when implemented.

TABLE 22.1 Blank Failure Modes and Effects Analysis (FMEA) Form

Process Step	Potential Failure Mode	Potential Cause of Failure Mode	Effects of Potential Failure Mode	Current Controls	Occurrence— Cause	Detectability of Failure Mode	Severity of Effect from Failure Mode	RPN	Corrective Action

TABLE 22.2 FMEA Ranking Scales for Occurrence, Detection, and Severity

Rank	Occurrence	Detection	Severity
	Probability that the cause will occur and lead to the failure mode	Probability that the failure mode will be detected before resulting in the end effect	Seriousness of the end effect when it occurs
1	Remote probability	Always	No effect
2	Low probability	High likelihood	Minor effect
3	Low probability	High likelihood	Minor effect
4	Moderate probability	Moderate likelihood	Moderate effect
5	Moderate probability	Moderate likelihood	Moderate effect
6	Moderate probability	Moderate likelihood	Moderate effect
7	High probability	Low likelihood	Serious effect
8	High probability	Low likelihood	Serious effect
9	Very high probability	Very low likelihood	Injury
10	100% probable	Never	Death

Summary

The end result of the FMEA should be a more robust and consistent process that is more likely to produce optimal outcomes and reduce the likelihood of hazardous outcomes. An effective QM program should focus on improving the entire process and preventing failures. Equipment parameters must still be considered, however, the FMEA should identify those parameters that are most critical and statistical techniques should be used to monitor those parameters.

References

Fraass, B., K. Doppke, M. Hunt, G. Kutcher, G. Starkschall, R. Stern, and J. Van Dyke. 1998. American Association of Physicists in Medicine Radiation Therapy Committee Task Group 53: Quality assurance for clinical radiotherapy treatment planning. *Med. Phys.* 25 (10): 1773–1829.

NRC. 2009. United States Nuclear Regulatory Commission 2009. Available from http://www.nrc.gov.

Stamatis, D. H. 1995. *Failure Modes and Effects Analysis.* Milwaukee, WI: ASQ Quality Press.

23

New Paradigm for Quality Management in Radiation Therapy Based on Risk Analysis

Mohammed Saiful Huq
University of Pittsburgh
Cancer Institute

Introduction

One of the goals of quality management in radiation therapy is to gain high confidence that patients will receive the prescribed treatment correctly. To accomplish these goals, professional societies such as the American Association of Physicists in Medicine (AAPM), the European Society for Therapeutic Radiology and Oncology (ESTRO), and the International Atomic Energy Agency (IAEA) have published many quality assurance (QA), quality control (QC), and quality management (QM) guidance documents. In general, the recommendations provided in these documents have emphasized the performance of device-specific QA at the expense of process flow and protection of the patient against catastrophic errors. Their focus has been on determining the functional performance of radiotherapy equipment through measureable parameters with tolerances set at strict but achievable values. Underlying the recommendations of most of these documents is the idea that everything that can affect the patient treatment should be checked. However, these recommendations are too comprehensive to be routinely implemented in many clinics due to limitations in staffing and/or other resources. Furthermore, they do not provide adequate guidelines on how to optimally distribute resources for QA and QM activities to maximize the quality of patient care. The difficulty and complexity of the situation have worsened as new equipment and procedures have been introduced into routine use in the radiation therapy clinic. Three- (and four-) dimensional treatment planning, intensity modulated radiation therapy (IMRT), image-guided radiation therapy (IGRT), stereotactic radiosurgery (SRS), stereotactic body radiotherapy (SBRT), 4D motion management systems, new brachytherapy sources, and widespread use of high-dose-rate (HDR) brachytherapy are all examples of this. The number and sophistication of possible QM activities, tests, and measurements have increased significantly with the introduction of these new technologies in the clinics. However, the time taken by the professional societies to develop consensus-based recommendations is usually too long for the individual clinical physicist, who is under pressure to implement new therapeutic strategies as soon as new technologies become available in the clinic. In such increasingly common situations, the clinical physicist needs guidance on how to develop a quality management program that includes devices, processes, and procedures that are new to the field as a whole.

A rich body of literature on the science of assuring quality already exists in industry. A seminal symposium in 2007 (ASTRO and AAPM 2007) was an important, formal effort to bring the industrial hazard analysis and radiation therapy worlds together. Industrial engineers have developed many tools to assist them in customizing methods to assure the quality of products ranging from foods to airline travel. Among these tools are process maps, failure modes and effects analysis (FMEA), and fault tree analysis. Of these, FMEA is beginning to be used in healthcare environments for performing safety analysis and promoting improvements in patient care (Sheridan-Leos, Schulmeister, and Hartranft 2006; Duwe, Fuchs, and Hansen-Flaschen 2005; Wetterneck et al. 2006).

New approaches are being developed in radiation therapy practice to address the QM needs of both existing and emerging technologies with finite financial and human resources. These are based on risk assessment that reexamines the entire radiation therapy process, critical or less critical, and accounts not only for the devices used for radiation treatment and the individual steps employed in each process (and their consequences should they fail), but also for interactions between them. They reflect jointly the probability of a failure occurring and the severity should it occur. Thus, they allow a radiotherapy department to develop a systematic QM program that acknowledges its current level of technology and clinical practice, and balances patient safety and treatment quality with available resources.

Task Group 100 (TG100) (Huq et al. 2008, 2010) of the American Association of Physicists in Medicine has developed a risk-assessment-based framework for QM activities in radiation therapy. This framework is based on the use of: (1) process mapping; (2) FMEA (Stamatis 1995); (3) fault tree analysis; and (4) the creation of a quality management program that will mitigate the most important risks that were identified in the previous analyses. The following sections give a brief description of these tools with illustrations of how they can be applied to critical procedures and processes in radiation therapy such as IMRT.

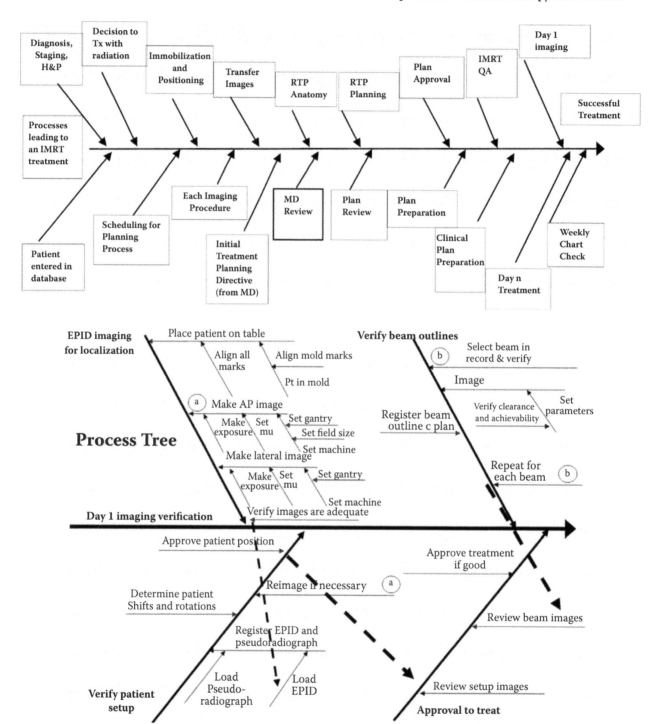

FIGURE 23.1 (a) An IMRT process tree. (From Huq et al. *Int. J. Radiat. Oncol. Biol. Phys.*, 71, 1 Suppl., S170–S1703, 2008. With permission.) (b) A process tree for the subprocess "day 1 imaging verification." (Courtesy of Bruce R. Thomadsen.)

Quality Management Tools

Process Map

The first step in the risk assessment approach is to delineate and then understand the processes to be evaluated. In this regard, a process map provides a convenient, visual illustration of the physical and temporal relationships between the different steps of a process that demonstrates the flow of these steps from process start to end. It shows how a process operates and focuses on what is done and how the different steps in the process are interrelated, not only on what is accomplished.

Figure 23.1a shows an example of a process map for IMRT treatment (Huq et al. 2008). Depending on the clinical practice, details of this tree will vary from one institution to another. The trunk, which takes the patient from entry into the radiation oncology system through end of treatment, runs across the center of the tree. The main boughs, representing the major subprocesses, emerge in approximately chronological order from the trunk. Each subprocess, in turn, can be subdivided into smaller steps. This is presented in the example of Figure 23.1b, which shows the various steps necessary for the day 1 image verification subprocess. For the treatment to be successfully executed, each step in all of the subprocesses must be performed correctly.

Failure Modes and Effect Analysis

After delineating the process, the next step is to assess the potential risks involved in that process. Although a number of methods exist for this assessment, TG100 uses the FMEA method. FMEA is a technique that is used to define, identify, and eliminate known and/or potential failures (failure modes), problems, and errors from a process before they reach the patient. A good

FMEA identifies known and potential failure modes, the causes and effects of each failure mode, and the likelihood that these failure modes will be detected and then prioritizes the identified failure modes according to a risk priority number (RPN) and provides for problem follow-up and corrective action. For example, at the first treatment day, a patient may be positioned at the wrong isocenter. Causes for this might include laser misalignment (at the treatment machine or at the simulator), therapist error or inattention (at treatment or at simulation), poor directions or setup documentation in the chart, and inadequate immobilization. Depending on such factors as the magnitude of the displacement from the planned isocenter, the treatment technique (IMRT, SRS, conformal, or large fields), the proximity of critical structures, and when the error is detected, the consequences can range from annoying to severe. For quantitative FMEA analysis, numerical values are assigned to three components for each cause of failure. O (occurrence) describes the probability that a particular failure mode will occur. S (severity) describes the severity of the effect resulting from the failure mode if it is not detected and corrected. D describes the probability that the failure will not be detected. RPN is then defined as a product of these three quantities, that is, $RPN = O \times S \times D$. TG100 uses numerical values for each of these quantities that range from 1 to 10. With the scales adopted by TG100, the RPN associated with a particular failure mode can range from 1 to 1000. The RPN values thus direct attention to failures that are most in need of QM, with higher values requiring immediate attention, and their component factors (O, S, D) help us see what features of the failure contribute most to the risk.

Table 23.1 is an example of the application of FMEA for the subprocess "RTP anatomy." When designing a QM program based on the RPN values, one will need to follow the guidelines given

TABLE 23.1　Example of the Application of FMEA for the Process RTP Anatomy for an IMRT Treatment

Process	Step	Potential Failure Modes	Potential Cause of Failure	Potential Effects of Failure	O	S	D	RPN
RTP Anatomy	Import images into RTP database	Wrong patient's images selected or imported	Human error Inadequate training and fatigue Standards/procedures/protocols (commissioning/acceptance testing) Defective materials (software or hardware) Inadequate design specifications Inadequate programming	Very wrong dose distribution Very wrong volume	4	8	5	160
RTP Anatomy	Import images into RTP database	Wrong imaging study (correct patient), viz.; wrong MR for target volume delineation	User error Standards, procedures, protocols Communication (poor) Inadequate training	Wrong dose distribution Wrong volume	5	7	6	210
RTP Anatomy	Import images into RTP database	File(s) corrupted	Defective materials (software or hardware) Inadequate maintenance Used incorrectly	Inconvenience (patient and staff)	3	3	2	18

in the section "Guidance for Designing a Quality Management Program in Radiation Therapy."

Fault Tree Analysis

The next step in the overall process is to evaluate the propagation of failures using a fault tree analysis. A fault tree complements a process tree and gives a visual representation of the propagation of a failure in the procedure and helps identify intervention strategies to mitigate the risks which have been identified.

Once the fault tree analysis has been completed, the final step in the process is to determine how best to avoid the faults and risks that have been identified. This analysis is then used to craft a quality management program.

Guidance for Designing a Quality Management Program in Radiation Therapy

The quality management tools described above can help design a QM program in radiation therapy. TG100 has performed an extensive FMEA and FTA for IMRT treatment. Based on this analysis, they have provided the following guidelines for designing a QM program for radiation oncology: (1) establish the goals of the program; (2) prioritize the potential failure modes based on RPN and severity values; (3) mark the highest-ranked steps on the process map; (4) mark the same highest-ranked steps on the fault tree; (5) select QM intervention placement; and (6) select appropriate quality management tools.

Summary

One of the most important components of the FMEA, FTA, and QM development processes is the involvement of all clinical groups (administrators, clinicians, nurses, dosimetrists, physicists, and therapists). Not every group needs to be involved in the analysis of each step of each process, but all need to be involved in the analysis of steps that are related to their clinical duties and for the procedure as a whole. Implementation of a risk analysis-based quality management program in radiation oncology would seem like a daunting task. However, once the basic principles are understood and the process is completed for one clinical area

or process, development of FMEA and FTA-based QM for other clinical areas or processes should be significantly more efficient.

Acknowledgments

The materials presented in this chapter resulted from the collaborative work with the members of TG100. The author would like to express his sincere thanks and gratitude to Benedick A. Fraass, Peter B. Dunscombe, John P. Gibbons Jr., Geoffrey S. Ibbott, Arno J. Mundt, Sasa Mutic, Jatinder R. Palta, Frank Rath, Bruce R. Thomadsen, Jeffrey F. Williamson, and Ellen D. Yorke for their contribution to this work.

References

American Society for Therapeutic Radiology and Oncology (ASTRO) and American Association of Physicists in Medicine (AAPM). 2007. Quality Assurance of Radiation Therapy and the Challenges of Advance Technologies. Dallas, Texas.

Duwe, B., B. D. Fuchs, and J. Hansen-Flaschen. 2005. Failure mode and effects analysis application to critical care medicine. *Crit. Care Clin*. 21 (1): 21–30, vii.

Huq, M. S., B. A. Fraass, P. B. Dunscombe, J. P. Gibbons Jr., G. S. Ibbott, P. M. Medin, A. Mundt et al. 2008. A method for evaluating quality assurance needs in radiation therapy. *Int. J. Radiat. Oncol. Biol. Phys*. 71 (1 Suppl): S170–S173.

Huq, S., B. Fraas, P. Dunscombe, J. Gibbons, G. Ibbott, A. Mundt, S. Mutic et al. 2010. Application of risk analysis methods to radiotherapy quality management: Report of Task Group 100 of the American Association of Physicists in Medicine (in press).

Sheridan-Leos, N., L. Schulmeister, and S. Hartranft. 2006. Failure mode and effect analysis: A technique to prevent chemotherapy errors. *Clin. J. Oncol. Nurs*. 10 (3): 393–398.

Stamatis, D. H. 1995. *Failure Modes and Effects Analysis*. Milwaukee, WI: ASQ Quality Press.

Wetterneck, T. B., K. A. Skibinski, T. L. Roberts, S. M. Kleppin, M. E. Schroeder, M. Enloe, S. S. Rough, A. S. Hundt, and P. Carayon. 2006. Using failure mode and effects analysis to plan implementation of smart i.v. pump technology. *Am. J. Health Syst. Pharm*. 63 (16): 1528–1538.

24

Risk Analysis and Control for Brachytherapy Treatments

Bruce Thomadsen
University of Wisconsin–Madison

Introduction

Of all the phases of brachytherapy, the treatment delivery presents the greatest risk. Each of the other phases, for example, calibration of source strength, treatment planning, and even implantation allow time for careful consideration and reconsideration and independent verification of parameters by highly trained staff. Treatment delivery forms an exception, although the reasons for the increased risk differ between most low-dose-rate (LDR) and high-dose-rate (HDR) treatments. While brachytherapy includes many different models for procedures, most cases fall into one of three major categories.

Permanent, Low-Dose-Rate Implants

The model for permanent implants currently applies mostly to prostate implants, although some breast and head and neck implants also use permanently implanted sources. Even for prostate implants, cases may be planned ahead or during the procedure. Either way, the implant environment induces stress with the number of persons involved, the patient under anesthesia, and time being of the essence. For cases where the plan was created based on ultrasound images performed some weeks ahead, and needles ordered loaded with the sources and spacers as per the plan, the greatest difficulty in the procedure room during the treatment becomes positioning the patient and the ultrasound transducer in the same position as when the planning images were obtained. Changes in the size of the patient's prostate complicate the alignment. During the source insertion, one problem involves assuring each needle is placed in the correct template hole and pushed to the correct depth, and another is ensuring that the thumb holding the stylet in place remains firmly in the same place during needle retraction.

Permanent prostate implants performed with intraoperative planning have their own times of high stress. Transducer alignment is simplified with no images to match, but the organ contouring and plan generation are performed with the whole staff waiting, a situation inherently stressful and leaving little time for reflection and quality management. Under these conditions, keeping the order of the sources and spacers straight while loading sources into needles for placement according to the plan is a challenge with little opportunity for verification. Using a delivery device such as the Mick applicator shifts the stress for bookkeeping from loading the needles to insertion and retraction distances and to the technique for use of the device.

Temporary Low-Dose-Rate Implants

Low dose-rate brachytherapy offers time and opportunity to reflect and check calculations and dosimetry, and to assure that the sources used were according to plan. However, the time of treatment proper, while not as stressful as with permanent implants, provides many avenues for errors. Most temporary LDR events occur during treatment. Many events follow failure to get the sources into the applicators, such as with Fletcher-style ovoids, or to get the source into the correct location in the applicator—to the end of an endobronchial tube with a tight bend, for example. Following insertion, nursing staff with little training in radiation therapy watch after the patient and the treatment appliance. Many events occur because a nurse adjusts an applicator not realizing that such actions affect the treatment.

High-Dose-Rate Brachytherapy

The potential for disastrous failures of the source drive during the delivery of HDR brachytherapy treatments has led to manufacturers installing multiple interlocks and mechanisms to prevent such problems, ameliorate injuries if a problem happened, and to warn the operator of anything unexpected. The operators receive considerable training in emergency procedures. All of this makes injury to the patient due to a mechanical failure during treatment unlikely. However, less dramatic but more likely errors exist. Examples include the use of transfer tubes of the wrong length or connection of the catheters or transfer tubes to the wrong channel of the treatment unit. If the patient has multiple plans, using the wrong plan for the treatment becomes possible.

Methods of Risk Analysis for Brachytherapy Treatments

Quality management (QM) seeks to achieve the desired quality of treatments. Of course, most facilities would claim to desire perfect treatments or treatments of the very highest quality. However, no facility gives *carte blanche* for resources assigned to the treatment process, so compromises in quality will always exist. Achieving quality in radiotherapy means maintaining the dose within some limits—often taken to be around ±5%—so that the dose distribution matches the target appropriately, with the desired gradients outside of the target to protect normal structures, and that no unexpected events injure the patient or compromise the treatment. Risk analysis attempts to identify actions or conditions that could result in failing to achieve the quality goals. Once identified, steps to prevent the failure could be put in place.

The first step in risk analysis entails understanding the process. One of the most common methods is making a process chart of some sort, such as a process map or a process diagram. Whatever method helps clarify the steps involved with the process and the interrelationship between them suffices. The mapping needs enough detail to provide a good understanding of the process without getting lost in the minutia. An example will help demonstrate some of these concepts. Figure 24.1 illustrates such a process map for a prostate implant with needles loaded before the procedure based on an imaging study performed 2 weeks

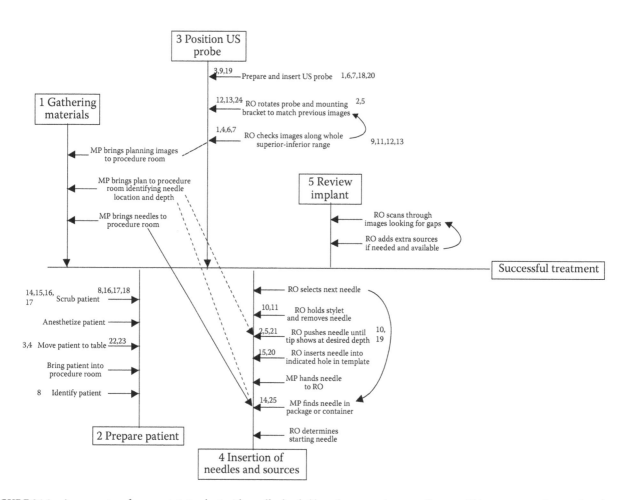

FIGURE 24.1 A process tree for a prostate implant with needles loaded based on a previous set of images. This mapping only considers the steps at the time of the implant proper.

earlier. Ignore the small numbers by each branch for the moment. This case was picked as about the simplest of the brachytherapy treatment steps, but we shall see that it offers good opportunities to see quality management development.

The risk analysis continues from the procedure map by identifying each possible way things could go wrong at each step. The most common tool for this step is failure modes and effects (FMEA), as described in Chapters 16, 22, 23, and 39. The values corresponding to scores for each of the three parameters require definition, and for this study, Table 24.1 gives the values. Generally, the occurrence (*O*) and detectability (*D*) values behave somewhat logarithmically. Table 24.2 lists the steps from the process map and their associated values for *O*, severity (*S*), *D*, and the risk priority number (RPN). Tables 24.3 and 24.4 give the steps from Table 24.2 sorted by the PRN ranking and the severity score, respectively. Table 24.4 sorts the steps within a severity value secondarily by RPN. The values for the parameters in Table 24.2 come solely from the author and only serve for this example. In practice, an FMEA should be developed by a team representing all the disciplines involved with a procedure so the values assigned have a broad basis and avoid the biases resulting from a single individual's opinions. In this example, the RPN values span a large range, from 576 to 0 (not counting failure mode 8). The severity scores also cover a range, but weighted toward the higher values, indicating that most failures can be significant. None of the failures carry a severity score between 2 and 4. Patient injury due to trauma or infection poses significant risks, but this example will confine the discussion to dosimetric failures.

Understanding how the failures in the FMEA propagate into a treatment failure also helps in designing methods to interrupt that propagation. Figure 24.2 plots the propagation of failures in a fault tree. The fault tree only considers failures that lead to dosimetry errors. In concept, any of the failures on the extreme

right-hand side lead directly to the effect on the left. The goal of QM design is to interrupt that flow.

Targeting the Potential Failures

QM includes several activities to eliminate the potential failures, such as:

- Quality design—Possibly the most effective method to prevent the propagation of a failure is to design the process to eliminate the potential failures. Some essential aspects of the design include:
  - Establishing standard procedures and protocols for the procedures so all involved know who performs what actions when. Standard protocols reduce the need for on-the-spot decisions, which open pathways for mistakes.
  - Establishing the lines of communication so that it becomes clear where information originates and how it is passed along. Forms often provide a simple, yet effective device to guide and control information flow.
  - Assuring that all the personnel receive training so they not only know their function but also understand how variations in how they perform their function affect the outcome. The training should contain some method to assess the student's understanding, such as a test.
  - Laying out the physical space to remove obstacles that might interfere with the work and positioning necessary objects, tools, and information in convenient, intuitive locations.

In many situations, managerial policies create situations that increase risk. A common example is understaffing,

TABLE 24.1 Description of Categories for Classification of *O*, *S*, *D*

Rank	Occurrence (*O*) Qualitative	Frequency	Severity (*S*) Qualitative	Categorization	Detectability (*D*) Estimated Probability of Failure Going Undetected (%)
1	Failure unlikely	1/10,000	No effect		0.01
2		2/10,000	Inconvenience	Inconvenience	0.2
3	Relatively few	5/10,000			0.5
4	failures	1/1000	Minor dosimetric error	Suboptimal plan or treatment	1.0
5		<0.2%	Limited toxicity or tumor underdose	Wrong dose, dose distribution, location, or volume	2.0
6	Occasional failures	<0.5%			5.0
7		<1%	Potentially serious toxicity or tumor underdose		10
8	Repeated failures	<2%			15
9		<5%	Possible very serious toxicity or tumor underdose	Very wrong dose, dose distribution, location,	20
10	Failures inevitable	>5%	Catastrophic	or volume	>20

Source: Huq et al., forthcoming, Application of risk analysis methods to radiotherapy quality nanagement: Report of Task Group 100 of the American Association of Physicists in Medicine. With permission.

TABLE 24.2 Table of the FMEA for the Example Process

Failure mode #	Major Processes	Step	Potential Failure Modes	Potential Causes of Failure	Potential Effects of Failure	O	S	D	RPN	Notes
1	1 Gathering materials	MP brings needles to procedure room	MP forgets needles	Other things on the mind, too many other things to bring	Delay as MP goes to get the needles	5	1	1	5	Assumes that the MP had ordered the needles at some time in the past
2	1 Gathering materials	MP brings plan to procedure room identifying needle location and depth	MP forgets documentation	Other things on the mind, too many other things to bring, did not notice those papers missing	Delay as MP goes to get the plan	5	1	1	5	
3	1 Gathering materials	MP brings planning images to procedure room	MP forgets images	Other things on the mind, too many other things to bring, did not notice the missing images	Delay as MP goes to get the plan	5	1	1	5	
4	2 Prepare patient	Identify patient	Patient implanted with wrong source pattern	Did not perform the time out (omission), performed the identification incorrectly (training failure)	Wrong dose distribution	3	9	6	162	
5	2 Prepare patient	Bring patient into procedure room	None			0	0	0	0	
6	2 Prepare patient	Move patient to table	Patient falls	Personnel did not help patient correctly (training failure)	Patient injury	3	9	10	270	The high score for detection reflects that once the bad decision on how to move the patient has been made, it is unlikely to be noted and changed before the patient falls
7	2 Prepare patient	Move patient to table	Patient falls	Personnel did not help patient correctly (no protocol)	Patient injury	3	9	10	270	
8	2 Prepare patient	Anesthetize patient	None relevant to this example							Leave this row for the anesthesiologists to work on
9	2 Prepare patient	Scrub patient	Patient not cleaned thoroughly	Training failure	Patient develops an infection	4	8	10	320	Infections can be life threatening. Not detectable after the scrub.
10	2 Prepare patient	Scrub patient	Patient not cleaned thoroughly	Inattention	Patient develops an infection	4	8	10	320	Infections can be life threatening. Not detectable after the scrub.
11	2 Prepare patient	Scrub patient	Patient not cleaned thoroughly	Rushed environment	Patient develops an infection	4	8	10	320	Infections can be life threatening. Not detectable after the scrub.
12	2 Prepare patient	Scrub patient	Patient not cleaned thoroughly	Poor performance	Patient develops an infection	5	8	10	400	Infections can be life threatening. Not detectable after the scrub.
13	3 Position US probe	Prepare and insert US probe	Probe not cleaned adequately	Poor performance, possibly rushed environment	Patient develops an infection	4	8	10	320	Assume that the person cleaning the probe has been trained.

#										
14	3 Position US probe	Prepare and insert US probe	Probe cover not correctly filled with water resulting in poor images	Training failure	Wrong dose location	9	9	7	567	With poor training, the likelihood of occurrence and of going undetected is high.
15	3 Position US probe	Prepare and insert US probe	Probe cover not correctly filled with water resulting in poor images	Inattention	Wrong dose location	6	9	4	216	
16	3 Position US probe	Prepare and insert US probe	Probe cover not correctly filled with water resulting in poor images	Poor performance	Wrong dose location	6	9	4	216	
17	3 Position US probe	Prepare and insert US probe	Template not seated properly	Inattention/Poor performance	Wrong dose location	8	7	7	392	
18	3 Position US probe	RO rotates probe and mounting bracket to match previous images	Prostate not aligned	Poor images (QA failure for imaging system)	Wrong dose location	5	9	8	360	Values contingent on the initial acceptance of the poor images.
19	3 Position US probe	RO rotates probe and mounting bracket to match previous images	Prostate not aligned	Poor performance	Wrong dose location	4	9	7	252	
20	3 Position US probe	RO rotates probe and mounting bracket to match previous images	Prostate not aligned	Changes in prostate gland	Wrong dose distribution	7	6	8	336	While detectable, there is little to be done to correct.
21	3 Position US probe	RO checks images along the whole superior–inferior range	Prostate not aligned	Training failure	Wrong dose location	8	8	8	512	Compared with the wrong location for probe alignment, the error in the dose distribution here likely would be less severe.
22	3 Position US probe	RO checks images along the whole superior–inferior range	Prostate not aligned	Inattention	Wrong dose location	7	8	8	448	
23	3 Position US probe	RO checks images along the whole superior–inferior range	Prostate not aligned	Poor performance	Wrong dose location	7	8	8	448	

(continued)

TABLE 24.2 Table of the FMEA for the Example Process (Continued)

Failure mode #	Major Processes	Step	Potential Failure Modes	Potential Causes of Failure	Potential Effects of Failure	O	S	D	RPN	Notes
24	3 Position US probe	RO checks images along the whole superior–inferior range	Prostate not aligned	Poor images (QA failure for imaging system)	Wrong dose location	8	8	9	576	Again, given poor image quality, detecting this failure would be difficult.
25	4 Insertion of needles and sources	RO determines starting needle	Poor selection of posterior needles at first obscuring later anterior locations	Training failure	Wrong dose distribution	3	5	8	120	Once selected, the probability of detection before insertion is low.
26	4 Insertion of needles and sources	RO determines starting needle	Poor selection of posterior needles at first obscuring later anterior locations	Inattention	Wrong dose distribution	3	5	8	120	Inappropriate selection due to inattention is not as likely due to habit.
27	4 Insertion of needles and sources	MP finds needle in package or container	MP selects the wrong needle	Inattention	Wrong dose distribution	5	6	8	240	The severity is likely only moderate, unless the erroneous needle places a source very close to the urethra or rectum. Detection for that needle error is low, but is likely to be found too late, when the needle used is actually called.
28	4 Insertion of needles and sources	MP finds needle in package or container	MP selects the wrong needle	Confusion between multiple packages	Wrong dose distribution	5	6	8	240	
29	4 Insertion of needles and sources	MP finds needle in package or container	MP selects the wrong needle	Poor demarcation of needles	Wrong dose distribution	4	6	8	192	
30	4 Insertion of needles and sources	MP finds needle in package or container	MP selects the wrong needle	MP fails to hear the correct needle identification	Wrong dose distribution	7	6	8	336	
31	4 Insertion of needles and sources	MP hands needle to RO	MP drops needle	Room layout makes transfer insecure	Wrong dose distribution	4	6	10	240	Assumes that the sources could not be reprocessed and loaded in a clean needle in time for use. If processing is available and performed, the effect is only delay. The frequency may be 1/10,000 needles or 1/1000 patients. Detection would be after the needle drops.
32	4 Insertion of needles and sources	MP hands needle to RO	MP drops needle	MP slips	Wrong dose distribution	4	6	10	240	

#	Step	Process	Failure mode	Cause	Effect	O	S	D	RPN	Comments
33	4 Insertion of needles and sources	MP hands needle to RO	MP stabs RO with needle	MP slips	Injury to RO, possible wrong dose distribution	2	5	10	100	The wrong dose distribution would be if the sources could not be reprocessed for the case. If the RO is unable to continue the case, a very wrong dose would result and the severity could be 9.0 for the patient.
34	4 Insertion of needles and sources	RO inserts needle into indicated hole in template	RO inserts needle into the wrong hole	Inattention	Wrong dose distribution	5	6	8	240	
35	4 Insertion of needles and sources	RO inserts needle into indicated hole in template	RO inserts needle into the wrong hole	Bad viewing conditions	Wrong dose distribution	7	6	8	336	
36	4 Insertion of needles and sources	RO inserts needle into indicated hole in template	RO inserts needle into the wrong hole	Confusion between holes	Wrong dose distribution	6	6	8	288	
37	4 Insertion of needles and sources	RO pushes needle until tip shows at desired depth	Needle at wrong depth	Training failure	Wrong dose distribution	8	8	9	576	Given poor training, the likelihood of occurrence is high.
38	4 Insertion of needles and sources	RO pushes needle until tip shows at desired depth	Needle at wrong depth	Inattention	Wrong dose distribution	6	6	8	288	
39	4 Insertion of needles and sources	RO pushes needle until tip shows at desired depth	Needle at wrong depth	Poor images (QA failure for imaging system)	Wrong dose distribution	8	7	9	504	
40	4 Insertion of needles and sources	RO pushes needle until tip shows at desired depth	Needle at wrong depth	Confusion between planes	Wrong dose distribution	5	6	8	240	
41	4 Insertion of needles and sources	RO holds stylet and removes needle	RO fails to hold stylet adequately	Poor performance	Wrong dose distribution	7	6	9	378	
42	4 Insertion of needles and sources	RO holds stylet and removes needle	RO pushes on stylet	Poor performance	Wrong dose distribution	7	6	9	378	

(continued)

TABLE 24.2 Table of the FMEA for the Example Process (Continued)

Failure mode #	Major Processes	Step	Potential Failure Modes	Potential Causes of Failure	Potential Effects of Failure	O	S	D	RPN	Notes
43	4 Insertion of needles and sources	RO selects next needle	Poor selection of posterior needles that later obscure more anterior locations	Training failure	Wrong dose distribution	3	5	8	120	
44	4 Insertion of needles and sources	RO selects next needle	Poor selection of posterior needles that later obscure more anterior locations	Inattention	Wrong dose distribution	3	5	8	120	
45	5 Review implant	RO scans through images looking for gaps	Missing gaps or not seeing sources	Poor images (either QA failure for imaging system or interference from the implanted sources)	Suboptimal dose distribution	5	6	8	240	Assumes that the source distribution did not follow the plan closely enough that coverage requires additional sources. Also assumes poor image quality.
46	5 Review implant	RO adds extra sources if needed and available	RO adds unnecessary sources	Poor images (either QA failure for imaging system or interference from the implanted sources)	Suboptimal dose distribution	5	6	8	240	Assumes that the source distribution followed the plan but images looked like there was a sizable gap. Also assumes poor image quality.
47	5 Review implant	RO adds extra sources if needed and available	RO adds unnecessary sources	Misinterpretation, training failure	Suboptimal dose distribution	5	6	8	240	
48	5 Review implant	RO adds extra sources if needed and available	RO fails to add needed source	Misinterpretation, training failure	Suboptimal dose distribution	5	6	8	240	

TABLE 24.3 List of Table 24.2, Sorted by Descending Values of the Risk Priority Number

Rank	Failure mode #	Major Processes	Step	Potential Failure Modes	Potential Causes of Failure	Potential Effects of Failure	O	S	D	RPN	Notes
1	24	3 Position US probe	RO checks images along the whole superior–inferior range	Prostate not aligned	Poor images (QA failure for imaging system)	Wrong dose location	8	8	9	576	Given poor image quality, detecting this failure would be difficult.
2	37	4 Insertion of needles and sources	RO pushes needle until tip shows at desired depth	Needle at wrong depth	Training failure	Wrong dose distribution	8	8	9	576	Given poor training, the likelihood of occurrence is high.
3	14	3 Position US probe	Prepare and insert US probe	Probe cover not correctly filled with water resulting in poor images	Training failure	Wrong dose location	9	9	7	567	With poor training, the likelihood of occurrence and of going undetected is high.
4	21	3 Position US probe	RO checks images along the whole superior–inferior range	Prostate not aligned	Training failure	Wrong dose location	8	8	8	512	Compared with the wrong location for probe alignment, the error in the dose distribution here likely would be less severe.
5	39	4 Insertion of needles and sources	RO pushes needle until tip shows at desired depth	Needle at wrong depth	Poor images (QA failure for imaging system)	Wrong dose distribution	8	7	9	504	
6	22	3 Position US probe	RO checks images along the whole superior–inferior range	Prostate not aligned	Inattention	Wrong dose location	7	8	8	448	
7	23	3 Position US probe	RO checks images along the whole superior–inferior range	Prostate not aligned	Poor performance	Wrong dose location	7	8	8	448	
8	12	2 Prepare patient	Scrub patient	Patient not cleaned thoroughly	Poor performance	Patient develops an infection	5	8	10	400	Infections can be life threatening. Not detectable after the scrub.
9	17	3 Position US probe	Prepare and insert US probe	Template not seated properly	Inattention/Poor performance	Wrong dose location	8	7	7	392	
10	41	4 Insertion of needles and sources	RO holds stylet and removes needle	RO fails to hold stylet adequately	Poor performance	Wrong dose distribution	7	6	9	378	
11	42	4 Insertion of needles and sources	RO holds stylet and removes needle	RO pushes on stylet	Poor performance	Wrong dose distribution	7	6	9	378	
12	18	3 Position US probe	RO rotates probe and mounting bracket to match previous images	Prostate not aligned	Poor images (QA failure for imaging system)	Wrong dose location	5	9	8	360	Values contingent on the initial acceptance of the poor images.

(continued)

TABLE 24.3 List of Table 24.2, Sorted by Descending Values of the Risk Priority Number (Continued)

Rank	Failure mode #	Major Processes	Step	Potential Failure Modes	Potential Causes of Failure	Potential Effects of Failure	O	S	D	RPN	Notes
13	20	3 Position US probe	RO rotates probe and mounting bracket to match previous images	Prostate not aligned	Changes in prostate gland	Wrong dose distribution	7	6	8	336	While detectable, there is little to be done to correct.
14	30	4 Insertion of needles and sources	MP finds needle in package or container	MP selects the wrong needle	MP fails to hear the correct needle identification	Wrong dose distribution	7	6	8	336	
15	35	4 Insertion of needles and sources	RO inserts needle into indicated hole in template	RO inserts needle into the wrong hole	Bad viewing conditions	Wrong dose distribution	7	6	8	336	
16	9	2 Prepare patient	Scrub patient	Patient not cleaned thoroughly	Training failure	Patient develops an infection	4	8	10	320	Infections can be life threatening. Not detectable after the scrub.
17	10	2 Prepare patient	Scrub patient	Patient not cleaned thoroughly	Inattention	Patient develops an infection	4	8	10	320	Infections can be life threatening. Not detectable after the scrub.
18	11	2 Prepare patient	Scrub patient	Patient not cleaned thoroughly	Rushed environment	Patient develops an infection	4	8	10	320	Infections can be life threatening. Not detectable after the scrub.
19	13	3 Position US probe	Prepare and insert US probe	Probe not cleaned adequately	Poor performance, possible rushed environment	Patient develops an infection	4	8	10	320	Assume that the person cleaning the probe has been trained.
20	36	4 Insertion of needles and sources	RO inserts needle into indicated hole in template	RO inserts needle into the wrong hole	Confusion between holes	Wrong dose distribution	6	6	8	288	
21	38	4 Insertion of needles and sources	RO pushes needle until tip shows at desired depth	Needle at wrong depth	Inattention	Wrong dose distribution	6	6	8	288	
22	6	2 Prepare patient	Move patient to table	Patient falls	Personnel did not help patient correctly (training failure)	Patient injury	3	9	10	270	The high score for detection reflects that once the bad decision has been made, it is unlikely to be noted and changed before the patient falls.
23	7	2 Prepare patient	Move patient to table	Patient falls	Personnel did not help patient correctly (no protocol)	Patient injury	3	9	10	270	
24	19	3 Position US probe	RO rotates probe and mounting bracket to match previous images	Prostate not aligned	Poor performance	Wrong dose location	4	9	7	252	

#	ID	Step	Action	Failure mode	Cause	Effect	O	S	D	RPN	Comments
25	27	4 Insertion of needles and sources	MP finds needle in package or container	MP selects the wrong needle	Inattention	Wrong dose distribution	5	6	8	240	The severity is likely only moderate, unless the erroneous needle places a source very close to the urethra or rectum. Detection for that needle error is low, but is likely to be found too late, when the needle used is actually called.
26	28	4 Insertion of needles and sources	MP finds needle in package or container	MP selects the wrong needle	Confusion between multiple packages	Wrong dose distribution	5	6	8	240	
27	31	4 Insertion of needles and sources	MP hands needle to RO	MP drops needle	Room layout makes transfer insecure	Wrong dose distribution	4	6	10	240	Assumes that the sources could not be reprocessed and loaded in a clean needle in time for use. If processing is available and performed, the effect is only delay. The frequency may be 1/10,000 needles or 1/1000 patients. Detection would be after the needle drops.
28	32	4 Insertion of needles and sources	MP hands needle to RO	MP drops needle	MP slips	Wrong dose distribution	4	6	10	240	
29	34	4 Insertion of needles and sources	RO inserts needle into indicated hole in template	RO inserts needle into the wrong hole	Inattention	Wrong dose distribution	5	6	8	240	
30	40	4 Insertion of needles and sources	RO pushes needle until tip shows at desired depth	Needle at wrong depth	Confusion between planes	Wrong dose distribution	5	6	8	240	
31	45	5 Review implant	RO scans through images looking for gaps	Missing gaps or not seeing sources	Poor images (either QA failure for imaging system or interference from the implanted sources)	Suboptimal dose distribution	5	6	8	240	Assumes that the source distribution did not follow the plan closely enough that coverage requires additional sources. Also assumes poor image quality.
32	46	5 Review implant	RO adds extra sources if needed and available	RO adds unnecessary sources	Poor images (either QA failure for imaging system or interference from the implanted sources)	Suboptimal dose distribution	5	6	8	240	Assumes that the source distribution followed the plan but images looked like there was a sizable gap. Also assumes poor image quality.
33	47	5 Review implant	RO adds extra sources if needed and available	RO adds unnecessary sources	Misinterpretation, training failure	Suboptimal dose distribution	5	6	8	240	
34	48	5 Review implant	RO adds extra sources if needed and available	RO fails to add needed source	Misinterpretation, training failure	Suboptimal dose distribution	5	6	8	240	

(*continued*)

TABLE 24.3 List of Table 24.2, Sorted by Descending Values of the Risk Priority Number (Continued)

Rank	Failure mode #	Major Processes	Step	Potential Failure Modes	Potential Causes of Failure	Potential Effects of Failure	O	S	D	RPN	Notes
35	15	3 Position US probe	Prepare and insert US probe	Probe cover not correctly filled with water resulting in poor images	Inattention	Wrong dose location	6	9	4	216	
36	16	3 Position US probe	Prepare and insert US probe	Probe cover not correctly filled with water resulting in poor images	Poor performance	Wrong dose location	6	9	4	216	
37	29	4 Insertion of needles and sources	MP finds needle in package or container	MP selects the wrong needle	Poor demarcation of needles	Wrong dose distribution	4	6	8	192	
38	4	2 Prepare patient	Identify patient	Patient implanted with wrong source pattern	Did not perform the time out (omission), performed the identification incorrectly (training failure)	Wrong dose distribution	3	9	6	162	
39	25	4 Insertion of needles and sources	RO determines starting needle	Poor selection of posterior needles at first obscuring later anterior locations	Training failure	Wrong dose distribution	3	5	8	120	Once selected, the probability of detection before insertion is low.
40	26	4 Insertion of needles and sources	RO determines starting needle	Poor selection of posterior needles at first obscuring later anterior locations	Inattention	Wrong dose distribution	3	5	8	120	Inappropriate selection due to inattention is not so likely due to habit.
41	43	4 Insertion of needles and sources	RO selects next needle	Poor selection of posterior needles that later obscure more anterior locations	Training failure	Wrong dose distribution	3	5	8	120	

#	Step	Action	Failure mode	Cause	Effect					Comments
42	4 Insertion of needles and sources	RO selects next needle	Poor selection of posterior needles that later obscure more anterior locations	Inattention	Wrong dose distribution	3	5	8	120	The wrong dose distribution would be if the sources could not be reprocessed for the case. If the RO is unable to continue the case, a very wrong dose would result and the severity could be 9.0 for the patient.
43	4 Insertion of needles and sources	MP hands needle to RO	MP stabs RO with needle	MP slips	Injury to RO, possible wrong dose distribution	2	5	10	100	
44	1 Gathering materials	MP brings needles to procedure room	MP forgets needles	Other things on their mind, too many other things to bring	Delay as MP goes to get the needles	5	1	1	5	Assumes that the MP had ordered the needles at some time in the past.
45	1 Gathering materials	MP brings plan to procedure room identifying needle location and depth	MP forgets documentation	Other things on their mind, too many other things to bring, did not notice those papers missing	Delay as MP goes to get the plan	5	1	1	5	
46	1 Gathering materials	MP brings planning images to procedure room	MP forgets images	Other things on their mind, too many other things to bring, did not notice the missing images	Delay as MP goes to get the plan	5	1	1	5	
47	2 Prepare patient	Bring patient into procedure room	None			0	0	0	0	
48	2 Prepare patient	Anesthetize patient	None relevant to this example							Leave this row for the anesthesiologists to work on.

TABLE 24.4   List of Table 24.2 Sorted by Descending Values of the Severity Score and by RPN

Rank	Failure mode #	Major Processes	Step	Potential Failure Modes	Potential Causes of Failure	Potential Effects of Failure	O	S	D	RPN	Notes
1	14	3 Position US probe	Prepare and insert US probe	Probe cover not correctly filled with water resulting in poor-quality images	Training failure	Wrong dose location	9	9	7	567	With poor training, the likelihood of occurrence and of going undetected is high.
2	18	3 Position US probe	RO rotates probe and mounting bracket to match previous images	Prostate not aligned	Poor-quality images (QA failure for imaging system)	Wrong dose location	5	9	8	360	Values contingent on the initial acceptance of the poor-quality images.
3	6	2 Prepare patient	Move patient to table	Patient falls	Personnel did not help patient correctly (training failure)	Patient injury	3	9	10	270	The high score for detection reflects that once the bad decision on how to move the patient has been made, it is unlikely to be noted and changed before the patient falls
4	7	2 Prepare patient	Move patient to table	Patient falls	Personnel did not help patient correctly (no protocol)	Patient injury	3	9	10	270	
5	19	3 Position US probe	RO rotates probe and mounting bracket to match previous images	Prostate not aligned	Poor performance	Wrong dose location	4	9	7	252	
6	15	3 Position US probe	Prepare and insert US probe	Probe cover not correctly filled with water resulting in poor-quality images	Inattention	Wrong dose location	6	9	4	216	
7	16	3 Position US probe	Prepare and insert US probe	Probe cover not correctly filled with water resulting in poor-quality images	Poor performance	Wrong dose location	6	9	4	216	
8	4	2 Prepare patient	Identify patient	Patient implanted with wrong source pattern	Did not perform the time out (omission), performed the identification incorrectly (training failure)	Wrong dose distribution	3	9	6	162	

#	ID	Step	Action	Failure mode	Cause	Effect	S	O	D	RPN	Comments
9	24	3 Position US probe	RO checks images along the whole superior–inferior range	Prostate not aligned	Poor-quality images (QA failure for imaging system)	Wrong dose location	8	8	9	576	Again, given poor image quality, detecting this failure would be difficult.
10	37	4 Insertion of needles and sources	RO pushes needle until tip shows at desired depth	Needle at wrong depth	Training failure	Wrong dose distribution	8	8	9	576	Given poor training, the likelihood of occurrence is high.
11	21	3 Position US probe	RO checks images along the whole superior–inferior range	Prostate not aligned	Training failure	Wrong dose location	8	8	8	512	Compared with the wrong location for probe alignment, the error in the dose distribution here likely would be less severe.
12	22	3 Position US probe	RO checks images along the whole superior–inferior range	Prostate not aligned	Inattention	Wrong dose location	7	8	8	448	
13	23	3 Position US probe	RO checks images along the whole superior–inferior range	Prostate not aligned	Poor performance	Wrong dose location	7	8	8	448	
14	12	2 Prepare patient	Scrub patient	Patient not cleaned thoroughly	Poor performance	Patient develops an infection	5	8	10	400	Infections can be life threatening. Not detectable after the scrub.
15	9	2 Prepare patient	Scrub patient	Patient not cleaned thoroughly	Training failure	Patient develops an infection	4	8	10	320	Infections can be life threatening. Not detectable after the scrub.
16	10	2 Prepare patient	Scrub patient	Patient not cleaned thoroughly	Inattention	Patient develops an infection	4	8	10	320	Infections can be life threatening. Not detectable after the scrub.
17	11	2 Prepare patient	Scrub patient	Patient not cleaned thoroughly	Rushed environment	Patient develops an infection	4	8	10	320	Infections can be life threatening. Not detectable after the scrub.
18	13	3 Position US probe	Prepare and insert US probe	Probe not cleaned adequately	Poor performance, possible rushed environment	Patient develops an infection	4	8	10	320	Assume that the person cleaning the probe has been trained.
19	39	4 Insertion of needles and sources	RO pushes needle until tip shows at desired depth	Needle at wrong depth	Poor-quality images (QA failure for imaging system)	Wrong dose distribution	8	7	9	504	
20	17	3 Position US probe	Prepare and insert US probe	Template not seated properly	Inattention/Poor performance	Wrong dose location	8	7	7	392	
21	41	4 Insertion of needles and sources	RO holds stylet and removes needle	RO fails to hold stylet adequately	Poor performance	Wrong dose distribution	7	6	9	378	
22	42	4 Insertion of needles and sources	RO holds stylet and removes needle	RO pushes on stylet	Poor performance	Wrong dose distribution	7	6	9	378	

(continued)

TABLE 24.4 List of Table 24.2 Sorted by Descending Values of the Severity Score and by RPN (Continued)

Rank	Failure mode #	Major Processes	Step	Potential Failure Modes	Potential Causes of Failure	Potential Effects of Failure	O	S	D	RPN	Notes
23	20	3 Position US probe	RO rotates probe and mounting bracket to match previous images	Prostate not aligned	Changes in prostate gland	Wrong dose distribution	7	6	8	336	While detectable, there is little to be done to correct.
24	30	4 Insertion of needles and sources	MP finds needle in package or container	MP selects the wrong needle	MP fails to hear the correct needle identification	Wrong dose distribution	7	6	8	336	
25	35	4 Insertion of needles and sources	RO inserts needle into indicated hole in template	RO inserts needle into the wrong hole	Bad viewing conditions	Wrong dose distribution	7	6	8	336	
26	36	4 Insertion of needles and sources	RO inserts needle into indicated hole in template	RO inserts needle into the wrong hole	Confusion between holes	Wrong dose distribution	6	6	8	288	
27	38	4 Insertion of needles and sources	RO pushes needle until tip shows at desired depth	Needle at wrong depth	Inattention	Wrong dose distribution	6	6	8	288	
28	27	4 Insertion of needles and sources	MP finds needle in package or container	MP selects the wrong needle	Inattention	Wrong dose distribution	5	6	8	240	The severity is likely only moderate, unless the erroneous needle places a source very close to the urethra or rectum. Detection for that needle error is low, but is likely to be found too late, when the needle used is actually called.
29	28	4 Insertion of needles and sources	MP finds needle in package or container	MP selects the wrong needle	Confusion between multiple packages	Wrong dose distribution	5	6	8	240	
30	31	4 Insertion of needles and sources	MP hands needle to RO	MP drops needle	Room layout makes transfer insecure	Wrong dose distribution	4	6	10	240	Assumes that the sources could not be reprocessed and loaded in a clean needle in time for use. If processing is available and performed, the effect is only delay. The frequency may be 1/10,000 needles or 1/1000 patients. Detection would be after the needle drops.
31	32	4 Insertion of needles and sources	MP hands needle to RO	MP drops needle	MP slips	Wrong dose distribution	4	6	10	240	
32	34	4 Insertion of needles and sources	RO inserts needle into indicated hole in template	RO inserts needle into the wrong hole	Inattention	Wrong dose distribution	5	6	8	240	
33	40	4 Insertion of needles and sources	RO pushes needle until tip shows at desired depth	Needle at wrong depth	Confusion between planes	Wrong dose distribution	5	6	8	240	

		Step	Action	Failure mode	Cause	Effect	O	S	D	RPN	Comments
34	45	5 Review implant	RO scans through images looking for gaps	Missing gaps or not seeing sources	Poor-quality images (either QA failure for imaging system or interference from the implanted sources)	Suboptimal dose distribution	5	6	8	240	Assumes that the source distribution did not follow the plan closely enough that coverage requires additional sources. Also assumes poor image quality.
35	46	5 Review implant	RO adds sources if needed and available	RO adds unnecessary sources	Poor-quality images (either QA failure for imaging system or interference from the implanted sources)	Suboptimal dose distribution	5	6	8	240	Assumes that the source distribution followed the plan but images looked like there was a sizable gap. Also assumes poor image quality.
36	47	5 Review implant	RO adds extra sources if needed and available	RO adds unnecessary sources	Misinterpretation, training failure	Suboptimal dose distribution	5	6	8	240	
37	48	5 Review implant	RO adds extra sources if needed and available	RO fails to add needed source	Misinterpretation, training failure	Suboptimal dose distribution	5	6	8	240	
38	29	4 Insertion of needles and sources	MP finds needle in package or container	MP selects the wrong needle	Poor demarcation of needles	Wrong dose distribution	4	6	8	192	
39	25	4 Insertion of needles and sources	RO determines starting needle	Poor selection of posterior needles at first obscuring later anterior locations	Training failure	Wrong dose distribution	3	5	8	120	Once selected, the probability of detection before insertion is low.
40	26	4 Insertion of needles and sources	RO determines starting needle	Poor selection of posterior needles at first obscuring later anterior locations	Inattention	Wrong dose distribution	3	5	8	120	Inappropriate selection due to inattention is not as likely due to habit.
41	43	4 Insertion of needles and sources	RO selects next needle	Poor selection of posterior needles that later obscure more anterior locations	Training failure	Wrong dose distribution	3	5	8	120	
42	44	4 Insertion of needles and sources	RO selects next needle	Poor selection of posterior needles that later obscure more anterior locations	Inattention	Wrong dose distribution	3	5	8	120	

(continued)

TABLE 24.4 List of Table 24.2 Sorted by Descending Values of the Severity Score and by RPN (Continued)

Rank	Failure mode #	Major Processes	Step	Potential Failure Modes	Potential Causes of Failure	Potential Effects of Failure	O	S	D	RPN	Notes
43	33	4 Insertion of needles and sources	MP hands needle to RO	MP stabs RO with needle	MP slips	Injury to RO, possible wrong dose distribution	2	5	10	100	The wrong dose distribution would be if the sources could not be reprocessed for the case. If the RO is unable to continue the case, a very wrong dose would result and the severity could be 9.0 for the patient.
44	1	1 Gathering materials	MP brings needles to procedure room	MP forgets needles	Other things on the mind, too many other things to bring	Delay as MP goes to get the needles	5	1	1	5	Assumes that the MP had ordered the needles at some time in the past.
45	2	1 Gathering materials	MP brings plan to procedure room identifying needle location and depth	MP forgets documentation	Other things on the mind, too many other things to bring, not notice those papers missing	Delay as MP goes to get the plan	5	1	1	5	
46	3	1 Gathering materials	MP brings planning images to procedure room	MP forgets images	Other things on the mind, too many other things to bring, did not notice the missing images	Delay as MP goes to get the plan	5	1	1	5	
47	5	2 Prepare patient	Bring patient into procedure room	None			0	0	0	0	
48	8	2 Prepare patient	Anesthetize patient	None relevant to this example							Leave this row for the anesthesiologists to work on.

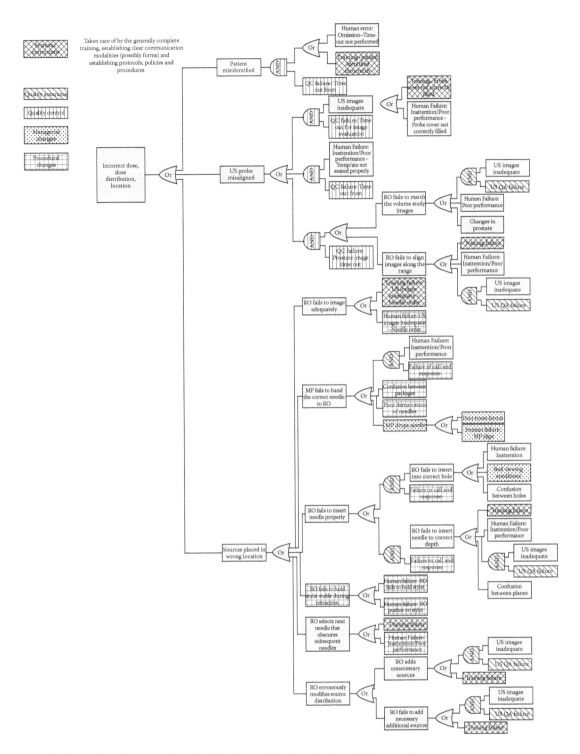

FIGURE 24.2   A fault tree for an incorrect dose distribution resulting from failures in the example.

which can lead to rushing during procedures, distractions caused by excessive demands, or failure to provide the necessary equipment or maintenance of equipment.

- Commissioning—While most medical physicists are familiar with commissioning equipment and instrumentation, processes also require commissioning. Commis-

sioning of a process consists of walking through the steps with the persons who will perform the various functions and seeing that the procedures as proposed flow smoothly and produce the desired outcomes. Part of commissioning a process includes commissioning all equipment used in the process. Commissioning both the

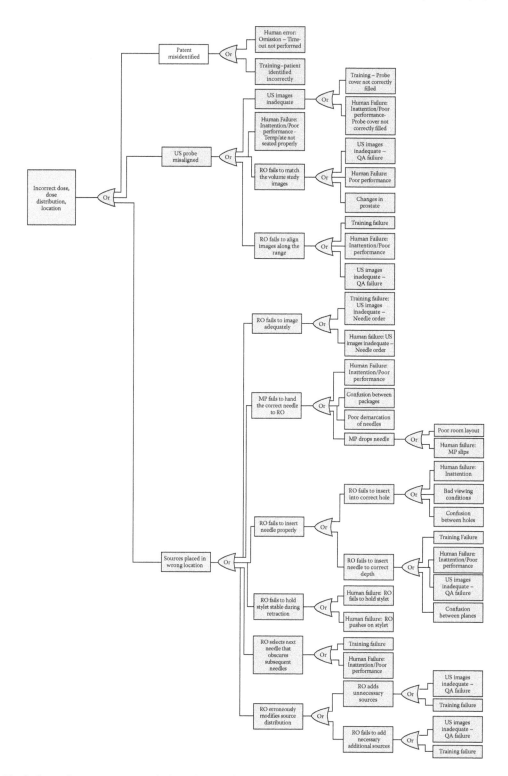

FIGURE 24.3 The fault tree from Figure 43.2 with the inclusion of quality management.

process as a whole and the equipment requires not only assuring that things operate as specified, but learning the limits of reliability, that is when things break down. Equipment should not only be tested following instructions, but also intentionally used incorrectly to learn how it responds.

- Quality control—Quality control (QC) acts to make sure the inputs to a process are correct (avoiding garbage in, garbage out.) As such, QC steps directly address the causes of failure at the extreme right side of the fault tree.
- Quality assurance—Quality assurance (QA) considers the output of a process and compares that with expectations

or standards to evaluate if it seems correct. QA mostly works toward the left side of the fault tree.

- Quality audits—During quality audits, an expert from outside the facility reviews the procedures used (process audit) and looks at patient records and logs (product audit). Very often, someone not familiar with how a given facility works can notice problems that those directly involved do not.
- Quality improvement—Gathering the information from the QC and QA, as well as the audits, reports, and analyses of any events that happened allows recognition of patterns that may indicate either where the existing QM does not work well or where the QM may have opened new possible failure modes. Periodically reviewing all this information and either redesigning parts of the process or changing some of the QM strengthens the quality program.

Our example assumes that the procedure is new, or that we are looking at it in a new light but with no information about past events or reports from auditors, thus the only approaches we have with which to work would be design, QA, QC, and commissioning.

The small numbers above the arrows in Figure 24.1 show the position in the process of the 25 steps with highest ranking RPN values in order, with 1 being the highest value. The small numbers by the text opposite the arrows denote the ranking of the steps according to severity, as in Table 24.3. This display calls attention to the parts of the procedure with the highest hazard, in this case moving the patient to the table, scrubbing the patient, all the steps in probe alignment and the insertion of the needle, and depositing the sources. This indicates where great care should be taken during the design of the procedure.

Figure 24.3 shows the fault tree from Figure 24.2, but with the addition of quality management. In this fault tree, quality management changes consist of systemic changes, QA, QC, managerial policy changes, or procedural changes. Many of the failures could result from inadequate training. Obviously, the persons involved need sufficient training in their functions to understand the ramifications of all possible problems and failures. Eliminating deficient training goes a long way toward preventing failures in the first place. Training would be a systemic change, shown by filling the boxes with cross hatch. Managerial changes, shown as dotted fill in this example, mostly entail environmental changes, such as correcting bad lighting and redesigning the procedural room to eliminate problems that would allow the medical physicist to drop a loaded needle in such a way that it could not be used.

Most of the QC and QA enter into the fault tree as the addition of AND gates, where a failure propagating requires a concomitant failure of the added QM step. For example, a new form would be used with the patient identification step to remind the staff to identify the patient and enter the method used. The time-out concept also finds use with the ultrasound probe alignment, enforcing a stop to check the work.

New procedures address some of the potential failures, for example having the physicist watch to make sure that the radiation oncologist selects the appropriate needle for insertion to avoid image distortion, and the addition of a call and response for verification of the needle identification on the medical physicist handing it to the radiation oncologist. Some of the new procedures require facility changes, for example, a flange added to the template that could slide to the end of the stylet could address failure of the radiation oncologist to hold the stylet in position during needle retraction.

The goal is to work through the listings of the FMEA until all potential failures that could cause significantly severe effects, or with a significant RPN, have protection from propagating to the end failure, in this example, a dosimetric failure. The definition of "significant" depends on a facility's risk adversity and resources available to provide for quality. In this example, an easy line for significance falls at line 44, where the severity drops from 5 to 1 and the RPN from 100 to 5. However, a facility may want to address the potential failures between 44 and 48 because they could affect the efficient operation of the department.

After performing the procedure for a while, reevaluation for the process should be performed. Better values could be assigned for the FMEA, and weaknesses discovered through events or near-misses could be addressed and the process improved.

Reference

American Association of Physicists in Medicine Task Group 100, S. Huq, B. Fraas, P. Dunscombe, J. Gibbons, G. Ibbott, A. Mundt, S. Mutic, J. Palta, F. Rath, B. Thomadsen, J. Williamson, and E. York. 2011. Application of Risk Analysis Methods to Radiotherapy Quality Management: Report of Task Group 100 of the American Association of Physicists in Medicine. http://www.aapm.org/pubs/reports/.

Methods to Assure
and Improve Quality

25

Medical Indicators of Quality: Terminology and Examples

Vincenza Viti
Istituto Superiore di Sanità (ISS)

Introduction

Health quality indicators are developed by diverse organizations for many different purposes. Health organizations and national and/or regional health authorities often develop and use them to survey activities. Professional colleges may design them to monitor the quality of delivered care, mainly within the framework of continuous quality improvement (CQI) projects. A health board can draw up indicators for internal monitoring of specific problems. Accreditation/certification systems and quality benchmarking activities usually agree on the indicators to be adopted in their processes. Given these large, diversified fields of application, knowing how to draw up and use general health quality indicators and how to design indicators for ad hoc purposes has become a preeminent necessity. The first part of this chapter deals with quality indicator terminology, while the second part illustrates examples of indicators that were developed specifically for radiotherapy.

Indicators as Measurements of Healthcare Quality

Performance monitoring is now a major process in the organization and professional management of health structures. Performance monitoring promotes continuous improvement in services, organization, and, ultimately, health service quality. It has led professionals to develop methodologies for measuring hospital performance which, in defining measurements, helps clarify objectives and provides the means not only for achieving its main objective, that is, deciding when changes are needed, but also for assessing any innovation that is introduced.

Quality measures incorporate items defined as clinical indicators. An indicator is an evaluation tool that serves to identify a problem, quantify it, and measure the success of intervention. One can define it as a measurable variable that is adopted to monitor a phenomenon in time, in space, or to estimate progress toward goals that should be reached.

When monitoring activities to improve health service quality, it is always helpful to define indicators and relative thresholds. Indicators should be used when starting monitoring within the framework of CQI projects and in specific audit activities. In both these instances, the indicator needs to be defined with a threshold (or standard) value. An objective to be reached or maintained can, at times, be considered an indicator without establishing a threshold as long as requirements are defined precisely.

Indicators are generally quantitative variables. They are expressed in numbers that stand for the value of a quantity, the frequency of an event, or the result of a scoring system. They can be also qualitative indices (e.g., sex, result of a diagnostic test). Bear in mind that an indicator should indicate if a problem exists or could potentially exist. Operators will then decide whether they need to take an action to improve quality.

Which Quality Can an Indicator Measure?

Indicators in healthcare are mainly used to measure professional quality and help operators provide correct diagnoses and treatments. However, managing quality and perceived quality should also be monitored, because these three dimensions of quality (professional, managing, and perceived) are not always clear-cut. For example, "waiting time," which is a widely used indicator in radiotherapy, is an overlapping composite of three qualities: (1) professional—the operator must consider appropriateness and priority of requests; (2) management—it is the responsibility of the organization to offer needed services; and (3) the patient's feelings—waiting for treatment must be considered a potential cause of avoidable stress. Finally, an indicator may serve for nationwide monitoring of social quality in terms of accessibility to, and equity of, medical care.

Indicator Quality Domains

There are different classifications of the quality domains that are monitored through indicators, but terms are often utilized outside the classification scheme that first coined them. Here is an outline of the main systems with some clarification to facilitate in the depth understanding.

Donabedian Classification

According to the Donabedian classification (Donabedian 1980), an indicator monitors one of three quality domains: structure, process, or outcome. Indicators of structure monitor the following:

- Infrastructure, equipment, and staffing in terms of number and qualifications
- System design—existence of MCQ programs, attention to equity, and continuity of services
- Availability of facilities

These measures precede patient interactions with physicians, nurses, and other health operators.

Indicators of processes monitor organizational processes and professional processes. Organizational processes include:

- Volume, that is, the number of patients
- Waiting lists
- Coordination and continuity
- Appropriateness (from a general point of view)
- Efficiency (which can also be measured in terms of use and complexity of resources)
- Support activities
- Education (basic and continuous)
- Managing activities, staff involvement, for example, existence and application of a reward system; implementation of MCQ programs, for example, internal and external audits

Indicators in this professional category monitor to what extent a healthcare operator competently and safely delivers appropriate, timely care in accordance, when possible, with international evidence and patient consensus. Examples include the following:

- Appropriateness, that is, convenience of a medical intervention for a specific patient
- Technical correctness
- Timeliness
- How operators dialogue with, and give information to, patients and caregivers

Outcome indicators signify a modification in health and include:

- Final health outcome, that is, improved functional status and quality of life in relation to pain and disability, survival, side effects, and complications
- Intermediate health outcome, that is, biological modifications, when associated with final health outcome
- Patient satisfaction with overall quality of care, which includes not only the final outcome, but everything the patient had to do to access the structure and within the structure itself

Joint Commission Classification

The Joint Commission classifies four types of indicators that define events to be assessed (JCAHO 1994, 1997): sentinel event, a rate-based indicator, outcome, and process.

A sentinel event is an unexpected occurrence involving death, serious physical or psychological injury, or the risk thereof. The phrase "or the risk thereof" includes any process variation, a recurrence of which would carry a significant chance of a serious adverse outcome. Such events are called "sentinel" because they signal the need for immediate investigation and response, see http://www.jointcommission.org/SentinelEvents/.

Rate-based indicators use data about events that are expressed as rates, ratios, or mean values for a sample population. Rate-based events are expected to occur with some frequency (Mainz 2003). Outcome and process measurements are defined in agreement with the Donabedian classification.

Besides this classification, the Joint Commission Classification specifies what an indicator should measure, including patient access to care, ease in obtaining what patients need when they need it, appropriateness, continuity, effectiveness, efficacy, patient perception, safety, and timeliness.

The American College of Radiology distinguishes between efficacy and effectiveness (American College of Radiology 1996). Efficacy is defined as "the degree to which the care received has the desired effect with a minimum effort, expense, or waste." Thus, the college refers to economic efficacy, that is, the ratio between outcomes and resources. Effectiveness is defined as "the degree to which care is provided in the correct manner." Thus, effectiveness considers the effectiveness of care in clinical practice.

The WHO Classification

A consensus was found by the WHO Regional Office for Europe around six key dimensions by choosing to emphasize the importance of patient care in acute care hospitals and performance areas that hospitals can impact (WHO 2003). Some of the performance areas that hospitals can impact include clinical effectiveness, patient centeredness, responsiveness to patients, production efficiency, safety for patients and staff, responsive governance (which includes access, continuity, and health), and promotion (equity). Clinical effectiveness includes both professional processes and outcomes where "technical quality, evidence-based practice and organization, health gain, outcome (individual and population)" are considered. Production efficiency, unlike the Joint Commission Classification, is the ratio between volume and resources. Patient centeredness includes all activities in response to the patient's needs, for example, confidentiality, client orientation (prompt attention, access to social support, quality basic amenities, choice of provider), patient satisfaction, and patient experience (dignity, confidentiality, autonomy, and communication).

Main Characteristics of an Indicator

An indicator should be reliable so that it has a low intra- and inter-observer variability. The indicator should be accurate, allowing data collection without systematic errors. Lastly, any indicator should be sensitive to changes and specific in terms of quality. Useful indicators are pertinent, scientifically robust (i.e., evidence-based), and able to influence decisions.

For the purposes of data collection, indicators should be easily understood, simple in conception, and reasonable in costs. Efforts to collect data should be timely and complete (i.e., the selected sample should provide a complete data set upon collection).

Defining an Indicator

When defining a new indicator, two elements are essential: (1) a working definition, which is crucial, and (2) the associated pilot study, which will test indicator reliability and real-life feasibility, including an examination of data collection difficulties. The essential points to consider when defining an indicator are threshold (or standard), confounding factors, data sources, and indicator checklists.

Threshold (or Standard)

An effective indicator needs a threshold (or standard). The threshold may be defined statistically in terms of indicator distribution. It may be based on international reports or on internal values (e.g., indicator distribution during the first year, which is increased annually). Although an indicator may lack a threshold when first defined, a threshold should be established as soon as you become aware of the state of things. A sentinel event can be considered as an indicator with zero threshold, as it signals a fault requiring immediate action.

Please note again the differences in meaning of the term "threshold." In quality improvement projects, "threshold" indicates the aim to be reached; in certification systems, "thresholds" must be reached to achieve certification.

In certification systems, an indicator together with its threshold is termed "a standard," that is, the criterion to ascertain data adherence. In this context, the "threshold" is part of a "standard." This terminology is sometimes adopted in healthcare quality.

Confounding Factors

Since data may reflect results that cannot be attributed to quality problems, potential confounding factors must be taken into account when defining an indicator. Confounding factors, or covariates, can sometimes be measured. For example, in considering prostate cancer treatment, the Rand group identified many covariates such as age, life expectancy, pretreatment PSA, clinical stage, Gleason grade, etc. (Litwin et al. 2000).

Data are stratified to exclude confounding factors. For example, the sample of patients is divided into different strata or groups and data are collected and analyzed separately for each group. Data may also be standardized (divided), for example, for age or sex. Data adjustment for these factors may be performed when comparing data from different centers, but must be used with caution because it may introduce data modifications, thus obscuring true results.

Results are also influenced by chance variability, which impacts mainly when using small data samples. When the data sample is large, a rare monitored event may be an expression of chance variability.

Data Sources

Databases for indicators are obtained from statistical and demographic data collections, systematic health data collections, clinical documents, and ad hoc data collections. If using previously collected data, information on data quality is needed. Before starting data extraction and/or collection, data collecting operators should be instructed about data uniformity in terms of information content and how the information was obtained.

Indicator Checklist

Table 25.1 reports a summary checklist that contains the main items that should be addressed when defining a quality indicator. Table 25.2 compares definitions for indicators of waiting times in radiotherapy as developed by Australian [Australian Council for Healthcare Standards (ACHS) and New Zealand College of Radiologists Faculty of Radiation Oncology 2009] and Italian (Cionini et al. 2007) groups of professionals. A waiting time threshold is included in the Australian indicator specifications. Therefore, according to the definition given in the section "Threshold (or Standard)," this appears as a standard.

Should Processes or Outcomes Be Measured?

As process indicators measure quality directly, there are many excellent examples of indicators being used to document and improve healthcare processes. Although quality indicators are less widely adopted for measuring outcome, international indications (European Commission—DG for Health and Consumers; Rubin, Pronovost, and Diette 2001) currently encourage operators to develop outcome indicators as they illustrate the overall

TABLE 25.1 Quality Indicator Checklist

Item	Explanations and Notes
Topic and name	What are you measuring?
Rationale	What is the rationale for measuring? What are the advantages? List evidence for those advantages.
Type of indicator	Is this indicator a structure, process, or outcome? Is this indicator from aggregate data or a sentinel event?
Numerator	The numerator is a parameter value. Which data are necessary to calculate it? What are the available data sources?
Denominator	The denominator is the reference population. Which data are necessary to calculate it? What are the available data sources?
Confounding factors	Are confounding factors present? If yes, define stratification or adjustment.
Stratification	What are the recommended categories for the application? May be used when confounding factors are present.
Threshold/standard	This is the reference value. Consider if it already exists or when it will be available. List evidences for the choice of threshold or standard.
Aspect of quality to be monitored	This can be professional, managing, and/or perceived.
Reliability and accuracy	How are these tested?
Data collection	What is the data type (population/sample) and time interval of data collection? Who is in charge of data collection? Necessary explanations should be given for uniform data collection.
Data analysis	Who is in charge of data analysis? Ideally, a statistician should be used to help in data interpretation.
Comparisons	Results should be monitored over time. Data comparison can be done among similar structures.
Use of the indicator	Specify the context (e.g., internal/external audit, CQI projects, etc.) for using the indicator. Who is in charge of the final utilization of the indicator? Who is in charge of continuous quality improvement projects?

TABLE 25.2 Comparison of Indicators for Radiotherapy Waiting Time

Name/Group	Waiting Time—Italian Indicator[a]	Waiting Time—Australian Indicator[b]
Topic	Treatment delay calculated as total mean waiting time (TWT). The total waiting time is called TWT and can be subdivided into three periods: WT1, WT2, and WT3.	Waiting time for radiotherapy.
Definitions	WT1: interval from the referral to the radiotherapy center to the initial prescription. WT2: interval from the initial prescription to the final prescription. WT3: interval from the final prescription to the beginning of treatment.	Waiting time is the interval from when the data treatment should commence and the first treatment is being delivered. Date ready for care is considered time zero.
Indicator dimension	Process and structure/mean value.	Rate based indicator/access to care.
Numerator	TWT is the sum of the waiting times (in days). WT1, WT2, and WT3 are waiting times of all patients considered in the study.	Total number of patients waiting more than 14 days from the date ready for care to the date of commencing radiotherapy.
Denominator	Number of treated patients.	Total number of patients commencing radiotherapy.
Recommended stratification	According to the treatment objective.	—
Deviate/exclude	Deviations from suggested values may be justified when medical treatments must precede radiotherapy (e.g., hormonotherapy for curative radiotherapy in prostate cancer and chemotherapy for postoperative radiotherapy in breast cancer).	Postoperative healing phase and/or post chemotherapy phase before which treatment should not commence. Any delay requested by the patient. Time for treating current morbidity. Delays outside the control of the radiotherapy department.
Further stratification	To obtain information on the three periods you may sum separately the three waiting times for all patients.	—
Total waiting time standard according to the recommended stratification	Curative/radical <30 days per patient. Palliative/symptomatic <10 days per patient. Preoperative adjuvant <15 days per patient. Postoperative adjuvant <60 days per patient.	Desired rate: low
Time period for data collection and frequency of analysis	At least 3 months, at least every 2 years.	1 week (specified in May or November)

[a] *Source:* Cionini et al., 2007, *Radiother. Oncol.*, 82, 191–200. With permission.

[b] *Source:* Australian Council on Healthcare Standards and Royal Australian (ACHS) and New Zealand College of Radiologists Faculty of Radiation Oncology, *Radiation Oncology Indicators Clinical Indicators User's Manual*, v. 3, ACHS, Sydney, 2009.

result of care. There are often, however, inherent difficulties. For example, a long period of time might need to elapse before the outcome of an event can be assessed or else an important event might be a rare event. Confounding factors that are difficult to measure but which may influence outcomes include nutrition, environment, lifestyle, and poverty (Mant 2001). These factors impact less in clinical trials because patients and controls are often carefully selected.

Experiences with Quality Indicators for Radiotherapy

Although indicator-related assessment of quality in radiotherapy is on the increase worldwide, a full set of indicators has not yet been drawn up by any international or national organization. Given the complexity of treatments in radiotherapy, the enormous effort that is required probably precludes designing suitable measures to control the entire radiotherapy process. Drawing up indicators for radiotherapy treatments for specific pathologies may often have seemed a better and wiser approach. Moreover, the need to adapt indicators to national and local realities sometimes hinders direct application of indicators developed elsewhere and so many validated indicators, besides being utilized as they stand, are modified for specific purposes or even adopted as models for designing indicators in the same field.

This report of quality measures in radiotherapy certainly does not present any definitive final statement and does not attempt to replace other prestigious studies that sometimes lack widespread diffusion because they are not easily accessed. Our aim is to show relevant examples of indicators that were developed in other conditions and that may be utilized to construct tools to measure care quality.

Even if indicators address different aspects of treatment, they are listed together when drawn up by national professional groups with a consensus on policy. Indicators for breast treatment derive from the pooled experience of several groups from different countries because a group of experts conducted a well-documented survey of indicators for this pathology. Finally, some recently developed indicators that are found in the international literature are referred to because the topic deserves particular emphasis or because they were drawn up by specific groups and organizations.

National Indicators

Australasian Indicators

Clinical Australian indicators constitute one of the longest professional indicator programs. First released in 1999, they are still an advanced system, forming part of the Healthcare Evaluation Program for the ACHS, which was developed in collaboration with professional colleges. In fact, radiotherapy indicators were developed by the Royal Australasian College of Radiologists. Indicators are updated from time to time and more indicators have been added to the original list; they currently total 10 indicators for radiotherapy. The latest version is n.3 (ACHS and New Zealand College of Radiologists Faculty of Radiation Oncology 2009).

Indicators for radiotherapy are listed in Table 25.3. The organization's manual (ACHS and New Zealand College of Radiologists Faculty of Radiation Oncology 2009) describes them in depth, citing sources of information and providing a series of forms for data collection. Complete indicator grids for some can be retrieved from the Agency for Healthcare Research and Quality (AHRQ) Web site (http://www.qualitymeasures.ahrq.gov).

Australian indicators follow the Joint Commission classification (see the section "Joint Commission Classification"), but specify rate-based events as access to care, efficiency, or patient consent. Table 25.3 reports this classification.

The 10 radiotherapy indicators refer to different themes and cancer pathologies. Eight are process indicators and two

TABLE 25.3   Australasian Indicators

Number	Topic of Monitoring	Dimension
1	Recorded informed consent	CP-RB, patient's consent
2	Waiting times longer than 14 days from the "ready for care" date to the date radiotherapy commences	CP-RB, patient's access
3	Clinical trial participation	CP-RB, patient's access
4	Treatment delay for postoperative radiotherapy for head and neck cancer	TP-RB, patient's access
5	Chemo-radiotherapy for curative cancer of the cervix	TP-RB, patient's access
6	Radiotherapy using multileaf collimators	TP-RB, patient's access
7	Curative radiotherapy using CT planning	TP-RB, patient's access
8	Letters on file to the referring doctor and general practitioner regarding radiotherapy	TP-RB, care efficiency
9	Complete follow-up and outcome for patients treated with radiotherapy for glottic cancer	OP-RB, care process
10	Complete follow-up for patients treated with radiotherapy for breast conservation	OP-RB, care process

Source: ACHS and New Zealand College of Radiologists Faculty of Radiation Oncology, *Radiation Oncology Indicators Clinical Indicators User's Manual,* v. 3, ACHS, Sydney, 2009.
Note: RB, rate-based; CP, consultation process; TP, treatment process; OP, outcome process.

encompass both outcome and process dimensions. Some of the process indicators refer to technical issues (indicators 6 and 7), while others (indicators 1 and 2) address the issue of patient care from a general point of view. These indicators were tested at length and now are currently employed in the framework of CQI programs. Insight into indicator-related promotion of changes derives from critical reports on data collections of these indicators (ACHS 2007; Gibberd et al. 2004). It is worth noting that the greatest change was shortening the time interval for commencing radiotherapy (waiting time) from 21 to 14 days after 2006 (indicator 2 of Table 25.3).

As already mentioned, waiting time thresholds are included in the indicator specifications. The Australians display a pragmatic attitude toward indicator values as they simply define a "desired rate," which might be high, low, or unspecified. Apparently their objective is to promote changes rather than monitor poor quality.

Italian Indicators (General)

A pilot study was undertaken in Italy with the aim of developing performance indicators for a typical radiotherapy center that could serve for a quality audit. A national multidisciplinary working group drew up a set of general indicators and tested them in a few Italian radiotherapy centers and medical physics services (Cionini et al. 2007; Working Group Continuous Quality Improvement in Radiotherapy). The final set of 13 indicators is listed in Table 25.4. They deal with general structural and/or operational features, health physics activities, accuracy,

and technical complexity of treatment. Some indicators, such as the patient's opinion survey or information in clinical records, can be used to monitor activities that are not specific to radiotherapy, while others, such as the indicator referring to fields per planned treatment volume, are directly related to the technical quality of treatment.

Indicators for Selected Pathologies

RAND Indicators

RAND Health, a division of the RAND Corporation, developed and tested quality assurance indicators for different pathologies including breast cancer, cervical cancer, colorectal cancer, lung cancer, prostate cancer, skin cancer, and cancer pain and palliation (Asch et al. 2000; Malin et al. 2000). Using the case-based approach, quality indicators for cancer were developed with the aim of covering the care continuum. Drafted on the basis of clinical literature, indicators were then reviewed by a panel of experts and endorsed by a consensus panel as appearing valid and feasible for further testing in a population-based sample. Table 25.5 reports the final list of indicators for radiotherapy.

Italian Indicators for Selected Pathologies

A larger multidisciplinary Italian group than the one that drew the general indicators for radiotherapy (see the section "Italian Indicators (General)") designed clinical indicators for some pathologies and tested them in 30 Italian radiotherapy centers (Progetto 2005). A set of six to seven indicators was defined for each selected pathology, which included gynecological and breast tumors, bone metastases, lung, prostate, rectum, and head-neck tumors. Table 25.6 reports the list of indicators. It is worth noting a few outcome indicators are on the list. An ongoing review process this year will allow indicators to be used by the general community.

Prostate Cancer

The RAND group attempted to establish specific indicators for assessing quality of care for patients with localized prostate carcinoma (Litwin et al. 2000). The team reviewed medical literature, interviewed experts in surgical and radiation treatment of prostate cancer, analyzed findings from focus groups with patients and spouses to understand what information patients needed, and, finally, considered the recommendations of an expert consensus panel. The outcome was a list of candidate indicators of quality of care for prostate cancer. Validity, feasibility, and appropriateness were the main parameters when assessing indicator acceptability. Table 25.7 reports candidate indicators for radiotherapy. They include measures of structure, process, and outcome, together with potentially confounding covariates that must be allowed for in quality-of-care studies.

The RAND report did not propose standards for recording compliance with each indicator and recommended performing

TABLE 25.4   List of Italian Indicators (General)

Number	Topic of Monitoring	Dimension
1	Patients per operator	S/P—staff efficiency
2	Patients per HEU	S/P—equipment efficiency
3	Treatment delay: mean waiting times	S/P—patient's access
4	Complete data in clinical records	P—care process
5	Questionnaires for patient's opinions	P—patient's opinion
6	Multidisciplinary approach	CP
7	Machine downtime for nonplanned maintenance	TP
8	Instrumentation for dosimetry and quality controls	TP
9	Equipment quality controls programs	TP
10	Treatment plans with CT	TP
11	Fields per PTV	TP
12	Shaped fields	TP
13	Portal verification	TP

Source: Cionini et al., *Radiotherapy and Oncology*, 82, 191–200, 2007.
Note: S, structure; P, process; CP, clinical process; TP, technical process.

TABLE 25.5 Quality Indicators Regarding Radiotherapy Recommended by RAND on the Basis of a Critical Review of Literature for Diagnosis and Treatment of Different Pathologies

	N. Indicator
Breast cancer	
Waiting times of radiotherapy for women treated with breast-conserving surgery	N. 6, p. 44
Colorectal cancer	
Patients with stage II and stage III rectal cancer should be offered:	N. 10, p. 108
• Postoperative radiotherapy to the pelvis	
• Preoperative radiotherapy to the pelvis	
• Preoperative radiotherapy with chemotherapy	
Non-small-cell lung cancer	
Patients with stage II nonsmall cell lung cancer with good performance status should be offered:	N.6, p. 168
• Thoractomy with surgical resection of the tumor	
• Radiotherapy to the thorax	
• Chemotherapy	
Patients with non-small-cell lung cancer who have metastases on MRI or CT of the brain should be offered:	N. 8, p. 169
• Radiotherapy to the brain	
• Surgical resection of the metastasis	
• Stereotactic radiosurgery	
Small-cell lung cancer	
Patients with limited small-cell lung cancer should be offered combined modality therapy with radiotherapy and chemotherapy.	N. 9, p. 170
Patients with small-cell lung cancer who have metastases on MRI or CT of the brain should be offered (unless they have received both previously):	N. 11, p. 170
• Radiotherapy to the brain	
• Chemotherapy	
• Stereotactic radiosurgery	
Patients with small-cell lung cancer who have bone pain and a corresponding positive radiography study should be offered (unless they have received both previously):	N. 12, p. 170
• Radiotherapy to the region	
• Chemotherapy	
Prostate cancer	
Men older than 60 years with minimal prostate cancer (stage 0/A1) should not be offered any of the following treatments:	N. 4, p. 211
• Bilateral orchiectomy	
• LHRH analogue	
• Antiandrogen	
• Radical prostatectomy	
• Radiotherapy	
Men younger than 65 years who do not have coronary artery disease or a second cancer should be offered radical prostatectomy or radiotherapy for localized prostate cancer (stages I&II/A2&B).	N. 5, p. 211
Prostate cancer patients with evidence of cord compression on MRI scan of the spine or CT myelogram should be offered one of the following:	N. 9, p. 214
• Radiotherapy to the spine	
• Decompressive lamunectomia	
Cancer pain and palliation	
Patients with painful bone metastases unresponsive to or intolerant of narcotic analgesia should be offered:	N.3, p. 233
• Radiotherapy to the sites of pain	
• Radioactive strontium therapy	

Source: Asch et al. *Quality of Care for Oncological Conditions and HIV. A Review of Literature and Quality Indicators.* Santa Monica, CA: RAND Health, 2000.

Note: Thresholds and other details on indicators, together with evidences from literature, are located in *Quality of Care for Oncological Conditions and HIV. A Review of Literature and Quality Indicators* (number and page in the second column).

TABLE 25.6   Italian Indicators for Selected Pathology (Tumor Site)

Pathology/Topic	1	2	3	4	5	6	7
Gyn	Multidisciplinary approach (f) P	Acute toxicity monitoring (f) O	Anaemia monitoring (f) P	Dose-volume histograms (f) P	Overall treatment time (v) P	Brachytherapy (f) P	
Breast	Complete staging (f) P	Modalities of dose to OAR evaluation (f) P	Set up verification (f) P	Multidisciplinary approach (f) P	Complete report anatomical data (f) P	Patient's satisfaction (v) O	
Lung	Written protocols (p) P	Necessary equipment (p) S	Complete staging (f) P	Multidisciplinary approach (f) P	Modalities of volume identification (f) P	Set up verification (f) P	Follow-up after radical treatment (f) P
Prostate	Infrastructure and methods (p) S	Complete Staging (f) P	Modalities of volume identification (f) P	Set up errors (f) P	Follow-up (f) P	Complete follow-up information (f) P	Rectum toxicity (f) O
Rectum	Multidisciplinary approach (f) P	Set up procedures (f) P	Set up verification (f) P	Dose-volume histograms (f) P	Acute toxicity (f) O	Quality of life monitoring (f) P	
H&N	Postoperative Waiting time (v) P/S	Waiting time (v) P/S	Modalities of volume identification (f) P	Multidisciplinary approach (f) P	Multidisciplinary protocols (p) P	Staging modalities (f) P	Breaks due to toxicity (f) P
Bone Met	Waiting time (v) S/P	Extent of illness evaluation (f) P	Fractionation (empirical/validated) (f) P	Multidisciplinary approach (f) P	Follow-up (f) P	Complete follow-up (f) P	

Source: Progetto Gruppo di lavora dell'Unita Operativa 1 del., Indicatori di qualita in radioterapia, Audit clinico su Indicatori di Qualita in Radioterapia selezionati per patologia p. Rapporti ISTISAN, 2005.

Note: Gyn, gynecological tumors; H&N, head and neck tumors; Bone Met, bone metastases. Topic indicates the variable under observation with respect to its frequency (f), value (v), or presence (p). P, process; O, outcome.

additional studies before accepting the RAND criteria as infrastructure for quality care assessments in patients with prostate carcinoma.

Subsequent studies (Miller et al. 2003; Penson 2008; Spencer et al. 2008) assessed the feasibility of measuring compliance with RAND quality indicators in a hospital setting and determined their sensitivity to changes in the care of patients with localized prostate carcinoma.

After establishing institution-specific standards, changes in compliance were detected in later data collections (Miller et al. 2003; Penson 2008; Spencer et al. 2008). These studies illustrated inconsistencies in prostate cancer care that are potential targets for quality improvement. In particular, explicitly reviewed medical records from 2775 men treated with radical prostatectomy or external-beam radiation therapy showed compliance ranged from 62.6% to 88.3% for the radiation technique (Spencer et al. 2008).

One of these studies (Miller et al. 2003) showed compliance with some candidate indicators of technical quality could not be measured because records were lacking and suggested that records of technical details of radiotherapy treatment will need to become routine. Furthermore, it was pointed out that pilot validity testing of indicators would require collecting and analyzing data from approximately 60–100 patients at each of 30 urology and 30 radiation oncology facilities.

Breast Cancer

The Agency for Healthcare Research and Quality at the University of Ottawa's Evidence-Based Practice Center (UOEPC) conducted a systematic review of the literature to survey the range of quality measures assessing breast cancer care in women and to identify specific parameters that could potentially affect their being suitable for wider use (Moher et al. 2004).

According to the authors, the main purpose of indicators is to highlight healthcare quality by identifying gaps in care. A total of 143 quality indicators were identified from 60 reports describing 58 studies that met eligibility criteria. In reply to the question of what quality of care measures were available to assess appropriate use and quality of radiotherapy treatment for breast cancer in women, 16 indicators for radiation therapy after breast-conserving surgery and postmastectomy were considered appropriate. Table 25.8 lists these 16 indicators; other details including sources of information can be found in the original report (Moher et al. 2004).

TABLE 25.7 Indicators by RAND Regarding Radiotherapy Treatment of Prostate Cancer

Topic	Dimensions		Covarietes
Availability of radiation oncology facilities	S	N. S9, p. 115	
Board certification of urologists and radiation oncologists	S	N. S12, p. 115	
Alternative treatment modalities (radical prostatectomy, radiation therapy—external beam, interstitial treatment, and expectant management) were presented to patient	P	N. P11, p. 116	
Patient was offered the opportunity to consult with a urologist or medical oncologist (if provider is radiation oncologist), or with a radiation oncologist or medical oncologist (if provider is urologist)	P	N. P13, p. 116	
Use of CT in conventional radiotherapy treatment planning	P	N. P16, p. 117	
Use of CT in conformal radiotherapy treatment planning	P	N. P20, p. 117	
Immobilization of patient during conventional radiotherapy	P	N. P23, p. 117	
Recommended doses (68–72 Gy isocenter [ICRU]) for conventional external beam radiation therapy	P	N. P34, p. 117	
Escalated doses (70–80 Gy ICRU) with conformal radiation therapy	P	N. P35, p. 118	
High energy (≥10MV) linear accelerator	P	N. P36, p. 118	
Failure after primary treatment by radiation therapy	O	N. O1, p. 119	Yes
Hospitalization (different diseases) following primary treatment by radiotherapy	O	N. O5, p. 119	Yes
Surgical treatment (different diseases) following primary treatment by radiotherapy	O	N. O6, p. 119	Yes
Medical (different diseases) following primary treatment by radiation therapy	O	N. O7, p. 119	Yes
Hospitalization (different diseases) following radical prostatectomy or radiation	O	N. O8, p. 119	Yes
Surgical treatment (different diseases) following radical prostatectomy or radiation	O	N. O9, p. 119	Yes
Medical treatment (different diseases) following radical prostatectomy or radiation	O	N. O10, p. 119	Yes
Patient assessment of different functioning following primary treatment by radiotherapy or radical prostatectomy	O	N. O11, p. 119	Yes
10-year clinical and/or biochemical disease-free survival following primary treatment by radiotherapy or radical prostatectomy	O	N. O13, p.119	
5-year clinical and/or biochemical disease-free survival following primary treatment by radiation therapy or radical prostatectomy	O	N. O14, p. 119	
Patient satisfaction with treatment choice	O	N. O17, p. 120	Yes
Patient satisfaction with continence	O	N. O21, p. 120	Yes
Patient satisfaction with potency	O	N. O22, p. 120	Yes

Source: Litwin et al., *Prostate Cancer Patient Outcomes and Choice of Providers: Development of an Infrastructure for Quality Assessment*, RAND, Santa Monica, CA.

Note: S, structure; P, process; O, outcome. Details can be found in the source at the indicated pages. "Yes" indicates when covariates are suggested among age, pretreatment PSA, clinical stage, Gleason grade, family history of prostate cancer, history of other cancer, comorbidity indicators, use of neoadjuvant or adjuvant hormone therapy, insurance, education, or income.

TABLE 25.8 Indicators for Radiotherapy Reported in the Review for Breast Cancer Treatment

Topic	Dimension	Reference
Radiotherapy after breast conserving surgery	P-appropriateness	p. 50–52[a]
Quality of radiotherapy after breast conserving surgery	TP	p. 53
Radiotherapy after mastectomy	P-appropriateness	p. 53
Radiotherapy via planning on a dedicated simulator	TP	p. 54
Homogeneous dose distribution	TP	p. 54
Use of wedges on tangent breast fields	TP	p. 54
Use of radiotherapy on axilla	P-appropriateness	p. 55
Use of parasternal radiotherapy	P-appropriateness	p. 55
Use of palliative radiotherapy	P-appropriateness	p. 55
Regional recurrence needing further surgery or rt	P-appropriateness	p. 55
Both tangent fields treated daily	TP	p. 55
Fractionation	CP	p. 54, p. 56
Electron beam breast radiation used	TP	p.55
Received enough information about surgery and radiotherapy	P-patient information	

Source: Moher et al., *Evid. Rep. Technol. Assess.*, 105, 1–8, 2004.

[a] Pages of the booklet where details can be found.

TABLE 25.9   Indicators by Healthcare Organizations Used to Monitor Access to Radiotherapy Treatments

Topic	Domain	Source
Radiation therapy for women under age 70 receiving breast-conserving surgery for breast cancer	Accountability	ASCO, NCCN, and the Commission on Cancer (COC), endorsed by the NQF (ASCO/NCCN)
Radiation therapy for patients under the age of 80 receiving surgical resection for rectal cancer	Surveillance	ASCO, NCCN, and the Commission on Cancer (COC) (ASCO/NCCN)
Radiation treatment for postoperative breast cancer patients	Access	Cancer Care Ontario (CCO) (Back et al. 2007)
Radiation treatment at some point during their illness	Access	Cancer Care Ontario (CCO) (Cancer Care Ontario)

Indicators for Accessibility to Radiation Therapy

Accessibility remains a key determinant of radiotherapy underuse, even in developed countries (Pagano et al. 2007). National and international societies and organizations emphasize the importance of timely access to radiotherapy treatments and insist radiotherapy capacity needs to be expanded. More than one indicator on accessibility to radiotherapy has been presented in previous paragraphs, but many other professional groups focus on monitoring waiting times.

Adjuvant radiotherapy after breast-conservation surgery is considered an indicator of quality of care for the majority of women with breast cancer (Hershman et al. 2008). ASCO-NCCN recently developed an ad hoc indicator for this pathology, together with an indicator for colon cancer, both of which are intended for application through cancer registries. Table 25.9 reports on these two indicators. Cancer Care Ontario (CCO), the provincial agency responsible for continually improving cancer services, adopted indicators of accessibility to radiotherapy treatments (Cancer Care Ontario 2008a, 2008b). These indicators are also reported in Table 25.9. Since 2006, Cancer Care Ontario reports not only whether radiation wait times are improving, but on how many patients are being treated in the recommended timeframe or targets, according to two intervals: (1) the time between being referred to being seen by a specialist; and (2) the time between being ready for treatment and receiving treatment (Cancer Care Ontario 2008a, 2008b). This procedure, similar to that performed with one of the Italian indicators (Table 25.2), aims to understand the nature of radiation therapy waits more precisely.

A recently developed criterion-based benchmark (CBB) was designed to give a reasonable estimate of the overall need for radiotherapy in breast cancer and to determine optimal utilization and accessibility of radiotherapy facilities (Kerba et al. 2007). The CBB was used to monitor treatments in Ontario and the USA (Kerba et al. 2007). When waiting times are longer, one of the factors contributing to longer waiting times was suggested to be the increased complexity of treatment involving 3-D conformal and intensity-modulated radiotherapy (Robinson et al. 2005).

Other Recent Clinical Indicators from Single Experiences

We want to mention some studies reporting quality monitoring of cancer treatments that are less addressed using quality control

indicators. One study established indices based on new clinical evidence for assessing the pattern of lung cancer care in the university-affiliated medical center of Taiwan (Chien and Lai 2006). The report indicated how indices were established and provided an example of the use of quality measures to monitor internal quality in radiotherapy treatments of lung cancer. Four of these indices refer to radiotherapy in the treatment of non-small-cell lung cancer and limited stage small-cell lung cancer.

In one of the first reports assessing a multidisciplinary approach in multiform glioblastoma the Department of Radiation Oncology at the Cancer Institute in Singapore retrospectively reviewed data from high-grade glioma patients using three quality indicators chosen by a multidisciplinary team involved in patient management (Back et al. 2007). Median survival and the potential impact of a multidisciplinary approach on care were monitored using these indicators: time from surgery procedure to starting radiotherapy, use of postoperative CT or MRI imaging, and adjuvant chemotherapy.

Summary

There could be some concerns about which measures should be selected for the purpose of monitoring the quality of radiotherapy treatments. In radiotherapy, as in any other disciplines, the first issue that needs to be addressed is the reason for collecting data on an indicator. Some indicators that monitor accessibility or resources may be very suitable for an external audit by the hospital management or by regional and/or national networks. When utilized for an internal audit, they might reveal shortcomings that need new resources, be used to create pressure, or stimulate proper motivation. This is the case, for example, with indicators of waiting times. There is general agreement that undue delay in treatment may adversely influence outcome and all professional groups recognize that waiting times in radiotherapy need to be measured, although different thresholds and formats are proposed, particularly for different pathologies. On the other hand, the waiting time indicator depends on many factors and the professional quality of the care is only one of them.

Whether process or outcome monitoring is more important in healthcare is still a matter of concern and open debate in many medical disciplines (Gabriele et al. 2006; Mant 2001). A process is generally easy to measure and straightforward to interpret. Radiotherapy process measurement aims at monitoring accessibility and effectiveness for patients, with the latter aspect often being addressed by means of technical indicators.

It is very important when choosing this latter type of indicator to remember that process measurements are a direct measure of the care quality if the process being measured is linked to outcome. Frequent updating due to rapid technological advances is another disadvantage of technical process indicators in radiotherapy.

In radiotherapy, some process indicators investigating accessibility (e.g., informed consent, participation in clinical trials, etc.) are shared with other medical disciplines, while others that can be used to verify improvement in the technical quality of treatments are very specific to radiotherapy (Tables 25.3, 25.4, 25.7, and 25.8). It is worth noting that at least three groups included irradiation with CT among technical indicators (Tables 25.3, 25.4, and 25.7) and other groups emphasized the importance of conformal treatments in updated care (Krupski et al. 2005).

Outcome indicators other than those referring to patient satisfaction are less frequently found among radiotherapy indicators. The scarcity of these indicators reflects the caution exerted by professionals who are aware of difficulties in data collection and potential confounding factors. On the other hand, monitoring follow-up of radiotherapy treatments was considered an integral part of care with radiation and many groups drew up indicators on this process (Tables 25.3, 25.6, and 25.7). As a long period of time might have to pass before some outcome events can be assessed, they are often monitored in clinical trials or retrospective analysis. In this case, one must bear in mind that treatment techniques may have changed considerably over the intervening period.

When defining indicators, the Joint Commission recommended a multidisciplinary approach involving personnel from more than one operative unit (JCAHO 1993). This holds true particularly for indicators in radiotherapy for two main reasons: (1) besides physicians, medical physicists and radiographers play major roles in radiotherapy treatments; and (2) a clinical multidisciplinary approach to care was indicated as important in radiotherapy treatments (Table 25.6) (Back et al. 2007). Therefore, participation of clinicians other than radiation oncologists, such as the prostate RAND group achieved (Litwin et al. 2000), is to be desired when developing indicators for radiotherapy.

The indicators reported in this section bear witness to a growing tendency to measure quality of care in radiotherapy, even in pathologies without curative intent. For example, different groups designed indicators for palliative care (Tables 25.5 and 25.6). Last, thresholds are not always given for indicators in radiotherapy, but results from data collections can provide the specifics for expanding indications and determining thresholds.

References

American College of Radiology. 1996. A Guide to Continuous Quality Improvement in Medical Imaging. Committee on Quality Assurance—Commission on Standards and Accreditation. Reston, VA: American College of Radiology.

Asch, S. M., E. A. Kerr, E. G. Hamilton, J. L. Reifel, and E. A. McGlynn, eds. 2000. *Quality of Care for Oncological Conditions and HIV. A Review of Literature and Quality Indicators.* Santa Monica, CA: RAND Health.

ASCO/NCCN. Quality Measures: Breast and Colorectal Cancers. Available from http://www.facs.org/cancer/qualitymeasures .html.

Australian Council on Healthcare Standards (ACHS). 2007. The Australian Council on Healthcare Standards Australasian clinical indicator report 2001–2007. In *Determining the Potential to Improve Quality of Care.* 9th ed. The Australian Council on Healthcare Standards (ACHS) Health Services Research Group, University of Newcastle.

Australian Council on Healthcare Standards (ACHS) and New Zealand College of Radiologists Faculty of Radiation Oncology. 2009. *Radiation Oncology Indicators Clinical Indicators User's Manual* version 3. Sydney: ACHS.

Back, M. F., E. L. Ang, W. H. Ng, S. J. See, C. C. Lim, L. L. Tay, and T. T. Yeo. 2007. Improvements in quality of care resulting from a formal multidisciplinary tumour clinic in the management of high-grade glioma. *Ann. Acad. Med. Singap.* 36 (5): 347–351.

Cancer Care Ontario. 2008a. Access to Radiation Treatment for Post-Operative Breast Cancer. Available from http://www .cancercare.on.ca/english/csqi2008/csqi-access-after-brs/.

Cancer Care Ontario. 2008b. Wait Times for Radiation Treatment. Available from www.cancercare.on.ca/english/csqi2008/ csqiaccess/csqi-radiation-utiliz/.

Chien, C. R., and M. S. Lai. 2006. Trends in the pattern of care for lung cancer and their correlation with new clinical evidence: Experiences in a university-affiliated medical center. *Am. J. Med. Qual.* 21 (6): 408–414.

Cionini, L., G. Gardani, P. Gabriele, S. Magri, P. L. Morosini, A. Rosi, and V. Viti. 2007. Quality indicators in radiotherapy. Italian Working Group general indicators. *Radiother. Oncol.* 82 (2): 191–200.

Donabedian, A. 1980. *The Definition of Quality and Approaches to Its Assessment, His Explorations in Quality Assessment and Monitoring.* Ann Arbor, MI: Health Administration Press.

European Commission—DG for Health and Consumers, Public Health Programme, EUPHORIC Project. Available from www.euphoric-project.eu.

Gabriele, P., G. Malinverni, C. Bona, M. Manfredi, E. Delmastro, M. Gatti, G. Penduzzu, B. Baiotto, and M. Stasi. 2006. Are quality indicators for radiotherapy useful in the evaluation of service efficacy in a new based radiotherapy institution? *Tumori* 92 (6): 496–502.

Gibberd, R., S. Hancock, P. Howley, and K. Richards. 2004. Using indicators to quantify the potential to improve the quality of health care. *Int. J. Qual. Health Care* 16 (Suppl. 1): i37–i43.

Gruppo di lavoro dell'Unità Operativa 1 del Progetto "Indicatori di qualità in radioterapia". 2005. Audit clinico su Indicatori di Qualità in Radioterapia selezionati per patologia. Rapporti ISTISAN 05/36.

Hershman, D. L., D. Buono, R. B. McBride, W. Y. Tsai, K. A. Joseph, V. R. Grann, and J. S. Jacobson. 2008. Surgeon characteristics and receipt of adjuvant radiotherapy in women with breast cancer. *J. Natl. Cancer Inst.* 100 (3): 199–206.

JCAHO. 1993. *The Measurement Mandate.* Oakbrook Terrace, IL: Joint Commission.

JCAHO. 1994. *A Guide to Establishing Programs for Assessing Outcomes in Clinical Settings.* Oakbrook Terrace, IL: Joint Commission.

JCAHO. 1997. *National Library of Healthcare Indicators.* Oakbrook Terrace, IL: Joint Commission.

Kerba, M., Q. Miao, J. Zhang-Salomons, and W. Mackillop. 2007. Defining the need for breast cancer radiotherapy in the general population: A criterion-based benchmarking approach. *Clin. Oncol. (R. Coll. Radiol.)* 19 (7): 481–489.

Krupski, T. L., J. Bergman, L. Kwan, and M. S. Litwin. 2005. Quality of prostate carcinoma care in a statewide public assistance program. *Cancer* 104 (5):985–992.

Litwin, M., M. Steinberg, J. Malin, J. Naitoh, K. McGuigan, R. Steinfeld, J. Adams, and R. Brook. 2000. *Prostate Cancer Patient Outcomes and Choice of Providers: Development of an Infrastructure for Quality Assessment.* Santa Monica, CA: RAND.

Mainz, J. 2003. Defining and classifying clinical indicators for quality improvement. *Int. J. Qual. Health Care* 15 (6): 523–530.

Malin, J. L., S. M. Asch, E. A. Kerr, and E. A. McGlynn. 2000. Evaluating the quality of cancer care: Development of cancer quality indicators for a global quality assessment tool. *Cancer* 88 (3): 701–707.

Mant, J. 2001. Process versus outcome indicators in the assessment of quality of health care. *Int. J. Qual. Health Care* 13 (6): 475–480.

Miller, D. C., M. S. Litwin, M. G. Sanda, J. E. Montie, R. L. Dunn, J. Resh, H. Sandler, and J. T. Wei. 2003. Use of quality indicators to evaluate the care of patients with localized prostate carcinoma. *Cancer* 97 (6): 1428–1435.

Moher, D., H. M. Schachter, V. Mamaladze, G. Lewin, L. Paszat, S. Verma, C. DeGrasse et al. 2004. Measuring the quality of breast cancer care in women. *Evid. Rep. Technol. Assess. (Summ.)* (105): 1–8.

Pagano, E., D. Di Cuonzo, C. Bona, I. Baldi, P. Gabriele, U. Ricardi, P. Rotta et al. 2007. Accessibility remains a major determinant of radiotherapy underutilization: A population based study. *Health Policy* 80: 483–491.

Penson, D. F. 2008. Assessing the quality of prostate cancer care. *Curr. Opin. Urol.* 18 (3): 297–302.

Robinson, D., T. Massey, E. Davies, R. H. Jack, A. Sehgal, and H. Moller. 2005. Waiting times for radiotherapy: Variation over time and between cancer networks in southeast England. *Br. J. Cancer* 92 (7): 1201–1208.

Rubin, H. R., P. Pronovost, and G. B. Diette. 2001. The advantages and disadvantages of process-based measures of health care quality. *Int. J. Qual. Health Care* 13 (6): 469–474.

Spencer, B. A., D. C. Miller, M. S. Litwin, J. D. Ritchey, A. K. Stewart, R. L. Dunn, E. G. Gay, H. M. Sandler, and J. T. Wei. 2008. Variations in quality of care for men with early-stage prostate cancer. *J. Clin. Oncol.* 26 (22): 3735–3742.

WHO. 2003. Measuring Hospital Performance to Improve the Quality of Care in Europe: A Need for Clarifying the Concepts and Defining the Main Dimensions—Report on a WHO Workshop, Barcelona, Spain.

Working Group Continuous Quality Improvement in Radiotherapy. General evaluation indicators for radiotherapy after a first clinical audit. Rapporti ISTISAN 05/43.

26

Medical Indicators of Quality: Structure, Process, and Outcome

Tony J. C. Wang
Columbia University

Benjamin A. Spencer
Columbia University

Introduction

Healthcare is undergoing a revolution in quality of care as healthcare policymakers and third-party payors create incentives to both improve healthcare and minimize its costs. The publication of *Crossing the Quality Chasm* by the Institute of Medicine in 2001 focused attention on deficits in our healthcare system and proposed a complete reengineering of the delivery of healthcare in the United States (Institute of Medicine (U.S.) Committee on Quality of Health Care in America 2001). Medicare's pay-for-performance program, "P4P," was launched in 2005 and links healthcare quality to reimbursement of providers. Hospitals scoring in the top 10% nationally are rewarded with an additional 2% payment on top of standard DRG charges (Centers for Medicare & Medicaid Services 2009). With these changes comes the need to measure, assess, report, and improve the quality of care. The development of the required tools is in its infancy, but efforts are underway by numerous stakeholders to develop such an infrastructure.

Quality of care is "the degree to which health services for individuals and populations increase desired health outcomes and are consistent with current professional knowledge" (Lohr, Donaldson, and Harris-Wehling 1992). Lessons from other industries that have practiced quality-of-care principles for much longer are increasingly being applied to healthcare (Adams et al. 2004; Frankel et al. 2005). The concept of continuous quality improvement comes from the Japanese "kaizen," an approach taken by their auto industry to make gradual small improvements toward perfection. Taking examples from the aviation industry, the specialty of anesthesiology has achieved Six Sigma status by minimizing medical errors to a frequency of less than 1 in 10^6 (Parker et al. 2007; Thiele, Huffmyer, and Nemergut 2008). The Leapfrog Group (2009) has publicized the use of evidence-based practices to improve the quality of healthcare delivery, such as

the administration of preoperative antibiotics within one hour prior to a surgical incision. These changes are gradually entering the field of radiotherapy and this chapter will provide an introduction to assessing quality of care in radiotherapy. Given the scope of this topic and our specialization in the treatment of prostate cancer, the authors will emphasize examples from the prostate cancer quality-of-care literature.

The quality-of-care paradigm was first described by Avedis Donabedian at the University of Michigan in 1966 (Donabedian 1966). He separated the delivery of healthcare services into three components: structure, process, and outcome (Brook, McGlynn, and Cleary 1996; Donabedian 1966, 1997). Structure of care refers to the equipment, resources, and provider experience needed to provide high-quality care. Process of care comprises interpersonal and technical aspects of the doctor-patient interaction. Outcomes of care refer to the results of medical treatment—overall or disease-specific survival and patient-reported outcomes, for instance (Donabedian 1988, 1992). Intrinsically, outcomes are of greater interest to us and have historically served as the main outcome measure in clinical studies. However, outcomes are limited in their usefulness for quality improvement because they cannot be modified directly. We can control structure and process and, therefore, finding deficits in structure and process indicator compliance can lead to identifying opportunities for interventions that will ultimately lead to improved outcomes.

A quality indicator is a performance standard considered representative of the likely quality of care. Quality of care studies and quality assurance activities utilize quality indicators to measure, evaluate, and report quality of care. Their use has the potential to lead to more uniform practice patterns and to improve the quality of care. Typically, quality assessment utilizes provider documentation of medical practice as a surrogate of the activities themselves. Therefore, poor documentation is equated with poor quality. Although this is a limitation of the methodology, medical record abstraction remains the standard for assessing quality of care (MacLean et al. 2006). Compliance rates with quality indicators have been used to create report cards to compare care among individual providers or hospitals. These report cards can be used internally for the purpose of quality assessment and improvement or publicly for consumers' benefits. Since 1991, New York has published statewide risk-adjusted mortality rates following coronary artery bypasssurgery (available online at http://www.health.state.ny.us/statistics/diseases/cardiovascular/). Concerns that such public reporting of surgeon-specific outcomes would lead to changes in surgeons' and hospitals' market share have proven unfounded (Jha and Epstein 2006). However, low-performing surgeons have a tendency to retire or relocate in the 2 years following a negative report (Jha and Epstein 2006). Such findings highlight the importance of accurate measuring tools and risk-adjustment with robust comorbidity indices.

The importance of quality-of-care assessment is magnified for highly-prevalent diseases such as prostate cancer, the malignancy with the highest incidence and second highest mortality among U.S. men (Jemal et al. 2004). Furthermore, the incidence of side effects following treatment, which include urinary and sexual dysfunction, is high and their impact on one's quality of life is significant (Blasko et al. 2002; D'Amico et al. 2000; Dillioglugil et al. 1997; Middleton et al. 1995; Shah et al. 2004). The RAND Corporation developed quality-of-care indicators for early-stage prostate cancer in order to begin to develop an infrastructure for quality assessment (Spencer et al. 2003). Given the lack of level I data (randomized clinical trials or meta-analyses of randomized clinical trials)—the usual source for initiating quality indicator development—in prostate cancer, the researchers used an alternate but well-established methodology previously used for developing clinical guidelines (Chassin, Rand Corporation and Commonwealth Fund 1986; Park et al. 1986). This methodology derives from the U.S. Army's Delphi Method and involves the use of experts to reach consensus on a controversial topic. For the prostate cancer indicator project, experts rated the validity and feasibility of 49 proposed quality indicators and 14 proposed covariates for risk-adjustment (Spencer et al. 2003). Among the proposed quality indicators, 22 were specific to radiation therapy and 14 were selected for pilot testing. The indicators are organized around the Donabedian paradigm of structure, process, and outcome to describe quality of care.

Structure Indicators

Important structural attributes for quality of care include clinician characteristics, organizational characteristics, patient characteristics, and community characteristics (Litwin et al. 2000). While the presence of certain structural characteristics may create the environment for high-quality care, it is not sufficient to ensure high-quality care. The optimal structural measures are those that have a positive influence on the process and outcomes of care (Brook et al. 1990). These measures are fixed characteristics that have advantages and disadvantages in

performing research. An advantage is that they do not require explicit review of the medical record-like process and outcome measures. However, they tend not to be specific to the patient, making them less useful when analyses are performed at the patient level, since compliance with an indicator would be the same for all patients treated at that institution during a given time period. An example of a structural measure is the tracking of treatment outcomes. Institutions that track their outcomes do not necessarily have better outcomes; however, awareness of its successes and failures usually leads to a process of continuous quality improvement (New York State Department of Health 2008, 2009). Organizations such as the Joint Commission on the Accreditation of Healthcare Organizations (JCAHO) have generally relied on structural measures in their accreditation procedures (Litwin et al. 2000).

The structural measure that has consistently demonstrated a strong association with improved clinical outcomes is the volume or number or cases treated by a particular physician or hospital (Begg et al. 2002; Grumbach et al. 1995; Hannan et al. 1997; Kitahata et al. 1996; Luft et al. 1990). High volume has been shown to be an important predictor of good quality care in the treatment of prostate cancer (Begg et al. 2002; Hu et al. 2003), lung cancer (Bach et al. 2001), and certain gastrointestinal cancers (Begg et al. 1998). Two studies pilot-tested the RAND indicators and identified five structure measures that were rated highly enough by an expert panel in both validity and feasibility to undergo further testing and consideration as quality indicators (Miller et al. 2007; Spencer et al. 2008). The structural indicators demonstrated consistently high compliance: availability of conformal therapy (89%), board-certified radiation oncologist (92%), and availability of psychological counseling (90%) (Spencer et al. 2003, 2008; Miller et al. 2007).

Process Indicators

Good process requires that medically appropriate decisions be made when diagnosing and treating the patient and that care be provided in a timely and skillful manner (Litwin et al. 2000). Assessing processes of care requires the development of quality indicators that apply to a particular type of patient in a specific clinical circumstance. These indicators are generally based on evidence in the medical literature and on current professional standards of care. Ideally, the evidence should be level I. The performance of physicians or health plans can be assessed by calculating rates of adherence to the indicators for a sample of patients. The National Quality Forum (NQF) is a nonprofit organization that aims to improve the quality of healthcare in several ways, including the endorsement of quality indicators for measuring and public reporting of quality of care (NQF 2009). Table 26.1 lists examples of process indicators established by the American Medical Association-convened Physician Consortium for Performance Improvement and endorsed by the NQF.

In the RAND study, 23 process indicators were endorsed by the expert panel, such as the documentation of the presence or absence of a family history of prostate cancer, baseline quality-

TABLE 26.1 National Voluntary Consensus Standards from the American Medical Association-Convened Physician Consortium for Performance Improvement for a Clinician-Level Cancer Care Candidate

	Description
Treatment summary documented and communicated	Percentage of patients with a diagnosis of cancer who have undergone BT or EBRT and who have a treatment summary report in the chart that was communicated to the physician(s) providing continuing care within 1 month of completing treatment
Normal tissue dose constraints	Percentage of patients with a diagnosis of cancer receiving 3D-CRT with documentation in their medical record that normal tissue dose constraints were established within 5 treatment days for a minimum of one tissue
Plan of care for pain	Percentage of visits for patients with a diagnosis of cancer currently receiving RT who report having pain with a documented plan of care to address pain
Three-dimensional radiotherapy	Percentage of patients with prostate cancer receiving EBRT who receive 3D-CRT or IMRT
Prostate cancer: avoidance of overuse measure—bone scan for staging low-risk patients	Percentage of patients with a diagnosis of prostate cancer at low risk of recurrence receiving BT or EBRT who did not have a bone scan performed at any time since diagnosis
Prostate cancer: adjuvant hormonal therapy for high-risk patients	Percentage of patients with a diagnosis of prostate cancer at high risk of recurrence receiving EBRT who were prescribed adjuvant hormonal therapy
Prostate cancer: initial evaluation	Percentage of patients with prostate cancer receiving BT or EBRT with documented evaluation of PSA, primary tumor (T) stage, and Gleason score prior to initiation of treatment
Prostate cancer: treatment options for patients with clinically localized disease	Percentage of patients with clinically localized prostate cancer receiving BT or EBRT who received counseling on, at a minimum, the following treatment options for clinically localized disease prior to initiation of treatment: active surveillance, BT, EBRT, and radical prostatectomy

Note: BT, brachytherapy; EBRT, external beam radiation therapy; 3D-CRT, three-dimensional conformal radiation therapy; IMRT, intensity-modulated radiation therapy; PSA, prostate-specific antigen.

of-life function, the performance of a digital rectal examination and a PSA blood test, communication with the patient's primary care physician, and referral by the radiation oncologist to a provider of an alternative therapy such as surgery (Spencer et al. 2003). Spencer found high levels of adherence among radiation oncologists to indicators for computed tomography planning (88%), high-energy accelerators (82%), and total dose ≥70 Gy (77%), but lower adherence to patient immobilization (66%) and rectal protection (63%) (Spencer et al. 2008; Miller et al. 2007). Another group investigated compliance with the RAND indicators using an administrative dataset, SEER–Medicare, and found similar rates of adherence to measures of computed tomography planning (85%) and the use of high-energy photons (75%) (Bekelman et al. 2007). However, rates of patient immobilization were higher (97%) (Bekelman et al. 2007).

Outcome Indicators

Given the prolonged natural history of prostate cancer, the success of treatment is best judged at several time points over the course of many years. Short-term outcomes include the incidence of radiation proctitis during treatment. Intermediate-term end-points include biochemical (PSA) failure, health-related quality of life, and patient satisfaction. Long-term end-points include overall and disease-free survival and health-related quality of life.

In the RAND study (Spencer et al. 2003), 16 outcome indicators were endorsed by the expert panel as potential indicators of quality, including: biochemical failure documented by PSA;

urinary, sexual, and bowel functioning after treatment reported by patients, not physicians, using a validated survey instrument; 10- and 15-year overall and disease-free survival; and patient-reported satisfaction with treatment choice, continence, and potency. An outcome in radiotherapy that is a measure of the quality of brachytherapy is postimplant dosimetry such as D90 (a dose that covers 90% of the prostate volume) and V100 (the fractional volume of the prostate that receives 100% of the prescription dose) (Robert Lee et al. 2005; Urbanic and Lee 2006).

Summary

Quality of care entered the national healthcare spotlight in the late 1990s with the publication of several seminal reports from the Institute of Medicine (Hewitt and Simone 1999). Since then, numerous organizations and professional societies have developed and endorsed evidence-based quality indicators and begun to develop systems for monitoring and reporting quality of care (Malin et al. 2006; Schneider et al. 2004; Institute of Medicine (U.S.) Committee on Quality of Health Care in America 2001; The Leapfrog Group 2009; National Quality Forum 2009). Voluntary consensus standards, for example, are shown in Table 26.1. Medicare's national "pay-for-performance" initiative rewards high quality care with bonus payments to top hospitals. Quality indicators can be used for various purposes, including maintenance of certification, quality improvement, accountability, and practice accreditation. Undoubtedly, the quality of care will become increasingly important in the healthcare system of the future.

References

Adams, R., P. Warner, B. Hubbard, and T. Goulding. 2004. Decreasing turnaround time between general surgery cases: A six sigma initiative. *J. Nurs. Adm.* 34 (3): 140–148.

Bach, P. B., L. D. Cramer, D. Schrag, R. J. Downey, S. E. Gelfand, and C. B. Begg. 2001. The influence of hospital volume on survival after resection for lung cancer. *N. Engl. J. Med.* 345 (3): 181–188.

Begg, C. B., L. D. Cramer, W. J. Hoskins, and M. F. Brennan. 1998. Impact of hospital volume on operative mortality for major cancer surgery. *JAMA* 280 (20): 1747–1751.

Begg, C. B., E. R. Riedel, P. B. Bach, M. W. Kattan, D. Schrag, J. L. Warren, and P. T. Scardino. 2002. Variations in morbidity after radical prostatectomy. *N. Engl. J. Med.* 346 (15): 1138–1144.

Bekelman, J. E., M. J. Zelefsky, T. L. Jang, E. M. Basch, and D. Schrag. 2007. Variation in adherence to external beam radiotherapy quality measures among elderly men with localized prostate cancer. *Int. J. Radiat. Oncol. Biol. Phys.* 69 (5): 1456–1466.

Blasko, J. C., T. Mate, J. E. Sylvester, P. D. Grimm, and W. Cavanagh. 2002. Brachytherapy for carcinoma of the prostate: Techniques, patient selection, and clinical outcomes. *Semin. Radiat. Oncol.* 12 (1): 81–94.

Brook, R. H., C. J. Kamberg, K. N. Lohr, G. A. Goldberg, E. B. Keeler, and J. P. Newhouse. 1990. Quality of ambulatory care. Epidemiology and comparison by insurance status and income. *Med. Care* 28 (5): 392–433.

Brook, R. H., E. A. McGlynn, and P. D. Cleary. 1996. Quality of health care: Part 2. Measuring quality of care. *N. Engl. J. Med.* 335 (13): 966–970.

Centers for Medicare & Medicaid Services. 2009. *Medicare "Pay for Performance (P4P)" Intiaitives,* http://www.cms.hhs.gov/apps/media/press/release.asp?counter=1343 (accessed December 18, 2009).

Chassin, M. R., RAND Corporation, and Commonwealth Fund. 1986. *Indications for Selected Rand Surgical Procedures: A Literature Review and Ratings of Appropriateness: Coronary Angiography.* Santa Monica, CA: Rand.

D'Amico, A. V., D. Schultz, L. Schneider, M. Hurwitz, P. W. Kantoff, and J. P. Richie. 2000. Comparing prostate specific antigen outcomes after different types of radiotherapy management of clinically localized prostate cancer highlights the importance of controlling for established prognostic factors. *J. Urol.* 163 (6): 1797–1801.

Dillioglugil, O., B. D. Leibman, N. S. Leibman, M. W. Kattan, A. L. Rosas, and P. T. Scardino. 1997. Risk factors for complications and morbidity after radical retropubic prostatectomy. *J. Urol.* 157 (5): 1760–1767.

Donabedian, A. 1966. Evaluating the quality of medical care. *Milbank Mem. Fund. Q.* 44 (3): Suppl:166–206.

Donabedian, A. 1988. The quality of care. How can it be assessed? *JAMA* 260 (12): 1743–1748.

Donabedian, A. 1992. The role of outcomes in quality assessment and assurance. *QRB Qual. Rev. Bull.* 18 (11): 356–360.

Donabedian, A. 1997. The quality of care. How can it be assessed? 1988. *Arch. Pathol. Lab. Med.* 121 (11): 1145–1150.

Frankel, H. L., W. B. Crede, J. E. Topal, S. A. Roumanis, M. W. Devlin, and A. B. Foley. 2005. Use of corporate Six Sigma performance-improvement strategies to reduce incidence of catheter-related bloodstream infections in a surgical ICU. *J. Am. Coll. Surg.* 201 (3): 349–358.

Grumbach, K., G. M. Anderson, H. S. Luft, L. L. Roos, and R. Brook. 1995. Regionalization of cardiac surgery in the United States and Canada. Geographic access, choice, and outcomes. *JAMA* 274 (16): 1282–1288.

Hannan, E. L., M. Racz, T. J. Ryan, B. D. McCallister, L. W. Johnson, D. T. Arani, A. D. Guerci, J. Sosa, and E. J. Topol. 1997. Coronary angioplasty volume–outcome relationships for hospitals and cardiologists. *JAMA* 277 (11): 892–898.

Hewitt, M., and J. V. Simone, eds. 1999. *Ensuring Quality Cancer Care.* Institute of Medicine and Commission on Life Sciences, National Research Council. Washington, DC: National Academy Press.

Hu, J. C., K. F. Gold, C. L. Pashos, S. S. Mehta, and M. S. Litwin. 2003. Role of surgeon volume in radical prostatectomy outcomes. *J. Clin. Oncol.* 21 (3): 401–405.

Institute of Medicine (U.S.). Committee on Quality of Health Care in America. 2001. *Crossing the Quality Chasm: A New Health System for the 21st Century.* Washington, DC: National Academy Press.

Jemal, A., R. C. Tiwari, T. Murray, A. Ghafoor, A. Samuels, E. Ward, E. J. Feuer, and M. J. Thun. 2004. Cancer statistics, 2004. *CA Cancer J. Clin.* 54 (1): 8–29.

Jha, A. K., and A. M. Epstein. 2006. The predictive accuracy of the New York State coronary artery bypass surgery report-card system. *Health Aff. (Millwood)* 25 (3): 844–855.

Kitahata, M. M., T. D. Koepsell, R. A. Deyo, C. L. Maxwell, W. T. Dodge, and E. H. Wagner. 1996. Physicians' experience with the acquired immunodeficiency syndrome as a factor in patients' survival. *N. Engl. J. Med.* 334 (11): 701–706.

Litwin, M., M. Steinberg, J. Malin, J. Naitoh, K. McGuigan, R. Steinfeld, J. Adams, and R. Brook. 2000. *Prostate Cancer Patient Outcomes and Choice of Providers: Development of an Infrastructure for Quality Assessment.* Santa Monica, CA: RAND.

Lohr, K. N., M. S. Donaldson, and J. Harris-Wehling. 1992. Medicare: A strategy for quality assurance, V: Quality of care in a changing health care environment. *QRB Qual. Rev. Bull.* 18 (4): 120–126.

Luft, H. S., D. W. Garnick, D. H. Mark, D. J. Peltzman, C. S. Phibbs, E. Lichtenberg, and S. J. McPhee. 1990. Does quality influence choice of hospital? *JAMA* 263 (21): 2899–2906.

MacLean, C. H., R. Louie, P. G. Shekelle, C. P. Roth, D. Saliba, T. Higashi, J. Adams et al. 2006. Comparison of administrative data and medical records to measure the quality of medical care provided to vulnerable older patients. *Med. Care* 44 (2): 141–148.

Malin, J. L., E. C. Schneider, A. M. Epstein, J. Adams, E. J. Emanuel, and K. L. Kahn. 2006. Results of the National Initiative for Cancer Care Quality: How can we improve the quality of cancer care in the United States? *J. Clin. Oncol.* 24 (4): 626–634.

Middleton, R. G., I. M. Thompson, M. S. Austenfeld, W. H. Cooner, R. J. Correa, R. P. Gibbons, H. C. Miller et al. 1995. Prostate Cancer Clinical Guidelines Panel Summary report on the management of clinically localized prostate cancer. The American Urological Association. *J. Urol.* 154 (6): 2144–2148.

Miller, D. C., B. A. Spencer, J. Ritchey, A. K. Stewart, R. L. Dunn, H. M. Sandler, J. T. Wei, and M. S. Litwin. 2007. Treatment choice and quality of care for men with localized prostate cancer. *Med. Care* 45 (5): 401–409.

National Quality Forum. 2009. National Quality Forum. Available from http://www.qualityforum.org (accessed 18 December 18, 2009).

New York State Department of Health. March 2008. *Adult Cardiac Surgery in New York State 2003–2005.* Albany, NY: New York State Department of Health.

New York State Department of Health. 2009. *Adult Cardiac Surgery in New York State 2004–2006.* Albany, NY: New York State Department of Health.

Park, R. E., A. Fink, R. H. Brook, M. R. Chassin, K. L. Kahn, N. J. Merrick, J. Kosecoff, and D. H. Solomon. 1986. Physician ratings of appropriate indications for six medical and surgical procedures. *Am. J. Public Health* 76 (7): 766–772.

Parker, B. M., J. M. Henderson, S. Vitagliano, B. G. Nair, J. Petre, W. G. Maurer, M. F. Roizen et al. 2007. Six sigma methodology can be used to improve adherence for antibiotic prophylaxis in patients undergoing noncardiac surgery. *Anesth. Analg.* 104 (1): 140–146.

Robert Lee, W., A. F. Deguzman, K. P. McMullen, and D. L. McCullough. 2005. Dosimetry and cancer control after low-dose-rate prostate brachytherapy. *Int. J. Radiat. Oncol. Biol. Phys.* 61 (1): 52–59.

Schneider, E. C., J. L. Malin, K. L. Kahn, E. J. Emanuel, and A. M. Epstein. 2004. Developing a system to assess the quality of cancer care: ASCO's national initiative on cancer care quality. *J. Clin. Oncol.* 22 (15): 2985–2991.

Shah, S. A., R. R. Cima, E. Benoit, E. L. Breen, and R. Bleday. 2004. Rectal complications after prostate brachytherapy. *Dis. Colon Rectum* 47 (9): 1487–1492.

Spencer, B. A., D. C. Miller, M. S. Litwin, J. D. Ritchey, A. K. Stewart, R. L. Dunn, E. G. Gay, H. M. Sandler, and J. T. Wei. 2008. Variations in quality of care for men with early-stage prostate cancer. *J. Clin. Oncol.* 26 (22): 3735–3742.

Spencer, B. A., M. Steinberg, J. Malin, J. Adams, and M. S. Litwin. 2003. Quality-of-care indicators for early-stage prostate cancer. *J. Clin. Oncol.* 21 (10): 1928–1936.

The Leapfrog Group. 2009. Factsheet: The Leapfrog Safe Practices Score Leap. http://www.leapfroggroup.org/media/file/FactSheet_SafePractices.pdf. Accessed 17 December 2009 [cited 17 December 2009].

Thiele, R. H., J. L. Huffmyer, and E. C. Nemergut. 2008. The "six sigma approach" to the operating room environment and infection. *Best Pract. Res. Clin. Anaesthesiol.* 22 (3): 537–552.

Urbanic, J. J., and W. R. Lee. 2006. Update on brachytherapy in localized prostate cancer: The importance of dosimetry. *Curr. Opin. Urol.* 16 (3): 157–161.

Role of Quality Audits: View from North America

Claudio H. Sibata
21st Century Oncology

Michael S. Gossman
Tri-State Regional Cancer Center

Introduction

Quality audits are emerging as an essential tool to achieve a state-of-the-art quality assurance (QA) program. In the United States, several institutions provide quality audits. This concept originated from the Radiological Physics Center (RPC), funded continuously since 1968 by the National Cancer Institute to provide for auditing dosimetric practices at institutions participating in NCI sponsored cooperative clinical trials (University of Texas MD Anderson Cancer Center 2008). The RPC is able to offer remote or in-depth site visits to assist and confirm treatment unit calibrations, employ thermoluminescent dosimetry, and offer benchmarks for radiosurgery and intensity modulated radiation therapy. Recently, the University of Texas MD Anderson Cancer Center, via their Radiation Dosimetry Services (RDS), began providing similar QA services on a fee-for-service basis. The American College of Radiology (ACR) (Hulick and Ascoli 2005; Ellerbroek et al. 2006) and the American College of Radiation Oncology (ACRO) (Dobelbower et al. 2001; Cotter and Dobelbower 2005) offer practice accreditation. However, site visits are an integral part of the audit with these organizations. They rely entirely on the RPC or RDS for radiation output confirmation.

For decades, peer review has been a cornerstone of QA for radiation oncology physicians (Johnstone et al. 1999; Amis 2004). Now, peer review has garnered strong momentum and formal guidance among medical physicists as outlined by the American Association of Physicists in Medicine (AAPM) in TG-103 (Halvorsen et al. 2005). It is critical that the QA program cover both the clinical as well as the physical aspects of treatment planning and delivery, as errors can occur from a myriad of sources in the clinical environment. Of possible errors, the most catastrophic are the errors made by a medical physicist (Figure 27.1). A quality audit addresses potential catastrophic errors and provides a comprehensive analysis with constructive criticisms to aid the supervising medical physicist in improving upon the overall QA program (British Institute of Radiology 1983, 1996; Hendee and Orton 2008).

Quality Audit Process

The quality audit process is complex. It involves every aspect of the radiation oncology department being audited. The first step in the audit process is to identify the department location, personnel, services provided (teletherapy, brachytherapy, and special procedures), equipment used for treatment and physics measurements including proper calibration, and proper licenses from the Nuclear Regulatory Commission (NRC 2009) or the agreement state. This information is usually provided to the accrediting entity in advance. An on-site review of the facility follows. A full day to review the QA program instituted by the Director of Medical Physics at the facility is typical. With frequent requests for data and responses to issues identified by the auditor, the presence of that supervising medical physicist must be maintained during the entire review period.

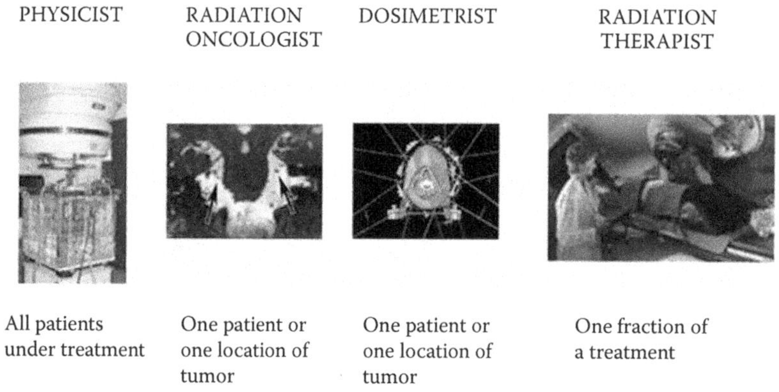

PHYSICIST	RADIATION ONCOLOGIST	DOSIMETRIST	RADIATION THERAPIST

| All patients under treatment | One patient or one location of tumor | One patient or one location of tumor | One fraction of a treatment |

FIGURE 27.1 Consequences of errors made by radiation oncology personnel.

Record Review

The second step is to review records related to therapy and simulation machine shielding. Included in this review are initial radiation surveys of the facility and personnel dosimetry records. It is common for direct questions to be posed to the chief physicist regarding how measurements were conducted. More complex discussions may include how skyshine, partial-ceiling shielding, typical barrier thicknesses, and neutron shielding were determined. Any reevaluation of shielding due to IMRT program development is also discussed.

Particular attention is paid to whether or not physics equipment is adequate for the services provided at the facility. A general list of equipment might include water-equivalent water phantoms, natural water phantoms, morning QA output devices, Accredited Dosimetry Calibration Laboratory (ADCL) referenced ionization chambers and electrometers, a complete cross-referenced backup dosimetry system, film scanners, in vivo dosimetry equipment, thermometers, barometers, survey meters, and other special equipment. At times, the site may be in need of additional equipment. In some situations, an explanation of the use of devices already present at the facility may result in a communication of their limitations. Recommendations on equipment may be presented in the written report to follow, and may benefit the facility in budgeting for its use in a timely manner as well as addressing correctable weaknesses in the program.

The reviewer generally reviews all records associated with the medical physics section and documents all of the special procedures performed at the facility. The clinical use of modalities involving dynamic wedges, intensity-modulated radiation therapy (IMRT), portal imaging, image-guided radiation therapy (IGRT), stereotactic radiosurgery, and respiratory gating are also studied. The most important factor for these procedures is the daily practice guidelines instituted from the QA program, encompassing each of these. Within such QA, it should be clearly observed by the reviewer that appropriate standards for QA are being met. Results from commissioning and periodic testing

should be in accordance with machine specifications as well as within criteria recommended by task groups of the AAPM. It is a common finding during on-site inspections that one or more of these special modalities is found to be unacceptable in daily use without having a single QA ran for it. In essence, it is the qualified medical physicists responsibility to commission each modality as well as to commission each plan.

A record and verify (R&V) system is an integral part of a clinical facility. If one is not in use, it will be strongly recommended. For sites with an R&V system in place, questions are posed that yield a response indicating how QA is involved with it. Simulated plan data should be exported directly from the planning system to the machine for treatment. Critically, if there is no direct networking to the R&V system, second checks should be conducted prior to the initial treatment. For sites where planned beam information is inserted manually; if a mistake is not found, then the possibility of treating a patient with the wrong parameters is more likely and may not be identified until the entire course has been completed. A good medical physics practice can employ a QA program that incorporates the R&V system to eliminate even the smallest of discrepancies. Results from an internal analysis are shown in Figure 27.2 for an accredited review site. In this case, the integration of an R&V system reduced simple charting errors by about 9% over a 20-month period.

External Beam Output Calibration

Verification of the protocol being used to calibrate accelerator output is an important component of a review. Significant errors have been documented over the years as a result of incorrect calibrations. The AAPM TG-51 formalism should be implemented (Almond et al. 1999). It is a recommended practice to verify such output calibrations by an independent organization. The RPC or RDS TLD mailing system is a good resource to consider for verification. Output should be determined for all clinical beams at least once annually. It should also be done immediately following significant accelerator repair or upgrade as well. Benchmark testing and cross-referencing other equipment are ideally

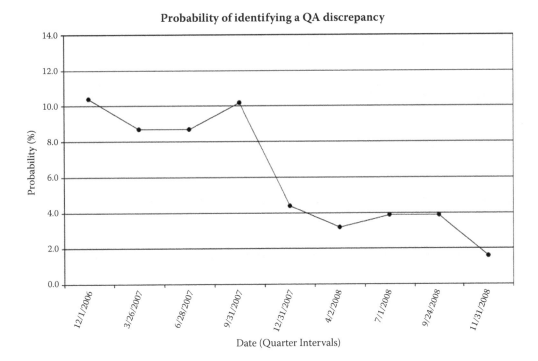

Probability of identifying a QA discrepancy

FIGURE 27.2 The use of quarterly statistical error tracking at one review site where a record and verify system was integrated has resulted in a reduction of simple charting errors by as much as 9% over a course of 1.8 years.

conducted immediately following calibration on the same day. Results from the RPC or RDS should be followed closely and inspected by the auditor. Protocol procedures should be reevaluated with repeated TLD exposures in instances when the RPC acceptance criterion is not met.

Calculation (Meter-Set) Checks

A third area of focus for on-site reviews in the auditing process is on second checks for the amount of time the machine is calculated to be on. Such a second check should be completed by another person or with independent calculation software within the first three treatments for a conventional course of radiotherapy. Short or hypo-fractionated courses should be completed in a much shorter time frame. Any in vivo measurements that are done or measurements conducted for special medical physics consultation should be conducted prior to or at the first treatment. Weekly continuing physics management checks should be conducted during the course of radiation delivery. Following the completion of the course prescription, final physics checks should be clearly indicated to the reviewer.

Comparison of the locally measured clinical data with published data is useful. ACRO compares the clinical data with data from the *British Journal of Radiology* (BJR) Supplement 25 (British Institute of Radiology 1996). Data found to exceed a review tolerance of ±3% may only be acceptable if annual scanning reports indicate that the measured data is consistent with the treatment planning system's commissioned data. A review of current machine beam data, annual scanning results, and commissioning physics reports are commonly asked for during accreditation review.

Brachytherapy

Brachytherapy services are also surveyed if provided by the facility. To begin the review in this area of the program, it is important to know who the radiation safety officer (RSO) is. This person should either be a radiation oncologist or a board certified medical physicist. It is encouraged that a medical physicist be appointed by the medical director to be the RSO, as superior radiation safety training has been received. The RSO should have independent authority (above all radiation oncologists and the medical director) to prohibit treatment if expected doses are questioned and if quality assurance has not been provided adequately. A radioactive material license should be available for examination. All afterloading devices, radionuclides, activity limits, authorized users, authorized medical physicists, and committee-approved procedure locations should be identifiable and updated. Issues pertaining to brachytherapy specifically, namely emergency procedures, assays, inventory, shipping and receiving methods, disposal, patient and room surveys, should all be properly documented upon review. A radiation safety committee should be clearly indicated and meet at least quarterly to oversee issues related to external beam radiation and brachytherapy. Communications of quarterly committee meetings and annual in-services should be available to the reviewer along with any reports from regulatory bodies such as the NRC, FDA, or agreement state. Past issues, incidents, or deviations from treatment prescription should have resulted in proper guidance for correction by the committee and the relevant documents should be available for inspection.

Policy and Procedure Documentation

There should be an active Policy and Procedure Manual in place that can be easily found by staff in the department. The procedures should be reviewed annually by the medical director and the chief medical physicist. The manual should include all clinical procedures related to simulation, treatment, and QA, along with the frequency of each test, training, chart round, peer review, and incident report. One manual located at each treatment machine is advisable. A program's maturity should be obvious to the reviewer. The credentials of staff members should be maintained and at the level stated as qualified by their professional society and governing statutes. Independent peer reviews should be conducted annually, both for the radiation oncologist's practice and for the chief medical physicist's program.

Patient Chart Reviews

The final component of a quality audit is the review of clinical charts. In order to determine consistency or recent changes, charts from both patients under treatment and completed patients should be reviewed. It is typical for a facility to be required to present twenty-five charts that are randomly selected by the auditor from facility ICD-9 code records. At a minimum, the following items should be included within each chart: a complete prescription, calculations for treatment time verification, and a daily record of treatment. Any additional charting records made mandatory by the accreditation organization are also closely inspected.

It is common to survey a clinic and find out that no medical event (or treatment variance) took place in the department. It is troublesome though that not a single treatment variance has occurred during that time. If no mistakes were identified, then it is very likely that the program was not adequately implemented. The auditor needs to ensure that the QA program is sound. Follow-up should be conducted in a pursuit to correct issues in a timely manner.

Another issue frequently observed is a weakness in the IMRT program. A QA program should be in place to second check a treatment planning system. For IMRT treatments, independent calculation software is not sufficient alone. Measurements should be conducted prior to the start of any treatment. Although one normally uses a test case to commission or check the system for IMRT capability, it is not proof enough that dynamic modalities are operating properly for each patient. It is a medical physics community common practice in the United States to compare measured isodose distributions from IMRT plans for every plan, in order to prove the proper transfer of leaf-motion files to the medical accelerator. This is fundamental to the quality of the IMRT program.

Discussion

The training of auditors is essential in order to insure that a meaningful audit of a program is provided. Determining the importance of correcting weaknesses in the audit is crucial. Looking for holes in the QA program, especially for causes of medical events, are more important than spending time on regulatory issues when time is short. QA program developers may also be spending a lot of time on less important issues when more useful tests could be conducted. This is especially true with the advent of R&V systems and the push for electronic charting.

Electronic charting is currently much more complicated than having written charts, although steadily improving. Many such systems cannot handle the information for each patient adequately. For example, frequently the treatment plan is scanned into the R&V system and a review of the plan and documentation is not done in the same area as would be found in a written chart. Although petty charting errors found with written charts are no longer found in chartless environments, such as an incorrect dose total or an error in counting the number of elapsed days, there are more places that need to be reviewed. It is common for paperless environments to have multiple computers running side-by-side in order to view all the components of the chart for a proper comparison.

A well-documented QA program can be easily audited. It seems that disorganization in documentation correlates with a poor QA program. Auditing a facility may be daunting due to the sheer complexities of the many services provided. Departments utilizing many different kinds of special procedures are the most troublesome ones. This may be the result of a lack of appropriate guidelines from the medical physics community. It may also be a direct result of poor supervision from the chief medical physicist and/or lack of appropriate staffing levels. Procedures should be properly assured of successful testing and management before being clinically implemented. It is for this reason that peer reviews for the chief medical physicist are required to be obtained annually and with high importance.

The director of medical physics should always have the sole responsibility for developing and supervising the QA program (Cotter and Dobelbower 2005; Halvorsen et al. 2005; Hendee and Orton 2008). It is important to weigh heavily on responsibilities of the medical physicist in general, when dealing with program oversights. As previously discussed, whenever a dosimetrist or radiation oncologist makes a mistake, it usually affects one patient. If a technologist makes a mistake, it normally affects one fraction of the treatment. If a physicist makes a mistake, it can affect all patients in the clinic during the time that the mistake is in place.

Mistakes can be avoided with proper training, continuation of proper development of a sound QA program, and quality audits through independent peer review. In general, facilities should adopt performance standards of task groups published by the AAPM for general direction. For auditors, this provides the foundation for evaluation regarding the facilities ability to construct an adequate medical physics program operating in accordance with the standards of the scientific field.

Summary

The AAPM has recently appointed a Peer Review Clearinghouse Subcommittee with a charge to provide the groundwork in providing AAPM members who desire independent peer review of their practice with access to a peer review service of reliable quality provided by vetted peers. Radiation oncology facilities may soon be permitted to request the services of the AAPM to handle professional medical physics site reviews directly. The review shall be conducted by way of a 1-day site visit, standardized by the AAPM through clear policies and procedures based on, but not restricted by, the recommendations in the TG-103 report, and based on the QA program of the chief medical physicist. The review shall not include independent measurement of clinical beam data, but could include a benchmark case if agreed by both parties.

It is important to note that the AAPM Peer Review Clearinghouse is currently in the design phase. It is expected to play an important role in making medical physics peer reviews easier to obtain, better conducted through AAPM supervision, and more informative to the site chief medical physicist. This effort should result in more interest in pursuing peer reviews in the medical physics community, better medical physics QA programs around the country, growth in accreditation awarded facilities, and, last but not least, improvement in patient care.

References

American College of Radiology (ACR). 2008. American College of Radiology (ACR) Radiation Oncology Accreditation Program Requirements 2008. Available from http://www.acr.org/accreditation/radiation/requirements.aspx.

Almond, P. R., P. J. Biggs, B. M. Coursey, W. F. Hanson, M. S. Huq, R. Nath, and D. W. Rogers. 1999. AAPM's TG-51 protocol for clinical reference dosimetry of high-energy photon and electron beams. *Med. Phys.* 26 (9): 1847–1870.

Amis, Jr., E. S. 2004. What is a good radiologist? *J. Am. Coll. Radiol.* 1 (3): 155.

British Institute of Radiology. Central axis depth dose data for use in radiotherapy. A survey of depth doses and related data measured in water or equivalent media. 1983. *Br. J. Radiol. Suppl.* 17: 1–147.

British Institute of Radiology. 1996. Central axis depth dose data for use in radiotherapy departments. *Br. J. Radiol. Suppl.* 25: 1–188.

Cotter, G. W., and R. R. Dobelbower Jr. 2005. Radiation oncology practice accreditation: The American College of Radiation Oncology, Practice Accreditation Program, guidelines and standards. *Crit. Rev. Oncol. Hematol.* 55 (2): 93–102.

Dobelbower, R. R., G. Cotter, P. J. Schilling, E. I. Parsai, and J. M. Carroll. 2001. Radiation oncology practice accreditation. *Rays* 26 (3): 191–198.

Ellerbroek, N.A., M. Brenner, P. Hulick, and T. Cushing. 2006. Practice accreditation for radiation oncology: Quality is reality. *J. Am. Coll. Radiol.* 3 (10): 787–792.

Halvorsen, P. H., I. J. Das, M. Fraser, D. J. Freedman, R. E. Rice 3rd, G. S. Ibbott, E. I. Parsai, T. T. Robin Jr., and B. R. Thomadsen. 2005. AAPM Task Group 103 report on peer review in clinical radiation oncology physics. *J. Appl. Clin. Med. Phys.* 6 (4): 50–64.

Hendee, W. R., and C. G. Orton, eds. 2008. *Controversies in Medical Physics: A Compendium of Point/Counterpoint Debates.* College Park, MD: AAPM.

Hulick, P. R., and F. A. Ascoli. 2005. Quality assurance in radiation oncology. *J. Am. Coll. Radiol.* 2 (7): 613–616.

Johnstone, P. A., D. C. Rohde, B. C. May, Y. P. Peng, and P. R. Hulick. 1999. Peer review and performance improvement in a radiation oncology clinic. *Qual. Manag. Health Care* 8 (1): 22–28.

U.S. Nuclear Regulatory Commission (NRC). 2009. United States Nuclear Regulatory Commission 2009. Available from http://www.nrc.gov.

University of Texas MD Anderson Cancer Center. 2008. *Radiological Physics Center History* 2008. Available from http://rpc.mdanderson.org/RPC/home.htm.

28

Role of Quality Audits: View from the IAEA

Joanna Izewska
*International Atomic
Energy Agency (IAEA)*

Eeva Salminen
*Finnish Radiation and Nuclear
Safety Authority (STUK)
and Turku University
Hospital, Finland.*

Introduction

Comprehensive quality assurance (QA) programs, or quality management systems, in radiotherapy include all components of radiation therapy practice (World Health Organization 1988). A wide range of QA recommendations and guidelines that describe procedures, tests, and tolerances for specific parts of the practice is available (Kutcher et al. 1994; Thwaites et al. 1995). In addition to maintaining the quality of patient treatment and the outcome at the required level, QA is also necessary to reduce the likelihood of dose misadministration, that is, reduce the likelihood of the actual delivered dose being substantially higher or lower than intended (International Commission on Radiological Protection 2000; International Atomic Energy Agency 2000). This is particularly important because radiotherapy is a potentially high-risk procedure.

The quality audit is recognized as an essential element of QA systems in radiotherapy. It is a method of checking that the quality of activities in a radiotherapy center adheres to the standards of good practice. The standards may be recommended nationally or internationally and should be derived from up-to-date evidence-based cancer management data.

The ultimate objective of the quality audit is quality improvement and the tool used is an assessment of a practice, or an activity, by an independent body. The quality audit is equivalent to peer-review or independent evaluation of the practice. The audit involves fact-finding and the interpretation of findings in the context of the evidence-based criteria for good practice. Deficiencies in structure, gaps in technology, or deviations in procedures will be identified by the auditors in the review process. This way the areas for improvement will be documented and a set of recommendations will be formulated for implementation by the center being audited. It is generally considered that the findings of

the audit and its outcome are confidential between the auditing body and the audited center.

It is worth mentioning that the quality audit in radiotherapy is not designed for regulatory purposes and the auditors have no power to enforce any actions based on their findings; they can only report their findings and give recommendations. The audit should be understood solely as an impartial source of advice on quality improvement (International Atomic Energy Agency 2007a). Therefore, it is the audited center that decides on any actions required for the implementation of the audit recommendations. A feedback system incorporated in the audit scheme should be in place in order to monitor the changes and to organize a reaudit when appropriate. With this approach, the auditing cycle will stimulate and promote continuous improvement for the benefit of the patient.

Scope and Focus

Quality audits are of a wide range of types and levels, either reviewing the whole radiotherapy practice (comprehensive audit) or selected, important parts of the practice (partial audit). A comprehensive audit, also called a clinical audit (Euratom Directive 97/43 1997), will cover the whole clinical pathway of the patient including all interconnected stages of radiotherapy. It addresses the three main elements of the practice: structure, process, and outcome. In contrast, a partial audit has a limited scope and only specific parts of the radiotherapy practice are reviewed. This may be a partial audit of structure (e.g., staffing levels and qualifications) or a process (e.g., a dosimetry audit checking the beam calibration in external beam radiotherapy). Another example of a partial audit is credentialing for entry into cooperative clinical research studies (Kron et al. 2002; Molineu et al. 2005), which examines the compliance of center's proce-

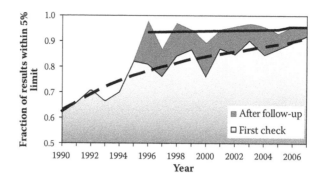

FIGURE 28.1 Fraction of TLD results within 5% acceptable limit in the IAEA/WHO TLD postal dose audit program.

dures with a specific clinical protocol for a selected group of patients.

Internal and external audits typically have different focuses and scopes, but they can complement each other. For example, an internal audit or a self-assessment may be used as a preparation for an external audit and to monitor the implementation of the audit recommendations. Also, an internal audit, rather than an external audit, would be more suitable for the review of the radiotherapy outcomes (especially in the international context). This is mostly due to the fact that outcome data reflect the past practice of the center, not the current practice that undergoes the audit.

Quality audits may be proactive, consisting of a review of ongoing procedures with the aim of improving the quality and preventing or reducing the probability of errors and accidents, or they may be reactive, that is, focused on the response to a suspected or reported incident. Examples of proactive and reactive quality audits are the IAEA/WHO TLD mailed dose program (Izewska et al. 2003) and onsite review visits to radiotherapy institutions by IAEA experts, respectively (International Atomic Energy Agency 2007b).

Dosimetry Audit

Audits of radiation dose have a long tradition (Aguirre et al. 2002; Izewska et al. 2003). Both onsite audit systems and mailed dosimetry programs exist in parallel. Typically, onsite audits review local dosimetry systems, test the dosimetric, electrical, mechanical, and safety parameters of radiotherapy equipment, test the treatment planning system, as well as review the clinical dosimetry records.

Many onsite review programs operate at a national level for a limited number of hospitals, whereas mailed systems provide cost effective audits on a larger scale, involving hundreds or thousands of radiotherapy facilities (Roue et al. 2004; Aguirre et al. 2002; Izewska, Svensson, and Ibbott 2002; Izewska and Thwaites 2002).

Typically, postal dose audit programs have a limited scope and are capable of providing verification of a few selected dose points or beam parameters. A four-level flexible audit system may be adapted for such audits (Izewska, Svensson, and Ibbott 2002):

- *Level 1.* Postal dose audits for photon beams in reference conditions (Izewska et al. 2003; Izewska and Thwaites 2002; Aguirre et al. 2002). This is the basic level, recommended for all radiotherapy centers and mandatory in several countries.
- *Level 2.* Postal dose audits for photon and electron beams in reference and nonreference conditions on the beam axis (Roue et al. 2004).
- *Level 3.* Audits for photon beams in reference and nonreference conditions off-axis and dose at depth on the beam axis for electron beams (Izewska et al. 2007).
- *Level 4.* Audits for photon and electron beams in semi- and anthropomorphic phantoms. This step is used to verify the dose distribution for more realistic treatment situations, such as breast, prostate, or lung (Gershkevitsh et al. 2008) or special treatment techniques, such as intensity-modulated radiation therapy (IMRT) of head and neck (Molineu et al. 2005).

Figure 28.1 illustrates the improvement in dosimetry practices in radiotherapy centers participating in the IAEA/WHO TLD postal dose audit program. After the regular follow-up of poor TLD results was introduced in 1996 by the IAEA/WHO, the fraction of acceptable results increased to about 0.96.

Comprehensive Audit

To optimize clinical outcomes, it is equally important that the clinical aspects as well as the physical and technical aspects of patient treatment are audited because, though essential for the radiotherapy process, accurate beam dosimetry and treatment planning alone cannot guarantee the required outcome of a patient's treatment. The comprehensive audit methodology is described by the IAEA (International Atomic Energy Agency 2007a) and an EC guidance document (European Commission). The IAEA audit methodology, also known as the Quality Assurance Team for Radiation Oncology (QUATRO) methodology, puts emphasis on radiotherapy structure and process rather than treatment outcome. It includes an assessment of infrastructure as well as of patient-related and equipment-related procedures involving radiation safety and patient protection aspects,

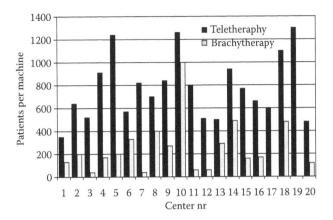

FIGURE 28.2 Patient throughput on radiotherapy machines in 20 radiotherapy centers of Central and Eastern Europe participating in QUATRO audits.

where appropriate. Staffing levels and professional training programs for radiation oncologists, medical radiation physicists, and radiation therapists are also reviewed.

The interpretation of audit results is made against the appropriate criteria of good evidence-based radiotherapy practice. As an example of such criteria, the IAEA has given a description of the design and implementation of a radiotherapy program regarding clinical, medical physics, radiation protection, and safety aspects (International Atomic Energy Agency 2008a, 2008b).

Since 2005, the IAEA has organized approximately 50 QUATRO audits in response to voluntary requests from its member states in Africa, Asia, Europe, and Latin America. The QUATRO audits included the assessment of the ability of centers to maintain the radiotherapy practice at the level corresponding to the best clinical practice in the specific economic setting of a given country. Gaps in technology, human resources, and procedures were identified and areas for improvement in current services were documented. Additionally, centers received advice for further development. Some centers have been acknowledged for operations at a high level of competence. Based on the audit results, it was possible for the IAEA to indentify specific areas and items needing improvement (weak links) in the different centers and address the common aspects such as, for example, staff training internationally. Figure 28.2 shows patient throughput on radiotherapy machines analyzed within the QUATRO process; equipment shortages in some centers have been addressed by the national governments following the QUATRO audits.

Summary

Radiation oncology requires a strong commitment to QA, including active participation of all staff directly involved in the radiotherapy process and supporting specialists. Using a regular audit system will bring continuous improvement through assessment and implementation of all those planned and systematic actions necessary to provide adequate confidence that the radiation treatment will satisfy the given requirements for quality in patient treatments. The analysis and utilization of the audit findings will help to set targets for improvement and the effects of any changes implemented on the practice should be monitored continuously.

References

Aguirre, J. F., R. Tailor, G. Ibbott, M. Stovall, and W. Hanson. 2002. Thermoluminescence dosimetry as a tool for the remote verification of output for radiotherapy beams: 25 years of experience. In *Standards and Codes of Practice in Medical Radiation Dosimetry: Proceedings of an International Symposium*, Vienna, Austria, November 25–28, 2002, 191–199. Vienna: IAEA.

Euratom Directive 97/43. 1997. On health protection of individuals against the dangers of ionising radiation in relation to medical exposures. *Official Journal of The European Communities* 1180:22.

European Commission. 1997. European guidance on clinical audit for medical exposure. Available from http://www .clinicalaudit.net.

Gershkevitsh, E., R. Schmidt, G. Velez, D. Miller, E. Korf, F. Yip, S. Wanwilairat, and S. Vatnitsky. 2008. Dosimetric verification of radiotherapy treatment planning systems: Results of IAEA pilot study. *Radiother. Oncol.* 89 (3): 338–346.

International Atomic Energy Agency. 2000. *Lessons Learned from Accidental Exposures in Radiotherapy*. In Safety Reports Series no. 17. Vienna: IAEA.

IAEA. 2007a. *Comprehensive Audits of Radiotherapy Practices: A Tool for Quality Improvement*. Quality Assurance Team for Radiation Oncology (QUATRO). Vienna: IAEA.

IAEA. 2007b. *On-Site Visits to Radiotherapy Centres: Medical Physics Procedures*. Quality Assurance Team For Radiation Oncology (QUATRO). Vienna: IAEA.

IAEA. 2008a. *Setting Up a Radiotherapy Programme: Clinical, Medical Physics, Radiation Protection and Safety Aspects*. Vienna: IAEA.

IAEA. 2008b. *Transition from 2-D Radiotherapy to 3-D Conformal and Intensity Modulated Radiotherapy*. Vienna: IAEA.

International Commission on Radiological Protection. 2000. *Prevention of Accidental Exposures to Patients Undergoing Radiation Therapy. ICRP Publication 86.* Oxford: ICRP.

Izewska, J., P. Andreo, S. Vatnitsky, and K. R. Shortt. 2003. The IAEA/WHO TLD postal dose quality audits for radiotherapy: A perspective of dosimetry practices at hospitals in developing countries. *Radiother. Oncol.* 69: 91–97.

Izewska, J., D. Georg, P. Bera, D. Thwaites, M. Arib, M. Saravi, K. Sergieva, K. Li, F. G. Yip, A. K. Mahant, and W. Bulski. 2007. A methodology for TLD postal dosimetry audit of high-energy radiotherapy photon beams in non-reference conditions. *Radiother. Oncol.* 84 (1): 67–74.

Izewska, J., H. Svensson, and G. Ibbott. 2002. Worldwide quality assurance networks for radiotherapy dosimetry. In *Standards and Codes of Practice in Medical Radiation Dosimetry: Proceedings of an International Symposium,* Vienna, Austria, November 25–28, 2002. Vienna: IAEA.

Izewska, J., and D. I. Thwaites. 2002. IAEA supported national thermoluminescence dosimetry audit networks for radiotherapy dosimetry. In *Standards and Codes of Practice in Medical Radiation Dosimetry: Proceedings of an International Symposium,* Vienna, Austria, November 25–28, 2002. Vienna: IAEA.

Kron, T., C. Hamilton, M. Roff, and J. Denham. 2002. Dosimetric intercomparison for two Australasian clinical trials using an anthropomorphic phantom. *Int. J. Radiat. Oncol. Biol. Phys.* 52 (2): 566–579.

Kutcher, G. J., L. Coia, M. Gillin, W. F. Hanson, S. Leibel, R. J. Morton, J. R. Palta et al. 1994. Comprehensive QA for radiation oncology: Report of AAPM Radiation Therapy Committee Task Group 40. *Med. Phys.* 21 (4): 581–618.

Molineu, A., D. S. Followill, P. A. Balter, W. F. Hanson, M. T. Gillin, M. S. Huq, A. Eisbruch, and G. S. Ibbott. 2005. Design and implementation of an anthropomorphic quality assurance phantom for intensity-modulated radiation therapy for the Radiation Therapy Oncology Group. *Int. J. Radiat. Oncol. Biol. Phys.* 63 (2): 577–583.

Roue, A., J. Van Dam, A. Dutreix, and H. Svensson. 2004. The EQUAL-ESTRO external quality control laboratory in France. *Cancer Radiother.* 8 Suppl 1: S44–S49.

Thwaites, D. I., P. Scalliet, J. W. Leer, and J. Overgaard. 1995. Quality assurance in radiotherapy, ESTRO recommendations. *Radiother. Oncol.* 35: 61–73.

World Health Organization. 1988. *Quality Assurance in Radiotherapy.* Geneva: WHO.

29

Peer Review:
Physician's View from Australia

Gerald B. Fogarty
Mater Hospital

Introduction

Peer review occurs when colleagues advise and comment on another's work. They may be invited to comment, as when asked for help, or an article is assessed for publication in a learned journal. They may comment anyway, as when questioning presented work at a clinical meeting, or when consulted in a medicolegal situation. Quality assurance (QA) ensures delivery of treatment that has been planned. There are different QA tools for different items that are being measured. Peer review, on the other hand, can be difficult to measure. A positive system for the delivery of peer review can have a beneficial impact on attitudes toward teamwork and patient care that lasts beyond treatment-related outcomes.

Overview of Peer Review

In the age of a multidisciplinary team approach to cancer therapy, the number and type of peers available has increased. For example, for an oncologist, a peer can be another oncologist or an oncology trainee. A peer could also be a person in radiation oncology but not an oncologist, for example, a medical physicist, radiation therapist, or nurse. There can also be peers outside radiation oncology, for example, a surgeon or medical oncologist. These different peers will approach and appraise work from their own viewpoint. The responsibility for incorporating suggested changes still remains the responsibility of the prescribing professional, although there is a move to spread the responsibility to include the rest of the team (Sidhom and Poulsen 2006).

Peer review aids reflection on judgment. An attitude of "expect the unexpected, from the unexpected" helps to capture all that may help to deliver the best quality care. The real benefit of peer review is realized when the professional to whom the comment is directed accepts the advice and incorporates it constructively into the case at hand and also into their ongoing professional formation. This way of acting is demanded of trainees, is essential for publication acceptance, and is a mark of greatness in a consultant. A conscientious professional will seek and invite peer review, will be grateful in receiving it, and will put it into effect.

Peer review aids reflection on motivations. Motivations include a desire to deliver quality service to patients, to fulfill employer expectations, to compete with and be compared favorably to other practitioners, and to avoid litigation. Involvement in peer review can be difficult. One can have a fear of criticism, fear of exposing a lack of knowledge and confidence in indications, and fear of having inappropriate techniques and quality of evidence that drives ones practice. Other considerations that work against peer review are the lack of time and resources, for example, a treatment waiting list that already shows that the department is overloaded. However, acknowledgment and documentation of these problems in the peer review process can lead to resolution.

Experience has shown that outcomes depend as much on the delivery of treatment as on its design. QA describes planned and systematic processes that ensure delivery of what has been promised. QA is more than quality control of procedures ensuring that certain performance standards are met. It also includes a "monitoring of outcome parameters and a vital feedback loop to implement remedial measures should outcome deficiencies be discovered" (Peters, Browning, and Potocsny 1991). QA tools differ depending on what is being measured. In medical treatments, the utility of QA tools are measured in what results in the best outcomes.

In radiation oncology, peer review as a QA tool has several specific challenges. There continues to be acceleration in the understanding of tumor biology, radiobiology, clinical evidence, technology, and innovation in treatment design and delivery.

There is increasing interdisciplinary interdependence. For example, on moving from two dimensional treatments to intensity-modulated radiotherapy (IMRT), the validation of treatment delivery has moved from a checking of a radiation portal on a plain x-ray film by a radiation oncologist, to significant interaction between physicists, therapists, and oncologists. Variability in type of practice is increasing; there are academic institutions through to community-based service providers. The workforce is changing, with a greater percentage of the radiation oncology workforce working on a part-time basis, but who all need the same benchmarks for delivery of quality care. The main variables in radiation oncology from the oncologists' point of view include the intent of treatment, in what defines the clinical target volumes (CTV), and the expansions of organs at risk (OAR) (Purdy 2004). In general, doses and fractionation schedules are fairly standard in radical treatment and are based on reasonable quality evidence. Continuing education, guidelines, and peer review are of great benefit in these variable areas. Radiation oncology is essentially a tertiary referral specialty that services a number of referring doctors (e.g., breast surgeons, hematologists, medical oncologists) that grows larger with increasing subspecialization. Fitting all these doctors into peer review presents a major challenge.

Guidelines are beginning to appear as the specialty attempts self-regulation. Guidelines go to great lengths to be inclusive and yet reasonable, advising doctors who deviate from them to ensure adequate documentation in the event of litigation (Hartford et al. 2009). Documentation in QA is essential and studies show there is room for improvement. Bekelman and Yahalom (2009) reported on the quality of radiotherapy reporting in randomized trials of lymphoma. They assessed 61 reports, looking for quality measures such as target volume, radiation prescription, the QA process used, and adherence to that process. Only 23 (38%) described the target volume. Of the 42 reports involving involved-field radiation therapy alone, only 8 (19%) adequately described the target volume. Thirteen reports specified the radiation therapy prescription point (21%). Twelve reports (20%) described using a QA process and 7 reports (11%) described adherence to the QA process. They concluded the quality of these studies was deficient, which is unfortunate as the treatment of lymphoma is based on this apparently high-level evidence. QA tools based on peer review may help to overcome these problems.

Implementation of Chart Rounds

Many initiatives have led to the development of QA tools based on peer review. An initiative put in place at the Head and Neck Unit at the Peter MacCallum Cancer Centre in Melbourne, Australia was the chart round (Fogarty 2001). This was encouraged by the work of Purdy et al. (1996). The aim of the chart round was to briefly review on a weekly basis all the current pertinent aspects of patient care of those undergoing radical radiation treatments for a head and neck malignancy. Participants included radiation oncology consultants, trainees, radiotherapy dosimetrists,

and treating radiation therapists on an "as available basis." The unit nurse, social worker, and dietician were also included. A form was designed in order to ensure that the chart round was directed, followed a format, and was succinct. A radiation oncologist, preferably the prescribing one, had the responsibility for filling out the form during the chart round discussion.

The number of head and neck patients undergoing treatment in the unit at any one time was between 30 and 45. Each case was discussed and, with brevity and effective leadership, the exercise could be completed within one and a half hours. The first week of treatment was seen as an opportunity to ensure that all relevant information for treatment design and prescription had been properly compiled. Each subsequent week, there was a check of whether the treatment verification had been accepted, whether planning for new techniques was on track, and on whether other items important to care had been completed, for example, to ensure if a continuation note from the Reaction Review Clinic had been written. The chart round was also a time each week when the unit nurse, dietician, and social worker could share information relevant to patient care. During the last week of treatment, it was a time to check whether a plan for follow-up had been made.

To see whether the chart round had an effect on patient care, selected items from the chart round list from the records of 47 consecutive patients were compared with 47 consecutive patients treated before the chart round was operational. Of a total of 354 items, 27 (7.6%) were detected at the chart round to have not been completed. Of these, 23 (6.5%) of these were then completed, leaving four (1%) of those detected undone. This compared favorably with 72 (20.7%) items that were not completed out of 347 in the pre-chart round period. The implementation of the chart round has raised the level of item completion from 80% to 99%.

The chart round has been an excellent feedback mechanism to help remedy deficiencies in items of patient care during the treatment process. Unexpected benefits in this area accrued from our chart round experiment. The best benefit that came from the chart round was an improvement in the culture of the unit, from a "blame game" to one of multidisciplinary cooperation in task completion. The chart round provides a forum for peer review in a timely fashion when all the unit radiation oncologists can assess the plans of others. The attitudes needed for a successful QA exercise like this are a desire to learn about unit mistakes, including one's own, and to improve in these areas. Openness to team input into one's treatment plans is also required. It has proven itself to be a valuable time for allied health personnel to contribute to and benefit from discussions as to why various medical and technical options have been utilized. The explanation of the rationale for plans has also provided an educational focus to the chart round. Adequate documentation of QA, like the chart round, during treatment delivery may become a criterion for departmental accreditation and/or personal recertification. Other units in the center have reported similar benefits from initiating chart rounds. In the skin/melanoma unit, an audit showed that the chart round helped to lift the completion rate to 98% (Fogarty and Ainslie 2005).

Studies on Peer Review

Other studies of peer review as a QA tool have been completed. Rosenthal et al. (2006) from the MD Anderson Cancer Center prospectively examined 134 consecutive patients needing radiation therapy for head and neck cancer who had preliminary radiation therapy plans. The study involved clinical peer review by up to four clinicians. This involved head and neck examination and review of the relevant imaging to confirm target localization. The oncologists were essentially blind to the fact that they were being reviewed. Peer review led to changes in treatment plans for 66% of patients. Major changes that may have significantly affected treatment outcomes in terms of cancer control and/or normal tissue toxicity were recommended in 11% of patients. Most changes involved target delineation based on physical examination findings. This practice may not be possible for community-based units or in units with resource problems. The authors suggested review of the treatment plans by at least one other specialist head and neck oncologist to ensure treatment success. Their study shows the importance of physical examination in peer review. This shows that it may not be possible to have an effective peer review of head and neck cancer by tele- or video-conferencing, that real-time, "same space" review is preferable to virtual alternatives.

Practice Audits

Brundage et al. (1999) reported on the development, structure, and implementation of a real-time clinical radiotherapy audit of the practice of radiation oncology in a regional cancer center. The audit reviewed a total of 3052 treatment plans in real-time over an 8-year period. Of these treatment plans, 124 (4.1%) were not approved by the audit due to apparent errors in radiation planning leading to 75 of these plans being modified prior to initiating treatment. An additional 110 (3.6% of all audited plans) were not approved due to deviations from radiotherapy treatment policy, and 22% of these plans were modified prior to initiating treatment. The remainder provided important feedback for continuous quality improvement of treatment policies. Their study was similar to our chart round study, in that there was a very low rate of errors (<1%) after the audit.

Summary

Peer review helps to reflect on real-time practice and is an important component of QA. Peer review provides an opportunity to question deeper motives—what motivates a unit to treat this patient in this way? It is the backbone of department protocol development. It helps insulate from subjective elements in treatment such as patient expectations, professional expectations, and even financial considerations—it challenges routine. Peer review studies also show that we need each other in the radiotherapy team—it really takes a unit with people of different outlooks to treat successfully.

Much peer review is happening. However as a science, peer review as a QA tool is in its infancy. Studies to date are single-institution and often single-unit within an academic department. Measuring its impact on outcomes may be difficult to do in a randomized setting, as current users are already convinced of its efficacy and may have lost the equipoise needed to allow some patients to be randomized to have no peer review. More studies are needed, and the design of them, especially across institutions, will be challenging.

References

Bekelman, J. E., and J. Yahalom. 2009. Quality of radiotherapy reporting in randomized controlled trials of Hodgkin's lymphoma and non-Hodgkin's lymphoma: A systematic review. *Int. J. Radiat. Oncol. Biol. Phys.* 73 (2): 492–498.

Brundage, M. D., P. F. Dixon, W. J. Mackillop, W. E. Shelley, C. R. Hayter, L. F. Paszat, Y. M. Youssef, J. M. Robins, A. McNamee, and A. Cornell. 1999. A real-time audit of radiation therapy in a regional cancer center. *Int. J. Radiat. Oncol. Biol. Phys.* 43 (1): 115–124.

Fogarty, G. B., and J. Ainslie. 2005. RE: Chart round in a skin radiotherapy unit. *Australas. Radiol.* 49 (6): 526–527.

Fogarty, G. B., C. Hornby, H. M. Ferguson, and L. J. Peters. 2001. quality assurance in a radiation oncology: The chart round experience. *Austral. Radiol.* 45: 189–194.

Hartford, A. C., M. G. Palisca, T. J. Eichler, D. C. Beyer, V. R. Devineni, G. S. Ibbott, B. Kavanagh et al. 2009. American Society for Therapeutic Radiology and Oncology (ASTRO) and American College of Radiology (ACR) practice guidelines for intensity-modulated radiation therapy (IMRT). *Int. J. Radiat. Oncol. Biol. Phys.* 73 (1): 9–14.

Peters, L. J., D. Browning, and A. S. Potocsny. 1991. Departmental support for a quality assurance program. In *Quality Assurance Radiotherapy Physics. Proceedings of an American College of Medical Physics Symposium*, ed. G. Starkschal and J. Horton. Madison, WI: Medical Physics Publishing.

Purdy, J. A. 2004. Current ICRU definitions of volumes: Limitations and future directions. *Semin. Radiat. Oncol.* 14 (1): 27–40.

Purdy, J. A., and C. A. Perez. 1996. Quality assurance in radiation oncology in the United States. *Rays* 21 (4): 505–540.

Rosenthal, D. I., J. A. Asper, J. L. Barker Jr., A. S. Garden, K. S. Chao, W. H. Morrison, R. S. Weber, and K. K. Ang. 2006. Importance of patient examination to clinical quality assurance in head and neck radiation oncology. *Head Neck* 28 (11): 967–973.

Sidhom, M. A., and M. G. Poulsen. 2006. Multidisciplinary care in oncology: Medicolegal implications of group decisions. *Lancet Oncol.* 7 (11): 951–954.

30

Peer Review: Physicist's View from North America

Per Halvorsen
Alliance Oncology LLC

Introduction

The medical profession pioneered the use of peer review long before other professions, with early Arabic medical texts referencing the process first described by Ishaq bin Ali al-Rahwi in his work titled *Ethics of the Physician* in the tenth century (Spier 2002). Although peer review of manuscripts did not evolve until the eighteenth century, a process for evaluating the practices of one's professional peers has existed in medicine for centuries, to ensure that basic standards of safety are followed.

More recently, peer review in radiotherapy has evolved into two broad categories: physicians' clinical practice, and physicists' technical programs.

Peer Review of Clinical Radiotherapy Practice

Chart rounds have been an integral part of most radiation oncology quality assurance programs for many decades (Fogarty et al. 2001). Analysis of the effectiveness of chart round peer reviews indicates that approximately 5% of cases are modified based on input from the chart round process (Horiot et al. 1993; Kane et al. 2008). Clearly, this form of peer review has proven useful as a quality assurance tool, and is likely to continue, with the format evolving to accommodate the new types of information in modern treatment charts (image-guided IMRT, image-based brachytherapy plans, SRS/SBRT plans, etc.).

A more comprehensive approach to chart-based peer review has been implemented in some centers, whereby every chart is peer reviewed prior to the first treatment session on an ongoing basis. Investigators have shown (Adams et al. 2009; Brundage et al. 1999) that such an approach is likely to uncover a larger number of discrepancies than the chart round format. A peer review of target volume delineation prior to the start of dosimetric treatment planning resulted in major adjustments to the target volume in 55% of the reviewed cases in one center's experience (Adams et al. 2009). This indicates that the most significant room for improvement in peer review of clinical radiotherapy practice may be early in the treatment process, rather than after a treatment technique has been implemented, as is the case with the conventional chart round format.

With the widespread participation in clinical trials, another form of clinical peer review has become common, whereby a peer from another institution evaluates the chart for compliance with clearly-defined criteria outlined in the trial's protocol. In some cases (Seegenschmiedt et al. 1993), more than 10% of cases require adjustment in order to ensure that a consistent standard is followed.

A large project was implemented in 1971 to provide clear data on radiation oncology practice patterns in the United States (Hanks 1984). The Patterns of Care studies, sponsored by the National Cancer Institute and administered by the American College of Radiology, proved highly valuable as a quality assessment tool and as a baseline for physician peer review (Coia and Hanks 1997).

As a natural extension of the Patterns of Care studies, the American College of Radiology implemented a radiation oncology practice accreditation program in 1986 (www.acr.org/accreditation/radiation.aspx). For its first decade, the program measured all practices against the Patterns of Care findings, using this data as the "standard of care." Just before 2000, the accreditation program was revised, scoring programs against the ACR Practice Guidelines and Technical Standards, which were by then quite comprehensive and more reflective of current practice patterns. The American College of Radiation Oncology operates a similar practice accreditation program (www.acro.org/Accreditation/index.cfm). Both programs provide a thorough peer review of an institution's clinical and technical programs in radiation oncology, benchmarked against well-defined standards and guidelines, and the peer review reports provide specific recommendations for improving the practice. Although the process is voluntary in most states, approximately 400 clinics in the United States participated in this form of peer review in 2009.

The Institute of Medicine, a division of the National Academy of Sciences sponsored by the National Institutes of Health, published a seminal publication in 2000 titled *To Err Is Human: Building a Safer Health System* (Kohn, Corrigan, and Donaldson 1999). The Institute of Medicine's report focused the medical community on the prevalence of avoidable medical errors and the need to incorporate process changes aimed at reducing the rate of errors. Peer review has been included as a core component of such error prevention programs.

Partly in response to the Institute of Medicine's recommendations, The American Board of Radiology recently introduced a maintenance of certification program, with one component of this program being a professional peer review activity to develop individualized goals for improving the quality of each physician's practice (Brennan et al. 2004; Kun et al. 2007; Miller 2005). All radiation oncology diplomats who obtained board certification after 1994 must complete this process on a regular basis in order to maintain their certification status.

Peer Review of Clinical Physics Programs

Peer review is a relatively recent component of clinical physics quality assurance programs. Although informal peer review has existed since the inception of the clinical radiation oncology physics profession, structured peer review was not commonly employed until the creation of the Radiological Physics center in 1968 (Hanson, Shalek, and Kennedy 1991). Funded continuously by the National Cancer Institute since its inception by the American Association of Physicists in Medicine, the center is overseen by the AAPM's Radiation Therapy Committee. Its charge is to assure the National Cancer Institute that institutions participating in cooperative clinical trials sponsored by the institute have the equipment, qualified personnel, and quality assurance procedures to deliver radiation doses that are comparable and consistent. The center accomplishes this through a combination of onsite audits, mailed thermoluminescent dosimeter evaluations, anthropomorphic phantom irradiations, and chart reviews. The center's findings have had a significant impact toward improving the consistency of machine calibrations and dose calculations in the United States (Gagnon et al. 1978). More recently, the center's use of anthropomorphic phantom irradiations has helped to strengthen the consistency and accuracy of beam modeling in modern treatment planning systems and dose delivery on modern intensity-modulated accelerators (Followill et al. 2007).

Outside the United States, the International Atomic Energy Agency operates an audit program consisting of mailed thermoluminescent dosimeters and onsite surveys, in a manner similar to the Radiological Physics Center (IAEA 2007). Thus, an impartial peer review of the core clinical physics responsibility—accurate calibration of accelerators' radiation output—is available in nearly all countries.

In the brachytherapy subspecialty, a peer review of 35 well-recognized brachytherapy physics programs identified considerable variability in treatment planning and dose calculation techniques (Prete et al. 1998). These findings in part helped to focus the AAPM's efforts toward stronger standardization of source activity specification and dose calculation formalisms in low-energy brachytherapy (DeWerd et al. 2004; Nath et al. 1997).

Despite clearly defined calibration protocols and dosimetry audit programs, errors can and do occur during the complex process of treatment planning and delivery, as evidenced by the International Atomic Energy Agency's recent publication (IAEA 2000). In response to these findings and the Institute of Medicine's recommendations (Kohn, Corrigan, and Donaldson 1999), the AAPM provided guidance for clinical radiation oncology physicists in solo practice (Halvorsen et al. 2003), with a central recommendation being a formal peer review at regular intervals. Following up on this report, the AAPM developed a process for radiation oncology clinical physics peer reviews (Halvorsen et al. 2005). This practice has gained broader application in recent years, with some large physics groups adopting formal peer reviews as an integral component of their quality assurance programs.

The American Board of Radiology's Maintenance of Certification program has been implemented for the Radiological Physics disciplines, and applies to all diplomates who achieved board certification since 2002 (Frey et al. 2007). One component of this program is quality improvement of the diplomate's clinical physics practice through an assessment of the overall practice environment. Among the five quality improvement categories are peer review and development of practice guidelines. The AAPM recently developed a hybrid peer review and self-assessment program, which has been qualified by the ABR as a society-based practice quality improvement activity.

Summary

From its inception in the early days of medicine to today, peer review has been an important component of ensuring quality and safety of medical procedures. As technological complexity has increased and public attention has focused on the prevalence of errors in medicine, peer reviews of clinical and physics practices have assumed an increased role in quality assurance programs within radiation oncology. The evidence continues to demonstrate that peer reviews can be an effective tool in a comprehensive quality assurance program.

References

Adams, R. D., S. Chang, K. Deschesne, S. Freeman, K. Karbowski, D. LaChappelle, D. E. Morris, B. F. Qaqish, J. L. Hubbs, and L. B. Marks. 2009. Quality assurance in clinical radiation therapy: A quantitative assessment of the utility of peer review in a multi-physician academic practice *Int. J. Radiat. Oncol. Biol. Phys.* 75 (S133).

Brennan, T. A., R. I. Horwitz, F. D. Duffy, C. K. Cassel, L. D. Goode, and R. S. Lipner. 2004. The role of physician specialty board certification status in the quality movement. *JAMA* 292 (9): 1038–1043.

Brundage, M. D., P. F. Dixon, W. J. Mackillop, W. E. Shelley, C. R. Hayter, L. F. Paszat, Y. M. Youssef, J. M. Robins, A. McNamee, and A. Cornell. 1999. A real-time audit of radiation therapy in a regional cancer center. *Int. J. Radiat. Oncol. Biol. Phys.* 43 (1): 115–124.

Coia, L. R., and G. E. Hanks. 1997. Quality assessment in the USA: How the patterns of care study has made a difference. *Semin. Radiat. Oncol.* 7 (2): 146–156.

DeWerd, L. A., M. S. Huq, I. J. Das, G. S. Ibbott, W. F. Hanson, T. W. Slowey, J. F. Williamson, and B. M. Coursey. 2004. Procedures for establishing and maintaining consistent air-kerma strength standards for low-energy, photon-emitting brachytherapy sources: Recommendations of the Calibration Laboratory Accreditation Subcommittee of the American Association of Physicists in Medicine. *Med. Phys.* 31 (3): 675–681.

Fogarty, G. B., C. Hornby, H. M. Ferguson, and L. J. Peters. 2001. Quality assurance in a radiation oncology unit: The chart round experience. *Australas. Radiol.* 45 (2): 189–194.

Followill, D. S., D. R. Evans, C. Cherry, A. Molineu, G. Fisher, W. F. Hanson, and G. S. Ibbott. 2007. Design, development, and implementation of the radiological physics center's pelvis and thorax anthropomorphic quality assurance phantoms. *Med. Phys.* 34 (6): 2070–2076.

Frey, G. D., G. S. Ibbott, R. L. Morin, B. R. Paliwal, S. R. Thomas, and J. Bosma. 2007. The American Board of Radiology Perspective on Maintenance of Certification: Part IV. Practice quality improvement in radiologic physics. *Med. Phys.* 34 (11): 4158–4163.

Gagnon, W. F., L. W. Berkley, P. Kennedy, W. F. Hanson, and R. J. Shalek. 1978. An analysis of discrepancies encountered by the AAPM radiological physics center. *Med. Phys.* 5 (6): 556–560.

Halvorsen, P. H., I. J. Das, M. Fraser, D. J. Freedman, R. E. Rice 3rd, G. S. Ibbott, E. I. Parsai, T. T. Robin Jr., and B. R. Thomadsen. 2005. AAPM Task Group 103 report on peer review in clinical radiation oncology physics. *J. Appl. Clin. Med. Phys.* 6 (4): 50–64.

Halvorsen, P. H., J. F. Dawson, M. W. Fraser, G. S. Ibbott, and B. R. Thomadsen. 2003. *The Solo Practice of Medical Physics in Radiation Oncology: Report of the Professional Information and Clinical Relations Committee Task Group #11.* New York, NY: American Institute of Physics.

Hanks, G. E. 1984. Future plans for quality assurance in radiation oncology in the United States. *Int. J. Radiat. Oncol. Biol. Phys.* 10 Suppl 1: 35–38.

Hanson, W. F., R. J. Shalek, and P. Kennedy. 1991. Dosimetry quality assurance in the United States from the experience of the Radiological Physics Center. In *Proceedings of the American College of Medical Physics Symposium*, ed. G. Starkschall and J. Horton. Madison, WI: Medical Physics Publishing.

Horiot, J. C., J. Bernier, K. A. Johansson, E. van der Schueren, and H. Bartelink. 1993. Minimum requirements for quality assurance in radiotherapy. *Radiother. Oncol.* 29: 103–104.

IAEA. 2000. *Lessons Learned from Accidental Exposures in Radiotherapy.* Vienna: International Atomic Energy Agency (IAEA).

IAEA. 2007. *Comprehensive Audits of Radiotherapy Practices: A Tool for Quality Improvement—QUATRO.* Vienna: International Atomic Energy Agency.

Kane, G. M., K. Kelly, J. Rockhill, U. Parvathaneni, S. Patel, J. Douglas, J. Liao, M. Phillips, and G. Laramore. 2008. Quality assurance of QA rounds: A prospective audit tracks practice performance. *Int. J. Radiat. Oncol. Biol. Phys.* 72: S483.

Kohn, L. T., J. M. Corrigan, and M. S. Donaldson, eds. 1999. *To Err Is Human: Building a Safer Health System.* Report of the Committee on Quality of Health Care in America, Institute of Medicine. Washington, DC: National Academy Press.

Kun, L. E., B. G. Haffty, J. Bosma, J. L. Strife, and R. R. Hattery. 2007. American Board of Radiology Maintenance of Certification—Part IV: Practice quality improvement for radiation oncology. *Int. J. Radiat. Oncol. Biol. Phys.* 68 (1): 7–12.

Miller, S. H. 2005. American Board of Medical Specialties and repositioning for excellence in lifelong learning: Maintenance of certification. *J. Contin. Educ. Health Prof.* 25 (3): 151–156.

Nath, R., L. L. Anderson, J. A. Meli, A. J. Olch, J. A. Stitt, and J. F. Williamson. 1997. Code of practice for brachytherapy physics: Report of the AAPM Radiation Therapy Committee Task Group No. 56. American Association of Physicists in Medicine. *Med. Phys.* 24 (10): 1557–1598.

Prete, J. J., B. R. Prestidge, W. S. Bice, J. L. Friedland, R. G. Stock, and P. D. Grimm. 1998. A survey of physics and dosimetry practice of permanent prostate brachytherapy in the United States. *Int. J. Radiat. Oncol. Biol. Phys.* 40 (4): 1001–1005.

Seegenschmiedt, M. H., W. Sauerbrei, R. Sauer, M. Schumacher, A. Schauer, H. F. Rauschecker, H. Dinkloh et al. 1993. Quality control review for radiotherapy of small breast cancer: Analysis of 708 patients in the GBSG I trial. German Breast Study Group (GBSG). *Strahlenther. Onkol.* 169 (6): 339–350.

Spier, R. 2002. The history of the peer-review process. *Trends Biotechnol.* 20 (8): 357–358.

31

Overview of Credentialing and Certification

Tomas Kron
Peter MacCallum Cancer Centre

Kwan Hoong Ng
University of Malaya

Introduction

Radiotherapy, like most cancer treatment services, is a complex and potentially dangerous process. It requires expensive and complicated equipment, highly skilled professionals, and, therefore, great attention to quality assurance. According to the International Standards Organisation (ISO), quality assurance (QA) consists of "all those planned and systematic actions necessary to provide confidence that a product or service will satisfy given requirements for quality" (ISO 2008). This approach has generally been taken up in radiation oncology (Leer et al. 1995).

The terms quality, quality assurance, accreditation, credentialing, and certification apply to all fields; however, in this chapter they will be specifically used in the context of radiation oncology, using medical physics services as an example. Unfortunately, it is often difficult for nonradiotherapy professionals to judge if a particular service is likely to be of high quality. This is where a need for accreditation, credentialing, and certification becomes obvious. It is essentially a peer review process that attests to all stakeholders that a service, process, equipment, or person is "up to the task."

In this chapter, after an attempt to define the terms, we look at elements and requirements for a credentialing and certification program using medical physics as an example. Consideration is given to stakeholders in the process and how credentialing impacts them. This is followed by a summary of how credentialing and certification affect quality in the provision of radiation oncology services as a whole. The chapter concludes with a look at the role of international organizations in the promotion of credentialing and, therefore, quality in radiation oncology.

Definitions

Unfortunately, the terms accreditation and credentialing are not used everywhere with the exact same meaning. We attempt to define them as illustrated in Figure 31.1. The process of credentialing is only meaningful in the context of a particular activity to which it will apply. As such, the first step is an analysis of requirements for the activity and the service needed. This would result in attributes that characterize the service or service provider. Ideally, this would be informed by evidence that these attributes indeed improve clinical outcomes.

Typical attributes are skills, knowledge, experience, and competencies. It is then necessary to develop a method that can test these attributes or at least a representative subset thereof. The assessment is against criteria that reflect the attributes, and the tests are typically developed by a group of expert peers who are familiar with the activities, the service needs, and the attributes. It is good practice to involve an institution with expertise in testing (such as a university or national testing authority) in the development of the test.

Accreditation and credentialing are the processes of assessing organizations, institutions, processes, or individuals against the defined criteria. Here, we use accreditation for the assessment of organizations and institutions, whereas credentialing will apply to individuals and processes, such as the capacity to participate in a particular clinical trial (Ibbott et al. 2008). In this chapter, the credentialing of professionals will be the focus.

Accreditation and credentialing have the same etymological root: "credere" is Latin and translates as "trust, believe," according to http://www.etymonline.com (accessed May 7, 2009). This

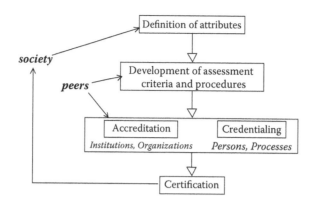

FIGURE 31.1   Relation between credentialing and certification.

illustrates the role of credentialing as a method to establish trust between two parties, here, the professional and society. The *Oxford English Dictionary* defines credential as "evidence of a person's achievements, trustworthiness, etc., usu. in the form of certificates, references, etc." (*Australian Pocket Oxford Dictionary* 2007).

There are different models of how credentialing can be administered. It can be managed by a professional organization that also sets the standards and criteria against which an assessment is performed. This is common practice in the case of medical professions where a college admits persons who have passed the examination process as members or fellows (see http://www.rcr.ac.uk or http://www.ranzcr.edu.au/ for comparison). To avoid conflict of interests, it is possible for the professional organization to set up an independent entity that administers all credentialing and accreditation matters. An example is the Canadian College of Physicists in Medicine (CCPM), which operates independently of the Canadian Organization of Medical Physicists (COMP), despite sharing many other resources (http://www.medphys.ca/). Alternatively, an entirely independent body performs the assessments, usually on a fee-for-service basis. In the case of radiation oncology professionals, this is not common as

there are too few professionals to be examined to make this a commercially viable option.

Certification refers to the award of a designation that provides evidence of the knowledge, skills, or competencies that were assessed. Certification is most commonly awarded by a professional board that is independent of the professional organization itself (Pusey et al. 2005; Thomas, Hendee, and Paliwal 2005; Brady, del Regato, and Levitt 1980). A good example is the board certification for medical physicists by the American Board of Radiology (ABR) in the United States (http://theabr.org/ic/ic_rp_landing.html). Certification is the information that members of society can expect and ask to see when requiring a particular service.

Elements of Credentialing of Radiation Oncology Professionals

Credentialing must examine a large number of aspects for each of the members of the radiation oncology treatment team. This is illustrated in Table 31.1, which shows the typical workflow in a radiotherapy department. Of relevance for the credentialing

TABLE 31.1   Analysis of Workflow and Functions of Staff in Radiotherapy

	Steps in the Radiotherapy Process	Staff Involved	Communication Pathways
1	Clinical evaluation	RO, multidisciplinary team	
2	Therapeutic decision	RO	
3	Patient immobilization	RT	
4	Imaging for treatment planning	RO, MP, RT	
5	Target and critical structures localization	RO, RT	
6	Treatment planning	MP, RT, RO	
7	Simulation/verification of treatment plan	RT, MP, RO	
8	Treatment delivery	RT	
9	Evaluation during treatment	RT, RO	
10	Follow-up after treatment	RO, multidisciplinary team	
	Performance of equipment	MP, engineer	

Source: IAEA, *Setting Up a Radiotherapy Programme STI/PUP 1296*, International Atomic Energy Agency, Vienna, 2008.
Note: RO, radiation oncologist/clinician; MP, medical physicist; RT, radiation therapist/technologist/RTT.

of staff is the involvement in the various aspects of work and the need to communicate effectively with others.

There are typically three elements that need to be addressed in a credentialing process that assesses a professional:

1. Theoretical knowledge, such as that gained in a university degree. This allows professionals to understand the processes and competently communicate with others. Curricula for this are widely available from organizations such as the American Association of Physicist in Medicine (AAPM 2002) and used by universities. This theoretical knowledge is easy to document and assess.
2. Practical/clinical skills and experience that can only be acquired in practical work under the supervision and guidance of an experienced colleague. In the past, this has often consisted of on-the-job or road learning. However, it has now been recognized that clinical training is an essential part of professional training for all professionals, including medical physicists. Clinical training guidelines are usually provided by professional organizations (AAPM 2006).
3. General professionalism as demonstrated in communication skills, work ethos, and general ethical behavior. This is more difficult to assess; however, many credentialing schemes include a requirement for a scientific presentation or other documentation of effective communication of one's work (e.g., reports or publications).

Requirement for Certification and Registration

As it is usually impossible for someone not familiar with the field to judge the credentialing process and its outcomes, it is important to summarize them in an easy, accessible form. This is where certification becomes useful.

In professional areas where public safety is at stake and strict control of professional activities is necessary, it is common to register professionals. This is the case for most medical professions, and also increasingly common for other professionals in radiation oncology, such as medical physicists. Registration is also often linked to the professional holding a license granted by a regulatory authority representing the interests of society. This system of licensure has been successfully used in many professional fields, for example, in radiation protection (IAEA 1996).

Figure 31.2 illustrates the process based on the scheme shown in Figure 31.1. Requirements for registration would typically include certification and, therefore, credentialing by a professional body. Being on a register typically also requires the professional to abide by a code of ethics (AAPM 2009). Governments may also allow other avenues for persons to be listed in a register. In particular, overseas qualifications and certification may need to be considered as being appropriate for at least temporary inclusion of a professional on a register in order to manage workforce issues in fields where qualified staff is difficult to get.

Stakeholders

There are four major groups of stakeholders in the credentialing and certification process:

1. Society at large and, particularly, patients who require the services—They have to trust the professionals and institutions that are involved in their diagnosis and treatment (and may need to have criteria to select them in the first place). Unfortunately, it is difficult to ascertain the qualifications and skills of professionals and patients have to rely on certifications to identify suitably qualified individuals and institutions.
2. Regulatory authorities, government, and insurers—They need a mechanism to identify persons and institutions who can provide appropriate services in order to inform the public and provide financial support. A register is a suitable method to identify persons with certain skills, and placement of individuals on a register requires (ongoing) credentialing/certification and a commitment to appropriate ethical behavior.
3. Health organizations, employers, and administrators—They use the skills and abilities of professionals to provide a service to patients (Street and Thomson 2008). As such, they rely on a method to identify suitable professionals. Credentialing also provides a tool to determine remuneration and status within an organization.
4. Professionals and practitioners—They are involved as the persons being credentialed and certified, as well as the ones who develop the schemes to do so. As such, credentialing is essentially a structured peer review process.

Quality and Credentialing and Certification

Credentialing and certification obviously identify persons and institutions as being able to provide services of an acceptable quality. However, the very nature of needing credentialing/certification results in an improvement of the quality of a service. There are direct and indirect influences on quality in radiation oncology. Several of them are illustrated in Figure 31.2:

1. The definition of important attributes of professionals or organizations is an opportunity to reflect on services and determine which are important and which are not. The process is likely to yield a syllabus and list of required competencies.
2. The assessment process will require candidates to prepare and, thus, improve their knowledge and skills. Some credentialing processes are linked to an education program that will provide further improvements in staff expertise and quality.
3. Certification and registration are normally only for a limited period. In fields that rely on technology and are

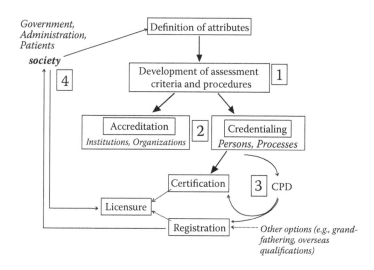

FIGURE 31.2 System of registration for professionals. Opportunities for quality improvement in the context of credentialing and certification are identified by numbers that are discussed in more detail in the text.

subjected to rapid change, continuing professional development (CPD) is essential (Street and Thomson 2008; Perkins and Kron 2007). It is common to make participation in CPD activities a requirement for renewal of certification and registration. Although CPD is an essential component of maintenance of certification (MOC), it also involves performance in practice and professional standing (Thomas, Hendee, and Paliwal 2005), all important aspects of quality services (Kun et al. 2007; Holmboe et al. 2008).

4. The role of society is of particular importance in terms of quality. By making certification compulsory, society indicates to professionals (and persons affected by cancer) that qualifications and expertise matter. As the needs of society and patients define the criteria for credentialing, there is also a gain in quality through a clearer definition of needs and expectations.

In addition to these obvious drivers of improved quality, there are a few more subtle ways in which credentialing and certification can improve the quality of radiotherapy services. Improving the status of credentialed professionals is one of the most important aspects of this.

Improving the Status of Professionals in Medical Physics

Although the status of medical practitioners is generally high, this is not necessarily the case for other professionals in radiation oncology. In particular, medical physicists with limited direct patient contact and, therefore, limited visibility, often have difficulties being recognized. Credentialing provides administrators with a list of required competencies and the ability to identify suitable individuals for a given task easily. As such, it conveys status because it provides professionals with an independent assessment

of their ability to perform a task. It also gives confidence to administrators that they have hired the right person and possibly even creates a bottleneck that makes certified individuals more difficult to find and, therefore, more valuable for employers.

In many cases, certification provides the basis for increasing salaries as demonstrated by the AAPM salary surveys of medical physicists. They consistently identify approximately 20% higher salaries for certified medical physicists compared to their noncertified peers. This, in turn, will attract better graduates to the professions and improve the overall quality of services to patients.

Role of International Organizations

Setting up a credentialing program is not a simple process. It requires organizational infrastructure, expertise, and a sufficient number of professionals who are willing to donate time and effort to setting up the procedures. In many countries, this is not easily achievable for some of the smaller professional groups in the workforce, and international organizations such as the World Health Organization (WHO) or the International Atomic Energy Agency (IAEA) have an important role to play (Zaidi 2008). These organizations also provide education through workshops that bring together groups of individuals (Coffey et al. 2004).

In particular, the IAEA has expended considerable resources in the past 10 years to develop recommendations for setting up a radiotherapy program (IAEA 2008). Although the recent Report 1296 does not specifically mention credentialing, it specifies what constitutes qualified staff. For example, "Medical physicists practicing in radiotherapy (or radiation oncology) must be qualified as physicists with academic studies in medical physics (typically at the postgraduate level) and clinical training in radiotherapy physics" (IAEA 2008). This is identical to the elements for credentialing discussed above. The IAEA continues to clarify that ... "[clinical] training should preferably be approved by a suitable professional body, i.e., a Board that will issue a clinical certification" (IAEA 2008).

The IAEA also offers a comprehensive audit program for radiotherapy services (IAEA 2007) resulting in recommendations for a center and possibly a commendation as a center of competence. In line with other IAEA documentation (IAEA 2008), the audit verifies that qualified staff is available and ascertains the type of professional certification of the staff. In addition, the audit determines availability and participation of individuals in professional education.

Finally, international organizations can help to secure resources required to educate professionals to a suitable level for credentialing and certification. WHO, IAEA, and the International Union against Cancer (UICC, http://www.uicc.org/) have such educational programs. Most recently, the IAEA has also started to get industry involved in this process. The Programme of Action for Cancer Therapy (PACT, http://cancer.iaea.org/documents/impactTOR.pdf) provides for a number of initiatives that include knowledge transfer and improvement of status for professionals.

Summary

Credentialing is the process of independent verification of competencies, skills, and qualifications of professionals or organizations, which often results in a formal certification. In radiation oncology, credentialing is usually a peer review process that plays an important part in quality systems, as it assures that training and capabilities match the requirements for a given task. It also provides individuals and organizations with extra status that creates trust between patient and provider and helps to attract and maintain the best staff. Credentialing and certification are inherently methods of quality improvement, because they require verifiable skills and participation in ongoing training and CPD.

Credentialing and certification are important parts of any quality program in radiation oncology. They provide incentives for individuals to improve their practice and increase the status of professionals who have undergone the process. Certification also constitutes a valuable link between professionals and society as it allows the latter to identify qualified persons with ease.

References

American Association of Physicists in Medicine (AAPM). 2002. *Academic Program Recommendations for Graduate Degrees in Medical Physics*. Madison, WI: AAPM.

AAPM. 2006. *Essentials and Guidelines for Hospital-Based Medical Physics Residency Training Programs: Report of the Subcommittee on Residency Training and Promotion of the Education and Training of Medical Physics Committee of the AAPM Education Council*. Madison: AAPM.

AAPM. 2009. *Code of Ethics for the American Association of Physicists in Medicine: Report of AAPM Task Group 109*. Madison, WI: AAPM.

Brady, L. W., J. A. del Regato, and S. H. Levitt. 1980. The board examinations in therapeutic radiology. *Int. J. Radiat. Oncol. Biol. Phys.* 6 (7): 951–952.

Coffey, M., J. Degerfalt, A. Osztavics, J. van Hedel, and G. Vandevelde. 2004. Revised European core curriculum for RTs. *Radiother. Oncol.* 70 (2): 137–158.

Holmboe, E. S., Y. Wang, T. P. Meehan, J. P. Tate, S. Y. Ho, K. S. Starkey, and R. S. Lipner. 2008. Association between maintenance of certification examination scores and quality of care for medicare beneficiaries. *Arch. Intern. Med.* 168 (13): 1396–1403.

International Atomic Energy Agency (IAEA). 1996. *International Basic Safety Standards for Protection against Ionizing Radiation and for the Safety of Radiation Sources*. Vienna: IAEA.

IAEA. 2007. *Comprehensive Audits of Radiotheraphy Practices: A Tool for Quality Improvement—QUATRO*. Vienna: IAEA.

IAEA. 2008. *Setting Up a Radiotherapy Programme STI/PUP 1296*. Vienna: IAEA.

Ibbott, G. S., D. S. Followill, H. A. Molineu, J. R. Lowenstein, P. E. Alvarez, and J. E. Roll. 2008. Challenges in credentialing institutions and participants in advanced technology multi-institutional clinical trials. *Int. J. Radiat. Oncol. Biol. Phys.* 71 (1 Suppl): S71–S75.

International Standards Organisation (ISO). 2008. *ISO 9000: 2000 Family of Standards on Quality Management*. Geneva: ISO.

Kun, L. E., B. G. Haffty, J. Bosma, J. L. Strife, and R. R. Hattery. 2007. American Board of Radiology Maintenance of Certification—Part IV: Practice quality improvement for radiation oncology. *Int. J. Radiat. Oncol. Biol. Phys.* 68 (1): 7–12.

Leer, J. W., R. Corver, J. J. Kraus, J. C. vd Togt, and O. J. Buruma. 1995. A quality assurance system based on ISO standards: Experience in a radiotherapy department. *Radiother. Oncol.* 35 (1):75–81.

Oxford University Press. 2007. *Australian Pocket Oxford Dictionary 2007*. Melbourne: Oxford University Press.

Perkins, A., and T. Kron. 2007. Continuing professional development needs of Australian radiation oncology medical physicists—An analysis of applications for CPD funding. *Australas. Phys. Eng. Sci. Med.* 30 (3): 226–232.

Pusey, D., L. Smith, E. M. Zeman, and R. Adams. 2005. A history and overview of the certification exam for medical dosimetrists. *Med. Dosim.* 30 (2): 92–96.

Street, M., and K. Thomson. 2008. Credentialing for radiology. *Biomed. Imag. Intervention J. (biij)* 4 (1).

Thomas, S. R., W. R. Hendee, and B. R. Paliwal. 2005. The American Board of Radiology Maintenance of Certification (MOC) program in radiologic physics. *Med. Phys.* 32 (1): 263–267.

Zaidi, H. 2008. Medical physicis in developing countries: Looking for a better world. *Biomed. Imag. Intervention J. (biij)* 4.

32

Approach to Radiation Oncology Practice Accreditation

Gregory W. Cotter
The University of South
Alabama College of Medicine

Introduction

Accreditation of medical services is based on the need or desire to demonstrate an identified level of medical care. This need may be driven by a variety of reasons. Some governmental agencies may require accreditation for licensure or other regulatory purposes. Governmental and private third-party payers for healthcare services may desire to reference payment to standardized care. Also, a medical practice may desire to seek accreditation to demonstrate an identified level of care to its patients or community.

History of the ACRO PAP

The American College of Radiation Oncology Practice Accreditation Program (ACRO PAP) was initiated in 1996 as a service to ACRO members. The goal of the ACRO PAP is to provide a method of assessing quality in the practice of radiation oncology. Quality of care is defined as the degree to which health services for individuals and populations increase the likelihood of desired health outcomes and are consistent with current professional knowledge (Lohr 1990).

Inherent in this definition is the public health aspect of medical care. This is an important aspect because substantial amounts of healthcare services are funded through governmental sources in the United States and other countries today. The reality in this acknowledgment is that medical care can only be delivered to a population within the economic limits of the funding source.

Accreditation Assessment Methods

The method of assessment of quality of care is important to the value of the recognition or accreditation at the completion of the process. In this regard, the methods of assessment should provide a reasonable review of aspects relevant to patient care. The Institute of Medicine (IOM) and the National Research Council established the National Cancer Policy Board (NCPB) in 1997 (Hewitt and Simone 1999). One of the questions tasked to the NCPB was to address the question, "What is quality cancer care and how is it measured?" The NCPB outlined a method for quality assessment of cancer care. "Quality assessment is the measurement of quality by expert judgment (implicit review) or by systematic reference to objective standards (explicit review)." Although implicit review is used as an accrediting method, its supplementation by explicit review allows for a broader and perhaps more meaningful assessment to be performed. The ACRO PAP uses both types of review.

The explicit review method provides a systematic approach to quality assessment and incorporates three dimensions of assessment. The three dimensions of assessment are structure, process, and outcome. "Structural quality" refers to health system characteristics, "process quality" refers to what the provider does, and "outcome quality" refers to the patient's ultimate health. Structural quality alone, although often easy to assess, is not an adequate measure of quality of care. Outcome quality is the best measure of quality; however, adequate outcome data are often lacking for a variety of reasons. As such, process quality is often used as a surrogate or proxy for assessing quality of care.

Process quality relies on technical quality and interpersonal quality. "Technical process" can be measured according to appropriateness criteria, practice guidelines, or professional standards. Evidence-based practice guidelines for patient care are often useful for assessment, particularly when they are linked to a defined outcome. "Interpersonal quality" refers to whether the care is provided to the patient in a humane manner. Interpersonal quality is often evaluated using patient surveys. Assessment of interpersonal quality may include issues such as whether the patient was provided with sufficient information to make an informed decision regarding his/her medical care. In the radiation oncology setting, the patient comes in contact with a variety of personnel providing care. Assessment of the aspects of nonphysician care may be appropriate and useful to the practice.

Standards in Radiation Oncology

Initiatives to define standards for radiation oncology treatment date back to at least the 1950s. In 1950, the National Cancer Institute of Canada published a booklet entitled *Minimum Standards of Radiation Therapy Centres* (Walton 1973). This was followed in 1957 with a revision entitled *Standards for Radiation Therapy Centers Recommended by the National Cancer Institute of Canada.*

In the United States, the Committee for Radiation Therapy Studies (CRTS) was formed in 1959 through the combined efforts of the National Cancer Institute (Wilson et al. 1995) and the Radiation Study Section (Kramer 1973). In 1967, the CRTS submitted a paper to the National Cancer Institute (NCI) in which the requirements for major and satellite cancer centers were outlined. In 1968, the CRTS submitted another report to the NCI entitled *A Prospect for Radiation Therapy in the United States.* This report, also referred to as the "blue book," described in some detail the current practice of radiation therapy as well as staffing and facility requirements. The "blue book" was periodically updated and was last published in 1991. It is now out of print.

Other initiatives have strived to develop guidelines or standards for medical practice including radiation therapy. Debate continues as to how best to characterize those elements that form the basis of current mainstream medical practice. Although the terms "guideline" and "standard" are often used interchangeably, there are significant differences between them. The term "guideline" is defined as a principle by which to determine a course of action (Webster's New World Dictionary 1990). The term "standard" is defined as something established for use as a rule or basis of comparison in measuring quality. In essence, a guideline points us in the right direction, whereas a standard is our ultimate goal.

Both terms have utility in trying to define acceptable practice parameters and are not mutually exclusive. Another useful concept is "scope of practice." This term perhaps begins the effort to define acceptable practice parameters by limiting or defining the realm of medical practice that one is focusing on. The process of first defining the scope of practice and next developing or identifying meaningful practice guidelines in the defined area of

interest hopefully leads to the establishment of acceptable standards for medical care, in this case for radiation therapy.

Different methods of developing guidelines and standards in medical practice have been used. Two common methods are expert opinion and consensus. Neither method, unto itself, ensures that the guideline or standard developed will, in fact, achieve the desired level of credibility. Another method involves surveying the practice patterns of physicians as performed in the Patterns of Care Study supported by the National Institutes of Health (Kramer 1977; Kramer and Herring 1976). Although information of the latter type is a useful profile of current practice, it does not, unto itself, establish scientifically proven standards of care. For this reason, The Patterns of Care Study incorporated Donabedian's model of quality of care assessment to try to compare practice patterns to quality of care endpoints (Donabedian 1966).

To improve credibility in guideline or standard development, some authors have incorporated evidence-based guidance into their process. The term evidence-based medicine first appeared in the medical literature in 1991 (Guyatt 1991). Clinical practice guidelines (CPGs) are defined as systematically developed statements aimed to assist in healthcare decisions made by clinicians, policymakers, and patients for specific clinical circumstances (Browman et al. 1995). This concept was developed by clinical epidemiologists at McMaster University in Canada beginning in the 1970s (Guyatt and Rennie 2002). It led to the formation of the international, evidenced-based Medicine Working Group, whose work was subsequently published in the *Journal of the American Medical Association* (Evidence-Based Working Group 1992). Today, various groups have published evidence-based guidelines and standards in the area of oncology and more specifically in the field of radiation oncology.

Governmental Requirements

In addition to the guidelines and standards in the field of radiation oncology, governmental regulations play an important role in defining acceptable practices in the field of medicine today. Examples of this are regulations applying to health and safety, persons with disabilities, patient privacy, and safety codes.

Current Status of ACRO PAP

Today, radiation oncology can be practiced in a variety of settings. These settings include the cancer center model, either hospital-based, freestanding, or independent of other specialties. The ACRO PAP recognizes the variety of settings of radiation oncology and also the combined modality approach used to treat many patients. As such, the ACRO PAP includes not only a review of the aspects of radiation oncology practice, but also a review of pharmacologic adjunctive and supportive therapy and oncologic imaging. The ACRO PAP process consists of the following steps:

- Application process. The practice seeking ACRO PAP accreditation will initially communicate their interest to the ACRO. The ACRO will then send an application form to the practice. Once the ACRO receives the application form and fee, the ACRO PAP office will be contacted to initiate the accreditation process.

- Survey process. The ACRO PAP Office will then send online data entry instructions to the practice. In addition, specific data will be requested from the practice for review. Requested data include personnel, facility and equipment information, physics data, and patient treatment information, including images related to treatment.
- Review process. The survey information and practice data are then turned over to the assigned reviewers. The reviewers may request additional information of the practice. After completion of the review, the reviewers return their findings and recommendation(s) to the Practice Accreditation Committee. Once the initial off-site review is completed, a site verification visit is performed.
- Facility accreditation. After completion of the review process, the ACRO PAP will inform the practice of its findings. If the practice meets the requirements of the ACRO PAP, accreditation will be issued by the ACRO. Accreditation will normally be for a period of 3 years.

If minor issues are noted, the practice may receive a provisional (probationary-type) accreditation with a defined time to address the issue(s). Once the issues are resolved by the practice, then full accreditation may be received. If major issues are identified during the ACRO PAP review process, accreditation may be deferred, allowing the practice to more fully address the issues. As the identified issues are resolved, the practice may then move forward through the remaining parts of the review process. During these steps in the ACRO PAP process, the following aspects of the practice are reviewed.

Practice Demographics

During the ACRO PAP review, demographics of the practice will be examined to help define the nature of the patients treated and the services offered. Requested demographic aspects of the practice include all relevant aspects of how the practice is functioning (i.e., equipment type and number, type of diseases treated, etc.).

Process of Radiation Therapy

The process of radiation therapy treatment consists of a series of steps (Figure 32.1). In the case of external beam radiation therapy, these steps typically follow in a logical order. When brachytherapy is used, the sequence is similar but may be more or less complicated depending on the specific type of treatment.

The process of radiation treatment within the practice will be evaluated for appropriateness of care. A semirandom sample of patient care medical records will be requested for off-site review and additional patient medical records will be evaluated at the time of the on-site review. Items to be checked include: consultation, informed consent, and treatment planning. Also included are aspects of simulation, dose calculation, treatment aids, radiation treatment delivery, and treatment verification. Radiation treatment management, follow-up medical care, and

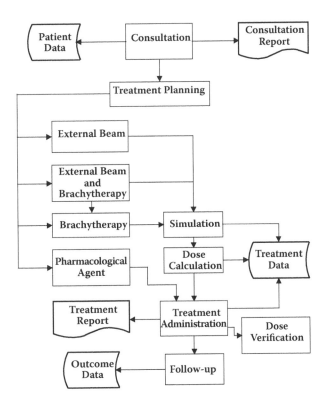

FIGURE 32.1 The process of radiation therapy.

clinical performance measures are included as part of the medical review. Similarly, all aspects of brachytherapy are included where appropriate for a given practice.

Facilities

During the practice review, the facilities are scrutinized to determine if patient care is being given in a reasonable manner consistent with applicable laws, regulations, and standards. Aspects of facility review include the following: parking availability, accessibility, waiting area(s), reception/business areas, restrooms, examination rooms, simulation areas, treatment planning/physics/dosimetry areas, megavoltage treatment room(s), treatment aid fabrication areas, and offices.

The practice should demonstrate compliance with the applicable rules of the Americans with Disabilities Act (ADA), the Health Insurance Portability and Accountability Act of 1996 (HIPAA), Occupational Safety and Health Administration (OSHA), and local fire codes.

Radiation Therapy Personnel

As noted above, the process of radiation therapy consists of a series of steps and involves a number of different individuals. Each practice should establish a staffing program consistent with patient care, administration, research, and other responsibilities. It is recognized that talent, training, and work preferences may vary from individual to individual. It is appropriate to factor these aspects into the staffing program. General staffing recommendations are outlined in Table 32.1.

Radiation Therapy Equipment

Today, there are various types of megavoltage radiation therapy units used in clinical treatment. The equipment for the general practice of radiation oncology should include the following: megavoltage radiation therapy equipment for external beam therapy, electron beam or superficial x-ray equipment suitable for treatment of superficial (e.g., skin) lesions, simulator capable of duplicating the treatment setups (fluoroscopic simulation, CT, or CT-Sim capability is desirable), brachytherapy equipment for intracavitary and interstitial treatment, computer dosimetry equipment, physics calibration devices for all equipment, and treatment aides.

Radiation Therapy Physics

The radiation therapy physics programs should include a radiation safety program, radiation room surveys, radiologic equipment licensure/registration, brachytherapy licensure/registration, a radiation exposure monitoring program, major equipment operating procedures, major equipment records, radiation safety and quality assurance (QA) procedures, morning and monthly QA procedures for all radiotherapeutic or radiologic equipment, as well as annual calibrations for all radiotherapeutic equipment. Furthermore, each practice must demonstrate a dosimetry reference for physics calibration purposes. Physics calibration equipment, treatment planning, and treatment quality assurance should be demonstrated. Similar aspects must be displayed for a brachytherapy program, as well as any other special procedures such as intensity modulated radiation therapy (IMRT), stereotactic radiosurgery (SRS), total body irradiation (TBI), etc.

Oncologic Imaging

A variety of imaging modalities are used in the care of oncologic patients. Those imaging modalities common to radiation oncology practices include: ultrasound, computerized tomography (CT), CT-Simulation (CT-Sim), magnetic resonance imaging (Thomadsen et al. 2003), and positron emission tomography

TABLE 32.1 Staffing per Annual Number of New Patients, 8 Hours per Day, Five Days per Week

Personnel	Staffing Level
Radiation oncologists	1 per 200–300
Medical physicists	1 per 300–400 (25% IMRT)
Dosimetrists	1 per 200–250 (25% IMRT)
Nurses (RNs)	1 per 200–300
Radiation therapists (RTTs)	1 per 100–150 (minimum of 2, 25% IMRT)
Simulation staff	1 per 200–250
Brachytherapy staff	As needed
Clerical staff	At least 1 per 200 patients
Treatment aides	As needed
Maintenance/service staff	By contract or 1 per 3–4 megavoltage units, CT, PET/CT, or MRI units
Dietician	
Physical or rehabilitation staff	As needed
Social worker	As needed

(PET and PET/CT). Those practices that provide oncologic imaging should have adequate facilities to safely care for patients.

Pharmacologic Adjunctive and Supportive Therapy

The use of pharmacologic modifying, adjuvant, and supportive agents in conjunction with radiation treatment is well established. Furthermore, research continues in this area with the primary goal being improvement in patient care. A wide variety of pharmacologic modifying, adjunctive, and supportive agents are available for general clinical use or use in the investigational setting. These agents can be categorized as follows: radiation protectors, radiation sensitizers, hormonal agents, cytotoxic agents, photosensitizers, immunologic agents, biologic agents, molecular therapeutic agents, antiemetics, pain medications, and general medications. Those practices that administer pharmacologic modifying, adjuvant, and supportive agents should have adequate facilities to safely care for patients.

Continuous Quality Improvement Program

The practice shall have a continuous quality improvement (CQI) plan. This may be combined with the radiation safety program. The following items should be included in a CQI program: chart review, general practice review, physics review, dose discrepancy review, new procedure review, incident report review, morbidity and mortality review, outcome studies review, physician peer review, and record maintenance and data collection.

Safety Program

The provision of a safe environment for patients, staff, and the public is essential for each practice. The practice shall demonstrate that it provides safety measures.

Education Program

Continuing medical education (CME) programs are required for physicians and physicists as well as the physics, dosimetry, nursing, and radiation therapy technology staffs. This program shall include access to information as appropriate for each individual's responsibilities that are pertinent to safe operation of all equipment within the facility.

Summary

In summary, the ACRO PAP covers all aspects of a radiotherapy practice. A review of a practice is comprehensive and time-consuming, but plays a crucial role in ensuring quality.

References

Browman, G. P., M. N. Levine, E. A. Mohide, R. S. Hayward, K. I. Pritchard, A. Gafni, and A. Laupacis. 1995. The practice guidelines development cycle: A conceptual tool for practice guidelines development and implementation. *J. Clin. Oncol.* 13 (2): 502–512.

Donabedian, A. 1966. Evaluating the quality of medical care. *Milbank Mem. Fund Q.* 44 (3) (Suppl): 166–206.

Evidence-Based Medicine Working Group. 1992. Evidence-based medicine. A new approach to teaching the practice of medicine. *JAMA* 268 (17): 2420–2425.

Guyatt, G., and D. Rennie, eds. 2002. *User's Guide to the Medical Literature: A Manual for Evidence-Based Clinical Practice.* The Evidence-Based Medicine Working Group. Chicago, IL: AMA Publisher.

Guyatt, G. H. 1991. Evidence-based medicine. *ACP Journal Club* 114: A16.

Hewitt, M., and J. V. Simone, eds. 1999. *Ensuring quality cancer care. Institute of Medicine and Commission on Life Sciences, National Research Council.* Washington, DC: National Academy Press.

Institute of Medicine (IOM). 1999. *To Err Is Human: Building a Safer Health System.* Washington, DC: National Academy Press.

Kramer, S. 1973. Radiation therapy and the cancer center. In *Frontiers of Radiation Therapy and Oncology*, ed. J. M. Vaeth. Baltimore, MD: University Park Press.

Kramer, S. 1977. The study of the patterns of cancer care in radiation therapy. *Cancer* 39 (2 Suppl): 780–787.

Kramer, S., and D. F. Herring. 1976. The patterns of care study: A nationwide evaluation of the practice of radiation therapy in cancer management. *Int. J. Radiat. Oncol. Biol. Phys.* 1 (11–12): 1231–1236.

Lohr, K., ed. 1990. *Medicare: A Strategy for Quality Assurance*, vol. 1. *Committee to Design a Strategy for Quality Review and Assurance in Medicine, Institute of Medicine.* Washington, DC: National Academy Press.

Thomadsen, B., S. W. Lin, P. Laemmrich, T. Waller, A. Cheng, B. Caldwell, R. Rankin, and J. Stitt. 2003. Analysis of treatment delivery errors in brachytherapy using formal risk analysis techniques. *Int. J. Radiat. Oncol. Biol. Phys.* 57 (5): 1492–1508.

Walton, R. J. 1973. Cancer centres. In *Frontiers of Radiation Therapy and Oncology*, ed. J. M. Vaeth. Baltimore, MD: University Park Press.

Webster's New World Dictionary. 3rd ed. 1990. New York, NY: Warner Books.

Wilson, R. M., W. B. Runciman, R. W. Gibberd, B. T. Harrison, L. Newby, and J. D. Hamilton. 1995. The quality in Australian health care study. *Med. J. Aust.* 163 (9): 458–471.

Clinical Trials: Credentialing

Geoffrey S. Ibbott
University of Texas

Introduction

The U.S. healthcare system has been criticized publicly recently for its high cost and for the variations in quality observed among facilities and among services available to different patients. Numerous organizations including Congress, the Joint Commission, and the American Board of Medical Specialties (ABMS) have proposed changes intended to improve quality and control costs. This, of course, is not the first time that recommendations have been made for improvements in U.S. healthcare. In the early 1900s, a professor at the Carnegie Foundation, named Abraham Flexner, studied professional education, including the medical profession. He published what is now known as the *Flexner Report* on medical education in the United States and Canada. The report contained the following statement condemning the quality of medical education and the broad disparity in quality of the services provided:

> We have indeed in America medical practitioners not inferior to the best elsewhere; but there is probably no other country in the world in which there is so great a distance and so fatal a difference between the best, the average, and the worst. (Flexner 1910)

A report published in 2000 by the Institute of Medicine highlighted the frequency of patient deaths from medical errors and alerted the U.S. public to the need for quality assurance (QA) (Kohn, Corrigan, and Donaldson 1999). Unlike some specialties, however, QA is well established in radiation therapy, resulting, at least in part, from the field's quantitative nature.

In recent years, a lot of media attention has been given to several significant radiation therapy errors (Donaldson 2008; Dunscombe, Lau, and Silverthorne 2008; Rogers 2009). Possible reasons for such errors include the introduction of, and increasing dependence on, advanced technologies (Barthelemy-Brichant et al. 1999; Macklis, Meier, and Weinhous 1998; Huang et al. 2005; Perrow 1999). At least one study suggested that the introduction of advanced technology equipment has permitted errors to occur that might otherwise have been detected (Patton, Gaffney, and Moeller 2003). There are also indications that the demands of advanced technologies on department resources have drawn resources from simpler or basic functions (Dunscombe, Lau, and Silverthorne 2008; Marks et al. 2007).

In the United States today, efforts are underway to improve and standardize the qualifications of physicians and medical physicists. Member boards of the ABMS, including the American Board of Radiology (ABR), have introduced changes to their examination procedures and have implemented programs to assure maintenance of certification (MOC). Beginning in 2014, in response to a resolution passed by the American Association of Physicists in Medicine (AAPM), the ABR will require that medical physicist candidates be enrolled in, or have graduated from, a residency program accredited by the Commission on Accreditation of Medical Physics Educational Programs (CAMPEP). This requirement will make medical physics education more equivalent to medical education and will assure that physics training is consistent with that required by other boards of the ABMS (Herman et al. 2008).

Terminology

In the United States, the terms *credentialing, certification*, and *accreditation* are understood to have quite specific meanings. *Credentialing* simply refers to the awarding of a *credential*, and can be applied to a number of activities. Credentialing is often required for participation in radiation therapy clinical trials, as will be discussed later. *Certification* refers to individuals (e.g., physicians and medical physicists) and recognizes the achievement of a qualification. *Accreditation* refers to organizations and is a mechanism for recognizing compliance with established standards.

Providers of Credentialing Programs

As was suggested earlier, the ABR plays an important role in establishing the qualifications of radiologists, radiation oncologists, and medical physicists. Today, ABR certification is recognized as an important credential for physicists with clinical responsibilities. The ABR recognizes that the quality of education and training is a key indicator of the likelihood of eventual certification and requires that physician candidates receive their training in accredited programs. For radiologists and radiation oncologists, accreditation of the training program by the Accreditation Council for Graduate Medical Education is required. For medical physicists, accreditation of the training program by CAMPEP will soon be required.

Purpose of Credentialing for Radiation Therapy Clinical Trials

Cooperative groups, such as the Radiation Therapy Oncology Group (RTOG), conduct U.S. national clinical trials that involve the use of radiation therapy through the National Cancer Institute (NCI). In preparation for such a trial, the cooperative group prepares a protocol to define the goals of the trial, the rationale for its design, and the details of the treatment procedure to be followed.

Institutions that deliver radiation therapy to patients in NCI-sponsored clinical trials are expected to participate in the Radiological Physics Center (RPC)'s various auditing programs. These include annual monitoring of treatment machine output with thermoluminescent dosimeters (TLDs) (Ibbott et al. 2008); on-site dosimetry reviews of radiation beam parameters, calculation methods, and QA procedures; and treatment record reviews. At least five organizations conduct regular remote audits of treatment machine output calibration with mailed dosimeters. In addition to the RPC, such programs are conducted by

Radiation Dosimetry Services (also located at M. D. Anderson Cancer Center), the International Atomic Energy Agency (IAEA) based in Vienna, the European Quality Assurance Laboratory (EQUAL) located in Paris, and the Section of Outreach Radiation Oncology and Physics at the National Cancer Center of Japan, in Tokyo. Of these five centers, the RPC has the largest program and monitors all of the institutions (1768 institutions, as of late 2009) that participate in clinical trials, both within the United States and internationally. The RPC initiated its TLD program for photon beams in 1977 (Kirby et al. 1986; Kirby, Hanson, and Johnston 1992). In 1982 electron beams were included and, in 2007, audits of proton beams were initiated.

An on-site audit has been recommended by several organizations, including the AAPM and the IAEA (Halvorsen et al. 2005; IAEA 2007). An independent audit is especially important for solo practitioners, but is a valuable exercise for all practicing clinical medical physicists. According to the AAPM, it need not be extensive but should address key activities such as basic calibrations, the overall QA program, and documentation. The RPC's on-site dosimetry audit is, in fact, comprehensive and includes the collection of a large amount of beam data, the comparison of measurements and dose calculations by the institution, the review of treatment records to determine the consistent use of dosimetry methods, and a review of QA documentation.

Clinical trials requiring the use of advanced technologies, such as intensity-modulated radiation therapy (IMRT) and prostate brachytherapy, are often considered sufficiently challenging that institutions are required to demonstrate their ability to use these technologies before being permitted to register patients. This process is called credentialing and may consist of several steps.

The RPC participates in the credentialing process for a number of clinical trials through several cooperative groups. In many cases, the RPC collaborates with one or more of the three other NCI-funded QA offices: the RTOG headquarters dosimetry group, the Quality Assurance Review Center (QARC), or the Image-Guided Therapy QA Center (ITC). The four QA offices together form the Advanced Technology Consortium (ATC), for which funds are provided through a separate NCI grant. The ATC allows the four QA offices to collaborate where practical, to share data through several methods, and to avoid duplication of effort whenever possible.

Components of a Credentialing Program for Clinical Trials

Credentialing for cooperative group clinical trials generally involves most or all of the procedures (see Table 33.1) described in the following subsections.

Previous Patients Treated with Technique

Institutions must demonstrate that they are familiar with the advanced technology being used by the trial and attest to having used it to treat at least some minimum number of patients.

TABLE 33.1 Steps in Credentialing for Clinical Trials by the RPC

1. Attestation that previous patients have been treated with the technique
2. Completion of facility questionnaire
3. Completion of knowledge assessment questionnaire
4. Submission of a treatment planning benchmark or irradiation of an anthropomorphic phantom
5. Demonstration of ability to submit treatment planning data digitally
6. Review of dosimetry system and QA procedures
7. Clinical review by a radiation oncologist

Facility Questionnaire

This questionnaire asks institutions to describe relevant aspects of their treatment planning and delivery equipment, their QA procedures, and, in some cases, the personnel who will be participating in protocol patient treatments. The RPC has recently developed a Web-based questionnaire that will allow an institution to retrieve data previously entered and provide updates on a regular basis. This will reduce the effort required for participation in a new clinical trial.

Knowledge Assessment Questionnaire

The physician is asked to take a simple open-book quiz to indicate that he or she is familiar with the protocol and its requirements. Again, this questionnaire is provided via a Web page.

Treatment Planning Benchmark or Anthropomorphic Phantom

For the more complex technologies, the institution may be required to submit a treatment plan generated for a standardized geometry or CT data set, or simulate, plan, and treat an anthropomorphic phantom. If the protocol requires a benchmark treatment plan, the RPC reviews the institution's plan and recalculates the doses at key locations to evaluate the accuracy of the planning system. When anthropomorphic phantoms are used, the RPC compares the delivered dose to the institution's plan to determine the agreement. See below for further details.

Digital Data Submission

Many clinical trials involving advanced technology radiation therapy are now requiring institutions to submit the treatment plans for protocol patients digitally to the ITC. The plans performed for irradiation of the RPC's anthropomorphic phantoms also must be submitted digitally.

QA and Dosimetry Review

The RPC reviews the institution's QA and dosimetry procedures and records.

Clinical Review by Radiation Oncologist

In some cases, the protocol requires that the institution submit representative treatment plans performed for patients treated previously using the technology being tested by the protocol, and techniques at least similar to those required by the protocol. The plans are reviewed by the study chair or a radiation oncologist who works with the RPC to ensure that they conform to the intentions of the study chair or designee.

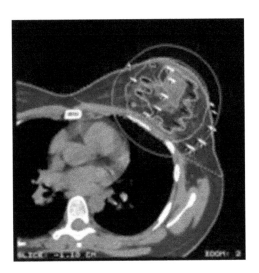

FIGURE 33.1 A treatment planning benchmark test developed by the RPC for credentialing institutions participating in a partial breast irradiation protocol.

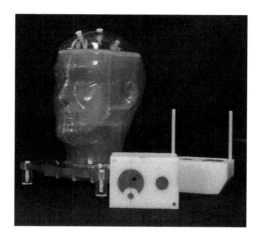

FIGURE 33.2   A phantom developed by the RPC to simulate the head and neck region of a patient. The phantom contains imageable structures and dosimeters to evaluate the delivered dose.

Important Features of a Credentialing Program

Independence

A key aspect of a credentialing program is independence, that is, the procedures conducted to assure the quality and accuracy of the product or process (in this case, the delivery of radiation therapy) must be independent of the product or process itself. Failure to establish independence can lead to the risk that the credentialing program merely mimics the performance of the parameter being measured, masking any error.

Reliability

The credentialing program must reliably test the capability of concern. For example, credentialing of IMRT is most dependably performed through the use of an anthropomorphic phantom. This procedure requires the institution to conduct almost all of the functions that would be undertaken to treat a patient with IMRT: preparation, CT simulation, treatment planning, positioning, and treatment delivery. Ideally, the institution would perform all of the procedures in exactly the same manner as would be done for a human patient; the regular staff would perform each step, the institution's data management and electronic medical record systems would be used, and the standard QA procedures would be followed. Because the phantom is as unfamiliar to the staff as a patient, the procedure prevents shortcuts from being taken. And because the phantom is anthropomorphic, the institution's standard procedures are more relevant than they might be if a rectangular or geometric phantom were to be used.

A treatment planning benchmark is a valuable test of the institution's treatment planning capabilities. However, it does not examine whether the institution can deliver the plans they generate. Likewise, a credentialing program that allows the institution to use their own equipment and simply report the results can be influenced by the quality of the measurement equipment, its independence from the quantity being measured, and the actual measurement procedures.

Results of the RPC's Credentialing Procedures

Several steps in a typical credentialing process are intended either to collect data, or to encourage the institution's staff to become properly prepared to participate in the protocol. For example, the attestation of use of the technology, the facility questionnaire, and the clinical reviews of patient records are steps that might require more than one attempt before the required information is completed, but rarely prevent participation in a trial.

TABLE 33.2   Passing Rates for Four of the RPC's Anthropomorphic Phantoms

Site	Institutions	Irradiations	Pass Rate (%)
H&N	472	631	75
Pelvis	108	130	82
Lung	67	77	71
Liver	15	18	50

Source: Ibbott, G., *Clinical Dosimetry Measurements in Radiotherapy*, Medical Physics, Madison, WI, 2009.
Note: The criteria for agreement were 7% and 4 mm for all phantoms except the lung phantom, for which the criteria were 5% and 5 mm.

On the other hand, the treatment planning benchmark test and the anthropomorphic phantom irradiation sometimes reveal deficiencies or misunderstandings in the institution's use of the technology. To the extent possible, the RPC reviews with the institution their procedures for calculation, QA, and delivery to determine the cause(s) of any discrepancy. When underlying causes are found and corrected, the benchmark is repeated or a repeat phantom test is scheduled.

Results of the Treatment Planning Benchmark Tests

An example of a treatment planning benchmark is shown in Figure 33.1. Institutions downloaded a CT data set of a breast patient from an FTP site maintained by the RPC and generated a treatment plan that complied with the requirements of the joint RTOG 0413/NSABP B-39 protocol. The treatment plan was submitted digitally to the ITC and made available to the RPC for evaluation. The target volume contours and DVHs were reviewed, and an independent calculation of dose to the target was performed. When benchmark cases failed to meet the criteria, the RPC contacted the institution, explained the discrepancies, and worked with the institution to resolve them.

Results of the Anthropomorphic Phantom Tests

During the period 2001 to 2008, the RPC mailed IMRT head-and-neck phantoms to 472 distinct institutions (see Figure 33.2 and Table 33.2) (Ibbott 2009; Molineu et al. 2005). A total of 631 irradiations were analyzed. Of these, 473 irradiations, or 75%, successfully met the irradiation criteria. More than 350 institutions failed to meet the irradiation criteria on the first attempt and had to repeat the phantom irradiation. Of those failing to meet the accreditation criteria, the majority failed only the dose criterion. The remaining unsuccessful irradiations failed the distance-to-agreement (DTA) criterion or both the dose and DTA criteria. Many institutions irradiated the phantom multiple times because they wanted to improve their initial irradiation results, test different treatment planning system algorithms, or test different treatment delivery systems.

The most common treatment-planning systems (TPSs) used to plan the irradiations of the phantom were the Phillips Pinnacle and Varian Eclipse systems. The pass rates for these two TPSs were approximately 73% and 85%, respectively. The difference is believed to be due to difficulties in modeling the penumbra at the ends of rounded multi-leaf collimator leaves (Cadman et al. 2002).

Summary

The RPC has compiled data showing that credentialing can help institutions comply with the requirements of a cooperative group clinical protocol. Phantom irradiations have been demonstrated

to exercise an institution's procedures for planning and delivering advanced external beam techniques (Ibbott et al. 2008; Molineu et al. 2005). Similarly, RPC data indicate that a rapid review of patient treatment records or planning procedures can improve compliance with clinical trials (Ibbott et al. 2007).

Acknowledgments

The Radiological Physics Center is supported by grants CA 10953 and CA 81647 from the National Cancer Institute.

References

Barthelemy-Brichant, N., J. Sabatier, W. Dewé, A. Albert, and J.-M. Deneufbourg. 1999. Evaluation of frequency and type of errors detected by a computerized record and verify system during radiation treatment. *Radiother. Oncol.* 53 (2): 149–154.

Cadman, P., R. Bassalow, N. P. Sidhu, G. Ibbott, and A. Nelson. 2002. Dosimetric considerations for validation of a sequential IMRT process with a commercial treatment planning system. *Phys. Med. Biol.* 47 (16): 3001–3010.

Donaldson, Sir Liam. 2008. *Towards Safer Radiotherapy*. The Royal College of Radiologists, Society and College of Radiographers, Institute of Physics and Engineering in Medicine, National Patient Safety Agency, British Institute of Radiology. London: The Royal College of Radiologists.

Dunscombe, P., H. Lau, and S. Silverthorne. 2008. The Ottawa Orthvoltage Incident: Report of the Panel of Experts Convened by Cancer Care Ontario. Available from http://www.ottawahospital.on.ca/about/reports/orthovoltage-cco-e.pdf (accessed 8/6/2010).

Flexner, A. 1910. *Medical Education in the United States and Canada*. New York, NY: Carnegie Foundation for Higher Education.

Halvorsen, P. H., I. J. Das, M. Fraser, D. J. Freedman, R. E. Rice 3rd, G. S. Ibbott, E. I. Parsai, T. T. Robin Jr., and B. R. Thomadsen. 2005. AAPM Task Group 103 report on peer review in clinical radiation oncology physics. *J. Appl. Clin. Med. Phys.* 6 (4): 50–64.

Herman, M. et al. 2008. *Alternative Clinical Medical Physics Training Pathways for Medical Physicists: Report of AAPM Task Group 133*. College Park, MD: AAPM.

Huang, G., G. Medlam, J. Lee, S. Billingsley, J. P. Bissonnette, J. Ringash, G. Kane, and D. C. Hodgson. 2005. Error in the delivery of radiation therapy: Results of a quality assurance review. *Int. J. Radiat. Oncol. Biol. Phys.* 61 (5): 1590–1595.

IAEA. 2007. *Comprehensive Audits of Radiotherapy Practices: A Tool for Quality Improvement—QUATRO*. Vienna: International Atomic Energy Agency.

Ibbott, G. 2009. QA for clinical dosimetry with emphasis on clinical trials. In *Clinical Dosimetry Measurements in Radiotherapy*, ed. D. W. O. Rogers and J. Cygler. Madison, WI: Medical Physics Publishing.

Ibbott, G. S., D. S. Followill, H. A. Molineu, J. R. Lowenstein, P. E. Alvarez, and J. E. Roll. 2008. Challenges in credentialing institutions and participants in advanced technology multi-institutional clinical trials. *Int. J. Radiat. Oncol. Biol. Phys.* 71 (1 Suppl): S71–S75.

Ibbott, G. S., W. F. Hanson, E. O'Meara, R. R. Kuske, D. Arthur, R. Rabinovitch, J. White, R. M. Wilenzick, I. Harris, and R. C. Tailor. 2007. Dose specification and quality assurance of radiation therapy oncology group protocol 95-17: A cooperative group study of iridium-192 breast implants as sole therapy. *Int. J. Radiat. Oncol. Biol. Phys.* 69 (5): 1572–1578.

Kirby, T. H., W. F. Hanson, R. J. Gastorf, C. H. Chu, and R. J. Shalek. 1986. Mailable TLD system for photon and electron therapy beams. *Int. J. Radiat. Oncol. Biol. Phys.* 12 (2): 261–265.

Kirby, T. H., W. F. Hanson, and D. A. Johnston. 1992. Uncertainty analysis of absorbed dose calculations from thermoluminescensce dosimeters. *Med. Phys.* 19: 1427–1433.

Kohn, L. T., J. M. Corrigan, and M. S. Donaldson, eds. 1999. *To Err Is Human: Building a Safer Health System.* The Institute of Medicine. Washington, DC: National Academy Press.

Macklis, R. M., T. Meier, and M. S. Weinhous. 1998. Error rates in clinical radiotherapy. *J. Clin. Oncol.* 16 (2): 551–556.

Marks, L. B., K. L. Light, J. L. Hubbs, D. L. Georgas, E. L. Jones, M. C. Wright, C. G. Willett, and F. F. Yin. 2007. The impact of advanced technologies on treatment deviations in radiation treatment delivery. *Int. J. Radiat. Oncol. Biol. Phys.* 69 (5): 1579–1586.

Molineu, A., D. S. Followill, P. A. Balter, W. F. Hanson, M. T. Gillin, M. S. Huq, A. Eisbruch, and G. S. Ibbott. 2005. Design and implementation of an anthropomorphic quality assurance phantom for intensity-modulated radiation therapy for the Radiation Therapy Oncology Group. *Int. J. Radiat. Oncol. Biol. Phys.* 63 (2): 577–583.

Patton, G. A., D. K. Gaffney, and J. H. Moeller. 2003. Facilitation of radiotherapeutic error by computerized record and verify systems. *Int. J. Radiat. Oncol. Biol. Phys.* 56 (1): 50–57.

Perrow, C. 1999. *Normal Accidents: Living with High Risk Technologies.* Princeton, NJ: Princeton University Press.

Rogers, L. 2009. Over 200 hurt or killed by botched radiation. *The Times (London) Online* 2009. Available from http://www.timesonline.co.uk/tol/news/uk/article711360.ece?token=null&offset=12&page=1 (accessed 3/1/2009).

Clinical Trials: Quality Assurance

James A. Purdy
UC Davis Medical Center

Philip Poortmans
Institute Verbeeten

Edwin Aird
Mount Vernon Hospital

Introduction

It is well understood that in order to reliably interpret the results of a clinical trial and to project its results to daily practice, the design of the protocol, the credentialing and quality assurance (QA) procedures used, and the evaluation of the data are of utmost importance. Credentialing and QA procedures must be designed so that they are able to reliably evaluate the ability of institutions to use the technology dictated by the trial. Other reasons for credentialing and QA are to identify flaws in trial design and to assess if deviations from the guidelines of the protocol are caused by misinterpretation of the trial prescriptions or by the inability to comply with the required procedures. It is the authors' strong conviction that clinical trials must be conducted such that participating patients are treated as stated per protocol, within the established tolerance levels, in order for the results to be interpreted meaningfully.

Radiation therapy treatment planning and delivery processes have changed dramatically since the introduction of three-dimensional treatment planning in the 1980s and continue to change in response to the implementation of new technologies (Purdy 1996, 2007). Computed tomography (CT) simulation and three-dimensional (3D) radiation treatment planning systems (3DTPS) provide the tools for carrying out 3D conformal radiation therapy (3DCRT), which is now widely used in clinical trials as well as in daily practice. Intensity modulated radiation therapy (IMRT) techniques provide conformal target volume coverage for even the most complex shapes, and conformal

avoidance of specific sensitive normal structures (Webb 2000) and its use in clinical trials is ever increasing. Already, National Cancer Institute (NCI) guidelines for use of IMRT in clinical trials have been developed (Palta et al. 2004). The highly conformal doses provided by IMRT and increased use of smaller safety margins to decrease the dose to the surrounding normal tissues have focused attention on the need to better account for both intrafraction and interfraction variations in patient positioning and internal organ shapes and locations, which has helped spur the development of treatment machines with integrated planar and volumetric imaging capabilities (Jaffray et al. 2002). In addition, improved tumor and normal organ definition are possible thanks to continuing advances in both anatomical and functional imaging. Advances in all of these technologies are occurring at a fast pace and are pushing radiation oncology toward what is referred to as image-guided radiation therapy (IGRT) (Bortfeld et al. 2006). Already, IGRT is being used for stereotactic body radiation therapy (SBRT) clinical trials (RTOG0236 2002) and its use in future protocol advanced technology processes such as adaptive radiation therapy (ART) are underway (Martinez et al. 2001).

All of these advanced technologies present credentialing and QA challenges for cooperative groups and clinical trial QA centers. It is also apparent that clinical trials are becoming much more international and, thus, harmonization of clinical trials credentialing and QA procedures/criteria is critical. This chapter will address these issues, focusing primarily on the United States and European clinical trial QA experience.

Brief History of Quality Assurance for Clinical Trials

U.S. Clinical Trials QA History

In the late 1960s and early 1970s, several organizations were created and eventually funded by the NCI that continue to play a major role in QA of clinical trials that use radiation therapy. Only the major QA offices that are currently active will be reported on here. The Radiological Physics Center (RPC) (under the direction of Dr. Robert Shalek at M. D. Anderson Cancer Center in Houston) was established by the NCI in 1968 to help ensure the correctness of physical data entered into clinical trials, provide feedback to institutions for correction of errors, and make instructive findings generally available (Golden et al. 1972; Grant et al. 1976; Gagnon et al. 1978). In that same time frame, the Radiation Therapy Oncology Group (RTOG) was organized (under the direction of Dr. Simon Kramer at Thomas Jefferson University in Philadelphia) for the purpose of conducting radiation therapy research and cooperative clinical investigations and received its initial NCI funding in 1971. The Quality Assurance Review Center (QARC) was created in the late 1970s under the direction of Dr. Arvin Glicksman and his work with the Cancer and Acute Leukemia Group B (CALGB) and obtained NCI funding beginning in 1980 to provide radiation therapy quality assurance services for multiple cooperative groups (Glicksman et al. 1981). In 1984, Perez et al. (1984) provided a description of the comprehensive QA programs typically in place at that time for U.S. cooperative groups' clinical trials involving radiation therapy. In 2004, the *AAPM Report 86* was published and provided information on QA for clinical trials specifically directed at the clinical physicist (AAPM Report 86 2004). More recently, Purdy (2008) provided a report on QA issues faced in conducting multi-institutional clinical trials using advanced technology and Ibbott et al. (2008) described the challenges the RPC has observed in credentialing institutions and participants in such trials.

Modern clinical trial QA efforts for U.S. protocols using advanced technologies can be traced back to the pioneering efforts of the Washington University 3D QA Center, which is now referred to as the Image-guided Therapy QA Center (ITC), and the Radiation Therapy Oncology Group (RTOG), which partnered in 1994 to provide QA for the 3D Oncology Group (3DOG) to conduct a prostate dose escalation study using 3DCRT (Purdy 1996; Purdy et al. 1998). The ITC was charged with developing a credentialing process and an informatics infrastructure that allowed institutions to submit (in a digital format) the volumetric imaging and treatment planning data for patients registered to this trial, and that provided a centralized QA review process. To that end, the ITC developed a successful format for electronic transfer and archival storage of volumetric treatment planning digital data called the RTOG Data Exchange (RDE) (Bosch et al. 1997; Harms, Bosch, and Purdy 1997). More than a thousand patients were enrolled in 3DOG/RTOG 9406 and multiple manuscripts have been published and will continue to be published using this historic data archive (Michalski et al. 2003, 2005). There is no doubt that 3DOG/RTOG 9406 revolutionized the way cooperative group clinical trial QA could be performed and the way data could be submitted, reviewed, and archived.

In 1999, the NCI funded (based on the success of the 9406 effort) the Advanced Technology Consortium for Clinical Trials QA (ATC) (http://atc.wustl.edu/), a consortium of U.S. QA centers that includes the ITC, RTOG, RPC, and QARC. The mission of the ATC is to facilitate the conduct of NCI-sponsored advanced technology radiation therapy clinical trials. This effort includes radiation therapy clinical trial QA, image and radiation therapy digital data management, and clinical trial research and developmental efforts. The ATC has strongly advocated the concept that advanced medical informatics can facilitate education, collaboration, and peer review, as well as provide an environment in which clinical investigators can receive, share, and analyze volumetric multimodality treatment planning and verification (TPV) digital data. The ATC's stated goal is to improve the standards of care in the management of cancer by improving the quality of clinical trials medicine. The ATC to date has archived more than 9000 protocol case digital data sets and now provides support to multiple clinical trial groups in the United States, Europe, and Japan.

European Clinical Trials QA History

A review of the QA program of the European Organization for Research and Treatment of Cancer Radiation Oncology Group (EORTC-ROG) has been reported by Poortmans et al. (2005) and is briefly presented here.

In 1982, the EORTC-ROG activated its QA program. Initially, 17 member centers were visited by a team of expert radiation oncologists and radiation physicists from January 1982 to December 1984. The evaluation included three steps: an evaluation of the infrastructure, staffing levels, mechanical and dosimetric integrity of treatment units, and a check of the data in the clinical and radiotherapy charts. Large variations were observed in the workload as expressed in the number of patients treated per year per radiation oncologist, radiation physicist, and radiation technologist. Also, there was a wide variation in the number of treatment machines that existed with major problems for five of the 17 centers, making it difficult to comply with the requirements of the EORTC-ROG (Horiot et al. 1986). The dosimetric evaluation showed deviations for a number of scanning electron beams, and for the flatness and the symmetry of several x-ray as well as electron beams (Johansson et al. 1986). No major deviations related to absorbed dose calibration or calculation in an anatomical phantom (containing a tonsillar tumor and a homolateral subdigastric node) were found. This investigation led to the development of the concept of what was later named the "dummy run" procedure (similar to the U.S. "dry run"): investigators were asked to treat a dummy patient (initially the Alderson anatomical phantom) as if it was a true patient entered in an active protocol (in that case, EORTC protocol No. 22791 comparing conventional versus twice a day fractionation for oropharyngeal cancers). Although the dose delivered was correct on the central axis in all cases, large differences in treatment planning were observed resulting in about one-third of the cases having insufficient or excessive irradiated volumes, exposing the "patient" to either increased failure or complication rates (Johansson, Horiot, and van der Schueren 1987). The evaluation of mechanical checks of megavoltage units and simulators showed that, in general, the deviations observed for accelerators and simulators were smaller than for cobalt units, possibly related to the advanced age (up to 20 years) of some of the latter units (Van Dam et al. 1993).

Technology Requirements

Institutions planning to participate in advanced technology clinical trials using 3DCRT, IMRT, SBRT, brachytherapy, and heavy particle beams must be able to submit the volumetric treatment planning and verification data electronically to a QA center. Film and other forms of hardcopy data are simply not adequate for evaluating target volumes and 3D dose distributions. In addition, dose volume histogram (DVH) data alone are not sufficient as there is no spatial information provided and the dose fractionation information is lost. Clearly, variation in dose distributions throughout an organ may lead to different expectations of toxicity for some organs. In addition, the volumetric treatment planning database resulting from a digital data submission process (that can be linked to clinical outcomes) will be a significant resource for secondary analyses and development of robust dose-response models.

Today, digital data exchange is rapidly becoming a nonissue for conducting clinical trials, as the RDE and DICOM RT export feature has been implemented on most treatment planning systems. As a result, there are several software (or multicomponent) systems that have been developed that can receive these digital data objects and have features to facilitate individual case review. As previously noted, the ITC pioneered the process of clinical trial case electronic treatment planning data collection and review. This ATC (ITC) informatics infrastructure is now referred to as the QuASA2R (Quality Assurance Submission, Archive, Analysis, and Review) system and provides an environment in which institutions can submit and ITC can receive, share, and analyze protocol-specific volumetric multimodality imaging/treatment planning/verification (ITPV) digital data (Purdy et al. 1996, 1998; Bosch et al. 2000; Purdy et al. 2006; Bosch et al. 2006, 2007). The system is fully RTOG- and DICOM-compliant and supports data submission that includes CT images, RT structures, beam geometry, 3D dose distributions, DVH, digital reconstructed radiographs (DRR), and verification images. Moreover, it provides fully web-based remote review. Another software system developed at Washington University called the Computational Environment for Radiotherapy Research (CERR) (Deasy, Blanco, and Clark 2003) has been integrated into the QuASA2R system, as well as QARC's clinical trials workflow and database system called MAX (FitzGerald et al. 2008). CERR provides multiaxis review of images and many other features that enhance the data review process. Until recently, CERR had been limited to local data review, but both QARC and ITC have implemented remote terminal services, which enable remote review.

Other systems now being used by clinical trial groups outside the United States include the Visualization & Organization of Data for Cancer Analysis (VODCA) system developed by Stefano Gianolini and Daniele Henggeler and the SWAN system developed by Martin Ebert and colleagues. The SWAN system is used to support Trans-Tasman Radiation Oncology Group (TROG) clinical trials (Ebert et al. 2008). The VODCA system is now available commercially from Medical Software Solutions, Switzerland (http://www.vodca.ch) and has recently been implemented at the EORTC headquarters. It provides similar review features to those previously discussed here. In the near future, VODCA will be further developed (in cooperation with other EORTC research groups) into an imaging platform that can be used in trials using modalities other than radiotherapy. The authors' agree that without such a platform, it is not possible to perform case QA at an internationally acceptable level within the EORTC. Such a system offers the opportunity to perform modern high-level QA by and within the EORTC-ROG, whereas until now this had to be outsourced to commercial QA providers. The possibility to perform trial-specific QA by itself might also lead to more competitive quotations to tender. Over the coming years, the EORTC-ROG will continue to cooperate on a case by case basis with QA providers, depending on the requirements for QA, the workload, and the availability of funding.

Finally, on a historical note, it should be mentioned that, in 2004, EqualEstro was founded thanks to an EU grant to the European Society for Therapeutic Radiology and Oncology (ESTRO) with the mission to provide QA solutions for all existing radiotherapy techniques. Later, it was reorganized as an independent provider for QA for clinical trials in radiation oncology (http://www.equalestro.org). At present, it is supporting a number of clinical trials in Europe.

Protocol Requirements

Every protocol for a clinical trial in which radiotherapy is involved should be approved by a QART team before submission to a clinical trial's office. This not only includes the checking of the planning/delivery protocol itself, but the QART team will also assist in the writing of the requirements for the trial-specific QA as well. The 3DOG/RTOG 9406 protocol led the way in changing how protocols should be written in terms of specifying the target volume and dose prescription, credentialing requirements, and QA processes. Instead of specifying radiation therapy fields based on anatomical guidelines, *ICRU Report 50* concepts were adopted and have been used in all subsequent protocols supported by the ATC (United States). This approach requires the use of volumetric imaging studies to define the volume of known tumors (GTV), suspected microscopic spread (CTV), and the volume necessary to account for setup variations, as well as organ and patient motion (PTV) (ICRU 1993). More recently, clinical protocols have embraced the use of *ICRU Report 62* concepts such as the internal margin (IM) around the CTV to define the Internal Target Volume (ITV), set-up margin (SM) to define the PTV, and a margin around the organs at risk (OAR) to create the planning organ-at-risk volume (PRV) to account for spatial uncertainties (ICRU 1999; Purdy 2004). The protocol should provide a rationale for the choice of margins to be used. It should also clearly define the OAR and/or PRV that are required to be delineated and provide clear guidelines for their contouring. The GTV, CTV, PTV, OAR, PRV, and skin contours should be, as a rule, delineated on all slices of the 3D volumetric imaging study in which each structure exists. The PRV does present a problem in some cases, for example, the instance of the rectum where the length over which it is delineated makes a huge difference on the DVH evaluation, necessitating clear delineation guidance to be spelled out in the protocol.

The protocol dose specification and dose inhomogeneity allowed throughout the PTV must be made explicitly clear as there are several prescription methods in use that do not concur completely with the ICRU guidelines, including specifying the prescription dose as (1) the minimum dose to the target volume, (2) the dose at or near the center of the tumor, or (3) the maximum dose within the target volume. This issue is clearly an important one, as there can be a significant difference in the resulting dose distributions. An example—but not more than that—of a clear prescription for 3DCRT is "the prescription dose is the minimum dose to the planning target volume. The maximum dose shall not exceed the prescription dose by more than 7%. The reported doses shall include the dose to the ICRU reference point as well as the maximum dose, minimum (prescription) dose, and mean dose to PTV."

For protocols that allow IMRT, the prescription specification is even more complicated and, as yet, there are no international guidelines or recommendations published by the ICRU. Purdy (2002, 2004) has pointed out the challenges that IMRT poses for the ICRU 50/62 methodology. More recently, Das et al. reported on the substantial variations in IMRT prescribed and delivered doses that exist among medical institutions, raising concerns about the validity of comparing clinical outcomes for IMRT (Das et al. 2008; Willins and Kachnic 2008). Cooperative groups/ QA Centers such as RTOG and the ATC have certainly made a concerted effort to provide clear dose prescription/volume specification and the NCI's IMRT Collaborative Working Group, and more recently ASTRO, have published recommendations for documenting IMRT treatments (IMRTCWG 2001; Holmes et al. 2009). That said, there still remains a clear need to work toward harmonizing IMRT prescription specifications, recording of what dose distribution was actually delivered, and reporting of pertinent IMRT information. The ICRU is in the process of completing a new report addressing IMRT that hopefully will provide guidelines/recommendations for meeting this challenge.

Advanced 3D dose calculation algorithms such as the convolution/superposition (Ahnesjö and Aspradakis 1999) are now available on most treatment planning systems and it is the

authors' recommendation that a heterogeneity-corrected dose distribution (generated using these types of algorithms) be used for all protocols using 3DCRT and IMRT. If both 3DCRT and IMRT modalities are allowed by the protocol, the dose heterogeneity for the PTV should be similar for both modalities; however, this may be difficult to achieve for a very complex PTV and, thus, protocols may need to limit the choice of treatment modalities allowed. Also, dose-volume tolerance constraints for each OAR/PRV situated within the irradiated volume should be defined and participants should be encouraged (and in some cases required) to remain within the protocol-specified limits. In some cases, dose constraints to OARs might be in conflict with the required dose within the PTV. This necessitates that clear guidelines be included in the protocol regarding the prioritization of the conflicting parameters and, if necessary, the inclusion of supplementary constraints, for example, constraints on the CTV in addition to the PTV. A DVH that is sometimes useful in plan evaluation QA is one for "nondelineated tissue" (NDT). This is defined as all tissue contained within the skin, but which is not otherwise within any other structure or target volume.

Protocols utilizing all types of external beam technologies should have specific procedures in place to ensure correct, reproducible positioning of the patient/target volume. Minimally, orthogonal (AP and lateral) DRR and corresponding orthogonal weekly portal images (film or electronic) should be required. Progressively more treatment machines with integrated CT features are available for use in clinical trials and, thus, protocol developers should anticipate using this modality for treatment localization and verification.

More recently, the use of IMRT was approved by the NCI for clinical trials involving intrathoracic lesions, in which significant heterogeneities are encountered and tumor mobility is likely (see http://atc.wustl.edu/home/NCI/NCI_IMRT_Guidelines.html). The new guidelines require that correction for heterogeneities and accounting for target motion be explicitly addressed in the design of the protocol. Also, the QA Centers must now establish explicit criteria for evaluating the acceptability of the dose calculation algorithms that can be used in the protocol.

Advanced technology protocols should explicitly require that the treatment machine MU setting generated by the TPS be checked independently before the patient's first treatment. For complex modalities such as IMRT and SBRT, individual patient specific QA measurement processes may be required.

Credentialing Requirements

Before participating in a cooperative group trial, institutions are typically credentialed at a level appropriate to the technology required by the trial. The concepts pioneered by ITC and RTOG with the 3DOG/RTOG 9406 protocol remain valid today. This process, adapted as well by the EORTC-ROG, includes: (1) a facility questionnaire that documents the institution's technical capabilities and identifies the critical treatment team individuals; (2) an external reference dosimetry audit (ERDA); and (3) a series of tests that are protocol-modality–specific and might include, depending of the complexity of the radiotherapy used in the trial under question, (a) an electronic data submission test, (b) a dry-run (or dummy run) test demonstrating the understanding of the protocol treatment planning and data submission requirements (or a rapid review of the first submitted case, with additional focus on the volume contouring), (c) a phantom planning-dosimeter test, and (d) other tests such as an IGRT test.

With regard to credentialing, it should be noted that a series of published exchanges between Back et al. (2008) and Williams et al. (2007, 2008) clearly point out that those involved in carrying out credentialing of clinical trials need to tread cautiously. Although the aim is to ensure the quality and integrity of the scientific data, achieving that aim may require consideration of factors that go beyond science and is very much dependent on healthy collaborations between radiotherapy centers (Williams et al. 2008). Bridier et al. (2000) have pointed out that this is particularly problematic for postal phantoms for external audits and that a number of special precautions have to be taken. Relevant information must be gathered in a small number of easily performed set-ups and appropriate dosimeters must be used. It is essential to make instructions as clear as possible and to keep the time necessary for the institution's staff to plan, set up, and irradiate the phantom as efficient as possible. If these criteria cannot be met, a large number of failures may be seen that are in fact "false positives," that is, they do not reflect the ability of the institution to plan and deliver the required dose to the protocol patient, but rather reflect the ability to pass a flawed test design. Such tests will clearly have a negative impact on protocol case accrual and may even diminish institutions' willingness to participate in clinical trials.

Facility Questionnaire

Before being approved for participation in any clinical trial, certain documentation requirements must be met. This is provided via a Facility Questionnaire. The information contained in the questionnaire typically includes: contact information for the responsible individuals (physician, physicist, dosimetrist/radiation technologist, and data manager) who will be working on the studies to enable QA Center personnel to contact the appropriate individual as needed in the qualification and participation review process, a list of the therapy machines to be used for the treatment of patients enrolled in the study, infrastructure available for data acquisition/imaging for treatment preparation, documentation about the image-based planning system that will be used, and other pertinent information depending on the protocol and modality used, such as patient motion management systems.

External Reference Dosimetry Audit

To verify that all centers within a clinical trial are delivering the same dose, it is vital to perform a dosimetry audit by an expert center that has direct links to a primary dosimetry laboratory (which also has links to primary laboratories throughout the world). This audit has become known as ERDA. The process by which this is accomplished varies between the United States and Europe at two major points: the provider and the frequency.

United States ERDA Experience

The RPC initiated a mailed thermoluminescent dosimetry (TLD) program for machine output checks for photon beams in 1977 and electron beams in 1982 (Kirby et al. 1986; Kirby, Hanson, and Johnson 1992). Monitoring of proton beams began in 2007. For a significant period, institutions participating in NCI-sponsored clinical trials underwent the RPC ERDA (mailed TLD check) every 6 months. This ERDA frequency was eventually changed to an annual check, but unfortunately, there are no subsequent peer-reviewed published reports analyzing the data to validate this yearly ERDA requirement.

European ERDA Experience

In 1987, the EORTC-ROG initiated a mailed TLD program for machine output check. This revealed a few large deviations (>7%) between the dose measured and the dose stated by the centers (Hansson and Johansson 1991; Hansson et al. 1993). Based on this, corrections were made by the participating centers. Subsequent mailings resulted in a decrease of the measured deviations and of the standard deviation. Correlating information from the TLD measurements with the onsite visits led to a further decrease in the beam output deviations (Hansson et al. 1993). In 1993, a report on the mailed TLD results of radiotherapy centers in Europe showed that the large majority of the beams (23/25) with deviations >3% were from centers that had not participated in external audits in the 5 years preceding the audit (Dutreix et al. 1993). Based on the outcome of these QA procedures, a document with minimum requirements for QA in radiotherapy departments was published (Horiot, Bernier, and Johansson 1993). The mailed TLD dosimetry program of the EORTC-ROG was later taken over by ESTRO (Derreumaux et al. 1995; Ferreira et al. 2000). Although funding ended for this program, it continues to be available for a fee through EqualEstro. In 2006, ERDA was made mandatory for all EORTC members before participation in RT trials. Based on the previously referenced reports, the ERDA is required to be performed only on a 2-year basis, for both electrons (if applicable for the trials in which centers participate) and photons. The ERDA can be performed by any provider, as long as the primary standard is traceable.

It should be noted that the U.S. requirement for an annual ERDA instead of the EORTC requirement of at least every 2 years is one of the two major differences in the clinical trial QA programs between the United States and Europe. Both agree that the institution must complete a successful ERDA prior to their participation in a clinical RT trial.

Phantom Test Measurements

New modalities (e.g., for IMRT protocols) may require additional credentialing tests such as a phantom dosimetry check.

United States Phantom/Benchmark Experience

The RPC has developed a series of postal anthropomorphic phantoms that contain dosimeters (film and TLD) that are intended to provide a comprehensive test of an institution's ability to image, plan, and treat a patient using IMRT or SBRT techniques (Molineu et al. 2005). Roughly 30% of institutions failed to deliver a dose distribution to the head and neck (H&N) phantom that agrees with their own treatment plan to within 7% or 4 mm (Ibbott et al. 2008; Ibbott 2009). Somewhat higher pass rates have been seen with the other RPC phantoms (Ibbott et al. 2008).

QARC has taken a different approach for cooperative group clinical trial credentialing. Rather than postal phantoms, they have developed a series of benchmarks that institutions are required to pass, but uses the institution's own phantoms, dosimetry systems, and treatment planning systems. For example, they developed an IMRT benchmark, which was reported on by ATC and does not require a postal phantom (Palta et al. 2004).

Another important point is that in the United States, the RTOG now requires an institution to obtain recredentialing when important elements of an institution's process change, for example, changing from standard multileaf collimator IMRT dose delivery to tomotherapy or volume arc therapy IMRT (VMAT or RapidArc).

European Phantom Experience

In 2000, Bridier et al. (2000) published a comparative description of several multipurpose phantoms (MPP) for external audits of photon beams. These MPP have seen relatively little use in clinical trial credentialing in Europe. More recently, Swinnen et al. reported on the use of a MPP called OPERA (Operational Phantom for European Radiotherapy Audits), which was designed to check the dosimetric aspects of complex treatments by comparing the planned versus the measured dose distribution as well as the online treatment portal imaging verification (Swinnen, Verstraete, and Huyskens 2002). OPERA was adapted for IMRT and used in five key centers in Europe to establish the principle and functionality for IMRT credentialing. The estimated cost of such complex dosimetry has been, however, found to be high. It is therefore expected that participating centers may have to bear at least part of the costs involved, unless sufficient funding can be found from other sources.

Dry-Run/Dummy Run Treatment Plan Test

Prior to being approved for participation in a clinical trial using advanced technology, each institution must successfully complete a dry run test. In the reports of the EORTC, this test is called a dummy run—see article by van Tienhoven et al. (1991)—which is intended to ensure that an institution preparing for participation in a particular clinical trial has both the technical capability and an appropriate understanding of the requirements of the protocol (tumor and target volumes, organs-at-risk, and dose prescription/heterogeneity requirements). Each protocol may

have its own dry run test that must be successfully completed to qualify for participation in a particular study.

Target Volume Delineation

Treatment planning based on modern imaging techniques including CT, MRI, and PET-CT has introduced a new level of uncertainty and variation. In the distant past, radiotherapy fields were often based on bony landmarks without added blocking to make the beam more conformal to the target volume. Whereas, these beams were large enough to cover the actual target, this approach limited the dose that could be prescribed due to normal tissue constraints and toxicity, and when it was attempted to make the radiotherapy beams more conformal, part of the target volume was often missed (Pilepich, Prasad, and Perez 1982). With the broad introduction of the use of imaging for 3D volume delineation and treatment planning in the 1990s, several new challenges were introduced:

- What the definition of the actual target volume to be treated is (for several sites, this is uncertain)
- What imaging technique should be used for a specific target volume
- What margins are needed from GTV to CTV to account for microscopic tumor spread
- What margins are needed from CTV to PTV to take into account set-up uncertainties, internal organ movement/shape change from day to day, and internal organ movement during a treatment session

Multiple reports confirmed the fear of many: that the variance in target volume delineation has become the largest remaining source of variation in clinical radiotherapy (Wu et al. 2005). Both U.S. and European groups have made serious efforts to assist in minimizing this source of uncertainty and variance by authoring a number of papers dedicated to the subject of delineation of the target volumes, as well as of the organs at risk (Michalski, Lawton et al. 2009; Boehmer et al. 2006; Poortmans et al. 2007; Hall et al. 2008; Lawton, Michalski, El-Naqa, Buyyounouski et al. 2009; Lawton, Michalski, El-Naqa, Kuban et al. 2009; Michalski, Roach et al. 2009). These reports are, in general, related to clinical trials under preparation so that later participants can be given a tool to decrease interobserver variability.

It should be noted that in most recent EORTC trials, a dummy run is used not only to perform treatment planning but also to perform volume delineation, which is then compared to a compromise based on the volume delineation done by a number of experts in the specific field. In several RTOG trials, a rapid review approach is used in which the first accrued patient treatment is held until the principal investigator reviews and approves their submitted volumes. Also, training sessions are organized and results presented and discussed at the various groups' meetings.

As margins from CTV to PTV depend on the magnitude of several uncertainties, including not only GTV/CTV delineation, but also patient positioning set-up errors and set-up corrections, the actual margin that should be used is related to both delineation and set-up variations and might therefore very well be different from one participating institution to another. This is not being done robustly in current clinical trials and how to take this into account in future trials is yet another challenge for clinical trial QA.

Individual Case QA Review Process

Although a dry run is based on a fictitious patient and can be done before actual patient accrual, an individual case review (ICR) provides the opportunity to evaluate in a very detailed way all parts of the complete patient file that might be relevant to the trial under evaluation (Bolla et al. 1995). In some circumstances, patient-, tumor-, and (non-radiation) treatment-related data must be evaluated to check the eligibility of the randomized patients and to compare the data reported on the case report forms (CRF) to the ones in the actual patient file. For many cooperative groups, both in the United States and Europe, this kind of evaluation is typically performed on the occasion of the group's meeting, where participants are invited to bring the required data with them. Alternatively, hard copies are sent after removing identifying information to the QA office for evaluation and comparison with the case report forms. Nowadays, in trials involving radiotherapy, the focus is put on the RT information for which digital data submission is required, as demonstrated by 3DOG/RTOG 9406.

To ensure a maximal effect of the individual case review, it should be performed as early as possible during patient accrual, eventually including the agreement to continue only after acceptance of the ICR. After a limited number of ICRs per center, a predefined number of patients can be evaluated, ranging from a low percentage to all randomized patients. In some specific trials, data on treatment planning are sent to the QA office for rapid evaluation, with treatment execution only after acceptance and, if necessary, correction.

The ICR review process pioneered by the ATC (ITC) is now clearly divided between the ITC and the cooperative groups. The ITC is responsible for digital data integrity QA (DDIQA), which is a review for completeness of protocol-required elements, format of data, spatial registration, possible data corruption, and recalculation of all DVHs. The cooperative group is responsible for protocol compliance QA (PCQA), which includes review of target volume and organ-at-risk contours, as well as protocol dose prescription and dose heterogeneity compliance. PCQA is performed by the cooperative-group-designated reviewer using a system such as QuASAR2's web-based Remote Review Tool (RRT). When a case is ready for review, the ITC notifies the PCQA reviewer who is responsible for the rest of the review process. This clear division of the QA review process has made it more efficient for both the ITC and the cooperative group to keep track of the status of the protocols for QA reports and data quality reports and allows the cooperative group to request delinquent data from the participating institution in a more efficient manner.

Data Integrity QA Review

Digital data submitted to a QA center undergo what is now referred to as a "digital data integrity QA (DDIQA)" review (Purdy et al. 2006; Straube et al. 2006). Submitted data are checked for completeness and consistency. Ensuring completeness of protocol-required elements, format of data, possible data corruption, uniformity in OAR/TV contour names, and recalculation of DVHs has been shown to be an important element of the overall protocol QA review process.

Experience has shown that submitted DVHs lack consistency due to algorithmic differences among treatment planning systems (Straube et al. 2005). Thus, recalculation of DVHs is necessary for consistent correlation of dosimetry with outcomes. The ITC DDIQA experience in accepting, processing, and reviewing digital data submissions for support (QA and analysis) of advanced technology protocols for the past 13 years has shown that approximately 25% of the cases require intervention, often requiring iterative communications with personnel at the submitting institution (Purdy et al. 2006). The ITC has implemented software tools to increase efficiency of the DDIQA process, but experience has shown that completely automated data DDIQA is not yet realistic, even for the foreseeable future.

At the EORTC headquarters, all data submitted on the case report form are verified using a double data entry procedure and, more recently, by direct data capture through the Internet. After verification for consistency and completeness, queries are sent to the participating centers to check, correct, and complete, if needed. For future trials in which the VODCA platform will be used, detailed data will be collected in a manner similar to the ATC (ITC) approach.

Protocol Compliance QA Review

The QA methodology pioneered by the 3DOG/RTOG 9406 study has proven most effective for advanced technology protocols, that is, review of the target volume and OAR contours and dose/dose heterogeneity compliance. The current process used for RTOG-ATC–supported protocols is that contours are reviewed by the study chairs and dosimetry compliance is reviewed by either RTOG HQ dosimetrist staff or by RPC dosimetry staff (for brachytherapy cases).

For the EORTC protocols, the paper reviewing procedures are now gradually being replaced by digital submission and review, either in collaboration with other organizations (including ATC) or using the VODCA platform. Most of the reviewing is done by the Quality Assurance in Radiation Therapy team members.

Discussion

In the past, several problems concerning clinical trial protocol inconsistencies and numerous aspects of routine clinical practice have been noted. Potential systematic protocol deviations, possibly leading to false negative/positive results, were often detected. Special attention must be given to the variations in the delineation of target volumes and to the adherence to ICRU recommendations for dose prescription and reporting. The possible influence of deviations in treatment delivery on the end results of studies has been estimated and this fact has emphasized the immense value of QA in clinical trials (Poortmans et al. 2001).

As a rule, a QA program is implemented for every clinical trial utilizing any form of radiation technology. To be effective, the QA program must be implemented before protocol activation. It should also include a direct interactive feedback procedure. The main aspect of QA is not only to check protocol compliance or to detect an inconsistency, but also to give advice aimed at improving protocol compliance and ensuring that corrections are made as soon as possible. Thus, a QA structure should be friendly, informal, and confidential. Therefore, published data from a clinical trial QA procedure should always be in an anonymized format. Such QA programs that have been implemented by the ATC, as well as by the EORTC, have generally been well-accepted by all participants, even though it adds to the burden of busy clinics. However, it is mandatory that QA centers constantly review the credentialing level that has been implemented for a particular trial and ascertain that the level is appropriate.

Clearly there is a cost associated with clinical trials quality assurance. In the United States, a large part of the cost is provided by the NCI. However, some of it is provided by the participating institutions, particularly the time spent by investigators (radiation oncologists and radiation physicists). In addition, some part is sometimes funded by the pharmaceutical industry when relevant to the protocol. In Europe, the question of funding is not fully solved yet.

New technologies continue to be developed and implemented in radiation oncology at a progressively higher rate. As these new technologies make their way into clinical trials, they pose new challenges for cooperative groups and for the ATC. Current challenges for clinical trials QA include the following: (1) multimodality imaging (including, but not limited to PET, MRI, MRS) used for target definition has to be checked for data import and for subsequent image coregistration; (2) data submission and QA evaluation of IGRT, including EPID, MV and kV Cone Beam CT, Helical Tomotherapy MV CT, US; (3) QA review of the accuracy and quality of the institution's motion management methodology; (4) heterogeneous dose calculations including QA evaluation criteria; (5) proton beam therapy; (6) ATC-compliant data export for stereotactic specialized treatment systems (e.g., Elekta Gamma Knife and Accuray CyberKnife); and (7) new processes such as adaptive radiation therapy requiring deformable registration QA tools. It is essential that radiation oncologists and medical physicists interested in clinical trials monitor the progress and help contribute to the solution in solving the credentialing and QA review issues associated with these new technologies.

Summary

The level of appropriate quality assurance in clinical trials is highly dependent on the complexity of the trial. Quality assurance should be comprehensive and its effects should also extend to the treatment of all patients. By doing this, efficient quality assurance in clinical trials will improve the overall quality of treatment for most patients at an acceptable cost.

The ATC (particularly the ITC) has been a leading pioneer in the development of electronic data exchange and software for clinical trials QA. This effort has greatly benefited treatment planning system vendors in developing and verifying implementation of digital data export. The informatics infrastructure developed (and continually updated) by the ITC in support of the 3DOG/RTOG 9406 clinical trial has provided a robust methodology for digital data submission, archiving, and QA review of volumetric RT data objects and has been the enabling technology for RTOG to uniquely conduct 3DCRT, IMRT, SBRT, HDR, and prostate brachytherapy clinical trials that require volumetric digital data submission.

The amount of digital data that may need to be transferred and stored (and thus the available information) to the QA center for new advanced technology protocols will clearly continue to increase in the future. IGRT kilovoltage cone-beam computed tomography (CBCT) images or megavoltage (MV) helical CT images can now be acquired on a daily basis during a patient's treatment course, adding tens of Mbytes to Gbytes of imaging data that must stored. Also, there is an increase in the amount of diagnostic imaging data (CT, MR, US, PET, etc.) that needs to be submitted. Although the data storage issue can be addressed with tiered data storage technology, the QA of submitted imaging data, including registration and fusion, is very much a developmental effort at this time.

The ATC has developed mechanisms and tools for the collection and QA review of images and RT data using DICOM (RT) and RTOG Data Exchange that are available worldwide now and working reasonably well. Progress will continue to be made toward improved interoperability by DICOM WG-7 and by the IHE initiative shared by healthcare professionals and industry to improve the way computer systems in healthcare share information. This enables the use of established standards such as DICOM and HL7 to address specific clinical needs in support of optimal patient care. Systems developed in accordance with IHE communicate with one another better, are easier to implement, and enable care providers to use information more effectively (http://www.ihe.net/). Concerning radiation oncology, the IHE-RO Interoperability Profiles in the 2007 (RT Objects) and 2008 (Image Registration) Technical Frameworks will help make DICOM export from commercial treatment planning systems more consistent and will facilitate the use of commercial software for clinical trials QA.

Notwithstanding the progress that has been made there are still some other technology issues that deserve mention. For instance, some effort is still needed for QA centers' software to facilitate coordinated QA review of planning CT and verification images (CBCT, MVCT, etc.), as well as pre- and post-treatment diagnostic images. And, of course, case QA review tools (including deformable registration) for IGRT and adaptive RT protocols are needed for future protocols utilizing this modality.

However, the most immediate and pressing concern for clinical trials is harmonizing the credentialing/QA methodology and criteria for the clinical trials protocols worldwide in view of the increasing intergroup cooperation. Currently, the frequencies of the ERDA and IMRT credentialing appear to be the major challenges for attaining harmonization. The methodology for IGRT credentialing/QA is still in its infancy without an agreement on the methodology to be used. Also, agreement on the nomenclature to be used for target volumes and OAR would be of great benefit for improving efficiency of clinical trials QA, as the largest remaining variance in current radiotherapy lays not in the technical aspects, but in the proper delineation of volumes for treatment planning.

References

AAPM. 2004. *Report 86. Quality Assurance for Clinical Trials: A Primer for Physicists.* New York, NY: American Institute of Physics, Inc.

Ahnesjö, A., and M. M. Aspradakis. 1999. Dose calculations for external photon beams in radiotherapy. *Phys. Med. Biol.* 44: R99–R155.

Back, M., L. Oliver, R. Bromley, and T. Eade. 2008. Multicentre quality assurance of intensity-modulated radiotherapy planning: Beware the benchmarker (letter to the editor). *J. Med. Imaging Radiat. Oncol.* 52: 197.

Boehmer, D., P. Maingon, P. Poortmans, M. Barond, R. Miralbelle, V. Remouchampsf, C. Scraseg, A. Bossih, and M. Bollai. 2006. Guidelines for primary radiotherapy of patients with prostate cancer. *Radiother. Oncol.* 79: 259–269.

Bolla, M., H. Bartelink, G. Garavaglia, D. Gonzalez, J. C. Horiot, K. A. Johansson, G. van Tienhoven, K. Vantongelen, and M. van Glabbeke. 1995. EORTC guidelines for writing protocols for clinical trials of radiotherapy. *Radiother. Oncol.* 36 (1): 1–8.

Bortfeld, T., R. Schmidt-Ullrich, W. De Neve, and D. E. Wazer. 2006. *Image-Guided IMRT.* Berlin: Springer.

Bosch, W., J. Matthews, W. Straube, and J. Purdy. 2007. QuASAR: Quality assurance submission, analysis, and review system for advanced technology clinical trials in radiation therapy. Paper read at XVth International Conference on the Use of Computers in Radiation Therapy, June 4–7, 2007, Toronto, Canada.

Bosch, W., J. Matthews, K. Ulin, M. Urie, J. Yorty, W. Straube, T. J. FitzGerald, and J. A. Purdy. 2006. Implementation of ATC Method 1 for clinical trials data review at the Quality Assurance Review Center (abstract). *Med. Phys.* 33 (6): 2109.

Bosch, W. R., W. B. Harms Sr., J. W. Matthews, and J. A. Purdy. 2000. Database infrastructure for multi-institutional clinical trials in 3D conformal radiotherapy and prostate brachytherapy. Paper read at XIII International Conference on the Use of Computers in Radiation Therapy, Heidelberg, Germany.

Bosch, W. R., T. L. Lakanen, M. G. Kahn, W. B. Harms Sr., and J. A. Purdy. 1997. An image/clinical database for multi-institutional clinical trials in 3D conformal radiation therapy. Paper read at XII International Conference on the Use of Computers in Radiation Therapy, May, 1997, Salt Lake City, UT.

Bridier, A., H. Nyström, I. Ferreira, I. Gomola, and D. Huyskens. 2000. A comparative description of three multipurpose phantoms (MPP) for external audits of photon beams in radiotherapy: the water MPP, the Umeå MPP and the EC MPP. *Radiother. Oncol.* 55 (3): 285–293.

Das, I. J., C.-W. Chang, K. L. Chopra, R. K. Mitra, S. P. Srivastava, and E. Glatstein. 2008. Intensity-modulated radiation therapy dose prescription, recording, and delivery: Patterns of variability among institutions and treatment planning systems. *JNCI* 100 (5): 300–307.

Deasy, J. O., A. I. Blanco, and V. H. Clark. 2003. CERR: A computational environment for radiotherapy research. *Med. Phys.* 30 (5): 979–985.

Derreumaux, S., J. Chavaudra, A. Bridier, V. Rossetti, and A. Dutreix. 1995. A European quality assurance network for radiotherapy: Dose measurement procedure. *Phys. Med. Biol.* 40 (7): 1191–1208.

Dutreix, A., E. van der Schueren, S. Derreumaux, and J. Chavaudra. 1993. Preliminary results of a quality assurance network for radiotherapy centres in Europe. *Radiother. Oncol.* 29 (2): 97–101.

Ebert, M. A., A. Haworth, R. Kearvell, B. Hooton, R. E. Coleman, N. Spry, S. Bydder, and D. Joseph. 2008. Detailed review and analysis of complex radiotherapy clinical trial planning data: Evaluation and initial experience with the SWAN software system. *Radiother. Oncol.* 86: 200–210.

Ferreira, I. H., A. Dutreix, A Bridier, J. Chavaudra, and H. Svensson. 2000. The ESTRO-QUALity assurance network (EQUAL). *Radiother. Oncol.* 55 (3): 273–284.

FitzGerald, T. J., K. White, J. Saltz, A. Sharma, E. Siegel, M. Urie, K. Ulin et al. 2008. Development of a queriable database for oncology outcome analysis. In *Cured II—LENT Cancer Survivorship Research and Education. Late Effects on Normal Tissues*, ed. P. Rubin, L. S. Constine, L. B. Marks, and P. Okunieff. New York, NY: Springer.

Gagnon, W., L. W. Berkley, P. Kennedy, W. F. Hanson, and R. J. Shalek. 1978. An analysis of discrepancies encountered by the AAPM Radiological Physics Center. *Med. Phys.* 5 (6): 556–560.

Glicksman, A. S., L. E. Reinstein, D. McShan, and F. Laurie. 1981. Radiotherapy quality assurance program in a cooperative group. *Int. J. Radiat. Oncol. Biol. Phys.* 7: 1561–1568.

Golden, R., J. Cundiff, W. H. Grant III, and R. J. Shalek. 1972. A review of the activities of the AAPM Radiological Physics Center in interinstitutional trials involving radiation therapy. *Cancer* 29: 1468–1472.

Grant, III, W., J. Cundiff, W. Hanson, W. Gagnon, and R. Shalek. 1976. Calibration instrumentation used by the AAPM Radiological Physics Center. *Med. Phys.* 3 (5): 353–354.

Hall, W. H., M. Guiou, N. Y. Lee, A. Dublin, S. Narayan, S. Vijayakumar, J. A. Purdy, and A. M. Chen. 2008. Development and validation of a standardized method for contouring the brachial plexus: Preliminary dosimetric analysis among patients treated with IMRT for head-and-neck cancer. *Int. J. Radiat. Oncol. Biol. Phys.* 72 (5): 1362–1367.

Hansson, U., and K. A. Johansson. 1991. Quality audit of radiotherapy with EORTC mailed in water TL-dosimetry. *Radiother. Oncol.* 20 (3): 191–196.

Hansson, U., K. A. Johansson, J. C. Horiot, and J. Bernier. 1993. Mailed TL dosimetry programme for machine output check and clinical application in the EORTC radiotherapy group. *Radiother. Oncol.* 29 (2): 85–90.

Harms, Sr., W. B., W. R. Bosch, and J. A. Purdy. 1997. An interim digital data exchange standard for multi-institutional 3D conformal radiation therapy trials. Paper read at XII International Conference on the Use of Computers in Radiation Therapy, May, 1997, Salt Lake City, UT.

Holmes, T., R. Das, D. A. Low, F.-F. Yin, J. Balter, J. Palta, and P. Eifel. 2009. American Society of Radiation Oncology recommendations for documenting intensity-modulated radiation therapy treatments. *Int. J. Radiat. Oncol. Biol. Phys.* 74 (5): 1311–1318.

Horiot, J. C., K. A. Johansson, D. G. Gonzalez, E. van der Schueren, W. van den Bogaert, and G. Notter. 1986. Quality assurance control in the EORTC cooperative group of radiotherapy: 1. Assessment of radiotherapy staff and equipment. European Organization for Research and Treatment of Cancer. *Radiother. Oncol.* 6 (4): 275–284.

Horiot, J. C., J. Bernier, and K. A. Johansson, E. van der Schueren, and H. Bartelink. 1993. Minimum requirements for quality assurance in radiotherapy. *Radiother. Oncol.* 29: 103–104

Ibbott, G. S., D. S. Followill, A. Molineu, J. R. Lowenstein, P. E. Alvarez, and J. E. Roll. 2008. Challenges in credentialing institutions and participants in advanced technology multi-institutional clinical trials. *Int. J. Radiat. Oncol. Biol. Phys.* 71 (1): S71–S75.

Ibbott, G. S. 2009. QA for clinical dosimetry, with emphasis on clinical trials. In *Clinical Dosimetry Measurements in Radiotherapy*, ed. D. W. O. Rogers and J. E. Cygler. Madison, WI: Medical Physics Publishing.

ICRU. 1993. *ICRU Report 50: Prescribing, Recording, and Reporting Photon Beam Therapy.* Bethesda, MD: ICRU.

ICRU. 1999. *International Commission on Radiation Units and Measurements.* Prescribing, recording, and reporting photon beam therapy. Supplement to Report 50. Washington, DC: ICRU.

IMRTCWG. 2001. NCI IMRT Collaborative Working Group: Intensity modulated radiation therapy: Current status and issues of interest. *Int. J. Radiat. Oncol. Biol. Phys.* 51 (4): 880–914.

Jaffray, D. A., J. H. Siewerdsen, J. W. Wong, and A. A. Martinez. 2002. Flat-panel cone-beam computed tomography for image-guided radiation therapy. *Int. J. Radiat. Oncol. Biol. Phys.* 53 (5): 1337–1349.

Johansson, K. A., J. C. Horiot, J. Van Dam, D. Lepinoy, I. Sentenac, and G. Sernbo. 1986. Quality assurance control in the EORTC cooperative group of radiotherapy. 2. Dosimetric intercomparison. *Radiother. Oncol.* 7 (3): 269–279.

Johansson, K. A., J. C. Horiot, and E. van der Schueren. 1987. Quality assurance control in the EORTC cooperative group of radiotherapy: 3. Intercomparison in an anatomical phantom. *Radiother. Oncol.* 9 (4): 289–298.

Kirby, T. H., W. F. Hanson, R. J. Gastorf, C. H. Chu, and R. J. Shalek. 1986. Mailable TLD system for photon and electron therapy beams. *Int. J. Radiat. Oncol. Biol. Phys.* 12 (2): 261–265.

Kirby, T. H., W. F. Hanson, and D. A. Johnson. 1992. Uncertainty analysis of absorbed dose calculations from thermoluminescence dosimeters. *Med. Phys.* 19 (6): 1427–1433.

Lawton, C. A. F., J. Michalski, I. El-Naqa, M. K. Buyyounouski, W. R. Lee, C. Menard, E. O'Meara, S. A. Rosenthal, M. Ritter, and M. Seider. 2009. RTOG GU radiation oncology specialists reach consensus on pelvic lymph node volumes for high-risk prostate cancer. *Int. J. Radiat. Oncol. Biol. Phys.* 74 (2): 383–387.

Lawton, C. A. F., J. Michalski, I. El-Naqa, D. Kuban, W. R. Lee, S. A. Rosenthal, A. Zietman et al. 2009. Variation in the definition of clinical target volumes for pelvic nodal conformal radiation therapy for prostate cancer. *Int. J. Radiat. Oncol. Biol. Phys.* 74 (2): 377–382.

Martinez, A. A., D. Yan, D. Lockman, D. Brabbins, K. Kota, M. Sharpe, D. A. Jaffray, F. Vicini, and J. Wong. 2001. Improvement in dose escalation using the process of adaptive radiotherapy combined with three-dimensional conformal or intensity-modulated beams for prostate cancer. *Int. J. Radiat. Oncol. Biol. Phys.* 50 (5): 1226–1234.

Michalski, J. M., K. Winter, J. A. Purdy, R. B. Wilder, C. A. Perez, M. Roach, M. B. Parliament et al. 2003. Preliminary evaluation of low-grade toxicity with conformal radiation therapy for prostate cancer on RTOG 9406 dose levels I and II. *Int. J. Radiat. Oncol. Biol. Phys.* 56 (1): 192–198.

Michalski, J. M., C. Lawton, I. El Naqa, M. Ritter, E. O'Meara, M. J. Seider, W. R. Lee et al. 2009. Development of RTOG consensus guidelines for the definition of the clinical target volume for postoperative conformal radiation therapy for prostate cancer. *Int. J. Radiat. Oncol. Biol. Phys.* 76 (2): 361–368.

Michalski, J. M., M. Roach III, G. Merrick, M. S. Anscher, D. C. Beyer, C. A. Lawton, W. R. Lee et al. 2009. ACR Appropriateness Criteria® on external beam radiation therapy treatment planning for clinically localized prostate cancer: Expert panel on radiation oncology—prostate. *Int. J. Radiat. Oncol. Biol. Phys.* 74 (3): 667–672.

Michalski, J. M., K. Winter, J. A. Purdy, M. Parliament, H. Wong, C. A. Perez, M. Roach, W. Bosch, and J. D. Cox. 2005. Toxicity after three-dimensional radiotherapy for prostate cancer on RTOG 9406 dose Level V. *Int. J. Radiat. Oncol. Biol. Phys.* 62 (3): 706–713.

Molineu, A., D. S. Followill, P. A. Balter, W. F. Hanson, M. T. Gillin, M. S. Huq, A. Eisbruch, and G. S. Ibbott. 2005. Design and implementation of an anthropomorphic quality assurance phantom for intensity-modulated radiation therapy for the Radiation Therapy Oncology Group. *Int. J. Radiat. Oncol. Biol. Phys.* 63 (2): 577–583.

Palta, J. R., J. A. Deye, G. S. Ibbott, J. A. Purdy, and M. M. Urie. 2004. Credentialing of institutions for IMRT in clinical trials. *Int. J. Radiat. Oncol. Biol. Phys.* 59 (4): 1257–1259; author reply 1259–1261.

Perez, C. A., P. Gardner, and G. P. Glasgow. 1984. Radiotherapy quality assurance in clinical trials. *Int. J. Radiat. Oncol. Biol. Phys.* 10 (1): 119.

Pilepich, M. V., S. C. Prasad, and C. A. Perez. 1982. Computed tomography in definitive radiotherapy of prostatic carcinoma: Part 2. definition of target volume. *Int. J. Radiat. Oncol. Biol. Phys.* 8: 235.

Poortmans, P., A. Bossi, K. Vandeputte, M. Bosset, R. Miralbell, P. Maingon, D. Boehmer et al. 2007. Guidelines for target volume definition in post-operative radiotherapy for prostate cancer, on behalf of the EORTC Radiation Oncology Group. *Radiother. Oncol.* 82: 121–127.

Poortmans, P. M., J. L. Venselaar, H. Struikmans, C. W. Hurkmans, J. B. Davis, D. Huyskens, G. van Tienhoven, V. Vlaun et al. 2001. The potential impact of treatment variations on the results of radiotherapy of the internal mammary lymph node chain: A quality-assurance report on the dummy run of EORTC Phase III randomized trial 22922/10925 in Stage I–III breast cancer(1). *Int. J. Radiat. Oncol. Biol. Phys.* 49 (5): 1399–1408.

Poortmans P. M., J. B. Davis, F. Ataman, J. Bernier, and J. C. Horiot. 2005. The quality assurance programme of the Radiotherapy Group of the European Organisation for Research and Treatment of Cancer: Past, present and future. *Eur. J. Surg. Oncol.* 31: 667–674.

Purdy, J. A. 1996. 3-D radiation treatment planning: A new era. In *Frontiers of Radiation therapy and Oncology. 3-D Conformal Radiotherapy: A New Era in the Irradiation of Cancer*, ed. J. L. Meyer and J. A. Purdy. Basel: Karger.

Purdy, J. A. 2002. Dose-volume specification: New challenges with intensity-modulated radiation therapy. *Semin. Radiat. Oncol.* 12 (3): 199–209.

Purdy, J. A. 2004. Current ICRU definitions of volumes: Limitations and future directions. *Semin. Radiat. Oncol.* 14 (1): 27–40.

Purdy, J. A. 2007. From new frontiers to new standards of practice: Advances in radiotherapy planning and delivery. In *IMRT, IGRT, SBRT—Advances in the Treatment, Planning and Delivery of Radiotherapy*, ed. J. L. Meyer, J. A. Purdy, B. D. Kavanagh, and R. Timmerman. Basel: S. Karger AG.

Purdy, J. A. 2008. Quality assurance issues in conducting multi-institutional advanced technology clinical trials. *Int. J. Radiat. Oncol. Biol. Phys.* 71 (1): S66–S70.

Purdy, J. A., W. B. Harms, J. Michalski, and J. D. Cox. 1996. Multi-institutional clinical trials: 3-D conformal radiotherapy quality assurance. Guidelines in an NCI/RTOG study evalu-

ating dose escalation in prostate cancer radiotherapy. *Front. Radiat. Ther. Oncol.* 29: 255–263.

Purdy, J. A., W. R. Bosch, W. L. Straube, J. W. Matthews, R. J. Haynes, J. M. Michalski, E. M. Martin, K. Winter, W. J. Curran Jr., and J. D. Cox. 2006. A review of the activities of the ITC in support of RTOG advanced technology clinical trials (abstract). *Int. J. Radiat. Oncol. Biol. Phys.* 66 (3): S134.

Purdy, J. A., W. B. Harms, J. M. Michalski, and W. R. Bosch. 1998. Initial experience with quality assurance of multi-institutional 3D radiotherapy clinical trials. *Strahlenther. Onkol.* 174 (Supplement II): 40–42.

RTOG0236. 2002. A Phase II Trial of Stereotactic Body Radiation Therapy (SBRT) in the Treatment of Patients with Medically Inoperable Stage I/II Non-Small Cell Lung Cancer.

Straube, W., W. Bosch, J. Matthews, R. Haynes, and J. Purdy. 2006. Digital data integrity QA for multi-institutional clinical trials (Abstract). *Med. Phys.* 33 (6): 2087.

Straube, W., J. Matthews, W. Bosch, and J. A. Purdy. 2005. DVH analysis: Consequences for quality assurance of multi-institutional clinical trials (Abstract). *Med. Phys.* 32 (6): 2021.

Swinnen, A., J. Verstraete, and D. P. Huyskens. 2002. The use of a multipurpose phantom for mailed dosimetry checks of therapeutic photon beams: 'OPERA' (operational phantom for external radiotherapy audit). *Radiother. Oncol.* 64 (3): 317–326.

Van Dam, J., K. A. Johansson, A. Bridier, G. Sernbo, and U. Hansson. 1993. EORTC radiotherapy group quality assurance: Mechanical checks and beam alignments of megavoltage equipment. *Radiother. Oncol.* 29 (2): 91–96.

van Tienhoven, G., N. A. van Bree, B. J. Mijnheer, and H. Bartelink. 1991. Quality assurance of the EORTC trial 22881/10882: "Assessment of the role of the booster dose in breast conserving therapy": The dummy run. EORTC Radiotherapy Cooperative Group. *Radiother Oncol* 22 (4): 290–298.

Webb, S. 2000. *Intensity-Modulated Radiation Therapy*. Bristol: Institute of Physics Publishing.

Williams, M. J., M. J. Bailey, D. Forstner, and P. E. Metcalfe. 2007. Multicentre quality assurance of intensity-modulated radiation therapy plans: A precursor to clinical trials. *Australas. Radiol.* 51: 472–479.

Williams, M. J., M. J. Bailey, D. Forstner, and P. E. Metcalfe. 2008. Reply: Multicentre quality assurance of intensity-modulated radiotherapy planning: Beware the benchmarker (letter to the editor). *J. Med. Imaging Radiat. Oncol.* 52: 303.

Willins, J., and L Kachnic. 2008. Clinically relevant standards for intensity-modulated radiation therapy dose prescription. *JNCI* 100 (5): 288–290.

Wu, D. H., N. A. Mayr, Y. Karatas, R. Karatas, M. Adli, S. M. Edwards, J. D. Wolff, A. Movahed, J. F. Montebello, and W. T. C. Yuh. 2005. Interobserver variation in cervical cancer tumor delineation for image-based radiotherapy planning among and within different specialties. *J. Appl. Clin. Med. Phys.* 6 (4): 106–110.

35

Vendor's Role in Quality Improvement

Ken Cashon
.decimal, Inc.

Chris Warner
.decimal, Inc.

Richard Sweat
.decimal, Inc.

Introduction

The main role of a vendor is to create new hardware and/or software products. This also includes products that facilitate quality control measurements or analysis of those measurements. One difficulty in implementing new hardware or software technologies is that the end user (customer) is largely left on their own to determine the appropriate clinical implementation of a new product. The vendor readily provides the "how-to" regarding their product, but they rarely, if ever, provide quantitative information on whether an implementation is optimal or less than optimal. Furthermore, it is very difficult, if not impossible, to determine what constitutes an optimal implementation for a given product. This leaves each institution more-or-less to their own internal experimentation procedures to commission and implement a product. Differences in acceptance testing, commissioning, and implementation naturally lead to variation in interinstitutional performance, for example, the accuracy of dose delivery (measured dose vs. planned dose) varies across institutions. This type of variation has been shown to exist for intensity modulated radiation therapy (IMRT) quality control (QC) measurements across different institutions (Pawlicki et al. 2008). There is a wide range of potential sources of variation in IMRT programs—linear accelerator mechanicals, treatment planning system (TPS) dose algorithm capabilities, TPS beam modeling quality, measurement equipment and methods, and dose quality assurance (QA) evaluation strategies, to name a few. It can be difficult to isolate and mitigate individual sources of errors that lead to problems, such as two sites with the same delivery and TPS equipment showing different levels of performance. Seminal publications have outlined tolerances for commissioning and QA of planning systems (Van Dyk et al. 1993), but to date, there are still no required standards regarding the methods used to assay the system.

This chapter describes one vendor's approach in working with the customer to improve quality. The reasons for taking a new approach, as well as some potential benefits, are discussed.

Background

The vendor (.decimal, Inc.) constructs physical compensators, among other things, for IMRT delivery. In general, the approach to improve quality works by the following six steps:

1. An IMRT plan is approved for patient treatment at an institution.
2. The optimized plan is exported and sent to the vendor.
3. The physical compensators are designed from the optimized fluence maps.
4. The compensators are manufactured and shipped to the institution.
5. Patient-specific IMRT QA is performed at the institution.
6. The patient is treated.

A major hurdle to do this successfully is that the vendor has to ensure that their compensators are being made accurately and precisely. To this end, the vendor had their own internal quality assurance steps to ensure that the optimized plan arrives correctly from the institution and the compensators are made correctly. Then, the final check on the correctness of the compensators is done by the physicist at the institution as part of their routine patient-specific IMRT QA program. Originally, it was thought that this process would be sufficient to ensure quality treatments. Early on and from time-to-time, the customer would obtain results from their patient-specific IMRT QA that indicated a potential problem with a compensator. Using the Six Sigma methodology (DMAIC—define, measure, analyze, improve, control—described in other chapters of this book), the vendor continued to improve their processes in making the compensators. At some point, no matter how much the vendor improved the construction of their compensators, roughly the same numbers of potential problems were being identified by different users. It was at this point that the vendor realized that in order to further improve their product from the customer's point of view, they could not just concentrate on the process of creating the compensators, but they must collaborate directly with the customer on the implementation and use of their system to make compensators for IMRT treatments.

Vendor Collaboration with the Customer

The vendor decided to apply the Six Sigma methodology to assisting the customer with implementing their product. The following describes what resulted from that initiative. The major objectives of the vendor's approach were to: (1) standardize the process of compensator implementation; (2) provide consistent, objective performance measures regarding each customer's ability to accurately deliver the optimal planned dose; (3) give feedback to identify and remove errors in using the compensators; and (4) allow each customer to see how their performance compares to the performance of their (anonymous) peers to understand if they should improve and to what potential level. In order to achieve these objectives, the following elements were applied:

- Employ well-defined standard measurements so that results are familiar to the customers.
- Generate quantitative metrics so that results are objective.
- Control all variables of the system that are controllable (treatment planning and dose delivery) in order to focus on distinct parts of the system and definitively identify sources of error and/or variation.
- Accrue results for a large number of institutions so that each customer can view their performance versus the population of institutions.

Does Plan Equal Actual?

The compensators require both the dose planning and delivery systems. Ultimately, the question that the vendor needed to answer is: Does the planned dose equal the actual dose? The main point is that it is not sufficient for the vendor to make a high-quality product; the vendor must provide support to ensure their product is implemented optimally. The flowchart shown in Figure 35.1 describes how the vendor achieved this in practice. A collection of standard equipment is used at each new institution. The equipment (shown in Figure 35.2) used is readily available off-the-shelf from other vendors.

Using the standard equipment is imperative, as the outputs of different equipment systems vary, and any such variation would make objective conclusions about the implementation more difficult. If different 2D measurement arrays were used, differences in the detector resolution and accuracy (along with variations in software analysis methodologies employed by each array) would prohibit meaningful direct intercomparisons of data. For example, a 2%/2-mm passing criteria for array A using detector B and calculation method C is not interchangeable with array X using detector Y and calculation method Z.

Assess Mechanicals

There are some basic mechanical prerequisites for all modes of complex dose delivery (IMRT, rotational, particle therapy, etc.). Verification that the customer's equipment meets specifications is done by the physicist at the institution. Specifically, these tests include:

- *Coincidence of compensator tray/machine crosshairs.* Since compensators are mounted to trays and positioned for each treatment beam, the position of the compensator in the beam path is tested with a tray/machine crosshair apparatus.
- *Collimator rotation.* Since each compensator is mounted to a collimator assembly, errors in the collimator isocenter position could cause errors with the compensator delivery. A collimator walk-around test is performed to determine if the crosshairs drift when rotating the collimator.
- *Treatment table validity.* Conventional tests are performed to assess table sag, rigidity, and positional accuracy.
- *Radiation isocenter.* Typical annual physics tests are used for the assessment of the radiation isocenter.

Assess Basics of the TPS Dose Algorithm/Beam Model

The TPS algorithm's ability to model the dose delivered through solid materials is verified. Specifically, field size effects (scatter from the compensator) and beam hardening are assessed. To measure the TPS dose algorithm's performance, absolute dose rates are measured for open fields over a range of field sizes and depths, followed by measurements of the same fields but delivered through uniform slabs of the compensator alloys. Then, the TPS is used to simulate those measured fields and the absolute calculated dose rates are compared to the actual measured dose rates. Variation with field size indicates how the TPS will perform regarding modeling variable scatter from the compensator, while variation with alloy slab thickness indicates how well the TPS models beam hardening.

At this point, if the measured versus calculated point doses do not agree for these simple arrangements (typically, differences less than 2% are required), the customer needs to make some adjustments to their beam model until their TPS calculations achieves at least a 2% agreement with measurement.

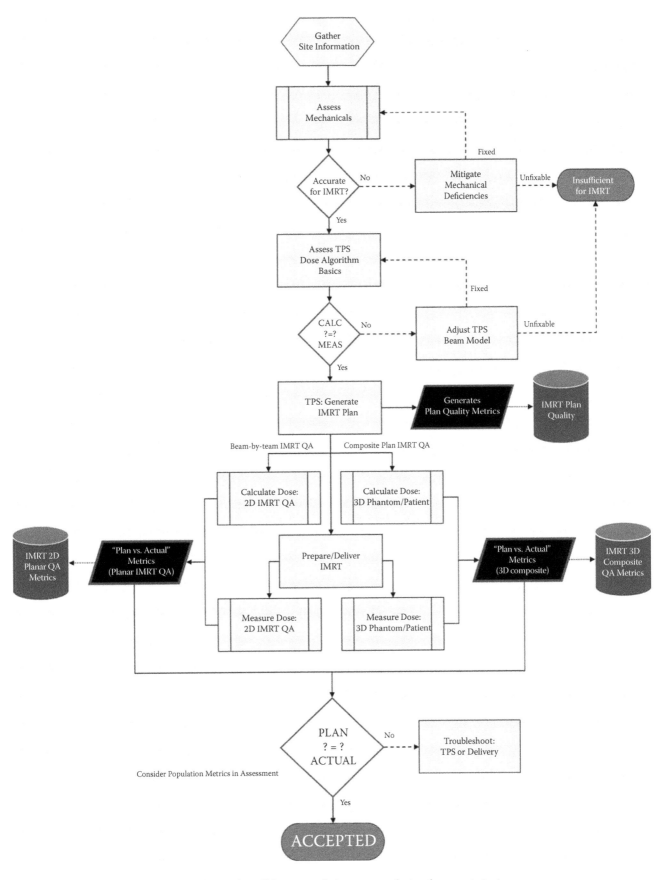

FIGURE 35.1 Flowchart describing how the vendor collaborates with the customer during the commissioning process.

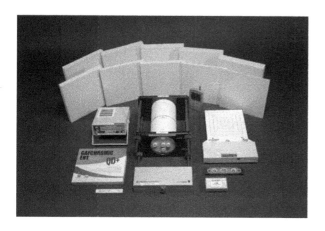

FIGURE 35.2 Collection of standard equipment, available off the shelf from other vendors, used at each new institution for commissioning.

End-to-End Testing

The customer then generates a seven-field IMRT plan designed on a standard head/neck phantom (CIRS, Norfolk, VA) with challenging predefined regions-of-interest (ROI) and dose objectives. Dose objectives are specified for each ROI (targets and sensitive structures). The treatment planner optimizes the seven beams to achieve all dose objectives.

A plan quality metric (a single quantitative metric from 0 to 100, with 100 being the best) is calculated as the sum of individual ROI plan metrics. The ROI-specific performance metrics are functions of the dose/DVH statistics from the treatment plan. Some ROI performance metrics are "all or none," such as that for the Target_1 getting ≥70 Gy or the serial organ at risk (OAR) getting <45 Gy, while other ROI performance metrics scale with a penalty levied the farther the dose objective strays from the goal. The details of the ROI metrics and composite plan metrics are detailed in Table 35.1.

Upon completing the phantom treatment plan, individual IMRT QA dose planes are calculated using a conventional IMRT QA process, that is, calculating each intensity-modulated beam separately, incident on a flat-water phantom at a set distance and depth in phantom, which will be used later for the beam-by-beam analysis.

The phantom is set up and a dose fraction is delivered. The internal dose is measured in two ways: (1) a single point dose at the isocenter, measured using an Exradin A16 Microchamber and MAX 4000 Electrometer (Standard Imaging, Middleton, WI); and (2) an axial dose plane, measured with Gafchromic EBT film (International Specialty Products, Wayne, NJ). These point doses and dose planes are compared directly to the calculated patient plan.

Beam-by-beam IMRT QA dose is measured with the Map-CHECK diode array (Sun Nuclear Corporation, Melbourne, FL). The diodes provide superior dose measurements for complex intensity-modulation due to their small detector size (0.8 × 0.8 mm active area in the BEV), which reduces the volume averaging effect of detectors with larger BEV area. These per-beam measurements are compared to the IMRT QA planes calculated by the TPS.

The dose difference and distance-to-agreement (DTA) pass/fail rates for a range of criteria are collected. All raw data are preserved so that any metrics can be generated in the future if a new metric is determined to be a better indicator of process performance than the ones currently used. Once the customer-specific performance metrics are generated, they can be compared to the population distributions of the same metrics, allowing the customer to decide if their site/equipment/TPS is performing accurately enough. This approach allows the constant monitoring

TABLE 35.1 Plan Quality Metric

Region of Interest (ROI)	Plan Goal	ROI Performance Metric	
Target 1	V70 Gy ≥ 95%	25	if V70 ≥ 95%
		0	if V70 < 95%
Target 1	ROI_{Max} < 73.5 Gy	5	if ROI_{Max} ≤ 73.5 Gy
		$5 - 2 \cdot (ROI_{Max} - 73.5)$	if ROI_{Max} > 73.5 Gy
		0	if ROI_{Max} > 76.0 Gy
Target 2	V50 Gy ≥ 95%	20	if V50 ≥ 95%
		$20 - 2 \cdot (95 - V50)$	if 85% < V50 < 95%
		0	if V50 ≤ 85%
Parallel 1	V22 Gy < 25%	10	if V22 ≤ 25%
		$10 - (V22 - 25)$	if 25% < V22 < 35%
		0	if V22 ≥ 35%
Parallel 2	V35 Gy < 25%	10	if V22 ≤ 35%
		$10 - (V22 - 35)$	if 35% < V22 < 45%
		0	if V22 ≥ 45%
Serial	ROI_{Max} < 45 Gy	25	if ROI_{Max} < 45 Gy
		0	if ROI_{Max} ≥ 45 Gy
Entire volume (global max)	D_{Max} < 77 Gy	5	if D_{Max} ≤ 77 Gy
		$5 - 2 \cdot (D_{Max} - 77)$	if D_{Max} > 77 Gy
		0	if D_{Max} > 79.5 Gy

Note: Composite plan metric = Σ (ROI performance metrics); perfect score = 100.0.

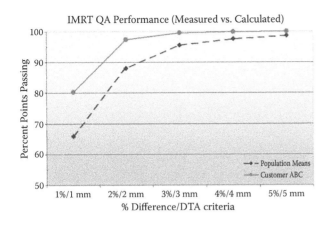

FIGURE 35.3 Beam-by-beam IMRT QA results for "Customer ABC" are plotted (for various IMRT QA criteria) versus the population means for those same criteria.

of variations across institutions, which can help identify best practices.

Key Performance Metrics

Examples of site-specific performance metrics versus population distributions are shown in Figures 35.3–35.5.

In Figure 35.3, the beam-by-beam IMRT QA results for "Customer ABC" are plotted (for various IMRT QA criteria) versus the population means for those same criteria. This is an example of a customer site whose performance exceeds the population averages.

Figure 35.4 shows data acquired for a new customer who was commissioning five linear accelerators. It shows where that customer's performance fell within a population distribution of a metric called delivery capability index (DCI), which is calculated

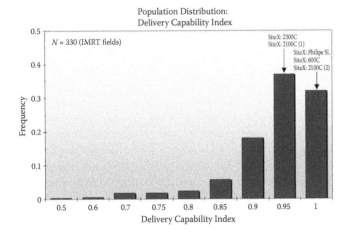

FIGURE 35.4 Data acquired for a new customer commissioning five linear accelerators. The customer's performance fell within a population distribution of a metric called the delivery capability index (DCI), where a perfect DCI performance is DCI = 1.00.

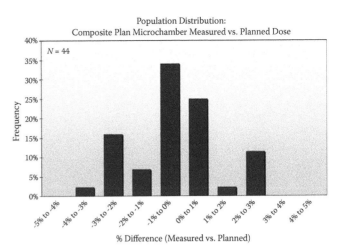

FIGURE 35.5 The variation in measured versus planned absolute point dose at the isocenter for the composite treatment delivered to the phantom.

as the average of the fractional passing rates in beam-by-beam analysis for 1%/1 mm DTA through 5%/5 mm DTA (perfect DCI performance is DCI = 1.00). All five linear accelerators performed well. It is interesting to note that the make and model of the linac and MLC made little difference in IMRT performance. In fact, some of the older models were on par or superior to much newer models.

Figure 35.5 shows the variation in measured versus planned absolute point dose at the isocenter for the composite treatment delivered to the phantom. The distribution of composite plan quality metrics is shown in Figure 35.6. Most of the calculated plans achieve very high performance scores for the test IMRT plan. Some of the lower plan quality scores (plan quality metric <85) were the result of misuse of optimization parameters in the TPS and, in some cases, due to poor beam models.

Figure 35.7 illustrates an outlier in beam-by-beam IMRT QA results for one particular institution. This institution's overall

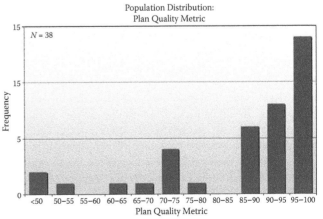

FIGURE 35.6 The distribution of composite plan quality metrics.

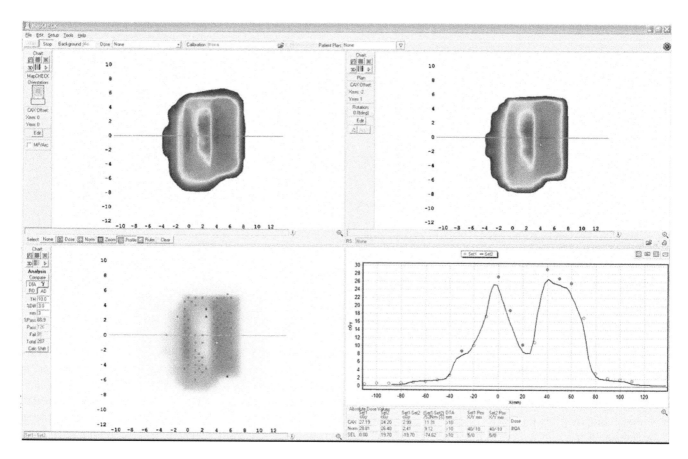

FIGURE 35.7 An outlier in beam-by-beam IMRT QA results for one particular institution. This institution's overall delivery capability index was measured to be 0.599, falling well into the underperforming tail of the population distribution shown in Figure 35.4.

delivery capability index was measured to be 0.599, falling well into the underperforming tail of the population distribution shown in Figure 35.4.

Summary

This commissioning and implementation approach is meant to close the gap between the treatment plan and actual delivered dose for customers using .decimal's compensator-based IMRT. The main point is that a new approach for product implementation was required in order to realize the full product potential. Assisting the customer with product implementation was good for business. However, it also has the potential to help the field in the following way. The vendor has information that an individual customer does not have, namely, a database of product implementation metrics. This would provide the information for any single institution to know whether their implementation is great, good, average, or poor. It is also useful for institutions to benchmark their current IMRT delivery capabilities against a database, which could potentially improve the standard of IMRT delivery.

In radiation therapy, we concentrate primarily on the performance means but tend to accept a lot of variation. For example, there may be a lot of variation in Physician A's contouring of the target versus Physician B's, or there is variation of the performance of Machine C from plan-to-plan or day-to-day, or there is variation from institution to institution concerning the per-plan prescriptions (Das et al. 2008) and QA measures (Nelms and Simon 2007). Two aspects of optimal quality are: (1) being on target and (2) minimal variation about the target. If the mean performance of a product or service deviates from the target, then there is poor quality. If there is variability in the performance, then there is poor quality. This approach to quality is widely accepted in other fields (Taguchi 1986) and being adopted in healthcare (Carey and Lloyd 1995). Working toward greater accuracy (on target) while working toward greater precision (minimal variation) continues to improve quality. This requires a mindset of *continual improvement* rather than a conventional mindset of *good enough* (i.e., within specifications).

Beyond radiation dose planning and delivery, there are other areas of radiation oncology that are well-suited for quantitative performance assessments and comparisons with population performance distributions. These areas include: target delineation methods (physician to physician, institution to institution, etc.); critical organ delineation (including, and especially, commercial autosegmentation algorithms); image-guided radiation therapy (IGRT); patient immobilization methods; and organ motion and gating techniques. An approach like this may also help in reducing variability in the submitted data for clinical trials.

Another benefit to generating these results it that, after a critical mass of results have been collected over many sites, it becomes obvious which errors can be attributed to imperfections in beam modeling (which can be repaired) and what degree of error is inherent in the imperfect TPS methods. The plan quality metric may be important for two reasons: (1) by requiring challenging plan goals, the complexity of beam modulation is ensured institution-to-institution; and (2) a database could prove valuable in assessing the capabilities of TPS optimization algorithms (TPS vs. TPS and/or optimization parameters vs. optimization parameters). A database such as this carries vital information to help the TPS vendors improve their software products and providing feedback to suppliers is a hallmark of a good quality system.

References

Carey, R. G., and R. C. Lloyd. 1995. *Measuring Quality and Improvement in Healthcare.* Milwaukee, WI: Quality Press.

Das, I. J., C. W. Cheng, K. L. Chopra, R. K. Mitra, S. P. Srivastava, and E. Glatstein. 2008. Intensity-modulated radiation therapy dose prescription, recording, and delivery: Patterns of variability among institutions and treatment planning systems. *J. Natl. Cancer Inst.* 100 (5): 300–307.

Nelms, B. E., and J. A. Simon. 2007. A survey on IMRT QA analysis. *J. Appl. Clin. Med. Phys.* 8 (3): 76–90.

Pawlicki, T., S. Yoo, L. E. Court, S. K. McMillan, R. K. Rice, J. D. Russell, J. M. Pacyniak et al. 2008. Process control analysis of IMRT QA: Implications for clinical trials. *Phys. Med. Biol.* 53 (18): 5193–5205.

Taguchi, G. 1986. *Introduction to Quality Engineering: Designing Quality into Products and Processes.* Tokyo: Asian Productivity Organization.

Van Dyk, J., R. B. Barnett, J. E. Cygler, and P. C. Shragge. 1993. Commissioning and quality assurance of treatment planning computers. *Int. J. Radiat. Oncol. Biol. Phys.* 26 (2): 261–273.

IV

People and Quality

36

Role of Leadership

Nina Kowalczyk
Ohio State University

Ann Scheck McAlearney
Ohio State University

Introduction

While many quality indicators within radiation oncology are quite visible, departmental success is often based less on technical performance indicators than on the quality of leadership. Yet investment in the development of leaders in medicine, and especially in radiology, appears to remain an underrated priority (Gunderman 2004). Although a definite imperative may exist to improve quality and efficiency, executing this through effective management remains a challenge because of the lack of leadership training. Similar to the circumstances in many clinical areas (Guthrie 1999), in many radiology and radiation oncology departments, an individual's expertise as a radiologist or radiation oncologist is greatly out of proportion to his or her expertise as a leader (Gunderman 2004).

Leadership and Development

Significant change in the clinical setting can only be achieved through the cooperation and engagement of clinicians, as the daily decisions of physicians, nurses, physicists, dosimetrists, therapists, and other staff greatly impact the care experienced by patients and other customers (Ham 2003). Clinical leaders from professional backgrounds can have a significant impact on improving health services (Eagle et al. 2003; Berwick 1994), but all too often clinicians recognize the need to cultivate their leadership skills only after assuming leadership roles. Therefore, it is vital that clinicians reflect on leadership at every stage of professional development (Gunderman 2004; Souba 2006).

The Association of American Medical Colleges suggests that healthcare quality in the twenty-first century requires a paradigm shift away from autonomy toward teamwork and clinical leadership in healthcare (Shine 2002). One major factor that differentiates the best performing organizations in terms of quality and efficiency is an organizational culture that embraces the engagement of clinical leaders in formal leadership development programs. Leadership development programs provide education and training for both new and experienced leaders (McAlearney 2008a, 2008b; Hernez-Broome and Hughes 2004; Casebow 2006) and have been shown to have measurable effects on organizational culture (Schein 2002) and organizational climate (Moxnes and Eilersten 1991). Further, leadership development programs provide opportunities for managing organizational change and limiting resource use while improving quality and efficiency in a healthcare setting (Xirasagar, Samuels, and Stoskopf 2005; Xirasagar, Samuels, and Curtin 2006; Stoller 2008).

At this point in time, there is clearly an essential need for strong leaders in medicine (McAlearney 2006; Schwartz and Pogge 2000), and leadership development activities and programs can help. Improving the leadership skills of all members of the healthcare team increases the competencies and capabilities of the physicians, nurses, therapists, dosimetrists, physicists, and other staff members within a radiation oncology department. In addition, healthcare quality is improved when the physicians and staff function as a medical team to jointly engage in problem-solving (Shine 2002). Table 36.1 provides examples of activities that can be supported within a healthcare organization committed to helping enhance the leadership capacity of its physicians and other clinicians.

Organization- or System-Level Leadership

A greater impact is obtained when the leadership education and development is provided at an enterprise or system level allowing for a strong organizational focus (McAlearney 2008a). Offering internal leadership conferences or online educational opportunities in leadership are successful approaches to educating and training personnel at all levels in the organization. These educational approaches, utilizing centralized, organization-wide

TABLE 36.1   Activities to Increase Organizational Focus on Quality and Efficiency

Department-Level Activities

- Form a journal club or book club
  ○ Assign a journal article or book focused on leadership and hold a discussion about that reading (e.g., Institute of Medicine report; article on Toyota Production Systems/Lean, etc.)
  ○ Follow up club meetings with reflection exercises
  ○ Invite participants to chair one session and select reading
- Develop a mealtime series focused on leadership topics related to quality and efficiency
  ○ Invite internal or external experts for focused discussion
  ○ Follow-up discussions with recommended readings
- Focus a regular part of staff meetings on quality and efficiency topics
  ○ Create opportunities for clinicians to get involved in investigating issues involving quality of care, patient safety, departmental efficiency, and so forth
  ○ Consider new ways to communicate about quality of care and efficiency
- Schedule monthly brainstorming sessions within the department on targeted topics
  ○ Focus sessions on issues such as how to proceed more rapidly with quality improvement processes, how to reduce waste in key areas, etc.
  ○ Hold lunch sessions with invited internal or external experts for focused discussion

Organization-Level Activities

- Have executives host chats to encourage communication with senior leadership
  ○ Focus conversations on topics related to quality of care and efficiency
  ○ Consider more formal executive forums on these topics
- Utilize existing education and training opportunities
  ○ Direct physicians, employees, and employee groups to existing educational and training opportunities to enhance development in targeted areas
- Consider revision of performance evaluation process
  ○ Establish linkages between individual performance and organizational performance in areas related to quality of care and efficiency
  ○ Hold executive leadership accountable for these metrics
- Develop a recognition program to reward creativity, innovation, and appropriate risk-taking in areas of quality of care, efficiency, and so forth
  ○ Consider appropriate forum for communicating about such initiatives to encourage discussion and transparency around these issues
- Foster multidisciplinary communication among clinicians and administrators to increase organization-wide knowledge about both perspectives in the context of quality of care and efficiency
  ○ Explore opportunities such as job shadowing, multidisciplinary committees, etc.

Source: McAlearney, A. S., *J. Healthc. Manag.,* 53(5), 319–331, discussion, 331–332; McAlearney, A.S. *Advances in Health Care Management,* 7, 205–231.

TABLE 36.2   Using Leadership Development Activities to Improve Quality and Efficiency in Healthcare Organizations

Leadership Development Goals	Potential Leadership Development Activities
Improve the caliber of the workforce	- Offer education and development programming to enhance leadership skills and capabilities of the workforce, including clinicians - Focus employee and clinician training in specific areas related to cost reduction, quality of care, etc. - Consider requiring structured follow-up after education and development activities in order to permit participants to apply new skills
Reduce employee turnover and turnover-related expenses	- Link performance evaluation process for managers and leaders to employee satisfaction (e.g., survey results), emphasizing importance of employee retention and satisfaction - Offer leadership skills training for clinicians to improve likelihood of leadership success when clinicians are promoted to management or leadership positions
Promote organizational culture that emphasizes quality of care and efficiency	- Investigate opportunities within the organization to emphasize quality of care and efficiency in visible ways such as revision of mission, vision, or values statement - Emphasize topics of quality of care and efficiency in regular communications
Focus attention on organizational strategic priorities	- Explicitly link leadership development activities and training to achievement of the organization's strategic goals (e.g., if a goal is to improve quality of care, include training in quality improvement methods) - Design and deliver targeted education and development modules, courses, and programs around topics related to quality of care, operational efficiency, etc. - Tie individual performance evaluation metrics to organizational goals around improving quality of care, reducing costs, improving patient satisfaction, and so forth

Source: McAlearney, A. S., *J. Healthc. Manag.,* 53(5), 319–331, discussion, 331–332.

leadership development programs, allow professionals to network and learn from one another. System-level leadership development programs improve quality and efficiency in a consistent manner and create opportunities for the application of new skills in the actual context of the work environment (McAlearney 2008a). They also allow the leadership development programs to be custom-made for a particular institution, permitting the training to be focused on specific organizational priorities (McAlearney et al. 2005; Emans et al. 2008). Moreover, developing leaders within an organization allows for internal promotion of qualified personnel throughout the organization, thus eliminating the need to hire senior leaders from the outside. Promoting from within an organization increases commitment to the organization and encourages a shared organizational vision. Table 36.2 highlights the opportunities of leadership development activities and programs to improve quality of care and efficiency in healthcare organizations.

Organizational Change

Most clinicians are motivated by the ability to offer high-quality patient care in a courteous and timely manner. The potential to eliminate errors and improve patient safety most commonly occurs through the systematic evaluation and improvement of existing organizational processes to identify barriers to service quality. In most instances, the medical staff are trying to do the best they can to provide excellent care, but they are often burdened by inefficient processes. The clinicians who deal with the processes on a daily basis are generally very effective at identifying causes of the inefficiencies and are able to propose viable corrective actions. Allowing clinical teams to be actively engaged in organizational change to improve quality is much more effective than having a change imposed upon them (Ham 2003). Therefore, if the clinical team members are active participants in efforts to redesign health services in ways that eliminate barriers to providing high-quality care, those team members are more likely to feel valued by the organization and be committed to the necessary change. This, in turn, can increase staff motivation and job satisfaction, which not only increases the quality of care provided to patients, but can also translate into cost savings through reduced turnover rates and higher levels of commitment to the organization (McAlearney 2008a). According to the U.S. Department of Labor (U.S. Department of Labor 2009), the estimated cost of replacing a management-level healthcare worker is approximately one-third of the associated annual salary. High turnover rates cost healthcare institutions large sums of money. Therefore, implementing practices aimed to reduce employee turnover and increase employee commitment to the organization can help those organizations save thousands of dollars annually.

Summary

Without a doubt, the role of effective leadership is crucial in ensuring quality and efficiency in healthcare. Regardless of the organizational approach to leadership development, improving the quality of clinical leaders has the capability to improve care quality in an efficient and effective manner. Whether an organization chooses to implement an enterprise-level leadership development program to strategically focus the training on organizational priorities or chooses to engage clinicians in a general leadership development program, the potential outcomes include a higher caliber workforce, increased employee satisfaction, increased organizational commitment, and decreased employee turnover. All of these outcomes, in turn, can aid in efforts to achieve the ultimate outcome of delivering high-quality patient care and service in radiology and radiation oncology departments.

References

Berwick, D. M. 1994. Eleven worthy aims for clinical leadership of health system reform. *JAMA* 272 (10): 797–802.

Casebow, P. 2006. Learning: How to get the right blend. *Train. Manag. Dev. Methods* 20 (5): 107–113.

Eagle, K. A., A. J. Garson Jr., G. A. Beller, and C. Sennett. 2003. Closing the gap between science and practice: The need for professional leadership. *Health Aff. (Millwood)* 22 (2): 196–201.

Emans, S. J., C. T. Goldberg, M. E. Milstein, and J. Dobriner. 2008. Creating a faculty development office in an academic pediatric hospital: Challenges and successes. *Pediatrics* 121 (2): 390–401.

Gunderman, R. B. 2004. Why radiology leaders fail. *J. Am. Coll. Radiol.* 1 (5): 359–360.

Guthrie, M. B. 1999. Challenges in developing physician leadership and management. *Front. Health Serv. Manage* 15 (4): 3–26.

Ham, C. 2003. Improving the performance of health services: The role of clinical leadership. *Lancet* 361 (9373): 1978–1980.

Hernez-Broome, G., and R. Hughes. 2004. Leadership development: Past, present, and future. *Hum. Resour. Plann.* 27 (1): 24–32.

McAlearney, A. S., D. Fisher, K. Heiser, D. Robbins, and K. Kelleher. 2005. Developing effective physician leaders: Changing cultures and transforming organizations. *Hosp. Top.* 83 (2): 11–18.

McAlearney, A. S. 2006. Leadership development in health care organizations: A qualitative analysis. *J. Organ. Behav.* 27 (7): 967–982.

McAlearney, A. S. 2008a. Using leadership development programs to improve quality and efficiency in healthcare. *J. Healthc. Manag.* 53 (5): 319–331; discussion 331–332.

McAlearney, A. S. 2008b. Improving patient safety through organizational development: Considering the opportunities. *Adv. Health Care Manag.* 7: 205–231.

Moxnes, P., and D. Eilersten. 1991. The Influence of management training upon organizational climate: An exploratory study. *J. Organ. Behav.* 25: 399–411.

Schwartz, R. W., and C. Pogge. 2000. Physician leadership: Essential skills in a changing environment. *Am. J. Surg.* 180 (3): 187–192.

Shine, K. I. 2002. Health care quality and how to achieve it. *Acad. Med.* 77 (1): 91–99.

Souba, W. W. 2006. The inward journey of leadership. *J. Surg. Res.* 131 (2): 159–167.

Stoller, J. K. 2008. Developing physician-leaders: Key competencies and available programs. *J. Health Adm. Educ.* 25 (4): 307–328.

U.S. Department of Labor. 2009. Available from www.bls.gov/jlt (accessed 9/2/09).

Xirasagar, S., M. E. Samuels, and T. F. Curtin. 2006. Management training of physician executives, their leadership style, and care management performance: An empirical study. *Am. J. Manag. Care* 12 (2): 101–108.

Xirasagar, S., M. E. Samuels, and C. H. Stoskopf. 2005. Physician leadership styles and effectiveness: An empirical study. *Med. Care. Res. Rev.* 62 (6): 720–740.

37

Human Factors Engineering: Overview

A. Joy Rivera-Rodriguez
University of Wisconsin

Ben-Tzion Karsh
University of Wisconsin

Introduction

In its 2000 report, the Institute of Medicine (IOM) made the public aware of the seriousness of patient safety, reporting that 44,000 to 98,000 individuals die each year in hospitals from medical errors (Institute of Medicine 2000). Radiotherapy (RT), specifically, has had its share of significant events that have caused harm to patients. One of the most publicized series of events was related to the Therac-25 radiation machine; six patients received massive overdoses of radiation (Leveson and Turner 1993). There have been other less publicized, but just as critical, events. Ostrom et al. (1996) examined 35 misadministration events reported to the Nuclear Regulatory Commission (NRC) in 1992 and described seven of them in detail. Four of the seven events were wrong site administrations and the other three events were wrong dose administrations. Direct causes of the events included poor organizational policy and procedures, lack of radiation safety officer oversight, lack of training and experience, poor supervision, decision errors, poor communication, and hardware failures. Yeung et al. (2005) studied 624 incidents that were reported using an incident reporting system from November 1992 to December 2002. More than 40% of the incidents (263) were due to errors in documentation related to data transfer or inadequate communication. Another 40% of the incidents (252) were due to errors in patient set-up. Thomadsen et al. (2003) analyzed 134 events of brachytherapy misadministrations. After performing a fault tree analysis, they found that 52% of errors were found in four process steps: (1) selection of the sources to place into the applicator; (2) loading of sources into the applicator; (3) using the required units when entering data into the computer; and (4) fixing the sources in the applicator or fixing the applicator in the patient. More recently, Thomadsen (2008) reported on two more events related to radiation machines, the Omnitron Event and the Stationary Multileaf Collimator Event.

To deal with events such as those described, radiotherapy has turned to quality assurance (QA). QA is performed by medical physicists (MPs) to ensure that all imaging devices are functioning properly, that the data they produce are reliable, that all software used to develop treatment planning programs are working correctly, that all equipment is performing the way they were designed to perform, and that all procedures for delivering radiation therapy are executed with the utmost safety and quality (Goitein 2008).

Although QA has continued to evolve with the new technology in RT, instances in which patients receive the wrong dose, receive the right dose to the wrong site, or that the wrong patient is treated still exist (Thomadsen 2008). This could be because organizational QA resources target the wrong part of the process, never catching errors that indeed exist (Thomadsen 2008). Another reason for the lack of patient safety could be that the system in which the MP works and performs QA on is poorly designed. That is, it is possible that the cognitive tasks MPs must perform during QA, such as sensing or perceiving malfunctioning devices, problem solving, etc., are not supported by the systems in which they work. If this is the case, despite MPs best efforts, errors will be likely.

A MP is relied upon by the rest of the radiation therapy team and the patient to be the gatekeeper of quality, so that the patient is treated appropriately. They are therefore expected to be accurate 100% of the time; anything less could lead to death. However, the achievement of 100% accuracy is *only in part* determined by the MPs' skills and knowledge, and the QA they perform. Their accuracy is actually determined by their *interaction* with the work system in which they work. A work system is comprised of system *elements* such as the physical environment (e.g., layout, lighting, noise), organization (e.g., culture, team members, structure), tasks (e.g., QA), people (e.g., patient, MP, oncologist, dosimetrist), and tools/technologies (e.g., checklists (paper or electronic), information displays, machines, information technologies) (Carayon et al. 2006). If any of those elements are not designed to support the cognitive or physical work of MPs generally, or during QA specifically, their performance will suffer.

The scientific discipline that studies the interaction between people and the systems in which they work is called human factors engineering and ergonomics (HFE). HFE scientists discover and apply information about human cognitive and physical abilities and limitations to the design of tools, machines, systems, tasks, and environments for productive, accurate, safe, and effective human use (Eastman Kodak Company 2004; Kroemer, Kroemer, and Kroemer-Elbert 2001; Salvendy 1997; Sanders and McCormick 1993; Wickens et al. 2004). HFE designs and interventions have improved operator performance and safety in aviation, manufacturing, nuclear power, process control, surface transportation, rail, air traffic control, service, construction, agriculture, and even healthcare. Little has been written about applications of HFE to RT specifically; Thomadsen is an exception (Thomadsen 2008; Thomadsen et al. 2003), but a growing amount of HFE research has been applied to many other facets of healthcare delivery, including surgery, pediatrics, emergency medicine, anesthesiology, critical care, home care, medical device design and use, health information technology, medication safety, transitions of care, and patient transport (Carayon 2006; Carayon, Schultz, and Hundt 2004; Cook 2006; Escoto, Karsh, and Beasley 2006; Fairbanks, Bisantz, and Sunm 2007; Gaba, Howard, and Small 1995; Hazlehurst, McMullen, and Gorman 2007; Koppel et al. 2008; Lin, Vicente, and Doyle 2001; Patterson, Cook, and Render 2002; Rogers and Fisk 2002; Scanlon, Karsh, and Densmore 2006; Wears and Perry 2007; Weinger, Gardner-Bonneau, and Wiklund 2009; Wiegmann et al. 2007). The remainder of this chapter provides a brief introduction to HFE thinking and discusses how to apply HFE thinking to RT.

Human Factors Engineering

Karsh, Holden, Alper and Or (2006) provided an introduction to human factors engineering for healthcare delivery in which they provided a figure explaining HFE conceptually (see Figure 37.1). The figure shows that the people such as patients, MPs, oncologists, radiation therapists, and dosimetrists all operate in a work system comprised of multiple interacting levels. Depending on how the elements of the work system are designed, they may or may not support the many different types of performance required by, in this case, people involved in RT. The types of performance required for RT are many. Physical types of performance include manipulating or adjusting knobs, dials and mice, and typing. Cognitive types of performance include sensing and perceiving information, sensing and perceiving physical object manipulation, searching for information, remembering all varieties of relevant information, planning, communicating, problem-solving, learning, and many others. There are certainly many sociobehavioral types of required performance as well.

Philosophically, HFE scientists and practitioners believe that when human performance needs to be improved, the first step is analyzing the system in which people work in order to discover how to redesign the system to improve performance. Similarly, HFE scientists and practitioners believe that when errors or poor quality occur, the first obligation is to, again, study the system to figure out what promoted the unwanted outcomes. This is in stark contrast to the "person-centered approach" (Reason 2000), common in healthcare, which is a belief that performance can only be improved through more training and education (though, training is a very important way to improve performance, and there exists an HFE science to well-designed training (Salas and Cannon-Bowers 2001)) or that when errors or poor quality occur, the only remedy is blaming, shaming, and retraining. The reason that HFE focuses on system design, and not blaming, shaming, and retraining is that, "We cannot change the human condition, but we can change the conditions under which humans work" (Reason 2000).

Other hallmarks of HFE thinking include (Sanders and McCormick 1993; Wickens et al. 2004):

- *Systems (e.g., machines or hospitals) need to be designed for and to work with people.* Examples of the opposite of this are if machine switches or dials are not intuitive, if lighting makes reading difficult, or if MPs are required to keep a lot of information in memory.
- *Systems must be designed to accommodate the range of users.* If the range of MPs includes those that are short and tall, have excellent and poor vision, and are novices and experts, then the systems they use or work in must accommodate *all* of that variation, otherwise the systems are poorly designed.
- *How systems are designed will influence human behavior and therefore system performance.* Even if a hospital insists they care about safety, if there is a manager that rewards subordinates for speed and production more than safety, the subordinates will be motivated to work around safety rules. If the layout of a room is inefficient for the required tasks, the people who work in the room will be motivated to find shortcuts in their work, potentially having a negative impact on quality and safety. Even visual displays of data affect behavior because some displays provide direct perception of the problem, while others require mental integration of discrete data.
- *Design needs to be evidence-based, not "common sense" or designer-driven.* Despite widespread belief, people, including subject matter experts (e.g., expert MP) are *not good at identifying* what kinds of things would help them perform better (Andre and Wickens 1995). In fact, people routinely report wanting things that do not improve their performance. It is thus critical to rely on existing HFE design standards or experimental testing to ensure designs will be good for performance.
- *All design must take into account the system of use.* Just because something, such as a piece of equipment or

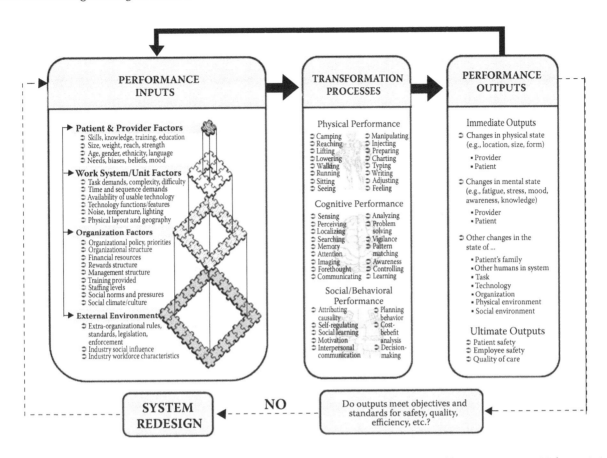

FIGURE 37.1 A human factors engineering model of performance. (From Karsh et al., *Qual. Saf. Health Care*, 15, I59–I65. With permission.)

information technology, works in one hospital or department, does not mean it will work in another. The reason is that if the other system elements (see Figure 37.1) are different, the context of use will be different, and so the technology might not be as effective. It is therefore critical to conduct HFE development methodologies such as user-centered design or cognitive work analysis (Fairbanks, Bisantz, and Sunm 2007; Norman and Draper 1986; Jacko et al. 2002; Vicente 1999), some of which have been used successfully in healthcare delivery settings (Coble et al. 1997; Gosbee and Gosbee 2005; Kushniruk 2002; Sawyer 1996; Fairbanks, Bisantz, and Sunm 2007).

As previously stated, Ostrom et al. (1996) analyzed seven misadministration events. During their investigation, they took a more HFE approach and determined that although some of the events were caused by decision errors, all the events occurred due to some variation of poor organizational policies and procedures. Additionally, they also found that due to poor or lack of proper organizational policies and procedures, in some of the events, the MPs lacked the proper training and experience necessary for the specific tasks they were performing. In some of those same events, there was no radiation safety officer present and proper supervision was also lacking. Thus, if those MPs needed help due to their lack of training and experience and the ambiguous organizational policies and procedures, there was no

one there to provide them with the needed support. From a HFE perspective, although decision errors were involved in some of the events, the focus of investigation and redesign would be on the causes of those errors, some of which might have included the organizational policies and procedures and the lack of training and support staff provided to the MPs.

There are likely many other examples of system designs that do not support the performance needs of the members of the RT team. For example, interruptions during documenting activities can cause erroneous reporting or can cause MPs to forget to report something. Poor lighting could lead to inappropriate machine inputs when trying to program a treatment plan into a machine, possibly causing a MP to inadvertently press the wrong button, which could lead to an overdose or underdose in treatment. Additionally, poorly designed technology, like the Therac-25 (discussed briefly above), can also lead to errors in programming and administration, without MPs having much control over the situation.

Summary

A system should be designed to support the physical and cognitive performance of the people within it (i.e., medical physicists, dosimetrists, nurses, patients, etc.). HFE methods can be used to identify those performance requirements and HFE design principles can be used to redesign systems to better support the users and their work. In the case of RT safety, HFE designs can directly improve performance and safety of RT team members and patients, and can also help ensure that important QA functions can be executed successfully.

The take-away points from this chapter are:

- Performance, efficiency, quality, and safety are the result of the *interaction* between people and the system in which they work. Human factors engineering helps you understand that interaction so that you can better design systems to improve performance, efficiency, quality, and safety (Eastman Kodak Company 2004; Kroemer, Kroemer, and Kroemer-Elbert 2001; Salvendy 1997; Sanders and McCormick 1993; Wickens et al. 2004).
- Human factors engineers and ergonomists have developed tools, standards, guidelines, and principles for improving human performance (e.g., safety and quality) (Federal Aviation Administration 2009; Karwowski 2005; Nielsen 1993; Salvendy 1997; Stanton et al. 2005).

References

Andre, A. D., and C. D. Wickens. 1995. When users want whats not best for them. *Ergon. Des.* 3 (4): 10–14.

Carayon, P., ed. 2006. *Handbook of Human Factors and Ergonomics in Patient Safety*. Mahwah, NJ: Lawrence Erlbaum Associates.

Carayon, P., A. S. Hundt, B. Karsh, A. P. Gurses, C. J. Alvarado, M. Smith, and P. F. Brennan. 2006. Work system design for patient safety: The SEIPS model. *Qual. Saf. Health Care* 15 (Suppl. I): i50–i58.

Carayon, P., K. Schultz, and A. S. Hundt. 2004. Wrong site surgery in outpatient settings: The case for a human factors system analysis of the outpatient surgery process. *Jt. Comm. J. Qual. Saf.* 20 (7): 405–410.

Coble, J. M., J. Karat, M. J. Orland, and M. G. Kahn. 1997. Iterative usability testing: Ensuring a usable clinical workstation. *J. Am. Med. Inform. Assoc.* 1997: 744–748.

Cook, R. I. 2006. Resilience and resilience engineering for health care. *Ann. Clin. Lab. Sci.* 36 (2): 232–232.

Eastman Kodak Company. 2004. *Ergonomic Design for People at Work*, 2nd ed. Hoboken, NJ: John Wiley and Sons.

Escoto, K. H., B. T. Karsh, and J. W. Beasley. 2006. Multiple user considerations and their implications in medical error reporting system design. *Hum. Factors* 48 (1): 48–58.

Fairbanks, R. J., A. M. Bisantz, and M. Sunm. 2007. Emergency department communication links and patterns. *Ann. Emerg. Med.* 50 (4): 396–406.

Federal Aviation Administration. 2009. *Human Factors Workbench*. Available from http://www.hf.faa.gov/Portal/default.aspx (accesed March 9 2009).

Gaba, D. M., S. K. Howard, and S. D. Small. 1995. Situation awareness in anesthesiology. *Hum. Factors* 37 (1): 20–31.

Goitein, M. 2008. *Radiation Oncology: A Physicist's-Eye View*. Heidelberg: Springer.

Gosbee, J. W., and L. L. Gosbee. 2005. *Using Human Factors Engineering to Improve Patient Safety*. Oakbrook Terrace, IL: Joint Commission Resources.

Hazlehurst, B., C. K. McMullen, and P. N. Gorman. 2007. Distributed cognition in the heart room: How situation awareness arises from coordinated communications during cardiac surgery. *J. Biomed. Inform.* 40 (5): 539–551.

Institute of Medicine. 2000. *To Err Is Human: Building a Safer Health System*. Washington, DC: National Academy Press.

Jacko, J. A., G. Salvendy, F. Sainfort, V. K. Emery, D. Akoumianakis, V. G. Duffy, J. Ellison et al. 2002. Intranets and organizational learning: A research and development agenda. *Int. J. Hum. Comput. Interact.* 14 (1): 93–130.

Karsh, B. T., R. J. Holden, S. J. Alper, and C. K. L. Or. 2006. A human factors engineering paradigm for patient safety: Designing to support the performance of the healthcare professional. *Qual. Saf. Health Care* 15: I59–I65.

Karwowski, W. 2005. *Handbook of Standards and Guidelines in Ergonomics and Human Factors*. Mahwah, NJ: Lawrence Erlbaum Associates.

Koppel, R., T. Wetterneck, J. L. Telles, and B. T. Karsh. 2008. Workarounds to barcode medication administration systems: Their occurrences, causes, and threats to patient safety. *J. Am. Med. Inform. Assoc.* 15 (4): 408–423.

Kroemer, K. H. E., H. B. Kroemer, and K. E. Kroemer-Elbert. 2001. *Ergonomics: How to Design for Ease Efficiency*, 2nd ed. Englewood Cliffs, NJ: Prentice Hall.

Kushniruk, A. 2002. Evaluation in the design of health information systems: Application of approaches emerging from usability engineering. *Comput. Biol. Med.* 32 (3): 141–149.

Leveson, N. G., and C. S. Turner. 1993. An investigation of the Therac-25 accidents. *Computer* 26 (7): 18–41.

Lin, L., K. J. Vicente, and D. J. Doyle. 2001. Patient safety, potential adverse drug events, and medical device design: A human factors engineering approach. *J. Biomed. Inform.* 34 (4): 274–284.

Nielsen, J. 1993. *Usability Engineering*. Boston, MA: Academic Press.

Norman, D. A., and S. W. Draper, eds. 1986. *User Centered System Design: New Perspectives on Human–Computer Interaction*. Boca Raton, FL: CRC.

Ostrom, L. T., P. Rathbun, R. Cumberlin, J. Horton, R. Castorf, and T. J. Leahy. 1996. Lessons learned from investigations of therapy misadministration events. *Int. J. Radiat. Oncol. Biol. Phys.* 34 (1): 227–234.

Ostrom, L. T., P. Rathbun, R. Cumberlin, J. Horton, R. Castorf, and T. J. Leahy. 1996. Lessons learned from investigations of therapy misadministration events. *Int. J. Radiat. Oncol. Biol. Phys.* 34 (1): 227–234.

Patterson, E. S., R. I. Cook, and M. L. Render. 2002. Improving patient safety by identifying side effects from introducing bar coding in medication administration. *J. Am. Med. Inform. Assoc.* 9 (5): 540–553.

Reason, J. 2000. Human error: Models and management. *Br. Med. J.* 320: 768–770.

Rogers, W. A., and A. D. Fisk. 2002. *Human Factors Interventions for the Health Care of Older Adults.* Mahwah, NJ: Lawrence Erlbaum Associates.

Salas, E., and J. A. Cannon-Bowers. 2001. The science of training: A decade of progress. *Annu. Rev. Psychol.* 52: 471–499.

Salvendy, G., ed. 1997. *Handbook of Human Factors and Ergonomics*, 2nd ed. New York, NY: John Wiley and Sons.

Sanders, M. S., and E. J. McCormick. 1993. *Human Factors in Engineering and Design*, 7th ed. New York, NY: McGraw-Hill Science.

Sawyer, D. 1996. Do It by Design: An Introduction to Human Factors in Medical Devices. US Food and Drug Administration. Available from http://www.fda.gov/cdrh/humanfactors/ (accessed May 17, 2007).

Scanlon, M. C., B. Karsh, and E. Densmore. 2006. Human factors and pediatric patient safety. *Pediatr. Clin. North Am.* 53: 1105–1119.

Stanton, N. A., P. M. Salmon, G. H. Walker, C. Baber, and D. P. Jenkins. 2005. *Human Factors Methods.* Aldershot: Ashgate.

Thomadsen, B. 2008. Critique of traditional quality assurance paradigm. *Int. J. Radiat. Oncol. Biol. Phys.* 71 (1 Suppl.): S166–S169.

Thomadsen, B., S. W. Lin, P. Laemmrich, T. Waller, A. Cheng, B. Caldwell, R. Rankin, and J. Stitt. 2003. Analysis of treatment delivery errors in brachytherapy using formal risk analysis techniques. *Int. J. Radiat. Oncol. Biol. Phys.* 57 (5): 1492–1508.

Vicente, K. J. 1999. *Cognitive Work Analysis.* Mahwah, NJ: Lawrence Erlbaum Associates.

Wears, R. L., and S. J. Perry. 2007. Status boards in accident and emergency departments: Support for shared cognition. *Theor. Issues Ergon. Sci.* 8 (5): 371–380.

Weinger, M., D. Gardner-Bonneau, and M. E. Wiklund, eds. 2009. *Handbook of Human Factors in Medical Device Design.* Boca Raton, FL: CRC Press.

Wickens, C. D., J. D. Lee, Y. Liu, and S. E. Becker. 2004. *An Introduction to Human Factors Engineering*, 2nd ed. Upper Saddle River: Pearson Prentice Hall.

Wiegmann, D. A., A. W. El Bardissi, J. A. Dearani, R. C. Daly, and T. M. Sundt. 2007. Disruptions in surgical flow and their relationship to surgical errors: An exploratory investigation. *Surgery* 142: 658–665.

Yeung, T. K., K. Bortolotto, S. Cosby, M. Hoar, and E. Lederer. 2005. Quality assurance in radiotherapy: Evaluation of errors and incidents recorded over a 10 year period. *Radiother. Oncol.* 74 (3): 283–291.

38

Human Factors Engineering: Radiotherapy Application

Enda F. Fallon
National University of Ireland Galway

Liam Chadwick
National University of Ireland Galway

Wil J. van der Putten
Galway University Hospitals and National University of Ireland Galway

Introduction

Since early 2000, significant investment in radiotherapy has taken place in Ireland. The requirements for this investment were based on poorer prognoses for patients with cancer in Ireland, compared to peer countries in Europe and elsewhere. One of the first major investments was the development of a new publicly funded radiotherapy facility in Galway, on the west coast of Ireland. Up till then, no radiotherapy had been provided outside the main urban centers of Dublin and Cork. The Galway center was new and based in a hospital where no prior radiotherapy existed. Consequently, the center could take advantage of the most modern technologies available.

The Galway center consists of three Linear Accelerators, CT and conventional simulators, 3D-treatment planning, HDR and orthovoltage systems, and a fully electronic patient management system. The specification of the equipment for the Galway Center was performed in the normal fashion with technical specifications and costs considered in a standard "value for money" assessment using a conventional option appraisal methodology (van der Putten et al. 2001). This placed great value on technical and service components of the system. During the evaluation it had already become evident that the patient management and system control software (a record and verify system) was critical to the operation of the center. Although it had been envisaged that a fully electronic patient record would be safer due to absence of transcription errors, no systematic assessment of the system and subcomponents was made using system safety methodologies. Despite the sophistication of the technology, no attempt was made to apply a systems engineering approach to the overall development of the center. Similarly, there was little systematic consideration of risk or human and organizational issues in this regard.

Following on from the commencement of clinical operations, the Center of Occupational Health & Safety Engineering and Ergonomics in National University Ireland, Galway was asked to perform a risk assessment associated with the use of a fully electronic, film- and paper-less radiotherapy management system in a busy acute hospital that still uses the conventional paper patient records. This analysis was performed using standard human factor engineering (HFE) tools. The outcome of the analysis has been reported elsewhere (Fallon, Chadwick, and van der Putten 2009). The purpose of this chapter is to illustrate the practical issues involved in applying tools from HFE to risk assessment in a working radiotherapy department.

Radiotherapy as a System

Although at first glance, errors in radiotherapy can be attributed to human failure, it has become abundantly clear that a large number of adverse events in radiotherapy can be attributed to system failures. This is demonstrated well in the report published by the Royal College of Radiologists in 2008, *Towards Safer Radiotherapy*, which outlines many of the contributing factors related to incidents, adverse events, and accidents in radiotherapy treatments (The Royal College of Radiologists et al. 2008). These include:

- Hierarchical department structure
- Changes in the treatment process
- Over-reliance on automated procedures
- Poor design and documentation of procedures
- Poor communication and lack of team work
- Lack of training and competence issues for complex treatment modalities

Radiotherapy can be regarded as a very much "process-driven" method of provision of care. In this it is very similar to an engineering process system and as such is quite unique in healthcare. The following attributes of engineering processes are readily identifiable in radiotherapy. Radiotherapy is unique in medicine in that it is the one medical discipline that would be impossible to practice without the support of physics. The strong influence of this discipline on this aspect of healthcare has led to systems of work that rely heavily on those found in engineering. This is true especially as medical physics, a branch of applied physics, can be considered closely related to engineering (Ihde 1997). Radiotherapy, relying as it does on medical physics, can thus be considered to be similar to engineering disciplines.

The radiotherapy process has very predictable patient flows, based on a well-defined and predictable demand. It is characterized by highly educated and well-qualified staff who operate and supervise the treatment process. In this they are supported by advanced technology systems comprising complex machines, interacting software, and control systems.

The modern radiotherapy treatment process includes a variety of computerized subsystems that cover all aspects of the treatment process, each with system-critical functionality. Examples of this are electronic patient records, image transfer and storage systems, radiotherapy treatment simulation and planning systems, treatment administration, recording and verification systems, and independent dose verification systems. The adoption of multiple computer systems has been driven by the need to reduce many of the familiar patient safety issues inherent in the operation of complex sociotechnical systems, that is, reducing the potential for human errors in activities such as data checking and manual data entry. However, this adoption makes the understanding of the systems' interactions and dependencies more opaque to the "sharp-end" users (Graber 2004). It also results in a more tightly-coupled treatment process where the risk of error propagation is increased and the opportunity of error recovery is dependent on the built-in process safety checks and the knowledge of the treatment staff. It places additional accountability for successful patient treatment onto the already highly pressured staff. Examples of reports where the computer systems were directly implicated in system error are those of Patton, Gaffney, and Moeller (2003) and Barthelemy-Brichant et al. (1999).

Practical Risk Assessment in Radiotherapy

As stated above, a risk assessment was undertaken of the radiotherapy system installed in Galway University Hospital. The purpose of the assessment was to determine:

- The extent to which there is risk to patient safety in the management of patient medical record using both soft-copy and hardcopy media.

- The extent to which the current information systems support legislative, accreditation, and audit standards and requirements, and also best practice.

The purpose of that risk assessment is quite specific to the department and to its workflow. The purpose of this chapter is to provide details on the practical ways to implement such a risk assessment. It will become clear that each department wishing to undertake such an exercise will have to carefully consider its own workflows and processes. There is no generic, "cookie cutter" approach to this and each analysis will have to be specific to the department involved.

Workflow and Process Mapping

As stated above, one of the characteristics of the radiotherapy process is that it has a well-defined workflow. This will be subtly different for each individual radiotherapy department and therefore it is important to spend a significant effort on the collection of data in order to gain an understanding of the radiotherapy treatment process and to model the patient and information flow within it.

The determination of critical patient and data flow through the department is essential. The tool used to develop this is IDEF0 (IEEE 1998). This is, in essence, a graphical modeling tool that has been used extensively to model flows in engineering systems, which, it has been argued, are similar to the radiation treatment process. Figure 38.1 shows one of the main building blocks used in the modeling of the system.

A detailed description of IDEF0 can be found in (IEEE 1998). One advantage of using a standardized modeling tool is that the workflow and processes can be interpreted without ambiguity. Initial data should be collected through a series of interviews with the various stakeholders in the RT department. Of course, any process flow maps that have been created previously by the clinical users can also be used. The interviews should be semiformal in nature and structured around a number of key-anchors/issues such as stakeholder function within the treatment process, information uploaded into the patient file, information required from the patient file, how systems are used during the

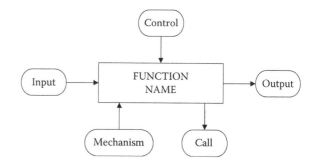

FIGURE 38.1 Main building block of IDEF0 modelling illustrating some of the definitions used to define the modelling syntax of the IDEF0 system. (From IEEE Std 1320.1-1998. With permission.)

treatment process, and concerns regarding the systems used. This can be facilitated by process maps that may be drawn up by the local service manager; however, as an alternative, patient flows such as those reported in *Towards Safer Radiotherapy* can also be used. It should be noted that process maps drawn up by clinical staff are typically more quality-oriented and, in general, will not emphasize risk assessment. Stakeholders include radiation oncologists, radiation therapists, medical physicists, and nursing and administrative staff.

Key words from the interviews are then used as key inputs in the modeling process. This is through a systematic process of collect data, build the model, and validate. This process is then to be iterated until overall agreement is achieved with all stakeholders about its appropriateness. Figure 38.2 shows part of the full IDEF0 map for the radiotherapy department in Galway University Hospital. Once the IDEF0 model has been agreed on by all stakeholders, a hazard analysis can be commenced. The IDEF0 model allows for the identification of processes which interact with patient data and data flows. The process map identifies critical paths and the influence of one sub-unit on others.

The IDEF0 model also identifies clearly who is responsible for a particular task.

Risk Assessment

Following the preparation of a detailed process map that is agreed to by all stakeholders, a detailed risk assessment of the identified processes can then be performed using standard risk assessment tools. An example of one such tool is that developed by the Irish Health Service Executive (HSE) (Hughes 2008). This is based on the Australian National Standard AS/NZS 4360 on Risk Management (Australian National Standard 2004). This follows a standard format in which a combination of frequency or likelihood of occurrence of each hazardous event and the severity associated with its outcome are considered. It considers each identified hazard process in terms of eight different risk categories (Hughes 2008):

1. Injury
2. Patient experience
3. Compliance with standards

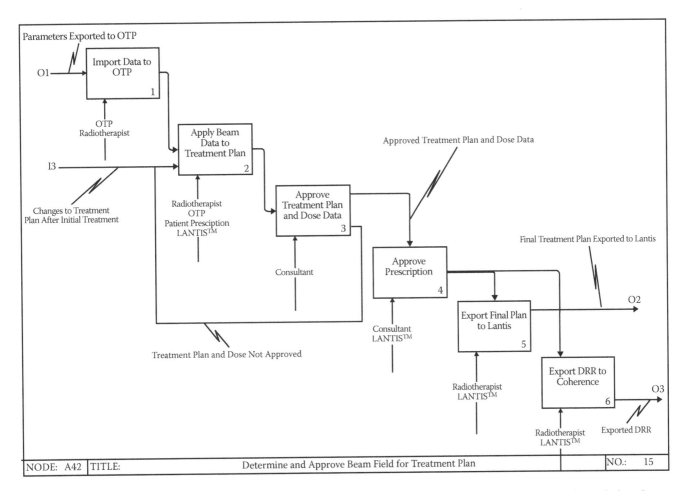

FIGURE 38.2 Part of IDEF0 process map for the radiotherapy department in Galway University Hospitals. The section shown deals with processes involved in preparing and approving radiotherapy treatment plans. The map shows inputs and outputs, nature and role of staff required, and constraints (such as availability of consultant).

4. Objective/projects
5. Business continuity
6. Adverse publicity/reputation
7. Financial loss
8. Environment

The tool defines five levels for the likelihood of occurrence of a hazard: rare/remote, unlikely, possible, likely, and almost certain. Similarly, five levels of severity of occurrence are defined specifically for each of the eight risk categories: negligible, minor, moderate, major, and extreme. Each of the individual processes in the IDEF0 map can now be analyzed using the risk model and the output of the analysis is a set of risk scores (1–25) for the eight risk categories identified above. A total risk score can then be calculated for every process by aggregating each of these eight risk category scores. Therefore, the maximum permissible score for any risk is 200. This assumes that each risk category is weighted equally. It should, however, be noted that, through stakeholder discussions, each of these individual categories can be assigned a different weighting. This is really up to individual departments and the environments they work in. It should also

be noted that different risk models can be adopted depending on local preferences, legal considerations, etc.

Analysis of Risk Factors

Having determined the risk scores for each of the processes, these now need to be analyzed further. There are a number of ways of doing this, but one way of performing this analysis is through a team consisting of key stakeholders such as

- Risk facilitator
- Radiotherapy services manager
- Lead physicist for the radiotherapy department
- Clinical nurse manager
- Radiotherapy administration manager

Others can, of course, be added depending on local circumstances.

The analyzed processes can be arranged in order of priority based on their total risk score. Using an agreed cut-off score, a subset of the processes is then selected for further analysis. This cut-off score is, of course, rather arbitrary. This reflects, to some

Failure Mode: First Evaluate failure mode before determining potential causes	Potential Causes	Scoring			Decision Tree Analysis				Action Type (Control, Accept, Eliminate)	Actions or Rationale for Stopping	Outcome Measure
		Severity	Probability	Hazard Score	Single Point Weakness?	Existing Control Measure ?	Detectability	Proceed?			
A425 (Export Final Plan on LANTIS™) - ii **Wrong Plan Exported**		Major	Remote	3	Yes	No	No	Yes			
	A425ii A Dosimetrist fatigued	Major	Uncommon	6	Yes	No	No	Yes	Control	Preferably the approved plan should be distinctly recognisable, more so than the current method, i.e. a small 'A' beside the plan name. Once a plan has been approved all other plans for that patient should be deleted/removed from the file.	It is possible to export an unapproved plan from OTP to LANTIS™. However this should be picked up by physics plan check or by pre-treatment radiotherapists.
	A425ii B Dosimetrist distracted/too busy	Major	Uncommon	6	Yes	No	No	Yes	Control	Preferably the approved plan should be distinctly recognisable, more so than the current method, i.e. a small 'A' beside the plan name. Once a plan has been approved all other plans for that patient should be deleted/removed from the file.	Due to the high workload of the radiotherapists and the operating environment it is possible for radiotherapists to be under pressure to complete work.
	A425ii C Dosimetrist picks wrong plan for export, i.e. more than 1 plan exists for the patients treatment	Major	Remote	3	Yes	No	No	Yes	Control	Preferably the approved plan should be distinctly recognisable, more so than the current method, i.e. a small 'A' beside the plan name. Once a plan has been approved all other plans for that patient should be deleted/removed from the file.	If two or more plans are generated for one patient it is possible that the wrong plan can be exported

FIGURE 38.3 Example of an output table from an HFMEA process related to the IDEF0 diagram in Figure 38.2. Failure mode examined: export of wrong treatment plan to record and verify system (Siemens Lantis™ system).

degree, the subjective nature of risk perception. The use of a multidisciplinary team approach in this is essential to ensure, as far as possible, absence of bias. It can not be stated *a priori* what the percentage of processes is that should be subject to further analysis. In a fully electronic department, this might be up to 20% of the total number of processes; in a department that still utilizes substantial numbers of paper files, it may be significantly higher due to the large amount of physical transcription that occurs and that is known to be error-prone.

In order to understand the consequences of the individual processes and to determine how to prevent possible adverse events, the technique of health failure mode and effect analysis (HFMEA™) can be applied (Wetterneck et al. 2006; Senders 2004; DeRosier et al. 2002). The technique was developed specifically for use in the healthcare sector as a "systematic approach to identify product and process problems before they occur." HFMEA is a team-based analysis technique completed under the guidance of a facilitator.

A proactive approach to risk management (through a FMEA-type analysis) is now part of the accreditation standards as set by the Joint Commission on Accreditation of Healthcare Organizations (JCAHO) in Standard LD.5.2 (see www.jointcommission.org for further details). An example of such a proactive approach can be found in the work of Stockwell and Slonim (2006). The same standard also requires the proactive analysis of "at least one high-risk process" each year by hospitals as part of accreditation requirements (Senders 2004).

The technique considers all possible failures of a process and considers the consequences of such an occurrence. Failures are then considered in terms of the categories "eliminate," "control," or "accept." Corrective actions are then determined by the analysis team. The participatory nature of HFMEA lends itself well to an intensive multidisciplinary process such as radiotherapy. Using the subset that was determined in the previous step of the process, HFMEA is then completed for each of the processes in this set in order to gain a greater understanding of the consequences of their effects and to determine how to defend against potential adverse events associated with them.

Figure 38.3 shows a typical output table from one small part of the HFMEA analysis. The table shown relates to the risk of a wrong treatment plan being sent to the record and verify system. Note that this would be a risk factor particular to a full electronic department where it is not always immediately obvious which version of the electronic treatment plan is the current one. It is easy to have multiple versions of old and new plans relating to the same patient in circulation if measures to prevent this are not in place.

In practice, HFMEA is conducted through an interactive team exercise. This starts with a presentation of the process in order to inform the participants. This is followed by a group meeting to analyze the first process in depth. At the end of the process, the large group of participants is reassembled to discuss and agree on the final HFMEA results. It should be clear that HFMEA is a very labor-intensive process and can be difficult to apply in a busy radiotherapy department. Due to the time requirements of the analysis and the time pressure on the participants, further processes can be analyzed in smaller subgroups of two to three people. Alternately, a more formal process such as the Delphi process (Browne 1968) could be used.

Summary

Risk assessment in a complex environment such as radiotherapy is a difficult and labor-intensive process. The systematic approach outlined in this chapter enables this process to be conducted as efficiently as possible. One advantage of preparing process maps using IDEF0 is that changes in the system—introducing new modalities, for example—can easily be accommodated and then the effect of this on the overall risk profile of the organization can be assessed in a relatively straightforward manner. A major advantage then is that the process outlined here allows a proactive approach to be taken to risk management. This process does not have to rely on postincident corrections.

References

Australian National Standard. 2004. AS/NZS 4360. *Risk Management*.

Barthelemy-Brichant, N., J. Sabatier, W. Dewé, A. Albert, and J.-M. Deneufbourg. 1999. Evaluation of frequency and type of errors detected by a computerized record and verify system during radiation treatment. *Radiother. Oncol.* 53 (2): 149–154.

Browne, B. B. 1968. Delphi process: A methodology used for the elicitation of opnions of experts. In *Rand Corporation Report P-3925*. Santa Monica, CA: Rand Corporation.

DeRosier, J., E. Stalhandske, J. P. Bagian, and T. Nudell. 2002. Using health care failure mode and effect analysis: The VA National Center for Patient Safety's prospective risk analysis system. *Jt. Comm. J. Qual. Improv.* 28 (5): 248–267, 209.

Fallon, E. F., L. Chadwick, and W. J. van der Putten. 2009. Learning from risk assessment in radiotherapy. In *13th International Conference on Human-Computer Interaction*. Town and Country Resort & Convention Center, San Diego, CA, USA. Lecture Notes in Computer Science (LNCS), Vol. 5620: 502–511.

Graber, M. 2004. The safety of computer-based medication systems. *Arch. Intern. Med.* 164 (3): 339–340.

Hughes, S. 2008. *'Your Service, Your Say' The Policy and Procedures for the Management of Consumer Feedback to Include Comments, Compliments and Complaints in the Health Service Executive (HSE)*. Dublin: Health Service Executive.

IEEE. 1998. IEEE standard for functional modeling language—syntax and semantics for IDEF0. IEEE Std 1320.1-1998.

Ihde, D. 1997. The structure of technology knowledge. *Int. J. Technol. Des. Educ.* 7 (1): 73–79.

Patton, G. A., D. K. Gaffney, and J. H. Moeller. 2003. Facilitation of radiotherapeutic error by computerized record and verify systems. *Int. J. Radiat. Oncol. Biol. Phys.* 56 (1): 50–57.

Senders, J. W. 2004. FMEA and RCA: The mantras of modern risk management. *Qual. Saf. Health Care* 13 (4): 249–450.

Stockwell, D. C., and A. D. Slonim. 2006. Quality and safety in the intensive care unit. *J. Intensive Care Med.* 21 (4): 199–210.

The Royal College of Radiologists, Society and College of Radiographers, Institute of Physics and Engineering in Medicine, National Patient Safety Agency, and British Institute of Radiology. 2008. *Towards Safer Radiotherapy*. London: The Royal College of Radiologists.

van der Putten, Wil J., B. McLean, D. Hollywood, Y. Davidson, K. Clancy, J. Folan, and W. Higgins. 2001. Any colour as long as it's black.... Equipment purchasing for radiotherapy. In *European Congress of Medical Physics and Clinical Engineering*. 12–15 September, 2001, Belfast, Northern Ireland.

Wetterneck, T. B., K. A. Skibinski, T. L. Roberts, S. M. Kleppin, M. E. Schroeder, M. Enloe, S. S. Rough, A. S. Hundt, and P. Carayon. 2006. Using failure mode and effects analysis to plan implementation of smart i.v. pump technology. *Am. J. Health Syst. Pharm.* 63 (16): 1528–1538.

39

Human Factors Engineering: Case Study

Edmond W. Israelski
Abbott

William H. Muto
Abbott

Introduction

Human factors engineering (HFE) is a design discipline that includes data about human capabilities and limitations, as well as methods from the behavioral sciences and engineering to make medical products safe, effective, efficient, and usable.

The flowchart in Figure 39.1 summarizes the systematic and scientific methods that are part of HFE. At the core, HFE is user-centered design, where users are the central focus from the beginning of design. The HFE process starts with a very thorough understanding of the context of use, that is, the task flows, the user profiles, and the use environment. The process proceeds to quantification of "use-error" risks and usability to iterative design using prototypes and simulations. Designs are also evaluated for usability using analytical techniques such as expert reviews and cognitive walkthroughs (as part of use-error risk analysis), followed by quantitative task-based usability testing. Usability testing is done one person at a time, where users' performances are observed and recorded as they perform essential and critical tasks. Usability testing is done early, as the design is iterated, and then at the end to validate that usability objectives are met and risk is reduced as much as practicable.

Throughout the HFE process, user data is obtained to iterate both the design and the formal risk analysis as appropriate. The end result should lead to a medical device that is effective, efficient, safe, and satisfying for its users, since they have been at the forefront during the design process.

Use of Error Risk Analysis

The Therac-25 linear accelerator is a well-known, but no longer marketed, medical device that was the inspiration for Steve Casey's book *Set Phasers on Stun* (Casey 1998). In the real-

life example from which the book gets its title, several cancer patients are injured or killed by overdoses of radiation from a computer-controlled radiation therapy system, the Therac-25. The tragedy resulted from a combination of (what can be interpreted as) operator use errors, software program errors, and a lack of redundant hardware design. An in-depth description of the Therac-25 case is given by Leveson and Turner (1993).

Risk analysis or hazard analysis has been used as an engineering tool for many years to identify system risks and control system modes of failure. These tools and related methods have more recently been applied to understanding "use-errors" made with medical devices. Use-errors are defined as a pattern of predictable human errors that can be attributable to inadequate or improper design. Use-errors can be predicted through analytical task walkthrough techniques and via empirically based usability testing.

Among the most widely used of the risk analysis tools are failure modes and effects analysis (FMEA) and fault tree analysis (FTA). FMEA is a "bottom-up" design evaluation technique used to define, identify, and eliminate known or potential failures, problems, and errors from the system. It has three components that help define and prioritize such failures or faults: (1) occurrence (frequency of failure); (2) severity (seriousness of the hazard resulting from the failure); and (3) selection of controls to mitigate the failure before it has an adverse effect. Each of these components can be assigned a numerical value by the investigative group. These are used to compute a risk priority number (RPN), which is used as a single measure to assess priorities. A variation on this quantitative RPN approach is a qualitative one where the components are assigned a descriptive categorical adjective.

For human-factors-related applications, the FMEA process is enabled by first performing a task analysis by identifying the

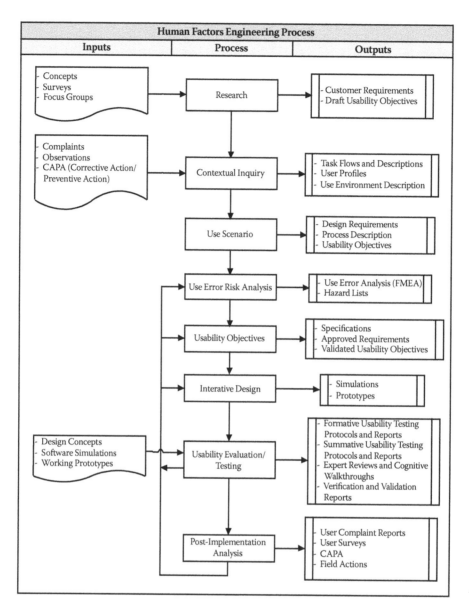

FIGURE 39.1 Human factors engineering process.

tasks, sequences, decisions, etc. Use-errors are then defined by identifying possible deviations from the expected or optimal performance of each of the tasks performed in the task analysis. The frequency of the error is then estimated, usually employing qualitative estimates, with either historical empirical data, experimental data, or, in most cases, informed judgment of a group of stakeholders. The severity of the failure is rated by seriousness of the resulting hazard and subsequent harm to the patient or user (often the worst case scenario of what can happen). When all of these values are estimated, they are combined into a risk index or risk acceptability level that is used to assess the overall resulting risk level. Changes designed to lower the risk index are fed back into the analysis to the remaining risk (residual risk).

Fault tree analysis (FTA) is a "top-down" analytical tool used to determine overall system reliability and safety (Stamatis 1995). It is a graphical representation of basic failures that are combined with Boolean logical operators (e.g., AND and OR), leading to the consequences of a system failure. For human factors applications, fault trees can be useful tools for determining the effects of human errors on overall system reliability and safety. The combination of events (using logical "gates") allows a design or evaluation team to estimate the probability of identified failures, including human errors, on the overall system failure probabilities.

When to Use FTA versus FMEA

Both techniques have their strengths and weaknesses. Because of its tabular format, FMEA is a straightforward method for documenting failures in terms of a "risk index," explicitly documenting the applied mitigations. Because of their bottom-up nature, however, it is often difficult to assess how a given fault will contribute to a system level fault or hazard (consequence of a fault). FTA is a good way of assessing overall system performance and the relationship of faults to higher-level failures. FTA is also better for quantitative analysis of system effects. For these reasons we believe that both techniques are complementary and should both be used. It makes good sense to use the output of each analysis to check the completeness of the other.

Also, these analytical techniques are best done iteratively, starting early and throughout the product development cycle.

Standards and FDA Guidance on Risk Analysis

FDA human factors guidance is very clear on the importance of performing use-error-focused risk analysis (Kaye and Crowley 2000). The guidance urges manufacturers to perform both analytical (e.g., risk analysis) and empirical analyses (e.g., usability testing) during product development to identify and control for use-errors. ISO standards also recommend the use of risk analysis in the development of medical devices. The current standard, ISO 14971:2007 *Medical Devices—Application of Risk Management to Medical Devices* (ISO 2007), describes the FMEA and FTA processes in general. The concept of human factors or use-error focused risk analysis is also endorsed in ANSI/AAMI HE-74:2001—*Human Factors Design Process For Medical Devices* (ANSI/AAMI HE 2001), which is a best practices process standard for human factors.

Background on the Therac-25 Case

The Therac-25 case has been written about and discussed in the literature in some detail (Leveson 1995; Israelski and Muto 2003; Leveson and Turner 1993). The device, after recalls and a number of field corrections, was eventually withdrawn from the market. The machine was driven by a minicomputer and used a command line user interface (employing typed commands). Between June, 1985 and January, 1987, there were six cases of overexposure associated with the Therac-25; one such case was on March 21, 1986 in Tyler, Texas. Prior to this case, the Therac-25 at that institution had been in use for 2 years and treated about 500 patients without incident. On this day, a male patient was scheduled for his ninth radiation treatment after a tumor had been previously removed from his back. He was to receive a 1.8 Gy from the Therac-25, using a 22-MeV electron beam. The operator initially incorrectly entered the command "X," which configured the machine for x-ray delivery. The operator quickly realized the error and hit the up arrow key to move the cursor to the beam selection field and then typed the correct command "E." Then, she entered "B" to begin treatment. After a moment, the machine shut down and the console displayed the message "Malfunction 54" and "treatment pause," which indicated a problem of low priority. Since this had repeatedly happened in the past, she hit the "P" key to proceed with the treatment. The operator was a longtime employee and had become very efficient in running the Therac-25.

Unknown to the operator, an 8-s timer was activated in the software by this irregular command sequence, causing any edits made before the 8 s elapsed to be ignored. The 8-s delay was due to setting the electron beam steering magnets. Any edits made to the Therac-25 console during those 8 s were not recognized by the necessary software subroutines. So the operator thought that the edit, showing the intended selection ("E"), was successful when it was not. After the command "B" for beam on, the system delivered the high-current electron beam that is used in x-ray mode (with a flattening filter). The patient felt like he received an electric shock on his back. Again, the cryptic error message appeared and the screen displayed no unusual amounts of delivered radiation, so the operator proceeded with another overdose. The repeated overdoses resulted in the patient receiving between 160 and 250 Gy instead of the intended 1.8 Gy. The patient died about 5 months later from these massive overdoses, and was heard to say before his passing, "Captain Kirk must have forgotten to set phasers on stun" (Casey 1998).

Table 39.1 shows excerpts from a theoretical retrospective use-error FMEA created by the authors. Tables 39.2 and 39.3 are the hypothetical decision tables used to enter values in the FMEA. In typical process instructions, procedures are given on how to assign the use-error probabilities in Table 39.2, that is, probable, occasional, rare, and improbable. Guidance would also be offered on assessing hazard severity, for example, "major" would be a hazard that is very likely to cause death or serious injury. Table 39.2 gives the risk level category as an input to Table 39.3 for determining risk acceptability. Likewise, instructions would be given on how to choose categories in Table 39.3 on the effectiveness of controls or design mitigations on the final risk acceptability. Figure 39.2 shows an excerpt from a theoretical fault tree analysis of the Therac-25 when use errors are included.

Considerations in Applying FMEA and FTA

It has been our experience that it is best to convene all of the critical organizational stakeholders to kick off the creation of FTAs and FMEAs. In a medical device company, the critical stakeholders include: human factors, medical affairs, quality assurance, and development. Other stakeholders can include: marketing, regulatory, documentation, and training. The first meeting gets everyone on board the process and, in a few hours, a start can be made on creating the fault trees and the FMEA matrix, assuming that a task analysis has been done. The completion of the analyses can be done in smaller groups or divided up among the group. When these steps are completed, the drafts are reviewed first by individuals and then by the group. The difficulty is to get consensus on the ratings among the group members. One approach may be to use Delphi-like techniques after doing calibration or leveling exercises for the process of estimating parameters needed in the analysis such as fault probabilities, hazard severities, and mitigation success. Good sources of data include usability tests, field complaints of similar products, and

TABLE 39.1 Theoretical Partial FMEA for the Therac-25 Radiation Therapy System

Task	Hazard	Use Errors	Use Error Prob	Hazard Severity	Risk Level	Method of Control	Effectiveness of Control on Risk Level	Risk Acceptability
				Generic Tasks				
Turn on device	Delay in therapy	Fail to hold power button for at least 2 s	Rare	Min	Low	Training and instructions note that power button must be held for 2 s	Little to no impact	Acceptable
				Programming Beam Modes				
Entering mode command—"E"	Overdose if mistake not edited	Mistyping "X" for high-current x-ray mode	Occ	Maj	Hi	Review screen display shows "X" instead of "E"	Reduces to medium	Acceptable with justification
	Overdose if mistake not edited	1. Mistyping "X" for x-ray mode 2. Miskeying down arrow to edit 3. Typing "E" to change to electron beam, but no change	Occ	Maj	Hi	1. System responds with short error tone and no display change 2. Review screen display shows "X" instead of "E"	Reduces to medium	Acceptable with justification
	Single overdose of radiation	1. Mistyping "X" for x-ray mode 2. Using up arrow to move to mode field 3. Typing in "E" for electron beam mode within an 8-s timing window 4. Press "B" for beam on	Occ	Maj	Hi	System responds with cryptic error message "Malfunction 54"	Little to no impact	Unacceptable
	Multiple overdoses to patient	1. Same sequence as above 2. Press "P" to proceed with second dose	Occ	Maj	Hi	System responds with cryptic error message "Malfunction 54" up to five times	Little to no impact	Unacceptable

Source: Occ, occasional; Maj, major; Min, minimal; Hi, high.

literature reviews. When consensus is achieved, it is important to get a signoff agreement from the main stakeholders and to put these analyses into a risk management file for the project.

These activities can be done in a few days or weeks, but more complex user interfaces may take longer to reach complete agreement. Again, these analyses are dynamic activities and should be done iteratively throughout the development cycle. The early analyses are important in making preliminary user interface design decisions. Later analyses examine how effective these design controls and mitigations were on reducing risk levels.

The Therac-25 incidents involved a combination of technical failures (software and possibly hardware) combined with human behavior resulting in catastrophic radiation overdoses. According to Leveson (1995), AECL, the manufacturer of the

TABLE 39.2 Factors Required for Deriving Risk Level

Probability of Use Error	Hazard Severity		
	Major	Moderate	Minimal
Probable	High	High	Medium
Occasional	High	Medium	Low
Rare	Medium	Low	Low
Improbable	Low	Low	Low

TABLE 39.3 Factors Required for Deriving Risk Acceptability

Effectiveness of Control	Risk Level		
	High	Medium	Low
No change to risk level	Unacceptable	Acceptable with justification	Acceptable
Risk lowered to medium or low	Acceptable with justification	Acceptable	Acceptable
Risk eliminated	Acceptable	Acceptable	Acceptable

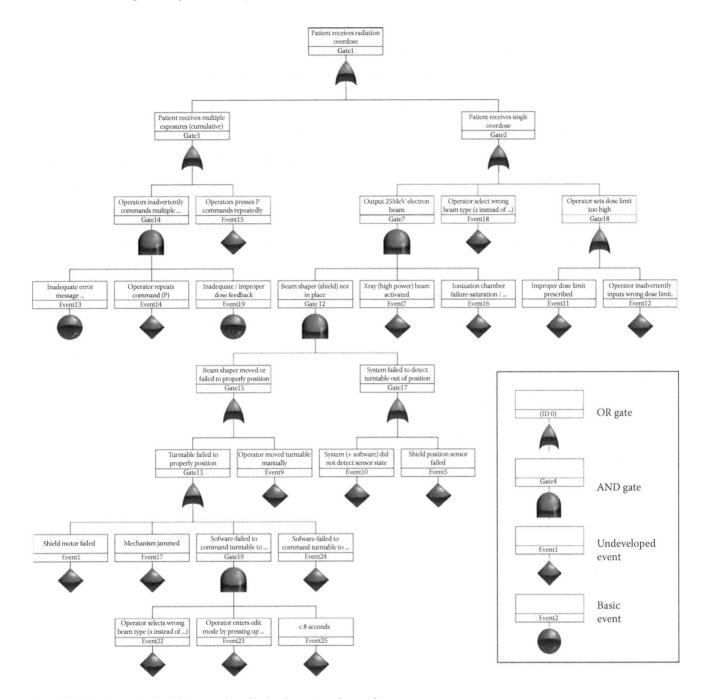

FIGURE 39.2 Example of a fault tree analysis for the Therac-25 radiation therapy system.

Therac-25, had conducted hazard analyses in the form of fault tree analyses. If so, what went wrong? Why were these hazards not anticipated and eliminated through design? We do know that use-error-focused FMEAs were not common in 1985. Although we may never know all the decisions and rationales that led to the resultant design defects, the authors believe that the inclusion of use-error risk analysis could have affected the outcome and effectiveness of any risk analyses.

Summary

The methods of both FMEA and FTA are valuable for systematically evaluating a design's potential for inducing use errors. When done early in the development cycle, potential faults and their resulting hazards are identifiable and much easier to mitigate with error-reducing designs. Design should be the primary method of control, followed by documentation and training as secondary methods. The hazard analysis methods not only make good business sense, but they are required by FDA guidelines and international standards for best practice human factors. Perhaps if these use-error-based risk analysis tools had been in widespread use at the time, the Therac-25 design deficiencies might have been eliminated.

References

American National Standards Institute (ANSI)/Association for the Advancement of Medical Instrumentation (AAMI) HE. 2001. *Human Factors Design Process for Medical Devices.* Arlington, VA: AAMI HE-74:2001.

Casey, S. M. 1998. *Set Phasers on Stun: And Other True Tales of Design, Technology, and Human Error.* Santa Barbara, CA: Aegean.

International Standards Organization (ISO). 2007. *Medical Devices—Application of Risk Management to Medical Devices.* Geneva: ISO 14971:2007.

Israelski, E. W., and W. H. Muto. 2003. Use-error focused risk analsis for medical devices: A case study of the Therac-25 Radiation Therapy System. In *Proceedings of the 47th Annual Meeting of the Human Factors and Ergonomics Society.* Santa Monica, CA: Human Factors and Ergonomics Society.

Kaye, R., and J. Crowley. 2000. *Medical Device Use Safety: Incorporating Human Factors Engineering into Risk Management.* Washington, DC: U.S. Food and Drug Administration.

Leveson, N. 1995. *Safeware: System Safety and Computers.* Reading, MA: Addison-Wesley Publishing Company.

Leveson, N. G., and C. S. Turner. 1993. An investigation of the Therac-25 accidents. *Computer* 26 (7):18–41.

Stamatis, D. H. 1995. *Failure Modes and Effects Analysis.* Milwaukee, WI: ASQ Quality Press.

40

Changing Role of the Radiation Oncologist

Vincenzo Valentini
Università Cattolica S.Cuore

Vincenzo Frascino
Università Cattolica S.Cuore

Marco Marchetti
Università Cattolica S.Cuore

Introduction

Radiation oncology is a medical specialty that has greatly developed over the past century. In recent years, its rich tradition of clinical care and evidence-based practice has become increasingly influenced by technological advances (Zietman 2008). Nowadays, quality in medical practice for a radiation oncologist has to face these two issues: clinical practice and technology.

From the 1960s to the 1980s, the clinical practice of radiation oncology was refined and mastered through a greater understanding of radiobiology and physics, through great attention to morbidity, and through the publication of clinical results. Radiation oncologists established their pivotal role in cancer care from diagnosis to death. They would admit patients for treatment, follow patients through the course of their disease, and offer palliative and terminal care. Radiation oncologists in many parts of the world, particularly the United Kingdom, Canada, and Australia, took on the responsibility of giving chemotherapy and remain "complete oncologists" to this day (Zietman 2008).

Changing Role of the Radiation Oncologist

The past two decades have brought tremendous changes to the practice of radiation oncology. There has been a huge focus on technology, and responsibility for clinical care has been shared with, and sometime abdicated to, other specialties. The radiation oncologist's efforts were focused on driving the image-guided delivery of a single physical therapy. This is something that has not yet been documented in any longitudinal study but is something that most practitioners would recognize (Zietman 2008).

As Antony Zietman stated recently, a "radiation ONCOLOGIST" risks becoming a "RADIATION oncologist" (Zietman 2008). He identified the following major drivers in this possible shift of practice:

- *The seduction of technology.* Neil Postman wrote in 1993 about a societal condition he termed "technopoly," which is a "state of culture and a state of mind which consists of the deification of technology. This means that the culture seeks its authorization in technology, seeks its satisfaction in technology, and takes its orders from technology" (Postman 1993).

- *The process of reimbursement that values the technical over the clinical* (as in the United States and other countries). A radiation oncologist could have an attractive incentive to use IMRT over three-dimensional radiation therapy and to use proton beam therapy, if available, over other less technologically advanced modalities. Little incentive to use combination with radiopharmaceuticals and even less incentive to perform inpatient care or outpatient follow-up can further support the strong interest in the technological aspects of radiation oncology.

- *The pressure from patients and vendors.* Patients are increasingly knowledgeable about their disease, its treatment, and their options. It is difficult for the radiation oncologist to resist multiple requests for the new technology being used by competing centers. Many hospitals, vendors, and disease-specific Web sites commonly promote new technologies, reflecting the advocacy of patients and their support groups.

- *The emergence of medical oncology as a specialty.* Radiation oncologists are increasingly pushed to the world of new technology and patient care has been progressively handed off to medical oncology. This trend can create the perception that the true oncologist is of the medical oncologist variety and can lead to the perception that the radiation oncologist as merely a deliverer of radiation; an interventional radiologist available to treat the patient with one particular therapy but out of the loop in terms of decision-making and aftercare, even when a multidisciplinary approach to patient management is practiced.

Even though these risks can threaten the identity and the perception of the radiation oncologist in daily practice, radiation oncologists are still perceived as having a holistic clinical outlook, seeing patients at every stage of their disease from curative to the most palliative. Radiation oncologists see themselves as discriminating users of their modality, selectively picking the arrows in their quiver and artfully manipulating dose, dose rate, and fractionation to improve outcome. Radiation oncologists supported by radiobiological knowledge and the evidence-based outcome of their practice can play a leading role in the multidisciplinary approach to the identification of guidelines, to the single patient treatment program prescription, and to clinical research perspectives.

Challenges Facing Radiation Oncologists

There are other challenges that can also engage the identity of the radiation oncologist if we examine the nature of that craft. The initial evaluation of the patient for therapy may be a formality that follows a decision already made by a medical oncologist or surgeon outside a multidisciplinary meeting. The simulation, drawing of volumes, and treatment planning is now largely performed by technicians and dosimetrists, but is increasingly automated and will increasingly be outsourced.

Overall, these challenges can place the radiation oncologist in a complex position. To manage all these complexities, the radiation oncologist has to enhance both clinical actions and skill profile. A reliable tool to accomplish this can be found in the research of quality in the medical practice. Quality for the radiation oncologist means to clarify the different components of the clinical decision and to supervise the steps with the proper methodology, which will bring the optimal outcomes of treatment.

A clinical decision represents the interaction between the knowledge and attitudes of the physician, the expectations of the patient, and the capability of the health system. With different levels of awareness and responsibility in their professional life, the radiation oncologist bases clinical decisions on information from different sources: scientific evidence, physician and patient preferences, and external rules and constraints as shown in Figure 40.1 (Mulrow, Cook, and Davidoff 1998).

Quality in this part of the radiation oncology practice is represented by the clear definition of the interaction between the different backgrounds. It involves ethical and framework landscapes. The quality is related to the transparency of the most hidden background of each clinical choice, which has to be agreed upon and signed by the patient.

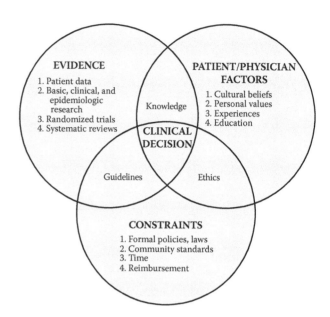

FIGURE 40.1 Factors that enter into clinical decisions. (From Mulrow, C. D., Cook, D. J., and Davidoff, F., *Systematic Reviews: Synthesis of Best Evidence for Healthcare Decisions,* American College of Physicians, Philadelphia, PA, 1998. With permission.)

Quality for the Radiation Oncologist

Quality for the radiation oncologist also means a guarantee that the agreement between the physician and the patient will arrive at the optimal outcome by supervising each clinical and technical component of the whole process. Currently, the methodology to guarantee the appropriateness of this second aspect of the quality of the radiation oncology medical practice is well-established, even if the development of the technology offers new pitfalls to identify and to overcome. Different agencies can offer guidelines and independent accreditation programs to radiation oncologists. Accomplishing accreditation for a program is very challenging, primarily because of the amount of time and changes in the organization required, but there are many related benefits. What is more relevant behind a quality assurance program is the educational opportunity for the team. The quality assurance program should promote the awareness of the relevance of details in each step of the radiation therapy procedure, the importance of the verifiability of daily practice to enhance its reliability, and the significance of the contribution of each single member of the staff and of the patients to the perspective of optimization of any procedure. The quality governance of the delivery process not only provides some protection from any legal issue, but it also increases the consciousness of the overall reliability of the radiation practice for the team, in the hospital, and in the community.

Quality and Outcomes

Quality requires an online overview of the outcomes. Outcomes could be perceived differently by the radiation oncologist, the patient, and the provider (health system). Actual involvement of all the actors in the identification of sustainable outcomes from the clinical decision to treatment can optimize patient flow through the system. Furthermore, knowledge of the average outcomes of their own practice allows them to identify how far their practice is from the benchmark outcomes. This evaluation allows for planning the changes according to the detected differences. It requires a finalized system to record reliable data and to integrate the data with the governance of the overall program for each patient.

The accomplishment of a shared clinical decision, of a certified proper therapy delivery, and of an actual outcome evaluation fosters quality in the medical practice of radiation oncologists and supports the enhancement of clinical actions and skill profiles.

Summary

As Antony Zietman reminded us, radiation is a superb, flexible, well-understood, organ-sparing, and cost-effective therapy (Zietman 2008). In this complex historical time, clinical and technological issues challenge radiation oncologists. The research of global quality could represent a further complexity, but it is the best tool to transform it in a feasible project to give perspective and a chance to further improve our discipline.

References

Mulrow, C. D., D. J. Cook, and F. Davidoff. 1998. Systematic reviews: Critical links in the great chain of evidence. In *Systematic Reviews: Synthesis of Best Evidence for Health Care Decisions*. Philadelphia, PA: American College of Physicians.

Postman, N. 1993. *Technopoly: The Surrender of Culture to Technology,* 1st Vintage Books ed. New York, NY: Vintage Books.

Zietman, A. 2008. The future of radiation oncology: The evolution, diversification, and survival of the specialty. *Semin. Radiat. Oncol.* 18 (3): 207–213.

41

Changing Role of the Medical Physicist

Brendan McClean
St Luke's Hospital

Introduction

The radiotherapy medical physicist forms an essential part of a multidisciplinary team of radiation oncologists, technologists, computer scientists, and other healthcare professionals providing a radiotherapy service. Medical physicists possess training and expertise in the areas of physics, mathematics, technology, and biology and are actively involved in the whole radiotherapy patient pathway, from imaging to dose delivery. The catalog of core tasks of a medical physicist in routine clinical service has been outlined in a number of publications (AAPM 1993; Belletti et al. 1996). Moreover, medical physicists are often trained in research methodology, allowing them to apply their knowledge of physical processes and technology to clinical problems in radiotherapy and to provide essential scientific input to the development of new treatment approaches and treatment optimization.

The importance of the role of medical physics has been recognized and embedded in the European Union's directives concerning radiation safety standards (European Commission 1996) and medical radiation exposures (European Commission 1997). Articles in these have given a statutory requirement for physicists to be involved in the medical uses of ionizing radiation. For example, in Article 6 of Euratom 97/43:

> In radiotherapeutic practices, a medical physics expert shall be closely involved. In standardized therapeutical nuclear medicine practices and in diagnostic nuclear medicine practices, a medical physics expert shall be available.

A medical physics expert is identified in Article 2 of the same directive as:

> An expert in radiation physics or radiation technology applied to exposure, within the scope of this Directive, whose training and competence to act is recognized by the competent authorities; and who, as appropriate, acts or gives advice on patient dosimetry, on the development and use of complex techniques and equipment, on optimization, on quality assurance, including quality control, and on other matters relating to radiation protection, concerning exposure within the scope of this Directive.

Together with the development of the role of medical physics, there is an ongoing effort to accept medical physics as a profession and scientific discipline. This has been articulated informally on a number of occasions (e.g., Ervin Podgorsak, *AAPM Newsletter*, November/December 2007) and more formally through the Malaga Declaration from the European Federation of Medical Physics (EFOMP 2006), in which the EFOMP and the International Organization of Medical Physics (IOMP) have lobbied for medical physics to become a regulated profession in all member states in response to the EU Directive 2005/36/EC on the recognition of professional qualifications.

Education and Training in Medical Physics

The acknowledged role of medical physics in the European Directives, acceptance of the need for medical physics to be recognized as a profession, and the central role of the medical physicist in increasingly complex radiation-based clinical procedures have given impetus to the discussions of education and training requirements in medical physics. There is a clear need for the medical physicist to demonstrate sufficient knowledge and competence to participate in patient care in a similar way to other professions involved in clinical practice. Appropriate training, coupled with a mechanism to ensure high standards of training, is central to the development of the profession and to provide both accountability and transparency to peers and the public. As such, training programs for medical physicists have to be constructed and accredited to ensure agreed standards are being met and should lead to certification and periodic recertification to demonstrate an adherence to quality standards and practice.

There are a number of accepted pathways to become a medical physicist (AAPM 2008). However, entry to medical physics education and training is usually through a university degree in physics, engineering, or equivalent physical science, usually followed by a graduate degree (MSc or PhD). The undergraduate and graduate education leads to a demonstrated acquired knowledge in areas of fundamental physics, advanced mathematics, and advanced atomic and nuclear physics. It is important to complement this undergraduate and graduate education with a training program involving further education and hospital-based clinical training. A suitable professional body to ensure consistency with and adherence to agreed high standards normally approves these training programs. The importance of this further training is recognized, for example, by the International Atomic Energy Authority (IAEA), which states that a person acquiring a university degree in medical physics without the required hospital training cannot be considered to be a clinically qualified medical physicist. In Europe, the graduate education is followed by 2 to 4 years of further training under the supervision of an experienced radiotherapy physicist to produce a qualified medical physicist (QMP). The QMP can then undertake further specialized training in a subspecialty (e.g., radiotherapy physics) for a further 2 years to qualify as a specialized medical physicist. Within the EU, as defined in the Medical Exposure Directive, "in relation to medical exposure," the medical physics expert is equivalent to the specialist medical physicist. Further developments in relation to the requirements for SMP are outlined in EFOMP Policy Statement No. 12 (Eudaldo and Kjeld 2010).

While the above indicates an idealized training route, in practice there is a wide variation between countries. In North America at present, there is a mix of accredited and nonaccredited graduate and training programs. In Europe, recent surveys (Eudaldo and Olsen 2008) indicate that while many countries have national societies for medical physics through which training programs are organized, there is a variation on the interpretation of the terms QMP and MPE, along with significant variations in the length and content of training programs. There are continuing attempts by EFOMP and the European Society for Therapeutic Radiology and Oncology (ESTRO) to develop the core curriculum for medical physics education and training and in setting up accredited training programs to provide appropriate clinical and research training in a consistent manner. In North America, the Commission on Accreditation of Medical Physics Education Programs (CAMPEP) sets the standards for medical physics education programs and controls these standards through a process of review and accreditation. While there are no equivalent European-wide organizations like CAMPEP, most training programs follow EFOMP/ESTRO curriculum guidelines and some either offer accreditation of programs or provision for limited official training institutes.

While there are variations in the content of educational and training programs in medical physics across countries, there is a similar core content which comprises three main components: didactics, clinical placement, and research. The core subjects for the didactic and placement components can be usefully divided into fundamental knowledge and applied knowledge and consist of some variation of the following subjects:

General fundamental knowledge

- Fundamentals of human anatomy and physiology
- Fundamentals of oncology
- Principles and applications of radiobiology and molecular biology
- Review of radiation physics
- Principles of quality management
- Statistical methods
- Quality management in radiotherapy
- Information and communication technology
- Organization, management, and ethical issues in healthcare
- General safety principles in the medical environment
- Health technology assessment

Applied knowledge and skills

- Dosimetry
- Principles of medical imaging and image handling
- External beam radiotherapy
- Brachytherapy
- Particle therapy
- Unsealed source therapy
- Radiation protection for ionizing radiation
- Mathematical modeling in radiation oncology
- Uncertainty in radiotherapy

Research Project
A short, focused research project with a topic relevant for radiotherapy physics and practice.

Certification and Registration of Medical Physicists

While the key objectives of a training program are to provide comprehensive training in the core competencies required to deliver a high standard of radiation oncology physics at an operational and clinical level, another objective is to prepare trainees for certification or registration and a professional career in radiation oncology. Certification or registration of medical physicists provides a demonstrated, transparent mechanism for identifying competence in the field. It is important, given the role and responsibility level of the medical physicist outlined earlier, that medical physicists are subject to the same processes as other medical subspecialties, which typically involve a certification step. At this point, the mechanism for certifying medical physics is not as strong as that for clinicians and others.

Certification usually refers to a process involving written and oral examinations. In the United States, the certification exams are conducted by the American Board of Radiology (ABR), which is a specialty board of the American Board of Medical Specialties (ABMS). Recent resolutions by the ABR require exam candidates to have completed a CAMPEP-accredited graduate program or a CAMPEP-accredited residency in medical physics. A similar mechanism for certification exists in Canada, where exams are administered by the Canadian College of Physicists in Medicine (CCPM). Registration is an alternative approach to demonstrate the competency of an individual in the diagnosis and treatment of patients. For example, the Institute of Physics and Engineering (IPEM), the medical physics professional body in the UK, accredits UK graduate programs and medical physics training programs. These training programs involve the construction of a portfolio of evidence of proven competency in specific areas of medical physics, with an oral examination of the advanced portfolio. Once successful in the examination, the trainee then joins the Register of Clinical Scientists, which is administered by the UK Health Professional Council. A successful examination certifies the trainee to practice unsupervised.

Many European countries have similar registration or certification processes (EFOMP 2001). While a CAMPEP equivalent body does not exist in Europe at this point, there is recognition that a European-wide accreditation and certification process is a means to achieve better harmonization across the EU member states, and preliminary discussions along these lines are taking place between EFOMP and ESTRO.

Following certification or registration, there is a requirement for continuing education and maintenance of this certification (Kun et al. 2005). For this reason, many certificates are time-limited, with certificates being reissued only upon evidence of the individual having completed a predetermined amount of further development and education. There are many courses in medical physics related areas (e.g., preconference courses, ESTRO courses, AAPM summer schools) to allow the certified medical physicist to retain the knowledge and skill set required for continuing certification or registration.

Importance of Certification of Medical Physicists

The key role of the medical physicist in all aspects of the radiotherapy process requires them to possess a high level of scientific knowledge and clinical competency. For accountability and transparency, and in keeping with other medical professions, these competencies have to be demonstrated through a certification or registration process. The path to certification is increasingly through the mechanism of accredited training programs, which although they vary in detail between countries, contain a very similar core of skills and competencies required for practicing medical physics. The nature of the training and education programs has to be broad and flexible to accommodate new technologies and techniques and should be kept under continual review to reflect the changing needs and demands of the profession. Greater collaboration with clinical and radiobiology colleagues on emerging trends within their areas of interest will help form strategic developments in training of medical physicists and allow the medical physics community to provide leadership in emerging technologies.

High common standards of training and education, together with common guidelines on certification, accreditation, and audit, will allow for improved consistency and quality of training and enable greater freedom of movement of qualified physics personnel across national boundaries (Starkschall 2009). Such an approach will also provide for greater confidence of the public in the medical physics profession.

References

AAPM. 1993. *Report 38: Role of a Physicist in Radiation Oncology*. New York, NY: American Association of Physicists in Medicine.

AAPM. 2008. *Report 133: Alternative Clinical Training Pathways for Medical Physics*. New York, NY: American Institute of Physics, Inc.

Belletti, S., A. Dutreix, G. Garavaglia, H. Gfirtner, J. Haywood, K. A. Jessen, I. L. Lamm et al. 1996. Quality assurance in radiotherapy: The importance of medical physics staffing levels. Recommendations from an ESTRO/EFOMP joint task group. *Radiother. Oncol.* 41 (1): 89–94.

EFOMP. 2001. Policy Statement No. 10. Recommended guidelines on national schemes for continuing professional development of medical physicists. *Phys. Med.* 17 (2).

EFOMP. Malaga declaration—EFOMP's position on medical physics in Europe 2006. Available from http://www.efomp.org/.

Eudaldo, T., and O. Kjeld. 2010 The present status of medical physics education and training in Europe. New perspectives and EFOMP Recommendations. *Phys. Med.* 26 (1): 1–5.

Eudaldo, T., and K. Olsen. 2008. The present status of medical physics education and training in Europe: An EFOMP survey. *Phys. Med.* 24 (1): 3–20.

European Commission. 1996. Euratom Directive 96/29. Laying down basic safety standards for the protection of the health of workers and the general public against the dangers arising from ionising radiation.

European Commission. 1997. Euratom Directive 97/43. On health protection of individuals against the dangers of ionising radiation in relation to medical exposures. *Off. J. Eur. Communities* 1180: 22.

Kun, L. E., K. Ang, B. Erickson, J. Harris, R. Hoppe, S. Leibel, L. Davis, and R. Hattery. 2005. Maintenance of certification for radiation oncology. *Int. J. Radiat. Oncol. Biol. Phys.* 62 (2): 303–308.

Starkschall, G. 2009. Editorial: International certification of medical physicists. *J. Appl. Clin. Med. Phys.* 10 (1): 3006.

Staffing for Quality: Overview

Julian Malicki
University of Medical Sciences
Greater Poland Cancer Centre

Introduction

The rapid development of science and technology has affected radiotherapy practice. Although the general concept—the use of photon and electron beams—has remained the same, substantial progress has been made in delivering a dose to a tumor more accurately.

The relationship between the quality of radiotherapy and staffing involves adequate staffing levels, clearly defined roles and responsibilities, and channels of communication between the staff. A problem arises when the above statements have to be quantified. Reliable analysis requires us to consider the following: the role of the radiotherapy department in the health delivery system in different contexts: organizational, social, economic, and geographic, which includes the case and procedure mix, type of clinical duties, research and teaching, the department size and hours of operation, the reimbursement system, local habits, and economic strength.

Does a Universal Standard in Staffing Exist and Can It Be Quantified?

Finite resources require that, in every radiotherapy department, the quality of services provided must be balanced against the number of services provided. Hospital administrators look at the required staff from a different perspective than clinical practitioners and scientists. An approach from detailed to generalization often leads to a number of posts (also known as positions) that are not acceptable from an economic point of view. Another approach is to survey representative departments and, based on that survey, draw a staffing model. Such studies (Slotman et al. 2005) usually show a large diversity in figures but reveal a common tendency that the more sophisticated are the procedures used, the more numerous are the staff needed in all or in certain categories (Miles et al. 2005; Budiharto et al. 2008). Indicators comprising the denominator and desired value of quantified activity allow one to evaluate staff performance and predict future needs (Cionini et al. 2007; Holmberg and McClean 2003).

No universally accepted model exists that accounts for all factors that occur worldwide and that is regularly reviewed against novel technologies. Performed studies revealed diversity in both guidelines and practice. Therefore, a common standard probably can be descriptive only while figures might differ in different places.

Relationship between Complexity of Equipment and Staff

Some aspects associated with technological development reduce workload, while others increase it, and the impact differs in staff groups. The increase in certain groups' workloads may be followed by a decrease in others (Miles et al. 2005). Many technological innovations improve machines so they require less outside maintenance. Self-checks are performed automatically and reduce procedures that personnel have to carry out, thus reducing the workload. On the other hand, novel techniques (IGRT, adaptive therapy) require constant development of quality procedures. The aim of quality procedures is to assure that in clinical practice, a better dose distribution is delivered to the patient in comparison to one without those novelties; and if any malpractice or error occurs, it will be detected and compensated.

Recruitment and Training

Recruitment should consider both primary qualifications and continuing education. Requirements on the levels of qualification of different groups of staff exist at the national level for most of the personnel involved. If a particular group is not regulated by official means, appropriate training must be organized at the departmental level (Leer et al. 1998). Training on new equipment and techniques as well as for emergencies and major disasters should be available. Departmental, institutional, or external professional meetings should be regular activities. Orientation programs help new staff while staff performance evaluations promote continuing professional development.

Staff Groups

Three major groups of staff contribute to radiotherapy: radiation oncologists, medical physicists, and radiation technologists (RTTs). Evaluation of staff performance is required. This can be accomplished by a single figure that accounts for all factors including science and technology development but, as a complex model, it is still only an estimate. Surveys in high-resource countries show a quite stable staffing situation during the past 10 years, with the exception of Japan where significant changes both in staffing and number of patients treated were noted (Teshima et al. 2008). Data for low-resource countries are incomplete.

Radiation Oncologists

Radiation oncologists (radiotherapists) are physicians trained in radiotherapy and oncology. The job is well defined and regulated in most countries. Most studies have revealed a guideline of 200–250 patients per year per radiation oncologist, but there can be significant deviations from practice to practice.

In Europe, the ESTRO project, QUARTS, revealed a large diversity in guidelines for the number of radiation oncologists needed. Less than half of European countries (45%) had guidelines, which were usually expressed as numbers of patients per radiation oncologist. It varied for university and nonuniversity centers and was influenced by other tasks in the field of oncology, such as administration of chemotherapy. The suggested figure from this study indicated 250 patients per 1 radiation oncologist for nonuniversity and no-chemotherapy units. When

other tasks are in place (e.g., chemotherapy), this figure should be proportionally lower (Slotman et al. 2005). The European EORTC centers' data for 2007 revealed a mean workload of 258 patients per radiation oncologist (range: 99–480) (Budiharto et al. 2008). The studies also set the limit of patients, that can be managed at the same time under the auspices of one radiation oncologist, from 15 to 35. In 1995, a comparative analysis between the United States and Japan showed differences in pattern and practice. In the United States, the guidelines were 200–250 (Parker et al. 1991), with an average figure in clinical practice of 256. In Japan, the average figure was 172 in 1995 and 247 in 2007. However, the ratio of total newly diagnosed patients treated with radiotherapy in Japan increased from 15% in 1995 to 25% in 2007, while in the United States it was quite stable and near 50%, which should be taken into account (Teshima et al. 1996). For data in countries with fewer resources, it should be considered that only part of the population has access to radiotherapy. Therefore, for example, the number of 933 qualified radiation oncologists against the population of 516.7 million in Latin America (Zubizarreta, Poitevin, and Levin 2004) shows a lack of service rather than severely inadequate staffing levels in the existing institutions. Data for several African and Asian countries are incomplete and further study is recommended in order to reveal the existing situation and more detailed guidelines (IAEA 1999; Sharma et al. 2008). Interestingly, as reported by Sharma et al. (2008) in 54% of centers across Africa, an average of 350 patients were treated per radiation oncologist, which is over 35% more than average for Europe (258). On the other hand, among some European EORTC centers, a much larger number of patients per radiation oncologist were treated (99–480) than is the case on average across Africa (Budiharto et al. 2008).

Physicists

Medical physicists in several countries are identified as an independent specialty. Usually their training includes a specific curriculum after a masters or doctoral degree in physics. Due to shortages in some countries, specialists from related fields (The Royal College of Radiologists et al. 2008) are eligible for this job; thus the job definition may not include a degree in physics but may describe the required expertise in radiation physics or radiation technology and a way of recognizing the competence (example in the Medical Exposure Directive 97/43 published in 1997 by European Commission).

The ESTRO/EFOMP joint group proposed a detailed model of calculating the recommended staffing level in medical physics for clinical routine (Belletti et al. 1996); however, the increasing complexity of radiotherapy might modify the outcome. If a model exists, it makes sense to check the results from surveys against the model, which is not easy as publications do not provide details needed to run the model. A rough estimate shows that even in high-resource countries, the workload of a physicist is higher than recommended: 450–500 patients per year from a European survey (Slotman et al. 2005) and 400 from the model estimate (Belletti et al. 1996) corresponding to survey conditions (brachytherapy excluded).

Studies in various regions revealed differences in guidelines and in practice for staffing, even more significant than for radiation oncologists. In the United States, physicists are members of radiotherapy departments while in Europe they usually have separate medical physics units. These two organizational models are copied worldwide. The staffing standard is usually expressed in relation to the number of machines or in relation to the number of new patients: 1 physicist per linear accelerator or 1 per 450–500 patients in Europe (Slotman et al. 2005). European EORTC centers disclosed a mean workload of 426 patients per physicist (range: 124–827) (Budiharto et al. 2008). A similar workload was noted in the United States, while indirectly estimated data for Japan showed approximately 1200 in 1995 and 1630 patients per physicist in 2007. It is unclear whether this was associated with a low fraction of patients treated with radiotherapy (15% and 25%, respectively) (Teshima et al. 1996, 2008) or if it indicated understaffing. A diverse staffing situation exists in Latin America. Out of 357 medical physicists there, 241 had a degree specifically in medical physics (Argentina 60, Brazil 176, and Chile 5) (Zubizarreta, Poitevin, and Levin 2004). This study shows, similar to data for radiation oncologists, an insufficient number of physicists in relation to the population, but does not show the staffing level in particular institutions.

Shortages of radiotherapy physicists are visible in various studies (Leetz et al. 2003; Slotman et al. 2005; Teshima et al. 2008; Zubizarreta, Poitevin, and Levin 2004) regardless of the economic strength of the world region, although different reasons contribute to the shortage to different extents.

The job of a dosimetrist varies significantly between countries. These persons are competent in basic dosimetric procedures (beam calibration, treatment planning) and they do not constitute an independent professional group but may be recruited from RTTs. In Central and East European countries, this job is usually done by physicists (Leer et al. 1998; Slotman et al. 2005).

Radiotherapy Technologists

Radiation technologists constitute an independent professional group in many countries (Leer et al. 1998). Historically, their staffing levels were determined using the numbers of posts per linear accelerator, simulator, etc. The increasing complexity of procedures and demand for patient comfort affect the method of counting the number of RTTs. The staffing models were reviewed by Coffey et al. (2004), with an indication of complex and descriptive approaches. It includes the educational standard that links personal responsibility for accurate preparation and administration of a course of radiation therapy and the subsequent monitoring of the patient while treatment is delivered (Coffey et al. 2004).

National guidelines for technologists vary and are dependent on local habits and complexity of work (available in 19/44 European countries) (Slotman et al. 2005). A change will come due to the demand to bring RTTs' training up to university levels, both bachelors and masters (Malicki 2005). Initiatives such as the Inter-University European program EMPIRION (joint diploma master course of Hogeschool Inholland–School of Health and Institute of Advanced Studies and Applied Research, Medizinische Universitat Wien, Medizinische Fakultat Karl Gustav Carus der Technische Universitat Dresden, Istituto Politecnico de Lisboa–Escola Seperior de Tecnologia da Saude de Lisboa and Uniwersytet Medyczny Poznan) contribute to job recognition and elevation. In European EORTC centers, a great diversity in numbers of patients treated by an RTT annually was revealed, with a mean of 107 but within a range of 34–734 (Budiharto et al. 2008). This indicates the need for additional parameters in order to quantify the job workload for this group. In Latin America, about 2300 radiotherapy technologists work, but the requirements for qualification are diverse (Zubizarreta, Poitevin, and Levin 2004).

Interactions between Various Groups of Staff

Overlaps in duties between radiation oncologists, physicists, and RTTs occur. Certain activities can be done by different staff groups and surveys reveal these facts. Sometimes a formal borderline between group tasks is moved by the existing practice due to staff shortages or habits. In all cases, if not regulated, an institutional policy should exist.

If radiation oncologists administer chemotherapy, the staffing standard should take that into account. In Europe, in 19.5% of countries, the majority of radiation oncologists administer chemotherapy, while in 39% of European countries, less than 25% of radiation oncologists administer chemotherapy. Treatment planning can be done by physicists, RTTs, dosimetrists, or radiation oncologists. In Europe, for high-resource countries, RTTs/dosimetrists and physicists perform treatment planning in equal shares of 42.8%; for medium- and low-resource countries, physicists do treatment planning in 85% of the countries. Quality assurance in Europe is done in 66.7% of cases by physicists in high-resource countries and by physicists in 90% of medium- and low-resource countries (Slotman et al. 2005).

Larger departments usually employ a quality officer. This person can be recruited from major staff groups or have a degree in another specialty (e.g., public health). A quality officer can be situated directly in the radiotherapy department or can serve it from outside. Nurses can contribute to optimizing patient care and therapeutic outcomes and they, in some countries, overlap in certain duties with RTTs and radiation oncologists. In Central and Eastern Europe, a historical underemployment of secretarial and medical record staff adds more administrative duties to radiation oncologists and physicists and should be considered.

Research and Teaching

If the mission of the institution includes teaching and research activities, there should be a policy that identifies the staff allocated for these activities. Usually in the departments that participate in university education programs and run courses, teaching is integrated with clinical duties. It is obvious that persons with teaching obligations cannot perform a quantity of clinical duties equal to those who do not teach. No simple formula exists to count the portions of clinical and teaching work but, depending on the type of employment, the proper compensation has to be foreseen for persons with both obligations, otherwise it compromises the quality. A similar approach should be taken with research, through compensating and promoting conceptual work and research projects.

Students of different specialties require unequal staff supervision. The medical student undergraduate curriculum does not contain extensive radiotherapy classes. However, if the radiotherapy unit belongs to a university hospital, the staff may be involved in teaching clinical subjects other than radiotherapy. Physics undergraduate curricula are differently focused, some of them on medical radiation physics and then various types of classes: practical or hands-on courses might substantially increase staff obligations.

Radiotherapy technologist (Izewska et al. 2003) undergraduate training usually includes an extensive onsite training program in a radiotherapy facility during regular working hours. Supervising those students and providing necessary explanations adds additional duties and substantially reduces the efficacy of the treatment performance if not balanced by additional staff members.

Role, Size, and Hours of Operation

Staffing should consider the role of the department in the catchment area. A sole radiotherapy provider or the leading provider needs better staffing (more numerous posts, staff skilled in all relevant procedures and able to work overtime hours in case of unforeseen needs) than is needed for a provider that contributes only partly to the radiotherapy service.

Small departments encounter problems in ensuring quality due to a limited number of staff in all categories, especially radiation oncologists and physicists. Small departments are more likely to fall below critical staff competence when normal absences such as illnesses, conferences, and holidays occur, but it is also more difficult to sustain double-checking of important procedures and provide ongoing training for sophisticated procedures to all staff members (Belletti et al. 1996; Dunscombe, Roberts, and Walker 1999). Fortunately for quality, a department's size and economic incentive go together as the cost of the radiotherapy procedure decreases with the increasing size of the institution (Dunscombe, Roberts, and Walker 1999). Additional tasks undertaken for quality procedures do not necessarily affect the unitary cost of radiotherapy (Malicki et al. 2009). Interestingly, in various countries, the size of radiotherapy centers vary regardless of the level of resources in a country, partly for historical reasons. Large groups of personnel may benefit from synergies and therefore safe and equal quality work can be done with a lower proportionate staffing level.

Working hours vary. In lower-resource regions with insufficient equipment and a restricted facility base, the way to treat more patients is to extend the hours of operation and/or increase the number of patients treated per hour. In high-resource countries, extended hours are observed as well and they may impact the quality as adequate staffing might not be present during overtime hours. Two solutions occur: (1) to work current facilities more intensively, which always increases staff overtime or shift work; and (2) facility expansion through purchase of additional accelerators and new staff recruitment. The labor cost dominates the other cost components (more than 50%) (Dunscombe, Roberts, and Walker 1999), which target cost-effectiveness over labor; thus careful analysis of a good quality–quantity balance is needed.

Impact of Safety Regulations Associated with Ionizing Radiation

Unlike the other medical disciplines, staff qualifications and post numbers in radiotherapy are of interest for the safety regulations of various radiation protection authorities. In some countries, the minimum number of staff members is cited in radiation protection laws and regulations. The professional and scientific societies are encouraged to approach the problem and give guidelines.

Brachytherapy

Questions concerning staff apply to brachytherapy as well. A major difference is that only 30%–50% of radiotherapy centers provide brachytherapy. Far fewer studies on staffing models have been published (Belletti et al. 1996) and few authors have evaluated facilities for brachytherapy in comparison to external beam radiotherapy (Guedea et al. 2007, 2008). In Europe, out of more than 1000 radiotherapy centers, only half provided brachytherapy and it is only an estimate considering the response rate to the PCBE survey (Guedea et al. 2008). Organizational differences make setting staffing standards more difficult, as the brachytherapy department may be part of the radiotherapy department or a separate unit. Consequently, staff may share obligations in brachytherapy and external beam therapy.

The workload of staff involved in brachytherapy in Europe was 11.4–20.2 hours per week for radiation oncologists, 10.2–17.2 hours per week for physicists, and 15.7–17.7 hours per week for RTTs, depending on the resource status of the country. Another figure revealed was the mean number of radiation oncologists per center performing radiotherapy: 3.0–4.5 (for detailed figures, see Guedea et al. 2008). Notable differences in brachytherapy treatment were observed, especially in the workload of radiation oncologists and physicists (number of patients per specialist), in a group of the 10 most recent members of the European Union (EU), and in a group of 14 other non-EU European countries, in comparison to the group comprising the 15 original member countries of the EU, plus four others according to economic wealth (Guedea et al. 2007, 2008).

Summary

The complexity of the procedures has been constantly increasing, which requires constant professional development of staff. The increasing cost of treatment enforces cost-effectiveness, which contradicts the staffing needs, as labor has become the major cost component in radiotherapy.

The annual patient load per center in many countries in Europe, North America, Australia, and Japan has increased during the past two decades. In the main staff categories—radiation oncologists and physicists—a limited reduction in absolute numbers of patients treated annually is noted, but it does not counterbalance the increasing demand of workload due to more sophisticated procedures applied, and it has led to considerable staff shortages in these categories. A more complex situation exists in the case of RTTs, for whom professional competencies are defined with much larger diversity. In many regions, including the United States, severe shortages of RTTs were noted. A way to overcome staff shortages is to strengthen the educational standard of the largest group, RTTs, who could take on more complex duties and responsibilities.

Surveys made worldwide revealed similar job descriptions, guidelines, and practices regarding workload among high-resource countries and more diverse ones in medium- and low-resource regions. In low-resource regions, the practice is more difficult to compare and a standard is more difficult to set, as only part of the population is treated. The average figure for these regions may produce a much worse result than the real staffing level in particular institutions.

No universally accepted and regularly updated staffing model (from detail to generalization) exists. However, it is advisable for a survey to refer to a model and collect the data that are needed to run the model. Thus, survey results could be checked against model results, which would support reliable comparative analysis.

References

Belletti, S., A. Dutreix, G. Garavaglia, H. Gfirtner, J. Haywood, K. A. Jessen, I. L. Lamm et al. 1996. Quality assurance in radiotherapy: The importance of medical physics staffing levels. Recommendations from an ESTRO/EFOMP joint task group. *Radiother. Oncol.* 41 (1): 89–94.

Budiharto, T., E. Musat, P. Poortmans, C. Hurkmans, A. Monti, R. Bar-Deroma, Z. Bernstein et al. 2008. Profile of European radiotherapy departments contributing to the EORTC Radiation Oncology Group (ROG) in the 21st century. *Radiother. Oncol.* 88 (3): 403–410.

Cionini, L., G. Gardani, P. Gabriele, S. Magri, P. L. Morosini, A. Rosi, and V. Viti. 2007. Quality indicators in radiotherapy. Italian Working Group General Indicators. *Radiother. Oncol.* 82: 191–200.

Coffey, M., J. Degerfalt, A. Osztavics, J. van Hedel, and G. Vandevelde. 2004. Revised European core curriculum for RTs. *Radiother. Oncol.* 70 (2): 137–158.

Dunscombe, P., G. Roberts, and J. Walker. 1999. The cost of radiotherapy as a function of facility size and hours of operation. *Br. J. Radiol.* 72 (858): 598–603.

Guedea, F., V. Montse, M. Ventura, A. Polo, J. Skowronek, J. Malicki, W. Bulski et al. 2007. Patterns of care for brachytherapy in Europe (PC BE) in Spain and Poland: Comparative results. *Rep. Pract. Oncol. Radiother.* 12: 39–45.

Guedea, F., M. Ventura, J. J. Mazeron, J. L. Torrecilla, P. Bilbao, and J. M. Borras. 2008. Patterns of care for brachytherapy in Europe: Facilities and resources in brachytherapy in the European area. *Brachytherapy* 7 (3): 223–230.

Holmberg, O., and B. McClean. 2003. A method of predicting workload and staffing level for radiotherapy treatment planning as plan complexity changes. *Clin. Oncol. (R. Coll. Radiol.)* 15 (6): 359–363.

IAEA. 1999. Proceedings of the co-ordination meeting of AFRA11-12 (RAF/6/014). In *African Regional Co-operative Agreement for Research, Development and Training (AFRA)*.

Izewska, J., P. Andreo, S. Vatnitsky, and K. R. Shortt. 2003. The IAEA/WHO TLD postal dose quality audits for radiotherapy: A perspective of dosimetry practices at hospitals in developing countries. *Radiother. Oncol.* 69: 91–97.

Leer, J. W. H., A. L. McKenzie, P. Scalliet, and D. I. Thwaites. 1998. Practical guidelines for the implementation of a quality system in radiotherapy ESTRO physics for clinical radiotherapy. In *ESTRO Quality Assurance Committee sponsored by "Europe Against Cancer," Physics for Clinical Radiotherapy, Booklet no. 4.*

Leetz, H. K., H. H. Eipper, H. Gfirtner, P. Schneider, and K. Welker. 2003. Staffing levels in medical radiation physics in radiation therapy in Germany. Summary of a questionnaire. *Strahlenther. Onkol.* 179 (10): 721–726.

Malicki, J. 2005. A career pathway for radiation therapists. Does it really exist?: In regard to Kresl et al. (*Int. J. Radiat. Oncol. Biol. Phys.* 2004;60:8–14). *Int. J. Radiat. Oncol. Biol. Phys.* 62 (1): 292; author reply 293.

Malicki, J., M. Litoborski, M. Bogusz-Czerniewicz, and A. Swiezewski. 2009. Cost-effectiveness of the modifications in the quality assurance system in radiotherapy in the example of in-vivo dosimetry. *Phys. Med.* 25 (4): 201–206.

Miles, E. A., C. H. Clark, M. T. Urbano, M. Bidmead, D. P. Dearnaley, K. J. Harrington, R. A'Hern, and C. M. Nutting. 2005. The impact of introducing intensity modulated radiotherapy into routine clinical practice. *Radiother. Oncol.* 77 (3): 241–246.

Parker, R. G., C. R. Bogardus, G. E. Hank, C. G. Orton, and M. Rotman. 1991. *Radiation Oncology in Integrated Cancer Management. Report of the Inter-Society Council for Radiation Oncology.* Merifield, VA: American College of Radiology Publications ISCRO.

Sharma, V., P. M. Gaye, S. A. Wahab, N. Ndlovu, T. Ngoma, V. Vanderpuye, A. Sowunmi, J. Kigula-Mugambe, and B. Jeremic. 2008. Patterns of practice of palliative radiotherapy in Africa, Part 1: Bone and brain metastases. *Int. J. Radiat. Oncol. Biol. Phys.* 70 (4): 1195–1201.

Slotman, B. J., B. Cottier, S. M. Bentzen, G. Heeren, Y. Lievens, and W. van den Bogaert. 2005. Overview of national guidelines for infrastructure and staffing of radiotherapy. ESTRO-QUARTS: Work package 1. *Radiother. Oncol.* 75 (3): 349–354.

Teshima, T., H. Numasaki, H. Shibuya, M. Nishio, H. Ikeda, H. Ito, K. Sekiguchi et al. 2008. Japanese structure survey of radiation oncology in 2005 based on institutional stratification of Patterns of Care Study. *Int. J. Radiat. Oncol. Biol. Phys.* 72 (1): 144–152.

Teshima, T., J. B. Owen, G. E. Hanks, S. Sato, H. Tsunemoto, and T. Inoue. 1996. A comparison of the structure of radiation oncology in the United States and Japan. *Int. J. Radiat. Oncol. Biol. Phys.* 34 (1): 235–242.

The Royal College of Radiologists, Society and College of Radiographers, Institute of Physics and Engineering in Medicine, National Patient Safety Agency, and British Institute of Radiology. 2008. *Towards Safer Radiotherapy.* London: The Royal College of Radiologists.

Zubizarreta, E. H., A. Poitevin, and C. V. Levin. 2004. Overview of radiotherapy resources in Latin America: A survey by the International Atomic Energy Agency (IAEA). *Radiother. Oncol.* 73 (1): 97–100.

Staffing for Quality: Physics

Michael D. Mills
University of Louisville

Introduction

Operating a first-class medical physics program requires justification of staffing allocations for qualified medical physicists (QMPs), and support staff such as physics assistants and medical dosimetrists. While appropriate staffing is necessary, it is not a sufficient condition to provide quality services. The purpose of this chapter is to provide direction on to how to equate clinical needs with the staffing levels and competencies required. This chapter will:

- Review why appropriate staffing is important to the practice of radiation oncology
- Look at different models based on data published and available to the radiation oncology community
- Estimate appropriate staffing (QMPs and support staff) for a radiation oncology physics section
- Apply these models to a variety of settings: academic practice, community hospitals, and freestanding cancer centers

When a physicist provides a staffing model for consideration by administration, that person often refers to recommendations published by the American College of Radiology (ACR 2008), and measured median standards found within the Abt Studies for Radiation Oncology Physics Services (Abt Associates 1995, 2003, 2008). However, this approach is open to criticisms: "These published median values are from several years ago; how long does it take our staff to accomplish routine and special physics procedures, today?" "Do current procedures and processes let our physicists and support staff members accomplish work more efficiently than these published standards?" "Do we need a physics section with so many years of combined experience, or will less experienced and less expensive employees be able to handle the workload?"

A proper staffing model should begin with what is required to provide the best possible treatment to the individual patients served by the clinic. The specific tasks include the expertise to commission appropriate new technologies in a timely manner, to provide consultant information to radiation oncologists during the treatment planning and delivery processes, to ensure that the technical aspects of the facility operate as efficiently as possible, and to ensure errors are minimal in number and magnitude. In addition, staff physicists are responsible for ensuring the safety of all patients, staff, and visitors from hazardous radiation and for fulfilling all state radiation regulatory requirements. The information in this article should enable the reader to construct an employment matrix or staffing justification grid to equate the clinical needs to the quantity and quality of staffing required.

Methods

The scope of professionals included in the staffing model is: qualified medical physicists (MS and PhD level), medical dosimetrists (or junior medical physicists), and physics assistants or technologists (including IMRT QA technologists). The scope of work includes clinical duties and support, educational teaching, research, and administration. Clinical practice factors impacting clinical medical physics staffing include:

At the main facility

- The number of patients treated annually
- The number of radiation oncologists
- The number and procedure mix of billed routine physics procedures
- The number and procedure mix of billed special medical physics procedures

At satellite facilities

- Parameters listed above for main facility
- Travel time
- Additional contracted efforts

For the United States, major sources of staffing information include the American College of Radiology (ACR 2008), The Abt Study of Medical Physicist Work Values for Radiation Oncology Physics Services: Round I, Round II, and Round III (Abt Associates 1995, 2003, 2008), and articles by Herman (Herman, Mills, and Gillin 2003), Mills (Mills et al. 2000; Mills 2005), and Klein (2009). Additional methods to assess physics staffing levels were conducted by the European Society for Therapeutic Radiology and Oncology (ESTRO) (Belletti et al. 1996), and the European Federation of Organizations for Medical Physics (EFOMP) (Eudaldo et al. 2004).

Review of Abt Studies I, II, and III

Abt Associates, Inc. is one of the nation's most respected medical economics consulting organizations; it has primary offices in Bethesda, MD, and Cambridge, MA. The AAPM and ACMP contracted for the Abt I and Abt II studies in 1995 and 2002; the AAPM contracted for the Abt III study in 2007. Abt I was limited to work associated with routine physics procedures; Abt II and III measured medical physicist work for both routine and special medical physics procedures. Abt III included such contemporary procedures as intensity modulated radiation therapy/image guided radiation therapy, high dose-rate afterloading brachytherapy, and stereotactic body radiation therapy. The benchmark procedure for all three Abt studies was CPT Code 77336—Continuing Medical Physics Consultation. The input information to the analysis for the Abt studies included:

- Collection of time estimates for services by CPT code
- Collection of intensity estimates for each service relative to the baseline service (77336)

TABLE 43.1 Median QMP Time Estimates in Hours by CPT Code

77295 Simulation 3D	1.18	77331 Sp Dosimetry	2.06
77300 Bas Dos Calc	0.55	77332 S Tx Device	0.13
77301 IMRT Tx Plan	4.53	77333 I Tx Device	0.34
77305 S Isodose	0.69	77334 C Tx Device	0.24
77310 I Isodose	0.78	77336 Cont MP Consult	1.00
77315 C Isodose	0.98	77370 Spec MP Consult	3.45
77321 Tele Port Plan	1.07	77781 HDR 1–4	2.70
77326 S Br Isodose	2.52	77782 HDR 5–8	3.79
77327 I Br Isodose	2.70	77783 HDR 9–12	4.70
77328 C Br Isodose	4.78	77783 HDR >12	3.43

Source: Abt Associates, Inc., Round III—Final Report: Abt Study of Medical Physicist Work Values for Radiation Oncology Physics Services, available from http://www.aapm.org/pubs/reports/ABTIIIReport.pdf, 2008. With permission.

TABLE 43.2 Median Support Staff Time Estimates in Hours by CPT Code

77295 Simulation 3D	2.75	77331 Sp Dosimetry	0.63
77300 Bas Dos Calc	0.5	77332 S Tx Device	0.50
77301 IMRT Tx Plan	4.25	77333 I Tx Device	0.50
77305 S Isodose	1.00	77334 C Tx Device	1.00
77310 I Isodose	1.50	77336 Cont MP Consult	N/A
77315 C Isodose	2.00	77370 Spec MP Consult	N/A
77321 Tele Port Plan	2.00	77781 HDR 1–4	0.50
77326 S Br Isodose	1.00	77782 HDR 5–8	0.88
77327 I Br Isodose	1.67	77783 HDR 9–12	1.00
77328 C Br Isodose	2.00	77783 HDR > 12	1.13

Source: Abt Associates, Inc., Round III—Final Report: Abt Study of Medical Physicist Work Values for Radiation Oncology Physics Services, available from http://www.aapm.org/pubs/reports/ABTIIIReport.pdf, 2008. With permission.

- Collection of service-mix data (annual number of procedures by service type)
- Analysis of the survey data to develop QMP and staff work estimates by service or procedure

A first step in the use of the Abt data is to customize the information to the unique clinical setting, determine the time to accomplish defined tasks, multiply by the measured or anticipated number of such tasks, and sum the results. The Abt data includes in its matrix development time for special procedures. However, it does not include time for meetings, vacations, sick days, etc. In addition, time for teaching, research, or administration is not included. The user may wish to account for these time categories according to local practices.

It may be desirable to perform the time analysis by mapping the level of experience or competency needed to perform the task. In an academic setting, this may be according to faculty appointment or administrative assignment. Special procedures such as stereotactic body radiation therapy, respiratory gating, or image guided radiation therapy may require experienced staff, while staff without a faculty appointment may be responsible for routine clinical support and quality assurance. A physics resident may also be included in a staffing model as this individual may provide real work support for the clinic, of course at a reduced FTE and usually at the level of a nonfaculty staff physicist. Table 43.1 shows the median QMP time estimates in hours by CPT code, as listed in the Abt III Report.

The data from Abt III is considered the most recent and most applicable to contemporary practice. In comparison with Abt II, it was found that commissioning times are reported to be a little shorter in 2007, compared with 2002, possibly reflecting greater familiarity with IMRT commissioning. Procedural times are a little shorter for IMRT procedures, a little longer for brachytherapy procedures, but not significantly different in either case. The Special Medical Physics Consultation was measured as 4.00 hours in 1995, 5.60 hours in 2002, and 3.45 hours in 2007. Support staff times were also measured and reported in

TABLE 43.3 Median Qualified Medical Physics Staffing Statistics

Physics Staffing	Overall	Private Hospital	Medical School University	Medical Physics Consulting Group	Physician Group
Patients/QMP	304	368	220	464	260

Source: Abt Associates, Inc., Round III—Final Report: Abt Study of Medical Physicist Work Values for Radiation Oncology Physics Services, available from http://www.aapm.org/pubs/reports/ABTIIIReport.pdf, 2008. With permission.

Note: Patients = total annual number of brachytherapy and teletherapy patients treated in a calendar year.

Abt III to distinguish this time from that of qualified medical physicists. Table 43.2 shows the median support staff time estimates in hours by CPT code as listed in the Abt III Report.

Work is considered as the product of time and intensity. Work as a concept will be dealt with only briefly in this report as most administrators are interested in the procedural time, not the intensity product associated with work. To measure work, a common representative procedure is selected and used as a benchmark with intensity = 1.0. The Abt studies all used 77336 as the benchmark and assigned it an intensity of 1.0. Respondents assigned all other procedures an intensity using 77336 as a reference. Qualified medical physicist work tables are reported in Abt I, II, and III Reports. All three studies indicate similar work values; indicating that QMP work has not changed very much over this 12-year period.

Results

To defend the effort to provide physics services at an institution:

- Determine the number and type of physics services your institution provides annually
- Use the median service mix and the median times per procedure in the 2007 Abt Report (Tables A8.2 and A5.4, respectively) to calculate the median procedure-hours provided by a medical physicist—for this report, the annualized service hours for the median FTE physicist is 2700 hours
- Use this information to show the service-hours provided by your program with reference to a national median standard

Of course, this does not mean the median FTE-qualified medical physicist works 2700 hours per year performing clinical procedures. It is well-known that time estimations in absolute hours by individuals performing services overestimates that time by approximately 30–50%. The reason the median medical physicist does not work 2700 hours per year is because of this overestimation, which is a characteristic of time magnitude estimation. Note that it is equally well known that systems engineers with stopwatches underestimate service times by up to 30–50%. The workweek reported by the AAPM Professional Information Survey indicates medical physicists work between 2200 and 2400 hours per year. This includes clinical, administrative, research, and teaching duties. Only the first of these are measured using the Abt information. Bear in mind that the Abt data includes the commissioning and acceptance time as well as quality assurance time associated with each procedure, so there is no need to include this time separately. The usefulness of this data is that it gives medical physicists a national work benchmark standard of comparison for the service mix at their respective venues of employment. Simply divide the total number of physics procedure service-mix hours for your institution by 2700 to give approximately the number of FTE physicists required for clinical support. Then, perform a quick reality check by using additional information found within the Abt III Report. Table 43.3 shows the median qualified medical physics staffing statistics from the Abt III Report and Table 43.4 shows the median support staffing statistics from the Abt III Report.

An example of this type of calculation based on the Abt II Report may be found in a previous article by Mills (2005). Notwithstanding, factors that may add complexity to a staffing model include:

- Achievement or progress toward ABR Certification in Therapeutic Radiological Physics
- Licensure (in states with a licensure requirement)
- Years of experience
- Experience and qualification in special procedures

TABLE 43.4 Median Support Staffing Statistics

Support Staffing	Overall	Private Hospital	Medical School University	Medical Physics Consulting Group	Physician Group
Medical Dosimetrist or Junior Physicist/QMP	1.5	1.5	0.6	1.15	0.87
Physics Assistant/QMP	0	0	0.1	0	0.3

Source: Abt Associates, Inc., Round III—Final Report: Abt Study of Medical Physicist Work Values for Radiation Oncology Physics Services, available from http://www.aapm.org/pubs/reports/ABTIIIReport.pdf, 2008. With permission.

Some additional factors needed to provide quality medical physics services include the acquisition of adequate equipment and support staff. Equipment includes treatment planning computers, dosimetry laboratory and quality assurance equipment, and information science support in data management an computing resources. It is important to incorporate proactive strategic planning in managing a medical physics section. The physics manager must provide clear direction, defend an adequate budget, and provide for time to fill vacant positions and acquire equipment. Staffing and recruiting should be planned, accounting for vacant positions and orientation time of new staff. It is prudent to assume a 12-month recruiting and orientation period for any open position. It is reasonable to assume a typical position will be occupied for 10 years. Therefore, an appropriate justification is:

$$\text{Number of positions needed} = \text{number of positions occupied, } X$$
$$(43.1)$$

Failure to proactively plan impacts clinical, teaching, and research efforts. Clinical implementation of new technology may be delayed, time to teach and mentor may be diminished, and research productivity may suffer. All of this could negatively impact institutional and section morale. An excellent example of a staffing model including many of these complexity issues based on Abt II is the recent investigation by Klein (2009).

Summary

We describe a time-driven method for justifying physics staffing based on Abt III data. The data can be used in different ways to estimate appropriate medical physics staffing for your institution. The methodology can be extended to an academic setting or, on a smaller scale, to a community hospital or freestanding center. This method equates the clinical needs with the quantity of staffing. The method is easily adaptable when changes to the clinical environment occur, such as an increase in IMRT or IGRT applications.

Abt-based estimates are based on median staffing, and medical physics staffing at your center may be more or less, depending on the scope of the radiation oncology practice, the experience and ability of the staff, and other factors. Abt data is a valuable resource for a baseline from which to justify appropriate staffing for your practice.

References

Abt Associates, Inc. 1995. The Abt Study of Medical Physicist Work Values for Radiation Oncology Physics Services. Cambridge, MA: Abt Associates, Inc.

Abt Associates, Inc. 2003. Round II—Final Report: Abt Study of Medical Physicist Work Values for Radiation Oncology Physics Services. June 2003. Available from http://www.aapm.org/pubs/reports/ABTReport.pdf.

Abt Associates, Inc. 2008. Round III—Final Report: Abt Study of Medical Physicist Work Values for Radiation Oncology Physics Services. March 2008. Available from http://www.aapm.org/pubs/reports/ABTIIIReport.pdf.

American College of Radiology (ACR). 2008. American College of Radiology (ACR) Radiation Oncology Accreditation Program Requirements 2008. Available from http://www.acr.org/accreditation/radiation/requirements.aspx.

Belletti, S., A. Dutreix, G. Garavaglia, H. Gfirtner, J. Haywood, K. A. Jessen, I. L. Lamm et al. 1996. Quality assurance in radiotherapy: The importance of medical physics staffing levels. Recommendations from an ESTRO/EFOMP joint task group. *Radiother. Oncol.* 41 (1): 89–94.

Eudaldo, T., H. Huizenga, I. L. Lamm, A. McKenzie, F. Milano, W. Schlegel, D. Thwaites, and G. Heeren. 2004. Guidelines for education and training of medical physicists in radiotherapy. Recommendations from an ESTRO/EFOMP working group. *Radiother. Oncol.* 70 (2): 125–135.

Herman, M. G., M. D. Mills, and M. T. Gillin. 2003. Reimbursement versus effort in medical physics practice in radiation oncology. *J. Appl. Clin. Med. Phys.* 4 (2): 179–187.

Klein, E. E. 2009. A grid to facilitate physics staffing justification. *J. Appl. Clin. Med. Phys.* 11 (1): 2987.

Mills, M. D. 2005. Analysis and practical use: The Abt study of medical physicist work values for radiation oncology physics services—Round II. *J. Am. Coll. Radiol.* 2: 782–789.

Mills, M. D., W. J. Spanos, B. O. Jose, B. A. Kelly, and J. P. Brill. 2000. Preparing a cost analysis for the section of medical physics—Guidelines and methods. *J. Appl. Clin. Med. Phys.* 1 (2): 76–85.

Role of Training

Kin Yin Cheung
Prince of Wales Hospital

Introduction

Quality assurance (QA) in radiotherapy is aimed at ensuring patient safety and that all processes and procedures involved in and leading to the treatment of a patient are within tolerances of a predefined set of quality standards. The ultimate objective of the measure is to prevent systematic errors and minimize the frequency and magnitude of random errors so as to improve treatment outcomes, that is, a better cure rate with a lower treatment-induced complication rate. It is important that every staff member involved in the radiotherapy process is fully aware of the potential for an accident and the need to strictly follow the preventive measures implemented to prevent any sentinel event from occurring. QA, if properly implemented, can be an effective preventive measure. The concept of QA has been an important aspect of the development of radiotherapy starting from the beginning of x-ray therapy.

The concept has evolved from simple equipment functionality checks to a comprehensive multidisciplinary process (AAPM Report 46 1994) and, in recent years, to specialized procedures dedicated to special treatment techniques such as IMRT (AAPM Report 82 2003). Despite the efforts and investments made by the radiotherapy community in implementing QA in the clinics, serious sentinel events have still frequently occurred and been reported over the years (ICRP 2000; IAEA 2000). According to the IAEA and ICRP, the lack of qualified and well-trained staff and inadequate QA procedures were identified as two of the main reasons that contributed to the occurrence of the events. The most important component of the entire radiotherapy process is qualified personnel with appropriate educational background and specialized training, including training in QA.

Role of Training

Staff training is one of the most important components of the entire radiotherapy program. Quite often, staff training is neglected or inadequately provided for a variety of reasons, including lack of funding or qualified staff. It has been reported that many of the accidental exposures have occurred because of the lack of well-trained staff (IAEA 2000). It is therefore vital that all the staff involved with any of the radiation therapy processes have the necessary educational background and specialized training. Staff training itself should be treated as an important part of the QA process. The provision of training, including training on QA, is one of the basic prerequisites for effective and efficient implementation of a QA system and, hence, an effective and safe radiotherapy service.

Training Requirements

The radiotherapy community has paid a high price in learning from bad experiences of clinical incidents that all the staff involved with radiation therapy must be properly trained and qualified for their specific jobs and responsibilities. That is to say, every staff member should have the educational background and have received the professional training relevant to their tasks. Radiotherapy is a multidisciplinary process involving different teams of staff from different disciplines and the use of a range of sophisticated and high precision, but potentially hazardous, equipment. For this reason, the implementation of an effective QA system covering every aspect of the radiotherapy service is essential. The QA system and the procedures involved should be under regular review and should be improved or upgraded whenever a new radiotherapy technique or equipment technology is introduced or there is a change in the treatment process. A corresponding training and development program for members of staff is essential in order to ensure that the intended quality standard of treatment can be achieved and sustained.

Who Should Receive Training on QA?

Effective implementation of a radiotherapy QA program requires the commitment and support of every staff member. It is important that every staff member understands the concept, principles, and structure of the department QA system and appreciates the need for strict implementation of such a system consistently. QA in radiotherapy is a multidisciplinary process involving radiation oncologists, medical physicists, radiation therapists, nurses, clerical and other supporting staff such as maintenance engineers (Thwaites et al. 1995). Each staff member should be trained to be competent in performing their duties in the radiotherapy service, including their own role and responsibilities in QA, in order to ensure service quality and patient safety. Because of the complexity of the work process involved and sophistication of the equipment used by them, dedicated, but specialized, training on different aspects of QA should be provided for the radiation oncologists, medical physicists, radiation therapists, nurses, and other supporting staff, including maintenance engineers. The training is particularly important when introducing new equipment technology.

Basic QA Training

Apart from being professionally competent to practice in their respective disciplines, every staff member should have received basic training on QA, including an understanding and appreciation of the following:

- The concept and rationale for a comprehensive QA program
- Details of the QA organization structure, including roles and responsibilities of the individual QA teams, lines of communication, etc.
- Principles of physical and operational aspects of radiotherapy equipment QA and quality control (QC)
- Appropriateness and limitation of the QA program and QC measures

Useful reference materials on basic QA training, such as guidelines and recommendations on QA in specific radiotherapy processes are published by and freely available through the Web sites of national and international organizations such as IAEA, WHO, ASTRO, AAPM, and ESTRO.

Equipment QA Training

Equipment QA covers the life cycle of the equipment starting from preparation of equipment specifications for procurement, acceptance testing, commissioning, to clinical operation and end of life decommissioning of the equipment. Training on equipment QA is crucial for medical physicists in the design

and proper implementation of an effective equipment QA program. Training courses organized at annual scientific meetings or dedicated workshops and symposia on QA of radiotherapy equipment are regularly organized by national and international organizations such as ASTRO, AAPM, ESTRO, and IAEA. Freely accessible resources on equipment QA in the form of guidelines and protocols are available from the Web sites of these organizations.

Training on New QA Tools, Techniques, and Concepts

Radiotherapy has undergone major advances over the past 10 years as a result of new developments in equipment technologies and treatment techniques, such as intensity-modulated radiotherapy (IMRT) and image-guided radiotherapy (IGRT). It is recognized that these new treatment modalities require more intensive and extensive QA than currently practiced. Some of the current QA guidance is incomplete or out of date and can no longer address all the quality and safety issues adequately. New approaches to quality and safety management are needed (Williamson et al. 2008; Thomadsen 2008; Huq et al. 2008). When planning the installation of new treatment equipment and the implementation of new treatment techniques or modalities, hospital management, and the chief radiation oncologist in particular, should include staff training in the budget. The training should include clinical and physics training as well as QA training. The efficiency and effectiveness of a QC procedure quite often is limited by unavailability of the right QA tools and techniques. Staff members should be able to keep up with the advances in QA equipment technology and techniques. The training on new QA approaches may be accomplished by sending staff to dedicated training courses or workshops or visit centers where such expertise is available. Investment in new equipment technology without investment in training at the same time can be dangerous.

Process- or Technique-Specific QA Training

A common limitation in the current practice of QA is that it is mainly focused on equipment performance and functionality and that fewer resources or emphasis are put on processes and procedures and human factors in performing the procedures. QC of radiotherapy processes and procedures is becoming more important due to the increasing complexity of the techniques and processes used in modern radiotherapy. It has been reported that operational and procedural errors were some of the major causes of accidents in radiotherapy (ICRP 2000; IAEA 1998). Radiotherapy treatment planning and dose calculation, dosimetry calibration, and commissioning of treatment machines are typical high-risk processes. Training on procedural QC should be provided to the respective staff members responsible for the clinical, physical, clerical, and engineering aspects of the radiotherapy process. For this reason, more investment on procedural QC, as suggested by Thomadsen (2008), may be beneficial.

Training on Safety Culture, Attitude, and Alertness

Provision of quality and safe radiotherapy service requires the effective implementation of a sound QA program which in turn requires the commitment and support of every staff member. Staff attitude and awareness of, and alertness to, procedural errors and mistakes play a very important role in preventing accidents. Specially designed training or exercises aimed at increasing staff awareness and alertness should be given to specific groups of staff. For instance, staff should be trained to have a sense of likeliness and reasonableness. In the case of treatment planning, staff should have a sense of similarity and consistency in the magnitude of MU calculations for the treatment plans of different patients treated by the same technique for the same disease. Any abnormality observed in the magnitude of the MU should trigger an investigative action by the staff before accepting a plan. Likewise, staff operating a treatment machine can be made more alert to the appropriateness or reasonableness of the treatment parameter settings on the machine.

Learning from Mistakes and Incidents

As in other medical specialties, one cannot totally prevent errors and mistakes in radiation therapy, although a zero accident rate is a goal. The most important issue is that radiation oncologists, medical physicists, radiation therapists, and other members of staff can learn from their own mistakes and those of others, especially the bitter experiences of major accidents, and implement appropriate QA measures to prevent similar incidents from happening again. Staff members, team leaders in particular, should review the reports of all cases of major radiotherapy incidents and understand the causes for such incidents and the recommended preventive measures made by investigation experts. The knowledge learned and information obtained should be valuable for development and implementation of appropriate preventive measures in their systems of work.

Quality Management Training

Effective implementation of a QA system in radiotherapy needs quality management on all the components of the individual RT processes. For instance, in the treatment planning process, a series of task-specific components is involved, ranging from image acquisition, image registration and contouring, dose planning, and optimization to treatment plan approval and verification. Quality management (QM) is applied to ensure that the QA system can address all the components of the process, not just QA of treatment planning equipment. Total quality management (TQM) is applied to address all the RT processes and their respective components. TQM should also address management issues such as risk and cost-effectiveness analysis, QA structure design and optimization, resources planning and management, appropriateness and compatibility in quality standards, the communication system between staff groups or teams, record

and documentation systems, monitoring and control systems, optimization systems, and development and improvement systems. Training of staff, especially the team leaders, on quality management is important for implementing and maintaining a cost-effective comprehensive QA system.

Continuing Education and Training

All critical staff, including radiation oncologists, medical physicists, and radiation therapists, should be aware of the limitations of a QA program and QC procedures and the need for continuing system improvement. This includes development of new and better QA concepts, knowledge, and techniques, especially when new equipment or new techniques are introduced into the clinic. Each staff should therefore be subject to an appropriate continuing education and training program in order to be able to keep track with rapid development in equipment technology. The exact training needs should be compatible with local equipment facility and service provision and can be quite different between centers.

Summary

Staff training plays an important role in maintaining treatment quality and preventing accidents in radiotherapy. The provision is an essential part of a QA system. Every member of staff should be trained and qualified for specific jobs and responsibilities. Training on equipment-specific QA and process-specific QC is essential for frontline staff. Supervisory staff and team leaders should also receive appropriate training on quality management, which can help improve their knowledge and skills in planning, implementation, and management of an effective QA system. A continual training and development program should be in place and appropriate corresponding resources such as manpower, facilities and materials should be made available to support the program.

References

AAPM Report 46. 1994. *Comprehensive QA for Radiation Oncology.* (Reprinted from *Medical Physics* 21 (4)). New York, NY: American Institute of Physics, Inc.

AAPM Report 82. 2003. Guidance document on delivery, treatment planning, and clinical implementation of IMRT: Report of the IMRT subcommittee of the AAPM radiation therapy committee. *Med. Phys.* 30 (8).

Huq, M. S., B. A. Fraass, P. B. Dunscombe, J. P. Gibbons Jr., G. S. Ibbott, P. M. Medin, A. Mundt et al. 2008. A method for evaluating quality assurance needs in radiation therapy. *Int. J. Radiat. Oncol. Biol. Phys.* 71 (1 Suppl.): S170–S173.

IAEA. 1998. *Accidental Overexposure of Radiotherapy Pin San José, Costa Rica.* Vienna: International Atomic Energy Agency (IAEA).

IAEA. 2000. *Lessons Learned from Accidental Exposures in Radiotherapy.* Vienna: International Atomic Energy Agency (IAEA).

ICRP. 2000. *Prevention of Accidental Exposure to Patients Undergoing Radiation Therapy.* Oxford: International Commission on Radiological Protection.

Thomadsen, B. 2008. Critique of traditional quality assurance paradigm. *Int. J. Radiat. Oncol. Biol. Phys.* 71 (1 Suppl): S166–S169.

Thwaites, D., P. Scalliet, J. W. Leer, and J. Overgaard. 1995. Quality assurance in radiotherapy. European Society for Therapeutic Radiology and Oncology Advisory Report to the Commission of the European Union for the 'Europe Against Cancer Programme.' *Radiother. Oncol.* 35 (1): 61–73.

Williamson, J. F., P. B. Dunscombe, M. B. Sharpe, B. R. Thomadsen, J. A. Purdy, and J. A. Deye. 2008. Quality assurance needs for modern image-based radiotherapy: Recommendations from 2007 interorganizational symposium on "quality assurance of radiation therapy: Challenges of advanced technology." *Int. J. Radiat. Oncol. Biol. Phys.* 71 (1 Suppl.): S2–S12.

Practical Aspects of Training

Ben Mijnheer
The Netherlands Cancer Institute

Introduction

The role of training all staff members cannot be overemphasized in a rapidly evolving field such as radiotherapy. It is evident that a radiotherapeutic treatment requires well-trained professionals for all steps involved in a patient treatment. Inadequate training may not only result in a nonoptimal treatment, important risks for patients occur if the users of radiotherapy equipment are insufficiently trained. This statement is based on numerous examples available in the literature as discussed recently, for instance, by Fraass (2008), Shafiq et al. (2009), and in the International Commission on Radiological Protection (ICRP) Report (2010). Careful analysis showed that all staff members, including radiation oncologists, medical physicists, and radiation technologists and therapists can be involved in the occurrence of accidents. It is therefore of utmost importance that staff members are aware of the caveats of that part of a patient treatment that they are performing. In other words, each staff member should have adequate training in those aspects of the treatment procedure for which they are responsible.

The role of training in avoiding accidents has recently also been emphasized in an editorial (Klein 2009) reviewing the issues discussed during a special symposium devoted to quality assurance of radiation therapy: *The Challenges of Advanced Technologies* (Williamson et al. 2008). If an institution has inadequate staffing in terms of quantity and quality, there will undoubtedly be a higher propensity of errors. During that symposium it was emphasized that the education, training, and continuing education of the radiation oncology team members are of critical importance to a QA program. In the past some of these programs, for instance, the clinical physics and dosimetrist/ treatment planner training, have been the weakest links with most programs relying on informally supervised or self-guided "on-the-job" training. Obviously adaptation of these types of training programs to the needs of modern radiotherapy was needed.

Training of All Staff Members

All professionals working in radiotherapy departments have experienced a basic training program before they are allowed to start with actual patient treatment. In principle, the level and content of the training program of the treatment team should be appropriate for the type of treatments performed in that particular institute. However, primary training programs are generally designed at the local or national level and may not always include the latest technological developments, or may not always be very relevant to the clinical practice in a specific institution.

Furthermore large differences exist in the curricula of training programs in different countries. For instance, a recent survey showed that the total length of medical physics training programs in Europe ranges from 2.5 to 9 years (Eudaldo and Olsen 2008). The same study indicated that in some European countries it is not even mandatory to hold a diploma or license to work as a medical physicist. Most likely this situation is also valid for the training of medical physicists in other countries, while the training of radiation oncologists and radiation technologists/therapists may show similar variation in primary education programs.

In addition to these primary education programs, training in daily clinical practice is of great importance for optimal patient treatment and the avoidance of accidents. On-the-job training, supervised by experienced radiotherapy (RT) professionals, in addition to a basic university education, is an approach often applied for staff members of a radiotherapy department. In this way, useful practical experience can be added to a (limited) formal training program resulting in competent staff.

Many errors described by Fraass (2008) and Shafiq et al. (2009) and in the ICRP Report (2010) occurred because of misunderstandings about the procedures by the staff involved. There are numerous ways to optimize information transfer in a radiotherapy department. These programs should not be restricted to training in the use of (new) equipment, but be much broader and

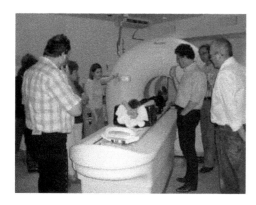

FIGURE 45.1 Demonstration of the use of a PET-CT scanner to RT professionals.

include interstaff discussions and continuing medical education activities such as demonstrations of newly installed equipment (see Figure 45.1).

An important approach to improving the quality of patient treatment in a radiotherapy department is the reporting of incidents. Careful analysis of incidents provides useful information concerning the safety of patient treatment in a specific radiotherapy department. For several years, organizations have existed where incidents can be reported on a voluntary basis (ROSIS 2008). The reports published by these organizations, besides providing detailed information about the cause of a specific error, pay a lot of attention to the lessons that can be learned from the incidents as, for example, in the report of the ICRP (2010). Incident reporting procedures have been implemented in many radiotherapy institutions and are even obligatory in some countries. A key constraint for the functioning of such a system is that (near)-accidents should be treated confidentially. It is crucial that all professionals understand that this is not a tool to monitor their weak points, but is meant to help make the treatment more accurate. In order to optimally profit from the analysis of incidents, staff members should be informed about incidents and discuss possible solutions, both formally in incident analysis groups and informally, if an open safety culture exists. The next step should be the incorporation of the various aspects related to a specific incident into departmental training programs.

A lot of information about the use of new equipment can be obtained at congresses, courses, and from the manufacturer and the literature. RT professionals should understand the procedures well and get confidence in applying new equipment in clinical practice. For those purposes informal visits to institutions having the same or similar equipment is a good procedure for easy exchange of knowledge between RT professionals. In this way, experts from RT departments already using that equipment for some time may share their experience with those starting in that field. Insight into the motivation behind solved problems or the reason for specific choices are of great importance in guaranteeing the safe introduction of these new techniques.

Training in the Use of New Equipment

The past decade has been characterized by a "technology explosion" in radiotherapy. The various and rapid developments in the field of three-dimensional imaging, treatment planning, and treatment delivery have led to more accurate and optimal RT treatments. Intensity-modulated radiation therapy (IMRT) is an example of such an advanced technological development and has been, or will soon be, implemented in many RT clinics. RT professionals need the necessary knowledge and skills in the field of state-of-the-art treatment planning, delivery, and quality assurance (QA) of IMRT. In addition to learning the theoretical background of IMRT from textbooks and at courses, interactive teaching sessions, focused on situations in practice, should be an important part of the training in various aspects of IMRT (see Figure 45.2). Experienced users of the systems and representatives of the companies should guide the potential users of the new equipment.

The introduction of new technology in a radiotherapy department brings new challenges to the radiation oncology team. It is often necessary to develop disease- and site-specific strategies and to carefully evaluate them with appropriate clinical studies for validation. Implementation of a new treatment modality depends on the individual resources and expertise within a department and should incorporate a multidisciplinary approach where radiation oncologists, physicists, and radiotherapy technologists are well integrated. Education and training are therefore critical issues for the safe implementation of new technology and should be well-thought out before the new tools are installed.

From the analysis of a number of recent accidents (and incidents) occurring in radiotherapy institutions after the introduction of new technology, it became evident that a considerable number of these accidents were due to insufficient training of the responsible staff (ICRP 2010). For instance, in one of the cases of recent accidents in France, the treatment planner made an error in the use of the treatment planning system, whereas in another accident, a physicist did not measure the output of small fields correctly (Derreumaux et al. 2008). In both cases,

FIGURE 45.2 Hands-on training in IMRT treatment planning.

additional training in the use of the new option of the treatment planning system, or the dosimetry of stereotactic radiosurgery beams, might have avoided these errors.

Although it is obvious that institutions must ensure properly trained staff before using new equipment, it is often difficult to determine when the training is sufficient to start actual patient treatment. Therefore, at the initial phase of using new equipment, it is important that various types of QA approaches are applied. For instance, in vivo dosimetry and external audits might pick up a number of errors before they cause substantial damage to patients. Furthermore, audits might identify learning needs (gaps in knowledge and skills) and recommend participation in additional training programs.

Web-Based Training

For basic training as well as for continuous medical education activities, online access to textbooks, journals, courses, or other resources is a prerequisite in a modern radiotherapy department. Generally a vast amount of information can be found on a specific topic using the Web and all staff members should therefore be trained in literature-searching skills and have the possibility of formal continuing professional development activities. For instance, the major sources of medical physics continuing education specifically for AAPM members are the summer school and continuing education courses at the annual meeting and their availability through the virtual library. Easy access to these tools should therefore be made available in their department for all AAPM members.

Web-based training and e-learning are nowadays part of many educational and training programs. An increasing number of Web-based educational modules relevant to the education and training of radiation oncologists, medical physicists, and radiation technologists are already available or under development. Web-based training offers several advantages over traditional instructor-led training. It is available at convenient times for the student, can be customized and tailored to the needs of a specific group, and it provides a tool to relieve mentors from the task of developing and administering a considerable amount of didactic training to their students. Web-based training can be delivered anywhere and may therefore also be very valuable at places where education and training resources are limited. Web-based modules should, however, be viewed as supplements to, and not full replacements for, classroom contacts between student and teacher or mentor. Classroom time should, for instance, always be used for skills-type training that requires some type of face-to-face interaction.

Several types of Web-based training are available for education and training of RT professionals. In its most simple form, teaching material in the form of a textbook can be downloaded from a Web site, the *Handbook for Teachers and Students on Radiation Oncology Physics*, for example (Podgoršak 2005). This handbook also includes a large set of PowerPoint slides, which can be used for lecturing and is used as the foundation of several teaching programs. More sophisticated Web-based tools include tests that will measure the learner's understanding of the material. Some form of interactivity such as exercises and case studies should make the information more personally relevant. Also, the opportunity to include details as an option, given as links to additional information, makes Web-based training an attractive alternative to classroom training. Real-time interactive remote education using the Web can resemble classroom-type education, where video and audio signals of the lecturer are transmitted to a remote classroom and shown on a television set, and the class can pose questions back to the lecturer. A particular advantage of this mode of tele-education is that specialized expertise can easily become accessible in otherwise impractical situations. Many vendors of radiotherapy equipment are now also offering Web-based training courses. These classes may vary from equipment-specific modules to more general teaching material related to basic training in the use of their equipment. Many companies also provide Web-based seminars, so-called "webinars," in which experts can be asked questions about their products in an interactive setting.

Furthermore, modern radiotherapy departments have created protocols for many parts of the various steps involved in the treatment of a patient. Obvious steps are clinical guidelines for tumor specific sites, planning objectives, and QA programs of equipment. Web sites are the most common places where protocols are stored and can be consulted. These protocols may vary in their detailed description of the whole process, but may have links to more specific information. By including these protocols in education and training programs, staff members may become aware of the various procedures applied in their department for the optimal treatment of a patient with a specific disease.

In addition to a Web site having practical information about patient treatment, it might also be useful to have a type of Web-based newsletter to inform staff members of recent events in the department. Such a newsletter may include information about the implementation of new equipment and also discuss incidents that happened recently in the department. All information should be written in a way that it is of interest to all staff members and contribute to their continuous medical education.

References

Derreumaux, S., C. Etard, C. Huet, F. Trompier, I. Clairand, J. F. Bottollier-Depois, B. Aubert, and P. Gourmelon. 2008. Lessons from recent accidents in radiation therapy in France. *Radiat. Prot. Dosim.* 131 (1): 130–135.

Eudaldo, T., and K. Olsen. 2008. The present status of medical physics education and training in Europe: An EFOMP survey. *Phys. Med.* 24 (1): 3–20.

Fraass, B. A. 2008. Errors in radiotherapy: Motivation for development of new radiotherapy quality assurance paradigms. *Int. J. Radiat. Oncol. Biol. Phys.* 71 (1 Suppl.): S162–S165.

ICRP. 2010. *Preventing Accidental Exposures from New External Beam Radiation Therapy Technologies.* ICRP Publication 112. Available from http://www.elsevierhealth.com.

Klein, E. E. 2009. Balancing the evolution of radiotherapy quality assurance: In reference to Ford et al. *Int. J. Radiat. Oncol. Biol. Phys.* 74 (3): 664–666.

Podgoršak, E. B., ed. 2005. *Radiation Oncology Physics: A Handbook for Teachers and Students.* Vienna: International Atomic Energy Agency.

ROSIS. 2008. Radiation Oncology Safety Information System (ROSIS). Available from http://www.rosis.info.

Shafiq, J., M. Barton, D. Noble, C. Lemer, and L. J. Donaldson. 2009. An international review of patient safety measures in radiotherapy practice. *Radiother. Oncol.* 92 (1): 15–21.

Williamson, J. F., P. B. Dunscombe, M. B. Sharpe, B. R. Thomadsen, J. A. Purdy, and J. A. Deye. 2008. Quality assurance needs for modern image-based radiotherapy: recommendations from 2007 interorganizational symposium on "Quality Assurance of Radiation Therapy: Challenges of Advanced Technology." *Int. J. Radiat. Oncol. Biol. Phys.* 71 (1 Suppl.): S2–S12.

V

Quality Assurance in Radiotherapy

46

CT Simulation

Performance Objectives and Criteria • QA of Emergency and Safety Systems • QA of Optical
Systems • QA of Mechanical Systems • QA of Imaging System • QA of Networking and
Archiving • QA of Simulation Process

Kathy Mah
*Odette Cancer Centre and
Sunnybrook Health Sciences Centre*

Parminder S. Basran
*BC Cancer Agency-Vancouver
Island Centre*

Introduction

The past two decades have observed significant developments in x-ray computed tomography (CT) technology and advances in CT-based simulation. Cone-beam detector technology, efficient reconstruction algorithms, subsecond x-ray rotation times with slip-ring technology, wider bores, larger reconstruction fields of view, and the steady progression of the Digital Imaging and Communications in Medicine (DICOM) and DICOM radiotherapy (RT) standards have brought about the era of 3D and 4D CT-based radiation therapy simulation. This chapter describes the important aspects of quality assurance, acceptance, and commissioning of CT simulators, many of which are discussed in the American Association of Physicist in Medicine (AAPM) Task Group (TG) TG-2 (AAPM Report 39 1993), TG-66 (Mutic et al. 2003), and other published reports (Mah 2007).

Acceptance and Commissioning of CT Simulation Systems

Acceptance testing involves radiation safety measurements, assessing the accuracy, precision, and reproducibility of mechanical systems, testing the congruence of laser localization systems with the mechanical and imaging isocenter, quantifying image quality, and testing other ancillary localization and patient monitoring devices. These measurements validate whether CT performance meets vendor specifications and the clinical demands

of CT simulation. Commissioning involves specific measurements that permit CT images to be used for virtual simulation and dose calculations. Commissioning would include testing data transfer and integrity, establishing site-specific imaging protocols, measuring a relative electron density curve for dose calculation purposes, and setting quality assurance procedures prior to clinical release.

Radiation Safety and Safety Systems

Important design considerations in assessing radiation safety for CT simulators should include: (1) the number of patients per year; (2) the expected scan protocols, accounting for the number of scans, kVp, and mAs setting; (3) utility of 4DCT acquisitions and the subsequent increase in scattered radiation fluence; (4) scatter radiation dose maps from newer cone-beam CT systems; and (5) shielding requirements.

In general, the workload for a busy radiotherapy CT simulator should not exceed 20,000 mA min/week. It is not uncommon for CT simulator suites to require additional shielding when single-slice systems are replaced with multislice systems. For all CT installations, radiation surveys must be performed prior to acceptance testing in order to validate the shielding design and construction, not only for local regulatory requirements, but also to ensure that the doses to staff and the public are kept as low as reasonably achievable. In most jurisdictions, both annual equivalent dose and instantaneous dose rates must fall within regulatory limits. These can be measured with calibrated large volume ion chambers or radiation survey devices. Radiation surveys should be conducted any time there are structural alterations to the CT suite. Shielding calculations should be reassessed if there are changes in occupancy factors in surrounding areas.

Additional safety considerations would include the location and functionality of emergency off switches, door-interlocks, circuit breakers and uninterruptible power supplies, emergency patient supplies (e.g., defibrillators, medical gases, suction), and intercom systems.

Mechanical Systems

The mechanical systems must be tested for accuracy and precision as they directly impact the spatial location of the CT images, subsequent 3D reconstructions, and ultimately the accuracy of the "simulation" of treatment. Most mechanical tests do not require specialized phantoms and can be accomplished with the aid of high precision levels and rulers.

The accuracy of mechanical read-outs for couch vertical and longitudinal positioning should be quantified against true spatial positioning of the couch in CT helical, axial, and scout modes, as well as verified against the couch index (or z coordinate) in the reconstructed images. The level of the couch, both parallel and orthogonal, to the scan axis must also be verified. All mechanical tests should be performed with and without the maximum permissible couch-load, and over durations typical of a CT simulation session. Quantifying couch reproducibility and accuracy is also important when absolute isocenter marking is performed.

Many commercial radiation treatment planning (RTP) systems assume that the CT images are truly vertical and cannot accept CT images that are not aligned in the transaxial plane. Once the CT gantry angle of zero has been verified to be in the true vertical position, it may be beneficial to disable the gantry-tilt and lock the CT gantry position. For some techniques such as radiosurgery, oblique CT slices for planning may be preferred. In this case, true spatial gantry angles must be validated against gantry tilt readouts.

CT Lasers Localization Systems

Externally mounted lasers facilitate the placement of fiducial marks, with, for example, small metallic ball bearings placed on the patient in order to provide a reference plane for beam placement. Typically, these lasers are mounted a fixed distance from the CT imaging plane in order to allow physical access to the patient to facilitate patient alignment, placement of fiducial marks on the skin, or preparation of immobilization devices. Lasers should be checked for congruence with the axis of rotation, true vertical and horizontal planes, and linearity. The latter must be validated for all translatable lasers across its range of motion. Lasers mounted within the gantry may be used as

a qualitative guide for patient positioning, but are generally not accurate or stable enough for quantitative use in treatment planning.

Ultimately, the laser and mechanical systems must be tested for congruence with the CT imaging plane and isocenter. This congruence will ensure that patient set-up using the CT-based reference marks will reproduce the anatomy and the planned radiation isocenter correctly on the treatment unit. Testing of this congruence can be conducted with commercial or customized phantoms. Figure 46.1a shows a precisely-machined phantom of known dimensions that provides orthogonal etchings for laser alignment and fiducial markers spaced 50 mm apart. Imaging such a phantom, as shown in Figure 46.1b, can confirm laser coincidence with $x = 0$ and $y = 0$ imaging axes, couch longitudinal accuracy, and spatial integrity as well as linearity of the image.

Imaging Systems

Basic image quality tests as defined by AAPM TG-2 should be undertaken in order to establish baseline image quality. Tests are conducted to ensure that the image quality meets the vendor's specifications and also to provide baseline values for quality assurance purposes. Many image quality tests may be performed by the vendor prior to shipment of the CT scanner with an automated procedure using the vendor's QA phantom. In a single session, these automated measurements can determine linearity, noise, high and low contrast, sensitometry, point-spread function, and, subsequently, the modulation transfer function. Repeating these automated tests after onsite installation is useful to ensure that image quality metrics are retained after shipment. However, image quality testing should also be performed manually with an independent phantom such as the Catphan® (Phantom Laboratory, Salem, NY) not only to validate the automated test results, but also to gauge the image quality under various clinical scanning conditions.

Low- and high-contrast resolution, noise, and sensitometry, as well as spatial integrity, have the greatest impact on treatment planning and should be thoroughly checked. Some tests recommended by TG-2, such as invasive measurement of the x-ray tube voltage, may prove difficult to implement and users are cautioned to check with the manufacturer before attempting such tests.

For radiation therapy, a large bore opening is often necessary to facilitate the patient treatment position including all immobilization devices. In order to encompass the entire patient and any devices that may intercept a treatment beam in the reconstructed images, vendors offer large scan field of views (FOV) and even larger display FOV. The former is limited by the x-ray fan beam width, while the latter is calculated by extrapolation (Anoop and Rajgopal 2009). Image quality and CT number accuracy should be measured using various FOVs and within different regions of the maximum display FOV since reconstruction artifacts can degrade image quality and affect CT simulator performance.

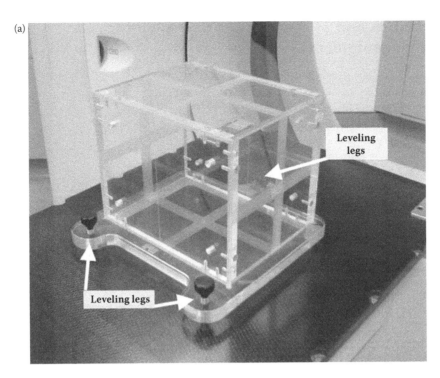

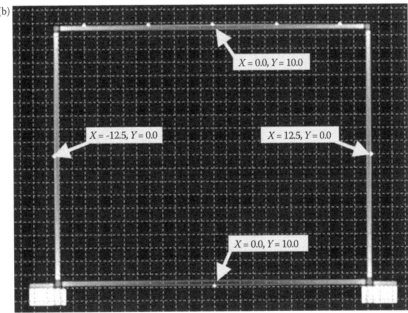

FIGURE 46.1 Custom laser-CT alignment phantom made at the Sunnybrook's Odette Cancer Centre. (a) The phantom is precisely machined with outer dimensions of 25.0 cm wide by 20.0 cm high. It consists of three leveling legs, orthogonally etched lines on each surface for laser alignment; embedded steel ball bearings along the $x = 0$ and $y = 0$ axes and at 5-cm intervals along the top surface. (b) The phantom is moved into the bore by digital longitudinal readout to confirm laser-to-image plane distance and imaged. The image is used to confirm (1) congruence of the lasers with the principal axes of the image plane; (2) the correct phantom size; (3) the correct image rotation (i.e., 0°); and (4) spatial linearity as visualized by the 5.0 cm separation of the anterior ball bearings.

Ancillary Localization and Patient Monitoring Equipment

In addition to external lasers, supplementary localization and patient monitoring systems may be used. Respiratory bellow devices (Kubo et al. 2000), ultrasound imaging systems (Serago et al. 2002), and electromagnetic tracking systems (Willoughby et al. 2006) have recently emerged as planning aides for motion assessment and localization. These systems should be checked to ensure that they: (1) function as expected and within specifications; (2) do not introduce systematic errors in the localization; (3) can be moved away from the patient quickly in the event of an emergency; and (4) do not inhibit staff access to the patient. Generally, such ancillary devices should have a separate acceptance testing procedure provided by the manufacturer. For each of these subsystems, quality assurance protocols should be established prior to clinical release. Automated contrast injectors should also be tested for functionality and safe operation.

Relative Electron Density Quantification

RTP systems require a conversion from CT number, or Hounsfield unit, to relative electron density (CT-RED) or physical density to be used in dose calculations accounting for inhomogeneities. Slice thickness, scan mode, and current settings affect the noise in the CT-RED datasets, whereas the tube potential and reconstruction filter will affect the sensitometry of the reconstructed data. The impact of tube potential and reconstruction filter on CT number conversion are shown in Figure 46.2a and b, respectively. Due to increased photoelectric effect at lower energies, the slope of CT-RED curves can change significantly for lower energies (90–100 kVp), with less pronounced changes in the mid- to high-energy ranges (120–140 kVp), as shown in Figure 46.2a. While it is important to characterize the CT-RED curves for different energies, clinical CT simulation does not generally use lower peak kilovoltage for helical scanning. Different reconstruction filters can introduce variations in the sensitometry estimates by as much as 5%, as illustrated in Figure 46.2b. Note that a 5% difference in sensitometry corresponds to a similar error in mass attenuation coefficient but not in dose (Henson and Fox 1984). The corresponding error in pathlength and kerma diminishes as the treatment energy increases, but increases with depth (Ahnesjo 1989). Overall, the CT-RED curve depends on the detector technology, the reconstruction filters, and algorithms on a particular scanner and therefore should be measured for every scanner.

Quality Assurance of CT Systems and the Simulation Process

The purpose of a quality assurance (QA) program is to assure that operational standards that were considered acceptable at the time of installation continue to be maintained, as closely as

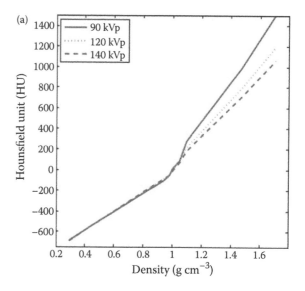

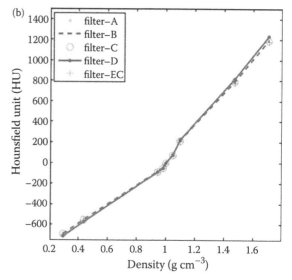

FIGURE 46.2 (a) CT number (or Hounsfield units, HU) to density conversion as a function of kVp. There is little change in this relationship for tissues of densities 1.1 g cm⁻³ or less. Helical CT scanning involves the mid- to high-range kVp of 120 to 140. (b) CT number to density conversion as a function of reconstruction filters. Data generated using filters available on the Big Bore Brilliance CT (Philips Medical Systems, Inc., Cleveland, OH).

possible, over the life of the unit. QA tests are performed periodically and involve partial or full repeats of acceptance and commissioning tests. For CT simulation, tests are required for optical, mechanical, imaging, and safety systems as well as network transference and communication. Essentially, QA testing is the periodic repetition of a subset of acceptance and commissioning tests. Table 46.1 is one example of a QA program for a CT simulator.

TABLE 46.1 Quality Control Tests for CT Simulators

	Test	Tolerance	Action
Daily			
DS1	Door interlock	Functional	
DS2	Beam status indicators	Functional	
DS3	Emergency off buttons (alternate daily)	Functional	
DS4	Lasers: parallel to scan plane	1°	2°
DS5	Lasers: orthogonality	1°	2°
DS6	Lasers: position from scan plane	1 mm	2 mm
DS7	Couch level: lateral and longitudinal	0.5°	1°
DS8	Couch and laser motions	1 mm	2 mm
DS9	CT number accuracy of water—mean	0 ± 3 HU	0 ± 5 HU
DS10	Image noise	5 HU	10 HU
DS11	Field uniformity of water	5 HU	10 HU
DS12	Simulated planning	1 mm	2 mm
Monthly			
MS1	Lasers: parallel to scan plane	1 mm	2 mm
MS2	Lasers: orthogonality	1°	2°
MS3	Lasers: position from scan plane	1 mm	2 mm
MS4	Lasers: linearity of translatable lasers	1 mm	2 mm
MS5	Couch level: lateral and longitudinal	0.5°	1°
MS6	Couch motions: vertical and longitudinal	1 mm	2 mm
MS7	Gantry tilt	1°	2°
Semiannually			
SS1	Slice localization from pilot	0.5	1
SS2	CT number accuracy of water—mean	0 ± 3 HU	0 ± 5 HU
SS3	CT number accuracy of other material—mean	[a]	
SS4	Field uniformity of water—standard deviation	5 HU	10 HU
SS5	Low contrast resolution	10 at 0.3%	[b]
SS6	High contrast resolution (5% MTF)	5 lp/cm	[c]
SS7	Slice thickness (sensitivity profile)	0.5	1
Annually			
AS1	Radiation dose (CTDI)	5%	10%
AS2	Independent quality control review	Complete	

Source: CAPCA Ref. 3.
[a] CT number accuracy of other materials will depend on the material and its uniformity.
[b] Low-contrast resolution will depend on the scan technique.
[c] High contrast resolution tolerance and action level will depend on the scan technique used.

Performance Objectives and Criteria

Performance objectives can fall into several categories:

1. Functionality: For some systems, the performance evaluation involves functionality—either it is working or not. Emergency stops, door interlocks, and beam-on indicators are examples of such systems.
2. Reproducibility: When reproducibility is the criterion, a system must perform within a set tolerance level about the acceptance specification or actions must be taken. Laser location, field uniformity, and sensitivity profiles are examples in which daily fluctuations are normal but the parameter must be maintained within a certain tolerance for acceptable performance.
3. Accuracy: Some systems require a test result that is accurate to a defined value. CT number accuracy and couch *z*-location are two examples that, if not accurate, could adversely affect the dose calculations and geometric integrity of the reconstructed images, respectively.
4. Data transfer and validation: The transfer of data must be accurate and efficient, whether by computer networking or human processes.

QA of Emergency and Safety Systems

Emergency stops, door interlocks, and beam-on indicators should be tested on a daily basis for functionality. Use of the CT should be suspended if any of these systems is not functioning.

TABLE 46.2   Example of Daily QA for CT Simulation Process

Task	Description
Set up phantom for CT scan and CT couch position readout check (*Use a different portion of the couch each day*)	• On the CT couch, position a mechanically-leveled, cylindrical water phantom with three markers at the exact location of the external lasers set at 0,0,0. • Phantom should also have additional markers at random locations. • Move couch using the digital longitudinal couch readout by the distance between the laser plane and the scan plane.
Laser/CT scan isocenter check	• Scan the phantom using 1.5- to 2-mm thick images, ensuring that the markers fall at the center of a transaxial image. • Verify that the markers appear on the $x = 0$ and $y = 0$ image axes.
Coincidence/image rotation check	• Verify that the three triangulation markers appear on the same image slice.
CT number accuracy, noise, and uniformity	• Measure the CT number and standard deviation of the water in five different regions of the transaxial image.
Data transfer and isocenter marking, and RTP beam placement check	• Transfer the CT images to the RTP system. Plan a beam, placing the isocenter onto the image of a random marker on the phantom.
Absolute localization, laser linearity, and couch position check	• Set the lasers and couch according to the parameters given by the RTP system for absolute patient marking. • Verify that the laser intersection coincides with the random marker used for the isocenter within 1 mm.

QA of Optical Systems

In-room lasers must be tested for spatial location and angulations. Spatially, the reference zero position must coincide with the scan axis of rotation at a set distance longitudinally beyond the scan plane (e.g., 500 mm from scan plane). The programmed position and true position of all translatable lasers must be validated to ensure correct patient reference marking.

QA of Mechanical Systems

The mechanical system of a CT simulator consists of the gantry and the couch. Most RTP systems accept only true transaxial images and therefore, the gantry angle and, subsequently, the scan plane must be vertical to within a degree. The patient support couch should be level throughout its range of motion. Couch readouts, both vertical and longitudinal, must be verified against its true position. The latter is critical, as it dictates the alignment of CT images along the superior–inferior plane in 3D reconstructions. Mechanicals systems should be tested daily.

QA of Imaging System

To ensure accuracy of CT number and consistency in localization, basic image quality parameters should be tested. Essential parameters such as CT number accuracy, field uniformity, and noise should be routinely compared with benchmark values. Low-contrast and high-contrast resolution and slice sensitivity also require testing.

QA of Networking and Archiving

CT systems require access to several databases, some internally within the CT imaging system, and others externally to image servers accessed by the RTP system and digital imaging archive systems. Simple network tests that check for data transferability

and integrity should be conducted after system and software upgrades as well as power failures.

QA of Simulation Process

On a daily basis, the entire CT simulation process should be verified to ensure all subsystems are operating within specifications and all transfer processes are intact. An example of such is given in Table 46.2. Daily QA must strike a compromise between the number of QA tests performed and amount of time allocated.

Summary

CT simulation is the cornerstone of all radiation treatment planning and delivery, permitting the development of 3D conformal radiation therapy (3D-CRT), intensity modulated radiation therapy (IMRT), and image guided radiation therapy (IGRT). The viability and efficacy of advanced treatment techniques depends on the accuracy and precision of the CT simulation process. Meticulous detail in acceptance testing, commissioning, and ongoing QA of CT simulators will minimize important sources of systematic error and allow for precise and accurate treatment planning and delivery.

References

AAPM Report 39. 1993. *Specification and Acceptance Testing of Computed Tomography Scanners*. New York, NY: American Institute of Physics, Inc.

Ahnesjo, A. 1989. Collapsed cone convolution of radiant energy for photon dose calculation in heterogeneous media. *Med. Phys.* 16 (4): 577–592.

Anoop, K. P., and K. Rajgopal. 2009. Image reconstruction with laterally truncated projections in helical cone-beam CT: Linear prediction based projection completion techniques. *Comput. Med. Imaging Graph* 33 (4): 283–294.

Henson, P. W., and R. A. Fox. 1984. The electron density of bone for inhomogeneity correction in radiotherapy planning using CT numbers. *Phys. Med. Biol.* 29 (4): 351–359.

Kubo, H. D., P. M. Len, S. Minohara, and H. Mostafavi. 2000. Breathing-synchronized radiotherapy program at the University of California Davis Cancer Center. *Med. Phys.* 27 (2): 346–353.

Mah, K. 2007. Standards for quality control at Canadian Radiation Treatment Centres: CT Simulators. Canadian Association of Provincial Cancer Agencies (CAPCA), Task Group, P. Dunscombe (Chair), C. Arsenault, J. P. Bissonnette, G. Mawko, and J. Seuntjens.

Mutic, S., J. R. Palta, E. K. Butker, I. J. Das, M. S. Huq, L. N. Loo, B. J. Salter, C. H. McCollough, and J. Van Dyk. 2003. Quality assurance for computed-tomography simulators and the computed-tomography-simulation process: Report of the AAPM Radiation Therapy Committee Task Group No. 66. *Med. Phys.* 30 (10): 2762–2792.

Serago, C. F., S. J. Chungbin, S. J. Buskirk, G. A. Ezzell, A. C. Collie, and S. A. Vora. 2002. Initial experience with ultrasound localization for positioning prostate cancer patients for external beam radiotherapy. *Int. J. Radiat. Oncol. Biol. Phys.* 53 (5): 1130–1138.

Willoughby, T. R., P. A. Kupelian, J. Pouliot, K. Shinohara, M. Aubin, M. Roach 3rd, L. L. Skrumeda et al. 2006. Target localization and real-time tracking using the Calypso 4D localization system in patients with localized prostate cancer. *Int. J. Radiat. Oncol. Biol. Phys.* 65 (2): 528–534.

47

MRI and MRS Simulation

Lili Chen
Fox Chase Cancer Center

Introduction

Radiotherapy (RT) is an effective local-regional treatment modality for cancers of all stages. Image-guided radiation therapy (IGRT) enables the precise delivery of a tumoricidal dose to the treatment volume to allow for dose escalation and hypofractionation, while sparing the nearby critical organs and structures. Magnetic resonance imaging (MRI) has been widely used for target and critical structure delineation in radiation therapy treatment planning due to its superior soft tissue contrast over computed tomography (CT) (Khoo et al. 2000; Buyyounouski et al. 2004). Studies demonstrated that it is possible to perform treatment planning dose calculation directly on MRI for intensity-modulated radiation therapy (IMRT) of prostate cancer (Chen, Price, Wang et al. 2004; Chen et al. 2007) and 3D conformal radiotherapy for the brain (Beavis et al. 1998) using commercially available treatment planning systems. Magnetic resonance spectroscopy (MRS) has recently garnered great interest in radiation oncology in the field of target definition for radiotherapy treatment planning, and for evaluation of response and recurrence. Over the past two decades, MRS has moved from being a basic research tool into routine clinical use (Payne and Leach 2006). MRS is able to detect signals from low molecular weight metabolites such as choline and creatine that are present at concentrations of a few mM in tissue. Spectra may be acquired from single voxels, or from a 2D or 3D array of voxels using spectroscopic imaging (Payne and Leach 2006).

Currently, the most identified sites using MRS clinically in RT are the brain and prostate. The correlation between CT, MRI, MRS, and biopsy is being evaluated for prostate cancer (Hossain et al. 2008). A study on MRS of the malignant prostate gland after RT was conducted (Menard et al. 2001). They reported that the sensitivity and specificity of MRS in identifying a malignant

biopsy were 88.9% and 92%, respectively, based on 116 tissue specimens from 35 patients, 18–36 months after external beam RT. The contrast achieved with MRS, based on tissue biochemistry, therefore provides a promising tool for identifying tumor extent and regions of high metabolic activity. MRS may become an essential tool for treatment planning and treatment assessment in advanced RT treatments (Payne and Leach 2006).

This chapter focuses on the quality assurance (QA) issues for the use of MRI and MRS in radiation therapy treatment simulation and briefly describes the QA methods, procedures, and phantoms as implemented at Fox Chase Cancer Center (FCCC). The specific details of QA and commissioning procedures for MR scanners and imaging parameters will be covered Chapter 68.

Accepting Testing and Commissioning for MR Simulators

Immediately after the installation of an MR simulator, and prior to its first clinical use, a set of acceptance tests should be performed on the equipment by a medical physicist. The acceptance testing procedure is an extremely thorough survey of the equipment and its capabilities, designed to confirm that the device was installed and set up properly and meets all vendor and industry performance specifications. The intention of the acceptance testing is not to summarily accept or reject the MR scanner and the associated software system, but rather to determine whether there are any performance items that should be addressed with the manufacturer and/or installer before final acceptance of the entire system. The physicist should work to address any such items in a collegial fashion with the installers and service engineers in order to ensure that the MR simulation system meets all manufacturer specifications.

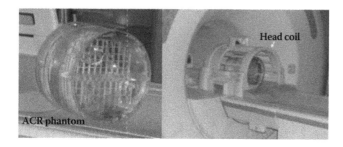

FIGURE 47.1 The ACR phantom, which is used to test geometric accuracy, high-contrast spatial resolution, slice thickness accuracy, slice position accuracy, image intensity uniformity, percent signal ghosting, low-contrast object detectability, and interslice RF interference.

An MR scanner designated for radiation therapy treatment simulation can be tested according to the methods and criteria established by the American College of Radiology (ACR) MRI Accreditation Program. An ACR phantom (Figure 47.1) can be used to test geometric accuracy, high-contrast spatial resolution, slice thickness accuracy, slice position accuracy, image intensity uniformity, percent signal ghosting, low-contrast object detectability, and interslice RF interference.

Detailed procedures for phantom positioning and MR scanning, together with the test worksheet and the testing criterion can be found at the ACR Web site (see the MRI testing & QC form at www.acr.org). In addition, the following items should also be checked during the acceptance testing: fringe field survey, magnetic field homogeneity, eddy currents, coils SNR (signal to noise ratio), cryogen consumption, patient/console intercom system, table stop buttons, emergency stop buttons, table docking, vertical movement, alignment lights and lasers, and the ventilation system.

Routine Quality Assurance

Routine QA procedures for an MR simulator are designed to detect changes in system performance relative to an established baseline. American Association of Physicists in Medicine (AAPM) *Report Number 28* (Price et al. 1990) provides detailed descriptions of quality assurance methods and phantoms for MRI. This document also includes recommendations for acceptable MRI phantom materials, phantom designs, and analysis procedures. Specific image parameters include resonance frequency, signal-to-noise, image uniformity, spatial linearity, spatial resolution, slice thickness, slice position/separation, and phase-related image artifacts. The ACR phantom and test criteria (see www.acr.org) can also be used to design the QA program for an MRI/MRS simulator.

Daily MR Machine QA

Daily MR QA procedures can be performed by a trained MR technologist. It takes approximately 20 minutes to complete the MR QA procedures including phantom setup, MR scanning, and

FIGURE 47.2 An example of the phantom used to check the three triangulation lasers

data recording. In-house or vendor provided QA phantoms and scanning protocols can be used to improve efficiency and effectiveness. The items to be checked include magnetic field stability (central frequency), signal-to-noise ratio (SNR), artifact inspection, and the isocenter position. The daily QA acceptance criteria can be determined based on the commissioning data.

Specific to radiotherapy applications, a set of three triangulation lasers (center and laterals) identical to those used on clinical accelerators are installed for patient positioning. The synchronized trackable lasers (laterals) ensure the anterior-posterior (AP) positions relative to the isocenter directly. The three triangulation lasers need to be checked daily. Figure 47.2 shows an example of the phantom used to check the lasers. The acceptance criterion for laser alignments is <2 mm, which is determined based on the calibration for clinical accelerators. In addition, the temperature in the MR scanning room needs to be checked for normal functioning of the ventilation system.

Weekly MR Machine QA

According to standards established by the ACR MRI Accreditation Program, the following items should be tested weekly: the MR table, console, central frequency (Hz), TX gain/attenuation (dB), phantom distance (both length and diameter), processor sensitometry, phase stability, magnetic field homogeneity, SNR in orthogonal planes, and image uniformity. The testing results should be compared with the criteria listed in the worksheet recommended by ACR (see form at www.acr.org).

Preventive Maintenance

A preventive maintenance (PM) program should be developed for both quality control and safety control. PM can be performed by either an in-house trained engineer who is qualified to work on the manufacture's equipment or service personnel from the manufacturer. The quality control checks may include items such as alignment light, coils, patch update, large volume shim, eddy current class, coherent noise, SNR, and spike noise while the safety control checks may include oxygen monitor operation, patient blower and filter, patient table, pneumatic patient alert system, and magnet rundown unit. A PM procedure is performed every 2 months at the radiation oncology department of FCCC.

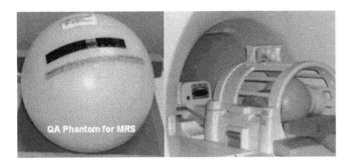

FIGURE 47.3 The QA protocol for an MRS scan uses a GE MRS QA phantom (sphere containing chemical solutions) shown in this figure.

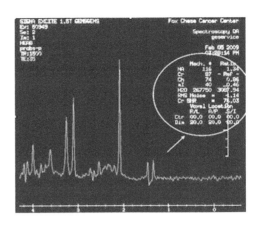

FIGURE 47.4 The numerical spectra of NA, Cr, Ch, mI, RMS Noise, and Cr SNR from the displayed image of the MRS QA scan using the GE built-in protocol.

Routine QA for MRS

Depending on the use frequency, daily and weekly QA procedures may be performed for clinical MRS simulation. A QA protocol was developed for brain and prostate simulation at FCCC, which is only applied when an MRS scan is needed. Figure 47.3 shows a GE MRS QA phantom that is a sphere containing chemical solutions. The QA phantom is placed in the MR scanning room for temperature equilibrium. The room temperature is recorded from the temperature strip on the phantom. The recorded temperature from the phantom needs to be entered in the MR protocol before scanning. It takes approximately 6 minutes to complete the MRS QA procedure including phantom setup, MR scanning, and data recording. The MRS scan is performed using the GE built-in protocol and the numerical spectra of NA, Cr, Ch, mI, RMS Noise, and Cr SNR can be recorded from the displayed image (Figure 47.4). Our QA acceptance criteria were determined based on the measurements over 20 days under normal performance of the MR unit.

Issues Related to MRI/MRS Simulation

MR Simulator

An MR simulator consists of an MR scanner and the associated simulation software for radiation therapy applications. The latter has functionalities and features similar to those of a CT simulator. CT and MR simulators are becoming widely used and are gradually replacing conventional simulators in radiation oncology centers. The 0.23 Tesla open MR unit (Philips Medical Systems, Cleveland, OH) was the first MR simulator reported (Mah, Steckner, Hanlon et al. 2002; Krempien et al. 2002; Mah, Steckner, Palacio et al. 2002; Chen and Price 2003). As shown in Figure 47.5, the scanner consists of two poles, each approximately 1 meter in diameter. The separation between the two poles is 47 cm. This is adequate to scan large patients (up to 400 lb). The

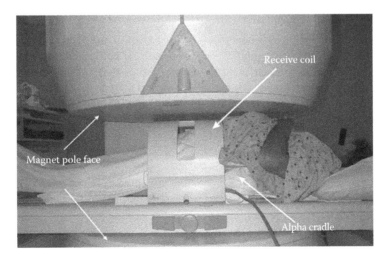

FIGURE 47.5 An open MR scanner consisting of two poles, each approximately 1 meter in diameter. The separation between the two poles is 47 cm, which is adequate to scan large patients (up to 400 pounds).

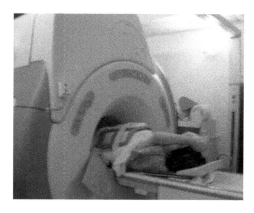

FIGURE 47.6 The GE 1.5 T MR simulator installed at Fox Chase Cancer Center, which is used for radiotherapy treatment planning and MR guidance for high-intensity focused ultrasound therapy.

MRI scanning table can be moved in orthogonal planes along a set of rails mounted on the floor and on an orthogonal set of rails built in the couch. Vertical adjustments can be accomplished with the flat table tops inserted beneath the patient. The flat table tops are made of special foam, which is stiff but light and MR compatible.

The advantages of using a low-field open magnet unit include convenience for patient setup and low maintenance cost compared to a high magnetic field scanner. An open MR scanner is also less claustrophobic, which is important to many patients. However, the high-field MR scanner can provide high-quality MR images, as well as MRS information for target determination and treatment assessment in advanced radiation therapy research. Figure 47.6 shows a GE 1.5 T MR simulator installed at Fox Chase Cancer Center, which has been used routinely in radiotherapy treatment planning. It is also used for MR guidance on high-intensity focused ultrasound surgery (Chen, Ma et al. 2009) and enhancement of drug delivery (Chen, Mu et al. 2009).

Target Delineation

For optimal target delineation for radiotherapy treatment planning, patients are scanned with both CT and MR. Both CT and MR scans must be performed on a flat, nonpadded scanning table with patients immobilized in their treatment positions. The MR-CT images can be fused either using a chamfer-matching algorithm according to the bony landmarks or using the maximization of mutual information so that the images are automatically fused to baseline images. Soft-tissue structures must be checked on both CT and MR images to avoid deformation effects between CT and MR scans. Image fusion and structure delineation tools are usually available in the simulation or treatment planning systems as provided by the vendor. On the fused images, MR is superb for delineating soft tissue structures (radiotherapy treatment targets and critical organs) while CT is useful in contouring bony structures, although it is best to use the two together as complementary modalities.

Image Distortion

Image distortion arises from both system-related effects and object-induced effects. System-related distortion is a result of inhomogeneities in the main magnetic and gradient fields. The geometric distortions increase with distance from the magnet isocenter and are more pronounced at the edges of the radial field. Maximum distortion errors of 2 mm for a field of view (FOV) with a radius of less than 20 cm have been reported (Moerland et al. 1995; Hill et al. 1994). In general, most of the targets and critical structures are within a 20-cm radius of the isocenter. The distortion for internal anatomy relevant to radiotherapy planning will be <2 mm. Routine MR QA should include phantom tests to ensure the functionality of the geometric distortion algorithm.

Object-induced effects are the result of both a chemical shift and susceptibility effects due to the difference in resonance between fat and water and the difference at tissue–air interfaces causing spatial misregistrations. Spatial distortions in MR images vary with field strength and with the image acquisition protocol. Chemical shift artifacts and susceptibility distortion can cause significant spatial misregistrations at high fields; their impact on the MR data at lower fields is substantially reduced. This means that at fields below approximately 0.5 T, imaging sequences providing sufficient signal to noise can be defined such that the geometric distortions due to either of these object-related effects can be kept below about 1–2 pixels. This is achieved by defining a lower limit for the bandwidth of the readout gradient during image acquisition. Studies have shown that, in vivo with 0.2 T using bandwidth readout gradient >100 Hz/pixel in the frequency direction, there is negligible artifact detected (Chen, Price, Nguyen et al 2004; Chen , Price, Wang et al. 2004 ; Fransson, Andreo, and Potter 2001). It was suggested that image protocols with high gradient bandwidths should be used to reduce the spatial distortions in MR images.

MRI Alone in Place of CT for Treatment Planning

Fused MR-CT images have been widely accepted as a practical approach for both accurate anatomical delineation (using MRI data) and dose calculation (using CT data). However, it would be ideal if MRI could be used alone for radiotherapy treatment planning (i.e., MRI-based treatment planning). The fusion process introduces additional errors due to difficulties in coordinating the CT and MR images and significant discordance caused by soft tissue deformation between image scans. Furthermore, MRI-based treatment planning will avoid redundant CT imaging sessions, which in turn will avoid unnecessary radiation exposure to the patient. It also saves patient, staff, and machine time.

Several perceived challenges of using MRI for radiotherapy planning have precluded its widespread use in this area. The challenges include: (1) lack of electron density information that is needed for heterogeneity corrections in dose calculation; (2)

image distortion that affects patient external contour determination and, therefore, introduces dose calculation error; and (3) lack of bony structures to derive effective digitally reconstructed radiographs (DRR) for patient setup. Studies have been carried out to explore the efficacy of MRI-based treatment planning for radiotherapy (Guo 1998; Mizowaki et al. 2000; Mah, Steckner, Hanlon et al. 2002; Mah, Steckner, Palacio et al. 2002; Chen, Price, Nguyen et al. 2004; Chen, Price, Wang et al. 2004; Chen et al. 2006, 2007; Lee et al. 2003). It has been shown that MRI clearly meets the dosimetry accuracy for prostate 3DCRT and IMRT (Chen, Price, Nguyen et al. 2004; Chen, Price, Wang et al. 2004; Lee et al. 2003) and 3DCRT for brain tumors (Beavis et al. 1998) using a large bandwidth sequence and bulk-assigned images with commercially available treatment planning systems. It is clear that before MRI-based patient geometry can be used for treatment planning, image distortions must be quantified and corrected. The development of a higher quality, low distortion MR sequence will allow regular practice of this technique. Currently, the identified sites using MRI-based treatment planning clinically in RT are for brain and prostate cancers (Beavis et al. 1998; Chen, Price, Nguyen et al. 2004; Chen, Price, Wang et al. 2004).

References

Beavis, A. W., P. Gibbs, R. A. Dealey, and V. J. Whitton. 1998. Radiotherapy treatment planning of brain tumours using MRI alone. *Br. J. Radiol.* 71 (845): 544–548.

Buyyounouski, M. K., E. M. Horwitz, R. G. Uzzo, R. A. Price, S. W. McNeeley, D. Azizi, A. L. Hanlon, B. N. Milestone, and A. Pollack. 2004. The radiation doses to erectile tissues defined with magnetic resonance imaging after intensity-modulated radiation therapy or iodine-125 brachytherapy. *Int. J. Radiat. Oncol. Biol. Phys.* 59 (5): 1383–1391.

Chen, L., C. M. Ma, T. Richardson, G. M. Freedman, and A. Konski. 2009. Treatment of bone metastasis using MR guided focused ultrasound. *Med. Phys.* 36: 2486.

Chen, L., Z. Mu, P. Hachem, C. M. Ma, and A. Pollack. 2009. Enhancement of drug delivery in prostate tumor in vivo using MR guided focused ultrasound (MRgHIFU). Paper read at World Congress on Medical Physics and Biomedical Engineering, September 7–12, 2009, Munich, Germany.

Chen, L., T. B. Nguyen, E. Jones, Z. Chen, W. Luo, L. Wang, R. A. Price Jr., A. Pollack, and C. M. Ma. 2007. Magnetic resonance-based treatment planning for prostate intensity-modulated radiotherapy: Creation of digitally reconstructed radiographs. *Int. J. Radiat. Oncol. Biol. Phys.* 68 (3): 903–911.

Chen, L., R. Price, J. Li, L. Wang, L. Qin, and C. Ma. Evaluation of MRI-based treatment planning for prostate cancer using the AcQPlan system. *Proc. Med. Phys.*, 30 (6): 1507.

Chen, L., R. A. Price Jr., T. B. Nguyen, L. Wang, J. S. Li, L. Qin, M. Ding, E. Palacio, C. M. Ma, and A. Pollack. 2004. Dosimetric evaluation of MRI-based treatment planning for prostate cancer. *Phys. Med. Biol.* 49 (22): 5157–5170.

Chen, L., R. A. Price Jr., L. Wang, J. Li, L. Qin, S. McNeeley, C. M. Ma, G. M. Freedman, and A. Pollack. 2004. MRI-based treatment planning for radiotherapy: Dosimetric verification for prostate IMRT. *Int. J. Radiat. Oncol. Biol. Phys.* 60 (2): 636–647.

Chen, Z., C. M. Ma, K. Paskalev, J. Li, J. Yang, T. Richardson, L. Palacio, X. Xu, and L. Chen. 2006. Investigation of MR image distortion for radiotherapy treatment planning of prostate cancer. *Phys. Med. Biol.* 51 (6): 1393–1403.

Fransson, A., P. Andreo, and R. Potter. 2001. Aspects of MR image distortions in radiotherapy treatment planning. *Strahlenther. Onkol.* 177 (2): 59–73.

Guo, W. Y. 1998. Application of MR in stereotactic radiosurgery. *J. Magn. Reson. Imaging* 8 (2): 415–420.

Hill, D. L., D. J. Hawkes, M. J. Gleeson, T. C. Cox, A. J. Strong, W. L. Wong, C. F. Ruff et al. 1994. Accurate frameless registration of MR and CT images of the head: Applications in planning surgery and radiation therapy. *Radiology* 191 (2): 447–454.

Hossain, M., L. Chen, M. Buyyounouski, B. Milestone, T. Richardson, and C. M. Ma. 2008. The role of MRS in radiation therapy: Correlation between T2-weighted MRI, biopsy and MRS. *Med. Phys.* 35: 2721.

Khoo, V. S., E. J. Adams, F. Saran, J. L. Bedford, J. R. Perks, A. P. Warrington, and M. Brada. 2000. A comparison of clinical target volumes determined by CT and MRI for the radiotherapy planning of base of skull meningiomas. *Int. J. Radiat. Oncol. Biol. Phys.* 46 (5): 1309–1317.

Krempien, R. C., K. Schubert, D. Zierhut, M. C. Steckner, M. Treiber, W. Harms, U. Mende, D. Latz, M. Wannenmacher, and F. Wenz. 2002. Open low-field magnetic resonance imaging in radiation therapy treatment planning. *Int. J. Radiat. Oncol. Biol. Phys.* 53 (5): 1350–1360.

Lee, Y. K., M. Bollet, G. Charles-Edwards, M. A. Flower, M. O. Leach, H. McNair, E. Moore, C. Rowbottom, and S. Webb. 2003. Radiotherapy treatment planning of prostate cancer using magnetic resonance imaging alone. *Radiother. Oncol.* 66 (2): 203–216.

Mah, D., M. Steckner, A. Hanlon, G. Freedman, B. Milestone, R. Mitra, H. Shukla, B. Movsas, E. Horwitz, P. P. Vaisanen, and G. E. Hanks. 2002. MRI simulation: Effect of gradient distortions on three-dimensional prostate cancer plans. *Int. J. Radiat. Oncol. Biol. Phys.* 53 (3): 757–765.

Mah, D., M. Steckner, E. Palacio, R. Mitra, T. Richardson, and G. E. Hanks. 2002. Characteristics and quality assurance of a dedicated open 0.23 T MRI for radiation therapy simulation. *Med. Phys.* 29 (11): 2541–2547.

Menard, C., I. C. Smith, R. L. Somorjai, L. Leboldus, R. Patel, C. Littman, S. J. Robertson, and T. Bezabeh. 2001. Magnetic resonance spectroscopy of the malignant prostate gland after radiotherapy: A histopathologic study of diagnostic validity. *Int. J. Radiat. Oncol. Biol. Phys.* 50 (2): 317–323.

Mizowaki, T., Y. Nagata, K. Okajima, M. Kokubo, Y. Negoro, N. Araki, and M. Hiraoka. 2000. Reproducibility of geometric distortion in magnetic resonance imaging based on phantom studies. *Radiother. Oncol.* 57 (2): 237–242.

Moerland, M. A., R. Beersma, R. Bhagwandien, H. K. Wijrdeman, and C. J. Bakker. 1995. Analysis and correction of geometric distortions in 1.5 T magnetic resonance images for use in radiotherapy treatment planning. *Phys. Med. Biol.* 40 (10): 1651–1654.

Payne, G. S., and M. O. Leach. 2006. Applications of magnetic resonance spectroscopy in radiotherapy treatment planning. *Br. J. Radiol.* 79 Spec. No. 1: S16–S26.

Price, R. R., L. Axel, T. Morgan, R. Newman, W. Perman, N. Schneiders, M. Selikson, M. Wood, and S. R. Thomas. 1990. Quality assurance methods and phantoms for magnetic resonance imaging: Report of AAPM nuclear magnetic resonance Task Group No. 1. *Med. Phys.* 17 (2): 287–295.

48

Image Registration, Fusion, and Segmentation

Michael B. Sharpe
Princess Margaret Hospital

Kristy K. Brock
Princess Margaret Hospital

Introduction

Medical imaging technologies have a central role in radiotherapy treatment planning, in guiding treatment, and in assessing treatment response. The localization of targets and normal anatomy for treatment planning relies on a variety of medical imaging modalities, including computed tomography (CT), magnetic resonance (MR), and positron-emission tomography (PET). Image-guided radiotherapy (IGRT) delivery controls treatment accuracy and precision with frequent reference to the treatment plan. Multiple modalities are also used for the assessment of treatment response in patient follow-up exams, with comparison with diagnostic and prior follow-up scans.

When a variety of imaging modalities are used within the context of a procedure, or when the required imaging studies are performed at different points in time, a geometric registration is required to establish a correspondence between imaging studies and to support geometric targeting in image guided radiation therapy (IGRT) (Kessler 2006; Dawson and Sharpe 2006; Brock 2007). There is a growing need for image registration and segmentation of anatomical structures in modern radiotherapy and a range of technologies is available to fulfill these requirements (Dawson and Sharpe 2006). This chapter provides a brief review of the role of registration and related technologies in radiotherapy processes. Common and emerging technologies are summarized, current quality assurance (QA) practices pertaining to image registration are reviewed, and implications in clinical process design are discussed.

Image Registration

Formally, registration is an operation that aligns images to achieve geometric correspondence, while fusion is a visualization procedure, mapping the complementary data in each modality (Kessler 2006). In practice, these two functions are closely linked and referred to almost interchangeably. A transformation model is used to establish the spatial relationship between images. Rigid body transforms, for example, employ linear translations and rotations to register image coordinates. The transformation model can be optimized with the aid of a registration metric. Registration metrics are an objective description of the degree of similarity between images as they are aligned using the registration model and will be discussed in the next section. Rigid registration models are manipulated manually or automatically, with appropriate tools to achieve a visual match or with a registration metric.

A general classification of transformation models has been described (Sharpe and Brock 2008). Registration algorithms operate principally on the basis of a geometric correspondence, but some formulations are based on an underlying physical model, such as viscous fluid flow or the biomechanical properties of organs and tissues. Rigid transforms are an example of purely geometric algorithms and currently are predominant in clinical applications. Achieving an accurate correspondence with rigid body registration can be difficult, however, when images are acquired with different modalities located at different geographic locations, or at different points in time. The

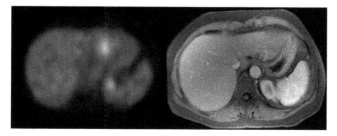

FIGURE 48.1 Advances in image registration will improve the integration of treatment planning and image-guidance for target localization to permit a more accurate estimation of the actual dose delivered and establish a formal link with follow-up imaging.

confident application of rigid registration is confounded even further when images encode different biophysical information, such as PET and MR (as depicted in Figure 48.1), and in the presence of nonlinear deformations associated with treatment response, weight change, and variation of organ position and volume between exams (Langen and Jones 2001; Hawkins et al. 2006; Ghilezan et al. 2005; Nichol et al. 2007). Deformable registration can address some of these challenges via affine transformations and polynomial (e.g., spline) basis functions, or using physical viscous-elastic models and biomechanical finite element modeling approaches based on the physical malleability of tissue (Kessler 2006; Brock 2007; Yan, Jaffray, and Wong 1999; Zhang et al. 2004; Brock et al. 2005; Chi, Liang, and Yan 2006).

Registration Metrics

Registration metrics are objective anatomical surrogates that express the degree of numerical similarity between images. Transformation models can be optimized by minimizing the difference between image sets compared using such a metric. Registration metrics are formulated using the distance between corresponding anatomical structures identified as points, line-segments, a segmented polygon describing the surface of a tumor or adjacent organ, or by voxel-based gray-scale intensities. Appropriate surrogates account for the clinical objectives in feature alignment in 2D and 3D images, and reflect the capabilities of the underlying transformation model (Kessler et al. 1991; Balter, Pelizzari, and Chen 1992; Van Herk and Kooy 1994; Gilhuijs et al. 1995; Gilhuijs, van de Ven, and Van Herk 1996; Ploeger et al. 2003). Points and surfaces can be delineated manually or automatically using intensity gradients or more advanced image processing for surface detection (Pekar, McNutt, and Kaus 2004).

Points or organ surfaces are used widely and successfully, but are time-consuming to segment manually and may ultimately limit registration accuracy in some situations, in particular when images present different biophysical properties. A target that is visualized poorly on CT may be more grossly visible on another modality, such as MR or PET studies. Surrogates based on gray-scale intensities are an attractive basis for registration metrics because they encompass more information and reduce manual

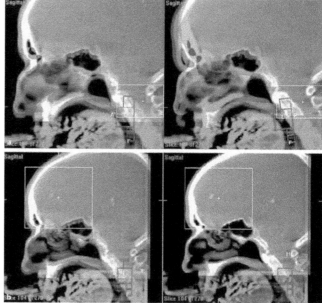

FIGURE 48.2 (a) Image alignment using a clip box positioned at the skull base, near the neck. (b) Registration of the same images using a clip box placed on the front skull. Matching regions appear in gray when overlaying green and purple color scales combine with high similarity.

intervention. Gray-scale similarity is calculated by several common methods, with mutual information (MI) emerging as a leading approach (Kessler 2006; Brock 2007). It is important to note that automated registration using an intensity-based metric examines correspondence over the entire imaging space, unless efforts are made to direct the focus to areas of clinical importance (Kessler 2006). If substantial deformation occurs between the two images (e.g., neck flexure) the global alignment results may by unacceptable. A geometric region of interest, such as a box or organ contour, can be used to define relevant information for an automated algorithm to focus only on the intensities in a region deemed important, as shown for the brain or head and neck in Figure 48.2.

Image Registration, Fusion, and Organ Segmentation

Image fusion creates a visual map of the registered secondary images and any associated target segmentations to the reference planning CT using the optimized transformation model. Combining or "fusing" image intensities allows simultaneous visualization of each imaging modality using a variety of display techniques (Kessler 2006; Brock 2007). Such techniques also support the visualization of 3D dose distributions mapped across imaging modalities.

The previous section alluded to the formulation of registration metrics from contoured targets and organ surfaces. Organs and target contours have other important roles in planning, guidance, evaluating dose distributions, and outcomes assessment.

Until recently, organ delineation involved manual outlining on transverse images. In some limited cases, tools that follow intensity gradients allowed semiautomated segmentation, but manual contouring continues to persist as a tedious and time-consuming task. In the past 5 years, 3D organ models have emerged as a feasible approach to automated segmentation and a means of reducing contouring time and effort (Pekar, McNutt, and Kaus 2004; Pizer et al. 2005).

Image Registration in Radiation Therapy

Figure 48.3 illustrates the processes involved in the delivery of radiation therapy. Two large light-gray boxes delineate planning from the steps involved in treatment delivery. This illustrates the separation of these events in time. Registration in planning establishes correspondence between imaging modalities for patient modeling. With each return treatment visit, the heavy black arrows indicate reproduction of the planned setup (black boxes) by registration of planning and treatment images (white boxes).

Treatment Planning

Modern systems support CT-guided planning augmented by an arbitrary number of images from additional modalities. Ideal supplemental exams are concordant with the planning CT, in both geometry and time. In practice, these modalities are employed at different time points and geographic locations. Because variation in body position, organ filling, and other changes introduce geometric uncertainties, registration by rigid

body models should be evaluated carefully to achieve acceptable consistency between modalities. The planning CT is considered the primary model of the patient over the course of therapy. Differences seen in the secondary modalities are minimized during registration to build a self-consistent static model. If the secondary images were acquired to aid in the delineation of the GTV, for example, the visible tumor or tumor-bearing organ would be the focus for aligning with the primary image.

"4D-CT" is a unique form of serial imaging that is used to assess organ movement due to breathing over a short time interval (Keall et al. 2004). 4DCT images reconstruct different breathing phases and tend to be registered inherently when acquired in the same scanning session as the primary CT exam. This is also the case for other forms of temporal imaging, which are emerging with the recognition of the importance of vascular architecture and angiogenesis to tumor growth (Miller et al. 2005).

Treatment Delivery

For each fraction of IGRT delivery, the patient is aligned nominally to the machine isocenter. To position the target, images are acquired and registered to the planning CT and the setup adjusted to fall within a defined tolerance interval. This process has mainly involved matching megavoltage (MV) portal radiographs with a digitally reconstructed radiograph (DRR) generated from the planning CT. Corrections are implemented by simple couch translations or manipulation of the patient's position on the couch (Balter, Pelizzari, and Chen 1992).

The advent of cone-beam computed tomography (CBCT) and its integration with treatment delivery have made it possible to

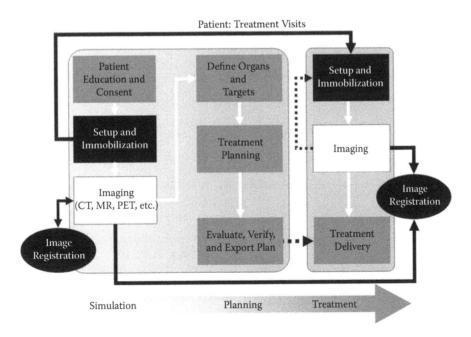

FIGURE 48.3 The clinical processes involved in the delivery of radiation therapy. The two large light-gray boxes delineate planning and delivery. With each return treatment visit, the heavy black arrows indicate the requirement to reproduce the planned treatment setup (black boxes) and to verify the patient setup by quantitative comparison of planning and treatment images (white boxes).

routinely guide treatment using images of soft-tissue (Purdie et al. 2007; Sharpe et al. 2006; Jaffray 2005; Letourneau et al. 2005). A registration of CBCT images with the planning CT supports the evaluation of the target location relative to the machine isocenter and the position of normal tissues that lie adjacent and peripheral to the target. This increases the quality of treatment delivery by ensuring that goals for target coverage and normal tissue sparing are translated properly from the planning context to the delivery of each treatment fraction.

Guidance differs from the planning context because images are not expected to be perfectly concordant over time. Instead of a static geometry, the images from each treatment session represent a temporal instance of the patient geometry incorporating residual setup variation, small shifts in the relative position of organs and targets, and other anatomical changes. Residual setup errors and deformation are undesirable, but they are anticipated and managed using PTV margins, for example. The availability of soft-tissue imaging for IGRT has spawned an area of active research and development to understand and mitigate the clinical consequences of residual uncertainties.

Adaptive Treatment Planning

Current applications of IGRT increase targeting accuracy and minimize setup uncertainties through online evaluation using rigid registration. Soft-tissue images also provide a record of residual variations and anatomical changes over the treatment course. Experience gained with MV portal imaging has led to strategies that include off-line components for determining statistical trends in systematic and random errors (Van Herk 2004; Yan et al. 2005). Statistics generated with offline strategies support the assessment and adjustment of PTV margins to exploit the precision inherent to an individual's treatment course (Lujan et al. 1999; Craig et al. 2001).

Routine IGRT leads naturally to the need to formally "serialize" information, that is, to arrange images in chronological

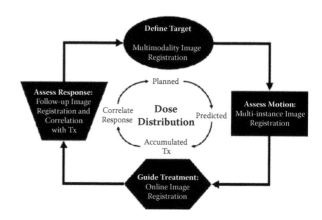

FIGURE 48.5 Emerging developments in image registration are creating a cycle of care in which images link treatment planning, image-guided treatment delivery, and adaptation to patient follow-up exams. Emerging developments will improve outcomes analysis by establishing a more quantitative relationship between images acquired for diagnosis, therapy, and follow-up.

order with registration to the planning context. The transformation model then supports computation and mapping of dose distributions representing individual treatments in the planning reference image, as illustrated in Figure 48.4. Deformable registration and anatomical structure mapping will provide a better estimate of anatomical changes (Δ) during and subsequent to therapy, as well as accumulation of the actual dose delivered (Σ) (Schaly et al. 2004; Mohan et al. 2005). Formal dose accumulation may improve the prediction of response compared to nominal planned distributions. When this process is managed actively during a course of therapy, adaptation to accumulating anatomical changes will be possible.

Outcomes Assessment

Figure 48.5 illustrates how emerging developments in image registration are creating a cycle of care in which images link treatment planning, image-guided treatment delivery, and adaptation to patient follow-up exams. Emerging developments will improve outcomes analysis by establishing a more quantitative relationship between images acquired for diagnosis, therapy, and follow-up. This, in turn, will support a more accurate estimation of the actual dose delivered for clinical assessment of radiation therapy via patient follow-up exams.

Quality Assurance

Image registration in planning and delivery requires verification of the results. Presently, rigid registration is predominant in commercial software offered for treatment planning and target localization applications, although deformable registration is beginning to be offered. Several groups recommend guidelines for QA of registration in radiosurgery and general planning; many of which are appropriate for the IGRT context (Fraass et al.

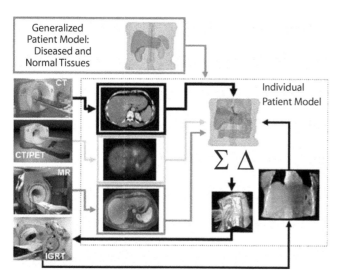

FIGURE 48.4 Image registration in the delivery of IGRT.

1998; Mutic et al. 2001; Yu et al. 2001; Moore, Liney, and Beavis 2004; Lavely et al. 2004; IAEA 2004). The International Atomic Energy Agency (IAEA), for example, emphasizes the appropriate use of software for the anatomical site and the number of imaging modalities employed. They recommend assessment of: the technical principles in use, any bias to a particular modality, constraints imposed on image acquisition, support for 2D and/or 3D registration, the degree and behavior of automation, the type of model (i.e., rigid or deformable), and dependence on image acquisition parameters or reliance on segmented structures.

The IAEA and American Association of Physicists in Medicine (AAPM) also recommend combined objective and visual queues for evaluating registrations. Recently, the AAPM formed Task Group 132 to review techniques for image registration, to identify issues related to their clinical implementation, to determine the best methods to assess accuracy, and to outline issues related to acceptance and QA. A key recommendation of the task group will be the need to perform process-oriented "end-to-end" testing of image registration tasks, spanning network-connected systems for image acquisition, planning, and delivery. This could be achieved, for example, using a geometric or anthropomorphic phantom that allows assessment of the accuracy and consistency of geometry, numerical gray scale intensities, demographic information, etc.

There is a demand for objective metrics of registration quality within radiotherapy and image-guided surgery (Rucklidge 1996; Jannin, Grova, and Maurer 2006; Crum, Camara, and Hill 2006). Phantom testing can determine if algorithms reproduce known displacements or changes in orientation under varying conditions (Mutic et al. 2001; Moore, Liney, and Beavis 2004; Lavely et al. 2004). Phantoms also help to confirm basic performance metrics like geometric scale calibration and orientation and the limits of linearity, accuracy, and precision. However, phantom studies do not completely capture factors degrading registration algorithm performance, such as variations in slice thickness, resolution, distortion, noise, and patient movement. In practice, process QA is required to monitor and manage these factors, and to achieve consistent practice. But, phantoms do offer a "best-case" scenario for users to understand the behavior and limitations of the algorithms registration metric and transformation model.

Setup variation and deformation in individual patients are the most limiting factors in image registration. Fortunately, the cross-comparison of many common imaging modalities inherently includes redundant structures, which help in the visual validation of each patient-specific procedure (Brock et al. 2005; Castillo et al. 2009). This type of qualitative assessment of accuracy can be achieved through image overlay, side-by-side comparison, split-screens, checker-board displays, and a variety of other visualization techniques (Kessler 2006; Brock 2007).

Traditionally, QA of registration tools occurs within the planning context and depends on a limited number of "upstream" imaging devices (Fraass et al. 1998; IAEA 2004). Image registration for planning has been limited to a single event, involving few staff and a limited need for physicians to delegate decision-making. At the same time, delivery guided by radiographs has relied on separate guidelines for QA and clinical practice (Herman et al. 2001). Imaging data now accompanies planning data when it is transferred to treatment management systems and individual linear accelerators; where it serves as a reference image. Each delivery system, in turn, is capable of generating new datasets that are fed back into planning systems to support vector-based dose accumulation and assessment of anatomical changes. It is likely that images will be stored on externally administered data storage. In order to complete and evaluate routine volumetric image registration tasks, a stable infrastructure must be maintained. QA programs for equipment and software in the IGRT era must assure operability and stability spanning across multiple imaging platforms. QA practices must reflect the integration of planning and delivery across information management systems and network infrastructure. Software, knowledge, and decision-making are disseminated throughout, and potentially beyond, the radiation therapy department. Consequently, a strong emphasis on standard guidelines for consistent practice and communication of expected results are required to support a working culture that is vigilant to exceptions, and to reduce the burden of patient-specific QA.

Summary

The routine use of IGRT leads to the requirement to serialize and register each image set to the treatment planning context. The potential to assess anatomical change quantitatively and to perform accurate dose accumulation are some of the anticipated benefits of an integrated approach to IGRT. Radiation therapy supports a strong culture of QA using prescriptive tests for equipment, logical intervals and tolerances for acceptance, commissioning, and periodic QA (Kutcher et al. 1994). With regard to the specific aspects of image registration, these principles are maintained in recent literature. However, there is no uniform consensus to guide the development of a comprehensive and prescriptive QA program in this arena.

It is feasible to confirm the performance of imaging devices, registration software, and networked storage and retrieval using phantom studies, but there is a strong need to rely on visual checks to assure consistent and unified practice in individual patient cases. The provision of robustly tested infrastructure, documentation, and well-exercised clinical processes will increase confidence in patient-specific situations. To minimize the possibility of human or algorithm error, clinical image acquisition and registration processes should be maintained in an integrated fashion where possible, to assure it is logical, sequential, and reproducible (Sharpe and Brock 2008).

References

Balter, J. M., C. A. Pelizzari, and G. T. Chen. 1992. Correlation of projection radiographs in radiation therapy using open curve segments and points. *Med. Phys.* 19 (2): 329–334.

Brock, K. K. 2007. Image registration in intensity-modulated radiation therapy, image-guided radiation therapy and stereotactic body radiation therapy. In *Advances in the Treatment Planning and Delivery of Radiotherapy*, ed. C. R. Meyer. Basel, Switzerland: Karger.

Brock, K. K., M. B. Sharpe, L. A. Dawson, S. M. Kim, and D. A. Jaffray. 2005. Accuracy of finite element model-based multi-organ deformable image registration. *Med. Phys.* 32 (6): 1647–1659.

Castillo, R., E. Castillo, R. Guerra, V. E. Johnson, T. McPhail, A. K. Garg, and T. Guerrero. 2009. A framework for evaluation of deformable image registration spatial accuracy using large landmark point sets. *Phys. Med. Biol.* 54 (7): 1849–1870.

Chi, Y., J. Liang, and D. Yan. 2006. A material sensitivity study on the accuracy of deformable organ registration using linear biomechanical models. *Med. Phys.* 33 (2): 421–433.

Craig, T., J. Battista, V. Moiseenko, and J. Van Dyk. 2001. Considerations for the implementation of target volume protocols in radiation therapy. *Int. J. Radiat. Oncol. Biol. Phys.* 49 (1): 241–250.

Crum, W. R., O. Camara, and D. L. Hill. 2006. Generalized overlap measures for evaluation and validation in medical image analysis. *IEEE Trans. Med. Imaging* 25 (11): 1451–1461.

Dawson, L. A., and M. B. Sharpe. 2006. Image-guided radiotherapy: Rationale, benefits, and limitations. *Lancet Oncol.* 7 (10): 848–858.

Fraass, B., K. Doppke, M. Hunt, G. Kutcher, G. Starkschall, R. Stern, and J. Van Dyke. 1998. American Association of Physicists in Medicine Radiation Therapy Committee Task Group 53: Quality assurance for clinical radiotherapy treatment planning. *Med. Phys.* 25 (10): 1773–1829.

Ghilezan, M. J., D. A. Jaffray, J. H. Siewerdsen, M. Van Herk, A. Shetty, M. B. Sharpe, J. S. Zafar et al. 2005. Prostate gland motion assessed with cine-magnetic resonance imaging (cine-MRI). *Int. J. Radiat. Oncol. Biol. Phys.* 62 (2): 406–417.

Gilhuijs, K. G., A. Touw, M. Vanherk, and R. E. Vijlbrief. 1995. Optimization of automatic portal image analysis. *Med. Phys.* 22 (7): 1089–1099.

Gilhuijs, K. G., P. J. van de Ven, and M. Van Herk. 1996. Automatic three-dimensional inspection of patient setup in radiation therapy using portal images, simulator images, and computed tomography data. *Med. Phys.* 23 (3): 389–399.

Hawkins, M. A., K. K. Brock, C. Eccles, D. Moseley, D. Jaffray, and L. A. Dawson. 2006. Assessment of residual error in liver position using kV cone-beam computed tomography for liver cancer high-precision radiation therapy. *Int. J. Radiat. Oncol. Biol. Phys.* 66 (2): 610–619.

Herman, M. G., J. M. Balter, D. A. Jaffray, K. P. MacGee, P. Munro, S. Shalev, M. Van Herk, and J. W. Wong. 2001. Clinical use of electronic portal imaging: Report of AAPM Radiation Therapy Committee Task Group 58. *Med. Phys.* 28:712.

IAEA. 2004. *Commissioning and Quality Assurance of Computerized Planning Systems for Radiation Treatment of Cancer, Technical Report Series.* Vienna: International Atomic Energy Agency.

Jaffray, D. A. 2005. Emergent technologies for 3-dimensional image-guided radiation delivery. *Semin. Radiat. Oncol.* 15 (3): 208–216.

Jannin, P., C. Grova, and C. Maurer. 2006. Model for defining and reporting reference-based validation protocols in medical image processing. *Int. J. Comput. Assisted Radiol. Surg.* 1 (2): 63–73.

Keall, P. J., G. Starkschall, H. Shukla, K. M. Forster, V. Ortiz, C. W. Stevens, S. S. Vedam, R. George, T. Guerrero, and R. Mohan. 2004. Acquiring 4D thoracic CT scans using a multislice helical method. *Phys. Med. Biol.* 49 (10): 2053–2067.

Kessler, M. L. 2006. Image registration and data fusion in radiation therapy. *Br. J. Radiol.* 79, Spec. No. 1: S99–S108.

Kessler, M. L., S. Pitluck, P. Petti, and J. R. Castro. 1991. Integration of multimodality imaging data for radiotherapy treatment planning. *Int. J. Radiat. Oncol. Biol. Phys.* 21 (6): 1653–1667.

Kutcher, G. J., L. Coia, M. Gillin, W. F. Hanson, S. Leibel, R. J. Morton, J. R. Palta, J. A. Purdy et al. 1994. Comprehensive QA for radiation oncology: Report of AAPM Radiation Therapy Committee Task Group 40. *Med. Phys.* 21 (4): 581–618.

Langen, K. M., and D. T. Jones. 2001. Organ motion and its management. *Int. J. Radiat. Oncol. Biol. Phys.* 50 (1): 265–278.

Lavely, W. C., C. Scarfone, H. Cevikalp, R. Li, D. W. Byrne, A. J. Cmelak, B. Dawant, R. R. Price, D. E. Hallahan, and J. M. Fitzpatrick. 2004. Phantom validation of coregistration of PET and CT for image-guided radiotherapy. *Med. Phys.* 31 (5): 1083–1092.

Letourneau, D., J. W. Wong, M. Oldham, M. Gulam, L. Watt, D. A. Jaffray, J. H. Siewerdsen, and A. A. Martinez. 2005. Cone-beam-CT guided radiation therapy: Technical implementation. *Radiother. Oncol.* 75 (3): 279–286.

Lujan, A. E., R. K. Ten Haken, E. W. Larsen, and J. M. Balter. 1999. Quantization of setup uncertainties in 3-D dose calculations. *Med. Phys.* 26 (11): 2397–2402.

Miller, J. C., H. H. Pien, D. Sahani, A. G. Sorensen, and J. H. Thrall. 2005. Imaging angiogenesis: Applications and potential for drug development. *J. Natl. Cancer Inst.* 97 (3): 172–187.

Mohan, R., X. Zhang, H. Wang, Y. Kang, X. Wang, H. Liu, K. K. Ang, D. Kuban, and L. Dong. 2005. Use of deformed intensity distributions for on-line modification of image-guided IMRT to account for interfractional anatomic changes. *Int. J. Radiat. Oncol. Biol. Phys.* 61 (4): 1258–1266.

Moore, C. S., G. P. Liney, and A. W. Beavis. 2004. Quality assurance of registration of CT and MRI data sets for treatment planning of radiotherapy for head and neck cancers. *J. Appl. Clin. Med. Phys.* 5 (1): 25–35.

Mutic, S., J. F. Dempsey, W. R. Bosch, D. A. Low, R. E. Drzymala, K. S. Chao, S. M. Goddu, P. D. Cutler, and J. A. Purdy. 2001. Multimodality image registration quality assurance for conformal three-dimensional treatment planning. *Int. J. Radiat. Oncol. Biol. Phys.* 51 (1): 255–260.

Nichol, A. M., K. K. Brock, G. A. Lockwood, D. J. Moseley, T. Rosewall, P. R. Warde, C. N. Catton, and D. A. Jaffray. 2007. A magnetic resonance imaging study of prostate deformation relative to implanted gold fiducial markers. *Int. J. Radiat. Oncol. Biol. Phys.* 67 (1): 48–56.

Pekar, V., T. R. McNutt, and M. R. Kaus. 2004. Automated model-based organ delineation for radiotherapy planning in prostatic region. *Int. J. Radiat. Oncol. Biol. Phys.* 60 (3): 973–980.

Pizer, S. M., P. T. Fletcher, S. Joshi, A. G. Gash, J. Stough, A. Thall, G. Tracton, and E. L. Chaney. 2005. A method and software for segmentation of anatomic object ensembles by deformable m-reps. *Med. Phys.* 32 (5): 1335–1345.

Ploeger, L. S., A. Betgen, K. G. Gilhuijs, and M. Van Herk. 2003. Feasibility of geometrical verification of patient set-up using body contours and computed tomography data. *Radiother. Oncol.* 66 (2): 225–233.

Purdie, T. G., J. P. Bissonnette, K. Franks, A. Bezjak, D. Payne, F. Sie, M. B. Sharpe, and D. A. Jaffray. 2007. Cone-beam computed tomography for on-line image guidance of lung stereotactic radiotherapy: Localization, verification, and intrafraction tumor position. *Int. J. Radiat. Oncol. Biol. Phys.* 68 (1): 243–252.

Rucklidge, W. 1996. *Efficient Visual Recognition Using the Hausdorff Distance, Lecture Notes in Computer Science.* Berlin: Springer.

Schaly, B., J. A. Kempe, G. S. Bauman, J. J. Battista, and J. Van Dyk. 2004. Tracking the dose distribution in radiation therapy by accounting for variable anatomy. *Phys. Med. Biol.* 49 (5): 791–805.

Sharpe, M. B., D. J. Moseley, T. G. Purdie, M. Islam, J. H. Siewerdsen, and D. A. Jaffray. 2006. The stability of mechanical calibration for a kV cone beam computed tomography system integrated with linear accelerator. *Med. Phys.* 33 (1): 136–144.

Sharpe, M., and K. K. Brock. 2008. Quality assurance of serial 3D image registration, fusion, and segmentation. *Int. J. Radiat. Oncol. Biol. Phys.* 71 (1 Suppl): S33–S337.

Van Herk, M. 2004. Errors and margins in radiotherapy. *Semin. Radiat. Oncol.* 14 (1): 52–64.

Van Herk, M., and H. M. Kooy. 1994. Automatic three-dimensional correlation of CT-CT, CT-MRI, and CT-SPECT using chamfer matching. *Med. Phys.* 21 (7): 1163–1178.

Yan, D., D. A. Jaffray, and J. W. Wong. 1999. A model to accumulate fractionated dose in a deforming organ. *Int. J. Radiat. Oncol. Biol. Phys.* 44 (3): 665–675.

Yan, D., D. Lockman, A. Martinez, J. Wong, D. Brabbins, F. Vicini, J. Liang, and L. Kestin. 2005. Computed tomography guided management of interfractional patient variation. *Semin. Radiat. Oncol.* 15 (3): 168–179.

Yu, C., M. L. Apuzzo, C. S. Zee, and Z. Petrovich. 2001. A phantom study of the geometric accuracy of computed tomographic and magnetic resonance imaging stereotactic localization with the Leksell stereotactic system. *Neurosurgery* 48 (5): 1092–1098.

Zhang, T., N. P. Orton, T. R. Mackie, and B. R. Paliwal. 2004. Technical note: A novel boundary condition using contact elements for finite element based deformable image registration. *Med. Phys.* 31 (9): 2412–2415.

4D Simulation

Dirk Verellen
Medical Physics Radiotherapy

Introduction

One of the most critical steps in the radiotherapy process is the delineation of the tumor volume and the surrounding healthy tissues. In fact, with the introduction of in-room image-guided radiotherapy (IGRT) and its aim of reducing planning target volume (PTV) margins (ICRU 1999; Verellen et al. 2007), one can argue that the delineation process currently has the highest contribution to the geometric uncertainties encountered in radiotherapy. It has been shown in the literature that intra- and interobserver variation has become the weakest link in the target localization process as the imaging system is only as good as the skills of the reader who interprets the images (Engels et al. 2009; Verellen et al. 2009; Van de Steene et al. 1998; Steenbakkers et al. 2005; Van de Steene et al. 2002). Currently, the combination of CT, MRI, and PET is widely used to delineate the gross tumor volume (GTV), which consists of all clinically macroscopic disease based on what is visible on these imaging modalities (ICRU 1999). The clinical target volume (CTV) is created by adding a certain margin to account for microscopic disease and, finally, a safety margin is added for geometric uncertainties such as setup errors and tumor motion, to create the so-called PTV. Uncertainties in target delineation, historically, were largely covered by the generous PTV margins encountered in conventional radiotherapy. With the aim of maximizing outcome while minimizing complications, the latest developments in radiotherapy have been focusing on dose sculpting and reduction of PTV margins, the latter being realized by IGRT. The high-dose gradients encountered in dose sculpting and dose-painting-by-numbers (Ling et al. 2000) are unforgiving with respect to geographical miss, explaining the synergistic developments in IMRT and IGRT. These efforts are, of course, pointless without accurate knowledge of the target that is to be treated.

Until the end of the previous century, patient setup and the determination of treatment beams were mainly guided by using a treatment simulator (using offline diagnostic X-ray imaging) and

drawing skin marks on the patient's surface, consequently used to position the patient with respect to the treatment machine. The announcement in the early 1970s of the CT scanner in a clinical environment has been described as the greatest step forward in radiology since Roentgen's discovery, and led to the 1979 Nobel Price for Medicine jointly awarded to Hounsfield and Cormack (Ambrose and Hounsfield 1973). This development was immediately adopted in radiotherapy for application in computerized CT-based planning using 3D imaging of the tumor and surrounding tissues, with radiation dose patterns superimposed on these CT scans (applying dose calculation algorithms based on attenuation properties of different tissues derived from the CT numbers) and special algorithms employed to optimize therapy treatment. Therefore, it has also been claimed that the introduction of CT in the radiotherapy planning process was a major step in the development of this discipline. Both CT and magnetic resonance imaging (Lauterbeur 1980; Mansfield and Maudsley 1977) paved the way for 3D conformal radiotherapy (Webb 1997), intensity-modulated radiotherapy (NCI 2001; Brahme 1988; Ezzell et al. 2003; Webb 2001), and the understanding of dose–volume relationships. These developments enabled determination of the size and location of the tumor and the beams of radiation to be aimed in such a way as to maximize the tumor dose and minimize the dose to healthy normal tissues. IMRT enabled the delivery of radiotherapy shaped to the contours of the PTV with a high-dose volume enveloping the target, even in the case of complex geometry or invagination of healthy tissue ("dose sculpting"). The result has been improved tumor control with normal tissue morbidities being equivalent or less than conventional treatment (Hanks et al. 1998; Zelefsky, Leibel, Gaudin et al. 1998; Zelefsky, Leibel, Kutcher et al. 1998). These developments were largely based on anatomical or morphological data. With the introduction of quantitative functional data, for example, diffusion-weighted and dynamic contrast-enhanced MRI and MR spectroscopy (Wuthrich, Shulman, and Peisach 1968; Gambhir 2002; Langer et al. 2009), and PET (Gregoire et al. 2007), dose-painting-by-numbers allows for a more biologically driven treatment strategy, as information can be obtained about tumor characteristic properties—also coined biologically conformal radiation therapy or BCRT (Ling et al. 2000; Alber et al. 2003; Yang and Xing 2005). These developments shifted the prevailing philosophy of radiotherapy of aiming for as uniform or homogenous dose as possible, toward a deliberately planned nonuniform dose in accordance with a metric obtained from functional imaging data (Ling et al. 2000; Bentzen 2005; Xing et al. 2002).

Fourth Dimension in Radiotherapy

The intrinsic features and advantages of the different imaging modalities are covered extensively in Chapters 46, 47, and 68. In this chapter, we will concentrate on the temporal aspects of the image acquisition and how this might influence the delineation process and definition of PTV margins in the case of, for instance, lung tumors. Motion of lung tumors with respiration causes difficulties with CT and PET in that valuable information is lost when these scans are not correlated with respiration. It is important to understand the consequences of motion artifacts and evaluate the motion patterns in clinical practice. Consequently, this information can be used for motion management in radiotherapy either by defining appropriate safety margins (population-based or individualized to the patient), or by breathing-synchronized irradiation techniques such as gating or real-time tracking (tumor chasing). In the AAPM TG-76 (Keall et al. 2006) report, the following components have been identified as contributing to the overall geometric error and should be considered when designing CTV-PTV margins:

1. Inter- and intraobserver variations in GTV and CTV delineation
2. Motion artifacts in the CT scan and other imaging modalities that cause target delineation errors
3. Respiratory motion and heartbeat during delivery, which are periodic functions of time
4. Daily variations of respiratory motion
5. Variations caused by changing organ volumes
6. Tumor growth and shrinkage
7. Treatment-related anatomical changes, such as reductions in bronchiole obstructions and changes in atelectasis regions
8. Patient setup error

Points 1–4 will be the focus of this chapter. Apart from techniques to accurately describe motion, different methods are being developed for motion management during treatment such as forced shallow breathing by abdominal compression (Wulf et al. 2000; Lax et al. 1994) or breath-hold (Mah et al. 2000; Wong et al. 1999) motion-encompassing techniques (Caldwell et al. 2003) and breathing-synchronized techniques such as respiratory gating (Shirato, Shimizu, Kitamura et al. 2000; Shirato, Shimizu, Kunieda et al. 2000; Murphy et al. 2000; Verellen et al. 2006; Ohara et al. 1989) and real-time tracking (Murphy

1997; Keall et al. 2001). Motion-encompassing techniques refer to the idea of incorporating information on tumor motion into the treatment planning process, either by introducing patient individualized margins or by using this information in the optimization procedure (Guckenberger et al. 2007; Brock et al. 2003; Bortfeld, Jiang, and Rietzel 2004). Internal motion can be assessed by time-resolved imaging techniques such as "slow" CT scanning or PET, which due to its slow acquisition time offers information on tumor position probability (Caldwell et al. 2003; Lagerwaard et al. 2001; Bosmans et al. 2006). Alternatively, multiple fast CT acquisitions during the respiratory cycle can be used to describe the target's motion during the respiratory cycle. Respiratory motion, however, is irregular and it is impossible to generalize respiratory patterns obtained by observation prior to treatment. Organ motion causes an averaging or blurring of the static dose distribution along the path of motion, and for IMRT an additional motion artifact follows from a possible interplay between the motion of the leaves of the collimator and the component of target motion perpendicular to the beam (Verellen et al. 2006; Bortfeld, Jiang, and Rietzel 2004; Jiang et al. 2003). As a result, respiratory-synchronized techniques will most often offer the optimal solution. The breathing pattern itself is usually obtained from an external signal, allowing real-time observation (e.g., infrared reflective markers placed on the patient's surface, spirometers, or flexible bellows) and mechanical coupling with the tumor is often weak, resulting in complex relationships. Breathing-synchronized techniques need to establish a correlation between the real-time external breathing signal and the internal tumor motion and this correlation needs to be verified with regular intervals during the course of treatment. Respiratory gating involves the administration of radiation during certain intervals within a particular portion of the patient's breathing cycle, commonly referred to as the "gate." The choice of the gate width is a trade-off between minimizing motion within the gate and beam-on time, and breath-hold techniques are being introduced to optimize the duty cycle (Wong et al. 1999; Linthout et al. 2009). Some studies have shown the anatomic position of the tumor to be more reproducible at end respiration (Mageras et al. 2001). On the other hand, when breath-hold techniques are introduced at (moderate) deep inspiration, the relative damage to healthy lung tissue can be reduced and it seems to be more comfortable for the patient (Wong et al. 1999; Hanley et al. 1999). Another means of accommodating respiratory motion is repositioning the radiation beam dynamically so as to follow the tumor's changing position, referred to as real-time tumor tracking. The latter can be established by synchronizing the linear accelerator's collimating system with the target motion (Keall et al. 2001). With conventional MLCs, only one dimension can be compensated for efficiently and the beam collimating device needs to be aligned such that the leaf motion coincides with the major axis of the tumor motion. Compensation of 2D motion might be offered with the pan-and-tilt-like solution currently under development (Kamino et al. 2006) and full 3D compensation is realized with the linear accelerator mounted to a robotic arm (Murphy 1997; Murphy et al. 2002; Schweikard et al. 2000).

Compared to the gating technique, the tracking approach potentially offers higher delivery efficiency and less residual target motion provided there is no system latency in the beam adaptation (Schweikard et al. 2000). It should be noted that these tracking approaches rely on an accurate predictive model of the breathing motion to anticipate the future position of the tumor, which proved to be challenging (Murphy 2004). An excellent overview of the management of respiratory motion in radiotherapy is given by the AAPM Task Group 76 (Keall et al. 2006).

4D or Respiration Correlated CT

Before embarking on the 4D concept in CT scanning, it should be mentioned that in many cases fluoroscopy using conventional simulators can be used to quantify the displacements and evaluate reliable PTV margins in order to assess tumor motion. However, as CT simulation is gradually replacing conventional simulation, the latter will not be covered in this chapter. Moreover, assessment of motion is only one aspect of tumor motion, artifacts introduced in the image reconstruction are also to be considered. Basically there are three options possible for CT imaging that can include the entire range of tumor motion for respiration. Listed in order of increased workload they can be classified as: slow CT, inhalation and exhalation breath-hold CT, and 4D or respiration-correlated CT (RC-CT). It is important to understand that the breathing patterns and, hence, tumor motion will change over time (between simulation or imaging sessions and treatment sessions) and are inherently irreproducible. When measuring tumor motion, the motion should be observed over several breathing cycles.

With the previous generation of "slow" CT scanners, the image of the tumor is smeared out due to breathing motion. In this case, tumor volume and organ delineation will be extremely difficult and subject to intra- and interobserver variability. This technique yields a tumor-encompassing volume, with the limitation that the respiratory motion will change between imaging and treatment. Moreover, motion may cause localization errors and, in some cases, disappearance of small tumors that should be detectable. The fast, helical multislice CT scanners actually freeze the image of the tumor at one location at one particular moment in the breathing cycle, thus offering a more anatomically relevant representation of the tumor. As this is not necessarily the average tumor position, however, this fast acquisition might introduce large systematic errors with respect to beam–tumor alignment when used inappropriately. Several strategies have been investigated to solve this problem, such as inhalation and exhalation breath-hold techniques, respiratory gating (Mageras and Yorke 2004), and respiration-correlated or 4D CT (RC-CT) (Keall et al. 2006; Bosmans et al. 2006; Ford et al. 2003; Vedam et al. 2003). Inhalation and exhalation breath-hold CT scans can also be applied to obtain a tumor-encompassing volume. The advantage of this approach over the slow scanning technique mentioned above is that the blurring caused by motion is significantly reduced. Dose calculation, however, should be performed on the CT data set that is most appropriate for that particular

treatment and patient. One option could be to use a free breathing CT for dose calculation and using the inhalation and exhalation scans to determine the range of motion and more accurate tumor delineation.

4D CT or RC-CT is a relatively new technology, made possible by the introduction of faster CT scanners with multiple detector rows. Basically, it is an oversampled or low-pitch CT scan during which the respiration signal is recorded. The latter can be obtained with different methods, of which abdominal straps with a pressure sensor (Li et al. 2006; Bosmans et al. 2006; Lu et al. 2005), infrared markers placed on the patient's chest (Ford et al. 2003), and measuring airflow with thermocouples or spirometers in a mouth mask (Wilson et al. 2003; Lu et al. 2006; Wolthaus et al. 2005) are most common. Afterward, the CT images can be binned (sorted) according to the phase or amplitude of the external respiratory signal (phase-angle or amplitude sorting) (Abdelnour et al. 2007; Lu et al. 2006). Most commercially available systems are based on the phase of the external breathing signal; however, Lu et al. (2006) have observed that the relationship to internal motion seems to be strongest with the amplitude of the external signal. These and other investigators showed that images generated using amplitude sorting displayed smoother lung-diaphragm boundaries and minimal reconstruction artifacts compared to phase-angle sorted imaging (Lu et al. 2006; Vedam et al. 2001; Rietzel, Pan, and Chen 2005; Fitzpatrick et al. 2006). In principle, phase-angle sorting regards a shallow breath the same as a normal and deep breath. Image binning itself can be performed prospectively or retrospectively. In prospective techniques, acquisition is synchronized with the patient's breathing, and all the projections are acquired during the same respiratory phase by means of a trigger produced by real-time tracking of the respiratory signal (Badea and Gordon 2004). On the other hand, retrospective algorithms (Keall et al. 2004; Kleshneva, Muzik, and Alber 2006; Sonke et al. 2005; Low et al. 2003) do not require any trigger signal during the acquisition, although they are constrained by the need to acquire at least one complete respiratory cycle per slice to avoid any empty phase bins. Therefore, the acquisition protocol usually requires multiple frames from every projection angle, each one corresponding to a different point during the breathing cycle.

Instead of one CT data set in 4D CT or RC-CT, several data sets according to the number of bins are available for tumor delineation and treatment planning. A limitation of 4D CT is that it is affected by variations in respiratory patterns during acquisition. Based on these 4D-CT data sets, movies can be generated to visualize and quantify the tumor's motion. However, the reader should realize that a movie-loop generated from a 4D-CT scan is not representative of the patient's breathing during treatment, as it represents only a few breathing cycles acquired several days prior to treatment repeated in a continuous loop. Patients' breathing during scanning or treatment is irregular, and can be affected by the patient gradually becoming more comfortable during the course of treatment and response to treatment. Again, this emphasizes the importance of the imaging technique being chosen with regard to the radiotherapy technique that will be used to treat the patient (motion-encompassing technique, immobilization, gating, tracking, etc.).

Incorporating the 4D information into the treatment planning is possible in different ways, again based on the treatment technique. One option is to use only one phase of the respiration, for example, the mid-ventilation phase, as being the phase where the tumor is at its average position, in combination with a "margin recipe" (based on the extent of motion observed from the other data sets) to account for the motion (Wolthaus et al. 2006). Some investigators use maximum intensity projections (MIPs), which reflect the highest pixel value encountered from all CTs along the viewing ray for each pixel, giving rise to an artificial intensity display of the brightest object along each ray on the projection image (Underberg et al. 2005). Another option is to contour each phase separately, or alternatively, contour one phase and use deformable registration to obtain the contours in the other phases, and use the union of the contours obtained from these data sets to obtain a margin recipe. In the case of respiratory gated radiotherapy, one might decide to delineate the tumor volume in the treatment-phase-angle or amplitude-sorted data set only (provided a similar technique is used to obtain the external breathing signal during CT-scanning as well as at the time of the actual treatment). Again, it is important to recall the fact that the patient's breathing during CT scanning might not be representative of the breathing during treatment and the correlation between the external signal and internal tumor motion is prone to changes (irregular breathing, tumor response, base line shifts, etc.). Respiratory gated or tracking techniques thus require IGRT during treatment to validate and/or update the correlation between the external breathing signal and the internal tumor motion.

Respiratory Correlated PET-CT

As mentioned earlier, inter- and intraobserver variability in tumor delineation can be improved with the combination of PET and CT. As RC-CT has been shown to be beneficial in imaging moving tumors, the effect of respiration on PET and the possibility of respiration-correlated PET-CT (Boucher et al. 2004; Wolthaus et al. 2005) needs careful consideration in radiotherapy treatment planning. PET imaging is a slow imaging technique requiring several minutes to obtain a reasonable signal/noise ratio. Therefore, several respiration cycles are covered by the images, obviously blurring the objects. Basically motion artifacts in PET lead to two major effects: first, they affect the accuracy of quantification, scrambling the measured standard uptake value; second, the apparent lesion volume is overestimated. Needless to say the blurring artifact hampers possible registration with fast CT images. Caldwell et al. (2003) suggested actually using this information to derive an individualized internal target volume (ICRU 1999). The hypothesis here is that the respiration is taken into account in the imaging process and the visible lesion represents a probability distribution of the tumor's position. However, some arguments can be raised against this approach. PET images are extremely sensitive to window/level settings and some kind of automated delineation software should

be applied to define the PET-based GTV (Geets, Lee et al. 2007; Lee, Geets et al. 2008; Lee, Langen et al. 2008). The latter should be correlated with pathology data (Stroom et al. 2007; Gregoire et al. 2007; Geets, Tomsej et al. 2007), which is extremely difficult with blurred PET images. Moreover, the resulting intensity as observed in the images is a complicated combination of metabolic uptake heterogeneity and motion (typically yielding a high intensity at the average position and low intensity at the extreme positions). Accurate information on several parameters such as the maximal standardized uptake value (SUV_{max}) is hampered and SUV values are only reliable when no or little movement of the target is present (Nehmeh et al. 2002). The attenuation correction could be wrongly performed when combining the fast CT scan with the slow PET-scan. Finally, the information on intratumor heterogeneity in tracer uptake one wants to use for dose-painting-by-numbers is completely lost due to the respiration motion. For these reasons, 4D or respiration-correlated PET imaging has been developed similar to the RC-CT.

Motion artifacts can be reduced by gating PET images in correlation with respiration. Nehmeh et al. (2002) have shown that respiratory gating (phase-based prospective gating using a respiration signal derived from infrared reflective markers rigidly mounted on a lightweight plastic block placed on the patient's chest) can reduce the degrading effect of breathing motion on PET images, allowing a more complete recovery of the true counts within the lesion and a more accurate estimation of the lesion volume. Based on phantom studies, these investigators showed a dependence of the reduction in the smearing effect on lesion size, motion amplitude, and bin size. Respiration-correlated PET (RC-PET) yielded an increase in the signal-to-noise ratio, as well as a recovery of the SUV values. Application to a patient study demonstrated that the technique was successful in reducing the smearing effect and correcting the SUV (Nehmeh et al. 2002). In these phantom studies, however, the PET data was acquired in gated mode and binned prospectively into 10 fractions. This approach is sensitive to irregular breathing patterns as the bins have predefined time lengths. In an attempt to account for varying breathing frequencies, Nagel et al. (2006) have investigated RC-PET based on continuous acquisition of the data in list mode and retrospective phase binning. Retrospective binning offers an advantage in that the time length of the bins is determined individually for each period of the respiration. After binning the CT and uncorrected PET data into corresponding phases, the tumor and tissue positions on PET and CT match more closely, reducing motion artifacts introduced to the PET reconstruction with CT-based attenuation correction. Using a moving lollipop phantom with a 3.2 cm diameter sphere filled with ^{18}F-FDG they observed that with a standard non-RC-CT-PET the volume as measured in CT and PET data sets was underestimated by as much as 46% and overestimated with 370%, respectively. Volumes obtained from RC-CT-PET had average deviations of 1.9% (±4.8%) and 1.5% (±3.4%) from the actual volume, for the CT and PET derived volumes, respectively (Nagel et al. 2006). Images with non-RC attenuation correction showed clear misplacement of the maximum activity (which, in clinical practice, would result in a mislocalization of the tumor). However, these investigators acknowledged that the sphere was imaged in air with little occurrence of attenuation, which might overestimate the effects. It is interesting to note that a small phase shift could be observed in some experiments because the respiratory signals were recorded with different devices. Synchronization of both modalities (CT and PET) with a single device for respiration correlation is assumed to dispose of this phase difference. To overcome some of these problems the same group has performed a simulation with real patient data based on respiratory gated CT studies of five patients: conventional PET, nongated PET with gated CT, gated PET with nongated CT, and phase matched PET and CT (Hamill, Bosmans, and Dekker 2008). As expected, phase-matched gated PET and CT gave essentially perfect PET reconstructions in the simulation. Gating of the PET alone (nongated CT) gave the correct tumor shape but was not quantitative. Gating of the CT only (nongated PET) resulted in blurred tumors and was again not quantitative, which was also true for the conventional PET.

Summary

RC-CT offers more reliable information on tumor motion and, consequently, allows for internal margins to be determined for each patient individually. Moreover, imaging artifacts can be reduced enabling more accurate tumor and healthy tissue delineation. RC-CT used in conjunction with the irradiation technique (individualized margins and motion-encompassing techniques, respiratory gating, or tracking) is a necessity for accurate treatment planning. RC-CT and RC-PET SUV determination and quantification of tracer uptake is more reliable and, thus, better suited for use in tumor characterization and automated delineation. Many of these developments are still under investigation and should be approached with great caution prior to clinical implementation. Prospective binning versus retrospective binning and phase-angle versus amplitude binning are influenced differently by variations in the patient breathing pattern and should be considered carefully before applying a treatment strategy based on these data sets.

References

Abdelnour, A. F., S. A. Nehmeh, T. Pan, J. L. Humm, P. Vernon, H. Schoder, K. E. Rosenzweig et al. 2007. Phase and amplitude binning for 4D-CT imaging. *Phys. Med. Biol.* 52 (12): 3515–3529.

Alber, M., F. Paulsen, S. M. Eschmann, and H. J. Machulla. 2003. On biologically conformal boost dose optimization. *Phys. Med. Biol.* 48 (2): N31–N35.

Ambrose, J., and G. Hounsfield. 1973. Computerized transverse axial tomography. *Br. J. Radiol.* 46 (542): 148–149.

Badea, C., and R. Gordon. 2004. Experiments with the nonlinear and chaotic behaviour of the multiplicative algebraic reconstruction technique (MART) algorithm for computed tomography. *Phys. Med. Biol.* 49 (8): 1455–1474.

Bentzen, S. M. 2005. Theragnostic imaging for radiation oncology: Dose-painting by numbers. *Lancet Oncol.* 6 (2): 112–117.

Bortfeld, T., S. B. Jiang, and E. Rietzel. 2004. Effects of motion on the total dose distribution. *Semin. Radiat. Oncol.* 14 (1): 41–51.

Bosmans, G., J. Buijsen, A. Dekker, M. Velders, L. Boersma, D. De Ruysscher, A. Minken, and P. Lambin. 2006. An "in silico" clinical trial comparing free breathing, slow and respiration correlated computed tomography in lung cancer patients. *Radiother. Oncol.* 81 (1): 73–80.

Boucher, L., S. Rodrigue, R. Lecomte, and F. Benard. 2004. Respiratory gating for 3-dimensional PET of the thorax: Feasibility and initial results. *J. Nucl. Med.* 45 (2): 214–219.

Brahme, A. 1988. Optimization of stationary and moving beam radiation therapy techniques. *Radiother. Oncol.* 12 (2): 129–140.

Brock, K. K., D. L. McShan, R. K. Ten Haken, S. J. Hollister, L. A. Dawson, and J. M. Balter. 2003. Inclusion of organ deformation in dose calculations. *Med. Phys.* 30 (3): 290–295.

Caldwell, C. B., K. Mah, M. Skinner, and C. E. Danjoux. 2003. Can PET provide the 3D extent of tumor motion for individualized internal target volumes? A phantom study of the limitations of CT and the promise of PET. *Int. J. Radiat. Oncol. Biol. Phys.* 55 (5): 1381–1393.

Engels, B., G. Soete, D. Verellen, and G. Storme. 2009. Conformal arc radiotherapy for prostate cancer: increased biochemical failure in patients with distended rectum on the planning computed tomogram despite image guidance by implanted markers. *Int. J. Radiat. Oncol. Biol. Phys.* 74 (2): 388–391.

Ezzell, G. A., J. M. Galvin, D. Low, J. R. Palta, I. Rosen, M. B. Sharpe, P. Xia, Y. Xiao, L. Xing, and C. X. Yu. 2003. Guidance document on delivery, treatment planning, and clinical implementation of IMRT: Report of the IMRT Subcommittee of the AAPM Radiation Therapy Committee. *Med. Phys.* 30 (8): 2089–2115.

Fitzpatrick, M. J., G. Starkschall, J. A. Antolak, J. Fu, H. Shukla, P. J. Keall, P. Klahr, and R. Mohan. 2006. Displacement-based binning of time-dependent computed tomography image data sets. *Med. Phys.* 33 (1): 235–246.

Ford, E. C., G. S. Mageras, E. Yorke, and C. C. Ling. 2003. Respiration-correlated spiral CT: A method of measuring respiratory-induced anatomic motion for radiation treatment planning. *Med. Phys.* 30 (1): 88–97.

Gambhir, S. S. 2002. Molecular imaging of cancer with positron emission tomography. *Nat. Rev. Cancer* 2 (9): 683–693.

Geets, X., J. A. Lee, A. Bol, M. Lonneux, and V. Gregoire. 2007. A gradient-based method for segmenting FDG-PET images: Methodology and validation. *Eur. J. Nucl. Med. Mol. Imaging* 34 (9): 1427–1438.

Geets, X., M. Tomsej, J. A. Lee, T. Duprez, E. Coche, G. Cosnard, M. Lonneux, and V. Gregoire. 2007. Adaptive biological image-guided IMRT with anatomic and functional imaging in pharyngo-laryngeal tumors: Impact on target volume delineation and dose distribution using helical tomotherapy. *Radiother. Oncol.* 85 (1): 105–115.

Gregoire, V., K. Haustermans, X. Geets, S. Roels, and M. Lonneux. 2007. PET-based treatment planning in radiotherapy: A new standard? *J. Nucl. Med.* 48 Suppl 1: 68S–77S.

Guckenberger, M., J. Wilbert, T. Krieger, A. Richter, K. Baier, J. Meyer, and M. Flentje. 2007. Four-dimensional treatment planning for stereotactic body radiotherapy. *Int. J. Radiat. Oncol. Biol. Phys.* 69 (1): 276–285.

Hamill, J. J., G. Bosmans, and A. Dekker. 2008. Respiratory-gated CT as a tool for the simulation of breathing artifacts in PET and PET/CT. *Med. Phys.* 35 (2): 576–585.

Hanks, G. E., A. L. Hanlon, T. E. Schultheiss, W. H. Pinover, B. Movsas, B. E. Epstein, and M. A. Hunt. 1998. Dose escalation with 3D conformal treatment: Five year outcomes, treatment optimization, and future directions. *Int. J. Radiat. Oncol. Biol. Phys.* 41 (3): 501–510.

Hanley, J., M. M. Debois, D. Mah, G. S. Mageras, A. Raben, K. Rosenzweig, B. Mychalcza et al. 1999. Deep inspiration breath-hold technique for lung tumors: The potential value of target immobilization and reduced lung density in dose escalation. *Int. J. Radiat. Oncol. Biol. Phys.* 45 (3): 603–611.

ICRU. 1999. *ICRU Report 62. Prescribing, Recording and Reporting Photon Beam Therapy (Supplement to ICRU Report 50).* Bethesda, MD: International Commission on Radiation Units and Measurements.

Jiang, S. B., C. Pope, K. M. Al Jarrah, J. H. Kung, T. Bortfeld, and G. T. Chen. 2003. An experimental investigation on intrafractional organ motion effects in lung IMRT treatments. *Phys. Med. Biol.* 48 (12): 1773–1784.

Kamino, Y., K. Takayama, M. Kokubo, Y. Narita, E. Hirai, N. Kawawda, T. Mizowaki, Y. Nagata, T. Nishidai, and M. Hiraoka. 2006. Development of a four-dimensional image-guided radiotherapy system with a gimbaled X-ray head. *Int. J. Radiat. Oncol. Biol. Phys.* 66 (1): 271–278.

Keall, P. J., V. R. Kini, S. S. Vedam, and R. Mohan. 2001. Motion adaptive x-ray therapy: A feasibility study. *Phys. Med. Biol.* 46 (1): 1–10.

Keall, P. J., G. S. Mageras, J. M. Balter, R. S. Emery, K. M. Forster, S. B. Jiang, J. M. Kapatoes et al. 2006. The management of respiratory motion in radiation oncology report of AAPM Task Group 76. *Med. Phys.* 33 (10): 3874–3900.

Keall, P. J., G. Starkschall, H. Shukla, K. M. Forster, V. Ortiz, C. W. Stevens, S. S. Vedam, R. George, T. Guerrero, and R. Mohan. 2004. Acquiring 4D thoracic CT scans using a multislice helical method. *Phys. Med. Biol.* 49 (10): 2053–2067.

Kleshneva, T., J. Muzik, and M. Alber. 2006. An algorithm for automatic determination of the respiratory phases in four-dimensional computed tomography. *Phys. Med. Biol.* 51 (16): N269–N276.

Lagerwaard, F. J., J. R. Van Sornsen de Koste, M. R. Nijssen-Visser, R. H. Schuchhard-Schipper, S. S. Oei, A. Munne, and S. Senan. 2001. Multiple "slow" CT scans for incorporating lung tumor mobility in radiotherapy planning. *Int. J. Radiat. Oncol. Biol. Phys.* 51 (4): 932–937.

Langer, D. L., T. H. van der Kwast, A. J. Evans, J. Trachtenberg, B. C. Wilson, and M. A. Haider. 2009. Prostate cancer detection with multi-parametric MRI: Logistic regression analysis of quantitative T2, diffusion-weighted imaging, and dynamic contrast-enhanced MRI. *J. Magn. Reson. Imaging* 30 (2): 327–334.

Lauterbeur, P. C. 1980. Progress in n.m.r. zeugmatography imaging. *Philos. Trans. R. Soc. Lond. B Biol. Sci.* 289: 483–487.

Lax, I., H. Blomgren, I. Naslund, and R. Svanstrom. 1994. Stereotactic radiotherapy of malignancies in the abdomen. Methodological aspects. *Acta Oncol.* 33 (6): 677–683.

Lee, C., K. M. Langen, W. Lu, J. Haimerl, E. Schnarr, K. J. Ruchala, G. H. Olivera et al. 2008. Assessment of parotid gland dose changes during head and neck cancer radiotherapy using daily megavoltage computed tomography and deformable image registration. *Int. J. Radiat. Oncol. Biol. Phys.* 71 (5): 1563–1571.

Lee, J. A., X. Geets, V. Gregoire, and A. Bol. 2008. Edge-preserving filtering of images with low photon counts. *IEEE Trans. Pattern Anal. Mach. Intell.* 30 (6): 1014–1027.

Li, T., L. Xing, P. Munro, C. McGuinness, M. Chao, Y. Yang, B. Loo, and A. Koong. 2006. Four-dimensional cone-beam computed tomography using an on-board imager. *Med. Phys.* 33 (10): 3825–3833.

Ling, C. C., J. Humm, S. Larson, H. Amols, Z. Fuks, S. Leibel, and J. A. Koutcher. 2000. Towards multidimensional radiotherapy (MD-CRT): Biological imaging and biological conformality. *Int. J. Radiat. Oncol. Biol. Phys.* 47: 551–560.

Linthout, N., S. Bral, I. Van de Vondel, D. Verellen, K. Tournel, T. Gevaert, M. Duchateau, T. Reynders, and G. Storme. 2009. Treatment delivery time optimization of respiratory gated radiation therapy by application of audio-visual feedback. *Radiother. Oncol.* 91 (3): 330–335.

Low, D. A., M. Nystrom, E. Kalinin, P. Parikh, J. F. Dempsey, J. D. Bradley, S. Mutic et al. 2003. A method for the reconstruction of four-dimensional synchronized CT scans acquired during free breathing. *Med. Phys.* 30 (6): 1254–1263.

Lu, W., P. J. Parikh, I. M. El Naqa, M. M. Nystrom, J. P. Hubenschmidt, S. H. Wahab, S. Mutic et al. 2005. Quantitation of the reconstruction quality of a four-dimensional computed tomography process for lung cancer patients. *Med. Phys.* 32 (4): 890–901.

Lu, W., P. J. Parikh, J. P. Hubenschmidt, J. D. Bradley, and D. A. Low. 2006. A comparison between amplitude sorting and phase-angle sorting using external respiratory measurement for 4D CT. *Med. Phys.* 33 (8): 2964–2974.

Mageras, G. S., and E. Yorke. 2004. Deep inspiration breath hold and respiratory gating strategies for reducing organ motion in radiation treatment. *Semin. Radiat. Oncol.* 14 (1): 65–75.

Mageras, G. S., E. Yorke, K. Rosenzweig, L. Braban, E. Keatley, E. Ford, S. A. Leibel, and C. C. Ling. 2001. Fluoroscopic evaluation of diaphragmatic motion reduction with a respiratory gated radiotherapy system. *J. Appl. Clin. Med. Phys.* 2 (4): 191–200.

Mah, D., J. Hanley, K. E. Rosenzweig, E. Yorke, L. Braban, C. C. Ling, S. A. Leibel, and G. Mageras. 2000. Technical aspects of the deep inspiration breath-hold technique in the treatment of thoracic cancer. *Int. J. Radiat. Oncol. Biol. Phys.* 48 (4): 1175–1185.

Mansfield, P., and A. A. Maudsley. 1977. Medical imaging by NMR. *Br. J. Radiol.* 50 (591): 188–194.

Murphy, M. J. 1997. An automatic six-degree-of-freedom image registration algorithm for image-guided frameless stereotaxic radiosurgery. *Med. Phys.* 24 (6): 857–866.

Murphy, M. J. 2004. Tracking moving organs in real time. *Semin. Radiat. Oncol.* 14 (1): 91–100.

Murphy, M. J., J. R. Adler Jr., M. Bodduluri, J. Dooley, K. Forster, J. Hai, Q. Le, G. Luxton, D. Martin, and J. Poen. 2000. Image-guided radiosurgery for the spine and pancreas. *Comput. Aided Surg.* 5 (4): 278–288.

Murphy, M. J., D. Martin, R. Whyte, J. Hai, C. Ozhasoglu, and Q. T. Le. 2002. The effectiveness of breath-holding to stabilize lung and pancreas tumors during radiosurgery. *Int. J. Radiat. Oncol. Biol. Phys.* 53 (2): 475–482.

Nagel, C. C., G. Bosmans, A. L. Dekker, M. C. Ollers, D. K. De Ruysscher, P. Lambin, A. W. Minken, N. Lang, and K. P. Schafers. 2006. Phased attenuation correction in respiration correlated computed tomography/positron emitted tomography. *Med. Phys.* 33 (6): 1840–1847.

NCI. 2001. Intensity-modulated radiotherapy: Current status and issues of interest. *Int. J. Radiat. Oncol. Biol. Phys.* 51 (4): 880–914.

Nehmeh, S. A., Y. E. Erdi, C. C. Ling, K. E. Rosenzweig, H. Schoder, S. M. Larson, H. A. Macapinlac, O. D. Squire, and J. L. Humm. 2002. Effect of respiratory gating on quantifying PET images of lung cancer. *J. Nucl. Med.* 43 (7): 876–881.

Ohara, K., T. Okumura, M. Akisada, T. Inada, T. Mori, H. Yokota, and M. J. Calaguas. 1989. Irradiation synchronized with respiration gate. *Int. J. Radiat. Oncol. Biol. Phys.* 17 (4): 853–857.

International Commission on Radiation Units and Measurements (ICRU). 1999. *Prescribing, Recording and Reporting Photon Beam Therapy, Report 62.* Bethesda, MD: ICRU.

Rietzel, E., T. Pan, and G. T. Chen. 2005. Four-dimensional computed tomography: Image formation and clinical protocol. *Med. Phys.* 32 (4): 874–889.

Schweikard, A., G. Glosser, M. Bodduluri, M. J. Murphy, and J. R. Adler. 2000. Robotic motion compensation for respiratory movement during radiosurgery. *Comput. Aided Surg.* 5 (4): 263–277.

Shirato, H., S. Shimizu, K. Kitamura, T. Nishioka, K. Kagei, S. Hashimoto, H. Aoyama et al. 2000. Four-dimensional treatment planning and fluoroscopic real-time tumor tracking radiotherapy for moving tumor. *Int. J. Radiat. Oncol. Biol. Phys.* 48 (2): 435–442.

Shirato, H., S. Shimizu, T. Kunieda, K. Kitamura, M. van Herk, K. Kagei, T. Nishioka et al. 2000. Physical aspects of a real-time tumor-tracking system for gated radiotherapy. *Int. J. Radiat. Oncol. Biol. Phys.* 48 (4): 1187–1195.

Sonke, J. J., L. Zijp, P. Remeijer, and M. van Herk. 2005. Respiratory correlated cone beam CT. *Med. Phys.* 32 (4): 1176–1186.

Steenbakkers, R. J., J. C. Duppen, I. Fitton, K. E. Deurloo, L. Zijp, A. L. Uitterhoeve, P. T. Rodrigus et al. 2005. Observer varia-tion in target volume delineation of lung cancer related to radiation oncologist–computer interaction: A 'Big Brother' evaluation. *Radiother. Oncol.* 77 (2): 182–190.

Stroom, J., H. Blaauwgeers, A. van Baardwijk, L. Boersma, J. Lebesque, J. Theuws, R. J. van Suylen et al. 2007. Feasibility of pathology-correlated lung imaging for accurate target definition of lung tumors. *Int. J. Radiat. Oncol. Biol. Phys.* 69 (1): 267–275.

Underberg, R. W., F. J. Lagerwaard, B. J. Slotman, J. P. Cuijpers, and S. Senan. 2005. Use of maximum intensity projections (MIP) for target volume generation in 4DCT scans for lung cancer. *Int. J. Radiat. Oncol. Biol. Phys.* 63 (1): 253–260.

Van de Steene, J., N. Linthout, J. de Mey, V. Vinh-Hung, C. Claassens, M. Noppen, A. Bel, and G. Storme. 2002. Definition of gross tumor volume in lung cancer: Inter-observer variability. *Radiother. Oncol.* 62 (1): 37–49.

Van de Steene, J., F. Van den Heuvel, A. Bel, D. Verellen, J. De Mey, M. Noppen, M. De Beukeleer, and G. Storme. 1998. Electronic portal imaging with on-line correction of setup error in thoracic irradiation: Clinical evaluation. *Int. J. Radiat. Oncol. Biol. Phys.* 40 (4): 967–976.

Vedam, S. S., P. J. Keall, V. R. Kini, and R. Mohan. 2001. Determining parameters for respiration-gated radiotherapy. *Med. Phys.* 28 (10): 2139–2146.

Vedam, S. S., P. J. Keall, V. R. Kini, H. Mostafavi, H. P. Shukla, and R. Mohan. 2003. Acquiring a four-dimensional computed tomography dataset using an external respiratory signal. *Phys. Med. Biol.* 48 (1): 45–62.

Verellen, D., M. De Ridder, N. Linthout, K. Tournel, M. Duchateau, T. Gevaert, T. Reynders, T. Depuydt, and G. Storme. 2009. Impact of the interplay between advances in imaging and radiotherapy on clinical care. *Imaging Med.* 1 (2): 12.

Verellen, D., M. D. Ridder, N. Linthout, K. Tournel, G. Soete, and G. Storme. 2007. Innovations in image-guided radiotherapy. *Nat. Rev. Cancer* 7 (12): 949–960.

Verellen, D., K. Tournel, N. Linthout, G. Soete, T. Wauters, and G. Storme. 2006. Importing measured field fluences into the treatment planning system to validate a breathing synchronized DMLC-IMRT irradiation technique. *Radiother. Oncol.* 78 (3): 332–338.

Webb, S. 1997. The physics of conformal radiotherapy. In *Advances in Technology*, ed. R. F. Mould, C. G. Orton, J. A. E. Spaan, and J. G. Webster. London: Institute of Physics Publishing.

Webb, S. 2001. Intensity-modulated radiation therapy. In *Series in Medical Physics*, ed. by C. G. Orton, J. A. E. Spaan, and J. G. Webster. London: Institute of Physics Publishing.

Wilson, E. M., F. J. Williams, B. E. Lyn, J. W. Wong, and E. G. Aird. 2003. Validation of active breathing control in patients with non-small-cell lung cancer to be treated with CHARTWEL. *Int. J. Radiat. Oncol. Biol. Phys.* 57 (3): 864–874.

Wolthaus, J. W., C. Schneider, J. J. Sonke, M. van Herk, J. S. Belderbos, M. M. Rossi, J. V. Lebesque, and E. M. Damen. 2006. Mid-ventilation CT scan construction from four-dimensional respiration-correlated CT scans for radiotherapy planning of lung cancer patients. *Int. J. Radiat. Oncol. Biol. Phys.* 65 (5): 1560–1571.

Wolthaus, J. W., M. van Herk, S. H. Muller, J. S. Belderbos, J. V. Lebesque, J. A. de Bois, M. M. Rossi, and E. M. Damen. 2005. Fusion of respiration-correlated PET and CT scans: Correlated lung tumour motion in anatomical and functional scans. *Phys. Med. Biol.* 50 (7): 1569–1583.

Wong, J. W., M. B. Sharpe, D. A. Jaffray, V. R. Kini, J. M. Robertson, J. S. Stromberg, and A. A. Martinez. 1999. The use of active breathing control (ABC) to reduce margin for breathing motion. *Int. J. Radiat. Oncol. Biol. Phys.* 44 (4): 911–919.

Wulf, J., U. Hadinger, U. Oppitz, B. Olshausen, and M. Flentje. 2000. Stereotactic radiotherapy of extracranial targets: CT-simulation and accuracy of treatment in the stereotactic body frame. *Radiother. Oncol.* 57 (2): 225–236.

Wuthrich, K., R. G. Shulman, and J. Peisach. 1968. High-resolution proton magnetic resonance spectra of sperm whale cyanometmyoglobin. *Proc. Natl. Acad. Sci. U. S. A.* 60 (2): 373–380.

Xing, L., C. Cotrutz, S. Hunjan, A. L. Boyer, E. Adalsteinsson, and D. Spielman. 2002. Inverse planning for functional image-guided intensity-modulated radiation therapy. *Phys. Med. Biol.* 47 (20): 3567–3578.

Yang, Y., and L. Xing. 2005. Towards biologically conformal radiation therapy (BCRT): selective IMRT dose escalation under the guidance of spatial biology distribution. *Med. Phys.* 32 (6): 1473–1484.

Zelefsky, M. J., S. A. Leibel, P. B. Gaudin, G. J. Kutcher, N. E. Fleshner, E. S. Venkatramen, V. E. Reuter, W. R. Fair, C. C. Ling, and Z. Fuks. 1998. Dose escalation with three-dimensional conformal radiation therapy affects the outcome in prostate cancer. *Int. J. Radiat. Oncol. Biol. Phys.* 41 (3): 491–500.

Zelefsky, M. J., S. A. Leibel, G. J. Kutcher, and Z. Fuks. 1998. Three-dimensional conformal radiotherapy and dose escalation: Where do we stand? *Semin. Radiat. Oncol.* 8 (2): 107–114.

Treatment Planning Systems

Swamidas V. Jamema
Tata Memorial Hospital

Introduction

A treatment planning system (TPS) is a crucial component of the radiotherapy treatment planning process, especially with the advent of image-based three-dimensional conformal radiotherapy (3DCRT) and intensity modulated radiotherapy (IMRT) in recent years. Current generations of TPSs are based on complex algorithms for volume delineation, dose calculation, and plan evaluation. A high-level of confidence must be ensured at the time of implementation of a TPS for clinical use. Errors of different kinds, such as inaccuracies in algorithms or incorrect input of parameters, could produce dose results not accurate enough to be used in clinical practice. Hence, TPS quality assurance (QA) is essential to ensure accurate dose delivery and to minimize the possibility of treatment errors. In the past, lack of proper TPS QA procedures has led to some serious accidents. An ICRP report states that lack of understanding of the TPS and lack of appropriate commissioning are the major contributory factors for the accidents associated with TPSs (ICRP 2000). Unlike treatment delivery errors, which are usually random in nature, the errors from the TPS are more often systematic and can be avoided if the system is properly commissioned and followed by the implementation of a periodic QA program.

This chapter is written with the following three objectives: (1) to provide a literature survey of the reports published regarding commissioning and QA of a TPS; (2) to describe a brief summary of the guidelines for commissioning and QA of TPS (the procedure presented here is based on Technical Report Series 430 from a recent publication from the International Atomic Energy Agency (IAEA) (Van Dyk et al. 2004); and (3) to address the issues that will be important for QA and commissioning of TPS in the future.

Literature Review

Reports by McCullough and Krueger (1980), Dahlin et al. (1983), and the International Commission Radiation Unit's *Report 42* (1987) were the early documents describing QA of a TPS. Later, a report published by Van Dyk et al. (1993) was the first with a detailed description of QA and commissioning of TPS. In 1996, the Institute of Physics and Engineering in Medicine and Biology (1996) published a report that was updated in 1999. In 1997, the Swiss Society for Radiobiology and Medical Physics (1997) published recommendations for the quality control of treatment planning systems for teletherapy. In 1998, the American Association of Physicists in Medicine (AAPM) published *Task Group Report 53*, which is a comprehensive document comprising acceptance testing and commissioning of a TPS followed by *Task Group Report 65* in 2004—a detailed description of dose calculation algorithms for inhomogeneity corrections (Fraass et al. 1998; Papanikolaou et al. 2004). The International Electro Technical Commission specifications and safety requirements were published in a report (62083) that was specifically aimed at manufacturers for acceptance test procedures (IEC 2000).

In 2004, both the IAEA and the European Society of Therapeutic Radiation and Oncology (ESTRO) published reports on the commissioning and QA of TPS (Van Dyk et al. 2004; Mijnheer et al. 2004). In 2006, the Netherlands Commission of Radiation Dosimetry also produced a report on QA of TPSs, which consisted of practical examples of the specific QA tests to be carried out (Bruinvis et al. 2006). In 2007, the IAEA published a protocol for the specification and acceptance testing of a TPS. This report is unique compared to the previously mentioned reports in that part of the tests addresses the vendor of the TPS. It requires vendors to perform a series of type tests, a subset of which is performed at the user level to ensure that the software complies with the standards. A new report is under development from the IAEA that will address the users of simple TPSs, particularly in developing countries (Ibbott et al. 2007). Another report, AAPM's *Task Group Report 67* (Bayouth et al. 2005), describes the development of a series of benchmark tests for the validation of dose calculation algorithms of TPSs (Van Dyk 2008).

Commissioning and QA of TPS

A recent document from the IAEA, *TRS 430*, explicitly lists all the tests to be carried out for a comprehensive commissioning and QA program. The implementation of the QA program can be classified into acceptance tests, commissioning, periodic QA, and patient-specific QA. Acceptance tests verify the functionality of the system hardware and software components as well as its agreement with the specifications supplied by the manufacturer. Commissioning consists of dosimetric and nondosimetric tests. It involves the characterization of treatment machines in the TPS and tests the functionality of the dose calculation algorithm. Periodic QA checks the reproducibility of planning, in accordance with the commissioning, while patient-specific QA checks the treatment process as a whole. The commissioning process can broadly be classified into the following categories: patient data acquisition, anatomical modeling, dose calculation and clinical test cases, plan evaluation, hard copy output, and plan transfer. A brief description of the above parameters is encompassed in the following section.

Nondosimetric QA

A large number of reports and publications are available in the literature about the QA of TPS and the verification of dosimetric calculations (Bedford et al. 2003; Starkschall et al. 2000). However, little interest has been shown by the investigators about the nondosimetric QA aspects. Nondosimetric commissioning is as complex as dosimetric commissioning in a modern 3D TPS. *TRS 430* presents a detailed list of nondosimetric tests to be performed. They may be quite elaborate to perform and include several crucial parameters as follows: image acquisition and transfer, beam display, CT image reconstructions, multiplanar CT image reconstructions, digitally reconstructed radiographs, anatomical definition, automatic tools, auto contouring, auto margin, 3-D display, and dose volume histograms.

Patient Data Acquisition

CT scans play a crucial role in radiotherapy planning as they contain information regarding relative electron densities in the form of CT numbers. To achieve accurate dose calculations, it is essential to define a valid CT calibration curve for the CT scanner used for image acquisition. An increasing number of facilities with satellite centers, where planning is carried out on images from a different CT scanner, are evolving. Hence, if multiple CT scanners are used with the TPS, specific calibration curves need

to be configured for each scanner. Multimodality imaging such as PET-CT, MRI, and SPECT are being increasingly used and these warrant a stringent QA so that the images are accurately transferred to the TPS (Grosu et al. 2000). Parameters related to registration, resolution, and geometric distortion need special attention. The accuracy of reconstructed images has to be checked for various parameters such as scale, orientation, labeling, coordinates, and grey levels. There are several other parameters such as uniqueness of the patient, imaging protocol, orientation, geometric accuracy, and image distortion that need to be tested.

Anatomical Modeling

The current generation of TPSs uses highly sophisticated algorithms for automatic contouring, 3D object description, expansion of structures, Boolean operations, digitally reconstructed radiographs, coregistration/fusion of multimodality image series, and the coordinate system. The tests for anatomical descriptions are dependent on the TPS and the clinical practice, hence most of the reports concentrate on the issues that should be considered and why they are important, rather than describing the specific tests. However, the Netherlands Commission on Radiation Dosimetry and ESTRO reports provide practical examples of specific tests to be carried out for anatomical modeling (Bruinvis et al. 2006; Mijnheer et al. 2004). The onus lies on the user to develop a QA program suitable for TPS, taking into consideration both the capabilities and the limitations of the various algorithms used for anatomical modeling. Very few investigators have reported the results of nondosimetric tests (Wang et al. 2006; Jamema et al. 2008). It was reported that object expansion was found to be distorted for structures with sharp corners. Discrepancies were also found with the use of the capping option. The use of multiple image data sets for image registration/fusion is still a challenge, and is not addressed by any report explicitly. It was recommended by AAPM Task Group 53 that another task group be formed specifically to develop a report on quality assurance of registration techniques. For guidance, the relevant literature can be used for setting up a QA process for image registration/fusion (Petti et al. 1994).

Plan Evaluation

Dose volume histograms (DVH) are an important part of a modern TPS, as the outcome/toxicities of all radiotherapy treatments/trials may be directly related to the dose volume parameters. It is important for the user to understand the algorithms of the DVH and to perform the tests accordingly. An ESTRO booklet has given explicitly the tests to be performed to verify the accuracy of DVH calculation (Mijnheer et al. 2004). The following parameters need careful investigation: self-consistency of various types of DVHs, normalization, relative and absolute dose volume, bin size, DVHs of compound structures, resolution of a DVH with respect to sampling, and the dose volume statistics. It is also recommended to verify the accuracy of the DVH in both high- and low-dose gradient regions.

The current generation of TPSs provides estimates of the biological effects of a dose distribution based on calculations with normal tissue complication probability (NTCP) and tumor control probability (TCP) models. It is essential to verify that the calculated NTCP and TCP values from the TPS are in agreement with the model they are based on taking into consideration the uncertainties of these models.

Dose Calculation

Verification of accurate dose calculation is a primary task during the commissioning process. The objective of any commissioning process is to identify and minimize errors and uncertainties. With the available resources in the clinics it is almost impossible to carry out complete verification of the accuracy of dose calculation algorithms. It can, however, be established once through a benchmark process. In a yet to be published report, AAPM TG-67 will present a dataset that will include benchmark data for various accelerators and beam energies. However, it is mandatory to verify that the algorithms have been implemented in the TPS correctly. AAPM Task Group 23 (AAPM 1995) provides a set of verification measurements. Later, Venselaar and Welleweerd (2001) proposed a test package to include the new possibilities offered by modern radiation therapy equipment. IAEA *TRS 430, TG 53,* and other reports propose an extensive list of tests to be carried out.

To begin with, a commissioning plan should be created taking into consideration the type of algorithm to be tested and other treatment conditions that may be encountered in the clinic. The choice of detectors, radiation field analyzer and its functionality, accuracy of measurements for various situations, processing, and documentation of the data are the direct responsibilities of the physicist. In addition to dose calculation algorithm checks, parameters that describe the machine configuration such as beam parameters, geometry, field size, accessories, display, and normalization need to be verified. During commissioning, it is important to understand the principles of the model, its functionality, and limitations, as the modern TPSs use sophisticated and complex algorithms. The comparisons between measurement and calculation should be performed and analyzed applying well-defined criteria. A recommended set of criteria is discussed in detail elsewhere (Venselaar, Welleweerd, and Mijnheer 2001). The following tests are suggested to verify the functionality and the accuracy of dose calculation algorithms: square and rectangular fields, SSD dependence, large and small field sizes, beam modifiers, oblique incidence, inhomogeneity correction, buildup region behavior, asymmetrical collimator settings, MLC shaped fields, multiple beam isocenter, field matching, missing tissue and dose compensation, DRR, and display of portal images.

It is worth mentioning here that a pilot study of the IAEA protocol *Development of Procedures for QA for Dosimetry Calculation in Radiotherapy* consisted of clinical test cases for the commissioning process. The project covered 11 hospitals with 15 different TPSs, encompassing altogether 37 different combinations of algorithms and beam qualities. The results have demonstrated that proposed test cases, such as oblique incidence, irregular field, and blocking the field center not only help to assure the safe use of the TPS, but may also help to understand the limitations of the algorithm. Camargo et al. (2007) have found that the confidence limits obtained for the dosimetric tests were much smaller than acceptability criteria with simple configurations. However, larger difference were found for tests cases such as oblique incidence, oblique incidence off-axis, symmetric open fields, shaped fields, off-axis wedged fields. Hence, it has been suggested to reduce the number of dosimetric tests to five where large deviations were observed. Other investigators have reported deviations in the buildup region behavior (Jamema et al. 2008) and inhomogeneity calculation (Kappas and Rosenwald 1995). Hence a consensus has to be reached based on experience gained to reduce the number of tests without compromising the quality of the QA process.

The verification of inhomogeneity corrections has to be dealt with carefully because these are difficult to measure accurately. Ample literature is available, on various types of algorithms, techniques of measurement, and interpretation of data (Van Esch et al. 2006). Typically, doses within the inhomogeneity and at the interfaces have to be measured accurately. Many TPSs do not support large fields at all due the limitation on the maximum field size and SSD. If it is used in the clinics, it must be commissioned appropriately in the TPS. Relevant literature in this regard is available for reference (Podgorsak and Podgorsak 1999).

The final step in the dose calculation process is the verification of MU. Generally, the MU calculation is related to the calculation of the relative dose distribution, normalization, and relative weights of the beams. It is important to understand precisely the relationship between the MUs and the calculated dose distribution. Verification of MUs should be carried out for various clinical test cases such as open, tangential, wedged, blocked, MLC shaped, central axis blocked, and any other conditions to be encountered in the clinic.

TPS QA Issues in the Future

IMRT QA

QA of IMRT treatment planning is very complex as it involves inverse optimization and the dose distribution often has large dose gradients. Comprehensive additional commissioning and QA procedures are required to validate inverse planned IMRT. None of the publications on QA of TPS addressed the QA issues adequately for IMRT planning. However, general guidelines regarding QA of IMRT have been formulated by other investigators (Ezzell et al. 2003).

TPS predicted doses in IMRT cannot be checked by measurements or manual calculations using the same methodology as for forward planning. Hence, the accuracy of TPS calculations in IMRT dose plans has to be ensured on a patient-specific basis by measuring the absorbed dose of the plan using an ion chamber, often in a homogeneous phantom or using a 2D array, film, or gel dosimeter. Various software programs have been developed for the independent verification of monitor units (Georg et al. 2007). A question to be addressed here is that, while the TPS uses sophisticated algorithms as compared to independent software, what should be the acceptability criteria between the TPS calculated MU and that of the check software calculated MU. A reasonable amount of accuracy can be obtained if the independent MU verification software takes into consideration all the parameters that influence the dose, for example, scatter, transmission through the collimator and MLC, rounded leaf, tongue and groove effects, penumbra of beamlets, etc. Otherwise, large discrepancies are likely to result between the original monitor unit calculations done by the TPS and those done by the independent software. Efforts are ongoing to improve the accuracy level of independent software. In a recent publication, Azcona and Burguete (2008) describe an independent method based on a deconvolution technique for accurate modeling of pencil beam kernels that lead to accurate dose calculations in IMRT conditions.

Another issue with patient-specific IMRT QA is that a significant amount of machine time is being used for IMRT QA. The question that arises here is if independent calculation checks are consistent and reliable, can we discontinue with patient-specific QA? In a recent publication, Pawlicki et al. (2008) have investigated IMRT QA using statistical process control method to compare the processes of patient-specific measurements and the corresponding independent computer calculations. Based on their findings, it has been concluded that it may be appropriate to suspend patient-specific IMRT QA measurements for every patient in the place of independent computer calculations. However, the issue is still open for discussion and needs a careful evaluation until a national and international consensus guideline is developed.

Optimization with Inclusion of Motion Uncertainties

Another interesting development in IMRT optimization is the inclusion of intra and inter-fraction motion and other uncertainties in the optimization process. Complex probabilistic approaches are adopted to account for motion and various types of uncertainties in the optimization routines. Following this approach, the dose distribution depends on a set of random variables that parameterize the uncertainty, as does the objective function used to optimize the treatment plan (Unkelbach et al. 2009). The question that arises here is: Are the quality assurance procedures for such optimization routines that need to be developed as complex and challenging as the algorithm itself or could the approach of patient-specific QA remain appropriate as the conventional IMRT verification procedure?

Biological Optimization

Emerging medical imaging techniques with more information on structural, biological, and chemical variations within tumors have opened up ample research opportunities in developing optimal nonuniform biological dose distributions (Yang and Xing 2005). The approach is a mathematical model taking into account the radio-sensitivity of the tumor and the response of the tumor to irradiation. The challenge is to develop a QA program, before clinical implementation, considering the fact that these models are approximate and are based on radio-sensitivity parameters with large uncertainties with limited clinical evidence.

4D Dose Calculations

Estimation of the dose delivered during respiration with the use of 4D CT data is being evaluated by many researchers (Rietzel et al. 2005). The issue of 4D dose estimation involves the handling of an extensive amount of data, which is not always feasible in the clinic. To simplify the dose calculation process, various attempts such as fewer breathing phases, midventilation, and averaged 4DCT have been tried (Wolthaus et al. 2006). The challenge is that the estimation of dose be straightforward and robust, without compromising quality, and in less time. Similar approaches would be required for the development of the QA program of 4D dose estimation.

Adaptive Radiotherapy (ART)

The reoptimization of a dose distribution on images acquired on a daily basis for image guidance (ART) will soon become a routine practice in clinics (Yoo and Yin 2006). ART will play a vital role in reducing the margins and, hence, improving the quality of life for the patients. It is extremely important to implement a QA protocol to ensure that reoptimization of dose distributions are accurate. It is even more challenging to develop a QA protocol for ART as it is practiced in real-time.

Monte Carlo–Based TPS

The development of Monte Carlo (MC) codes optimized for radiotherapy calculations and the availability of faster computers will soon make MC-based calculations widely available in the clinic. MC dose calculation algorithms are known to be very accurate when used properly, especially in heterogeneous patient tissues. However, MC should be treated as any conventional dose algorithm during clinical commissioning (Chetty et al. 2007). AAPM TG 105 gives a detailed review of MC algorithms that includes the issues associated with the clinical implementation of MC-based TPS and specific quality assurance tests for experimental verification. The challenge is to develop the commissioning and QA procedures for MC-based TPSs, considering the measurement uncertainties involved in various measurement devices as they are used in different situations. Another difficulty with MC algorithms is how to handle the statistical uncertainty in the results, which is not part of deterministic models.

Administrative Issues

The design of an appropriate QA program depends strongly on the TPS's capabilities, limitations, and its clinical use. Although there are standard protocols available for the QA of a TPS, it's impossible to provide step-by-step guidance or tests to be carried for the implementation of a QA program specific to various TPSs with their wide range of applications. TPS QA procedures must be established based on each institution's own special situation. The increasing use of a modern TPS will require all physicists to begin educating themselves as well as the management and staff at their institution. It is important that an appropriately qualified medical physicist should be given the responsibility for the commissioning and implementation of a QA program to ensure its ongoing validity. There should be greater cooperation and understanding between the physicist, vendor, and imaging physicist to better address the modern TPS QA issues.

One of the challenges in QA of a TPS is to design a QA program such that it does not impose an unrealistic commitment of time and resources in a busy clinical center. The complexity of the current generation of TPSs has increased the commissioning and QA process dramatically. A clear understanding of each parameter to be tested, and possible errors caused by them, and its clinical impact has to be analyzed critically. There is a need to prioritize the QA process based on the potential failure of a parameter and the consequences of such a failure on the quality of patient care. A recent publication by Huq et al. (2008) provides guidelines on risk assessment approaches with emphasis on failure modes and effects analysis (FMEA) and an achievable QA program based on risk analysis. Task Group 100 of the AAPM is developing a framework for designing QM activities and, hence, allocating resources based on estimates of clinical outcome, risk assessment, and failure modes. Government and private entities committed to improved healthcare quality and safety should support research directed toward addressing QA of TPSs.

Summary

The complexities of the current generation of TPSs warrant a stringent QA and commissioning program. QA for a TPS that is individualized and institution-specific must be designed by the physicist based on the guidelines laid out in national and international documents. Currently, most of the approaches for QA and commissioning of a TPS increase the time and effort required by physicists, and routine QA testing is still quite time-intensive. Considerable research is needed to reduce the time required for QA processes; benchmarking measurements is one example in that process. The current approach of parameter- and device-centered QA may be complemented by process-centered approaches while designing a QA program. With the rapid technological advancements in radiation therapy, update of the QA of a TPS is the need of the hour to ensure optimization, effectiveness, and safe practice of these advanced techniques.

References

AAPM. 1995. Report 55: *Radiation Treatment Planning Dosimetry Verification, Radiation Therapy Committee Task Group #23*, ed. D. Miller. College Park, MD: American Institute of Physics.

Azcona, J. D., and J. Burguete. 2008. A system for intensity modulated dose plan verification based on an experimental pencil beam kernel obtained by deconvolution. *Med. Phys.* 35 (1): 248–259.

Bayouth J., D. Followill, B. Fraass et al. 2005. AAPM Radiation Therapy Committee Task Group 67: Benchmark datasets for photon beams. Unpublished data.

Bedford, J. L., P. J. Childs, V. Nordmark Hansen, M. A. Mosleh-Shirazi, F. Verhaegen, and A. P. Warrington. 2003. Commissioning and quality assurance of the Pinnacle(3) radiotherapy treatment planning system for external beam photons. *Br. J. Radiol.* 76 (903): 163–176.

Bruinvis, I. A. D, R. B. Keus, W. J. M. Lenglet, G. I. Meijer, B. J. Mijnheer, A. A. van Veld, J. L. M. Venselaar, J. Weleweerd, and E. Woudstra. 2006. *Quality Assurance of 3-D Treatment Planning Systems for External Photon and Electron Beams: Report 15 of The Netherlands Commission on Radiation Dosimetry.* Delft, The Netherlands: The Netherlands Commission on Radiation Dosimetry.

Camargo, P. R., L. N. Rodrigues, L. Furnari, and R. A. Rubo. 2007. Implementation of a quality assurance program for computerized treatment planning systems. *Med. Phys.* 34 (7): 2827–2836.

Chetty, I. J., B. Curran, J. E. Cygler, J. J. DeMarco, G. Ezzell, B. A. Faddegon, I. Kawrakow et al. 2007. Report of the AAPM Task Group No. 105: Issues associated with clinical implementation of Monte Carlo-based photon and electron external beam treatment planning. *Med. Phys.* 34 (12): 4818–4853.

Dahlin, H., I. L. Lamm, T. Landberg, S. Levernes, and N. Ulso. 1983. User requirements on CT-based computed dose planning systems in radiation therapy. *Acta Radiol. Oncol.* 22 (5): 397–415.

Ezzell, G. A., J. M. Galvin, D. Low, J. R. Palta, I. Rosen, M. B. Sharpe, P. Xia, Y. Xiao, L. Xing, and C. X. Yu. 2003. Guidance document on delivery, treatment planning, and clinical implementation of IMRT: Report of the IMRT Subcommittee of the AAPM Radiation Therapy Committee. *Med. Phys.* 30 (8): 2089–2115.

Fraass, B., K. Doppke, M. Hunt, G. Kutcher, G. Starkschall, R. Stern, and J. Van Dyke. 1998. American Association of Physicists in Medicine Radiation Therapy Committee Task Group 53: Quality assurance for clinical radiotherapy treatment planning. *Med. Phys.* 25 (10): 1773–1829.

Georg, D., M. Stock, B. Kroupa, J. Olofsson, T. Nyholm, A. Ahnesjo, and M. Karlsson. 2007. Patient-specific IMRT verification using independent fluence-based dose calculation software: Experimental benchmarking and initial clinical experience. *Phys. Med. Biol.* 52 (16): 4981–4992.

Grosu, A. L., W. Weber, H. J. Feldmann, B. Wuttke, P. Bartenstein, M. W. Gross, C. Lumenta, M. Schwaiger, and M. Molls. 2000. First experience with I-123-alpha-methyl-tyrosine spect in the 3-D radiation treatment planning of brain gliomas. *Int. J. Radiat. Oncol. Biol. Phys.* 47 (2): 517–526.

Huq, M. S., B. A. Fraass, P. B. Dunscombe, J. P. Gibbons Jr., G. S. Ibbott, P. M. Medin et al. 2008. A method for evaluating quality assurance needs in radiation therapy. *Int. J. Radiat. Oncol. Biol. Phys.* 71 (1 Suppl.): S170–S173.

Ibbott, G., J. Izewska, R. Schmidt, K. Shortt, J. Van Dyk, and S. Vatnitsky. 2007. Protocol for specification and acceptance testing of radiation treatment planning systems. In *IAEA TECDOC Series No. 1540*. Vienna: International Atomic Energy Agency.

International Commission on Radiological Protection (ICRP). 2000. *Prevention of Accidental Exposure to Patients Undergoing Radiation Therrapy*. Oxford: ICRP.

International Commission on Radiation Units and Measurements (ICRU). 1987. Use of computers in external beam radiotherapy procedures with high energy photons and electrons. In *ICRU Report 42*. Baltimore, MD: ICRU.

International Electrotechnical Commission (IEC). 2000. *Medical Electrical Equipment—Dose Area Product Meters (60580)*. Geneva: International Electrotechnical Commission.

Institute of Physics and Engineering in Medicine and Biology (IPEMB). 1996. A guide to commissioning and quality control of treatment planning systems. In *Report 68*. York, United Kingdom: IPEMB.

Jamema, S. V., R. R. Upreti, S. Sharma, and D. D. Deshpande. 2008. Commissioning and comprehensive quality assurance of commercial 3D treatment planning system using IAEA Technical Report Series-430. *Australas. Phys. Eng. Sci. Med.* 31 (3): 207–215.

Kappas, C., and J. C. Rosenwald. 1995. Quality control of inhomogeneity correction algorithms used in treatment planning systems. *Int. J. Radiat. Oncol. Biol. Phys.* 32 (3): 847–858.

McCullough, E. C., and A. M. Krueger. 1980. Performance evaluation of computerized treatment planning systems for radiotherapy: External photon beams. *Int. J. Radiat. Oncol. Biol. Phys.* 6 (11): 1599–1605.

Mijnheer, B., A. Olszewska, C. Fiorino, G. Hartmann, T. Knoos, J. Rosenwald, and H. Welleweerd. 2004. *Quality Assurance of Treatment Planning Systems: Practical Examples of Non-IMRT Photon Beams*. Brussels: European Society of Therapeutic Radiation Oncology (ESTRO).

Papanikolaou, N., J. Battista, A. Boyer, A. Kappas, E. Klein, T. Mackie, M. Sharpe, and J. Van Dyk. 2004. *AAPM Report 85: Tissue Inhomogeneity Corrections for Megavoltage Photon Beams. Report by Task Group 65 of the Radiation Therapy Committee of the American Association of Physicists in Medicine*. Madison, WI: Medical Physics Publishing.

Pawlicki, T., S. Yoo, L. E. Court, S. K. McMillan, R. K. Rice, J. D. Russell, J. M. Pacyniak et al. 2008. Moving from IMRT QA measurements toward independent computer calculations using control charts. *Radiother. Oncol.* 89 (3): 330–337.

Petti, P. L., M. L. Kessler, T. Fleming, and S. Pitluck. 1994. An automated image-registration technique based on multiple structure matching. *Med. Phys.* 21 (9): 1419–1426.

Podgorsak, E. B., and M. B. Podgorsak. 1999. Special techniques in radiotherapy. In *The Modern Technology of Radiation Oncology: A Compendium for Medical Physicists and Radiation Oncologists*, ed. J. Van Dyk. Madison, WI: Medical Physics Publishing.

Rietzel, E., G. T. Chen, N. C. Choi, and C. G. Willet. 2005. Four-dimensional image-based treatment planning: Target volume segmentation and dose calculation in the presence of respiratory motion. *Int. J. Radiat. Oncol. Biol. Phys.* 61 (5): 1535–1550.

SSRPM Report 7. 1997. *Quality Control of Treatment Planning Systems for Teletherapy*. Geneva: Swiss Society for Radiobiology and Medical Physics.

Starkschall, G., R. E. Steadham Jr., R. A. Popple, S. Ahmad, and I. I. Rosen. 2000. Beam-commissioning methodology for a three-dimensional convolution/superposition photon dose algorithm. *J. Appl. Clin. Med. Phys.* 1 (1): 8–27.

Unkelbach, J., T. Bortfeld, B. C. Martin, and M. Soukup. 2009. Reducing the sensitivity of IMPT treatment plans to setup errors and range uncertainties via probabilistic treatment planning. *Med. Phys.* 36 (1): 149–163.

Van Dyk, J. 2008. Quality assurance of radiation therapy planning systems: Current status and remaining challenges. *Int. J. Radiat. Oncol. Biol. Phys.* 71 (1 Suppl.): S23–S27.

Van Dyk, J., R. B. Barnett, J. E. Cygler, and P. C. Shragge. 1993. Commissioning and quality assurance of treatment planning computers. *Int. J. Radiat. Oncol. Biol. Phys.* 26 (2): 261–273.

Van Dyk, J., J. Rosenwald, B. Fraass, P. Andreo, J. Cramb, F. Ionescu-Farca, J. Izewska et al. 2004. *Commissioning and Quality Assurance of Computerized Planning Systems for Radiation Treatment of Cancer*. Vienna: International Atomic Energy Agency IAEA TRS-430.

Van Esch, A., L. Tillikainen, J. Pyykkonen, M. Tenhunen, H. Helminen, S. Siljamaki, J. Alakuijala, M. Paiusco, M. Lori, and D. P. Huyskens. 2006. Testing of the analytical anisotropic algorithm for photon dose calculation. *Med. Phys.* 33 (11): 4130–4148.

Venselaar, J., and H. Welleweerd. 2001. Application of a test package in an intercomparison of the photon dose calculation performance of treatment planning systems used in a clinical setting. *Radiother. Oncol.* 60 (2): 203–213.

Venselaar, J., H. Welleweerd, and B. Mijnheer. 2001. Tolerances for the accuracy of photon beam dose calculations of treatment planning systems. *Radiother. Oncol.* 60 (2): 191–201.

Wang, L., J. Li, K. Paskalev, P. Hoban, W. Luo, L. Chen, S. McNeeley, R. Price, and C. Ma. 2006. Commissioning and quality assurance of a commercial stereotactic treatment-planning system for extracranial IMRT. *J. Appl. Clin. Med. Phys.* 7 (1): 21–34.

Wolthaus, J. W., C. Schneider, J. J. Sonke, M. van Herk, J. S. Belderbos, M. M. Rossi, J. V. Lebesque, and E. M. Damen. 2006. Mid-ventilation CT scan construction from four-dimensional respiration-correlated CT scans for radiotherapy planning of lung cancer patients. *Int. J. Radiat. Oncol. Biol. Phys.* 65 (5): 1560–1571.

Yang, Y., and L. Xing. 2005. Towards biologically conformal radiation therapy (BCRT): selective IMRT dose escalation under the guidance of spatial biology distribution. *Med. Phys.* 32 (6): 1473–1484.

Yoo, S., and F. F. Yin. 2006. Dosimetric feasibility of cone-beam CT-based treatment planning compared to CT-based treatment planning. *Int. J. Radiat. Oncol. Biol. Phys.* 66 (5): 1553–1561.

External Beam Radiotherapy

Jatinder R. Palta
Davis Cancer Center

Introduction

External beam radiotherapy (EBRT) is reaching new pinnacles with continued advancement in treatment planning, accelerator and computer control technology, and treatment verification. These advances are attributed to newer innovations in areas that include volumetric imaging, optimized 3-D dose calculations and display, and computer-controlled linear accelerators (linacs) with image-guidance capabilities. As a result, there is a complete shift in the paradigm of the treatment delivery process. Historically, linear accelerators have been used to deliver radiation of uniform intensity through field apertures shaped by blocks. Now the emphasis is to shape the field apertures with a multileaf collimator system and vary the radiation intensities with dynamic motion of the collimator and gantry system to deliver conformal radiation to the target volume. Early investigators (Brahme 1988; Cormack 1987; Cormack and Cormack 1987) demonstrated that the use of nonuniform beam intensities to deliver dose distributions that conform to the target shape while sparing surrounding normal tissue optimally. Their work led to the development of intensity-modulated radiotherapy (IMRT), which is now widely used in clinical practice. The modern radiation-delivery systems capable of delivering IMRT are quite complex, with sophisticated computer controls that allow dynamic beam-delivery capabilities. These systems require an elaborate infrastructure to input complex treatment prescriptions and delivery parameters. It is therefore very important to pay attention to how a computer-controlled linear accelerator integrates with a facility-management system that includes networking of treatment planning, simulation, and treatment-verification equipment.

This chapter is focused primarily on identifying the key elements and issues that relate to commissioning and quality assurance (CQA) of IMRT, which is the most complex EBRT process. Readers interested in a more exhaustive review of delivery systems are referred to a landmark publication on this subject (Karzmark, Nunan, and Tanabe 1993). The American Association of Physicists in Medicine (AAPM) also has numerous task group reports that deal with quality assurance issues in various aspects of EBRT (report nos. 13, 24, 46, 47, 56, 62, 72, 82, 91, 95, 106, 119, and 142; http://aapm.org/pubs/reports/). Furthermore, International Electrotechnical Commission reports (60976 and 60977; http://webstore.iec.ch/webstore) are also very helpful in understanding performance guidelines and functional performance characteristics of medical electron accelerator relating to dynamic beam delivery techniques, image-guided radiotherapy, stereotactic radiotherapy/stereotactic radiosurgery, and the use of electronic imaging devices.

Current Paradigm of IMRT Commissioning and QA

The current paradigm for the commissioning and QA program for EBRT includes CQA of the treatment planning system, CQA of the delivery system, and patient-specific QA measurements. Although the treatment planning and delivery system is the same as that for both IMRT and conventional radiotherapy, IMRT has more parameters to coordinate and verify. Because of complex beam intensity modulation in IMRT, each treatment portal often includes many small, irregular, off-axis fields resulting in isodose distributions for each treatment plan that are more conformal than those from conventional radiotherapy. Therefore, these features impose a new and more stringent set of QA requirements for IMRT planning and delivery. The generic test procedures to validate dose calculation and delivery accuracy for both treatment planning and conformal delivery have to be customized for each type of IMRT planning and delivery strategy. The rationale for such an approach is that the overall accuracy of conformal delivery is dependent on the piecewise uncertainties in both the planning and delivery processes. The end user must have well-defined evaluation criteria for each element of the planning and delivery process. Such information can then potentially be used to determine a priori the accuracy of IMRT planning and delivery.

Process of IMRT Planning and Delivery

The process of IMRT from the CQA perspective in general includes patient immobilization, 3-D imaging, inverse planning, leaf sequencing, plan verification, patient setup verification, and treatment delivery (Purdy et. al. 2001; Ezzell et al. 2003; Galvin et al. 2004; Palta et al. 2004). IMRT requires more stringent tolerance limits for patient immobilization than 3-D conformal radiotherapy (3DCRT). This is because IMRT treatment delivery may take a longer time, thus increasing the potential for intrafraction patient motion. The computer optimization process in inverse planning relies on the accurate delineation of target volumes and critical structures and their spatial integrity relative to each other, and the definition of clinically appropriate planning objectives. The inverse planning algorithm determines optimal beam intensities that lead to the desired conformal dose distribution. However, these intensity distributions cannot be delivered directly by the IMRT delivery systems. They are first converted into an MLC leaf sequence. The leaf sequencing algorithms need to account for the mechanical limitations of the delivery system, beamlet size, leaf end leakage, leaf transmission, and leaf travel. Each one of these parameters has tolerance limits that impact the overall accuracy of the beam intensity delivery as planned. The nonintuitive nature of the beam intensity patterns makes it necessary to verify each IMRT treatment plan on a hybrid phantom. The sharp dose gradients in IMRT warrant much tighter tolerance limits in the verification of patient setup for treatment delivery. Finally, the accuracy of IMRT delivery as planned depends on the mechanical accuracy and integrity of the MLC system. IMRT places much greater mechanical demand on the MLC and can result in accelerated wear and tear of the system. Therefore, periodic QA test procedures with appropriate tolerance limits and action levels are crucial in the planning and delivery of IMRT. The most comprehensive guidance to date on IMRT QA, tolerance limits, and action levels for planning and delivery of IMRT can be found in the AAPM summer school proceedings (Palta and Mackie 2003). Finally, a recent International Atomic Energy Agency report (IAEA 2008) provides a comprehensive overview of required infrastructure and processes for the state-of-the-art EBRT.

IMRT Treatment Planning Systems

Task Group Reports from the AAPM (Kutcher et al. 1994; Fraass et al. 1998) recognized that the treatment-planning computer is a crucial component of the radiation treatment process. They discuss initial and ongoing testing, the importance of understanding the planning system documentation, and outline the treatment planning procedure and treatment planning QA for individual patients. All these issues are equally important in the IMRT treatment planning system QA. The specific test procedures and the tolerance limits described in these reports are not directly applicable to IMRT. IMRT treatment planning system QA includes but is not limited to the integrity of imaging data for planning, the calculation of relative dose distributions for modulated beams, heterogeneity corrections, leaf sequencing, and monitor unit (MU) calculation. Each institution in turn should follow the general principles of QA procedures and evaluation and must develop its own QA program that suits the local environment, ensures patient safety, and enables treatment planning and delivery accuracy. The amount of effort required in clinically implementing and maintaining an IMRT planning system must not be underestimated. IMRT treatment planning systems are also different in that the commissioning and QA process must include determining the effect of the input parameters on the optimized dose distribution. IMRT dose distributions are calculated by dividing each beam into smaller sections called *beamlets* that can have varying intensities. Because these beamlets are very small in size, typically 1×1 cm, a small error in the size of the beamlet can result in a large change in the radiation output (LoSasso, Chui, and Ling 2001). Accurate modeling of beam parameters such as transmission through the collimators, penumbra, and dose outside the field is more significant for IMRT. The accuracy with which the planned intensity distribution is reproduced by a delivery system depends on parameters such as collimator transmission, shape and size of the leaves, interleaf leakage, and mechanical limitations in the motion of the MLC. A recent AAPM report (Ezzell et al. 2009) has produced quantitative confidence limits as baseline expectation values for IMRT commissioning at 10 institutions using a suite of test cases. Data from this report can be used to establish criteria of acceptability of IMRT treatment planning systems.

IMRT Delivery Systems

The three most important characteristics of the MLC-based IMRT delivery system include mechanical integrity of the delivery system, precise spatial and temporal positioning of the MLC system, and radiation beam fidelity for small MUs. The mechanical demands on delivery system components, especially the MLC system, are more stringent for IMRT than 3DCRT. The wear and tear of the mechanical system is also accelerated by the same magnitude. Therefore, special QA tests are required in addition to what is necessary for 3DCRT using an MLC to ensure the delivery system continues to meet the functional performance specifications. The MLC performance characteristics requiring continuous monitoring include the following: the leaf position accuracy and reproducibility, the leaf gap width reproducibility, and the leaf speed accuracy.

Sharp dose gradients, which are typical of an IMRT delivery, mandate better mechanical accuracy of the delivery equipment to realize its full clinical potential. The tolerance limits for QA tests for an IMRT delivery system are described in a recent AAPM report (Klein, Hanley, and Bayouth 2009). IMRT delivery techniques have relatively small gaps between opposed leaves while the radiation is delivered at each gantry position. The radiation output for small gap widths is very sensitive to the size of the gap width, which changes the magnitude of the extrafocal radiation (LoSasso, Chui, and Ling 2001). In addition, leaves shield most regions most of the time during radiation delivery. Therefore, the delivered dose is very sensitive to the transmission through the leaves and the rounded leaf ends. Other factors impacting the accuracy of IMRT delivery with an MLC include leaf speed, dose rate of the linear accelerator, and the fidelity of the delivery control system for small MU at high-dose rates. The accuracy of dose output and beam stability for small MU cannot be overlooked in IMRT because a large fraction of the total MU for each IMRT field is delivered with field segments that have very small MU. IMRT beam delivery with small MU can be inaccurate, especially at higher dose rates (Li et al. 2003).

QA Requirements for Individual Patients

The QA for 3DCRT planning and delivery typically relies on the performance evaluation of individual parameters of the system only. It is not necessary to perform patient-specific QA except when a clinical situation warrants the monitoring of dose to a specific area of interest with in vivo dosimeters. This is always with an assumption that once the system is properly commissioned, the periodic QA checks of the subsystem will guarantee that all patients are treated with accuracy that is within the limits of established QA criteria. For IMRT, the traditional QA is not sufficient. It is very difficult to anticipate all likely problems in IMRT. There is little correlation between the MU and the delivered dose from each intensity-modulated field. Therefore, direct measurements are commonly made of a "hybrid plan," which is generated by applying the intensity-modulated field from a patient plan to a CT study of a geometric phantom. The computed dose distributions are then compared with the measured dose distributions with either a film or a diode array device. Often, an ion chamber is also used to measure the dose in a high-dose, low-gradient region in the phantom. One must recognize that the patient-specific QA is only a total system check. It does not tell anything about the accuracy with which the patient receives an IMRT treatment. The accuracy of the patient treatment is strongly dependent on the accuracy of patient positioning, internal organ motion, and the presence of heterogeneities.

Sources of Error in the EBRT Process

The EBRT process has multiple steps with a potential to incur small errors at each step along the way. Most of the treatment delivery errors occur for each treatment fraction and are classified as random errors. There are other errors such as organ motion that can occur both during imaging for treatment planning and during treatment delivery. The errors that occur only during treatment planning are classified as systematic errors. It should be noted that the systematic errors are the most significant in terms of undesirable clinical outcomes in EBRT. It is generally believed that the total error in clinical practice of EBRT has a normal distribution and that errors in each step can be added quadratically (van Herk et al. 2000). The total error can then be obtained by adding the standard deviations of each error in quadrature. It is quite obvious that the overall error is dominated by the error in a step with the largest magnitude. Therefore, every attempt should be made to reduce that error. Errors cannot be eliminated completely, and a method to accommodate the errors without compromising a positive clinical outcome is to select margins around the clinical target volumes and organs at risk judiciously. It is critically important that the tolerance limits and action levels for the QA test procedures for the different steps in the EBRT process preserve the rationale used in the selection of margins for each disease site.

The error analysis should be done for each disease site specifically because, as described earlier, the internal organ motion and target delineation uncertainties can vary from site to site and each one of these has a much greater impact on the overall uncertainty. One parameter that can impact the overall uncertainty the most is the target and critical structure delineation. Physician-to-physician variability in target delineation can also be significant. Having explicit delineation protocols, adequate training, and frequent consultations with the diagnostic imaging experts can reduce the uncertainties in target and critical structure delineation.

Achievable Accuracy and Goal of the EBRT CQA Process

The goal in any radiation therapy treatment is to deliver the prescribed radiation dose with less than 5% overall uncertainty. This requires the set tolerance limits for the EBRT planning and delivery system to be such that the overall uncertainty is within the acceptable limits. Determining the required and achievable accuracy for complex EBRT such as IMRT is very difficult because what is achievable on one IMRT system may not be so with another system. Therefore, the onus is on each institution to establish limits of accuracy of their IMRT process in their own local environment and for each disease site. In IMRT, individual intensity-modulated beams that vary in complexity from beam to beam create these high-dose gradients. It is sometimes difficult to have agreement at all points in a 3-D dose distribution. A disagreement between measured and calculated dose values at a few points does not necessarily lead to a negative overall result if other comparable points are well within the established tolerance limits. However, it is abundantly clear that functional performance specifications of the MLC system and its dosimetric characterization in the treatment planning system for IMRT need to be more stringent than for 3DCRT. Furthermore, the tolerance limit for the locus of gantry, collimator, and the table isocenter should be much tighter for the IMRT delivery systems than the delivery systems used for conventional radiation therapy. The tolerance limits on the output stability and beam symmetry for low-MU delivery should also be tighter for IMRT delivery than for conventional radiation therapy.

Interconnectivity and Interoperability of Devices for EBRT

The treatment planning and delivery parameters associated with 3DCRT and IMRT can be quite large. It is practically impossible to keep these records in paper form. The only viable possibility is to keep these records in electronic form. The proliferation of computer information systems in radiation oncology has made it easier to keep track of this large amount of patient-related data in an organized manner electronically in a facility management system. A very significant advantage of the facility management system is its ability to verify and automatically set treatment parameters before radiation delivery. This completely eliminates the possibility of wrong treatment delivery as long as the transcription/transfer of treatment parameters into the facility management system is correct. It is therefore very important that all components of the treatment-delivery system are fully integrated. The only way to have a seamless data exchange between two nodes is if both comply with standard data communication protocols. The electronic data from different sources in a radiation oncology facility should be IHE-RO compliant (http://www.ihe.net/Radiation_Oncology/).

Resource Requirements for EBRT CQA

To ensure that advanced EBRT services consistently meet the highest clinical standards, each institution must invest in a comprehensive CQA program for EBRT planning and delivery. Besides a large initial investment in the hardware and software, adequate qualified personnel resources are necessary for the initial commissioning and ongoing QA of complex EBRT systems. For example, current estimates of additional resources necessary for the implementation and maintenance of an IMRT program (40 IMRT patients of a total of 300 patients treated per year on a single machine) are 550 h, which includes the following: 100 additional hours for machine QA, 50 additional hours for treatment planning QA, 200 h for patient-specific QA, and 200 additional hours for IMRT treatment planning. It should also be noted that IMRT decreases daily throughput on the machine. The maximum machine workload is expected to go down from 32 to 27 patients per day (8-h shift). In addition, the machine uptime is expected to go down from 99% to 95% because of more wear and tear of delivery equipment hardware and complexity of the control software. Finally, the importance of investing resources in personnel training, development of policies and procedures, documentation of well-defined clinical workflow, and CQA protocols should not be underestimated.

Summary

The discipline of radiation oncology is changing rapidly with continued advancement in treatment planning, delivery, and treatment verification. The goal of these enhancements is to provide capabilities for more conformal radiation treatments and for potentially less morbidity. The complexity of EBRT almost mandates that the whole process be automated with appropriate safeguards and a comprehensive CQA program. It is not sufficient just to look at the dosimetric performance characteristics of the treatment planning and delivery system. Equal or more attention must be paid to the integration of all key components, staff training, developing policies and procedures, identifying potential failure modes, and developing strategies for error mitigation.

References

Brahme, A. 1988. Optimization of stationary and moving beam radiation therapy techniques. *Radiother. Oncol.* 12 (2): 129–140.

Cormack, A. M. 1987. A problem in rotation therapy with X rays. *Int. J. Radiat. Oncol. Biol. Phys.* 13 (4): 623–630.

Cormack, A. M., and R. A. Cormack. 1987. A problem in rotation therapy with X-rays: Dose distributions with an axis of symmetry. *Int. J. Radiat. Oncol. Biol. Phys.* 13 (12): 1921–1925.

Ezzell, G. A., J. W. Burmeister, N. Dogan, T. J. LoSasso, J. G. Mechalakos, D. Mihailidis, A. Molineu et al. 2009. IMRT commissioning: Multiple institution planning and dosimetry comparisons, a report from AAPM Task Group 119. *Med. Phys.* 36 (11): 5359–5373.

Ezzell, G. A., J. M. Galvin, D. Low, J. R. Palta, I. Rosen, M. B. Sharpe, P. Xia, Y. Xiao, L. Xing, and C. X. Yu. 2003. Guidance document on delivery, treatment planning, and clinical implementation of IMRT: Report of the IMRT Subcommittee of the AAPM Radiation Therapy Committee. *Med. Phys.* 30 (8): 2089–2115.

Fraass, B., K. Doppke, M. Hunt, G. Kutcher, G. Starkschall, R. Stern, and J. Van Dyke. 1998. American Association of Physicists in Medicine Radiation Therapy Committee Task Group 53: Quality assurance for clinical radiotherapy treatment planning. *Med. Phys.* 25 (10): 1773–1829.

Galvin, J. M., G. Ezzell, A. Eisbrauch, C. Yu, B. Butler, Y. Xiao, I. Rosen et al. 2004. Implementing IMRT in clinical practice: A joint document of the American Society for Therapeutic Radiology and Oncology and the American Association of Physicists in Medicine. *Int. J. Radiat. Oncol. Biol. Phys.* 58 (5): 1616–1634.

IAEA. 2008. Transitioning from 2-D Radiotherapy to 3-D conformal and intensity modulated radiotherapy, IAEA-TECDOC-1588. International Atomic Energy Agency.

Karzmark, C. J., C. S. Nunan, and E. Tanabe. 1993. *Medical Electron Accelerators*. New York, NY: McGraw-Hill, Inc., Health Professions Division.

Klein, E. E., J. Hanley, J. Bayouth, F. F. Yin, W. Simon, S. Dresser, C. Serrago et al. 2009. Task Group 142 report: Quality assurance of medical accelerators. *Med. Phys.* 36 (9): 4197–4212.

Kutcher, G. J., L. Coia, M. Gillin, W. F. Hanson, S. Leibel, R. J. Morton, J. R. Palta et al. 1994. Comprehensive QA for radiation oncology: Report of AAPM Radiation Therapy Committee Task Group 40. *Med. Phys.* 21 (4): 581–618.

Li, J. G., J. F. Dempsey, L. Ding, C. Liu, and J. R. Palta. 2003. Validation of dynamic MLC-controller log files using a two-dimensional diode array. *Med. Phys.* 30 (5): 799–805.

LoSasso, T., C. S. Chui, and C. C. Ling. 2001. Comprehensive quality assurance for the delivery of intensity modulated radiotherapy with a multileaf collimator used in the dynamic mode. *Med. Phys.* 28 (11): 2209–2219.

Palta, J. R., J. A. Deye, G. S. Ibbott, J. A. Purdy, and M. M. Urie. 2004. Credentialing of institutions for IMRT in clinical trials. *Int. J. Radiat. Oncol. Biol. Phys.* 59 (4): 1257–1259; author reply 1259–1261.

Palta, J. R., and T. R. Mackie, eds. 2003. *Intensity Modulated Radiation Therapy—The State of the Art.* Madison, WI: Medical Physics Publishing.

Purdy et al. 2001. Intensity Modulated Radiation Therapy Collaborative Working Group. Intensity-modulated radiotherapy: Current status and issues of interest. *Int. J. Radiat. Oncol. Biol. Phys.* 51: 880–914.

van Herk, M., P. Remeijer, C. Rasch, and J. V. Lebesque. 2000. The probability of correct target dosage: Dose-population histograms for deriving treatment margins in radiotherapy. *Int. J. Radiat. Oncol. Biol. Phys.* 47 (4): 1121–1135.

52

Linear Accelerator: Resource Analysis for Commissioning

Michael C. Schell
University of Rochester Medical Center

Robert J. Meiler
University of Rochester Medical Center

Introduction

The specific details of calibration and image acquisition, storage, and transfer to the treatment planning system are documented in the American Association of Physicists in Medicine (AAPM) task-group reports referenced in Table 52.1. Before detailing methods of error reduction, the selection process of the linear accelerator is fundamental to error reduction in treatment delivery. The following items are crucial to a system-successful linear accelerator deployment in radiation oncology. It is important not only to evaluate the linear accelerator and the properties that are desired but also to evaluate the ongoing support for the linear accelerator. The vendor must have reliable service contracts and representatives. The physicist should inquire as to whether the service representatives are well trained and competent and also whether they receive ongoing training on a regular basis. There are aspects of reliability in the linear accelerator selection. What are the flaws of the linear accelerator as it comes out of the factory before installation? Some vendors have a higher flaw rate that can be missed during the installation.

Another aspect of linear accelerator selection is the training for its use during or before installation. But even more crucial is the ability to obtain ongoing training as linear accelerator technology is upgraded and personnel turnover impacts the cohort of trained physicists in the facility. Simulation training (drill of procedures from start to completion) is the essential complement to continuing medical education.

In summary, we are looking at the aspects of a sustainable installation and the use of the linear accelerator over its lifetime. Is there support for ongoing training to offset the loss of skill with the passage of time? Are there funds to support the entire operation from installation through use year after year?

The Heinrich Triangle is pertinent to this endeavor. Heinrich studied approximately 500,000 accidents in the 1930s (Heinrich 1931). He discovered that only 10% were caused by unsafe working conditions. Eighty-eight percent of the accidents were precipitated by workers' unsafe actions. The ratio of fatal to minor to no-harm accidents was 1:29;300. The focus should be on drilling for proper behavior in the clinic.

- Determine the behaviors that minimize unsafe clinical conditions.
- Train all department personnel in these behaviors.
- Measure that personnel are indeed doing these behaviors.
- Promote appropriate behaviors with incentives.

Successful error reduction during linear accelerator installation depends critically on the design of the installation plan and testing or commissioning after acceptance testing with the vendor. This plan should include the review of the installation with the vendor's installer, the physicist, and also the vendor's service representative who will participate in the acceptance tests and commissioning. Frequently, vendors expect the physicists to check their work, which is necessary but not sufficient. Omission of the review of the installation by the service representative is

TABLE 52.1 WTF of Two Linear Accelerator Models

C-Arm Component	Time to Commission (Weeks)	Training Required	AAPM Task Group Number	Spiral Component	Time to Commission (Weeks)	Training Required
Linear accelerator	1.0	Vendor's machine acceptance course 1.0 weeks	142 (AAPM Report 142 2009) 45 (AAPM Report 47 1994) 51 (AAPM Report 67 1999) 106 (AAPM Report 106 2008)	Integrated unit	1.4	Vendor's machine acceptance course 0.6–1.0 weeks
MLC	0.4		50 (AAPM Report 72 2001)			
EPID	0.2		28 (AAPM Report 24 1987) 58 (AAPM Report 75 2001) 75 (AAPM Report 95 2007)			
CBCT	0.4	Linear accelerator vendor's kV imaging course	66 (AAPM Report 83 2003) 75 (AAPM Report 95 2007)			
kV imaging	0.4	1.0 weeks	95 (AAPM Report 95 2007)	Not available	n/a	n/a
Respiratory motion compensation	0.4	Device vendor's training 0.2–0.4 weeks	76 (AAPM Report 91 2006)	Not available	n/a	n/a

C-Arm Technique	Time to Commission (Weeks)	Training Required	AAPM Task Group Number	Spiral Technique	Time to Commission (Weeks)	Training Required
SRS/SBRT	1	Treatment planning system vendor's course 1 week	42 (AAPM Report 54 1995)	SRS	0.8 weeks	Vendor training course
IMRT	0.6	Treatment planning system vendor's course 1 week	(AAPM Report 82 2003)	Integral to the unit's operation	Included in basic commissioning time	Vendor's treatment planning course 1.0 weeks
IGRT	0.5	Linear accelerator vendor's kV imaging course and treatment planning system course. Possibly third-party course as well ~1.4 weeks total	76 (AAPM Report 91 2006)			

equivalent to a treatment plan beam calculated and checked by the same physicist. The lack of a second vendor representative reviewing the installation is a serious omission in quality control. It is simply placing all the responsibility for the vendor's review of the installation on the facility's medical physicists. This should be regarded as unacceptable. There is no excuse for a vendor omitting the review of the installation by either a second installer or the regional service representative.

Error Reduction Methods

Plan twice, commission once. This is a simplification of the fault hazard analysis method. Planning and designing linear accelerator commissioning will save time and avoid errors. The plan design depends on the linear accelerator configuration. The error reduction methodology does not. Consider the calibration of an x-ray beam. The TG-51 (Almond et al. 1999) measurements will be acquired to determine the output and beam energy. The

physicist would repeat the measurements to ensure that errors were avoided. The physicist will perform the TG-51 protocol calculations and check these calculations. The linear accelerator calibration will be adjusted in accordance with the TG-51 findings. The linear accelerator output will be remeasured to confirm the proper calibration. A second physicist will repeat the above process to confirm the first physicist's calibration. Finally, the linear accelerator calibration will be validated by measurement by a third party. This can be by an external audit such as the Radiation Dosimetry Services (http://www.mdanderson.org/education-and-research/resources-for-professionals/scientific-resources/core-facilities-and-services/radiation-dosimetry-services/index.html).

The above example could be the result of a fault hazard analysis. Application of fault hazard analysis should occur in (1) the design of the equipment selection, (2) the design of the equipment acceptance/commission, and (3) the design of clinical use of the equipment.

Error Reduction by Staffing Analysis

The Abt study (ABT 2008) is a nationwide survey of physics staffing as a function of patient and QA workload. This enables medical physicists to estimate the required staffing for the radiation oncology department's planned workload for the planned linear accelerator installation. Together with the work time function (WTF) linear accelerator work estimate and vendor training courses, the staffing requirements and the commissioning requirements can be methodically assessed.

The American College of Radiology provides radiation oncology audits that review the clinical and medical physics strengths and weaknesses. The ACR audit reviews the work of the radiation oncologists, medical physicists, radiation therapists, and treatment documentation. The ACR audit collects the data for analysis by ACR staff. The analysis includes comparison with ACR data nationwide as a function of patient load for academic and private practices. The ACR audit does not include dosimetric assessment of the treatment plan calculation, dose delivery, or linear accelerator calibration.

Error Reduction by External Audits

The Radiological Physics Center (RPC) is funded by the National Cancer Institute to audit dosimetry practices at institutions participating in National Cancer Institute cooperative clinical trials (RTOG, SWOG, ACOSOG, etc.). The AAPM Radiation Therapy Committee is the scientific advisory body to the RPC. The RPC quantitatively evaluates the treatment plan accuracy, the radiation dose delivery of these treatment plans, and the linear accelerator calibration.

External dosimetry audit by thermoluminescent dosimetry can be provided, for example, by the Radiation Dosimetry Services, which is housed administratively in the Section of Outreach Physics within the Department of Radiation Physics at M. D. Anderson Cancer Center. Thermoluminescent dosimetries are posted to the client department, irradiated, and measured to confirm the therapy beam calibration.

Evaluation

We introduce a WTF factor, which is a linear accelerator configuration-dependent estimate of the time to commission the linear accelerator (see Table 52.1).

This exercise is a method to formulate or assess the workload involved with a C-arm linear accelerator and tomotherapy commissioning starting with and normalized to 1.0 for the basic dual-energy linear accelerator. The basic linear accelerator configuration is two x-ray beams and five electron beams. We assume that the workload (WTF factor) for commissioning the basic unit is normalized to 1.0. The addition of electronic portal imaging increases the commissioning workload by an estimated 10%. Thus, a basic unit with electronic portal imaging has a relative workload of 1.1. Adding on the second level, we estimate that

the addition of a multileaf collimator and intensity-modulated radiotherapy (IMRT) adds a 40% increase relative to the basic workload. Thus, the WTF factor is now 1.5. Adding stereotactic radiosurgery and stereotactic body radiotherapy yields an increase of 30% relative to the commissioning of the basic linear accelerator. The 30% is confined to the commissioning of the localization apparatus and small field dosimetry. Thus, now the WTF is coming up to 1.8 relative to the basic linear accelerator. The addition of onboard imaging itself is a 40% increase in the basic linear accelerator commissioning workload, which brings us to a WTF equal to 2.2. Cone-beam CT is an estimated further 40% increase relative to the basic linear accelerator. Here we are essentially commissioning a computed tomography unit, which is attached to the treatment unit itself. The 40% increase reflects the work of CT alignment with the rotation axis, image quality, kVp tube testing, and DICOM export. Now the WTF is 2.6. The final addition to this exercise is 4D CT/gated beam therapy, which we estimate to be a 60% increase in the workload of commissioning a basic unit. We are now at a WTF factor of 3.2 times the typical workload for a stripped linear accelerator.

Although estimates on the relative increase may vary with each technological addition to the 1980s linear accelerator, the modern linear accelerator has evolved into a device that is extremely complex with an estimated 3.2 times the commissioning workload of the linear accelerator of 20 years ago. The message of this exercise is that error reduction is critically dependent on the proper estimate of the work involved and the time required for an error-free commissioning of the C-arm linear accelerator. This message must be made clear to the administration and the radiation oncologists involved in the planning process of equipment acquisition. Proper project planning must lay out all of the work involved and the time required to achieve these goals, and must determine how the revenues generated by the project will support ongoing quality assurance and ongoing training for the radiation oncologists, the physicists, the dosimetrists, and the radiation therapists that comprise the overall team.

Note that the total training for a complete C-arm linear accelerator is 5.8 weeks. The training courses add on with each added layer of technology of the C-arm. The C-arm linear accelerator planning system can have up to four photon dose calculation algorithms and two electron beam algorithms. This does not include the training for use and administration of the patient information system. The total physics training for the spiral therapy unit is 2.2 weeks. The spiral therapy unit has one photon energy and one photon dose calculation algorithm. The spiral therapy unit lacks beam gating but has one unified software system for treatment planning, the patient database, and linear accelerator control.

Equipment Selection

Establish the Total Cost of the New Technology

The total cost obviously includes the direct cost of the equipment and the equipment support cost after the warranty expires, including software and hardware support. However, it also includes training of existing staff as well as training of new personnel because of existing personnel turnover or when new positions are added. It must also include annual retraining of personnel to ensure competencies are maintained. Training is all too often neglected, which increases the risk of treatment errors. This risk increases with time as memories fade and as personnel depart.

Fit to Existing Equipment and Software

Before purchasing a unit, the medical physicist should review the acceptance procedure used by the manufacturer and assess whether it will meet his/her minimum quality standards. Alterations to the acceptance procedure should be made before signing a purchase agreement.

Supporting Equipment—Calibration, Redundancy, and Rehearsal

Ensure that the ionization chambers and electrometers have been calibrated within the past 2 years by an accredited dosimetry calibration laboratory, preferably alternating years. Rehearse/drill for the linear accelerator commissioning to maximize efficiency and determine that the equipment is functioning. Rehearse all steps of the commissioning.

Design of Equipment Acceptance

Obtain necessary and sufficient training for the new technology (Table 52.1 is a starting point). During acceptance testing, the installer, medical physicist, and field service engineer must all be present. Agreement by all three parties of successful completion of installation must be documented.

Optimizing data acquisition and data verification during acceptance testing must include both demonstration of operation (mechanical, dosimetric, image acquisition, gating, etc.) within acceptable tolerance and accurate and reliable data transfer, recording and verification, data review, storage, and archiving.

Select tests necessary and sufficient for patient safety and therapy. The installer, field service engineer, and medical physicist should have regular meetings to communicate, discuss, and document problems and solutions met during the installation process. These should include discussion of issues of meeting tolerances and reliability.

A complete "front-to-back" test of all systems (e.g., planning, imaging, treating measuring delivered dose to phantoms) before clinical use is necessary.

Design archival storage of testing records and perform external tests such as the RPC phantom tests, external thermoluminescent dosimetry tests, external physics review by RPC or ACR or independent medical physicists, and external calibration of dosimetry equipment by ADCL.

Data Acquisition

Check data samples during each step of the linear accelerator commission for errors/discrepancies. Resolve the discrepancies promptly and avoid wasting time during the commissioning. Have a second physicist review the data during commissioning.

Match beam data to extant data. If this is not the first linear accelerator in the facility, match the new linear accelerator beam data to the existing linear accelerator when possible. This reduces the commissioning time and minimizes beam data errors.

Design of the Clinical Use of the Equipment

Current clinical processes need to be evaluated. Redesign the current clinical process to incorporate the new technology as well as redesign the QA/QI procedures and timetable.

Users of new equipment need to be trained or retrained on the new equipment. Training of staff (RTT, medical physicists, physicians) must occur before clinical use (not concurrent with) and be sufficiently documented. Establish an annual ongoing training schedule of all practitioners before purchasing a unit. The medical physicist should review the acceptance procedure used by the manufacturer and assess whether it will meet their minimum quality standards. Alterations to the acceptance procedure should be made before signing a purchase agreement.

The installer, field service engineer, and medical physicist should have regular meetings to communicate, discuss, and document problems and solutions met during the installation process. These should include discussion of issues with meeting tolerances and reliability.

Acceptance testing must include both demonstration of operation (mechanical, dosimetric, image acquisition, gating, etc.) within acceptable tolerances as well as accurate and reliable data transfer, recording and verification, data review, storage, and archiving.

Training

According to the IAEA Safety Report Series No. 17 (IAEA 2000) and the Radiation Risk Profile by the WHO (2009), lack of training plays a key role in every misadministration in radiation oncology. Error reduction is greatly enhanced by training at all levels. It is assumed that the physicist fulfills the basic AAPM definition of a credentialed medical physicist and is board certified by the American Board of Medical Physics or the American Board of Radiology. Requisite training on all linear accelerator technology is an obvious requirement. Training courses for linear accelerator technology are delineated in Table 52.1 as a function of linear accelerator configuration.

One aspect of training is frequently overlooked. Ongoing training is essential to error reduction and a safe treatment environment. The medical field focuses on continuing medical education to ensure competence. Continuing medical education can be fulfilled by quizzes and journal reading and meeting attendance. Continuing medical education does not test the actual performance of professionals in the workplace in stressful conditions where procedures fail. Consider the training policy of a nuclear power plant. The training policies are those of the Ginna nuclear power plant in upstate New York. The technologists and engineers receive two full weeks of training each year. The training includes simulator training that tests the personnel for the proper response to various failure modes. The nuclear power plant operators train every sixth week for every possible operation. A professional fireman in Canandaigua, NY, receives 2 weeks training per year. In contrast to these training policies, how does any medical field fare? Training and testing of medical professionals in simulated conditions is at best amateurish. Ongoing training typically is 2 days per year for a few departmental clinicians rather than for the entire department. Weak or nonexistent ongoing training erodes the infrastructure of the department as memories fade and personnel turnover takes it toll.

After the Three Mile Island Nuclear Power Plant accident in 1979, the Kemeny Commission recommended that the U.S. nuclear energy industry act as a nuclear professional organization and regulate itself under the supervision of the Nuclear Regulatory Commission. The industry formed the Institute of Nuclear Power Operations to self-regulate nuclear power plant operations, including the training of personnel. The Institute of Nuclear Power Operations guided the nuclear industry to establish a comprehensive system of personnel training and qualifications. The National Academy for Nuclear Training integrated the training programs of the Institute of Nuclear Power Operations, the training efforts of the U.S. nuclear energy companies, and the activities of the National Nuclear Accrediting Board. The following is a summary of the National Academy of Nuclear Training objectives and criteria for training program accreditation, which are published as Academy Accreditation Documents. The Academy Accreditation Documents is shown below. It contains six training criteria:

1. Training for performance improvement
2. Management of training processes and resources
3. Initial training and qualification
4. Continuing training
5. Conduct of training and trainee evaluation
6. Training effectiveness evaluation

The AAPM defines the minimum qualifications of a clinical medical physicist as board certification by the American Board of Medical Physics or the American Board of Radiology (criterion 3). However, the training required for linear accelerator commissioning depends entirely on the linear accelerator configuration. Table 52.1 delineates the required training by linear accelerator configuration.

Summary

It was the best of data; it was the worst of data. The choice is yours. Successful error reduction depends on the factors reviewed here.

If the safety culture/environment permits the time and resources to plan and train well, the risk of errors and misadministrations can be minimized. Training—it is our opinion that a lack of ongoing training in the medical field plays a significant role in the 100,000 to 200,000 hospital deaths each year. This is equivalent to 33 to 66 9–11 attacks annually. Simulation training exercises are the only way to prove the process is effective without performing the experiment on the patient. Simulation training must be applied to radiation oncology procedures to prove that the staff is efficient and that the procedures are safe. These methodologies must be developed, applied, and analyzed for outcome. The following are the steps that should be taken for error reduction:

1. Clear documentation of procedures, job descriptions and responsibilities, and the chain of command.
2. Commission the linear accelerator(s) with teams of physicists. Double check the work of one by the other and alternate the roles of effectors and checker.
3. Clear communication between all disciplines: medical physics, oncologists, radiation therapy technologists, engineering, nurses. Maintain professional communication among the staff. Verbal and written communications must be clear and correct and free of impediments (reference 3, p. 74).
4. Ensure that both phases of training are sufficient. Provide training for new equipment and new procedures—Phase 1. Provide the necessary and sufficient ongoing training for all personnel. This includes drills for emergency procedures and drills for recovery from failure modes in treatments. Simulation training is necessary in linear accelerator commission and in all aspects of medical physics procedures and processes.
5. Maximize equipment redundancy and software compatibility. Preempt equipment failures through proactive equipment repairs—replace components on a routine basis. If water pumps fail every 5 to 6 years, replace off hours every 5 years.
6. Log all equipment maintenance and calibrations. Maintain the calibration schedule of the linear accelerators and calibration equipment.
7. Perform annual audits of the quality assurance procedures. Analyze the audits and QA procedures for improvement opportunities when technologies or practices change.
8. Emergency preparedness plans must be in place for accidents and unusual events. These plans must be drilled to the point where personnel respond immediately and are error-free. The drills must be documented clearly and reviewed periodically.

As new technologies continue to be incorporated into existing medical accelerators, and the complexity of these implementations increases, thorough training in their function and use becomes even more important. Very often, errors in installation or calibration of new equipment are subtle or nonintuitive. In these instances, training in, and understanding of, the equipment's design and operation are critical to the commissioning process's successful detection of errors.

The AAPM task group reports are a valuable resource for guidance in how to safely implement new technology. They offer background on the physical principals involved as well as suggested procedures and criteria that have been accepted by the medical physics community. Task-group reports are also extensively referenced, which facilitates the search for more detailed or more fundamental information when needed.

In some cases, further vendor-specific training is necessary. Often, control software, calibration procedures, and networking requirements are highly vendor-specific. Similarly, one vendor's design may impose restrictions that another has overcome. In short, as errors frequently hide in details, such training becomes essential.

Similarly, the treatment techniques such as IMRT and image-guided radiotherapy also require training to implement correctly. Again, the AAPM provides valuable task group reports on a broad range of subjects. But as with the implementation of the technology itself, the implementation of advanced treatment techniques often requires vendor-specific training. This is most obvious in cases such as treatment planning system software and plan quality assurance procedures.

When estimating the time necessary for commissioning a new treatment modality or technique, it is important to include the time needed for training.

A second important factor in estimating the time required to commission a medical accelerator is the degree of integration between the basic accelerator and its various components. As shown in the table, different accelerator designs have been more or less optimized for different treatment techniques.

Acknowledgments

I am indebted to Richard J. Watts of the Monroe County Office of Emergency Management, Rochester, NY. Mr. Watts directed training at the Ginna Nuclear Power Plant and provided the insight to nuclear power plant training.

References

AAPM Report 24. 1987. *Radiotherapy Portal Imaging Quality.* Madison, WI: Medical Physics Publishing.

AAPM Report 47. 1994. *AAPM Code of Practice for Radiotherapy Accelerators* (reprinted from *Medical Physics*, vol. 21, issue 7). Madison, WI: Medical Physics Publishing.

AAPM Report 54. 1995. *Stereotactic Radiosurgery.* Madison, WI: Medical Physics Publishing.

AAPM Report 67. 1999. *Protocol for Clinical Dosimetry of High-Energy Photon and Electron Beams* (reprinted from *Medical Physics*, vol. 26, issue 9). Madison, WI: Medical Physics Publishing.

AAPM Report 72. 2001. *Basic Applications of Multileaf Collimators.* Madison, WI: Medical Physics Publishing.

AAPM Report 75. 2001. *Clinical Use of Electronic Portal Imaging* (reprinted from *Medical Physics*, vol. 28, issue 5). Madison, WI: Medical Physics Publishing.

AAPM Report 82. 2003. *Guidance Document on Delivery, Treatment Planning, and Clinical Implementation of IMRT: Report of the IMRT Subcommittee of the AAPM Radiation Therapy Committee.* Madison, WI: Medical Physics Publishing.

AAPM Report 83. 2003. *Quality Assurance for Computed-Tomography Simulators and the Computed-Tomography-Simulation Process.* Madison, WI: Medical Physics Publishing.

AAPM Report 91. 2006. *The Management of Respiratory Motion in Radiation Oncology* (synopsis published in *Medical Physics*, vol. 33, issue 10, 27 pp). Madison, WI: Medical Physics Publishing.

AAPM Report 95. 2007. *The Management of Imaging Dose during Image-Guided Radiotherapy: Report of the AAPM Task Group 75* (*Medical Physics*, vol. 34, issue 10). Madison, WI: Medical Physics Publishing.

AAPM Report 106. 2008. *Accelerator Beam Data Commissioning Equipment and Procedures: Report of the TG-106 of the Therapy Physics Committee of the AAPM.* Madison, WI: Medical Physics Publishing.

AAPM Report 142. 2009. *Task Group 142 Report: Quality Assurance of Medical Accelerators* (*Medical Physics*, vol. 36, issue 9). Madison, WI: Medical Physics Publishing.

ABT. 2008. *The Abt Report on Medical Physics Work Values for Radiation Oncology Physics Services: Round III.* Bethesda, MD: Abt Associates, Inc.

Almond, P. R., P. J. Biggs, B. M. Coursey, W. F. Hanson, M. S. Huq, R. Nath, and D. W. Rogers. 1999. AAPM's TG-51 protocol for clinical reference dosimetry of high-energy photon and electron beams. *Med. Phys.* 26 (9): 1847–1870.

Heinrich, H. W. 1931. *Industrial Accident Prevention. A Scientific Approach.* 1st ed. New York, NY: McGraw-Hill Insurance Series.

IAEA. 2000. *Lessons Learned from Accidental Exposures in Radiotherapy.* Vienna: International Atomic Energy Agency (IAEA).

WHO. 2009. *Radiotherapy Risk Profile.* Geneva: World Health Organization.

Linear Accelerator: Implementation and Use

Jean-Pierre Bissonette
Princess Margaret Hospital

Introduction

The clinical use of medical linear accelerators for radiotherapy is a complex process involving highly skilled individuals who carry out a variety of interrelated activities with multiple hand-offs of responsibility and interact with intricate hardware and software. Modification of this process, through technology or practice generally involves the careful assessment of physical and human factors. Classically, the former involves the collection of beam data relevant for reliable clinical practice, whereas the latter focuses on staff training (Podgorsak, Metcalfe, and Van Dyk 1999). Special procedures, such as brain radiosurgery, total body irradiation, and intensity-modulated radiotherapy, are usually dealt with as additions to classic radiation therapy and are documented elsewhere in this book; each of these techniques therefore involves its own additional set of tests and measurements that are subsequently used to validate calculations from treatment planning systems, establish tables for secondary dose calculations, and establish baseline values for future quality assurance tests.

The rapid evolution of commercially available technologies for radiotherapy, in combination with the increased complexity of therapies, has required new guidelines for commissioning and quality assurance. Early adopters of such innovative approaches document and disseminate their knowledge in scholarly and academic papers that clinical users may find somewhat haphazard, overwhelming, or inconsistent; for example, parameters that may be of academic interest can turn out to have a low likelihood of failure, therefore draining limited QA resources. Still, such papers eventually lead to the production of commissioning

and QA guidelines recognized by professional organizations, but this process frequently takes several years after the novel technologies or practices have been introduced. In the absence of such guidelines, early adopters typically follow the spirit of documents such as the AAPM TG-40 and TG-45 reports in North America (Kutcher et al. 1994; Nath et al. 1994), with the premise that each of the technical components of a device or process requires testing, but allowing some provision to adjust test tolerances and frequency (Pawlicki and Whitaker 2008). However, Thomadsen (2008) reminds us that the main goal of QA is to ensure that patients receive the prescribed treatment correctly and that QA guidelines should be risk informed. This implies that not only radiotherapy equipment but also human interaction with radiotherapy equipment is subject to failure.

Analysis of organizations has shown that a large proportion of major change initiatives, including the successful implementation of new technologies or procedures, fail to realize their intended gains, most probably due to lack of human acceptance and adoption of a change. The successful implementation of new technologies or processes depends on the availability of (1) clear goal statements, (2) available expertise and skills, (3) proper incentives, (4) sufficient resources, and (5) an action plan (Kotter 1995). A team that is lacking clear goals and aims will quickly become confused, whereas lack of skills or resources will lead to anxiety and frustration. Similarly, lack of motivation and empowerment will slow down progress considerably, whereas the lack of an action plan will likely abort the project altogether. Still, in the clinical field, the accomplishment of a successful commissioning and QA plan may evolve differently from the action plan as surprises may arise.

General Framework for Implementation of Novel Technologies or Practices

In this section, a more comprehensive approach for commissioning and building QA programs for novel technologies or processes is presented. This approach presupposes that the clinical goals and aims have been clearly established, that a motivated team has been identified, and that all regulatory requirements have been met.

Acceptance Testing

Acceptance testing is a formal process where the customer, usually represented by a qualified medical physicist, and the vendor demonstrate that new equipment meets or exceeds the specifications detailed in the purchase order or in the tender (Kutcher et al. 1994; Nath et al. 1994). The first step of the acceptance process is to verify that all safety systems are operational; a radiation survey may be performed at this stage to comply with local radiation safety licensure regulations. Vendors typically provide an exhaustive acceptance testing procedure to be performed by both the customer and a representative of the vendor on the premises. Customers should ensure, at this stage, that any additional technical specifications mentioned in the tender or purchasing contract have been met; a common example of additional specifications is to ensure that the beam properties of a new accelerator match those of another accelerator within the department. A careful review of published descriptions or guidelines for acceptance testing may prompt customers to enrich the acceptance process with any additional technical specifications that are felt necessary to be met or exceeded by the vendor prior and enclose these additional specifications into the tender or purchase order. Data generated from acceptance testing can subsequently be used as baseline performance values for quality control tests or to verify performance of equipment after a major repair. When all tests are completed to the satisfaction of both parties, the final acceptance certificates are signed by representatives of the vendor and the customer, who thereupon accepts the new merchandise and its ownership.

Clinical Commissioning

Clinical commissioning occurs when all the steps needed to bring a system into clinical service are performed. Typically, for linear accelerators, commissioning involves measurements performed to specify, for a wide range of settings, geometric and dosimetric properties of beams that could be encountered clinically. Classically, this step involves the measurement of data required for creating beam models for computerized treatment planning systems and includes beam calibration, following nationally accepted protocols aimed to achieve high dosimetric accuracy (Almond et al. 1999; IAEA 2000). However, during clinical commissioning, not only is the minimal required data set measured

but additional beam measurements are also performed to test the accuracy of calculations for a range of settings for general use (Fraass et al. 1998) or for special techniques.

Once classic accelerator commissioning is completed, teams typically enter a phase where new or existing clinical processes are validated on the new equipment. This is a crucial moment when clinical teams can field-test preparedness to treat patients with new equipment or practices, preferably with an element of external validation. A common way to commission new clinical processes is by performing dry runs until saturation (i.e., the team does not learn anything new) is achieved. During the initial dry run, the team would perform the new process, beginning to end, using appropriate phantoms and stock images, (1) to identify all glitches, deficiencies, or any other issues; (2) to test the connectivity between all system elements; and (3) confirm that the system behaves in an accurate and reproducible manner. All clinical members, including therapists, radiation oncologists, and physicists should participate to ensure good communication at all levels and build confidence in the new process or technology. The dry run may also involve the irradiation of a phantom to measure dose at several points of interest. The Radiological Physics Centre provides a series of dosimetry phantoms that, after irradiation, can yield measured recorded doses to be compared with calculated doses from a treatment plan generated locally, thereby providing an external validation of the process. Once the development team is confident in the new process and has reached saturation, the last step is to educate the staff about the new learnings, and use these learnings to develop a sensible QA program, based on an initial assessment of the risks involved with the new procedure.

Periodic Quality Controls

In the absence of professionally endorsed QA guidelines, constructing a QA program to ensure the stable and reproducible performance of novel technologies or procedures may very well take several months, if not years. Early adopters will likely seek initial guidance from the manufacturer's customer acceptance procedure (Yoo et al. 2006), leading to the exhaustive and typical QA programs that drain so much of the clinical resources. Modern QA tools use statistical control charts to define appropriate test frequencies and tolerances based on variances of quality metrics over appropriate lengths of time (Pawlicki and Whitaker 2008). Another useful tool is risk analysis, where quality metrics are prioritized according to the severity, likelihood, and detectability of a process or device failure (Huq et al. 2008). Thus, a technical parameter, such as gantry or collimator angle indicator, that is unlikely to fail, or has low impact on the risk of inaccurately delivering radiation therapy to patients, should not be checked as frequently as a high-impact factor, such as beam output calibration. Performing such analyses requires time and careful analysis of potential QA data accrued over several months; this internal audit of QA data may lead to a documented, in-house protocol. It should be noted that published QA guidelines give users the freedom to adjust quality control test tolerances and frequencies according to internal review of quality control data.

Patient-Specific Quality Control

The clinical team responsible for commissioning a new protocol would be well advised to document how this new process will be applied for each patient undergoing this process. Examples of patient-specific quality control checks can be as simple as verifying patient identification to as complex as verifying fluences for intensity-modulated radiotherapy. Thus, ensuring that the appropriate treatment or imaging protocol is selected for a properly identified patient is an important action.

Audits

Modern QA uses audits to provide assessments that devices and processes are valid and reliable and to provide reasonable assurance that they are free of errors. Quality audits, which form the basis of the ISO 9001 quality system standard, are performed to ensure that an institution has clearly defined quality monitoring procedures that are effective and to demonstrate that the institution achieves the desired level of quality care. Internal audits can be used, for instance, to draft a QA program and to provide exhaustive analyses of those internal processes that employees have access to. External audits are useful to compare the performance of one institution with respect to another, each attempting to achieve similar goals.

Human Factors

The commissioning and QA programs for novel equipments or procedures involve a series of carefully chosen preparatory steps and studies that involve not only machine and beam parameters but also dissemination and communication of information and, equally important, adoption of new behaviors by colleagues and staff. A successful plan may include:

1. Clear enunciation and demonstration of the needs and goals, and their perceived benefits
2. Review of the technical specifications of commercially available radiotherapy equipment or measured data that are consistent with the enunciated goals
3. Availability of appropriate resources (time, space, funds), expertise, and administrative support
4. Identification of internal or external documents or guidelines for acceptance testing and commissioning for clinical use
5. Assign and empower a leader or a team of champions that is credible and motivated with proper incentives
6. Develop a strategy and an action plan to achieve the goals, including a training program to disseminate expertise and information and the development of a comprehensive QA program
7. Generate short-term successes, such as dry runs, to build on and ramp up the novel program
8. Anchor the new technology, approaches, and behaviors into the core values of the clinical team

Monitoring incident and QA reports would help assess how successfully the new technology or practice is anchored in clinical practice.

Application of the Framework: Image-Guided Radiotherapy

One of the recent advances in technology and practice of radiotherapy is image-guided radiotherapy (IGRT), where imaging devices are installed in radiotherapy treatment suites to direct radiation beams accurately using imaging information. This new modality can achieve several aims. First, IGRT increases the accuracy of radiotherapy by verifying the position of the patient, with respect to the treatment beam, immediately before irradiation. Second, users may use the enhanced geometric accuracy to review and, hopefully, reduce setup margins for PTV design, leading to reduced doses to organs at risk and perhaps enabling dose escalation. Finally, IGRT may also empower modification of the radiotherapy treatment plan as clinicians can respond to anatomical changes seen during a course of radiation. Secondary aims of IGRT might include replacing film or portal imaging to document positional accuracy, managing inter- and intrafractional motion of organs during radiotherapy, or measuring the actual efficacy of immobilization accessories. Thus, first-time users of this technology must ascertain which of these aims are desirable in their own clinical contexts and construct their commissioning and QA programs in alignment with these aims. The following paragraphs will summarize the commissioning process for IGRT, based on cone-beam CT (CBCT), at Princess Margaret Hospital, Toronto, Canada, with the original aims to ensure that high geometric accuracy is achieved and to replace portal imaging in the clinical practice.

Implementation of wide-scale IGRT was built on several short-term successes, starting with acceptance testing of the first CBCT device. Teams were built and allowed to develop their confidence in the IGRT process, using dry runs where a phantom was treated exactly like a patient, starting with a simulation CT scan through to treatment delivery. Such dry runs simulated the process in a multidisciplinary environment, helped identify and resolve issues, and developed expertise. A powerful motivation factor for the teams was the clinical research and development opportunities offered to clinical site groups by IGRT; for example, obtaining soft-tissue contrast on volumetric data sets using lower doses than portal imaging opened the possibility of frequent and accurate positioning of the patient at the onset of each treatment. Teams have tested and benchmarked CBCT with portal imaging guidance, initially using phantoms and dry runs (Sharpe et al. 2006), followed by patient studies (Hawkins et al. 2006; Purdie et al. 2006; Moseley et al. 2007; Zeng et al.

2009; White et al. 2007; Bissonnette et al. 2009). Often, these studies involved verification imaging to assess the accuracy of the couch shift required to place patients back in the treatment position; this helped build confidence in the accuracy of the IGRT process. Coordination meetings were instituted to share the findings from the various groups and to identify infrastructure issues, such as disk space and data archiving requirements. In parallel, a QA program was established, using guidance from the manufacturer and guidance from reports on CT imaging (Jaffray, Bissonnette, and Craig 2005).

Clinical patient studies were gradually implanted. Anatomical site groups were selected based on (1) research interests, (2) sufficient patient volume to allow the clinical team to learn without overloading it (about five patients on a treatment unit), (3) ease or difficulty of the imaging process, as influenced by visualization of soft tissues and mobility of internal organs, and (4) when gains in contrast offered with kilovoltage over portal imaging are immediately obvious (lung, pediatrics, brain). All professionals were involved in the development of site-specific IGRT techniques. Radiation oncologists defined the initial positional accuracy requirements, whereas physicists recommended appropriate imaging techniques while keeping radiation doses low. Therapists advised on proper immobilization techniques and performed image guidance; subsequently, physicists analyzed positional data and presented results back to the team to reassess setup margins, tolerances for residual positional errors, and opportunities for dose escalation. Administrators also participated in the process, allowing time for staff to learn the new process while making teams aware of the fiscal constraints that needed to be met once the new IGRT processes were stabilized. Each team fully documented site-specific image-guided processes, with clear statements of the accuracy requirements, that were used as a learning tool for staff.

Since the first few steps in 2003, the Princess Margaret Hospital has enabled volumetric CBCT-based IGRT on eight of its 16 linear accelerators. This form of IGRT has been firmly anchored in the institutional values, and daily CBCT IGRT is now part of routine practice for many anatomical sites: in 2008, 3532 patients were treated with this form of image guidance, yielding more than 36,000 volumetric CBCT data sets.

Summary

The framework presented in this chapter attempts to broaden the view of classic radiotherapy QA with elements of risk assessment and human factors engineering. The former is to help avoid or mitigate the risk of catastrophic error and maintain high quality of treatment. The latter ensures that the classic, device-centered approach to QA is complemented with a process-centered approach dealing implicitly with human interaction with the process; well-trained and motivated teams can recognize and mitigate errors and variations before they manifest themselves as actual incidents. Field testing under research protocols and dry runs are QA tools that can be used to test a system as a whole, from technical factors to process optimization accounting for human factors. Monitoring QA data and treatment variances is helpful in assessing how successful the new equipment or practice has been integrated by the clinical team.

References

Almond, P. R., P. J. Biggs, B. M. Coursey, W. F. Hanson, M. S. Huq, R. Nath, and D. W. Rogers. 1999. AAPM's TG-51 protocol for clinical reference dosimetry of high-energy photon and electron beams. *Med. Phys.* 26 (9): 1847–1870.

Bissonnette, J. P., T. G. Purdie, J. A. Higgins, W. Li, and A. Bezjak. 2009. Cone-beam computed tomographic image guidance for lung cancer radiation therapy. *Int. J. Radiat. Oncol. Biol. Phys.* 73 (3): 927–934.

Fraass, B., K. Doppke, M. Hunt, G. Kutcher, G. Starkschall, R. Stern, and J. Van Dyke. 1998. American Association of Physicists in Medicine Radiation Therapy Committee Task Group 53: Quality assurance for clinical radiotherapy treatment planning. *Med. Phys.* 25 (10): 1773–1829.

Hawkins, M. A., K. K. Brock, C. Eccles, D. Moseley, D. Jaffray, and L. A. Dawson. 2006. Assessment of residual error in liver position using kV cone-beam computed tomography for liver cancer high-precision radiation therapy. *Int. J. Radiat. Oncol. Biol. Phys.* 66 (2): 610–619.

Huq, M. S., B. A. Fraass, P. B. Dunscombe, J. P. Gibbons Jr., G. S. Ibbott, P. M. Medin, A. Mundt et al. 2008. A method for evaluating quality assurance needs in radiation therapy. *Int. J. Radiat. Oncol. Biol. Phys.* 71 (1 Suppl): S170–S173.

IAEA. 2000. Absorbed dose determination in external beam radiotherapy: An international code of practice for dosimetry based on standards of absorbed dose to water. In *Technical Reports Series No. 398*. Vienna: International Atomic Energy Agency (IAEA).

Jaffray, D. A., J.-P. Bissonnette, and T. Craig. 2005. X-ray imaging for verification and localization in radiation therapy. In *The Modern Technology of Radiation Oncology*, ed. J. Van Dyk. Madison, WI: Medical Physics Publishing.

Kotter, J. P. 1995. Leading change: Why transformation efforts fail. *Harv. Bus. Rev.* 73 (2): 59–67.

Kutcher, G. J., L. Coia, M. Gillin, W. F. Hanson, S. Leibel, R. J. Morton, J. R. Palta et al. 1994. Comprehensive QA for radiation oncology: Report of AAPM Radiation Therapy Committee Task Group 40. *Med. Phys.* 21 (4): 581–618.

Moseley, D. J., E. A. White, K. L. Wiltshire, T. Rosewall, M. B. Sharpe, J. H. Siewerdsen, J. P. Bissonnette et al. 2007. Comparison of localization performance with implanted fiducial markers and cone-beam computed tomography for on-line image-guided radiotherapy of the prostate. *Int. J. Radiat. Oncol. Biol. Phys.* 67 (3): 942–953.

Nath, R., P. J. Biggs, F. J. Bova, C. C. Ling, J. A. Purdy, J. van de Geijn, and M. S. Weinhous. 1994. AAPM code of practice for radiotherapy accelerators: Report of the AAPM Radiation Therapy Task Group No. 45. *Med. Phys.* 21: 1093–1121.

Pawlicki, T., and M. Whitaker. 2008. Variation and control of process behavior. *Int. J. Radiat. Oncol. Biol. Phys.* 71 (1 Suppl): S210–S214.

Podgorsak, E. B., P. Metcalfe, and J. Van Dyk. 1999. Medical accelerators. In *The Modern Technology of Radiation Oncology*, ed. J. Van Dyk. Madison, WI: Medical Physics Publishing.

Purdie, T. G., D. J. Moseley, J. P. Bissonnette, M. B. Sharpe, K. Franks, A. Bezjak, and D. A. Jaffray. 2006. Respiration correlated cone-beam computed tomography and 4DCT for evaluating target motion in Stereotactic Lung Radiation Therapy. *Acta Oncol.* 45 (7): 915–922.

Sharpe, M. B., D. J. Moseley, T. G. Purdie, M. Islam, J. H. Siewerdsen, and D. A. Jaffray. 2006. The stability of mechanical calibration for a kV cone beam computed tomography system integrated with linear accelerator. *Med. Phys.* 33 (1): 136–144.

Thomadsen, B. 2008. Critique of traditional quality assurance paradigm. *Int. J. Radiat. Oncol. Biol. Phys.* 71 (1 Suppl): S166–S169.

White, E. A., J. Cho, K. A. Vallis, M. B. Sharpe, G. Lee, H. Blackburn, T. Nageeti, C. McGibney, and D. A. Jaffray. 2007. Cone beam computed tomography guidance for setup of patients receiving accelerated partial breast irradiation. *Int. J. Radiat. Oncol. Biol. Phys.* 68 (2): 547–554.

Yoo, S., G. Y. Kim, R. Hammoud, E. Elder, T. Pawlicki, H. Guan, T. Fox, G. Luxton, F. F. Yin, and P. Munro. 2006. A quality assurance program for the on-board imagers. *Med. Phys.* 33 (11): 4431–4447.

Zeng, G. G., S. L. Breen, A. Bayley, E. White, H. Keller, L. Dawson, and D. A. Jaffray. 2009. A method to analyze the cord geometrical uncertainties during head and neck radiation therapy using cone beam CT. *Radiother. Oncol.* 90 (2): 228–230.

54

Cobalt: Implementation and Use

Cari Borrás
Federal University of Pernambuco

Introduction

According to the latest UNSCEAR Report (UNSCEAR 2008) for the period 1997–2007, the global use of radiation therapy increased to 5.1 million treatments from 4.7 million treatments in the period 1991–1996. About 4.7 million patients were treated with external beam radiation therapy, whereas 0.4 million were treated with brachytherapy. The availability of linear accelerators worldwide was about 1.6 machines per million people in the population, whereas the availability of x-ray machines and of cobalt units was about equal, at 0.4 per million people. In absolute numbers, UNSCEAR claims that in the period 1997–2007, the number of linear accelerators worldwide increased to about 9000 from about 5000 in the previous period. The 2009 IAEA Directory of Radiotherapy Centers shows the global distribution of high-energy machines, totaling 7177 clinical accelerators and 2505 Co-60 units (IAEA 2009). There are no numerical data on the global distribution of GammaKnife and GyroKnife systems, but it is known that most of the GammaKnife units have been installed in level I countries, where most of the linear accelerators function and where 24% of the population received 76% of the total radiation therapy treatments (UNSCEAR 2008). (Level I countries are countries where there is at least one physician for every 1000 people in the general population.) Gyro Knife units are mostly working in China, where they were first developed (Gammastar 2008).

Of great concern is this skewed distribution of radiotherapy units. There are some parts of the world, such as large regions of Africa and South East Asia, where there may be only one high-energy radiation therapy machine for 20–40 million people, and one machine may be used to treat more than 600 new patients per year. Many cancer patients have no access to radiotherapy services. Yet, increases in life expectancy mean there will be even a greater demand of radiation therapy, mostly consisting of conventional Co-60 units. And if cobalt-based tomotherapy units become commercially available, their use will be extended to both industrialized and developing countries.

Types of Cobalt-Based Radiotherapy Machines

There are three types of external beam radiotherapy units that use the radionuclide Co-60 (mean energy, 1.25 MeV; half-life, 5.27 years) as the source of radiation: conventional cobalt-60 units, the GammaKnife, and the GyroKnife. They all have cobalt-60 sources, either as pellets or as a casting, which are double encapsulated in stainless steel capsules and seal-welded by the manufacturer, who leak-tests them before inserting them in the housing. Similarities and differences among these systems related to quality assurance (QA) will be discussed in this chapter.

Conventional Co-60 Units

Since the technology was first developed in Canada in the late 1940s, both the source and the machine have undergone several designs (Johns and Cunningham 1983). Vertically mounted units, which moved the treatment head over a fixed patient table and allowed different surface-skin distances, were eventually replaced by isocentric machines, with source-axis distances first of 0.80 m and more recently of 1 m. Design specifications for equipment to be used in developing countries were articulated in 1993 (Borrás and Stovall 1995).

Source Characteristics

The outer capsule of the Co-60 source is a cylinder of about 50 mm in diameter, positioned in the treatment head with its circular end facing the patient. The radioactive source inside has an active diameter from 10 to 25 mm, with the consequent effect on the penumbra of the dose profile. Activities range from 185 to 370 TBq.

Beam Collimation

Several mechanisms have been developed over the years to collimate the beam to the area of interest and to control the irradiation process (Johns and Cunningham 1983). Current collimators consist of a set of multiplane or multivane (moving arc) heavy metal blocks that can be moved independently or in pairs to obtain a rectangular-shaped field. To minimize the penumbra, the collimator blocks can be shaped so that the inner surface of the blocks is always parallel to the radiation beam. The possibility of attaching multileaf collimators and even an accessory for tomotherapy has been explored (Dhanesar 2008).

Source Motion Mechanisms

To control the irradiation time, there are two source motion mechanisms currently in use. In one of them, the source is mounted on a rotating wheel that moves the source 180° toward (source "on") or away (source "off") from the collimator aperture. In another design, the source is mounted on a metal drawer that slides horizontally in (source "on") and out (source "off") an opening in the treatment head facing the collimator aperture.

GammaKnife

The GammaKnife device, developed in Sweden in the late 1960s, contains 201 Co-60 sources of approximately 1.1 TBq each, placed in a spherical array in a heavily shielded assembly. The machine was designed exclusively for noninvasive brain surgery. The 201 beam channels are focused to a single point at the center of the radiation unit—the focal distance is 403 mm. Rigid immobilization to prevent head movement is achieved by a lightweight stereotactic head frame fixed to the outer skull.

Source Characteristics and Beam Collimation

Each Co-60 source consists of 20 pellets, 1 mm in diameter and height, stacked on top of one another, and encased in a double-walled stainless steel capsule. Each source fits into a source-bushing assembly and has a precollimator and a collimator. The final collimation is accomplished with one of four collimator helmets with circular apertures that produce nominal 4-, 8-, 14-, or 18-mm-diameter fields at the focus. The apertures of individual collimators can be replaced with occlusive plugs to block the radiation. Each helmet is equipped with a pair of trunnions, which serve as the fixation points for the stereotactic frame in the right–left direction (Schell et al. 1995). To achieve positional accuracy, there is an automatic patient positioning system.

Gyro Knife

The Gyro Knife, developed in China in the late 1990s, is a rotating stereotactic radiotherapy and radiosurgery system that, using gyroscopic principles, allows for triple focusing of the radiation beams from a Co-60 source, mounted on two gyroscope structures that rotate synchronously in the vertical direction. The technology has the capacity for noninvasive whole-body stereotactic fractionated radiotherapy and single high-dose radiosurgery. There are four sets of collimators of diameters 5, 15, 25, and 50 mm for brain irradiation; body radiotherapy treatments are done with large fields. Patient positioning is facilitated by several three-dimensional positioning systems. The activity of the Co-60 is 240.5 TBq (Gammastar 2008).

Commissioning Considerations

Radiation Safety

Before equipment commissioning, it is necessary to verify the adequacy of the structural shielding and to perform a radiation safety survey of the facility. This includes testing the door interlocks, the source safe/fail mechanism, the radiation warning signs, the radiation monitor alarms, the emergency off switches, the beam on–off lights, the audio visual system, and all the collision and alignment microswitches present. The next steps are to check the integrity of the source encapsulation by wipe-testing the accessible areas and to measure the kerma rate around the source housing to ensure compliance with federal and state regulations.

Equipment Commissioning

Before performance parameters are tested and radiation characteristics are measured, the unit should be checked for mechanical integrity, mechanical stability, electrical integrity, and electrical safety, in accordance with manufacturer's specifications and national and/or local safety codes (Borrás 1997). The initial testing should verify the accuracy of readouts (scales, meters, and digital displays) and the proper functioning of all the equipment components and accessories. The dimensions of the mechanical and radiation isocenter have to be measured and their congruence has to be verified. The activity of the source(s) needs to be confirmed, and absorbed dose rates, as well as dose beam profiles, have to be determined for all the field sizes clinically available. Calibrated ionization chambers and electrometers, as well as suitable phantoms to be used for these measurements, are to follow dosimetry protocols (Almond et al. 1999; IAEA 2000a, 2000b). Corrections for source decay are to be applied during the commissioning of the machine and periodically thereafter. A monthly recalculation is recommended. The source transit time (timer error) needs to be accounted for in the determination of the absorbed dose rate, not only at the time of the initial calibration of the machine but also periodically thereafter, as changes over time may occur. In addition, the timer constancy, linearity, and accuracy need to be established. Because of their relatively short half-life, Co-60 sources require replacement every 3–6 years. If replaced less frequently, treatment times may need to be increased so that biologically effective doses that account for differences in fractionation schemes remain constant.

Conventional Co-60 Units

In addition to the tests listed in the preceding paragraph, it is necessary to perform mechanical and optical checks to verify gantry motions, including those of all moving parts, such as the collimator and detachable accessory trays, as well as motions of the patient support assembly. Radiation field symmetry needs to be tested to ensure the proper placement of the Co-60 source. The next step is to verify the alignment of the gantry, collimator, and table isocenters and the congruence between the radiation and the light fields. Beam modifiers such as wedges and compensators need to be measured and the corresponding beam profiles for all the fields and depths to be used clinically should be entered in the treatment planning computer together with the profiles corresponding to the open fields. Patient positioning devices and laser lights are to be tested and adjusted if needed.

GammaKnife

The commissioning of the gamma knife includes additional tests to those listed in a previous section. The pressure of the hydraulic system that opens and closes the door and moves the treatment table in and out of the unit needs to be checked. The accuracy of the trunnions' centricity, helmet alignment, and target localization need to be determined. These tests can be done with tools specifically designed for this purpose by the manufacturer. Absolute calibration should be carried out for the 18-mm collimator with the ionization chamber positioned at the convergence of the radiation beams, either in the center of an 800-mm-radius sphere of tissue equivalent material (Schell et al. 1995) or in a thimble-shaped water phantom with a 2-mm plastic wall to contain water (Drzymala, Wood, and Levy 2008). Dose profiles of each helmet (collimator) have to be measured individually while plugging the other 200 collimators. This can be

achieved using film. Relative helmet factors can be determined using diodes or thermoluminiscent dosimeters. The methodology and needed instrumentation are described in the AAPM Report 54 (Schell et al. 1995).

Quality Assurance Program

QA protocols should address facilities, equipment, and procedures, as discussed elsewhere. Regarding equipment, the QA program should repeat on a periodic basis the tests performed during commissioning and be closely coordinated with the maintenance program, as adherence to the set of maintenance steps recommended by the manufacturer is critical to ensure patient safety and prolong the life of the machine (Borrás 1997).

Conventional Co-60 Units

Table 1 of AAPM TG 40 (AAPM Report 46 1994) lists the frequencies and tolerances of the procedures that constitute a comprehensive QA program. A critical maintenance aspect for those units with a source motion mechanism that functions by compressed air is to check the air pressure periodically and adjust it as needed.

GammaKnife

AAPM Report 54 (Schell et al. 1995) and Mack et al. (2004) describe the methodology and instrumentation required to conduct a comprehensive QA program, which includes the parameters, tolerances, and measurement frequency recommended.

Safety Aspects

Patient Safety

Patient safety is integrated with the QA of the radiotherapy treatments. Like in other radiation therapy modalities, the goal of the QA program of cobalt-based treatments is to ensure the accurate delivery of the prescribed dose to the tumor in the patient and to minimize the dose to other tissues (IAEA 1996). The measures to ensure quality of a radiotherapy treatment inherently provide for patient safety and for the avoidance of accidental exposures (IAEA 1998). The most common overexposures have occurred because of beam miscalibrations (IAEA 2000a, 2000b; ICRP 2000), but patient fatalities due to mechanical problems have also occurred (AECL 1987). Patient safety has been the subject of recent comprehensive publications (The Royal College of Radiologists et al. 2008; WHO 2008).

Staff Safety

Staff exposure may be higher with machines with radioactive sources than with linear accelerators because the treatment head shielding does not block entirely the emitted radiation. This exposure will be even higher around some conventional cobalt-60 units that use depleted uranium collimators (to diminish the weight of the treatment head). Accidents have occurred where the source has been stuck in the "on" position. Emergency procedures should be developed, posted at the entrance of the room, and periodically tested.

Public Safety

Co-60 sources require replacement at least every 6 years. Manufacturers will remove the old sources when purchasing new ones, but if the facility decides to buy a linear accelerator or close the center, the disposal of Co-60 sources is very expensive. Disposal costs should be budgeted at the time the equipment is first purchased. Irresponsible disposal or abandonment of a Co-60 unit can result in severe radiation accidents such as the one that occurred in Ciudad Juarez, Mexico (CNSNS 1984).

References

AAPM Report 46. 1994. *Comprehensive QA for Radiation Oncology* (reprinted from *Medical Physics*, vol. 21, issue 4). New York, NY: American Institute of Physics, Inc.

AECL. 1987. *Safety Alert Letter for All El Dorado Cobalt Therapy Unit Users*. Ottawa, ON: Atomic Energy Canada Limited (AECL).

Almond, P. R., P. J. Biggs, B. M. Coursey, W. F. Hanson, M. S. Huq, R. Nath, and D. W. Rogers. 1999. AAPM's TG-51 protocol for clinical reference dosimetry of high-energy photon and electron beams. *Med. Phys.* 26 (9): 1847–1870.

Borrás, C., ed. 1997. *Organization, Development, Quality Control and Radiation Protection in Radiological Services—Imaging and Radiation Therapy*. Washington, DC: Pan American Health Organization/World Health Organization.

Borrás, C., and J. Stovall, eds. 1995. *Design Requirements for Megavoltage X-ray Machines for Cancer Treatment in Developing Countries: Report of an Advisory Group Consultation*. Los Alamos, NM: Los Alamos National Laboratory (LA-UR-95-4528).

CNSNS. 1984. *Accidente por Contaminación con Cobalto-60*. México, DF: Comisión Nacional de Seguridad Nuclear y Salvaguardias (CNSNS), Secretaría de Energía, Minas e Industria Paraestatal.

Dhanesar, S. K. 2008. Conformal Radiation Therapy with Cobalt-60 Tomotherapy 2008. http://hdl.handle.net/1974/1182 (accessed February 28, 2009).

Drzymala, R. E., R. C. Wood, and J. Levy. 2008. Calibration of the GammaKnife using a new phantom following the AAPM TG51 and TG21 protocols. *Med. Phys.* 35 (2): 514–521.

Gammastar. 2008. Radiotherapy Treatment Equipment. Gyro Knife γ-Series 2008. http://www.gammastar.com/en/zhiliao/y_series.html (accessed February 28, 2009).

IAEA. 1996. *International Basic Safety Standards for Protection against Ionizing Radiation and for the Safety of Radiation Sources*. Vienna: International Atomic Energy Agency (IAEA).

IAEA. 1998. *Design and Implementation of a Radiotherapy Programme: Clinical, Medical Physics, Radiation Protection and Safety Aspect*. Vienna: International Atomic Energy Agency.

IAEA. 2000a. Absorbed dose determination in external beam radiotherapy: An international code of practice for dosimetry based on standards of absorbed dose to water. In *Technical Reports Series No. 398*. Vienna: International Atomic Energy Agency (IAEA).

IAEA. 2000a. *Lessons Learned from Accidental Exposures in Radiotherapy*. Vienna: International Atomic Energy Agency (IAEA).

IAEA. 2009. Directory of Radiotherapy Centers. http://www-naweb.iaea.org/nahu/dirac/default.asp (accessed February 28, 2009).

ICRP. 2000. *Prevention of Accidental Exposure to Patients Undergoing Radiation Therapy*. Oxford: International Commission on Radiological Protection.

Johns, H. E., and J. R. Cunningham. 1983. *The Physics of Radiology*. 4th ed. Springfield, IL: Charles C Thomas.

Mack, A., S. G. Scheib, M. Rieker, D. Weltz, G. Mack, H. Czempiel, H. J. Kreiner, H. D. Boettcher, and V. Seifert. 2004. A system for quality assurance in radiosurgery. In *Radiosurgery*, ed. D. Kondziolka. Basel: Karger Vol 5, pp 236–246.

Schell, M. C., F. J. Bova, D. A. Larson, D. D. Leavitt, W. R. Lutz, E. B. Podgorsak, et al. 1995. *Stereotactic Radiosurgery: Report of Task Group 42 Radiation Therapy Committee*. Woodbury, NY: American Institute of Physics.

The Royal College of Radiologists, Society and College of Radiographers, Institute of Physics and Engineering in Medicine, National Patient Safety Agency, and British Institute of Radiology. 2008. *Towards Safer Radiotherapy*. London: The Royal College of Radiologists.

UNSCEAR. 2008. *Sources and Effects of Ionizing Radiation: UNSCEAR 2008 Report to the General Assembly, with Scientific Annexes*, New York, NY: United Nations: United Nations Scientific Committee on the Effects of Atomic Radiation.

WHO. 2008. Radiotherapy Risk Profile—Technical Manual by the World Health Organization. Geneva, Switzerland: World Health Organisation.

55

Superficial and Orthovoltage: Implementation and Use

Manuel A. Morales-Paliza
Vanderbilt University

Charles W. Coffey
Vanderbilt University

Introduction

Superficial and orthovoltage x-ray equipment continues to be used today in radiotherapy. Photon beams in the range from 40 to 300 kVp remain of interest in limited radiation oncology applications, such as for the treatment of some skin disorders and superficial lesions. In addition, low- and medium-energy photons are often used as a source of radiation for various applications in radiobiology research laboratories. The relative high dose rates achievable and the limited penetration of superficial and orthovoltage x-rays make them very appropriate for the irradiation of small animals and cells grown in culture media. For accurate dosimetry results in this often infrequently used modality, correct calibration and beam parameterization and adequate quality assurance of the equipment are of utmost importance; physicists should follow standard procedural protocols to guarantee proper equipment function and dose delivery.

Following the AAPM's TG-61 (Ma et al. 2001) protocol guidelines for kilovoltage x-ray beam dosimetry in the range of 40–300 kVp, there are two regions of clinical radiation oncology and radiobiological interest: (1) low-energy (or superficial) x-rays for tube potentials lower than or equal to 100 kVp and (2) medium-energy (or orthovoltage) x-rays for tube potentials higher than 100 kVp.

Scope of Use

Limited centers in the United States and a large number of centers around the world use kilovoltage x-ray machines to treat a wide variety of superficial disorders, including the treatment of nonneoplastic diseases and some superficial malignant lesions (Malinverni et al. 2002). Superficial and orthovoltage x-rays remain a significant modality choice for endocavitary irradiation of some rectal carcinomas, malignant skin disorders seen in patients with autoimmune-related diseases, and in limited radiation shielding environments using intraoperative radiation (Coffey 1992). Recent introduction of electronic brachytherapy (Smitt and Kirby 2007) is the proposed use of a miniature 50-kVp x-ray tube for radiotherapy of accelerated partial breast radiotherapy, vaginal carcinoma, and endocavitary irradiation of rectal carcinoma.

The use of small fixed treatment cones, limited penetration, historically known tumor and normal tissue responses, ease of use, and associated low equipment and maintenance costs make low-energy x-rays often the preferred treatment of choice for some patients with basal cell and squamous cell tumors of the skin. In addition, some malignant skin disorders associated with immune deficiency such as Kaposi's sarcoma can be treated with superficial and orthovoltage x-rays (Cooper and Fried 1987). Typically, 100-kVp x-rays delivered with 8–10 Gy as a single dose or 15–20 Gy over 1 week are given to treat Kaposi's sarcoma as an alternative to electron irradiation.

Superficial and orthovoltage x-rays have been used in intra-operative radiotherapy (Dobelbower and Abe 1989). Although electron beams deliver a more homogeneous dose distribution in large tumors and offer a better sparing of normal tissues distal to the tumor, use of a superficial/orthovoltage unit in intra-operative radiotherapy possesses the advantage of placing the unit within an operating room because of the relatively modest shielding requirements. In this way, transport of anesthetized patients from the operative suite to the radiotherapy linear accelerator can be avoided. Hence, superficial/orthovoltage radiotherapy may offer a cost-effective answer to the often cost-prohibitive financial concerns of linac-based intraoperative radiotherapy.

As an alternative to surgery, selected cases of rectal adeno-carcinomas can be treated using low-energy x-rays (at 50 kVp, called contact therapy) including a fiberoptic scope for guidance (Matar, Coffey, and Maruyama 1988). This endocavitary technique is an adaptation of the original method developed by Papillon (1975) in France, in which tumors—preferably adeno-carcinomas that have not penetrated the bowel wall and are within 10–12 cm of the anal verge—receive a very high local dose at a high dose rate while normal tissues are spared. Rusch et al. (2008) have recently proposed electronic brachytherapy as an alternative irradiation method for delivering a modified endocavitary Papillon technique.

Commissioning

To commission new kilovoltage equipment, several parameters and specifications need to be investigated. An appropriate selection of x-ray peak voltages (in peak kilovolts), tube currents (in milliamperes), and thicknesses of filtration to harden the beam are needed. In addition, cone sizes and associated SSDs need to be specified. With all these different parameters, the commissioning process evaluates the absolute dose, beam uniformity, beam quality, shutter/timer errors, and the inverse square factor.

Selection of Milliampere, Peak Kilovolt, and Filtration Combinations

Each kilovoltage x-ray unit provides specifications of different combinations of tube potentials (peak kilovolts), tube currents (milliamperes), and thicknesses of filtration [in half value layer (HVL)]. Each set will define a different energy beam quality. Tube currents in the range of 5–8 mA for low-energy and 10–20 mA for medium-energy x-rays are typically used, as well as typical HVL values that range between 1 and 8 mm aluminum for superficial x-rays and 1 and 4 mm copper for orthovoltage x-rays (Khan 2003).

Cone Sizes and Associated SSDs vs. Variable Collimators

Superficial x-ray treatments are usually given with the help of applicators or cones attachable to the diaphragm of the machine. In orthovoltage therapy, cones can be used to collimate the beam to the desired size as well; but additionally, a movable diaphragm, consisting of lead plates, allows a continuously adjustable field size. The SSD typically ranges between 15 and 20 cm for superficial treatments, whereas it is usually set at 50 cm in orthovoltage therapy (Khan 2003).

Beam Uniformity

The intensity of the radiation across the treatment field is an important concept to evaluate at any radiotherapy beam energy.

At low energies, the heel effect (uneven radiation distribution along the anode to cathode axis) may introduce a clinically significant nonuniform radiation distribution. To evaluate the uniformity of the beam, scanning ionization chambers or film densitometry methods may be used (Coffey 1992).

Beam Quality: Tube Potential, HVL, and Percentage Depth Dose

A measurement of HVL may be affected by the experimental setup, procedures, and the energy dependence of the particular dosimeters. In general, the manufacturers give target and filtration information, whereas other factors are not usually very well known. The AAPM's TG-61 protocol recommends separating the beam quality specification into two main stages: (1) obtaining the air-kerma calibration factor N_K from the standards laboratory by calibrating the chamber in terms of both the tube potential and HVL and (2) measuring the absorbed dose in the user beam by considering the HVL as the only quality metric.

Specific recommendations on how to setup the HVL measurements are provided in the TG-61 protocol. In general, the first HVL for an x-ray beam is defined as the thickness of an absorber that reduces the air-kerma rate to one half under conditions that minimize the number of scattered photons originating in the absorber that reach the detector, usually referred as "good geometry" conditions. Then, detectors with sufficient buildup thickness shall be used to minimize the effect of contaminant electrons. The beam diameter, as defined by the diaphragm, shall be 4 cm or less and the detector shall be placed at least 50 cm away from the attenuating material and the diaphragm. The monitor chamber for normalization of the ion chamber signal, if needed, shall be positioned such that its response is not affected by changing the attenuator thickness, as shown in Figure 1 of the AAPM's TG-61 protocol (Ma et al. 2001).

In addition, the beam quality specification may be given, in terms of percentage depth dose, in tissue or water. The percentage depth dose is defined as the ratio of the dose at depth relative to the dose at d_{max} (d_{max} is at the surface for low- and medium-energy x-rays). Experimental measurements of percentage depth dose can be made with the appropriate ionization chamber in a water phantom and the corresponding suitable absorbed-dose-to-water equation. Depending on the range of energy and depth, the AAPM's TG-61 protocol provides the equations for the absorbed dose to water as a function of parameters that can be obtained from the manufacturers and from the literature. The formalisms derived from these equations will be discussed in the section Outputs (TG-61).

Shutter/Timer Errors

Superficial and orthovoltage therapy machines generally show a "ramp-up region" allowing x-ray peak voltages and tube currents to reach the steady state. During this period the dose rate increases from zero to the steady-state output. This can generate a significant dose deficit, which can be compensated by adding a correction to the treatment time. This correction can be determined from the intercept of the dose (reading) versus time plot. Alternatively, the timer/shutter error or end effect, δt, can be calculated using a derived equation (Attix 1986).

Inverse Square

Provided the effective source position is known, the inverse-square relation can be used to test the accuracy of in-air measurements at different source-chamber distances. In general, the effective source position is different from the x-ray focal spot, and it can be determined by using measurements performed at different distances and extrapolating to the desired distance (Li, Salhani, and Ma 1997). This method is particularly useful for in-air calibration with closed-end cones when a cone-end air kerma measurement needs to be made (Ma et al. 2001).

Outputs (TG-61)

Historically, there have been several dosimetry protocols regarding calibration and specification of kilovoltage x-ray beams. In 2001, the AAPM Task Group 61, based on these previous protocols, established a set of recommendations that has become the standard protocol for kilovoltage x-ray dosimetry.

In-Air vs. In-Water

The AAPM's TG-61 protocol is based on in-air calibrated ionization chambers in terms of air kerma. Depending on where the point of interest is located, one of two methods can be used. For points near the surface, the absorbed dose to water at the surface of a water phantom can be calculated by in-air measurements (the "in-air" method). For points of interest at a depth and tube potentials greater than or equal to 100 kVp, the absorbed dose to water can be evaluated by in-water measurements at a depth of 2 cm (the "in-phantom" method). The protocol provides guidelines to determine the dose at other points in water and the dose at the surface of other materials of interest.

Chamber Selection

The TG-61 protocol recommends using air-filled ionization chambers for reference dosimetry in kilovoltage x-ray beams. The center of the sensitive air cavity of the chamber should be the effective point of measurement for both cylindrical and parallel-plate chambers. The in-air or in-phantom raw reading, M_{raw}, should be corrected for temperature–pressure (P_{TP}), ion recombination (P_{ion}), polarity (P_{pol}), and electrometer accuracy (P_{elec}), such that the fully corrected reading can be calculated as $M = M_{raw} P_{TP} P_{ion} P_{pol} P_{elec}$. Care should be taken in ensuring that chamber calibration factors do not change significantly between two calibration beam qualities (uncertainty less than 2%). Calibrated soft x-ray parallel-plate chambers with a thin entrance window are recommended for use for low-energy x-rays with tube potentials less than 70 kVp, whereas suitable

cylindrical chambers can be used with tube potentials greater than or equal to 70 kVp. To remove electron contamination and provide full buildup, thin low-Z (polyethylene or PMMA) foils or plates can be added to the entrance window. Table 1 of the AAPM's TG-61 protocol refers to the total wall thicknesses for various tube potentials (Ma et al. 2001). For instance, the total required wall thickness for a tube potential of 60 kVp for a thin-window plane-parallel chamber is 5.5 mg/cm² (ICRU 1984); then, for a chamber with window thickness of 2.5 mg/cm², a 3.0 mg/cm² of foil thickness needs to be added.

For medium-energy x-rays, cylindrical chambers with calibration factors changing with the beam quality by less than 3% should be used for reference dosimetry. In water, care should be taken in waterproofing the chamber (talcum powder and air gaps between the chamber and sleeve of no greater than 0.2 mm should be allowed). In air, cylindrical chambers do not require a buildup cap because they have adequate thimble thickness.

Chamber Calibration: Air-Kerma

The ionization chamber calibration involves the measurement of the air kerma free in air (K_{air}) in an appropriate x-ray beam in a standard laboratory reference beam quality. If K_{air} is the air kerma at the reference point in air for a given beam quality and M the corrected (see section "Chamber Selection") reading of an ionization chamber to be calibrated with its reference point at the same point, the air-kerma calibration factor N_K for this chamber at the specified beam quality is defined as $N_K = K_{air}/M$.

The air-kerma calibration factors must be traceable to national standards institutions such as the Accredited Dosimetry Calibration Laboratories, the National Institute for Standards and Technology, or the National Research Council of Canada, preferably for several x-ray beam qualities. Both HVL and tube potential shall be used to specify N_K. Then, several values of N_K corresponding to different combinations of HVLs and tube potentials should be provided to adequately evaluate the beam qualities for each machine.

"In-Air" Formalism

For low- and medium-energy x-rays (tube potentials between 40 and 300 kVp) and point of interest at the phantom surface ($z_{ref} = 0$), absorbed-dose-to-water measurements must be achieved free in air using a backscatter factor to account for phantom scattering. Under these conditions, the absorbed dose to water at the phantom surface is determined by

$$D_{w,z=0} = M N_K B_w P_{stem,air} \left[\left(\frac{\bar{\mu}_{en}}{\rho} \right)_{air}^{w} \right]_{air}, \qquad (55.1)$$

where M is the free-in-air corrected (see section "Chamber Selection") chamber reading, with the center of the sensitive air cavity of the ionization chamber placed at the measurement point ($z_{ref} = 0$); N_K the air-kerma calibration factor; B_w the backscatter factor accounting for the effect of the phantom scatter (which must include the effect of end plates in close-ended cones, if applicable); $P_{stem,air}$ the chamber stem correction factor that accounts for the change in photon scatter from the chamber stem between the calibration and measurement, due mainly to the change in field size (i.e., if the calibration and measurement were done using the same field size, this factor is equal to one); and $\left[\left(\bar{\mu}_{en}/\rho\right)_{air}^{w}\right]_{air}$ the ratio for water to air of the mean mass energy-absorption coefficients averaged over the incident photon spectrum. Tables with values for B_w, $P_{stem,air}$, and $\left[\left(\bar{\mu}_{en}/\rho\right)_{air}^{w}\right]_{air}$ can be found in the AAPM's TG-61 protocol.

Equation (55.1) stands under conditions of charged particle equilibrium and in the absence of electron contamination from the primary beam, which applies to both open and closed cones. If the measurement cannot be performed at the point of interest, it should be done at a point as close as possible to the point of interest; then, an inverse square correction can be used (see section "Inverse Square").

"In-Phantom" Formalism

For tube potentials greater than 100 kVp (medium-energy x-rays) and point of interest at depth in water, the measurement must be performed at the reference depth ($z_{ref} = 2$ cm) in a water phantom. The in-phantom formalism adopted a reference depth of 2 cm to ensure that enough buildup material exists in the upstream direction to cover the whole chamber and, at the same time, the ionization signal in the chamber is not too small as at deeper depths. Under these conditions, the absorbed dose to water at 2 cm reference depth in water for a 10×10 cm^2 field size at 100 cm SSD can be determined by

$$D_{w,z=2cm} = M N_K P_{Q,cham} P_{sheath} \left[\left(\frac{\bar{\mu}_{en}}{\rho}\right)_{air}^{w}\right]_{water}, \quad (55.2)$$

where M is the corrected (see "Chamber Selection") chamber reading, with the center of the air cavity placed at the reference depth ($z_{ref} = 2$ cm); N_K the air-kerma calibration factor; $P_{Q,cham}$ the overall chamber correction factor that accounts for the change in the chamber response due to the displacement of water by the ionization chamber and chamber stem, the change in the energy, and angular distribution of the photon beam in the phantom compared to that used for the calibration in air; P_{sheath} the correction for photon absorption and scattering in the waterproofing sleeve (if any); and $\left[\left(\bar{\mu}_{en}/\rho\right)_{air}^{w}\right]_{water}$ the ratio for water to air of the mean mass energy-absorption coefficients averaged over the photon spectrum. Tables with values for $P_{Q,cham}$, P_{sheath}, and $\left[\left(\bar{\mu}_{en}/\rho\right)_{air}^{w}\right]_{water}$ can be found in the AAPM's TG-61 protocol.

Other Considerations

In section "Chamber Selection," we mentioned that the in-air or in-phantom raw reading for the ionization chamber, M_{raw}, should be corrected for temperature–pressure (P_{TP}), ion recombination (P_{ion}), polarity (P_{pol}), and electrometer accuracy (P_{elec}). Measurements of P_{TP}, P_{ion}, and P_{pol} follow the same criteria as in the AAPM's TG-51 protocol for megavoltage energies (Almond et al. 1999). P_{elec} is determined at the Accredited Dosimetry Calibration Laboratories or the National Institute for Standards and Technology. Also, in previous sections, we dealt with shutter/timer errors (end effect) and the inverse-square factor, respectively. All these parameters need to be considered to determine the absorbed doses to water using any of the previous formalisms.

The AAPM's TG-61 protocol provides two easy-to-follow worksheets to calculate the dose to water and the dose rate with any of these formalisms. The dose rate is defined as the dose to water (Equations 55.1 or 55.2) divided by the sum of the total radiation time and the end-effect correction (timer error, δt).

Regulatory Issues and Routine Quality Assurance

Importantly, physicists should refer to applicable state and federal radiation regulations and guidelines governing the proper installation and use of low-energy radiotherapy x-ray units. These regulations and guidelines will include radiation safety requirements including, but not limited to, leakage radiation limits and safety interlocks [emergency stop, door interlock, treatment timer, x-ray warning lights, radiation area monitors, patient monitors (visual and audio)]. Also, the regulations and guidelines will refer to the type and frequency of quality assurance checks and measurements required, frequency of ion chamber calibration for absolute output measurements, qualifications of the physicist making/reviewing the measurements, and records keeping requirements. Additional machine manufacture and operational recommendations have been established by the International Electrotechnical Commission (IEC 601-2-8), which contains requirements for the safety of therapeutic x-ray generators operating with nominal tube voltages from 10 to 400 kVp (IEC 1987).

In addition to compliance with required state and federal regulations and guidelines and IEC recommendations, routine quality assurance procedures for superficial and orthovoltage x-ray equipment should be performed on a regular basis. Although AAPM's TG-40 (Kutcher et al. 1994) does not specifically list quality assurance procedures for low-energy x-ray radiotherapy units, selective quality assurance procedures for cobalt teletherapy units and linear accelerators can be modified and adapted for applications with low-energy x-ray beams. Two studies (Niroomand-Rad et al. 1987; Coffey 1992) have addressed the need and suggested procedures for the routine quality assurance of low energy x-ray units. Table 55.1 lists these procedures. AAPM's TG-152 Report (Thomadsen et al. 2009) addresses the

TABLE 55.1 Suggested Quality Assurance for Low- and Medium-Energy X-Rays

Frequency	Parameter
Yearly	HVL[a]
	Beam uniformity
	Light field vs. radiation field coincidence
	Dose reproducibility[a]
	Absolute dose (reference field size)[a]
	Cone (field) size correction factors[a]
	Optical distance indicator (if applicable)
	Inverse square law evaluation
	Timer accuracy and linearity
	Timer–shutter error evaluation [a]
Daily	Proper x-ray tube warm-up
	Treatment door interlock check
	X-ray warning light functionality
	Area radiation monitor(s) functionality
	Timer accuracy
	Dose output reproducibility[b]

[a] For each peak kilovolt, milliampere, and added filter combination.
[b] For one selected peak kilovolt, milliampere, and added filter combination.

safety, clinical operations, calibration, and routine quality assurance issues that are involved with the anticipated introduction of electronic brachytherapy methods within the radiotherapy community. In addition to the quality assurance issues discussed for low-energy x-ray units, electronic brachytherapy quality assurance will incorporate additional checks associated with the use of remote high dose rate afterloaders.

Radiation protection guidelines and facility shielding methodologies for low-energy x-ray units can be found in the National Council on Radiation Protection and Measurements Report No. 151 (NCRP 2005). Shielding equations and calculations of necessary lead or concrete are straightforward for low-energy x-rays; appropriate estimation of workload may be the most difficult shielding parameter for the medical physicist to assess. Specific shielding caveats include the retrofit of a low-energy x-ray unit into an existing space (e.g., a conventional radiotherapy simulator room), the replacement of a superficial x-ray unit (100 kVp) with an orthovoltage unit (300 kVp) into an existing space, and the introduction of electronic brachytherapy into a radiotherapy or operating room environment. With each scenario, the medical physicist needs to follow current state and federal radiation permissible radiation limits and low-energy x-ray unit safety regulations and guidelines.

References

Almond, P. R., P. J. Biggs, B. M. Coursey, W. F. Hanson, M. S. Huq, R. Nath, and D. W. Rogers. 1999. AAPM's TG-51 protocol for clinical reference dosimetry of high-energy photon and electron beams. *Med. Phys.* 26 (9): 1847–1870.

Attix, F. H. 1986. *Introduction to Radiological Physics and Radiation Dosimetry*. New York, NY: Wiley.

Coffey, C. 1992. Calibration of low energy x-ray units. In *Advances in Radiation Oncology Physics*, ed. J. Purdy. New York, NY: AAPM's Medical Physics Monograph 19: 148–180.

Cooper, J. S., and P. R. Fried. 1987. Defining the role of radiation therapy in the management of epidemic Kaposi's sarcoma. *Int. J. Radiat. Oncol. Biol. Phys.* 13 (1): 35–39.

Dobelbower, R. R., and Mitsuyuki Abe. 1989. *Intraoperative Radiation Therapy*. Boca Raton, FL: CRC Press.

ICRU. 1984. *Stopping Powers for Electrons and Positrons*. Bethesda, MD: International Commission on Radiation Units and Measurements.

IEC. 1987. *Specification for Therapeutic X-ray Generators*. Geneva: International Electrotechnical Commission.

Khan, F. M. 2003. *The Physics of Radiation Therapy*. Philadelphia, PA: Lippincott Williams & Wilkins.

Kutcher, G. J., L. Coia, M. Gillin, W. F. Hanson, S. Leibel, R. J. Morton, J. R. Palta et al. 1994. Comprehensive QA for radiation oncology: Report of AAPM Radiation Therapy Committee Task Group 40. *Med. Phys.* 21 (4): 581–618.

Li, X. A., D. Salhani, and C. M. Ma. 1997. Characteristics of orthovoltage x-ray therapy beams at extended SSD for applicators with end plates. *Phys. Med. Biol.* 42 (2): 357–370.

Ma, C. M., C. W. Coffey, L. A. DeWerd, C. Liu, R. Nath, S. M. Seltzer, and J. P. Seuntjens. 2001. AAPM protocol for 40–300 kV x-ray beam dosimetry in radiotherapy and radiobiology. *Med. Phys.* 28 (6): 868–893.

Malinverni, G., M. Stasi, B. Baiotto, C. Giordana, G. Scielzo, and P. Gabriele. 2002. Clinical application and dosimetric calibration procedure of the superficial and orthovoltage therapy unit Therapax DXT300. *Tumori* 88 (4): 331–337.

Matar, J. R., C. Coffey, and Y. Maruyama. 1988. Rectal carcinoma: Treatment with Papillon technique and fiberoptic-guided methods. *Radiology* 168 (2): 562–564.

NCRP. 2005. *Structural Shielding Design and Evaluation for Megavoltage X- and Gamma-ray Radiotherapy Facilities*. Berkeley, CA: National Council on Radiation Protection and Measurements.

Niroomand-Rad, A., M. T. Gillin, F. Lopez, and D. F. Grimm. 1987. Performance characteristics of an orthovoltage x-ray therapy machine. *Med. Phys.* 14 (5): 874–878.

Papillon, J. 1975. Intracavitary irradiation of early rectal cancer for cure. A series of 186 cases. *Cancer* 36 (2): 696–701.

Rusch, T., E. Klein, L. Kelley, R. Myerson, and S. Axelrod. 2008. Dosimetry of an x-ray endocavitary proctoscope adapted for use with the Axxent electronic brachytherapy system. *Med. Phys.* 35: 2851.

Smitt, M. C., and R. Kirby. 2007. Dose-volume characteristics of a 50-kV electronic brachytherapy source for intracavitary accelerated partial breast irradiation. *Brachytherapy* 6 (3): 207–211.

Thomadsen, B. R., P. J. Biggs, L. A. DeWard, C. W. Coffey, S. C. Tsao, M. S. Gossman, et al. 2009. *Report of AAPM Task Group 152: Model Regulations for Electronic Brachytherapy*. College Park, MD: American Association of Physicists in Medicine.

56

Computer-Controlled and Intensity-Modulated Radiotherapy

Benedick A. Fraass
University of Michigan

Introduction

Modern radiotherapy (RT) is complex, and current treatment delivery is a significant part of that complexity. Conformal RT, whether it consists of multiple fixed, shaped fields or intensity-modulated radiation therapy (IMRT) created by nonuniform intensity distributions delivered using segmental or dynamic multileaf collimator (MLC) fields, is usually delivered using a computer-controlled treatment machine. The planning and delivery process used for this delivery involves many steps and many individuals within the RT department.

As everyone in the RT community knows, hardware breaks, software always has bugs, processes mutate and devolve, and people make mistakes. Every RT clinic is susceptible to many kinds of errors, potentially including big errors that can cause patients or staff toxicity, injury, or even death. There are a series of reports demonstrating that computer-controlled treatment delivery has been associated with errors that cause injury and death [see reports on the Therac-25 accidents (Leveson and Turner 1993) and the more recent reports of IMRT-related accidents (Bogdanich 2010)]. Given the potential gravity of the errors that are possible, it is essential that treatment delivery systems and processes be made as safe and error-free as possible.

In an effort to help satisfy the goal of safe and error-free treatment delivery, the objectives of this chapter are to describe modern computer-controlled (and IMRT) treatment delivery systems and processes, to describe differences between those methods and more traditional methods, and to describe issues that require improved quality assurance (QA) methods appropriate for these new delivery methods. The interconnectedness of treatment planning, treatment preparation, and treatment delivery mean that it is important to consider the planning/delivery process as a whole, although this particular chapter will concentrate mainly on the delivery-related aspects of the process.

Given the significant differences in susceptibility to and types of errors that occur in modern computer-controlled delivery (compared to traditional delivery methods), it is essential that our quality management systems (including QA, quality control, and other techniques) be developed to optimize the efficiency and effectiveness of our efforts to ensure error-free computer-controlled treatment delivery. Various methods and techniques that can be valuable for a modern quality management program for computer-control treatment delivery will be briefly described. Finally, the need for more work toward application of modern risk-aware and process-oriented analysis and quality management program development will be summarized.

Transition from Traditional to Computer-Controlled Treatment Delivery

The process used for modern computer-controlled treatment planning and delivery has changed dramatically since the era of manually controlled machines and treatment delivery. That manual process, from simulation to 2-D treatment planning to paper treatment charts to manually controlled treatment delivery, is the basis for much of the QA that is currently used and recommended for the planning and delivery process (see, e.g., Kutcher et al. 1994). In this manual process, information was transferred from person to person, or to and from a paper chart, at each step, meaning that each step in the process was subject to transcription errors of one kind or another. Given all the transcription and look-up errors that are possible with this kind of manual and paper-based process, it is logical that nearly all the QA that is normally recommended is aimed at finding and correcting these transcription and manual activity errors. Many AAPM task group reports on QA (see, e.g., Kutcher et al. 1994) have been developed for this traditional process, as were many of the reports that guide our use of daily, weekly, monthly, and annual accelerator QA, weekly portal imaging for setup verification, treatment plan and monitor unit (MU) calculation checks, and weekly chart rounds, to name a few standard QA procedures. For computer-controlled treatment delivery, these procedures must be reevaluated in the context of the process used for modern treatment delivery.

During the past 10–15 years, most of the specific devices used in the delivery process have been replaced with computer-based technologies. Manually controlled treatment machines are now computer-controlled. CT simulators and 4-D CT have replaced x-ray simulators, diagnostic and megavoltage electronic portal images have replaced port films, MLCs have replaced focused blocks, and IMRT has replaced flat and wedged fields. As these computer-based technologies have come into use, more and more automation has become part of the delivery process.

In the late 1970s, "record and verify" (R/V) systems (Perry, Mantel, and Lefkofsky 1976; Kartha et al. 1977; Rosenbloom, Killick, and Bentley 1977; Sternick et al. 1979) were developed with the general goal of avoiding many of the transcription and look-up errors that plagued the traditional treatment process. Use of an R/V system provides a solution to transcription errors by using a computer to check the parameters used to treat each field. To accomplish this check, the R/V system needs to know the patient, their treatment plan parameters, and which field is selected for treatment. Unfortunately, each of these steps or

pieces of information again requires manual selection or entry of information by people, and so are also susceptible to error. In fact, numerous papers have shown that the R/V system "caused" an error rate of 15–25% (per patient) due to these kinds of entry and selection errors (Huang et al. 2005; Macklis, Meier, and Weinhous 1998; Patton, Gaffney, and Moeller 2003).

The solution to this "new" failure mode is to integrate the R/V system into the planning/delivery system because the R/V system will always provide the correct parameters. If this is always correct, then it should be safer to set the machine directly by control of the R/V system. However, this solution now means that the R/V system is no longer an independent check but rather is part of the integrated computer-controlled treatment delivery system. Modern RT treatment systems do not have R/V systems any longer; rather, they are integrated computer-controlled treatment delivery systems. The QA of these computer-controlled systems needs to be designed with that fact in mind rather than with the traditional manual treatment mind set.

Modern treatment with IMRT and computer-controlled treatment delivery is typically performed with a process similar to that shown in Figure 56.1. Virtually every step in this process is either new or significantly changed from those of the traditional process. Although this chapter will concentrate on treatment delivery, it should be noted that many aspects of treatment planning are quite a bit different from the traditional process, particularly when IMRT is used. The planning/delivery process also includes completely new steps, such as the sequencing of MLC leaf trajectories (for IMRT delivery), the planning of automated treatment delivery (treatment delivery planning), and the download process that transfers the plan information electronically from the treatment planning system to the treatment delivery

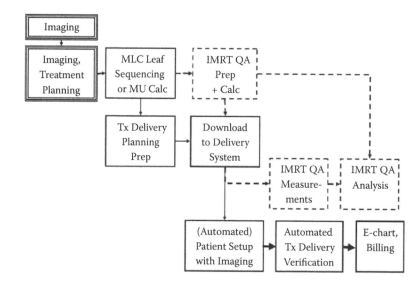

FIGURE 56.1 Modern computer-controlled IMRT planning/delivery process. Dashed boxes and lines show additional steps involved in typical patient-specific IMRT QA checks. Because this chapter concentrates on computer-controlled delivery issues, all of the processes involving in imaging and treatment planning are condensed into the first box at the top left.

system. Automated patient setup and imaging (with cone beam CT or other electronic imaging) and automated treatment delivery are also new features.

Clearly, this new kind of treatment delivery process requires new study and analysis to determine what kinds of errors or failure modes are possible. All of the QA associated with treatment delivery equipment, systems, and the delivery process should be reevaluated to make sure it is both efficient and effective for the new equipment, methods, and treatments that are now part of computer-controlled RT.

Basic Issues Affecting Quality and Safety for Computer-Controlled Treatment Delivery

In traditional RT (i.e., before extensive computer-controlled machinery arrived on the scene), it was relatively straightforward to determine the steps necessary to provide a safe and high-quality treatment to each patient. The treatment planning system needed to be commissioned; the treatment plan was checked by hand as it was transcribed into the paper treatment chart; the patient was set up daily on the machine using skin marks; patient position was documented using weekly port films; the treatment machine light field, calibration, and operation were checked (daily, monthly, annually); and everything involved in the patient's treatment progress was checked each week in chart rounds. Commissioning the treatment planning system was performed with straightforward dose versus data comparisons; the treatment machine was commissioned using simple film, ion chamber, and scanning system measurements; and the various external devices used during patient treatment (mainly immobilization devices) had little to no impact on how the treatment delivery or planning equipment was commissioned or used. Many different QA guidance documents (see, e.g., Kutcher et al. 1994) apply to treatment planning and delivery of this era.

The situation is dramatically different in current RT clinics. The complex and interconnected behavior of this new process leads to a plethora of new design and procedural issues that must be considered when designing a quality management program to ensure safe and high-quality patient treatment—even if this list concentrates just on issues that are related to the computer-controlled delivery aspects of the process:

- All modern treatment begins with a plan based on image data (CT, MR, PET, and/or other modalities), making the operation, calibration, imaging protocols, file formats, coordinate systems, and many other aspects of each imaging system an important QA issue.
- Patient position and immobilization, on each imaging system and through the process to the treatment delivery machine, must be consistent, documented, and indexed wherever possible, so that the relationship of the patient to the machinery is fixed or at least known.
- The transfer of the imaging and patient position information into treatment planning (and then treatment

delivery) is a crucial step because a problem here causes a systematic error in any planning or treatment that occurs downstream from the error. This problem is intensified by the fact that diagnostic radiological use of imaging, from nonflat imaging system tables to positioning to geometrical accuracy requirements to the standard viewpoint for image display (e.g., from the foot of the patient), is often quite different from the RT conventions for image display and use.

- Treatment planning for modern treatment involves a huge variety of QA issues, as described in treatment planning QA guidance documents (Fraass, Doppke et al. 1998; Van Dyk et al. 2004). However, most of these reports concentrate on the treatment planning issues themselves, not on things that are more directly tied to modern computer-controlled delivery of those plans. Especially for IMRT planning, in which inverse planning or optimization techniques are used to generate complex beam intensity distributions, the step in which actual machine delivery constraints are input into the planning process (during dose calculations, optimization, or MLC leaf sequencing) can be problematic in ways that can be hard to detect until the IMRT plan is run on the treatment machine. It is also a well-known fact that the IMRT leaf sequencing process can degrade the "optimized" IMRT plan, and many systems do not make it easy to evaluate the extent of that degradation.
- Even for non-IMRT plans, methods to calculate and particularly to check the number of MUs needed for each treatment field are dramatically affected by the kinds of fields treated with computer control. Multisegment fields often are composed of small, irregularly shaped, off-axis fields for which it is quite difficult to accurately calculate MUs using traditional methods, making it very hard to perform the second hand check that has been a staple of MU calculation QA for decades. The large number of smaller segments or strangely shaped fields also makes quite difficult the intuitive check of the number of MU used in each field that a treatment therapist is supposed to perform during each treatment session. Furthermore, the large number of MUs regularly involved in IMRT fields means that therapists are often used to ignoring whatever intuition they had about the sanity of MUs used for a given field or dose prescription.
- For IMRT plans, it is not practical to perform a hand-based check of the plan parameters and MUs—there are thousands of parameters to be verified. Patient-specific IMRT QA has typically been used to fill this big gap in the QA program (Boyer and Yu 1999), although there are widely varying methods and tolerances used for this step of the process. Some parts of the IMRT QA process are shown in Figure 56.1 using dotted lines, but much more discussion will occur later in this chapter.
- Once the plan is finalized, a completely new step in the process, treatment delivery planning (Kessler, McShan,

and Fraass 1995), is required because computer-controlled treatment delivery requires that all motions, trajectories, and actions by the delivery equipment be planned, added to the delivery script, and checked for correctness before the automated treatment is applied to the real patient. Although systems providing tools to perform this delivery planning and preparation have been developed and described (Kessler, McShan, and Fraass 1995), current commercial systems often do not support the graphical and procedural tools that make this easy and efficient. Treatment delivery planning is an important issue where more effort is needed.

- Rather than writing a plan in the paper chart, having the physician sign the prescription, and then carrying the chart out to the machine, computer-controlled treatment involves directly downloading the plan information into the treatment delivery system's database and preparing that plan for delivery to the patient. Even if the same vendor's software system is used for this process, there are numerous issues that must be analyzed from a QA standpoint, including how and when checks of the information and approvals by dosimetry, physician, and physicist are included in the process. Major and systematic errors in patient treatment are possible if this step is not correctly performed and verified (see, e.g., Bogdanich 2010; Fraass 2008a, 2008b).

- No longer is the treatment machine a large mechanical system driven by electrical motors and switches, or even by a hardware-based electronic control system. All current treatment delivery systems involve mechanical and electronic systems that are driven by software-based control systems that interface to or are integrated with software databases, computer-based user interfaces, and usually a department-wide RT information system (which may or may not be connected or integrated with the delivery system software and hardware). Most QA issues concerned with the treatment machine (and its control system) can no longer be addressed or even evaluated by simple machine tests and measurements.

- The use of automated setup techniques, image-guided RT based on cone beam CT, diagnostic and megavoltage imagers, or other guidance devices (e.g., radiofrequency beacons) are all setup methods that depend completely on the accuracy with which the desired patient localization information is transmitted through the entire planning and delivery process, from the initial imaging studies through to treatment. Making sure the patient setup and localization decisions are based on correct data, with tolerances of only a millimeter or two, is a completely different dynamic than the traditional methods based on lasers and skin marks, so QA of this process is also completely different from the weekly port film checks that used to suffice.

- How treatment delivery is performed and, in particular, verified, is significantly changed. Traditional delivery

techniques depended importantly on the therapist's ability to verify that the correct treatment was being performed. For computer-controlled delivery, QA issues include how to involve the therapists in the delivery, how to show the therapists what is going to happen next in an automated delivery, and how to demonstrate that automated and/or IMRT delivery is proceeding correctly, in a way that they can easily identify is correct for that patient, plan, and fraction.

- The documentation, recording, billing, and handling of other chart-related issues (e.g., prescriptions and changes or updates to prescriptions) are now almost entirely handled inside the computer-based delivery system and radiation oncology information system. This means the process for handling and quality assuring changes or updates, documentation of the actual expectation and results for each treatment, and other such chart-related issues also are completely different than the old paper-based methods. Errors made in the handling of prescriptions or prescription updates or changes are potential systematic errors that can lead to significant treatment errors and are also hard to identify.

- In most clinics, treatment delivery is performed through use of a series of different computer systems and software applications, as illustrated in Figure 56.2. Even if all the pieces of the system are supplied by a single vendor, these different tools or applications have interfaces, different data representations, different user interfaces, and other distinguishing characteristics. For each of these systems or applications, specific QA analysis and testing is important. However, consideration and testing of each piece of the process, in isolation, misses two very large issues. First, the interfaces between systems and applications are particularly problematic, as a source of both major and systematic errors or misunderstandings about meanings of variables and parameters, either of which can lead to big clinical problems. Second, there are often many paths through the complex set of procedural steps in this

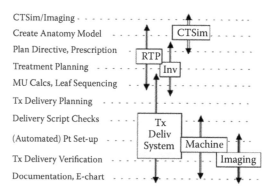

FIGURE 56.2 Different systems or applications typically involved in the various steps of the planning/delivery process.

process. Therefore, it is essential to consider the process and to use process-oriented QA rather than just using system testing to try to ensure safety.

Safe, high-quality computer-controlled treatment delivery depends on organized and effective approaches to address all of these issues.

Approaches to QA for Computer-Controlled Treatment Delivery

Study of errors and accidents involving computer-controlled RT machines (or other computer-controlled medical devices) can quickly increase one's awareness of the importance and seriousness of the QA issues that must be appropriately handled for safe use of computer-controlled treatment delivery. One of the most well-known accidents involved the Therac-25 accelerator system. During the mid-1980s, 11 machines were installed, and there were a total of six accidents involving massive overdoses of doses on the order of 40–50 Gy delivered in a few seconds, fatally injuring several patients. Although various aspects of the early computer-control system of this machine were directly involved in the accident, the investigations of the root causes of the problem clearly involved "a combination of organizational, managerial, technical, and sometimes sociological or political factors. Preventing accidents requires paying attention to all the root causes . . . ," as an investigation of these accidents concluded (Leveson and Turner 1993).

In reaction to these accidents, AAPM Task Group 35 was created to help provide improved QA guidance for use of computer-controlled delivery machines. Although published nearly 20 years ago (Purdy et al. 1994), this report provides instruction and guidance on many of the issues associated with computer-controlled accelerator use and is an important resource for users as QA for computer-controlled systems is developed. The report uses a classification scheme for potentially dangerous problems, describes how medical physicists should respond to potential safety hazards, and discusses training requirements and specific issues for software-controlled treatment machines. All of the recommendations in this report are still relevant to our current technologies. Other AAPM reports address additional technology that is involved in current treatment delivery methods, including TG142 (QA for accelerators) (Klein et al. 2009), the IMRT Guidance Document (Ezzell et al. 2003), and others.

Our own local experience with computer-controlled treatment delivery began with one of the first commercially available MLC-equipped computer-controlled accelerator systems, the Scanditronix Racetrack Microtron. Our group designed and implemented the Computer-Controlled Conformal Radiotherapy System to direct and control the computer control system of the microtron, as described in a series of papers (Kessler, McShan, and Fraass 1995; Fraass, McShan, Kessler et al. 1995; Fraass, McShan, Matrone et al. 1995; McShan et al. 1995). Through the experience of analyzing safety, control, and functionality needs of this computer-controlled system, we arrived at a series of points

that may be helpful to anyone commissioning or implementing the clinical use of a computer-controlled delivery system:

- The more you know about the design specifications, system requirements, and hazard analyses that have been performed for all parts of the treatment machine and its computer control system, the better you will be able to design and perform QA testing relevant to your clinic.
- Obtain relevant QA documentation from the vendor and use that documentation to design detailed tests of potential weak spots for your clinical process.
- Remember that there will always be bugs and design limitations in any software-based system.
- The treatment process, the design of the QA program, and the system commissioning methods should all be dependent on the design of the delivery system, the capabilities that will be used clinically, and the clinical goals and requirements of the treatments that will be performed.
- It is not possible to verify that every possible use of a system works correctly, so it is essential that clinical commissioning and QA efforts be tuned toward the clinically relevant uses of the system—and these are tied to the process used as much as to the specific technical features of the machinery that are used.

Analysis of delivery errors comparing a modern automated and computer-driven treatment process (with the Computer-Controlled Conformal Radiotherapy System) to a traditional manual delivery process showed that virtually all the random errors in delivery are avoided in the more automated process (Fraass et al. 1998).

However, more attention must be directed toward the possibility of systematic errors. A number of resources can be useful to help users identify, describe, and understand potential failure modes or problems with existing delivery systems or processes. The US Food and Drug Administration maintains a reporting system and database of such events called MAUDE (Manufacturer and User Facility Device Experience), which can be searched online. A more independent RT-specific reporting system and database called ROSIS (www.rosis.info) contains almost 900 error reports submitted from RT clinics around the world.

However, particularly in the United States, there is a major limitation to our ability to learn from errors, problems, and near misses that occur in other clinics. Various laws, concerns about medico/legal issues, and general marketing issues all tend to make it difficult to report, discuss, investigate, and document errors or problems. As demonstrated recently with errors reported in a series of newspaper articles (Bogdanich 2010), although rumors about the incident circulated in the field relatively soon after the incident, it was several years later before the real details became widely available (Fraass 2008a, 2008b). One of the most potentially useful actions that could help improve the safety of RT treatment would be the creation of a system for reporting, dissemination, and analysis of errors, problems, and near misses that would be available to all RT users. This system would help prevent the following issues:

- For each incident that we ever hear about, there are more that are unknown or not widely reported.
- To actually respond to any error that we hear about, it is essential to 1) understand the string of faults that combined to let the error happen and 2) whether related events have happened elsewhere.
- Typically, it is impossible to determine from current error reporting systems a detailed description of each problem, what the real issues or dangers are, and a detailed and authoritative description of how to address the issue.

Describing a complete QA program for computer-controlled treatment delivery is well beyond the scope of this chapter and is very dependent on the technology and clinical treatment delivery goals and techniques. Some brief suggestions that may help in the design of the local QA program for computer-controlled treatment delivery are as follows:

- To design the QA program for the clinic, one needs to understand what kinds of errors can occur in your own clinic and how to prevent them.
- There are many independent hardware/software systems and processes. Each system requires detailed QA, including testing, and on-going checks, but interfaces between systems are also particularly important.
- As systems become more connected, random errors become less important and systematic errors become more dangerous. This means that the more traditional QA methods, which were directed in large part toward transcription-based errors, should be replaced or enhanced with modern process-oriented risk-aware error analysis methods, like the on-going efforts of AAPM Task Group 100 (Huq et al. 2008).
- Given the many different systems involved in this process, attention to process flow and hand-offs between the various systems is a crucial part of the QA program.
- There should be a low probability of random transcription or manual errors, but systematic errors are much more dangerous, so the QA program must be reoriented toward these errors.
- Edits to planned or delivered data (including positioning) should be preplanned or documented, and unexpected differences should be investigated. The delivery system should be able to search for and then highlight changes as part of a routine QA procedure (see Fraass 2008b).
- As with most processes, users must have continual vigilance for changes caused by actions outside the standard and automated process because those deviations often lead toward errors.
- Treatment machine QA must be significantly different for these computer-controlled machines. The daily QA check can be significantly more detailed than traditional checks, using computer-controlled technology to perform automated

checks. Thompson et al. (1995) describes automated checks of computer-controlled motion of gantry, collimator, and table angles, table x,y,z, MLC shaping, and automated treatment of multiple segments, as well as much more complete dosimetry checks of the machine's dose delivery. Other things such as dynamic MLC leaf trajectories, collision avoidance software/hardware, motion-control devices like 4-D scanning, gating, tracking, and active breathing control should also be checked, although it is important to determine efficient and straightforward check methods.

Needed: Process-Oriented and Risk-Aware Quality Methods Applied to Computer-Controlled Treatment Delivery

To effectively and efficiently design and perform QA that will provide higher-quality and safer computer-controlled treatment delivery, modern risk-aware and process-oriented quality methods must be applied to the planning/delivery process used in RT. One of the first attempts to apply these kinds of approaches to RT is the effort by the AAPM Task Group 100, which is addressing the IMRT and High Dose Rate (HDR) brachytherapy processes. The method used by TG 100 (see Huq et al. 2008) includes the following general steps:

- Map the clinical process, without current QA steps, because we want to use the new analysis to place QA steps where they will be most efficient and effective
- Analyze how the process can fail, and what the effects of the failures will be, using failure modes and effects analysis
- Determine or assign frequency, severity, and detectability scores to each failure mode to help prioritize which failure modes are most important to mitigate
- Map how the faults propagate through the process using fault tree analysis
- Find efficient ways to minimize propagation of errors through the process by adding improved quality management, QA, and quality control steps into the process

Such an analysis should be performed within an individual clinic to most effectively address the particular clinical issues of that clinic. To assist in the creation of a process-oriented review of safety and quality issues for computer-controlled treatment delivery, the following identifies some of the issues that should be considered:

1. Leaf sequencing or MU calculation
   - How is the plan degradation by sequencing evaluated, limited, or corrected?
   - How does the process compensate for an inability to evaluate the plan quality after sequencing?
   - Where are the compromises in the sequencing process?

- What possible incompatibilities exist between planning, optimization, sequencing, and delivery systems?
- How does the process document compromise in plan quality?
- How is the correctness of the sequenced plan verified?

2. Treatment delivery planning
   - Is there a check of plan deliverability?
   - Are standard protocols for field naming and ordering, creation of imaging fields or processes, and definition of anatomy for positioning checks used?
   - How do you make sure that plan changes for deliverability do not cause other problems?
   - How does the system ensure that modified imaging fields do not become inconsistent with treatment fields?
   - How is the electronic chart (e-chart) documentation of the setup and treatment process checked for completeness and correctness?

3. Plan download to delivery system
   - How do you verify that all data transfer between the different systems involved in the delivery is correct, especially issues that can highlight differences in software capabilities and/or configurations?
   - What information is or can be lost, modified, or transformed as it is transferred into the delivery system?
   - What failure modes involve user error (e.g., choice of wrong method, plan, patient), and how are these possible failures mitigated?
   - What failure modes lead to inaccurate or incomplete transfer of information, and how are those issues mitigated?

4. Patient-specific IMRT QA
   - What are the main purposes of patient-specific IMRT QA? Some reasons include the nonintuitive dose and fluence distributions that are generated in IMRT planning; the fact that there are no hand-type MU calculation checks; that plans are too complicated to check by hand; that there is a need to verify which sequencing, MLC limits, plan transfers, and delivery are working correctly; and that dose calculations and optimization often have significant approximations plus limitations.
   - Does IMRT QA actually confirm the correctness of the patient's entire IMRT plan delivery script, or are there possibilities for errors in the patient's delivery plan that the IMRT QA will not pick up?
   - Is the creation of the IMRT QA plans and beams performed automatically or manually? What failures are possible in either of these methods?
   - What resolution is used for both measurements and calculations for IMRT QA, and what tolerances (geometric, dosimetric) are necessary for the clinical cases being evaluated? Are QA tolerances for analysis really adequate for patient IMRT? For example, do we really accept the tolerances that are often used for analysis of IMRT QA (3%/3 mm or even 5%/5 mm), when a 3-mm geometric difference could potentially lead to dosimetric differences on the order of 30%?
   - If the exact treatment machine MLC trajectory description is not used for patient-specific IMRT QA checks, how do you confirm that the two MLC trajectory descriptions are identical? How do you confirm that the IMRT QA plans are identical to that used for patient treatment?
   - What happens when the patient-specific QA for a plan is out of tolerance? How do you evaluate the clinical relevance of that difference?

5. Automated patient setup with imaging (i.e., image-guided RT)
   - How do you avoid selecting or presenting the wrong patient, plan, or imaging information?
   - What controls prevent incorrect application of setup instructions?
   - How is a misalignment or poor decision about image-guided RT setup avoided?
   - What control is there for wrong patient positioning?
   - How are the prescription, course, fraction, plan, and delivery process chosen and verified?

6. Automated treatment delivery and verification
   - What methods control for and check that plan or MLC trajectory descriptions for IMRT are not changed (or are different than expected) without user knowledge?
   - What conditions cause a direct failure of delivery, and what are the methods to avoid this kind of failure triggering other delivery, prescription, or other problems, including mishandling of machine faults or aborts?
   - What is the process to verify that incorrect recording of actual treatments performed (either electronic or manual) does not occur? What is the correction process when a problem is identified?

7. Electronic chart and billing
   - How are desired changes in prescription, plan, and delivery process documented, identified, and handled so that all members of the treatment team understand what is supposed to happen?
   - What kinds of electronic chart and billing documentation must be routinely verified, and what kinds of errors or problems are possible?

Summary

There has been some work investigating the planning/delivery process for IMRT and computer-controlled treatment, but much of the old traditional delivery process has simply been ported for use with computer-controlled treatment systems. QA procedures have changed even more slowly, and many QA steps appropriate for the old process have been taken over to the new process, often without much justification. New, more organized and radical approaches to QA for computer-controlled IMRT are necessary. Probably the best way to accomplish the needed improvements is to use modern risk-based process-oriented analysis to help design more effective QA techniques and methods.

Analysis of errors has shown that random transcription errors, which happen as humans transfer information, are no longer the most important issue. In our new computer-driven planning and delivery methods, systematic errors, which are much less common but potentially much more severe, are a major concern, especially in interfaces between systems and processes. In addition, automation removes much human scrutiny from the process, and we must find ways to reimplement human scrutiny into our automated planning and delivery process. With careful analysis and more sophisticated application of modern risk-aware analysis methods to our process, we will improve the safety and quality of RT treatment, an important goal for the next years.

References

Bogdanich, W. 2010. Radiation offers new cures, and ways to do harm. *New York Times*, January 24.

Boyer, A. L., and C. X. Yu. 1999. Intensity-modulated radiation therapy with dynamic multileaf collimators. *Semin. Radiat. Oncol.* 9 (1): 48–59.

Ezzell, G. A., J. M. Galvin, D. Low, J. R. Palta, I. Rosen, M. B. Sharpe, P. Xia, Y. Xiao, L. Xing, and C. X. Yu. 2003. Guidance document on delivery, treatment planning, and clinical implementation of IMRT: Report of the IMRT Subcommittee of the AAPM Radiation Therapy Committee. *Med. Phys.* 30 (8): 2089–2115.

Fraass, B. A. 2008a. Errors in radiotherapy: Motivation for development of new radiotherapy quality assurance paradigms. *Int. J. Radiat. Oncol. Biol. Phys.* 71 (1 Suppl.): S162–S165.

Fraass, B. A. 2008b. QA issues for computer-controlled treatment delivery: This is not your old R/V system any more! *Int. J. Radiat. Oncol. Biol. Phys.* 71 (1 Suppl.): S98–S102.

Fraass, B. A., K. L. Lash, G. M. Matrone, S. K. Volkman, D. L. McShan, M. L. Kessler, and A. S. Lichter. 1998. The impact of treatment complexity and computer-controlled delivery technology on treatment delivery errors. *Int. J. Radiat. Oncol. Biol. Phys.* 42 (3): 651–659.

Fraass, B. A., D. L. McShan, M. L. Kessler, G. M. Matrone, J. D. Lewis, and T. A. Weaver. 1995. A computer-controlled conformal radiotherapy system: I. Overview. *Int. J. Radiat. Oncol. Biol. Phys.* 33 (5): 1139–1157.

Fraass, B. A., D. L. McShan, G. M. Matrone, T. A. Weaver, J. D. Lewis, and M. L. Kessler. 1995. A computer-controlled conformal radiotherapy system: IV. Electronic chart. *Int. J. Radiat. Oncol. Biol. Phys.* 33 (5): 1181–1194.

Fraass, B., K. Doppke, M. Hunt, G. Kutcher, G. Starkschall, R. Stern, and J. Van Dyke. 1998. American Association of Physicists in Medicine Radiation Therapy Committee Task Group 53: Quality assurance for clinical radiotherapy treatment planning. *Med. Phys.* 25 (10): 1773–1829.

Huang, G., G. Medlam, J. Lee, S. Billingsley, J. P. Bissonnette, J. Ringash, G. Kane, and D. C. Hodgson. 2005. Error in the delivery of radiation therapy: Results of a quality assurance review. *Int. J. Radiat. Oncol. Biol. Phys.* 61 (5): 1590–1595.

Huq, M. S., B. A. Fraass, P. B. Dunscombe, J. P. Gibbons Jr., G. S. Ibbott, P. M. Medin, A. Mundt et al. 2008. A method for evaluating quality assurance needs in radiation therapy. *Int. J. Radiat. Oncol. Biol. Phys.* 71 (1 Suppl.): S170–S173.

Kartha, P. K., A. Chung-Bin, T. Wachtor, and F. R. Hendrickson. 1977. Accuracy in radiotherapy treatment. *Int. J. Radiat. Oncol. Biol. Phys.* 2 (7–8): 797–799.

Kessler, M. L., D. L. McShan, and B. A. Fraass. 1995. A computer-controlled conformal radiotherapy system: III. Graphical simulation and monitoring of treatment delivery. *Int. J. Radiat. Oncol. Biol. Phys.* 33 (5): 1173–1180.

Klein, E. E., J. Hanley, J. Bayouth et al. 2009. Task Group 142 report: Quality assurance of medical accelerators. *Med. Phys.* 36 (9): 4197–4212.

Kutcher, G. J., L. Coia, M. Gillin, W. F. Hanson, S. Leibel, R. J. Morton, J. R. Palta, J. A. Purdy, L. E. Reinstein, G. K. Svensson, M. Weller, and L. Wingfield. 1994. Comprehensive QA for radiation oncology: Report of AAPM Radiation Therapy Committee Task Group 40. *Med. Phys.*, 21 (4): 581–618.

Leveson, N. G., and C. S. Turner. 1993. An investigation of the Therac-25 accidents. *Computer* 26 (7): 18–41.

Macklis, R. M., T. Meier, and M. S. Weinhous. 1998. Error rates in clinical radiotherapy. *J. Clin. Oncol.* 16 (2): 551–556.

McShan, D. L., B. A. Fraass, M. L. Kessler, G. M. Matrone, J. D. Lewis, and T. A. Weaver. 1995. A computer-controlled conformal radiotherapy system: II. Sequence processor. *Int. J. Radiat. Oncol. Biol. Phys.* 33 (5): 1159–1172.

Patton, G. A., D. K. Gaffney, and J. H. Moeller. 2003. Facilitation of radiotherapeutic error by computerized record and verify systems. *Int. J. Radiat. Oncol. Biol. Phys.* 56 (1): 50–57.

Perry, H., J. Mantel, and M. M. Lefkofsky. 1976. A programmable calculator to acquire, verify and record radiation treatment parameters from a linear acceleration. *Int. J. Radiat. Oncol. Biol. Phys.* 1 (9–10): 1023–1026.

Purdy, J. A., P. J. Biggs, C. Bowers, E. Dally, W. Downs, B. A. Fraass, C. J. Karzmark et al. 1993. Medical accelerator safety considerations: Report of AAPM Radiation Therapy Committee Task Group No. 35. *Med. Phys.* 20 (4): 1261–1275.

Rosenbloom, M. E., L. J. Killick, and R. E. Bentley. 1977. Verification and recording of radiotherapy treatments using a small computer. *Br. J. Radiol.* 50 (597): 637–644.

Sternick, E. S., J. R. Berry, B. Curran, and S. A. Loomis. 1979. Real-time computer verification for radiation therapy treatment machines. *Radiology* 131 (1): 258–262.

Thompson, A. V., K. L. Lam, J. M. Balter, D. L. McShan, M. K. Martel, T. A. Weaver, B. A. Fraass, and R. K. Ten Haken. 1995. Mechanical and dosimetric quality control for computer controlled radiotherapy treatment equipment. *Med. Phys.* 22 (5): 563–566.

Van Dyk, J., J. Rosenwald, B. Fraass, P. Andreo, J. Cramb, F. Ionescu-Farca, J. Izewska et al. 2004. Commissioning and quality assurance of computerized planning systems for radiation treatment of cancer. International Atomic Energy Agency IAEA TRS-430.

57

Intensity-Modulated Volumetric Arc Radiotherapy

Stine Sofia Korreman
Rigshospitalet, University of Copenhagen

Introduction

The term "volumetric treatment" has been coined for treatment delivered in an arc motion of a linac gantry around the patient, while the delivery is being modulated by varying, for example, the dose rate, the gantry speed, and the multileaf collimator shape. These techniques will, in the following, also be referred to as modulated arc therapies. Note that helical tomotherapy will not be covered in this chapter, as the principles are so basically different from gantry-based arc therapy, and tomotherapy is covered in Chapter 75. Conformal arc therapy is considered more of a standard technique and also will not be covered specifically in this chapter.

There are specific issues of quality assurance (QA) related to these volumetric treatment techniques that apply only to these techniques because of their unique interplay of variables. In this chapter, issues in QA for volumetric treatment will be covered under five headings: equipment for arc therapy quality procedures, commissioning and acceptance testing, QA for the machine performance, QA for the delivery of patient-specific dose distributions, and QA for the treatment planning process.

Distinguishing Factors of QA for Volumetric Treatment

The QA of modulated arc therapy can be considered an addition to "standard" QA and should include procedures for features that are not otherwise tested in the QA programs related to other treatment techniques. The special needs for modulated arc therapy QA thus depend on the features that are unique to this treatment modality, including, as the key issue, obviously the feature of the rotating gantry during beam-on time.

Conventional gantry-based arc delivery is available as a conformal arc solution in several commercially available systems. At the time of the writing of this text, gantry-based modulated arc therapy is featured in two different commercially available solutions (RapidArc® by Varian Medical Systems, Inc., and VMAT by Elekta AB) as well as a number of home-made solutions (Duthoy et al. 2003; Otto 2008; Ulrich, Nill, and Oelfke 2007; Wang et al. 2008). The various solutions include different degrees of freedom of modulation, and the QA program should test these degrees of freedom and their interplay completely. However, as mentioned above, this should be built on top of, and considered an extension to, the already existing QA program. In this chapter, only QA for the parts specific to volumetric treatment will be covered, and it will be assumed that QA for standard use of the treatment unit is already covered. However, there is no clear-cut line between standard use and volumetric treatment delivery because there may be differences in what is considered "standard" use in different clinics and different installations (for instance, standard use in some clinics may include dynamic MLC IMRT, whereas for others it may include only static field delivery).

The schedules of the described QA procedures will be dependent on the needs in each installation, and can depend on inherent stability of the linear accelerator, frequency of use, measurement equipment used, and other issues.

Equipment for Arc Therapy QA Procedures

As a consequence of the gantry rotation being an essential part of arc therapy, it is also essential that equipment used in the QA of these techniques is suited to handle gantry rotations. The QA program will include measurements performed while the gantry is rotating with the beam irradiating from all angles of rotation, and the equipment must therefore exhibit little or no angular dependence. Several of the major vendors of measurement equipment offer phantoms with built-in measurement arrays exhibiting little angular dependence of measurement—this includes the OCTAVIUS with 2D-ARRAY seven29 (PTW Freiburg GmbH), the MatriXXEvolution with MULTICube (IBA Dosimetry), the Delta4® (ScandiDos AB), and the ArcCHECK (Sun Nuclear Corporation). All these equipments consist of one or more measurement array(s)—ion chambers or diodes (or film)—embedded in a solid phantom with rotational symmetry (see Figure 57.1).

An alternative to phantoms such as described above is to use the portal imager of the treatment machine that rotates with the gantry. This is inherently independent of gantry angle as the angle of incidence of the radiation does not change. Furthermore, the imager often has a very high spatial resolution and temporal resolution. An issue for this type of equipment is that rather than representing accumulated absorbed dose delivered by the plan, it measures beam fluence (potentially accumulated over the entire rotation). Some mathematical modeling (e.g., back projection) is needed to convert this measurement to a representation of absorbed dose accumulated in a patient over the arc rotation. There is presently no commercially available software to perform this conversion for arc therapies, but there are several noncommercial independent solutions (Mans et al. 2010; Nicolini

et al. 2008), and undoubtedly commercially available solutions underway.

A third approach is to attach a measurement array (or film) to the gantry, to allow it to rotate along with the gantry and thereby eliminate the need for rotational symmetry of the phantom, much as for the portal imager. This approach is currently being developed with two different scopes: attachment of a phantom with an embedded measurement array at isocenter distance and attachment of a transmission ion chamber array close to the gantry head immediately after the beam exit. The latter has the advantage of being potentially usable during actual patient treatment for online dosimetry.

Although the measurement of fluence profiles or absorbed dose as described above may be spatially and temporally resolved, one thing that is not resolved is the gantry angle as a function of time. An independent inclinometer is necessary for that, and this should be connected to the measurement equipment used, so that the measurements are synchronized with the gantry angle. Several commercially available systems for arc therapy measurements come with an independent inclinometer as part of the package.

Commissioning and Acceptance Testing

At the time of installation/upgrading/change of a volumetric delivery system, acceptance testing and commissioning must be done. Acceptance test guidelines are usually provided by the vendor in the case of commercially available systems and include verification of the system's capability of performing delivery of a prespecified sequence of delivery schemes. This would typically include import of a plan for treatment, delivery of that plan, interruption of delivery, and so on.

Commissioning issues specific to volumetric treatment delivery are, for instance, the interplay of gantry rotation and dose rate variation/MLC positioning. For RapidArc, a set of commissioning tests have been suggested (Ling et al. 2008) including a picket fence–type test during gantry rotation, and different combinations of sweeping leaf speeds and dose rates during gantry rotation at different speeds. Likewise for VMAT, a commissioning program has been suggested (Bedford and Warrington 2009) including flatness and symmetry tests during rotation, various dynamic MLC tests (as dynamic MLCs are not part of standard Elekta linac operation), test of accuracy of accumulated dose with simple rotational plans, and beam interruption tests. Detailed descriptions of these kinds of tests are given in other chapters.

As part of the commissioning process, running a number of actual patient plans and measuring absorbed dose in a phantom can be beneficial. These test the entire treatment planning to dose deposition process end to end. Although a comprehensive program of commissioning tests is planned, there may still be transfer and delivery issues that are not tested, particularly for the very complex cases of volumetric modulated arc treatments. Running a representative set of actual patient plans would give an indication as to whether such issues will arise in the clinical setting.

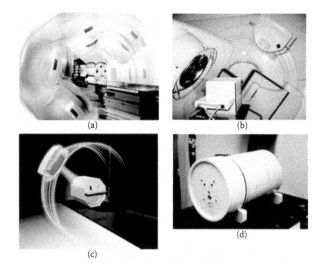

FIGURE 57.1 Four commercially available phantoms with no angular dependence and embedded detector arrays. (a) Delta4 with two orthogonal diode arrays (courtesy of ScandiDos AB); (b) MatriXXEvolution two-dimensional ion chamber array with MULTICube (courtesy of IBA Dosimetry); (c) OCTAVIUS with 2D-ARRAY seven29 ion chamber array (courtesy of PTW Freiburg GmbH); and (d) ArcCHECK with a helical diode array cylindrical around the central axis (courtesy of Sun Nuclear Corporation).

Machine-Specific QA

Machine-specific QA encompasses procedures to verify the performance of the equipment parts involved in modulated arc treatment delivery separately and in combination. The procedures for machine-specific QA can, to a large degree, be identical to or derived from the procedures used in commissioning. The procedures should include tests to investigate the extreme limits of behavior, as well as the moderate (more commonly used) behavior. The machine-specific QA procedures should be part of a QA program with schedules for daily, weekly, monthly, and yearly procedures.

For most of the machine-specific QA procedures, the portal imager is a very well-suited measurement tool. For machine-specific QA procedures, it is not so much an absorbed dose one wants to test but rather the performance with respect to the geometric and temporal behavior of the various parts of the machinery. As the portal imager will mostly have high temporal and spatial resolution, it serves this purpose well. It is possible to use phantoms with ion chambers or diode arrays as well or arrays attached directly to the gantry, and several vendors provide software suited for machine QA measurement analysis. However, the spatial resolution is poorer than that of the portal imager for all existing arrays. Film is a third choice, but there is no temporal resolution in the use of films.

Other tools include log reports from the linear accelerator, which may contain information such as actual gantry position and actual MLC leaf positions as a function of time, as well as

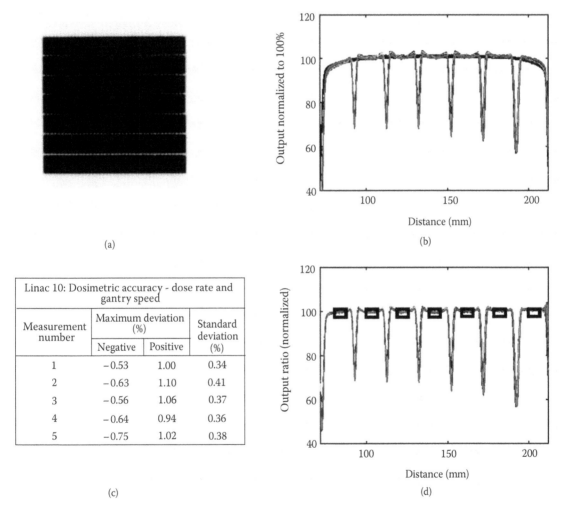

(a)

(b)

Linac 10: Dosimetric accuracy - dose rate and gantry speed			
Measurement number	Maximum deviation (%)		Standard deviation (%)
	Negative	Positive	
1	−0.53	1.00	0.34
2	−0.63	1.10	0.41
3	−0.56	1.06	0.37
4	−0.64	0.94	0.36
5	−0.75	1.02	0.38

(c)

(d)

FIGURE 57.2 Test of gantry speed variation interplay with dose rate variations. (a) An image from the portal imager showing seven segments of different combinations of gantry speed and dose rate obtained during arc delivery. (b) Vertical profiles of the detector outputs; each color corresponds to one MLC pair, and black is the output for an open beam. (d) The output profiles for the test image divided by the open beam image output. Each color corresponds to an MLC pair, and the squares indicate where analysis was performed, resulting in: (c) variations from mean output level are calculated for each MLC pair for each segment, and maximum deviations and standard deviations are shown. The five rows indicate five different deliveries of the test plan. Data and analysis courtesy of Anna Fredh, Department of Radiation Oncology, Rigshospitalet, Denmark. (Data from Fredh et al., *Radiother. Oncol.*, 94 (2), 195–198, 2010. With permission.)

dose rate, MU chamber readouts, and so on. It should be noted, however, that these files do not represent an independent measurement. One way of using these files may be for more detailed analysis of equipment behavior for specific error finding, if a measurement on the portal imager (or other detector) fails the specified tolerance levels.

For machine-specific QA in relation to volumetric modulated treatment delivery, the following aspects should be included.

First, in all the volumetric treatment techniques, the multileaf collimator leaves move dynamically during the rotation of the linear accelerator gantry, and QA procedures for dynamic MLCs must therefore be part of the machine-specific QA program. The leaf motion may be driven either by the gantry position or by the delivered dose. In installations where the dynamically moving MLCs are used for other treatment deliveries (such as sliding window IMRT), QA procedures for this may be already available. However, one should pay attention to whether the DMLCs are driven the same way for the different types of delivery. DMLC tests include sweeping window tests and picket fence pattern tests.

Second, the dynamic gantry is a key feature of the volumetric treatment techniques. The gantry rotation may be driven either by time or by the dose delivered. In installations where dynamic arc therapies are already in use (for instance, for dynamic conformal arc treatments), there may already be QA for this incorporated into standard QA procedures; however, these usually incorporate only fixed gantry speed, whereas the volumetric modulated arc therapies may involve varying gantry speeds. Gantry rotation speed may be checked in combination with the use of an independent inclinometer.

Third, varying dose rate is a feature of volumetric modulated therapies, however, in quite different ways for the different solutions. For RapidArc, the dose rate is continuously varying during beam-on time, whereas for the VMAT approach (in its present implementation), the dose rate can take on a number of discrete magnitudes. The varying dose rate is difficult to test independently because it is an integrated part of the arc delivery scheme and is not used as a parameter in other delivery methods.

The interplay between all of these variables is crucial to test. Test plans can be custom-made to run different combinations of variations, relevant to the specific setup of volumetric modulated therapy. The commissioning test plans mentioned in the previous section for both RapidArc and VMAT are examples of such customized test plans, suited for different setups, which may be used for scheduled QA as well as for commissioning. In Figure 57.2, the results of one test plan run for RapidArc machine QA using the portal imager for measurements (for a linear accelerator at the author's institution) are shown. Also shown are the results of analysis of the measurements using an in-house–developed MATLAB® application. The plan tests seven combinations of varying gantry rotation speeds and dose rates in one arc delivery, and the results are given in the table in terms of maximum deviations and standard deviations of the response from the normalized expected level for all of the combinations.

Patient-Specific QA

The intention of performing patient-specific QA is to ensure that each treatment plan can be/is delivered such that the dose deposited in the patient is identical to the dose as calculated and evaluated in the treatment planning system. The patient-specific QA procedure must be performed using a system independent of the treatment planning and delivery system. Also, the procedure should mimic the treatment situation as closely as possible. The more complex a treatment plan is, the more warranted the need for patient-specific QA is. Comprehensive commissioning and QA of all the subparts involved in the entire treatment chain could, however, render patient-specific QA obsolete. The scope of this text is not to discuss the necessity of patient-specific QA as such but rather to discuss the requirements of tools used for patient-specific QA for modulated arc therapies.

The important part of patient-specific QA is to measure the actual dose deposited in a phantom mimicking the patient or in the patient himself/herself. Phantoms with embedded dose measurement arrays (gantry angle–independent) are therefore quite well suited for the purpose of these types of measurements when they are performed off-line. The measured dose is then compared with the dose calculated in the treatment planning system (as delivered to the phantom, not the patient). Comparison is most often performed using a gamma evaluation, in which both the dose difference and the geometric difference between measured and calculated dose are considered (Low and Dempsey 2003). In Figure 57.3, an example is shown of measurements in a phantom with two orthogonal measurement arrays for a prostate volumetric modulated plan. Distributions are shown for dose differences at the measurement points, geometric distance to closest point of agreement, and for gamma values using 3 mm distance-to-agreement and 3% dose difference criteria. The pass or fail of a measurement will depend on tolerance levels for these three distributions decided in each clinic.

Film can be used in the same way as described above and has the advantage of a much higher spatial resolution than an ion chamber or diode array. This is relevant for volumetric modulated plans, as these plans may contain a high degree of modulation, which can only be measured with high-resolution equipment. Ion chambers have a limit with respect to spatial resolution given by their large volume, which in effect smears out measurements. Diodes can be small, and to a higher degree, give point measurements, so the smearing effect is avoided, but they are not very closely spaced, and some modulation may be "lost" between measurement points. The measurements shown in Figure 57.3 are for a diode array, and when looking in detail at the dose profiles, it may be suspected that there is some modulation between the measurement points. These phantom measurements should all be performed with rotation of the gantry. It is a common simplification of patient-specific QA measurements to collapse all delivery into gantry angle 0° (or 180°) for simplicity. This should not be done for volumetric modulated arc therapy because the gantry rotation is a key component of the treatment—for the RapidArc solution, the gantry position is the driving force for the DMLC positioning.

(a)

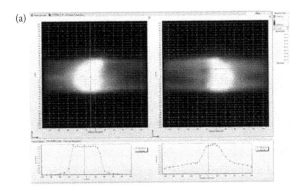

(b)

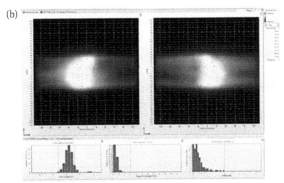

FIGURE 57.3 Data from measurements performed using the Delta4 phantom for a prostate RapidArc plan. (a) The output of detectors in the two orthogonal arrays are shown in the top; below are two output profiles through the arrays (at the positions of the dotted lines in the detector views). The lines indicate dose calculated in the treatment planning system, and the dots indicate measured dose. (b) For the same measurement, the lower plots show distributions of point dose deviations, distances to agreement, and gamma values (with 3% dose deviation, 3 mm distance-to-agreement). Above, in the detector views, the points with gamma values more than 1 are highlighted. Images in the figure courtesy of Lotte Fog, Department of Radiation Oncology, Rigshospitalet, Denmark.

Portal dosimetry is another option for patient-specific QA, which has the advantage of being useful with the patient both off and on the couch. With portal dosimetry, the response of the portal imager detector array can be accumulated over the treatment arc, beaming through air, a phantom, or the patient. In principle, this check is primarily a fluence check, which should be combined with an independent gantry angle measurement to check the fluence profile of the beam as a function of gantry angle. In this sense, it is a check that the beam delivery is performed correctly. If this technique is to be used to check the accumulated dose to the patient, a back-projection calculation is needed based on a model of the patient—the CT scan for treatment planning or possibly an online cone-beam CT scan of the patient.

For patient-specific QA, temporal resolution of detectors is not so relevant because the property one really wants to measure is the accumulated dose to the patient for a specific treatment. Nevertheless, temporal resolution can be used to track the

sources of any detected errors in a measurement. If an error in the accumulated dose can be traced to stemming from one or more specific arc segments, it may be determined what the cause of the delivery error was. This information may in turn be used to make a decision on how to proceed in the specific case.

QA for Arc Therapy Treatment Planning

The final issue requiring attention for volumetric treatment is that of the treatment planning process. There are many subparts in modulated volumetric treatment planning and most will not be covered here, as much of the necessary QA for modulated volumetric treatments will be already done as part of a standard QA program. Also for the treatment planning process, the procedures necessary for implementation of volumetric modulated arc therapy are an add-on to the existing QA program. Several of the standard procedures for treatment planning QA, such as consistency checks, should be just extended to include the case of plan optimization for volumetric treatment.

For volumetric modulated therapy, the following issues of the treatment planning process, however, require special attention:

- *Beam commissioning for small fields.* Often, there will be small MLC openings in the arc field at some gantry angles, and the correct modeling of small fields becomes more important than for other treatment techniques.
- *MLC settings such as transmission and leaf gap.* For the volumetric modulated therapies, the secondary collimator jaws may not move during the arc delivery but are open to cover the entire field size from all angles during the complete arc delivery (this is the case in the RapidArc implementation). This means that effects of MLC properties on the delivered dose will be larger than for other treatment techniques.
- *Couch modeling.* The delivery is being performed from all angles (potentially), which implies that the beam will go through all parts of the couch, and beam angles will most often not be chosen to avoid certain structures, such as rails. It therefore becomes more important to model couch properties correctly for optimization and dose calculation.
- *Dose calculation grid size.* The beam delivery may be highly modulated because of the complex combination of MLC, gantry, and dose rate variations, and it is important to verify to what extent these modulations can be seen for various magnitudes of dose calculation grids.

Summary

Volumetric modulated arc therapies are very complex techniques that require a high level of QA. The interplay of the many variables being used over a broad range of magnitudes implies that this is very demanding. Angular independence is a requirement, and measurement of high modulation should be possible preferably with a possibility of temporal resolution as well.

References

Bedford, J. L., and A. P. Warrington. 2009. Commissioning of volumetric modulated arc therapy (VMAT). *Int. J. Radiat. Oncol. Biol. Phys.* 73 (2): 537–545.

Duthoy, W., W. De Gersem, K. Vergote, M. Coghe, T. Boterberg, Y. De Deene, C. De Wagter, S. Van Belle, and W. De Neve. 2003. Whole abdominopelvic radiotherapy (WAPRT) using intensity-modulated arc therapy (IMAT): First clinical experience. *Int. J. Radiat. Oncol. Biol. Phys.* 57 (4): 1019–1032.

Fredh, A., S. S. Korreman, A. F. Munck, and P. Rosenschold. 2010. Automated analysis of images acquired with electronic portal imaging device during delivery of quality assurance plans for inversely optimized arc therapy. *Radiother. Oncol.* 94 (2): 195–198.

Ling, C. C., P. Zhang, Y. Archambault, J. Bocanek, G. Tang, and T. Losasso. 2008. Commissioning and quality assurance of RapidArc radiotherapy delivery system. *Int. J. Radiat. Oncol. Biol. Phys.* 72 (2): 575–581.

Low, D. A., and J. F. Dempsey. 2003. Evaluation of the gamma dose distribution comparison method. *Med. Phys.* 30 (9): 2455–2464.

Mans, A., P. Remeijer, I. Olaciregui-Ruiz, M. Wendling, J. J. Sonke, B. Mijnheer, M. van Herk, and J. C. Stroom. 2010. 3D Dosimetric verification of volumetric-modulated arc therapy by portal dosimetry. *Radiother. Oncol.* 94 (2): 181–187.

Nicolini, G., E. Vanetti, A. Clivio, A. Fogliata, S. Korreman, J. Bocanek, and L. Cozzi. 2008. The GLAaS algorithm for portal dosimetry and quality assurance of RapidArc, an intensity modulated rotational therapy. *Radiat. Oncol.* 3: 24.

Otto, K. 2008. Volumetric modulated arc therapy: IMRT in a single gantry arc. *Med. Phys.* 35 (1): 310–317.

Ulrich, S., S. Nill, and U. Oelfke. 2007. Development of an optimization concept for arc-modulated cone beam therapy. *Phys. Med. Biol.* 52 (14): 4099–4119.

Wang, C., S. Luan, G. Tang, D. Z. Chen, M. A. Earl, and C. X. Yu. 2008. Arc-modulated radiation therapy (AMRT): A single-arc form of intensity-modulated arc therapy. *Phys. Med. Biol.* 53 (22): 6291–6303.

58

Four-Dimensional Treatment

Wei Lu
University of Maryland

Lakshmi Santanam
Washington University
School of Medicine

Camille Noel
Washington University in St. Louis

Parag J. Parikh
Washington University
School of Medicine

Daniel A. Low
UCLA School of Medicine

Introduction

Respiratory motion can cause significant dose delivery errors in conformal radiation therapy for thoracic and upper abdominal tumors. Four-dimensional (4D) radiotherapy has been developed to partially account for respiratory motion during treatment planning and delivery. This chapter presents an overview of 4D treatment methods first, followed by detailed commissioning and quality assurance (QA) tests for an external respiratory gating system (Varian RPM). The focus of this chapter is on QA for the gated treatment delivery system. The critical patient-specific QA of motion management is addressed in Chapter 87.

Four-Dimensional Treatment Delivery

Keall (2004) defined 4D radiotherapy as "the explicit inclusion of the temporal changes in anatomy during the imaging, planning and delivery of radiotherapy." It consists of 4D CT imaging, 4D treatment planning, and 4D treatment delivery. The AAPM Task Group 76 provides an overview of various methods to account for respiratory motion in radiotherapy treatment delivery, along with their related QA issues (Keall et al. 2006). The methods can be grouped into a few categories: (a) generation of a motion-encompassing tumor volume; (b) respiratory gating method (using external respiration signal or internal fiducial markers); (c) breath-hold method (deep-inspiration breath-hold, active breathing control, and self-held breath-hold); (d) forced shallow breathing with abdominal compression; and (e) nearly real-time tumor tracking by repositioning the MLC (under development) or by repositioning a linear accelerator with a robotic arm (CyberKnife system, Accuray). For category (a), the term internal tumor volume (International Commission on Radiation Units and Measurements 1999) has been widely used to denote the union of the gross tumor volumes contoured in CT images

at multiple respiratory phases (Underberg et al. 2004, 2005a, 2005b) or in maximum intensity projection CT (Underberg et al. 2005a, 2005b; Bradley et al. 2006). The internal tumor volume encompasses the partial or the entire extent of the gross tumor volume motion, yielding a larger radiation field size and increased dose to normal tissue (Underberg et al. 2005a, 2005b). The use of the internal tumor volume does not require additional QA on the treatment delivery system. Respiratory gating with an external respiration signal has been implemented clinically and its benefit in reducing dose to normal tissue has been demonstrated (Underberg et al. 2005a, 2005b). A breath-hold method can be considered as a special case of respiratory gating. Forced shallow breathing requires patient-specific QA to verify that the motion is consistent. It does not require additional QA on the treatment delivery system. QA for real-time tumor tracking with the CyberKnife system will be addressed in Chapter 77. In this chapter, we will focus on QA for treatment delivery systems that use external respiratory gating for motion management. Such systems will likely be adopted widely in the near future.

Acceptance Testing, Commissioning, and QA for a Respiratory-Gated Treatment Delivery System

In this section, we provide details of acceptance testing, commissioning, and QA for a Varian Trilogy linear accelerator with Real Time Position Management Respiratory Gating System (RPM, Varian Medical Systems). We use the following terms to describe the possible situations: (1) stationary or stationary phantom: the phantom for dosimetric measurements is stationary; (2) moving or moving phantom: the phantom for dosimetric measurements is set to simulated respiratory motion; (3) gated delivery: the radiation beam is turned on only when the respiratory signal falls into the gating window, which typically is set to the end of exhalation; (4) nongated delivery: the beam is not turned on or off during the planned monitor units (MU).

RPM System

The RPM system uses a CCD camera to image infrared reflective markers attached to a marker block placed on the patient's abdomen or chest. The calculated marker trace represents the patient's surface displacement in the anterior–posterior direction, which serves as an external surrogate to the internal (tumor) respiratory motion. The user can then specify upper and lower gating thresholds on either the marker's amplitude trace or the phase trace (Lu et al. 2006). The beam is turned on only when the marker falls into a gating window defined by the upper and lower gating thresholds. A larger extension between the two thresholds produces a larger duty cycle, which is the ratio of beam-on time to the total beam delivery time. A detailed description of the RPM with illustrations can be found in the report of Vedam et al. (2001).

Acceptance Tests by the Vendor

The vendor's acceptance test of the RPM system (Varian Trilogy Acceptance Test Document) includes (1) verification that the camera can track a marker block using a motion phantom and verification that audio coaching works properly; (2) testing that the beam is held when the phantom motion exceeds the gating window; (3) verification that the length of the treatment changes accordingly with the extension of the gating thresholds or the duty cycle; (4) verification of a moving-gated delivery (at 30% duty cycle) by acquiring a film of a "BB" placed on the moving marker block and confirming that the observed blurring and total displacement of the BB is similar to that observed during the simulation; and (5) verification of a moving nongated delivery by repeating test (4) with a duty cycle of 100% and confirming that the results are similar to those in simulation.

Commissioning Tests by Physicists

Physicists perform commissioning tests for the RPM system to verify (1) the stationary-gated delivery has the same beam characteristics (output, percent depth dose, beam profiles) as the stationary-non-gated delivery and (2) the moving-gated delivery functions properly: (a) it delivers a dose distribution as predicted by the treatment plan assuming a stationary phantom and (b) it reduces the dosimetric errors caused by motion or it produces a dose distribution more similar to the stationary delivery than does the moving-non-gated delivery.

The following tests were performed at our institution when a linear accelerator (Varian Trilogy) with RPM was installed. For all simulated motion tests, we used a simple two-dimensional motion platform. The motion platform moves in a sinusoidal pattern in the superior–inferior (SI) direction with a synchronized surrogate axis that moves in the anterior–posterior direction. These tests were performed for all beam energies (6, 10,

and/or 18 MV) used for gating, although only illustrations for 6 MV are presented.

Comparison of Central Axis Output from Stationary-Gated and Stationary-Non-Gated Deliveries

An ionization chamber is placed in a stationary solid water phantom ($30 \times 20 \times 20$ cm³) to measure the central axis outputs. For gated delivery, the marker block is placed on the motion platform's surrogate axis that moves in the anterior–posterior direction. The infrared camera is focused on the marker block to monitor the simulated respiratory motion for generating a gating signal. Motion at various rates (12–20 cycles/min) can be tested to simulate different respiratory frequencies. The solid water phantom for dosimetric measurements remains stationary. Beams with various monitor units (5, 10, 25, 50, 100, 300, 500 MU) and dose rates (100, 300, 600 MU/min) are delivered to the phantom for gated and nongated deliveries, respectively. For nongated delivery, the gating window is opened to encompass the entire breathing wave (duty cycle = 100%). The planned MU is delivered without any beam interruptions. For gated delivery, the gating window is adjusted to encompass only a portion of the breathing wave (at the end of exhalation) and the planned MU is delivered with beam hold when the motion exceeds the gating window. A range of duty cycles can be tested for the gated delivery. For each gated delivery, output readings from the ion chamber are recorded and compared to those of the nongated delivery. The difference between the two was smaller than 1% for larger MU or less than the dose corresponding to 1 MU for smaller MU.

Comparison of Percent Depth Dose from Stationary-Gated and Stationary-Non-Gated Deliveries

The percent depth doses for beams with field sizes of 5×5 and 10×10 cm² are measured with similar setup to that described in the section "The RPM System." Gated and nongated ion chamber measurements are made in a $48 \times 48 \times 41$ cm³ stationary water phantom. Figure 58.1 shows that gated percent depth dose curves with 33% and 50% duty cycles match well with the percent depth dose curve from nongated delivery (within 1% beyond the buildup region).

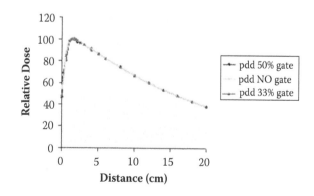

FIGURE 58.1 Comparisons of percent depth dose.

Comparison of Beam Profiles from Stationary-Gated and Stationary-Non-Gated Delivery

For beam profile comparisons, a high-resolution dosimeter (film or small volume ion chamber) is preferred over a low-resolution diode array (MapCheck or Profiler, Sun Nuclear). The setup is similar to that described in the section "Acceptance Tests by the Vendor," where small-volume ion chamber measurements are made in a stationary phantom. Figure 58.2 shows typical beam profiles using a 5×5-cm^2 field as an example. The gated beam profiles match well with those from nongated delivery at both d_{max} and 10 cm depth.

The above results demonstrate that the stationary-gated delivery does not significantly alter the beam characteristics compared with stationary-non-gated delivery. Ramsey, Cordrey, and Oliver (1999) reported similar results that in most of their gating sequences, the stationary-gated beam output, energy, flatness, and symmetry differed by less than 0.8% from the stationary-non-gated deliveries. These results are expected for the Varian RPM system because it holds off a beam by stopping the electron flow to the waveguide with a gridded electron gun (Ramsey, Cordrey, and Oliver 1999). Larger differences due to the end effect were reported for gating systems that power up and down the linear accelerator (Ohara et al. 1989).

Comparison of IMRT Dose Distribution from Stationary-Non-Gated, Moving-Gated, and Moving-Non-Gated Deliveries

A two-dimensional dosimeter (film or diode array such as MapCheck) set to motion can be used to study IMRT dose distributions for a gating system. In our commissioning, a MapCheck was used to measure the dose distribution field by field for an SMLC IMRT plan. The plan was first delivered to the stationary MapCheck. The MapCheck was then placed on the motion phantom and moved in the superior–inferior direction

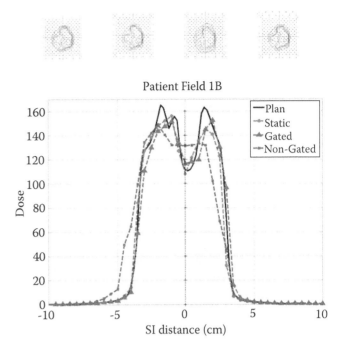

Patient Field 1B

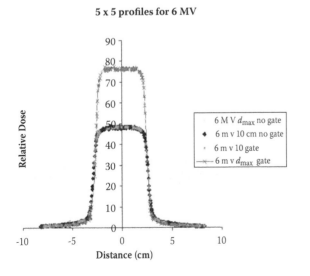

FIGURE 58.3 Planar dose distributions for (a) planned; (b) stationary-non-gated delivery; (c) moving-non-gated delivery for a 2-cm motion; and (d) moving-gated delivery for the same motion. (e) Center dose profiles along the SI direction.

with an amplitude of 2 cm and frequency of 15 cycles/min. The same IMRT plan was delivered with gating (duty cycle = 25%) and without gating to the moving MapCheck. An example of the measured dose distributions and center dose profile along the SI direction for one field is shown in Figure 58.3. For moving-non-gated delivery, motion-induced blurring is seen in the high dose gradient region (Figure 58.3c), and the dose profile is shifted in the SI direction (Figure 58.3e). For moving-gated delivery (Figure 58.3d), these motion effects were clearly reduced and the dose distribution matched better with the planned and stationary-non-gated deliveries.

In addition to single field studies, Hugo, Agazaryan, and Solberg (2002) performed composite studies on multiple-field IMRT plans. Films were placed in a phantom perpendicular to the motion direction (SI), on the central axis and off-axis. Their results show that on the central axis slice, both the moving-gated and moving-non-gated dose distributions match the stationary distribution well, due to the minimal amount of dose modulation in the direction of motion on this slice. However, on the off-axis slice, the dose modulation becomes high, nongated delivery fails to produce the planned dose distribution, whereas the gated delivery substantially improves that: the average γ value is reduced from 16.27 to 0.44.

Additional QA Considerations for Gated DMLC IMRT

For DMLC IMRT, it is known that a communication lag between the treatment console and MLC controller causes leaf lag and leads to dosimetric errors. The errors may be exacerbated by gated operation as illustrated by Duan et al. (2003). Their analysis for

FIGURE 58.2 Comparisons of beam profiles using small volume ion chamber.

sliding a 1-cm-wide DMLC leaf gap across a stationary film shows that gated delivery generates distinctive dosimetric artifacts: 5% cold and hot stripes at beam hold-off and beam resume, respectively. The magnitude of the artifact is proportional to the dose rate, the width of the artifact is proportional to the leaf speed, whereas the dose discrepancy within a small volume is proportional to the number of beam hold-offs triggered by the gating system. Hugo, Agazaryan, and Solberg (2002) also reported that their DMLC IMRT generates larger gated delivery errors than SMLC IMRT because of more complex implementation of the former. For gating with DMLC IMRT or dynamic arc, it is therefore important to test and choose appropriate treatment (e.g., a moderate dose rate of 300 MU/min) and gating parameters (e.g., a lower duty cycle for a faster breathing patient) to balance the treatment efficiency and gating dosimetric artifacts.

Daily QA and Frequency of QA

At our institution, a daily QA for gating is performed using the vendor-provided RPM phantom and marker block. It tests proper functioning of the gating system by (1) verifying the camera can track the marker block and (2) verifying the "radiation beam enable" light turns on and off as the marker block moves in and out of the gating window. A comprehensive QA should be performed after hardware or software changes that affect the respiratory gating. This may be part of annual QA for the linear accelerator.

Summary

In this chapter, commissioning and QA tests for an external respiratory gating system (Varian RPM) are presented. The results demonstrate that this gating system functions properly and does not cause significant differences in beam characteristics, although cautions should be used for gating DMLC IMRT.

New techniques are being developed, by both researchers and vendors, for (1) directly monitoring tumor position during a treatment with on-board kV images or implanted markers and (2) nearly real-time tumor tracking by dynamic MLCs or dynamic arc delivery. Additional QA of the treatment system will be required when these new techniques are introduced into clinical use. It is important to note that even with these advanced techniques, all 4D radiotherapy systems depend largely on a certain level of consistency in the patient's breathing. Respiratory coaching can be useful for improving the breathing consistency and the stability of the relationship between tumor motion and the external respiratory surrogate (Neicu et al. 2006). Patient-specific QA remains critical for motion management. The reader is referred to Chapter 87 and to the study of Jiang, Wolfgang, and Mageras (2008).

References

Bradley, J. D., A. N. Nofal, I. M. El Naqa, W. Lu, J. Liu, J. Hubenschmidt, D. A. Low, R. E. Drzymala, and D. Khullar. 2006. Comparison of helical, maximum intensity projection (MIP), and averaged intensity (AI) 4D CT imaging for stereotactic body radiation therapy (SBRT) planning in lung cancer. *Radiother. Oncol.* 81 (3): 264–268.

Duan, J., S. Shen, J. B. Fiveash, I. A. Brezovich, R. A. Popple, and P. N. Pareek. 2003. Dosimetric effect of respiration-gated beam on IMRT delivery. *Med. Phys.* 30 (8): 2241–2252.

Hugo, G. D., N. Agazaryan, and T. D. Solberg. 2002. An evaluation of gating window size, delivery method, and composite field dosimetry of respiratory-gated IMRT. *Med. Phys.* 29 (11): 2517–2525.

International Commission on Radiation Units and Measurements. 1999. *Prescribing, Recording, and Reporting Photon Beam Therapy. Supplement to Report 50.* Washington, DC: ICRU.

Jiang, S. B., J. Wolfgang, and G. S. Mageras. 2008. Quality assurance challenges for motion-adaptive radiation therapy: Gating, breath holding, and four-dimensional computed tomography. *Int. J. Radiat. Oncol. Biol. Phys.* 71 (1 Suppl): S103–S107.

Keall, P. 2004. 4-Dimensional computed tomography imaging and treatment planning. *Semin. Radiat. Oncol.* 14 (1): 81–90.

Keall, P. J., G. S. Mageras, J. M. Balter, R. S. Emery, K. M. Forster, S. B. Jiang, J. M. Kapatoes et al. 2006. The management of respiratory motion in radiation oncology report of AAPM Task Group 76. *Med. Phys.* 33 (10): 3874–3900.

Lu, W., P. J. Parikh, J. P. Hubenschmidt, J. D. Bradley, and D. A. Low. 2006. A comparison between amplitude sorting and phase-angle sorting using external respiratory measurement for 4D CT. *Med. Phys.* 33 (8): 2964–2974.

Neicu, T., R. Berbeco, J. Wolfgang, and S. B. Jiang. 2006. Synchronized moving aperture radiation therapy (SMART): Improvement of breathing pattern reproducibility using respiratory coaching. *Phys. Med. Biol.* 51 (3): 617–636.

Ohara, K., T. Okumura, M. Akisada, T. Inada, T. Mori, H. Yokota, and M. J. Calaguas. 1989. Irradiation synchronized with respiration gate. *Int. J. Radiat. Oncol. Biol. Phys.* 17 (4): 853–857.

Ramsey, C. R., I. L. Cordrey, and A. L. Oliver. 1999. A comparison of beam characteristics for gated and nongated clinical x-ray beams. *Med. Phys.* 26 (10): 2086–2091.

Underberg, R. W., F. J. Lagerwaard, J. P. Cuijpers, B. J. Slotman, J. R. van Sornsen de Koste, and S. Senan. 2004. Four-dimensional CT scans for treatment planning in stereotactic radiotherapy for stage I lung cancer. *Int. J. Radiat. Oncol. Biol. Phys.* 60 (4): 1283–1290.

Underberg, R. W., F. J. Lagerwaard, B. J. Slotman, J. P. Cuijpers, and S. Senan. 2005a. Benefit of respiration-gated stereotactic radiotherapy for stage I lung cancer: An analysis of 4DCT datasets. *Int. J. Radiat. Oncol. Biol. Phys.* 62 (2): 554–560.

Underberg, R. W., F. J. Lagerwaard, B. J. Slotman, J. P. Cuijpers, and S. Senan. 2005b. Use of maximum intensity projections (MIP) for target volume generation in 4DCT scans for lung cancer. *Int. J. Radiat. Oncol. Biol. Phys.* 63 (1): 253–260.

Vedam, S. S., P. J. Keall, V. R. Kini, and R. Mohan. 2001. Determining parameters for respiration-gated radiotherapy. *Med. Phys.* 28 (10): 2139–2146.

Combined Planning and Delivery Systems

Sonja Dieterich
Stanford University Hospital

Introduction

The rapidly increasing diversity in treatment delivery systems has resulted in systems in which the software and hardware are not separate modules of the treatment planning and delivery process. We define a "combined system" as radiation delivery system for which no single item within the "technical component" box of the "combined system" shown in Figure 59.1 could be replaced by technology from another vendor without requiring FDA clearance. Examples of combined systems include Tomotherapy, Gamma Knife, CyberKnife, and the various hadron therapy machines. In this chapter, R&V systems and backup issues will not be included in the discussion.

Special Challenges Arising from Combined Systems

The combined systems provide special challenges to the clinical user that are outlined in more detail in the paragraphs below. Generally, a combined system is developed in a more closed environment, which limits second checking through other parallel applications or interfaces with independent vendor software or devices.

Challenge of Using a Combined System

In a multivendor solution, the protocols on how systems interact are, by necessity, well defined and published. The design of a quality assurance (QA) program to monitor the correct handoff of information from system to system is relatively straightforward; failure modes and their effects can be determined based on this information. The situation is more challenging for combined systems. The interaction between system components typically is handled internally via the vendor-provided software. Computer-controlled delivery systems, although preventing manual entry errors, can also contribute to errors. The report of AAPM TG-35 (Purdy et al. 1993) discusses many aspects of QA of computer-controlled radiation delivery systems. The recent paper of Fraass (2008) makes a clear statement on the status of QA management at this time in our field, and deserves to be cited in full:

> The knowledge, documentation, and understanding of control system architecture, design, and implementation that a typical user has is quite limited, making effective testing difficult. TG-35 recommends that vendors provide reasons for changes, bug fix descriptions, modification details, operational changes, site-dependent and user-accessible data or software which may be affected, testing procedures, revised specifications, support documentation, operations manuals, and beta test results with any software installation or update. Unfortunately, little of this is typically available, although vendor adherence to TG-35 recommendations would make possible significantly more effective testing and use by the user.

Closed Feedback Loops

That a system does not throw an error does not mean it is working correctly. Software is indeed capable of taking the fork in the road. A medical physicist should take great care never to rely exclusively on software to catch malfunctions in system performance. Software in itself can be faulty. Therefore, all interactions between system components need to be rigorously tested.

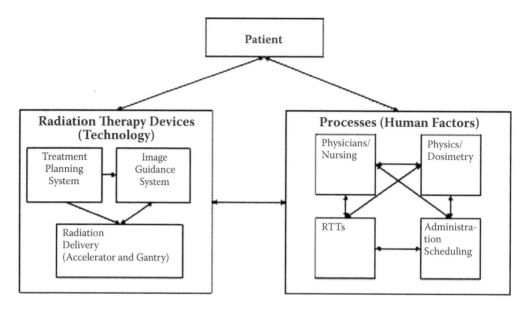

FIGURE 59.1   Definition of a combined system. None of the technical components can be exchanged by a similar component from another vendor.

Developing a failure modes and effects analysis (FMEA) for closed feedback loops requires documentation of the functionality of the closed system, including the design of the backup safety systems, to be available to the end user. Often, however, manufacturers decline to provide this information because of concerns about trade secrets.

Install Timelines and User Base

In the initial clinical adoption phase of equipment containing combined systems, there will be a relatively small user base. The increased variety of delivery system options will also lead to a reduced user base for each individual system. Communication and close collaboration between the early adopters of the combined system are essential to quickly generate intercomparison charts and QA protocols, as well as communication of software anomalies. Vendors providing QA tools often have not yet developed device-specific tools for closed systems, or it is not financially viable until the user base has reached a critical mass.

Commissioning

In the commissioning of a closed system, careful attention has to be paid to manufacturer's specifications and guidelines. Required input parameters may differ from other systems the medical physicist is familiar with. Beam data, for example, may have to be taken in TPR instead of PDD configuration.

In addition, technical design parameters may differ enough from existing technologies to require development of new approaches. An example of this situation is the available beam aperture for Tomotherapy or CyberKnife. Neither of these technologies allows for a 10 × 10-cm field size required by AAPM

TG-51 (Almond et al. 1999) for reference dosimetry. During the early years of clinical use of these devices, medical physicists had to develop and improve on the reference dosimetry protocols (Kawachi et al. 2008; Francescon, Cora, and Cavedon 2008; Thomas et al. 2005). Only recently, the IAEA working group on small field dosimetry published a conceptual paper (Alfonso et al. 2008) to define a new formalism for nonstandard reference beams. Lack of familiarity with these new techniques and/or detectors used for measurements is another concern for the medical physicist.

Clinical Use

Combined systems can also pose unexpected challenges for clinical use. Often, commercially available QA technology for clinical plan QA is not up-to-date with technical innovations. If, for example, a combined system uses a new tracking algorithm for patient setup and monitoring during the treatment, the medical physicist needs to have tools available to check the clinical accuracy and performance of this algorithm.

Similar challenges exist for adaptive planning algorithms or deformable registration-based planning. Both concepts are patient-specific, but a thorough concept of what a well-designed QA program should look like does not yet exist.

Contradictory approaches for clinical QA in combined systems have historically developed to a point where the path to standardization of clinical practice remains unclear. The classic example is whether patient-specific delivery QA is needed for SRS/SBRT delivered by Gamma Knife, CyberKnife and MLC-based delivery (Tomotherapy and linear accelerator-based). The combined systems that originated from SRS, Gamma Knife, and CyberKnife, chose a QA philosophy that focused on QA testing technical functionality, including monthly end-to-end tests for delivery QA. In combination with setup accuracy (either through using a patient frame or image-guidance during treatment), it is assumed that the delivery to the patient is accurate. The two combined systems using the same MLC technology as IMRT, Tomotherapy and Linac-based SRS, stayed with the IMRT tradition of patient-specific QA, which originated from the early implementation of MLC where there were no sophisticated real-time leaf position verification tools available yet. The added complication of different reimbursement patterns for patient-specific QA for each of the SRS delivery systems together with tradition has so far prevailed over a scientific study of whether patient-specific QA, technical QA, a combination of both, or a FMEA/statistical process control approach would be the safest QA program making optimum use of resources.

QA Approach for Combined Systems

Failure Modes and Effects Analysis

The best approach to designing an efficient QA/QM program for a combined system is to perform a rigorous FMEA analysis at system install, with updates at regular intervals (e.g., annually and after system changes/upgrades). The FMEA analysis requires investment of time before the physical installation, ideally starting after the purchase agreement. This requires also the commitment of the clinical team and should not be the sole responsibility of the medical physicist. Creating the FMEA flowcharts requires knowledge of potential failure modes and at least an educated guess at the failure probability. Both of these cannot be established well unless there is close cooperation and information exchange between the manufacturer and the clinical users. Therefore, the FMEA analysis and support for QA/QM program design need to be integrated into the vendor-provided training programs for the equipment.

Vendor input is also essential in gaining a good understanding of where the failure modes may be. Vendors need to provide, and customers (i.e., the clinical users) need to request in writing, documentation on system safety design and results of the safety testing done during development and release. All too often, the "Technology" box in Figure 59.1 remains a black box to the clinical user. This lack of information is the main obstacle to developing a good technical QA program.

Data Collection for FMEA

To generate an accurate FMEA analysis, understanding the probability of a failure in the absence of safeguards is necessary. A large database based on voluntary error reporting, such as ROSIS in Europe (http://www.clin.radfys.lu.se) or the FAA database on airline safety incidents in the United States, would be invaluable. Currently, the medical professions in the United States support and encourage root cause analysis of medical errors and near-miss events within the clinical setting. However, open disclosure and reporting to a national or international body may put the institution and/or reporting person at high risk for legal action. In the airline industry, only waiver of enforcement/waiver of disciplinary action for voluntary reporting, strict anonymization, and close collaboration between government, industry, and workforce representatives enabled the establishment of large databases quickly. Until the healthcare industry finds a similar pathway to support reporting on safety issues, the creation of a large enough database for the purpose of error prevention analysis, and therefore sound data on which to develop QM programs, will be extremely difficult.

Safeguards

After the FMEA analysis is completed, safeguards have to be set for events where the FMEA scores above a certain limit. Safeguards can be in the form of process (e.g., surgical time-outs on a SRS delivery system), technological (e.g., software or mechanical safeguards), or equipment QA.

Process safeguards are outside the scope of this chapter and are covered in Parts I and II. For software safeguards and mechanical safeguards, it is essential for the medical physicists to educate themselves and test the primary and secondary safeguards against machine failure. In a combined system, this may require close interaction with the manufacturer in regard to questions that are not covered in the standard user guides and system manuals. QA tests should be the tertiary safety layer to ensure functionality of the equipment in clinical practice. An error-free combined system is not necessarily a functioning system!

Summary

Combined systems provide convenience as well as special risks in clinical practice. The seamless integration of treatment planning and delivery avoids issues with data transfer across different vendor platforms. On the other hand, the seeming ease of use of a combined system can easily lead to blind trust, that is, insufficient testing of potential risk factors within the closed system environment. The clinical medical physicist has to rely much more on excellent vendor communication through system design manuals and timely customer technical bulletins to design a comprehensive, efficient QA program.

References

Alfonso, R., P. Andreo, R. Capote, M. S. Huq, W. Kilby, P. Kjall, T. R. Mackie et al. 2008. A new formalism for reference dosimetry of small and nonstandard fields. *Med. Phys.* 35 (11): 5179–5186.

Almond, P. R., P. J. Biggs, B. M. Coursey, W. F. Hanson, M. S. Huq, R. Nath, and D. W. Rogers. 1999. AAPM's TG-51 protocol for clinical reference dosimetry of high-energy photon and electron beams. *Med. Phys.* 26 (9): 1847–1870.

Fraass, B. A. 2008. Errors in radiotherapy: Motivation for development of new radiotherapy quality assurance paradigms. *Int. J. Radiat. Oncol. Biol. Phys.* 71 (1 Suppl): S162–S165.

Francescon, P., S. Cora, and C. Cavedon. 2008. Total scatter factors of small beams: A multidetector and Monte Carlo study. *Med. Phys.* 35 (2): 504–513.

Kawachi, T., H. Saitoh, M. Inoue, T. Katayose, A. Myojoyama, and K. Hatano. 2008. Reference dosimetry condition and beam quality correction factor for CyberKnife beam. *Med. Phys.* 35 (10): 4591–4598.

Purdy, J. A., P. J. Biggs, C. Bowers, E. Dally, W. Downs, B. A. Fraass, C. J. Karzmark, F. Khan, P. Morgan, R. Morton, et al. 1993. Medical accelerator safety considerations: Report of AAPM Radiation Therapy Committee Task Group No. 35. *Med. Phys.* 20 (4): 1261–1275.

Thomas, S. D., M. Mackenzie, D. W. Rogers, and B. G. Fallone. 2005. A Monte Carlo derived TG-51 equivalent calibration for helical tomotherapy. *Med. Phys.* 32 (5): 1346–1353.

60

Proton Radiotherapy

Richard L. Maughan
University of Pennsylvania

Introduction

As the use of proton radiation therapy expands and the new proton beam delivery techniques such as uniform scanning and pencil beam scanning are introduced and applied, there is a need to better document and disseminate quality control and quality assurance methods for this modality. There has been a significant increase in the number of hospital-based, commercially manufactured proton therapy facilities in the past 10 years and it is likely that this trend will continue in spite of the recent recession. However, there is relatively little published data on proton therapy quality control and quality assurance procedures. Although many aspects of proton therapy are similar to conventional photon and electron radiation therapy, there are significant differences that require specific proton beam quality control and assurance. Detailed discussion of the literature appears in the other proton therapy–related chapters in this publication. The purposes of this chapter are to provide an introduction to the topic of QA for proton beams and to outline efforts by organizations such as the Proton Therapy Cooperative Group (PTCOG), the International Atomic Energy Agency (IAEA), and the American Association of Physicists in Medicine (AAPM) to establish recommendations for proton radiation therapy QA. Another goal of this chapter is to identify how proton therapy systems and their application differ from conventional photon and electron beams and how this impacts QA procedures.

Role of AAPM, PTCOG, ICRU, and IAEA

Conventional photon and electron therapy are well established, and there is a wealth of literature on quality assurance. In the United States, the AAPM has published guidelines on many aspects of quality control and assurance for the conventional external beam therapy modalities. These guidelines result from the work of many AAPM-sponsored task groups, and reports on these activities are easily accessible in *Medical Physics* or online at the AAPM Web site (www.aapm.org).

It is only relatively recently that the AAPM formed a Working Group on Particle Beams (WGPB), and it is part of this group's mandate to create task groups to address issues related to particle beams. This group is focusing its attention on the needs of the proton therapy community because, in the United States, the emphasis has been on installing hospital-based proton therapy facilities rather than combined ^{12}C/proton centers. Of course, many of the issues related to proton therapy are directly relevant to ^{12}C ion therapy and other ions heavier than protons.

At the time of writing, the AAPM WGPB has three active task groups. One group is preparing a report on "Reference Dosimetry and Patient Specific Field Calibration for Passively Scattered Proton Therapy Beams" (TG156), another is dealing with "Clinical Commissioning of Proton Therapy Systems," (TG185), and the third is addressing "Nomenclature, Specifications, Safety Requirements and Acceptance Procedures for Proton Radiation Therapy Systems" (TG183). The first task group deals directly with an important aspect of proton therapy QA, whereas the other two have significant relevance to proton beam QA. A fourth task group, which is awaiting approval, is planned to give guidance on uncertainties and margins in proton therapy.

The WGPB intends to create task groups that will directly address QA procedures specific to proton beam therapy to supplement the information in the report of AAPM Task Group 40, "Comprehensive QA for Radiation Oncology: Report of AAPM Task group 40" (Kutcher et al. 1994) and in the pending TG-100 report, which should be published soon. The WGPB has delayed initiation of these task groups as proton radiation therapy is undergoing a rapid expansion and new techniques such as uniform scanning and pencil beam scanning are just being introduced in commercial systems. This increase in activity is expected to result in a significant increase in the available literature on proton beam QA that will aid the task groups in establishing standards and making recommendations.

In parallel with the AAPM WGPB, the PTCOG has established a publications committee, which is also mandated to create reports related to proton and heavier ion beam radiation therapy. There is a liaison between these two groups to minimize unnecessary duplication of effort. The PTCOG group, with its broader international and proton/ion therapy–focused membership base, will complement the AAPM efforts.

Until recently, the major sources of general information on proton therapy QA procedures have been found in the work of Moyers (1999) and in the International Commission on Radiation Units and Measurements (ICRU) Report 59, "Clinical Proton Dosimetry—Part I: Beam Production, Beam Delivery and Measurement of Absorbed Dose" (ICRU 1998). In 2008, the ICRU published an updated report on proton therapy entitled "Prescribing, Recording and Reporting Proton-Beam Therapy" (ICRU 2008). This report contains a chapter addressing QA for proton therapy but also addresses the issue of absolute dose calibration of proton beams. ICRU Report 78 reconciles absorbed dose calibration discrepancies between ICRU Report 59 and the IAEA technical report TRS-398 (IAEA 2004). The recently published *Proton and Charged Particle Radiotherapy*, edited by Delaney and Kooy, also includes information on proton QA (Maughan and Farr 2008).

Importance of System Specifications, Acceptance Testing and Commissioning in Defining a Quality Assurance Program for Proton Therapy

The capital costs of proton therapy equipment are high and, in general, this requires that a single accelerator, generally a cyclotron or a synchrotron, supply the proton beam to multiple treatment rooms, which may contain gantry delivery systems or fixed beam lines at various angles. Typically, room configurations in commercial proton therapy systems are of three types. (1) Fixed horizontal beams in which the beam delivery system or nozzle is positioned at the end of a fixed horizontal beam line and directed toward a treatment couch or treatment chair. Highly specialized dedicated treatment chairs are often used for treating uveal melanoma patients. (2) Fixed beam configurations at two angles. Two beam lines in one room generally at 0° (horizontal) and 60° or 90° above horizontal, sometimes with a nozzle that can be translated between the two angles and docked to one or the other of the beam lines. (3) A gantry treatment room in which the nozzle is mounted on a beam line in a gantry that rotates through 360°.

In the presently installed commercial systems, the number of treatment rooms varies between three and five; five being considered the maximum number of rooms that can be supported by one accelerator. Because of the high cost of these systems, there are presently several research initiatives aimed at developing proton accelerator systems that will fit into a single shielded room and that can be operated in a community hospital environment.

System Specifications and Acceptance Testing

Because proton therapy is a field that is not yet as mature as conventional radiation therapy and where the equipment and installation costs are considerably higher, it is still common practice for customers to enter into detailed and prolonged negotiations with the vendors when purchasing this equipment. Proton therapy systems are still in a state of development. The beam delivery systems are more complex than for conventional linear accelerators because the attraction of proton therapy is that the dose distribution can be accurately controlled in the third dimension (i.e., depth) by varying beam penetration. A variety of beam delivery options exists: single scattering, double scattering, uniform scanning, and modulated scanning (or pencil beam scanning). For the potential customer, these multiple options can be confusing. In addition, it is important for the customer to understand the beam specifications in detail to obtain a proton therapy system that meets their needs. Over time, it is expected that the beam and system specifications of each vendor's equipment will develop to be well defined and unambiguous. However, at the present time, there is some flexibility in defining system specifications during negotiations for these expensive and high-technology systems.

Once system specifications are defined and agreed on it is necessary to have a well-defined documentation process by which compliance with the specifications will be demonstrated by the vendor to the customer. In conventional x-ray systems, this is the acceptance procedure/test that occurs over a 1- to 2-week period after installation of the equipment. Linear accelerator vendors have well-established, documented acceptance procedures as they have sold many systems (numbered in the thousands) and the technology is mature and changes relatively slowly. Again, the situation is quite different with proton therapy systems where relatively few commercial systems have been sold and installed (numbered in the tens). Proton equipment and technology are still undergoing development, often based on the customer's requirements. The greater complexity of the proton therapy systems in comparison to conventional x-ray linear accelerators requires system installation to be extended over many months (up to 18–21 months for a large five room system). During this period a large number of verification and validation procedures are performed on the system components and the proton beam modalities. The complexity level of these verification and validation procedures increases steadily over the installation period. Many of the higher-level verification and validation procedures can be incorporated into the acceptance testing so that the final acceptance procedures can be accomplished in a more traditional series of tests at the end of installation without extending the acceptance period beyond 3 or 4 weeks. Hence, acceptance test procedures are not as well defined as for conventional therapy. For a complex proton therapy system, establishment of an agreed on acceptance test strategy may become part of the contract negotiations.

Commissioning in Proton Therapy

The final stage of installation is commissioning. This includes not only collecting the beam data needed for the treatment planning system but also customer-defined benchmarking of that system and many of the other components of the proton therapy system. These other components include the imaging system and patient positioning system. Not only is benchmarking a commissioning issue but personnel training in the use of this largely unfamiliar equipment is also necessary. The benchmarking and training program need to be considered as an element of the overall program of quality assurance if proton therapy treatment is to be implemented and delivered in a safe manner. Because most institutions are introducing proton therapy for the first time, some departmental infrastructure for reviewing the proposed new procedures should be created. This review process should include physicians, physicists, radiation therapy technologists, nurses, and administrators. An external review of proposed procedures by a group experienced in the delivery of proton therapy should also be considered.

Defining a Quality Assurance Program for Proton Therapy

The system specifications should serve as a basis for many of the medical physics-based quality assurance procedures because much of the QA process relies on ensuring the correct operation of the system within the originally defined specifications. Acceptance test procedures and commissioning benchmarking procedures should form the basis for a comprehensive set of quality assurance tests, which may be performed at appropriate regular intervals (daily, weekly, monthly, or annually). As system specifications may vary from installation to installation, particularly as the technology is developing rapidly, it is important that the medical physicists at each site pay close attention to the QA procedures required by their specific equipment. For instance, the facility at the University of Pennsylvania will be the first site to use a unique proton multileaf collimator, which has been developed as part of a joint agreement between Varian Medical Systems (Palo Alto, CA, USA), Ion Beams Applications, S.A. (Louvain-La-Neuve, Belgium), and the University of Pennsylvania. MLC QA procedures specific to proton therapy will need to be developed. This is just one example of many unique specifications, incorporated in much of the vendor-supplied equipment, which require special attention from a QA perspective.

Proton Radiation Therapy–Specific QA Requirements

Many elements of a proton therapy system are similar to those found in conventional photon and electron beam therapy. However, there are some important differences; for example,

in addition to accurately controlling the lateral dimensions of the beam through collimation, in proton therapy the depth of penetration of the beam and the length of the spread-out Bragg peak can be controlled with a comparable accuracy. This creates some unique QA problems in proton therapy. The introduction of pencil beam scanning presents further unique QA challenges. In pencil beam scanning, collimation is no longer used to control the lateral extent of the beam, but the current in the scanning magnets themselves determines the beam position.

Another major difference between proton therapy and conventional therapy QA at the present time is the scarcity of commercially available proton-specific QA tools. In conventional therapy there is a wide variety of tools available from an array of commercial vendors. Although many of these devices can be adapted for use with proton beams, they often require major modifications. Not all conventional radiation detectors are suited to quantitative proton therapy measurements. Diode detectors and film both have significant energy dependence, which in principle can be overcome, but only at the expense of laborious calibration measurements. Pencil-scanned beams present many problems if accurate dosimetry measurements are required throughout the scanned volume. The ionization chamber is the most reliable energy-independent detector for proton beams, but the point-by-point measurements that would be required to characterize a 3D scanned beam dose distribution would be prohibitively time-consuming. There is a need for a reliable, convenient-to-use, 3D tissue equivalent integrating dosimeter that would cover the largest scanned beams (potentially $30 \times 40 \times 30$ cm^3), allowing a dose distribution to be measured in a single scan or multiple uninterrupted scans. In practice, a smaller device could be useful for a large number of measurements. A further requirement would be that the analyzed scan data should be available within minutes of completing the scan. The best solution to this problem at the moment is a 2D ionization chamber array in a water-proof enclosure, which can be scanned in depth through a water phantom (Farr et al. 2008). Another 2D device uses a scintillator screen coupled to a calibrated CCD camera (Boon et al. 2000) to measure the lateral dose distribution of the beam. For percentage depth dose measurements, a multilayer ionization chamber has been developed at the University of Indiana (Farr et al. 2008).

Proton beam therapy is a modality that needs to be delivered with high precision, in three dimensions, similar to stereotactic radiosurgery, stereotactic radiation therapy, and IMRT with conventional photons. Beam positioning, gantry isocentricity, and couch accuracy are of the utmost importance. The large size of the gantry presents special problems, and in some systems, gantry isocentricity is achieved by making couch adjustments based on the known and highly reproducible flexing characteristics of the gantry as it moves through 360°. This solution to the isocentricity problem may require unique QA procedures in comparison to a linac gantry. Laser systems may be used for preliminary patient alignment, but in gantry systems, some of these lasers may be mounted on the gantry itself (in the side walls and on the face of the treatment nozzle).

Another difference between a conventional linac and a proton double-scattering beam delivery system is that the collimation and beam shaping system (aperture and compensator, or MLC and compensator) should be positioned as close to the surface of the patient as possible to obtain the best penumbra. This requires that the collimation/compensator be mounted on a translational stage that moves along the direction of the beam over a range of about 400–500 mm. Thus, a QC procedure that can monitor the mechanical stability of this translation over time is required.

Of course, imaging is an important part of any modern radiation therapy equipment. Orthogonal kV images have been a part of proton therapy for many years. In proton therapy, the importance of accurate positioning may be critical to achieving range accuracy. Conventional therapy has "leapfrogged" proton therapy in recent years with the introduction of sophisticated on-board imaging systems that include cone-beam CT systems. Only recently has there been interest in installing CBCT on the proton therapy gantry or in the treatment room. Because of geometrical constraints in some systems, it may not be possible to image the patient in the treatment position and, therefore, the couch will need to be translated to the treatment position after imaging. This is similar to the situation when a "CT-on-rails" is used in conventional therapy, but nonetheless, it presents a unique set of QA problems that must be solved. In fact, a CT in the treatment room solution for soft tissue imaging will be used at the Paul Scherrer Institute proton therapy facility in their second gantry room that is presently under construction (Pedroni et al. 2004).

Pretreatment imaging is an important part of any treatment planning process, but in proton therapy, there are additional special requirements. The proton range depends on the stopping power of protons in the tissues through which they pass. Proton stopping power depends critically on the exact atomic composition of those tissues, not on their average electron density. However, it is possible to calibrate CT Hounsfield units against stopping power, but this requires calibration of individual CT devices, adding yet another proton-specific QA procedure (Schneider, Pedroni, and Lomax 1996; Schaffner and Pedroni 1998).

Summary

Proton therapy equipment and proton therapy delivery techniques are sufficiently different from those used in conventional therapy that they pose some unique QA challenges for the medical physicist. Some aspects of the technology are still in their infancy (e.g., pencil beam scanning), and universally accepted QA procedures are still to be defined. In the immediate future, it is expected that there will be much activity in developing the required QA procedures. As more proton therapy centers have become operational in recent years, this process has already begun. The other chapters in this volume dealing with proton therapy discuss specific proton QA procedures in greater detail, giving insight into what has already been achieved and what needs to be accomplished in the future.

References

Boon, S. N., P. van Luijk, T. Bohringer, A. Coray, A. Lomax, E. Pedroni, B. Schaffner, and J. M. Schippers. 2000. Performance of a fluorescent screen and CCD camera as a two-dimensional dosimetry system for dynamic treatment techniques. *Med. Phys.* 27 (10): 2198–2208.

Farr, J. B., A. E. Mascia, W. C. Hsi, C. E. Allgower, F. Jesseph, A. N. Schreuder, M. Wolanski, D. F. Nichiporov, and V. Anferov. 2008. Clinical characterization of a proton beam continuous uniform scanning system with dose layer stacking. *Med. Phys.* 35 (11): 4945–4954.

IAEA. 2004. Absorbed dose determination in external beam radiotheraphy: An international code of practice for dosimetry based on standards of absorbed dose to water. In *IAEA Technical Report Series TRS-398*. Vienna: International Atomic Energy Agency.

ICRU. 1998. Clinical Proton Dosimetry Part 1: Beam Production, Beam Delivery and Measurement of Absorbed Dose. In *ICRU Report 59*. Bethesda, MD: International Commission on Radiological Units and Measurements.

ICRU. 2008. Prescribing, recording, and reporting proton-beam therapy. ICRU report 78. *J. ICRU* 7 (2).

Kutcher, G. J., L. Coia, M. Gillin, W. F. Hanson, S. Leibel, R. J. Morton, J. R. Palta et al. 1994. Comprehensive QA for radiation oncology: Report of AAPM Radiation Therapy Committee Task Group 40. *Med. Phys.* 21 (4): 581–618.

Maughan, R. L., and J. B. Farr. 2008. Quality assurance for proton therapy. In *Proton and Charged Particle Radiotherapy*, ed. T. F. Delaney and H. M. Kooy. Philadelphia, PA: Lippincott, Williams and Wilkins. pp 50–56.

Moyers, M. F. 1999. Proton therapy. In *The Modern Technology of Radiation Oncology*, ed. J. Van Dyk. Madison, WI: Medical Physics Publishing.

Pedroni, E., R. Bearpark, T. Bohringer, A. Coray, J. Duppich, S. Forss, D. George et al. 2004. The PSI Gantry 2: A second generation proton scanning gantry. *Z. Med. Phys.* 14 (1): 25–34.

Schaffner, B., and E. Pedroni. 1998. The precision of proton range calculations in proton radiotherapy treatment planning: Experimental verification of the relation between CT-HU and proton stopping power. *Phys. Med. Biol.* 43 (6): 1579–1592.

Schneider, U., E. Pedroni, and A. Lomax. 1996. The calibration of CT Hounsfield units for radiotherapy treatment planning. *Phys. Med. Biol.* 41 (1): 111–124.

61

Stereotactic Radiosurgery and Stereotactic Radiation Therapy

Timothy D. Solberg
*University of Texas Southwestern
Medical Center at Dallas*

Paul M. Medin
*University of Texas Southwestern
Medical Center at Dallas*

Introduction

The field of stereotactic radiosurgery was pioneered in 1951 by the Swedish neurosurgeon Lars Leksell when he applied the methodology of stereotactic surgery to the delivery of external beam irradiation (Leksell 1951). Leksell defined radiosurgery as "a single high dose of radiation, stereotactically directed to an intra-cranial region of interest." Throughout the 1960s and 1970s, the field progressed slowly, hampered primarily by the lack of optimal radiation sources. Leksell subsequently abandoned his use of low energy x-rays in favor of protons (Larsson et al. 1958). In 1954, John Lawrence and Cornelius Tobias, working at the Lawrence Berkeley Lab, began a clinical program for stereotactic irradiation using proton, and later, helium ion beams (Lawrence et al. 1958, 1962). Raymond Kjellberg and colleagues at the Harvard Cyclotron Lab treated patients with proton beam radiosurgery beginning in 1961 (Kjellberg et al. 1962). The field changed radically with Larsson's invention of the GammaKnife in 1968 (Larsson, Liden, and Sarby 1974). For the next two decades, however, stereotactic radiosurgery remained a highly specialized and relatively obscure procedure, practiced at only a handful of centers worldwide. With the development of linear accelerator–based radiosurgery beginning in the mid-1980s, dedicated, low-cost stereotactic radiosurgery apparatus provided many clinicians with both the capabilities and confidence to effectively treat both benign and malignant cranial disease (Betti and Derechinsky 1984; Colombo et al. 1985; Lutz, Winston, and Maleki 1988; Podgorsak et al. 1988; Winston and Lutz 1988; Friedman and Bova 1989).

In the past two decades, developments in radiosurgery technology have progressed at a torrid pace, with innovations including MR localization, image fusion, dedicated radiosurgery linacs, relocatable frames, micro-multileaf collimators, image-guided ("frameless") localization techniques, and application to extracranial tumor sites. Guidance for radiosurgery quality assurance has significantly lagged the technological development. The primary reference document today, AAPM Report No. 54 "Stereotactic Radiosurgery" (Schell et al. 1995), was published nearly 15 years ago and covers none of the more recent innovations noted above. Several relevant efforts originating within the American Association of Physicists in Medicine are presently ongoing and are expected to be completed and published in the near future. These include:

- Task Group 101—"Stereotactic Body Radiotherapy"
- Task Group 104—"KiloVoltage Localization in Therapy"
- Task Group 117—"Use of MRI Data in Treatment Planning and Stereotactic Procedures—Spatial Accuracy and Quality Control Procedures"
- Task Group 132—"Use of Image Registration and Data Fusion Algorithms and Techniques in Radiotherapy Treatment Planning"
- Task Group 135—"QA for Robotic Radiosurgery"
- Task Group 155—"Small Fields and Non-Equilibrium Condition Photon Beam Dosimetry"
- Task group 178—"Gamma Stereotactic Radiosurgery Dosimetry and Quality Assurance"

In the absence of a modern, comprehensive radiosurgery QA document, this chapter is intended to provide guidance and examples for assessing and minimizing geometric (localization) and dosimetric uncertainties in cranial stereotactic radiosurgery and stereotactic radiation therapy. Other chapters in this edition are dedicated to gamma- and robotic-based radiosurgery; thus, this chapter focuses primarily on linear accelerator (linac)-based stereotactic radiosurgery/stereotactic radiation therapy. Nevertheless, many of the procedures described here have a much more universal application.

Geometric Localization

There are many sources of localization uncertainty in stereotactic radiosurgery/stereotactic radiation therapy. Table II of the AAPM Report 54 lists several: isocenter alignment, stereotactic frame, image resolution among others, with an overall localization uncertainty of $\sigma = 2.4$ mm for all sources combined when 1.7-mm-thick CT slices are used (Schell et al. 1995). It is essential that uncertainties be evaluated and understood so that appropriate guidance can be provided with regard to planning target volume margins and other relevant clinical parameters. In general, these sources can be separated into those originating with the linac (which we refer to as *isocenter* uncertainty) and those originating from the radiosurgery localization process (such as imaging, positioning in the treatment room, and others).

Isocentric Accuracy in Stereotactic Radiosurgery

In cranial stereotactic radiosurgery, the "Winston–Lutz" test is the universally accepted methodology for assessing mechanical isocentric accuracy (Lutz, Winston, and Maleki 1988; Winston and Lutz 1988). Using the lasers as the isocenter reference, a small radiopaque sphere is positioned at the presumed linear accelerator isocenter. A series of beams-eye-view images are obtained at predetermined gantry, couch, and collimator angles within a clinically relevant range. By observing where the projection of the ball falls relative to the central axis of the beam, any offset from the presumed isocenter can be determined and, if needed, corrected. The process is shown in Figure 61.1. For radiosurgery systems that use circular collimators ("cones") in addition to a micro-MLC, it is important to repeat the Winston–Lutz evaluation for both field-defining devices, as the cones and MLC define independent beam axes.

Modern linac systems are capable of Winston–Lutz results averaging better than 0.75 mm over the full range of gantry/couch/collimator positions; on installation, it is possible to achieve results on the order of 0.5 mm. Nevertheless, there may be outliers at particular gantry/couch/collimator combinations. It is important to be aware of any outliers and to compensate for unacceptable uncertainties they may induce. One could, for example, simply avoid use of beam or arcs at the offending positions.

It is important to perform a Winston–Lutz assessment regularly, on a daily basis at a minimum, and preferably before each treatment. There are numerous reasons that a Winston–Lutz test can fail. The most common is that the lasers are easily bumped

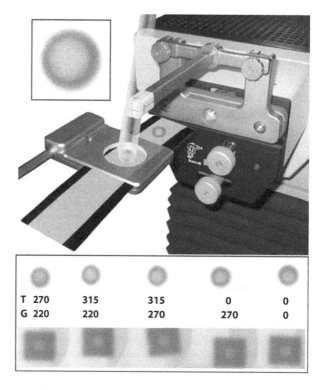

| T 270 | 315 | 315 | 0 | 0 |
| G 220 | 220 | 270 | 270 | 0 |

FIGURE 61.1 The Winston–Lutz test is performed by aligning a small radiopaque sphere to the presumed linac isocenter and subsequently obtaining a series of beams-eye radiographs (top) at various gantry, couch, and collimator combinations. The Winston–Lutz test should be performed for both cones and MLC (bottom). In the examples shown, a 5-mm ball is shown projected in circular fields 7.5 mm in diameter and in 12×12 mm square MLC fields.

or improperly aligned. As the lasers are the primary reference for patient localization (even in image-guided localization), accurate coincidence of the lasers with the linac isocenter is essential. Regular use of any linac, wear and tear over time, will also adversely impact isocentric accuracy. In general, the couch remains the weak mechanical link with regard to isocentric accuracy.

In addition, service to the MLC can alter both the MLC and cone axes; thus, a Winston–Lutz test is essential after any such action. Finally, on Novalis (BrainLAB AG, Feldkirchen, Germany) and Varian (Varian Medical Systems, Palo Alto, CA) treatment units, anomalies have also been observed after service to, or replacement of, the gantry/couch motor driver "AMC" printed circuit board (Advanced Motion controls, Camarillo, CA).

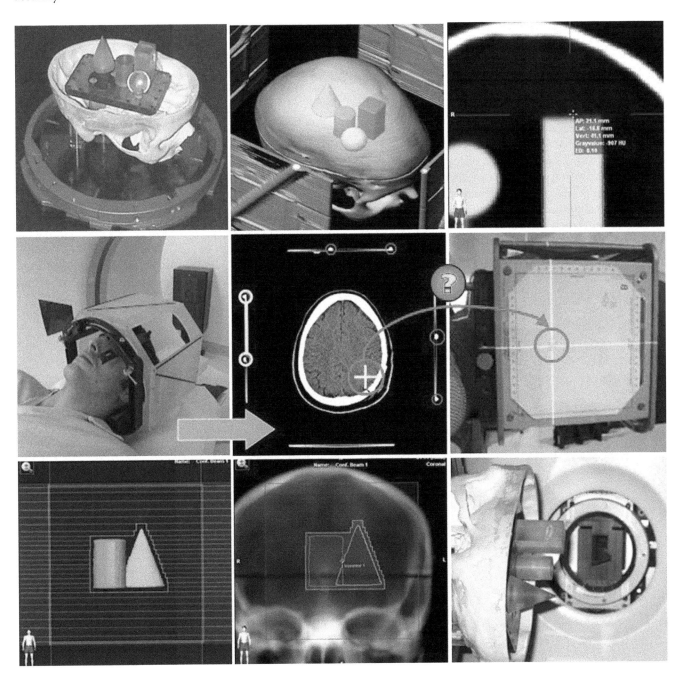

FIGURE 61.2 Ensuring accurate calculation and transfer of all mechanical parameters is of fundamental importance. By scanning the Radionics QMP Phantom (top left) and querying the objects' size and location, one can assess whether the stereotactic coordinates are correctly calculated by the treatment planning computer (top right). Similarly, the phantom can be used to assess proper transfer of these coordinates, and other mechanical parameters, to the treatment device (middle). For example, one can contour objects, place, and shape beams (bottom left/center), then position the phantom on the linac and determine (by projecting the shaped light field on the phantom; bottom right) whether all mechanical parameters are consistent with the planning system.

End-to-End Target Localization

The ability of the treatment planning system to correctly dictate all mechanical treatment parameters (e.g., stereotactic coordinates, gantry, couch and collimator angles, leaf positions) and translate these mechanical parameters to the treatment device is of fundamental importance to any radiosurgery system. The Radionics QMP Phantom (Integra Radionics, Inc., Burlington, MA) provides several desirable features for assessing many of these capabilities. The phantom, which contains four objects, each with a known location and volume within stereotactic space, can be used to evaluate accuracy of the localization process as well as consistency between treatment planning and delivery. This is illustrated in Figure 61.2.

An approach similar to the Winston–Lutz test using a "hidden target" can be used to assess "end-to-end" localization capabilities. The Lucy® 3D QA Phantom (Standard Imaging, Inc., Middletown, WI), designed to aid in evaluation of image transfer QA, dosimetry, and machine QA within the exact coordinate system of commercially available stereotactic head frames, contains a small radiopaque ball specifically for hidden target evaluation. Figure 61.3 shows a CT scan of the Lucy phantom and the radiopaque ball. An isocenter is placed at the center of the ball, and the phantom is positioned in the room using a conventional laser-template method. A series of beams-eye-view images (in this case using a 10 × 10 mm square field) are subsequently obtained at various gantry, couch, and collimator angles.

Hidden target tests can also be used to assess accuracy of image-guided localization methodologies. A process for assessing frameless (image-guided) localization techniques has been described previously (Solberg et al. 2008). Briefly, an anthropomorphic head phantom containing a 4-mm radiopaque sphere was fixed in a stereotactic head frame. A CT was obtained and the sphere was identified and defined as an isocenter. In the treatment room, stereoscopic localization x-rays were obtained (ExacTrac 6D, BrainLAB AG, Feldkirchen, Germany) and fused with digitally reconstructed radiographs, and the phantom was positioned accordingly. AP and lateral beams-eye-view films were obtained using a 10-mm circular collimator. The offset of the hidden sphere within the projection of the field was determined using commercial software (RIT113, Radiological Imaging Technology, Inc., Colorado Springs, CO). This process is shown in Figure 61.4. The procedure was repeated 50 times, yielding an average 3D displacement of 1.11 mm with a precision of 0.42 mm (1σ) (Table 61.1). Averages near zero in each direction are indicative of only random uncertainties. These results are similar to those reported by other groups (Verellen et al. 2003; Chang et al. 2003; Yu et al. 2004).

It is also useful to assess image-guided localization techniques in a clinical setting. As shown in Figure 61.5, patients

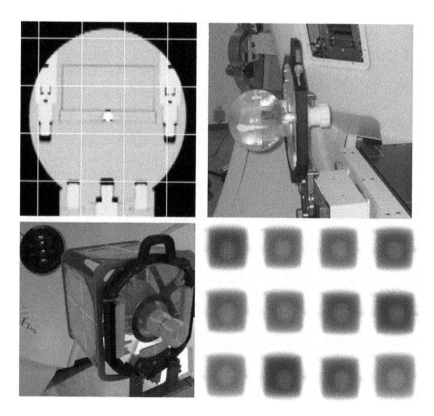

FIGURE 61.3 A hidden target test performed using the Lucy phantom. An isocenter is placed at the center of the ball (top left), the phantom is subsequently positioned using the room lasers (bottom left), and a series of images are obtained at various gantry and couch combinations (bottom right).

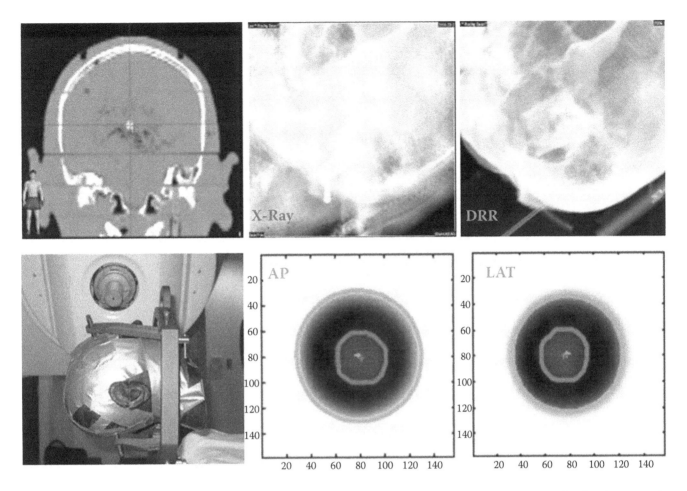

FIGURE 61.4 A hidden target assessment performed using frameless (image-guided) localization. An isocenter is placed at the center of a radiopaque ball inside an anthropomorphic head phantom (top left). Stereo x-rays are obtained and fused with the digitally reconstructed radiographs (top right) and the phantom is positioned accordingly. AP and lateral verification films are obtained (bottom right), and the offsets of the ball within the radiation field are determined.

were initially positioned using the conventional methodology of aligning to room lasers. Stereoscopic localization x-rays were subsequently obtained and fused with the patient digitally reconstructed radiographs to determine the difference between frame-based and frameless isocenter locations. Data were collected from 45 stereotactic radiosurgery cases and 565 fractions from 37 stereotactic radiation therapy cases (Table 61.1). A mean vector deviation of 1.00 mm ($\sigma = 0.55$ mm) was observed in patients for whom rigid fixation was used versus 2.36 mm ($\sigma = 1.32$ mm) for patients immobilized in a mask.

Dosimetric Localization

Dosimetric localization, the ability to accurately deliver the intended dose and make the desired dose distribution coincide with the desired point in space, is a critical element of radiosurgery. Ensuring dosimetric integrity relies on several critical elements: accurate characterization of photon beams, which can be particularly challenging for the small fields used in radiosurgery; commissioning of the resulting beam data and the treatment

planning system; and demonstrating end-to-end dosimetric capabilities in treatment delivery. In this section we emphasize practical aspects of radiosurgery dosimetry and point out some of the common pitfalls.

Beam Data Acquisition

The critical consideration in beam characterization is the choice of detector. Small fields necessitate small detectors to eliminate partial volume effects and to provide acceptable resolution in central and off-axis scans. Candidate detectors include small ionization chambers, semiconductor devices such as diodes and diamond detectors, and film, both radiographic as well as radiochromic. Excellent references on the characteristics and application of various detectors are provided by Pappas et al. (2008) and Ding, Duggan, and Coffey (2006).

Radiosurgery beams are typically characterized by three parameters: central axis profiles (percent depth dose [PDD] or tissue maximum ratio), off-axis profiles, and relative output factors. Radiosurgery beam data characterization will require

TABLE 61.1 Localization Offsets Determined Using an X-Ray–Guided Frameless Positioning System

	Anterior–Posterior	Lateral	Axial	3D Vector
Phantom (n = 50)				
Average (mm)	−0.06	−0.01	0.05	1.11
Standard deviation (mm)	0.56	0.32	0.82	0.42
Frame patients (n = 45)				
Average (mm)	−0.01	0.05	0.25	1.00
Standard deviation (mm)	0.66	0.56	0.71	0.55
Mask patients (n = 565)				
Average (mm)	0.17	0.17	0.47	2.36
Standard deviation (mm)	1.32	1.24	2.11	1.32

Note: Phantom measurements were performed using a radiopaque sphere "hidden" in an anthropomorphic head phantom. Patient measurements were compared with conventional radiosurgery positioning methodology.

the use of multiple detectors, depending on field size and type of measurement being performed. It should be noted that very small fields exhibit an absence of lateral electronic equilibrium on the central axis, a phenomenon that is exacerbated at higher energies. For a variety of reasons, including uncertainties in output factors, which fall off rapidly with field size, and beam perturbation by tissue heterogeneities, small field irradiation using high-energy photons should be avoided. Energies more than 10 MV were expressly prohibited in the original lung SBRT trial, RTOG 0236 (RTOG 0236 2002).

Central axis profiles are best obtained by scanning PDD in a motorized water phantom. A small volume (≤0.015 cm³) ionization chamber is the detector of choice and can be used for all fields larger than approximately 2 cm. At smaller field sizes, the smaller sensitive area/volume of a stereotactic diode can minimize the partial volume effects that may manifest with ionization chambers. The most difficult practical challenge may be in establishing detector alignment with the beam axis for very small fields. Fortunately, modern scanning systems have automated features that can help with this task. Small field PDD

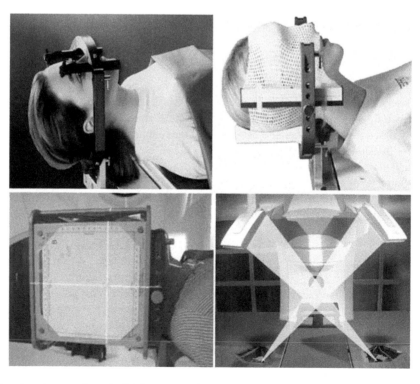

FIGURE 61.5 PDD (left) and total scatter factor (right) for a 6-MV photon beam measured as a function of square field size.

curves exhibit the familiar dependence on field size, specifically the slope of the distal portion of the curve becomes increasing shallow at larger field sizes. Therefore, by overlaying PDD curves for different field sizes on a single plot, errors in detector alignment become quite apparent. Figure 61.6 (left) shows PDD curves for six MV fields ranging from 5 × 5 to 150 × 150 mm^2. A secondary phenomenon that can be observed through careful measurement is a slight shift of the depth of maximum dose (D_{max}) toward shallower depths at the smallest field sizes. This is due to the absence of lateral electronic equilibrium, which is manifest at fields smaller than approximately 30 × 30 mm^2. For a 6-MV beam with a nominal D_{max} of 15 mm at a 100 × 100 mm^2 reference field, D_{max} may shift to 12 mm for a 5 × 5 mm^2 field.

An appropriately small detector is also required for accurate measurement of off-axis profiles because detectors with too large a volume will produce artificially broad profiles. Several authors have shown that even the small volume air ionization chambers, such as those intended for small field use, are not suitable for small field dosimetry (McKerracher and Thwaites 1999; Ding, Duggan, and Coffey 2006; Pappas et al. 2008; Fan et al. 2009). In addition, radiographic film can overrespond in the penumbra region of small photon beams, also contributing to a slight broadening. This effect can be reduced by the use of radiochromic film. For many physicists, the detector of choice in small field profile measurements is an appropriate diode (stereotactic or electron diode), oriented and scanned horizontally to maximize resolution.

The total scatter factor is defined as the ratio of dose for a given depth, field size, and source-to-surface distance to that for a reference depth, field size, and source-to-surface distance. In a general form, this can be written as:

$$S_{c,p} = \frac{D(r, d, \text{SSD})}{D(r_{ref}, d_{ref}, \text{SSD}_{ref})} \qquad (61.1)$$

Typically, the measured and reference depth and SSD are the same, so:

$$S_{c,p} = \frac{D(r, d, \text{SSD})}{D(r_{ref}, d, \text{SSD})} \qquad (61.2)$$

Determination of small field output factors is often the most challenging radiosurgery measurement that physicists will encounter. As output factors are directly correlated with the calculated dose, any error in measurement will correspond to an equivalent error in the delivered dose. There are three common errors in the measurement of output factors, all of which result in an underestimation of $S_{c,p}$ and a resulting overestimation of MU and dose: (1) failure to properly center the detector within the field, (2) the use of a detector that is too large, and (3) incorrect application of solid state detectors. A number of groups have investigated the relationship between different detector types and sizes and the resulting output factors (Heydarian, Hoban, and Beddoe 1996; McKerracher and Thwaites 1999; Zhu et al. 2000; Ding, Duggan, and Coffey 2006; Pappas et al. 2008; McKerracher and Thwaites 2008; Fan et al. 2009). For reasons relating to detector size, most investigators recommend a stereotactic diode as the detector of choice for measurement of small field output factors. Figure 61.6 (right) shows $S_{c,p}$ as a function of square field size for two commonly used detectors—a stereotactic diode (SFD, IBA Dosimetry, Louvain-la-Neuve, Belgium) and a 0.015-cm^3 PinPoint chamber (PTW, Freiburg, Germany). The SFD diode has a sensitive volume of 0.6 mm in diameter × 0.06 mm in thickness, ideal size characteristics for small field measurements. Ion chamber measurements were obtained with the PinPoint chamber oriented horizontally and vertically. For a 6 × 6 mm^2 field, the ion chamber clearly underestimates the true output factor at 12 × 12 mm^2 and higher; however, all techniques produce good agreement.

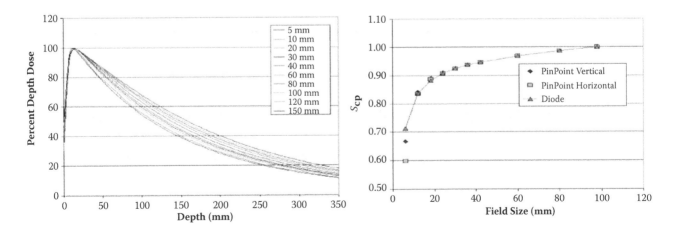

FIGURE 61.6 Measured PDD (left) and output factors for a 6-MV beam shown as a function of field size.

It is well known that the energy spectrum of the incident photons and that of the resulting secondary electrons changes as a function of field size. Using Monte Carlo methods, Heydarian, Hoban, and Beddoe (1996) have shown an increase in mean incident photon energy of more than 200 keV for a 5-mm-diameter beam relative to a 100-mm beam. This difference increases with depth in water. For nontissue equivalent detectors such as diodes, the energy difference results in an overresponse at small fields relative to the standard 100×100 mm² reference field. Fundamentally, if one relies solely on a diode for all output measurements, small field ratios will be incorrect. To circumvent this problem, the use of diodes should be limited to measurements in which field size differences produce little change in the energy spectrum. In practice then, scatter factors for small fields can be obtained by normalizing the diode response to an intermediate-size field at which an accurate ion chamber measurement can

be obtained. For example, the output factor for a 5-mm circular cone can be determined from:

$$S_{c,p}(r = 5\,\text{mm}) = \frac{D_{\text{diode}}(r = 5\text{mm}, d, \text{SSD})}{D_{\text{diode}}(r = 20\text{mm}, d, \text{SSD})}$$

$$\times \frac{D_{\text{pinpoint}}(r = 20\text{mm}, d, \text{SSD})}{D_{\text{pinpoint}}(r = r_{\text{ref}}, d, \text{SSD})} \quad (61.3)$$

End-to-End Dosimetric Commissioning

Once reliable beam data have been obtained, a thorough dosimetric commissioning of the radiosurgery treatment planning system must be performed. Validation of the TPS should begin by comparing calculation of the basic beam characteristics

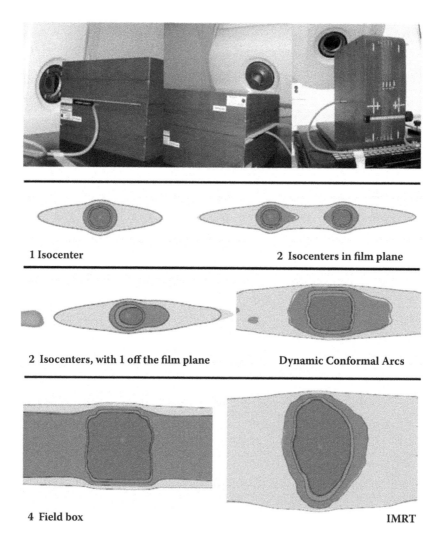

1 Isocenter   **2 Isocenters in film plane**

2 Isocenters, with 1 off the film plane   **Dynamic Conformal Arcs**

4 Field box   **IMRT**

FIGURE 61.7 Dosimetric commissioning of various radiosurgery delivery modalities and configurations using film and ionization chambers in solid water phantoms. Film dosimetry (colorwash) is shown with calculations from the treatment planning system superimposed (solid lines) for one and two isocenter combinations delivered in a series of circular collimated arcs, a four-field box shaped with MLC, a five-arc configuration delivered through five dynamically shaped arcs, and a seven-beam IMRT plan.

(PDD, profiles, output factors) with measured input data. Failure to reproduce beam data measured in simple geometry indicates a fundamental problem that must be resolved before pursuing evaluation of more complex arrangements. Following this, we recommend a systematic approach to assessing the various system capabilities, beginning with simple plans, accompanied by measurement in phantom, and working toward increasingly more sophisticated configurations. A combination of film (or other 2D detector array) and ionization chamber measurements is useful to verify both the absolute dose delivered to the target as well as the overall dose distribution. Figure 61.7 shows an example of film/ion chamber measurement in solid water phantoms (top). These procedures are common practice in most institutions. The resulting film dosimetry (colorwash) is shown with calculations from the treatment planning system superimposed (solid lines); comparison of the two distributions is facilitated through the use of dedicated software described by Agazaryan, Solberg, and DeMarco (2003). Dark green areas indicate regions where gamma exceeds 1 (Low et al. 1998). Small areas of disagreement are common, as the criteria adopted for comparison, with

scaling criteria of 3% dose and 3 mm distance criterion used, are quite stringent. Plan/irradiation configurations include one and two isocenter combinations delivered in a series of circular collimated arcs, a four-field box shaped with MLC, a five-arc configuration delivered through five dynamically shaped arcs, and a seven-beam IMRT plan. The absolute dosimetry for each configuration is also measured (in Figure 61.7 using a 0.015-cm^3 PinPoint chamber) and compared against calculation. It is important to adopt rigorous acceptance criteria, no more lax than ±3% in absolute dose, during commissioning of radiosurgery systems. In our experience, this is readily achievable.

Standard dosimetric procedures such as those described above are important but incomplete with regard to radiosurgery commissioning. It is essential that the process include combined end-to-end localization and dosimetric elements. This may present additional challenges and may require additional equipment. Three examples of end-to-end localization/dosimetry assessments are shown in Figure 61.8. In each, a phantom placed in a stereotactic head frame is subsequently imaged, localized, and planned based on the stereotactic coordinates, set up in the

FIGURE 61.8 End-to-end localization and dosimetric commissioning performed in a solid water spherical phantom, a BANG gel phantom (top/right), and the Lucy phantom (bottom).

treatment room, and irradiated in a manner identical to a clinical procedure. A spherical solid water phantom (Radiology Support Devices, Long Beach, CA) has interchangeable inserts to accommodate a variety of dosimeters (top left/center). A mailed dosimetry service using prepared BANG® gel phantoms (Medical Gel Dosimetry Systems, Inc., Madison, CT) is available commercially (top/right). The Lucy phantom (Standard Imaging, Inc.) facilitates QA of numerous imaging, localization, and dosimetry procedures, all within the exact coordinate system of commercially available stereotactic head frames (bottom). CT reconstructions from planning scans show inserts for a PinPoint chamber and a dose distribution based on MR-contoured objects for subsequent film dosimetry.

As a final note, end-to-end verification of localization and dosimetric parameters should always be performed through a record and verify system. In this manner, potential errors that may propagate through the electronic record can be detected and corrected long before they ever reach a patient.

Summary

Although the clinical field of stereotactic radiosurgery is nearly 60 years old, commissioning and quality assurance of radiosurgery systems remain challenging and time-consuming processes. The high dose nature of the discipline imposes rigorous requirements with regard to accuracy and precision. Specialized equipment is required to assess a broad range of characteristics and capabilities. Beam data acquisition is fraught with uncertainties, but these uncertainties can be minimized through proper detector selection and use. End-to-end assessment of all aspects of the process, localization as well as dosimetric, is essential. Finally, it is strongly recommended that new practitioners seek guidance from more experienced colleagues. Many serious and highly publicized mistakes have resulted from building new programs in a vacuum.

References

Agazaryan, N., T. D. Solberg, and J. J. DeMarco. 2003. Patient specific quality assurance for the delivery of intensity modulated radiotherapy. *J. Appl. Clin. Med. Phys.* 4 (1): 40–50.

Betti, O. O., and V. E. Derechinsky. 1984. Hyperselective encephalic irradiation with linear accelerator. *Acta Neurochir.* 33 (Suppl): 385–391.

Chang, S. D., W. Main, D. P. Martin, I. C. Gibbs, and M. P. Heilbrun. 2003. An analysis of the accuracy of the CyberKnife: A robotic frameless stereotactic radiosurgical system. *Neurosurgery* 52 (1): 140–146; discussion 146–147.

Colombo, F., A. Benedetti, F. Pozza, R. C. Avanzo, C. Marchetti, G. Chierego, and A. Zanardo. 1985. External stereotactic irradiation by linear accelerator. *Neurosurgery* 16 (2): 154–160.

Ding, G. X., D. M. Duggan, and C. W. Coffey. 2006. Commissioning stereotactic radiosurgery beams using both experimental and theoretical methods. *Phys. Med. Biol.* 51 (10): 2549–2566.

Fan, J., K. Paskalev, L. Wang, L. Jin, J. Li, A. Eldeeb, and C. Ma. 2009. Determination of output factors for stereotactic radiosurgery beams. *Med. Phys.* 36 (11): 5292–5300.

Friedman, W. A., and F. J. Bova. 1989. The University of Florida radiosurgery system. *Surg. Neurol.* 32 (5): 334–342.

Heydarian, M., P. W. Hoban, and A. H. Beddoe. 1996. A comparison of dosimetry techniques in stereotactic radiosurgery. *Phys. Med. Biol.* 41 (1): 93–110.

Kjellberg, R. N., A. M. Koehler, W. M. Preston, and W. H. Sweet. 1962. Stereotaxic instrument for use with the Bragg peak of a proton beam. *Confin. Neurol.* 22: 183–189.

Larsson, B., L. Leksell, B. Rexed, P. Sourander, W. Mair, and B. Andersson. 1958. The high-energy proton beam as a neurosurgical tool. *Nature* 182 (4644): 1222–1223.

Larsson, B., K. Liden, and B. Sarby. 1974. Irradiation of small structures through the intact skull. *Acta Radiol. Ther. Phys. Biol.* 13 (6): 512–534.

Lawrence, J. H., C. A. Tobias, J. L. Born, Combs Rk Mc, J. E. Roberts, H. O. Anger, B. V. Low-Beer, and C. B. Huggins. 1958. Pituitary irradiation with high-energy proton beams: A preliminary report. *Cancer Res.* 18 (2): 121–134.

Lawrence, J. H., C. A. Tobias, J. L. Born, C. C. Wang, and J. H. Linfoot. 1962. Heavy-particle irradiation in neoplastic and neurologic disease. *J. Neurosurg.* 19: 717–722.

Leksell, L. 1951. The stereotaxic method and radiosurgery of the brain. *Acta Chir. Scand.* 102 (4): 316–319.

Low, D. A., W. B. Harms, S. Mutic, and J. A. Purdy. 1998. A technique for the quantitative evaluation of dose distributions. *Med. Phys.* 25 (5): 656–661.

Lutz, W., K. R. Winston, and N. Maleki. 1988. A system for stereotactic radiosurgery with a linear accelerator. *Int. J. Radiat. Oncol. Biol. Phys.* 14 (2): 373–381.

McKerracher, C., and D. I. Thwaites. 1999. Assessment of new small-field detectors against standard-field detectors for practical stereotactic beam data acquisition. *Phys. Med. Biol.* 44 (9): 2143–2160.

McKerracher, C., and D. I. Thwaites. 2008. Phantom scatter factors for small MV photon fields. *Radiother. Oncol.* 86 (2): 272–275.

Pappas, E., T. G. Maris, F. Zacharopoulou, A. Papadakis, S. Manolopoulos, S. Green, and C. Wojnecki. 2008. Small SRS photon field profile dosimetry performed using a PinPoint air ion chamber, a diamond detector, a novel silicon-diode array (DOSI), and polymer gel dosimetry. Analysis and intercomparison. *Med. Phys.* 35 (10): 4640–4648.

Podgorsak, E. B., A. Olivier, M. Pla, P. Y. Lefebvre, and J. Hazel. 1988. Dynamic stereotactic radiosurgery. *Int. J. Radiat. Oncol. Biol. Phys.* 14 (1): 115–126.

RTOG 0236. 2002. *A Phase II Trial of Stereotactic Body Radiation Therapy (SBRT) in the Treatment of Patients with Medically Inoperable Stage I/II Non-Small Cell Lung Cancer.* Philadelphia, PA: Radiation Therapy Cooperative Group.

Schell, M. C., F. J. Bova, D. A. Larson, D. D. Leavitt, W. R. Lutz, E. B. Podgorsak, and A. Wu. 1995. Stereotactic Radiosurgery, Radiation Therapy Committee Task Group 42. In

American Association of Physicists in Medicine Report No. 54. Woodbury, NY: American Institute of Physics Publishing.

Solberg, T. D., P. M. Medin, J. Mullins, and S. Li. 2008. Quality assurance of immobilization and target localization systems for frameless stereotactic cranial and extracranial hypofractionated radiotherapy. *Int. J. Radiat. Oncol. Biol. Phys.* 71 (1 Suppl): S131–S135.

Verellen, D., G. Soete, N. Linthout, S. Van Acker, P. De Roover, V. Vinh-Hung, J. Van de Steene, and G. Storme. 2003. Quality assurance of a system for improved target localization and patient set-up that combines real-time infrared tracking and stereoscopic X-ray imaging. *Radiother. Oncol.* 67 (1): 129–141.

Winston, K. R., and W. Lutz. 1988. Linear accelerator as a neurosurgical tool for stereotactic radiosurgery. *Neurosurgery* 22 (3): 454–464.

Yu, C., W. Main, D. Taylor, G. Kuduvalli, M. L. Apuzzo, and J. R. Adler Jr. 2004. An anthropomorphic phantom study of the accuracy of Cyberknife spinal radiosurgery. *Neurosurgery* 55 (5): 1138–1149.

Zhu, X. R., J. J. Allen, J. Shi, and W. E. Simon. 2000. Total scatter factors and tissue maximum ratios for small radiosurgery fields: Comparison of diode detectors, a parallel-plate ion chamber, and radiographic film. *Med. Phys.* 27 (3): 472–477.

Total Body Irradiation

Amjad Hussain
Tom Baker Cancer Center

Jose Eduardo
Villarreal-Barajas
University of Calgary

Derek Brown
University of Calgary

Introduction

Total body irradiation (TBI) is used primarily for the purpose of suppressing the immune system before bone marrow transplantation but can also be used as a means of eradicating malignant cells (e.g., leukemias, lymphomas, and certain solid tumors) or cell populations with genetic disorders (e.g., thalasemia).

TBI treatments present unique challenges in the radiotherapy department. The need to treat patients with very large radiation fields may require that the patient be positioned far from the source of radiation. TBI is further complicated by the requirement that the lungs, and sometimes eyes, kidneys, and brain, may need to be partially shielded and that a uniform dose must be delivered to a nonuniform patient body. In addition, no commercially available device, or dose calculation algorithm, exists specifically for TBI, so treatment techniques and dose calculations are not standardized across different radiotherapy clinics.

Given so many complicating factors, a TBI program is challenging to commission and, once established, requires a stringent quality assurance (QA) program.

TBI Treatment Techniques

The choice of technique for delivering TBI is dependent on the availability of space in the treatment room, the length of time the treatment takes, staff expertise, and access to a backup in case of equipment malfunction.

The goal of any TBI technique is to deliver an accurate and uniform dose to the entire body to within ±10% of the prescribed dose (Sarfaraz et al. 2001; Lam et al. 1979; Khan 2010), including circulating malignant cells, the immune system, and skin. The technique should allow well-controlled and accurately localized partial shielding of the lung, kidney, eye, and brain. It should be reproducible, reliable, and comfortable for patients and staff, and should be designed such that it can be easily integrated into the routine treatment environment.

TBI techniques fall into two distinct categories: (1) stationary techniques that make use of radiation fields that are large enough that the entire patient is encompassed in a single field and (2) dynamic techniques that use radiation fields that are smaller than the patient, requiring that the beam be scanned across the entire patient body, by moving the patient, the beam, or both.

Stationary TBI Techniques

The first TBI treatment was reported in 1905 by Dessauer (1905). Since then, there have been significant developments aimed at increasing the accuracy of TBI using stationary radiation sources. Heublein (1932) reported one of the first dedicated TBI facilities in North America using deep x-ray therapy units. With the development of cobalt-60 units and linear accelerators, more penetrating radiation beams (up to 25 MV) were available for use in TBI treatments. The use of these sources/beams in combination with extended SSDs to accommodate the entire body of the patient is extensively reviewed in the AAPM Report No. 17 (TG-29) (Van Dyk et al. 1986).

The main advantages of extended SSD, stationary TBI techniques are simultaneous dose delivery to the entire body and circulating cells, and a comparatively short treatment time.

Dynamic TBI Techniques

Cunningham and Wright (1962), in the early 1960s, designed a rotating, ceiling-mounted cobalt-60 TBI system that could scan the entire patient body at 120 cm SSD. In 1985, Quast (1985) introduced a novel treatment concept for TBI, moving the patient under a vertical photon beam at shorter SSDs (more than 100 cm but less than 200 cm). These techniques improve the photon fluence uniformity and depth dose homogeneity. Combined arc therapy with dynamic output control for TBI was proposed by Hugtenburg et al. (1994). La Macchia et al. (2007) described a segmented intensity modulated technique to achieve a uniform total body dose using the Pinnacle™ treatment planning software. Two to three segmented MLC fields were created to cover the entire body.

Dynamic TBI techniques offer a number of advantages over stationary techniques. They can be used to produce large effective field sizes at shorter SSDs and, because only a small portion of the patient is treated at any one time, dynamic techniques offer more flexibility for improving dose homogeneity (Briot, Dutreix, and Bridier 1990). Furthermore, patient setup is simple and accurate, which improves patient comfort (Sarfaraz et al. 2001; Gerig et al. 1994). Provided the beam is continuously on and is delivered to the entire patient length, the possibility of missing circulating malignant cells is minimized. Dynamic techniques, however, are inherently complex and require a customized QA program that may consume more resources than traditional, stationary, extended SSD TBI techniques.

Recent Developments in TBI

In 2000, the first variable velocity translating bed TBI technique was reported by Chretien et al. (2000). The patient is positioned on a translating bed that moves under a static radiation beam. The velocity of the translating bed is modulated to account for variations in patient thickness, moving faster over thinner regions and slower over thicker regions. Recently, the same group (Lavallee et al. 2009) has described a similar variable velocity TBI technique using the Pinnacle treatment planning system to optimize dose homogeneity along the patient midline.

The idea of selective conformal total marrow irradiation and total marrow and lymphatic irradiation, also known as targeted TBI, was introduced by Schultheiss et al. in 2004 (Schultheiss et al. 2007). In targeted TBI, the dose is delivered primarily to the bone marrow and lymphatic system. Investigators predicted that, with this technique, the dose to bone marrow could be escalated to 20 Gy, while, at the same time, keeping doses to normal organs lower than in conventional TBI (Wong et al. 2006).

Hui et al. (2005) investigated the planning and delivery of TBI and targeted total marrow irradiation using PET/CT images and image-guided helical tomotherapy. A comparison of dose volume histograms showed that the organ-at-risk doses were reduced by 41–87% compared to conventional TBI (Schultheiss et al. 2007; Wong et al. 2009). In 2006, a pilot study was performed to investigate the feasibility of a linac-based approach to intensity-modulated total marrow irradiation at standard SSD using the Eclipse treatment planning software (Aydogan, Mundt, and Roeske 2006). Three intensity-modulated radiotherapy (IMRT) plans with different isocenters were developed to cover the head to mid-femur in a typical adult. With this technique, the doses to the organs at risk were reduced to 77–22% compared to conventional TBI. However, this IMRT total marrow irradiation approach presented significant challenges, including field size limitations, field junction hot and cold spots, and reproducibility of patient positioning.

TBI Commissioning

There is a wide variety of TBI techniques currently in use however; documentation and guidelines for the commissioning and QA of TBI are generally sparse, particularly for newer techniques. In 1986, TG-29 described general dosimetric and QA protocols for TBI (Van Dyk et al. 1986). The commissioning of any TBI technique should include, but is not limited to, absolute beam output calibration, the acquisition of central axis depth dose data, beam profiles, and attenuation data; all of which should be measured under the physical and technical conditions used for the specific TBI technique (Van Dyk et al. 1986; Podgorsak et al. 1985; Van Dyk 1987; Quast and Sack 2003).

In the following sections, we discuss three components of commissioning that are specific to TBI in the context of stationary and dynamic treatment techniques. Those components are

1. The setup used to produce a sufficiently large field size
2. The technique used to improve dose homogeneity
3. The system used to provide partial shielding of organs at risk

Producing a Sufficiently Large Field Size

Stationary TBI

Stationary TBI treatments are typically performed using a linear accelerator with the collimator opening set as large as possible (40 × 40 cm at 100 cm from the source) and the patient positioned at extended SSD (~400 cm). All manner of patient setups have been reported, including seated, semiseated, lying prone/supine, decubitus, and standing up (Mutyala et al. 2010; Cunningham and Wright 1962; Quast and Sack 2003; Van Dyk et al. 1986; Van Dyk 1999; Quast 1987; Harden et al. 2001). Depending on the chosen patient setup, two opposed beams are delivered either from the lateral directions or from the anterior/posterior directions.

In general, a single field size is used for all patients, and patients are always smaller than the field, so there is no need to measure collimator scatter factors. However, the patient size in the short axis, perpendicular to the central beam axis, is variable. For this reason, one should measure, and include in dose calculations, a phantom scatter factor that accounts for changes in central axis dose with patient size. Beam profiles should also be measured to determine the useful geometric beam size and to quantify beam flatness at extended SSD.

Dynamic TBI

Unlike stationary techniques, dynamic TBI treatments are delivered exclusively from the anterior/posterior directions with the patient lying in the supine and prone positions. Typically, the field size along the short axis of the patient is made large enough to encompass the largest patient and held constant for all patients, whereas the field size along the long axis of the patient can be varied. This means that both collimator scatter factors, to account for different field sizes along the long axis, and phantom scatter factors, to account for different patient sizes along both axes, must be measured and included in dose calculations.

Beam profiles in the direction of motion are of particular interest for dynamic techniques because the patient receives the dose from the entire profile, from the tail of the profile as the point of calculation enters the beam to the tail of the profile on exiting the beam. Accurate measurements in these regions of the profile are essential for precise dose calculations.

Improving Dose Homogeneity

As previously stated, the goal of TBI is to deliver a uniform dose to the entire body to within ±10% of the prescribed dose (Sarfaraz et al. 2001; Lam et al. 1979). This criterion can be difficult to achieve in the buildup region, especially at higher beam energies. Two common methods of increasing surface dose, used for both stationary and dynamic treatments, are beam spoilers and bolus material.

Bolus material simply provides an additional buildup layer, increasing dose at the surface of the patient. The surface dose is dependent on the material used as bolus, the thickness of the bolus, and the beam energy used for treatment.

Beam spoilers are thin polymethylmethacrylate sheets placed between the radiation source and the patient. Electrons liberated from the polymethylmethacrylate by high-energy photons provide the desired increase in dose at the surface of the patient. The surface dose is dependent on the thicknesses of polymethylmethacrylate used, the distance from the polymethylmethacrylate to the patient, and the photon energy used for treatment.

Stationary TBI

A common approach to improve dose homogeneity in stationary TBI is to use layered copper sheets to provide variable attenuation (Khan 2010; Van Dyk et al. 1986; Greig et al. 1998). Several more complex compensation schemes for stationary treatments have been described in the literature (Van Dyk et al. 1986; Lin, Chu, and Liu 2001; Bradley et al. 1998; Brix et al. 1984). Dose profiles, both with and without the compensator in place, must be measured to enable calibration of the compensator attenuation. The reproducibility of the compensator positioning must also be demonstrated and the effect of the compensator on scatter and surface dose should be measured.

Dynamic TBI

Dynamic treatments inherently offer more options for improving dose homogeneity because only a fraction of the patient is in the beam at any given time. One of the most ingenious systems for dose uniformity was reported by Shigeo et al. (2000) and Gallina, Rosati, and Rossi (2005). They developed a unique water-compensating filter, consisting of many small water-filled containers of variable sizes, which was positioned between the radiation source and the patient on a translating bed. This system could be used to compensate for variations in external body contours and was demonstrated to provide improved dose uniformity over the whole patient body. Chretien et al. (2000) have described in detail the use of variable speed translation, increasing the speed of translation over thinner sections of the patient, to achieve a more uniform dose distribution. Changing the field size during translation could offer similar improvements in dose homogeneity. Although these methods do very well at compensating for variations in thickness along the central, long axis of the patient, they do little to account for variations along the short axis of the patient. In an effort to improve dose homogeneity along both axes, we have recently developed an MLC-based technique that modulates the field shape dynamically as the patient is translated under the beam. Preliminary simulations performed using the Eclipse treatment planning system on a static anthropomorphic phantom (Rando®) at extended SSD (180 cm) demonstrate the feasibility of using such a technique to improve dose homogeneity along both axes of the patient (Brown 2010; Hussain, Villarreal-Barajas, and Brown 2010). Figure 62.1 compares dose homogeneity along the short axis of the patient using a static field shape and a dynamically modulated field shape.

For dynamic techniques, the speed at which the patient moves plays a major role in determining the dose delivered. Moreover, for treatment techniques that use variable speed, or variable field size or shape, the location at which these variations occur must be matched to patient anatomy. To prevent motion errors, mechanisms should be in place to continuously monitor the speed and position of the dynamic component of treatment. Ideally, those mechanisms should automatically turn the beam off and stop all motion upon detection of a mismatch between planned and monitored motion.

An additional complication is that the optimization of dose homogeneity requires the ability to predict the dose distribution within the patient. This is generally achieved using a treatment planning system, which must therefore also be commissioned and validated for the physical and technical conditions used for the specific TBI setup.

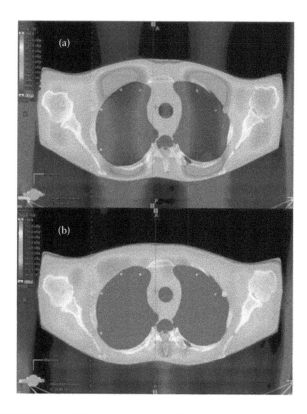

FIGURE 62.1 Calculated dose distributions for opposed photon beams with a static beam shape (a) and a dynamically modulated beam shape (b). Calculated dose is displayed in color wash, with lower doses in blue and higher doses in red. Dose homogeneity is greatly improved using dynamically modulated beams.

Partial Shielding of Organs at Risk

For most of the TBI treatments, partial shields are used as a means of reducing dose to the lungs, which would otherwise receive a much larger dose than prescribed. In rare cases it may also be necessary to shield the eyes, kidneys, or brain.

Stationary TBI

For stationary treatments delivered from the anterior/posterior directions, partial shields can be constructed from Cerrobend™ or lead-rubber sheets (Khan 2010; Van Dyk et al. 1986; el-Khatib and Valcourt 1989; Smith et al. 1996; Hussein and Kennelly 1996; Lavallee et al. 2008). The attenuation coefficients of these shields must be measured under the specific conditions that they will be used during TBI treatment. Some means of verifying the correct positioning of the shields before treatment must also be established. For stationary treatments at extended SSD, the change in beam divergence is minimal along the length of the block and does not change over the course of treatment, so divergent blocks can be used.

Techniques that make use of more complex compensation methods, such as the water compensating filter, can also be used to reduce the dose to the lungs.

Dynamic TBI

Cerrobend or lead-rubber sheets may also be used for dynamic treatment techniques. Their use is complicated by the fact that beam divergence along the length of the block changes over the course of treatment. The blocks must be constructed with this in mind, and some solution, be it rounded block edges or an expansion of the block length, must be made (Lavallee et al. 2008). One of the advantages of using dynamically modulated field shapes, discussed in the previous section, is that lung compensation is easily achieved without the need for additional shields (see Figure 62.1).

Quality Management

Given the unconventional nature of TBI, and the limited number of high-dose fractions, TBI techniques require a stringent and comprehensive QA program. It is the responsibility of the medical physicist to develop and implement the QA program. As part of this QA program, a subset of the data acquired for commissioning should be verified periodically. These tests should include machine output, PDD/TPR, beam profiles, and in vivo dosimeter calibration. If treatment planning software is used for TBI, the calculations performed should be verified under TBI conditions.

It is worth mentioning that despite the complexity of dose distributions, the TBI prescription is generally based on a single point dose calculation approach. Ideally, real-time in vivo dosimetric verification should be performed during TBI to ensure the accuracy of the dose delivered to the patient.

The "beam on" time for TBI is large enough that the dose rate may vary during treatment, thus affecting the average dose rate. In addition, a minor error in a single-fraction high-dose TBI has a greater impact on the treatment outcome than the same error would have in a standard dose fractionation scheme (20–30 fractions).

Failure Modes and Effects Analysis

Unlike the traditional QA approach where the main focus is on checks and measurements to achieve performance within specific tolerance values, the AAPM is developing a framework (TG-100) (Huq 2009) to promote the objective design of quality management activities. The TG-100 project is largely based on the concept of failure modes and effects analysis (FMEA) and risk assessment based on predicted clinical outcomes. FMEA describes the nature of the failures, estimates their probability at each step in the process map, and estimates the consequences of the failure and the likelihood that the failure will not be detected before impacting clinical treatment. Numerical values are assigned to provide a quantitative analysis. O (occurrence) describes the probability of the failure, S (severity) defines the severity of the effect, and D (detectability) represents the inability to detect the failure. TG-100 has developed numerical scales (1–10) for the O, S, and D indices to be used in radiotherapy-

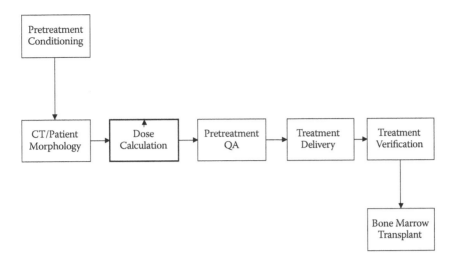

FIGURE 62.2 Process map of TBI treatment.

related scenarios. The product of these three numbers is the risk priority number that can be used to rank failure modes in order of importance.

FMEA Application to TBI

The focus of this section is the application of FMEA to the TBI process. The process is composed of the following steps: (1) development of TBI process map, (2) application of FMEA to the process map, and (3) design of a QA program based on the results of the FMEA analysis.

Development of a Process Map

The first step toward the implementation of FMEA is to describe the entire TBI process. Process maps can be used to define all of the major steps involved in the treatment, from initial contact with the patient to follow-up visits after treatment. Process maps help us to understand the complexity of the process and enable visualization of the relationship between the various steps during a course of treatment. Of interest for the current discussion are the more detailed aspects of the TBI treatment. An example of a process map for the major steps in a typical TBI treatment is shown in Figure 62.2.

Application of FMEA to TBI Process Map

The goal of the FMEA is to identify possible failure modes that could occur, the basic causes that precipitate these failures, and the associated clinical consequences. The potential causes of failure at each individual step can be visualized along the process map. There may be many different causes for a particular failure

TABLE 62.1 Sample FMEA for Dose Calculation in TBI

Step No.	Failure Mode	Failure Causes	Failure Effects	O	D	S	RPN	Actions to Reduce Occurrence of Failure
1	Dose per fraction	Not written correctly on prescription sheet Use incorrect dose per fraction during dose calculation	Wrong dose Dose too high: increase normal tissue effects Dose too low: reduce treatment efficacy	3	3	10	90	Prescription should be second checked Dose per fraction should be within local TBI dose guidelines
2	Dose calculation factors	Incorrect factors used in dose calculation	Wrong dose Dose too high: increase normal tissue effects Dose too low: reduce treatment efficacy	3	6	6	108	RTT and physicist should second check dose calculation
3	Patient morphology	Not measured correctly Not entered correctly in patient prep sheet Change between measurement and treatment dates	Wrong dose Dose too high: increase normal tissue effects Dose too low: reduce treatment efficacy	3	5	5	75	Morphological measurements should be second checked Measurements should be spot checked on day of treatment

Note: RPN, risk priority number.

mode at each step. Table 62.1 provides an example of an FMEA analysis for the dose calculation step, defined earlier in the TBI treatment process map.

Design of TBI QA Program

The results of the FMEA should be used to guide the development of the QA program. Most commonly, QA programs are divided into two main categories: infrastructure QA and process QA. Infrastructure QA deals with the physical equipment involved in the TBI process, including the linac, treatment planning system, simulator, mold room, and so on. The performance of the equipment should be checked periodically to ensure accuracy and consistency of function. The QA program should describe the procedures and checks in detail, the required frequency of these checks, the action levels, and the responsible personnel required to perform the checks (Kutcher et al. 1994). Process QA should be developed by the TBI team (consisting of a radiation oncologist, medical physicist, radiation therapist, and dosimetrist) and will require proper education and practical training of all team members. The responsible physicist should develop an effective and comprehensive QA program in consultation with all the involved personnel.

Summary

Most TBI treatment techniques make unconventional use of linacs that are designed to treat small tumors using conformal fields. Scatter conditions, beam flatness, equivalent field calculations, and target volumes are different from those normally encountered during standard radiotherapy procedures. This, coupled with the lack of standardization of TBI treatment techniques across different radiotherapy clinics and the recent development of a number of novel treatment methods, demands the implementation of a comprehensive, efficient, and stringent QA program. Process maps, which detail the treatment procedure, used in conjunction with FMEA to identify failure modes, provide an effective strategy for developing a comprehensive and effective QA program.

References

Aydogan, B., A. J. Mundt, and J. C. Roeske. 2006. Linac-based intensity modulated total marrow irradiation (IM-TMI). *Technol. Cancer Res. Treat.* 5 (5): 513–519.
Bradley, J., C. Reft, S. Goldman, C. Rubin, J. Nachman, R. Larson, and D. E. Hallahan. 1998. High-energy total body irradiation as preparation for bone marrow transplantation in leukemia patients: Treatment technique and related complications. *Int. J. Radiat. Oncol. Biol. Phys.* 40 (2): 391–396.
Briot, E., A. Dutreix, and A. Bridier. 1990. Dosimetry for total body irradiation. *Radiother. Oncol.* 18 (Suppl. 1): 16–29.
Brix, F., E. Duhmke, H. Gremmel, D. Hebbinghaus, and M. Jensen. 1984. Optimization of dose distribution in whole body irradiation by means of compensators. *Strahlentherapie* 160 (2): 108–113.
Brown, W. D. 2010. A novel translational total body irradiation technique. *J. Med. Devices* 4 (2).
Chretien, M., C. Cote, R. Blais, L. Brouard, L. Roy-Lacroix, M. Larochelle, R. Roy, and J. Pouliot. 2000. A variable speed translating couch technique for total body irradiation. *Med. Phys.* 27 (5): 1127–1130.
Cunningham, J. R., and D. J. Wright. 1962. A simple facility for whole body irradiation. *Radiology* 78: 941–949.
Dessauer, F. 1905. Eine Neue Anordnug zur Röntgenbestrahlung. *Arch. f. Phys. Med. U. Med. Techn.* (2): 218–223.
el-Khatib, E. E., and S. Valcourt. 1989. Calculation of lung shielding for total body irradiation. *Int. J. Radiat. Oncol. Biol. Phys.* 17 (5): 1099–1102.
Gallina, P., G. Rosati, and A. Rossi. 2005. Implementation of a water compensator for total body irradiation. *IEEE Trans. Biomed. Eng.* 52 (10): 1741–1747.
Gerig, L. H., J. Szanto, T. Bichay, and P. Genest. 1994. A translating-bed technique for total-body irradiation. *Phys. Med. Biol.* 39 (1): 19–35.
Greig, J. R., F. S. Harrington, R. W. Miller, N. G. Wersto, and P. Okunieff. 1998. Head and neck compensation for total body irradiation using opposed laterals. *Med. Dosim.* 23 (1): 11–14.
Harden, S. V., D. S. Routsis, A. R. Geater, S. J. Thomas, C. Coles, P. J. Taylor, R. E. Marcus, and M. V. Williams. 2001. Total body irradiation using a modified standing technique: A single institution 7 year experience. *Br. J. Radiol.* 74 (887): 1041–1047.
Heublein, S. C. 1932. Preliminary report on continuous irradiation of the entire body. *Radiology* 18: 1051–1062.
Hugtenburg, R. P., J. R. Turner, S. P. Baggarley, D. A. Pinchin, N. A. Oien, C. H. Atkinson, and R. N. Tremewan. 1994. Total-body irradiation on an isocentric linear accelerator: A radiation output compensation technique. *Phys. Med. Biol.* 39 (5): 783–793.
Hui, S. K., J. Kapatoes, J. Fowler, D. Henderson, G. Olivera, R. R. Manon, B. Gerbi, T. R. Mackie, and J. S. Welsh. 2005. Feasibility study of helical tomotherapy for total body or total marrow irradiation. *Med. Phys.* 32 (10): 3214–3224.
Huq, M., B. A. Fraass, P. B. Dunscombe, J. P. Gibbons Jr., G. S. Ibbott, P. M. Medin, A. Mundt et al. 2010. Methods for evaluating QA need in radiation therapy. *American Association of Physicists in Medicine Radiation Therapy Committee Task Group 100.*
Hussain, A., J. E Villarreal-Barajas, and W. D. Brown. 2010. Validation of Eclipse™ AAA algorith at extended SSD. *J. Appl. Med. Phys.* 11 (3).

Hussein, S., and G. M. Kennelly. 1996. Lung compensation in total body irradiation: A radiographic method. *Med. Phys.* 23 (3): 357–360.

Khan, F. M. 2010. *The Physics of Radiation Therapy.* 4th ed. Philadelphia, PA: Lippincott Williams & Wilkins.

Kutcher, G. J., L. Coia, M. Gillin, W. F. Hanson, S. Leibel, R. J. Morton, J. R. Palta et al. 1994. Comprehensive QA for radiation oncology: Report of AAPM Radiation Therapy Committee Task Group 40. *Med. Phys.* 21 (4): 581–618.

La Macchia, N. A., R. Tsang, M. van Prooijen, F. Cheung, R. Heaton, and A. Parent. 2007. The development of a segmented intensity modulated beam technique for total body irradiation prior to hematologic stem cell transplantation. *Int. J. Radiat. Oncol. Biol. Phys.* 69: S540.

Lam, W. C., B. A. Lindskoug, S. E. Order, and D. G. Grant. 1979. The dosimetry of 60Co total body irradiation. *Int. J. Radiat. Oncol. Biol. Phys.* 5 (6): 905–911.

Lavallee, M. C., S. Aubin, M. Chretien, M. Larochelle, and L. Beaulieu. 2008. Attenuator design for organs at risk in total body irradiation using a translation technique. *Med. Phys.* 35 (5): 1663–1669.

Lavallee, M. C., L. Gingras, M. Chretien, S. Aubin, C. Cote, and L. Beaulieu. 2009. Commissioning and evaluation of an extended SSD photon model for PINNACLE3: An application to total body irradiation. *Med. Phys.* 36 (8): 3844–3855.

Lin, J. P., T. C. Chu, and M. T. Liu. 2001. Dose compensation of the total body irradiation therapy. *Appl. Radiat. Isot.* 55 (5): 623–630.

Mutyala, S., A. Stewart, A. J. Khan, R. A. Cormack, D. O'farrell, D. Sugarbaker, and P. M. Devlin. 2010. Permanent iodine-125 interstitial planar seed brachytherapy for close or positive margins for thoracic malignancies. *Int. J. Radiat. Oncol. Biol. Phys.* 76 (4): 1114–1120.

Podgorsak, E. B., C. Pla, M. D. Evans, and M. Pla. 1985. The influence of phantom size on output, peak scatter factor, and percentage depth dose in large-field photon irradiation. *Med. Phys.* 12 (5): 639–645.

Quast, U. 1985. Physical treatment planning of total-body irradiation: Patient translation and beam-zone method. *Med. Phys.* 12 (5): 567–574.

Quast, U. 1987. Total body irradiation—Review of treatment techniques in Europe. *Radiother. Oncol.* 9 (2): 91–106.

Quast, U., and H. Sack (eds.) 2003. DEGRO/DGMP Leitlinie: Ganzkörperbe-Strahlung. www.DEGRO.de, DGMP-Bericht Nr. 18, Ganzkörperbe-Strahlung. www.DGMP.de. 2002. DGMP-Report No. 18, Guideline: Whole Body Radiotherapy. www.DGMP.de.

Sarfaraz, M., C. Yu, D. J. Chen, and L. Der. 2001. A translational couch technique for total body irradiation. *J. Appl. Clin. Med. Phys.* 2 (4): 201–209.

Schultheiss, T. E., J. Wong, A. Liu, G. Olivera, and G. Somlo. 2007. Image-guided total marrow and total lymphatic irradiation using helical tomotherapy. *Int. J. Radiat. Oncol. Biol. Phys.* 67 (4): 1259–1267.

Shigeo, A., I. Takashi, K. Gorou, K. Kouji, Y. Masanori, and A. Tsutomi. 2000. Study of water-compensating filter in total body irradiation (TBI). *Jpn. J. Radiol. Technol.* 56: 826–833.

Smith, C. L., W. K. Chu, M. R. Goede, D. Granville, and G. V. Dalrymple. 1996. An analysis of the elements essential in the development of a customized TBI program. *Med. Dosim.* 21 (2): 49–56; quiz 58–60.

Van Dyk, J. 1987. Dosimetry for total body irradiation. *Radiother. Oncol.* 9: 107–118.

Van Dyk, J. 1999. Special techniques in radiotherapy; total body irradiation with photon beams. In *The Modern Technology of Radiation Oncology: A Compendium for Medical Physics and Radiation Oncologists.* Wisconsin Medical Physics Publishing. pp 641–662.

Van Dyk, J., M. Galvin, G. Glasgow, and E. B. Podgorsak. 1986. *The Physical Aspects of Total and a Half Body Photon Irradiation,* AAPM Report 17. New York, NY: American Institute of Physics.

Wong, J. Y., A. Liu, T. Schultheiss, L. Popplewell, A. Stein, J. Rosenthal, M. Essensten, S. Forman, and G. Somlo. 2006. Targeted total marrow irradiation using three-dimensional image-guided tomographic intensity-modulated radiation therapy: An alternative to standard total body irradiation. *Biol. Blood Marrow Transplant.* 12 (3): 306–315.

Wong, J. Y., J. Rosenthal, A. Liu, T. Schultheiss, S. Forman, and G. Somlo. 2009. Image-guided total-marrow irradiation using helical tomotherapy in patients with multiple myeloma and acute leukemia undergoing hematopoietic cell transplantation. *Int. J. Radiat. Oncol. Biol. Phys.* 73 (1): 273–279.

63

High Dose Rate Brachytherapy

Jack L. M. Venselaar
Instituut Verbeeten

Introduction

In many national and international publications, recommendations are given on frequencies and tolerances of quality control procedures for brachytherapy. Some books and reports present an extensive description of the required procedures, but the simple repetition of their contents would be out of the scope of this chapter. Comprehensive overviews can, for example, be found in the books and book chapters by Thomadsen (2000) and Williamson et al. (1994). In this chapter, in general, concise descriptions of these procedures are given in combination with frequency tables that provide medical physicists with useful material to assist with their quality assurance (QA) tasks. The frequencies presented here are essentially taken from previously published AAPM and ESTRO documents, with a short discussion on the differences in the approaches expressed in their reports. At the end of the chapter, some developments in the paradigms of QA in radiation oncology are discussed. It should be further noted, however, that whenever any national set of (legal) requirements exists, these should be followed. One should be aware that in clinical practice, an increase in test frequency is also required when the stability of a system is suspect or when a specific patient treatment method demands a special accuracy.

Because the quality control procedures differ somewhat between high dose rate (HDR), pulsed dose rate (PDR), medium dose rate, and low dose rate brachytherapy, often these modalities are taken separately. The present chapter deals with HDR only, but any difference with PDR would be minor. In specific afterloader designs, PDR treatments can even be given on the same hardware as HDR, with the only difference that the machine is equipped with other or additional software to control the timer setting of the pulses.

The main reports from the AAPM that provide guidance for QA and the procedures necessary to maintain safety and efficacy in brachytherapy are (1) the report of TG-40 of 1994 on comprehensive QA for radiation oncology in which brachytherapy was included (Kutcher et al. 1994) and (2) the TG-56 report of 1997 that narrowed the focus to a code of practice for brachytherapy physics (Nath et al. 1997), whereas (3) the TG-64 report dealt only with permanent prostate brachytherapy (Yu et al. 1999; not discussed here any further). The dose calculation formalism for brachytherapy treatments that was developed in (4) the AAPM TG-43 report (Nath et al. 1995) and its update TG-43U1 (Rivard et al. 2004) has proven to be a great success in terms of its rapid and wide adoption and its robust ability to incorporate advancing technology. These reports provide a consistent set of recommendations regarding QA, and particularly with respect to calibration of sources for brachytherapy implants. A recent paper by Butler and Merrick (2008) deals with the clinical practice and QA challenges in modern brachytherapy sources and dosimetry and focuses primarily on low dose rate low-energy photon-emitting sources, but the rationale for the procedures described in their paper are similarly of importance for the high-activity and high-energy sources used in HDR brachytherapy.

This series of publications is not at all finished. Other task groups under the Brachytherapy Subcommittee (AAPM-BTSC) are still working very hard on further development of recommendations to complete the picture. New reports are forthcoming. The High-Energy Brachytherapy Source Dosimetry Working Group (HEBD) has the heavy duty to solve problems associated with the application of TG-43 to the higher-energy photon-emitting sources. For that purpose, a close cooperation with the Brachytherapy Source Registry Working Group (BSR) is essential. TG-143 focuses on Dosimetry for Elongated Brachytherapy Sources. TG-138 has had several meeting presentations on Brachytherapy Dose Evaluation Uncertainties and is finalizing the publication of a report. TG-167 on Investigational Brachytherapy Source Recommendations has also almost finalized its task, similar to TG-182 on Electronic Brachytherapy Quality Management. A new task group to investigate the future of dose calculation algorithms was installed as the TG-186 on Model-Based Dose Calculation Techniques for Advanced Dosimetry. A challenging series of publications will therefore be available soon.

Further interesting reading is found in a number of recent publications, of which the following can be mentioned in the present context:

- The view on past and current trends in brachytherapy physics by Thomadsen et al. (2008).
- The anniversary paper "Fifty Years of AAPM Involvement in Radiation Dosimetry" by Ibbott et al. (2008).
- The overview on "Brachytherapy Technology and Physics Practice Since 1950: A Half-Century of Progress" by Williamson (2006).
- "Current Brachytherapy Quality Assurance Guidance: Does It Meet The Challenges of Emerging Image-Guided Technologies?" by Williamson (2008).
- "Quality Assurance Issues for Computed Tomography-, Ultrasound-, and Magnetic Resonance Imaging-Guided Brachytherapy" by Cormack (2008).
- A Vision 2020 paper on the evolution of brachytherapy treatment planning was presented by Rivard, Venselaar, and Beaulieu (2009).

Technical Considerations for Afterloading Equipment

Afterloader systems are complex biomedical devices that are critical to the safe and efficacious practice of brachytherapy. Figure 63.1 shows five of these devices. Manufacturers must comply with the standards for these systems that have been developed by the IEC (http://www.iec.ch/). In addition to the general electrical standards, IEC 60601, two additional standards are relevant. The first is IEC 60601-2-8, "Particular requirements for the safety of therapeutic X-ray equipment operating in the range 10 kV to 1 MV." The second is IEC 60601-2-17, "Particular requirements for the safety of automatically-controlled brachytherapy afterloading equipment."

The Food and Drug Administration (FDA) is responsible for assuring the public in the United States of the safety of most types of medical devices and radiation-emitting devices. The Center for Devices and Radiological Health is the branch of the FDA responsible for the approval of all medical devices before they can be legally marketed in the United States, as well as for overseeing the manufacturing, performance, and safety of these devices. There are basically two groups of new products: those similar to existing ones and those that are not. For that reason, there are different FDA regulatory paths for these two classes of products to become legally marketed medical devices in the United States. The vendor must provide sufficient evaluation data that has been performed either by the vendor or by some independent investigators to receive approval and to market the device. The FDA's role is thus to assure the public that medical devices are safe and effective and therefore any new device, including a new or innovative brachytherapy source, is generally evaluated for approval as a medical device by the FDA based on its risk factor.

European legislation states that only medical devices, which have been certified with a *CE-marking* according to the Council Directive 93/42/EEC, may be sold/exported as commercial products on/to the European market (see: http://www.ce-marking

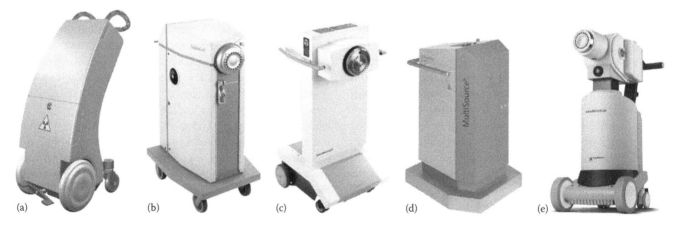

FIGURE 63.1 (a) Flexitron, Isodose Control; (b) Varisource iX, Varian; (c) GammaMed iX, Varian; (d) MultiSource, Bebig; (e) MicroSelectron HDR, Nucletron. (Courtesy of Isodose Control, Varian, Bebig, Nucletron.)

.org/directive-9342eec-medical-devices.html). Directive 2007/47/ec of the European Parliament and of the council of 5 September, 2007, which amended the Council Directive 93/42/EEC of 14 June, 1993, defines a medical device as: "any instrument, apparatus, appliance, software, material or other article, whether used alone or in combination, including the software intended by its manufacturer to be used specifically for diagnostic and/or therapeutic purposes and necessary for its proper application, intended by the manufacturer to be used for human beings. Devices are to be used for the purpose of: diagnosis, prevention, monitoring, treatment or alleviation of disease; diagnosis, monitoring, treatment, alleviation of or compensation for an injury or handicap; investigation, replacement or modification of the anatomy or of a physiological process; control of conception. This includes devices that do not achieve its principal intended action in or on the human body by pharmacological, immunological or metabolic means, but which may be assisted in its function by such means." Commercial equipment, which has been in use before July 14, 1998, will not have a CE certification. By this definition, the afterloading machine and each individual applicator, transfer tube, and connector must all be CE certified. On delivery of new equipment, the user should check that the product is CE certified. In some European countries, the national legislation is stricter and the requirement is extended in the sense that CE marking must be present for all medical devices, even if developed in-house and for in-house usage only.

An overview of technical specifications of currently available afterloading equipment (as per 2009) is presented in Table 63.1.

Safety and Physics Aspects of QC of HDR Afterloading Equipment

A description of quality control procedures for HDR afterloading equipment is separated in this chapter into the procedures of (1) safety systems and (2) physical parameters. Methods for calibration of the source and monitoring the source strength are only very briefly described in a separate section. In this chapter, we have assumed that the afterloader makes use of an HDR ^{192}Ir source. Application of other radionuclides is not discussed, although some other types are suggested as an alternative to iridium because of the specific physical properties such as a longer half-life (^{60}Co) or a lower mean energy of the emitted photons (^{169}Yb, ^{170}Tm). A quality management system is not affected by the choice of a specific radionuclide. Only the frequencies of the checks may be influenced to adapt to the source exchange periodicity.

Unintended radionuclide contamination is an important aspect that may not usually be checked. A check can be performed on the radionuclide composition of the delivered source. Any short-half-life contamination in a very newly produced source would show up with a deviating decay factor. Some users therefore repeat the initial calibration check on the ^{192}Ir source after 1 or 2 weeks and compare the result with the expected value. Some authorities such as the German Ministry for

Environment, Nature Conservation and Reactor Security even require officially that the source strength is measured to check the purity of the radionuclide after a period of 2 weeks from the first calibration. This is mainly focused on the use of ^{192}Ir-based afterloading systems where there exists the problem of possible contamination with ^{194}Ir, but the procedural rule is applicable to other radionuclides as well.

Calibration of Brachytherapy Sources

Calibration of brachytherapy sources at the hospital is an essential component of a well-designed QA program. The aim of the calibration is twofold: to ensure that the value entered into the treatment planning system agrees with the source calibration certificate to within a predetermined limit and to ensure traceability to international standards. The traceability is important because it simplifies national and international comparison of treatment results.

The recommended quantities to specify the source strength are the reference air-kerma rate $\dot{K}_R$ and air-kerma strength S_K, which are mutually related through the inverse-square law to the reference distance r_0 as $\dot{K}_R = S_K \cdot 1/r_0^2$. Note that the specification in terms of air-kerma strength S_K is adopted in the AAPM TG-43 formalism (Nath et al. 1995), whereas the reference air-kerma rate $\dot{K}_R$ is recommended by ICRU 38 and 58. This latter quantity is also used in the report from IAEA TecDoc 1274 (IAEA 2002). Other quantities for specifying source strength are now considered obsolete.

The calibration can essentially be done with a so-called in-air measurement technique or with the use of a well-type ionization chamber. Another method is to use a dedicated solid phantom for calibration purposes. In principle, any source can be calibrated with these methods, but there are some practical limitations. With in-air and solid phantom calibrations, the signals typically obtained when using too low activity (low dose rate) sources are very low and the final uncertainty in the air-kerma strength or reference air kerma rate may be unnecessarily high. For HDR sources, however, these methods may be considered.

For several reasons, the solid phantom measurement technique (see the literature for example on the so-called Meertens' and Krieger phantoms) is usually not recommended for a direct and traceable in-house HDR calibration procedure. The readings from ionization chamber measurements in such phantoms depend on the general design and on the material composition, influencing the scatter dose contribution and the degradation of the mean photon energy over the distance to the measurement volume. Nevertheless, these phantoms do have advantages in their stability and reproducibility and can thus be applied if validated against one of the other directly traceable methods. So, specifically as a *relative measurement technique*, solid phantoms are useful.

With in-air calibrations, the measured charge or current is strongly dependent on the measurement distance, and errors in the distance may yield large uncertainties in the

TABLE 63.1 Technical Specifications of Currently Available Afterloading Equipment (July 2009)

1. Vendor and product specification

	Nucletron B.V.	Eckert & Ziegler BEBIG GmbH	Varian Medical Systems Inc.	Varian Medical Systems Inc.	Isodose Control BV
Name of vendor	Nucletron B.V.	Eckert & Ziegler BEBIG GmbH	Varian Medical Systems Inc.	Varian Medical Systems Inc.	Isodose Control BV
Web site	www.nucletron.com	www.ibt-bebig.eu	www.varian.com	www.varian.com	www.isodosecontrol.com
Name of product(s)	microSelectron Digital	MultiSource: 20 channels GyneSource: with five channels but same specs as Multisource	Varisource iX	GammaMedplus iX; GammaMedplus iX 3/24 with 3 channels but same specs	Flexitron
Specify capability for use as HDR and/or PDR	HDR or PDR	HDR only	HDR only	iX model: HDR or PDR; iX 3/24 model: HDR only	HDR and PDR

2. Specifications of source or sources

Single- or dual-source capability	Single source	Single source	Single source	Single source	Dual source
Possible types of source (radionuclide), available now and/or under development	^{192}Ir	^{192}Ir, prod code IR2.A85-2 ^{60}Co, prod code Co0.A86	^{192}Ir	^{192}Ir	^{192}Ir
Maximum source strength for each possible radionuclide, with approval for marketing by authorities	57 mGy h^{-1} at 1 m (14 Ci, 518 GBq)	Co0.A86: 22.6 mGy h^{-1} at 1 m (2 Ci, 74 GBq) values ±10% IR2.A85-2: 40 mGy h^{-1} at 1 m (10 Ci, 370 GBq) values +30% −10%	44.8 mGy h^{-1} at 1 m (11 Ci, 407 GBq)	61.1 mGy h^{-1} at 1 m (15 Ci, 555 GBq) Local regulations may prohibit to install more than 10 Ci	2× 44.8 mGy h^{-1} at 1 m (2× 11 Ci, 2× 407 GBq). Max storage capacity 22 Ci ^{192}Ir; pending local regulations for max storage
Source dimensions for each of the source types (L = length, OD = outer diameter, in mm)	0.9 (OD) × 5 (L)	Co0.A86 1.0 (OD) × 5 (L) mm IR2.A85-2 0.9 (OD) × 5 (L)	0.59 (OD) × 5 (L)	HDR: 0.9 (OD) × 4.52 (L) PDR: 0.9 (OD) × 2.97 (L)	HDR: 0.85 (OD) × 4.6 (L) PDR: 0.85 (OD) × 4.6 (L)
Guarantee for the maximum number of source transfers or source cycles	25,000 cycles	Co0.A86 100,000 cycles IR2.A85-2 25,000 cycles	1000	5000	30,000 cycles

3. Specification of applicators

Outside diameter of the applicators (needles and or tubes, in mm)	Needles 1.3 mm, flexible 4F	Interstitial needles 1.5 mm (17 Ga), with both sources	18G (1.27 mm) needles constructed from robust thick wall tubing. 4.7F robust thick-walled catheters	17G (1.5 mm) needles. 5F catheters.	Needles 1.3 mm, flexibles 4F
Minimum curvature of an applicator (plastic loop radius, e.g., for an 180° curve, in mm)	15 mm for 6F catheter	10 mm (loop of 90°) 15 mm (loop of 180°) Both sources	17 mm	13 mm	17 mm

	Laser welded to ultraflexible drive wire	Laser welded	Embedded in the Nitinol (nickel-titanium) source drive wire	Laser welded to ultraflexible drive cable	Laser welded to ultra flexible drive wire, including weld protection
4. Source to cable attachment					
Method of source attachment to cable	Laser welded to ultraflexible drive wire	Laser welded	Embedded in the Nitinol (nickel-titanium) source drive wire	Laser welded to ultraflexible drive cable	Laser welded to ultra flexible drive wire, including weld protection
5. Source extension and movement					
Maximum source extension (in mm from the indexer)	1500	1500	1500	1300	1400
Speed of source movement, in seconds over maximum source extension	500 mm/s, typ. outdrive time 4 s for 1500 mm	300 mm/s, 5 s for 1500 mm	600 mm/s	630 mm/s	500 mm/s
6. Number of channels					
Number of hardware applicator channels	30	MultiSource model: 20 GyneSource model: 5	20	iX model: 24; iX 3/24 model: 3	40
Maximum number of channels that can be used in one plan/treatment	90	MultiSource model: 20 GyneSource model: 5	20	iX model: 24; iX 3/24 model: 3	40
7. Method of source movement					
Method of source movement	Forward stepping, 48 dwell positions with 2.5/5 or 10 mm step size	Source pulling (step backward)	Source pulling (step backward) 60 steps 2–99 mm step size in 1-mm increments	Source pulling (step backward) 60 steps 1–10 mm step size in 1-mm increments	Forward stepping, 401 dwell positions with any stepsize of multiples of 1 mm
Treatment window for defining possible steps and dwell positions (in mm)	475 mm	max 600 mm; step size selectable 1–10 mm in 1-mm increments max 100 dwell positions	700–1500 mm	710–1300 mm	400 mm
Method of counting dwell positions	From catheter tip to indexer	From catheter tip to indexer	From catheter tip to indexer	From catheter tip to indexer	From coupling to catheter tip
8. Source arrangement and dose calculation					
Source arrangements and dose calculations	Dwell times in 0.1-s increments, 0.1–999.9 s per dwell position	Dwell times maximum: 3600 s; minimum: 0.3 s; increments and precision: 1 ms	Dwell times in 0.1-s increments, 0.1–999.9 s at 60 positions	Dwell times in 0.1-s increments, 0.1–999.9 s at 60 positions	Dwell times in 0.1-s increments 0.1–999.9 s per dwell position

(continued)

TABLE 63.1   Technical Specifications of Currently Available Afterloading Equipment (July 2009) (Continued)

9. Safety features

Method of source retraction event of failure	Backup DC motor with friction clutch, backup batteries, manual hand crank	Independent backup-retraction motor powered by backup battery system and additional hand crank	Backup motor and independent backup drive assembly; backup battery and additional backup hand crank	Backup battery and additional backup hand crank	Backup DC motor with friction clutch, backup batteries, manual hand crank
Independent measurement system to detect safe source retraction	Triple safety check by optodetector, mechanical end-switch, and radiation detector	Radiation measurement and several photoelectric barriers	Onboard radiation detectors and mechanical switches	Onboard radiation detectors and mechanical switches	Onboard radiation detectors and proximity switches
Other safety features	Independent (hardware) secondary timer per dwell position, independent shaft encoders for motion detection and position verification, independent (software) verification of treatment record	Database for applicators and applicator geometry; automated calibration and guide tube verification; position check by camera; secondary timer; independent room monitoring	Wire length check, independent radiation detector, secondary timer and software watchdog checks, mechanical source home positional check, physical wire overtravel check, wire force feedback system, wire slippage check, plan data checksum verification	Independent radiation detector, secondary timer and software watchdog checks, mechanical source home positional check, applicator/ guide tube length check, plan data checksum verification	Independent (hardware) secondary timer per dwell position, independent shaftencoders for motion detection and position verification, independent torque measurement per drive, independent (software) verification of treatment record
What is the maximum level of leakage dose, in $\mu Gy\ h^{-1}$ at 0.1 m from any point of the surface of the machine, at maximum source strength loading?	$1.1\ \mu Gy\ h^{-1}$ at 0.1 m	According to IEC60601-2-17 Co0.A86: $<10\ \mu Gy\ h^{-1}$ at 1 m IR2.A85-2: $<1\ \mu Gy\ h^{-1}$ at 1 m	$6\ \mu Gy\ h^{-1}$ at 0.1 m for 11-Ci source	$4.5\ \mu Gy\ h^{-1}$ at 0.1 m for 15-Ci source	$<1\ \mu Gy\ h^{-1}$ at 1 m

10. Control unit and treatment planning system

Digital communication system	Separate control system with network connection. Extended graphical user interface with patient and library/QA plan database	PC-based, OS Windows XP or Vista, control PC networked to planning PC	All signals are digital	All signals are digital	Separate control system with network connection. Extended graphical user interface with patient and library/QA plan database
Remote access to system control unit	Yes	Yes, if Internet connection is available	No	No	Yes, if Internet connection is available

Dicom RT Plan import interface for treatment plan input data	Yes	At TPS side DICOM RT Plan export available, no import	At TPS side DICOM RT Plan export available	Yes
Dicom RT Report for R&V	Yes	Export DICOM RT Plan Dose and Structure at TPS	Export DICOM RT Plan Dose and Structure at TPS	Yes
TPS dedicated to vendor's afterloader or able to create plans for other makes of afterloading equipment	TPS dedicated to Nucletron afterloader	Export DICOM RT or Bebig proprietary data format	Export in DICOM RT or Varian proprietary format (for any radioactive source)	TPS able to plan for different afterloaders via DICOM RT Plan export
Dedicated TPS supporting all afterloader features and brachytherapy applications	Yes	Dedicated 3D TPS, all kinds of image modalities available. 2D x-ray images import either by x-ray scanner or DICOM. Fusion. Real-time online US prostate planning	Yes, supports 2D and 3D planning without limitations	Dedicated 3D TPS, all kinds of image modalities available. 2D x-ray images import either by x-ray scanner or DICOM. Fusion. Real-time online US prostate planning
11. Special features				
User ID and password protection	Yes, configurable permissions per user	Yes, protection by user ID and password, file protection by checksum	Yes, configurable permissions per user	Yes, protection by user ID and password, file protection by checksum
Details of available applicators:				
• MR/CT compatibility with applicators available	Yes, full range of carbon/plastic GYN applicators, interstitial plastic and titanium needles	Yes, full range of titanium or plastic (MR/CT compatible) for GYN; plastic needles and tubes for interstitial	Yes, full range	Yes, full range available in titanium (MR/CT compatible) for GYN; plastic needles for interstitial
• Capability to add interstitial needles to intracavitary GYN applicators	Yes, for ovoids and for ring	Yes, for ring techniques (work in progress); template used with cylinder applicator	Yes, for ring applicator	Yes, for ovoids and for ring
• Sterilization	All applicators (including CT/MR) can be steam-sterilized at 134°C or are delivered sterile (e.g., catheters)	Steam sterilization 134°C 3 bar for 5 min	All GYN applicators and majority of others can be steam sterilized	All applicators (including CT/MR) can be steam-sterilized at 134° C or are delivered sterile (e.g., catheters)
• Other features	Integrated in vivo dosimetry			

source calibration. To improve accuracy, several distances should be used with in-air measurements. In this procedure of HDR ^{192}Ir source calibration, for which no primary S_K standard exists, the AAPM endorses traceability to an interim secondary standard using an ion chamber with directly traceable ^{137}Cs and orthovoltage external-beam calibration coefficients, allowing a method in which a ^{192}Ir air-kerma calibration coefficient is estimated by interpolation. This approach, described by Goetsch et al. (1991), has become the standard of practice in North America. See, for further reading, also the details on the history of brachytherapy dosimetry in the work of Williamson (2006) and on the involvement of the AAPM in dosimetry recommendations during the past decades in the report of Ibbott et al. (2008). In the European area, several primary standards laboratories, among which PTB in Braunschweig, Germany, can provide the user with a calibration factor for their ion chamber for the ^{192}Ir source.

Normally for in-air calibrations, a low-scattering jig is used in which the source is kept at a distance of 5–15 cm from the ion chamber (Figure 63.2). With such distances it is required to apply a fluence nonuniformity correction factor to the readings, such as the ones calculated by Kondo and Randolph (1960), although the details of that publication were later criticized by Tölli et al. (1997). Depending on the measurement setup, a small room scatter correction factor needs to be determined and applied in the transformation of reading into air-kerma rate. For open ionization chambers, temperature and air pressure corrections are applied. A detailed description of this procedure is given in the IAEA TecDoc 1274 (IAEA 2002), which was later further completed with a number of directly accessible tables for parameters in a chapter in ESTRO Booklet #8 (Venselaar and Pérez-Calatayud 2004).

Most of the present recommendations rely on the use of a dedicated well-type ionization chamber. Although such chambers provide an easy, fast, and reliable method for source calibration, it must be borne in mind that in-air calibration is a more fundamental method. Still, the well-type chambers offer the best of practice: reliable, reproducible, and very easy to use. Several primary or secondary standards laboratory or ADCLs provide the users all over the world with a calibration factor for their instrument that is to be used with its accompanying electrometer and inserts. The procedure for in-house calibration is then simple and can be repeated at a few intervals in the lifetime of the local HDR afterloader source.

A general requirement is to use the available dosimetry equipment at each source installation before clinical treatments take place. In many institutions the procedure is repeated at the end of the lifetime of the source. In this way, the old and the new sources are checked in a short period, the stability of the measurement system is demonstrated, and confidence in the use of the correct decay factor is created. The instrument itself should be checked for linearity, leakage, and for consistency of the readings with a measurement at regular intervals with a long-lived source, for example, a ^{137}Cs source, once a year. Any deviation for the source under consideration larger than 3% must be

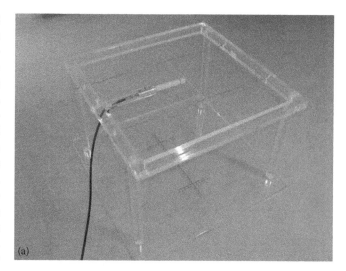

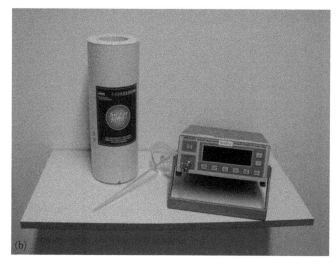

FIGURE 63.2 (a) Example of an in-air jig measurement setup. (b) Example of a well-type chamber with electrometer. (Courtesy of Institute Verbeeten.)

inspected by repetition of the measurement and/or with independent means.

In this context, it is worth mentioning that there is noticeably a renewed interest in the calibration issues regarding brachytherapy sources at the level of the European primary standard laboratories and, more specifically, in the development of a dose-to-water calibration standard. This goal was at the end of 2008 accepted as an explicit part of a new research activity in the EURAMET network of national measurement institutes. The iMERA joint research project on brachytherapy aims at developing a primary standard of absorbed dose to water not only for ^{192}Ir to be available at three national measurements institutes (NMIs) but similarly also for ^{125}I available at four national measurements institutes in a time frame of about 3 years (2009–2011). Regardless of whether this research is successful and how this advance is further implemented, there is still much work to be done before direct chamber calibrations for liquid water dose rates are available to clinical users. It is the responsibility of the international PSDL community to coordinate efforts to uniformly transition to the development of dose-to-water standards worldwide and to disseminate these standards jointly with the professional societies to the clinical end users of the sources. Coordination for this transition from reference air-kerma to dose to water across the entire field of brachytherapy is key. A distinct requirement from the clinical user's point of view is that such a new approach should at least give the same, but preferably, lower source calibration uncertainties.

Note that several vendors of afterloading equipment now provide their system with an internal radiation measurement device. This allows a simple daily verification (with maybe a somewhat larger uncertainty) of the strength of the source in the machine, thus avoiding rather useless daily repetition of a measurement.

QA of Imaging Techniques

As described in a Vision 2020 paper (Rivard, Venselaar, and Beaulieu 2009) and elsewhere, the use of x-ray, CT, and other 3D imaging modalities such as MR, US, and more recently PET, marks a distinct improvement in brachytherapy planning, to be considered as a departure from the surgical practice of brachytherapy. The step forward is made primarily due to direct availability of the radiological equipment for interventional procedures. The more expensive imaging modalities are nowadays readily available in every modest-sized hospital, whereas in many radiotherapy departments, 3D imaging has replaced the conventional RT simulator for patient setup. TRUS-guided permanent and HDR temporary implantation have become standard practice for prostate brachytherapy in North America and Europe. For other body sites, especially for gynecology, 3D imaging is now, or will be soon, included in the newest recommendations as a standard guidance for volume definition and dose prescription. A new generation of applicators in which a combination of intracavitary and interstitial techniques is pursued to increase the flexibility for source position optimization in the implant geometry is coming into the field, based on the use of carbon fiber or titanium as construction materials. Although MRI is only slowly entering the field, it is widely recognized that this technique readily adds to the quality of the treatments because of its ability to discriminate between different tissue types. Contouring will increasingly be based on this imaging technology. Functional MRI depends on its ability to demonstrate active parts of the tissue under consideration, allowing suborgan identification. If tissue contouring can be performed in this way, this would allow a spatially modulated dose deposition dose painting procedure, in which the full organ, for example, a prostate, is treated to the required dose but subparts are administered a much higher dose. There is a similar expectation for the use of PET and PET–CT imaging for identifying active tumor areas. One of the challenges in brachytherapy imaging is the further development of registration for different modalities and image sets made at different points in time. A key to this challenge is the need to address deformable image registration. In brachytherapy, not only organ movements or the increase or decrease in tumor volume but also the required transportation of a patient from one location to the other, for example, operating area to radiology department, may influence the relative position of the applicators and the organs.

Not only the imaging tools themselves but also QA of these imaging procedures must be adapted to the needs of radiation oncology. New QA protocols are required specifically for clinical applications in brachytherapy. Although the requirements of QA for both brachytherapy and imaging devices are well defined (Cormack 2008), image-guided brachytherapy has raised new issues. Image guidance in brachytherapy involves the transition from reference point dosimetry using films to volumetric imaging such as computed tomography, ultrasonography, and magnetic resonance imaging for treatment planning and guidance of applicator, needle, or seed placement. The QA of these devices might not reflect the conditions of use in brachytherapy or the requirements of brachytherapy treatment planning. The QA may not be the responsibility of the medical physicist involved in the brachytherapy procedures. According to Cormack, image interpretation both toward structure recognition and toward geometric accuracy becomes much more important with image-guided brachytherapy. This change toward image guidance has implications at the level of treatment, the process, and the field of brachytherapy as a whole. In Cormack's paper, the QA concerns are discussed that arise from brachytherapy procedures using ultrasound, computed tomography, and magnetic resonance imaging guidance, as well as problems associated with using imaging in an interventional setting, showing areas in which technical improvements are needed.

Williamson (2008) also raises the question if current BT recommendations meet the challenges of emerging image-guided technologies. For non-image-based brachytherapy, the AAPM Task Group reports 56 (Nath et al. 1997) and 59 (Kubo et al. 1998) provide reasonable guidance on procedure-specific process flow and QA. But more guidance is needed for intraoperative imaging systems and image-based planning systems, and improved guidance is needed even for established procedures such as ultrasound-guided prostate implants. Adaptive replanning in brachytherapy faces unsolved problems similar to that of image-guided adaptive external beam radiotherapy.

Williamson identifies the following unaddressed issues: (1) indications for physicist and/or dosimetrist attendance at volume studies and implant procedures, and delineation of critical QA duties; (2) processes for ensuring adequate TRUS probe positioning and image quality during the procedure; (3) verifying target volumes manually segmented from intraoperative or volume-study images; (4) review procedures for image-based treatment plans, including checks of manual dose-calculation and target localization accuracy; (5) improving operator performance by postprocedure implant quality assessment. He concludes with the observation that because of the complexity and variability of image-based brachytherapy procedures, risk-based guidance that includes clear guidelines for cost-effectively adapting general QA protocols to the specific clinical implementations is urgently needed. Finally, research directed toward quantifying random and systematic source positioning, target delineation, and dose-estimation errors is key.

QA protocols for image-based brachytherapy are not mature and need further elaboration. Joint efforts from radiodiagnostics and radiation oncology physics groups are needed. Training programs in medical physics with specialty brachytherapy may have to be reconsidered to include sufficient knowledge of the specific problems and issues in this challenging field.

QA of Treatment Planning in Brachytherapy

Treatment planning relies on (1) the proper reconstruction of the implant geometry, discussed in the previous section, and (2) accurate algorithms for calculation of the dose deposition from the sources in the medium. The evolution of treatment planning in brachytherapy was extensively discussed in the Vision 2020 paper in medical physics (Rivard, Venselaar, and Beaulieu 2009), describing the history, the conventional dose calculation approach, the present TG-43 and TG-43U1 status, and the ongoing developments in Monte Carlo and MC-based methodology. The latter evolution is aiming at improved accuracy of dose calculation in inhomogeneous media (shields, lung, lack of scatter situations). It is not the intention to go into any of these details here. It is assumed that the actual standard of the algorithms in the clinically applied treatment planning systems for brachytherapy is the TG-43 and TG-43U1 approach (Nath et al. 1995; Rivard et al. 2004). If this is not the case, at least the dose and/or dose rate in "along and away" tables found for the specific sources under the TG-43 formalism should be used as benchmark data under the assumption of homogeneous infinite-sized water media.

The other issue that should be considered as being "solved" is the worldwide accepted and implemented recommendation to use the air-kerma strength (or reference air-kerma rate) to specify the strength of the individual source. Other quantities must be considered obsolete, although it is very common to use the "activity" in conjunction with the strength definition. This is almost inevitable in view of, being the main reason, the ease of use in the clinic of the value of a 10-Ci ^{192}Ir source for the miniaturized HDR sources (and the order of 1 Ci for PDR).

QA of brachytherapy treatment planning should cover at least the following four issues, which are each discussed in the following text:

1. What exactly is the source used in the HDR afterloader (avoiding misinterpretation of terminology such as *new* or *classical*, or otherwise due to technological developments)?
2. Where do we find the (benchmark) data sets that belong to the individual source (dose rate constant; radial dose function; geometry function; anisotropy factor or function) in the TG-43 formalism?
3. How to validate the calculation by the TPS of the dose distribution around the individual source?
4. What else is there to check or validate: source decay, optimization, dose volume histogram calculation, and so on?

Issue 1

It is the responsibility of the medical physicist to validate and clearly describe in the documentation of the brachytherapy system which source is used clinically. These data should come from the vendor and the information should be unambiguous. Over time, afterloader types may have changed in design of the source with implications for the dosimetric data that should be used in the TPS to calculate the dose distribution. Any such change should be documented and any next step in the QA or commissioning process should be taken as if the system were a completely new system. Vendors have responsibility in the communication to their users for any such changes.

Issue 2

In the past decade, many papers have been published in the *Medical Physics* journal describing the required dosimetric data for, initially, ^{125}I and ^{103}Pd seeds. Later, such overviews of experimental and MC calculated data were presented also for the high-energy photon-emitting sources of ^{137}Cs and ^{192}Ir. The reader could be referred to those published articles, although it appears that the interpretation of the published data is, maybe, not univocal. The AAPM Brachytherapy Subcommittee in cooperation with the Radiological Physics Center maintains a database of validated and weighted recommended publications of the data sets of these sources. Thus, the reader can be referred to the Radiological Physics Center database at MD Anderson. The original papers can thus easily be traced. One step further is the use of the databases in which the full and unified formatted data are found and can be downloaded in an accessible format, for example, MS Excel. The following Internet references are available for this purpose:

- http://rpc.mdanderson.org/RPC/home.htm
- http://www.estro.org/estroactivities/Pages/TG43BTDOSMETRICPARAMETERS.aspx
- http://www.physics.carleton.ca/clrp/seed_database/

The work of the groups in setting up and maintaining these databases is very useful for the inexperienced user and therefore greatly appreciated. But to avoid possible confusion between data published in different databases, a joint effort of the Brachytherapy Subcommittee with the owners of the databases is required. So far, any differences that appeared between these databases seem to be minor and their use for validation purposes can be recommended. Especially the inclusion of along and away tables for the dose rates in a grid of points around the source makes it suitable for fast QA.

Issue 3

Previously, calculation of the dose or dose rate at a number of representative points around a source was recommended. Because of the burden of definition of these points at clinically relevant positions, the number of points to consider was kept limited. With, for example, Microsoft Excel, the medical physicist can relatively easily include the algorithm and data sets into a spreadsheet calculation procedure and do the calculation at many grid points. In this way, a 2D comparison can be constructed between these test calculations and the outcome of the treatment planning system. The recommendation remains to perform these comparisons at least at a number of clinically relevant points.

The validation should be done initially (commissioning) and with any changes in the use of the source. Some of the published recommendations state that the procedure should be repeated at regular intervals (e.g., annually).

A clinical dose distribution is essentially a summation of contributions from several or many sources or source positions. Because the computer is designed to perform summations, it is generally not necessary to perform dose calculations of many complex source arrangements. A few examples of dose distributions typical of the departmental practice are sufficient. These can be used not only for demonstration of stability of the system but also for training purposes of new people entering the brachytherapy staff.

Relative to the input data supplied and the algorithm assumed, the computer-assisted dose calculations should have a numerical accuracy of at least 2%. Systems using the TG-43 formalism generally perform much better except at points where extrapolation of the input data set is required.

Issue 4

Because the source decay is important in dose delivery, consistency in the use of the correct source decay constants must be validated. For example, for ^{192}Ir, a 1% per day change of source strength occurs. Because there may be several places where dose decay is calculated (on the desk of the medical physicist, in the TPS, and in the afterloader), small changes may occur if different values of the half-life are used.

In clinical practice, dose optimization routines are often used in the brachytherapy TPS. Often a TPS has more than one optimization strategy, which may be suited for different types of implants. Small changes in the geometry of implants will result in different outcomes of the optimization step such that individual dwell times are different, although the overall results may look the same. Often there is no one-answer solution to a specific problem. It is customary to use a few simple cases for training purpose of new staff members, which can also be used for clinical validation of new software versions. Inspection of both the general result of the optimized plan (e.g., if the preset conditions were met and using TRAK for overall comparison of different solutions) and the details of the outcome especially at the high-dose region is essential for understanding the optimization routine.

Accuracy of dose volume histogram calculation is highly dependent on the conditions under which the 3D image data set is collected. Slice thickness and slice distance have a major impact on the results, and therefore, imaging procedures should be taken from a detailed protocol. There are definitely differences between the volume calculations in the products of different vendors (Kirisits et al. 2007). But as the user normally has no choice other than the algorithm in his/her TPS, only initial studies on objects with well-known dimensions and volumes are reasonably easy to perform. Typical problems arise with inclusion of the end slices in a volume calculation.

For an independent check of an individual dose calculation, at least one critical point is recommended for each implant. Comparison with TRAK (in previous terms: an mg-hr concept) values or tables from established implant systems may also be used to check many clinical implants. It is then recommended to pursue an agreement within 15% between the independent check and the dose calculation. In many cases of standardized treatments, the agreement is often even better than 5%. If the treatment was custom-planned, the check should verify that the final result was consistent with the plan. The geometrical configuration of the reconstructed implant should be compared to the radiographs or CT scout views and an assessment of the dose distribution should be compared to the written prescription.

The AAPM report 46 (Kutcher et al. 1994) summarizes these verification steps for brachytherapy, which is here slightly adapted to be specifically for HDR applications.

Tests on Ancillary Equipment

Dosimetry Equipment

Because all HDR afterloaders make use of sources with relatively short half-lives, the calibration system, any handheld radiation monitor, survey detectors, or any other radiation detection device should be checked with an independent method. A long-lived source such as a ^{137}Cs source is very suitable for the purpose. Simple procedures with well-described measurement setup are sufficient to demonstrate the stability of the reading of the devices against such sources over the years.

Protective Means

Lead aprons and lead shields may be available at the department. Although the aprons do not provide any significant protection against the relatively high energy of the photons emitted by a ^{192}Ir source, they are useful in cases of lower-energy sources (less than 100 keV) and in x-ray procedures that may be applied for imaging. The aprons suffer from wear and tear over the years and should be inspected on an annual basis. A standardized procedure for lead aprons can be to inspect them on a therapy simulator table or a similar device. Lead shields are sometimes used as additional shields in treatment rooms in case the walls are not sufficiently thick. These shields are unlikely to change shielding properties, but visual inspection may show damaging from, for example, falling over. The same holds for lead slabs if mounted to walls. If mounted invisibly *in* a wall, a radiation survey with suitable equipment may be necessary to show that it is still in place and not torn off from its fastening bolts.

Frequencies and Tolerances

Frequencies and tolerances for the individual QC tests and items for HDR systems are listed in Table 63.2. The table contains two sets of recommendations, illustrating some differences in interpretation. The first set is taken from the ESTRO Booklet #8 (Venselaar and Pérez-Calatayud 2004), the second set from AAPM TG56 (Nath et al. 1997). The ESTRO table is rather concise, indicating the main aspects of QC and the associated frequencies, whereas the AAPM report is very valuable in that it is much more detailed and also contains more information on the methods for the QC tests and their alternatives. Therefore, the reader is referred to the original document.

In the ESTRO booklet, the recommendations for a given test frequency must be considered as a *minimal*, not as an *optimal test frequency*. An increase in the frequency of the test is

TABLE 63.2 Frequencies and Tolerances of Quality Control Tests for HDR Afterloading Equipment

Description	ESTRO Booklet #8		AAPM TG56 1997	
	Minimum Requirements		Minimum Requirements	
Safety Systems	Test Frequency	Action Level	Test Frequency	Action Level
Warning lights	Daily/3M[a]	—	Daily	—
Room monitor	Daily/3M[a]	—	Daily	—
Communication equipment	Daily/3M[a]	—	Daily	—
Emergency stop	3M	—	3M	—
Treatment interrupt	3M	—	3M	—
Door interlock	3M	—	Daily	—
Power loss	3M	—	3M	—
Applicator and catheter attachment	6M	—	Daily	—
Obstructed catheter	3M	—	3M	—
Integrity of transfer tubes and applicators	3M	—	3M	—
Timer termination	Daily	—	Daily	—
Contamination test on r.a. leak	A	—	6M (NRC)	—
Leakage radiation, survey around machine	A	—	3M	—
Emergency equipment	Daily/3M[a]	—	Daily	—
(forceps, emergency safe, survey meter)				
Practicing/training (emergency) procedures	A	—	A	—
Hand crank functioning	A	—	A	—
Handheld monitor	3M/A[b]	—	Daily	—
	Physical Parameters			
Source calibration	SE	>5%	Daily/3M	5% (d)/3% (3M)
Source positioning	Daily/3M[a]	>2 mm	Daily/3M	>1 mm (d)/>0.5 mm (3M)
Length of treatment tubes	A	>1 mm	3M	>1 mm
Irradiation timer	A	>1%	Daily	>2%
Date, time, and source strength in treatment unit	Daily	–	Daily	–
Transit time effect	A	–	A	–

Note: 3M, quarterly; A, annual; SE, source exchange.

[a] Daily checks are assumed to be an implicit part of normal operation. The department's policy determines whether a separate logbook of these daily checks should be kept. A "formal" check by the medical physicist should be performed at least at the lower indicated frequency, for example, quarterly.

[b] The lower frequency determines the interval to verify the proper function of the handheld monitor, for example, with a known source of radiation.

required whenever the stability of the system is suspect, or when the specific treatment method demands a special accuracy. The medical physicist should carefully consider which recommended test frequency is applicable in his or her clinical situation, considering:

- The likelihood of the occurrence of a malfunction
- The seriousness of the possible consequences of an unnoticed malfunction to patients and/or to the personnel
- The chances that, if a malfunction occurs, this will not be identified during normal treatment applications

For example, it is recommended to *formally* check the performance of the warning lights in the treatment room, the proper functioning of the room radiation detector, and the audio and/or video communication system for the patient only once every 3–4 months and then to record the results of the check in a logbook. The reason is that it can be assumed that a malfunction of any of these systems will be quickly detected by the radiation technologists during their routine work. Starting the treatment and signing the documents for that treatment may implicitly assume that the daily tests were performed and that the results were satisfactory, according to a department's written policy. Other departments may wish to develop special daily check forms to record and sign for the execution of these tests on satisfactory completion.

The AAPM report states that the tables of their recommendations should not be interpreted as a rigid prescription of tests and frequencies but rather as a starting point for developing a written QA program, individualized to the need of each institution. Nevertheless, because of differences in the liability issues between countries, the practical interpretation of the reports from the professional organization leaves less room for the medical physicist in the United States. The requirements must be strictly followed and any deviation from these recommendations must be well documented in the institution.

In addition, local authorities such as the NRC in the United States may have developed their own sets of guidelines, which then must be followed.

Please note that the daily tests should be executed on a routine basis before treating the first patient of the day, and only on days when patients are treated. For most of the tests in Table 63.2, a 3-month interval is suggested because this is usually the frequency with which HDR sources are replaced. Some departments may apply a 4-month interval instead, if source replacement takes place only three times annually. The quality control checks, which are performed quarterly or with a lower frequency, must be explicitly logged in a logbook, which is kept by the physicist.

Note further that the data, provided in this table as "action level," reflect the upper limit in clinical conditions. For an acceptance test, the design specifications must be compared. Often the design of the system is such that a much better performance can be obtained under reference conditions, such as positional checks on straight catheters with autoradiography or with a high-resolution video camera system on a check ruler. Larger deviations will be observed with curvatures in the catheters or in the applicators.

It is the task of the medical physicist to inspect the performance history of the system thoroughly, using the data in the logbook noted during the clinical lifetime of the equipment.

Applicators and Appliances

Acceptance Tests

It has to be verified after receiving a delivery of any equipment for brachytherapy that the delivered product is in accordance with what has been ordered. Whenever applicable, the instructions for use must be checked for availability and studied for their contents. One should check that all individual parts have been delivered. The mechanical properties of the applicators should be checked and verification should be made that the applicators and transfer tubes are of the correct length. Checks should be made to ensure that the product is functioning as described in the instructions for use. Any mechanical code, which is intended to force the connection of a specific applicator with a specific transfer tube, should be verified. Applicators for a specific irradiation device are not to be used with other irradiation devices.

When applicators are shielded, checks should be made to ensure that the positional shielding marker is correct. For applicators that are more complex to handle (e.g., Fletcher applicators), each user should become familiar with the use of the device by instruction and training without a patient (Figure 63.3).

Instructions for sterilization must be provided with the products. When selecting products, care must be taken to have the proper sterilization procedures available. Not all hospitals have facility for gas sterilization. In some cases, one might be obliged to find a nearby hospital that does gas sterilization and arrange a contract with them for the sterilization of the applicators. Many users therefore prefer to order presterilized products, such as plastic tubes for interstitial and endoluminal work.

A detailed discussion on the QA procedures of applicators and appliances can be found in the ESTRO Booklet #8 (Venselaar and Pérez-Calatayud 2004) and the book of Thomadsen (2000). Steps and procedures that can be used for verification of the correct functioning, as mentioned in this section, are summarized in Table 63.3.

Regular Tests of Applicators and Transfer Tubes

Because of usage and sterilization, the mechanical properties of the applicators may change. The mechanical integrity of all applicators and transfer tubes including the connectors should be regularly checked. It is especially necessary to check the length of reusable plastic applicators and (nonmetallic reinforced) transfer tubes because of the possibility of expansion or shrinkage over time. Depending on the type of afterloading machine,

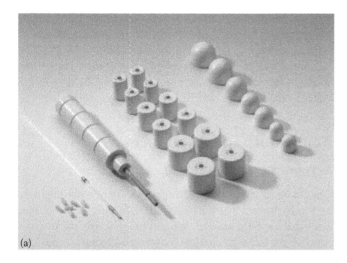

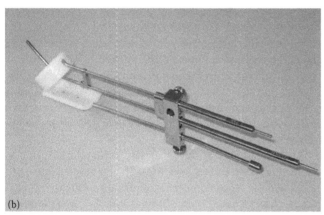

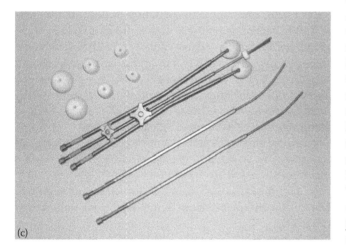

FIGURE 63.3 CT and MR compatible applicator sets. (a) Segmented vaginal set; (b) ring applicator set; (c) Fletcher-Suit-Delclos ovoids and intrauterine tandem set. (Courtesy of Varian.)

applicators and transfer tubes that are too short may lead to an overlap of several dwell positions if the advancing source hits the end of the applicator before it reaches the first programmed position if there is no properly functioning interlock system. The length of the applicator–transfer tube combination can simply

be measured with a wire of appropriate length, a check ruler, or a high-resolution camera system.

Contamination, Cleaning, and Sterilization

High-activity sources such as ^{192}Ir HDR sources should never be touched by hand for wipe tests performed for verification that the surface of the source is not contaminated with radioactivity. Usually the tip of a used catheter or transfer tube is measured in a sensitive detector (e.g., using an NaI crystal). If any signal is detected above background radiation, the contaminated parts must be checked in more detail and there should be no clinical use of the source and/or afterloader until the cause of the contamination is clarified. The radiation safety officer must be informed immediately. Any further spread of radioactive contamination must be avoided.

Cleaning and sterilization of reusable applicators and transfer tubes should be performed only with methods given by the manufacturer of the product. If there are no cleaning and sterilization methods described in the instructions for use, the manufacturer should be contacted unless it is possible to use a normal procedure, for example, for the cleaning and sterilization of metallic applicators. Applicators specified for single use by the manufacturer may not be reused.

Time for a New Paradigm for QA?

In a recent paper on QA in RT supplement to the *International Journal of Radiation Oncology, Biology, and Physics*, Thomadsen (2008) presents his critique of the traditional QA paradigm. After discussing a number of organizations that have contributed to the present status of a series of QA recommendations (see the "Introduction" of this chapter for many references), he questions the efficacy of the time spent in RT physics to go through the prescribed exhaustive lists of tests to perform, the frequency of the tests, and suggested tolerances. He refers to a few contributions of colleagues who, for example, showed a logbook of 460 pages formatted to facilitate compliance with the AAPM TG-40 report's recommendations for one linear accelerator for 1 year. His own book on QA procedures for brachytherapy alone counted 239 pages (Thomadsen 2000). Thus, "the conglomeration of all these reports and recommendations has left the clinical physicist with a rapidly expanded job just addressing QA. Although most of the reports and texts suggest that not every test is appropriate for every institution, the standard was set such that failure to follow all recommendations appeared negligent, certainly to many state radiation regulators."

Others have also questioned if all effort spent at QA is efficient and argued that it is sometimes inefficient at best and wasted at worst. Regardless of the efficacy of QA in general, it seems that at least some of the QA customarily performed in RT serves little purpose. Examples are easily given. Radiation safety measurements can demonstrate efficacy of shielding, but it is unlikely that most of the shielding characteristics change over time. So

TABLE 63.3 Steps and Procedures for Verification of the Correct Functioning of Applicators and Appliances

Article	Procedure
Integrity of applicator materials	Visual inspection of the applicators, depending on their use: before or after each treatment (such as in the case of reusable materials)
Fixation mechanisms	Check each fixation screw and mechanism for proper functioning: before and after each treatment
Shielding in applicators	Check for the presence and position of shields included in the applicator, at acceptance (radiography)
Source positioning	Autoradiography whenever applicable for verification of source position, at acceptance or when there is suspicion of (length) changes in the applicators
Identification of connecting mechanisms	Check the integrity of the applicator in relation to its use or to its connection to the afterloader, at acceptance
Sterilization procedures	Check for the presence of the instructions for sterilization and follow these instructions meticulously to avoid unintended damaging
Validity of dose distributions in relation to specific applicators	Carefully check the applicability of any dosimetrical "atlas" for precalculated and tabulated treatment times, at acceptance
Radioactive contamination	Careful handling with the applicators at detachment and check the tubes in a NaI crystal to detect possible leakage or contamination

Source: From Venselaar J. L. M. and Pérez-Calatayud J., *A Pratical Guide to Quality Control of Brachytherapy Equipment.* 2004.

why perform such checks repeatedly, for example, on an annual basis? If a ^{192}Ir source was calibrated and eventually contamination of possible short-half-life radionuclides during source production has been excluded, why repeat a full calibration procedure on a patient-to-patient basis? Maybe—if at all needed—a built-in detector serves the purpose to show that nobody changed the source? Changes in most applicators, often rigidly constructed or metallic, are unlikely to appear. So why do more than a visual inspection?

Just performing all the recommended QA might not provide protection against events. Instead, each facility should consider its own needs for addressing potentially hazardous situations and, particularly, target measures for control. This process of failure mode and effect analysis would also include assessment of the efficacy of any QA procedure in their setting. Such an assessment should demonstrate potentially hazardous situations, the frequencies with which they can occur, and the associated risks for patient, personnel, or environment. An analysis like that would best suit a facility's practice and provide the most efficacious guard for quality.

Risk analysis is in itself a time-consuming exercise. But the situations to describe and analyze are often very similar for different users. It is the author's opinion that, therefore, joint efforts from professional groups combined with the input from the vendors of the products under consideration are required. The AAPM and other national or international organizations have recognized this and as an example, AAPM TG-100 "Method for Evaluating QA Needs in Radiation Therapy" (M. Saiful Huq) was, in part, charged with the task to (see www.aapm.org) identify a structured systematic QA program approach that balances patient safety and quality versus resources commonly available and strikes a good balance between prescriptiveness and flexibility. It is explicitly stated that the task group members are also working on failure mode and effect analysis for HDR brachytherapy. Still, until this promising approach has been elaborated on the details of the specific fields of medical physics, established,

and received full support from the communities, the existing recommendation reports form the basis of this chapter.

References

Butler, W. M., and G. S. Merrick. 2008. Clinical practice and quality assurance challenges in modern brachytherapy sources and dosimetry. *Int. J. Radiat. Oncol. Biol. Phys.* 71 (1 Suppl.): S142–S146.

Cormack, R. A. 2008. Quality assurance issues for computed tomography-, ultrasound-, and magnetic resonance imaging-guided brachytherapy. *Int. J. Radiat. Oncol. Biol. Phys.* 71 (1 Suppl): S136–S141.

Goetsch, S. J., F. H. Attix, D. W. Pearson, and B. R. Thomadsen. 1991. Calibration of 192Ir highdose-rate afterloading system. *Med. Phys.* 18: 462–467.

IAEA. 2002. *Calibration of Photon and Beta Ray Sources Used in Brachytherapy. Guidelines on Standardized Procedures at Secondary Standards Dosimetry Laboratories (SSDLs) and Hospitals.* Vienna: International Atomic Energy Agency.

Ibbott, G., C. M. Ma, D. W. Rogers, S. M. Seltzer, and J. F. Williamson. 2008. Anniversary paper: Fifty years of AAPM involvement in radiation dosimetry. *Med. Phys.* 35 (4): 1418–1427.

Kirisits, C., F. A. Siebert, D. Baltas, M. De Brabandere, T. P. Hellebust, D. Berger, and J. Venselaar. 2007. Accuracy of volume and DVH parameters determined with different brachytherapy treatment planning systems. *Radiother. Oncol.* 84 (3): 290–297.

Kondo, S., and M. L. Randolph. 1960. Effect of finite size of ionization chambers on measurement of small photon sources *Radiat. Res.* 13: 37–60.

Kubo, H. D., G. P. Glasgow, T. D. Pethel, B. R. Thomadsen, and J. F. Williamson. 1998. High dose-rate brachytherapy treatment delivery: Report of the AAPM Radiation Therapy Committee Task Group No. 59. *Med. Phys.* 25 (4): 375–403.

Kutcher, G. J., L. Coia, M. Gillin, W. F. Hanson, S. Leibel, R. J. Morton, J. R. Palta et al. 1994. Comprehensive QA for radiation oncology: Report of AAPM Radiation Therapy Committee Task Group 40. *Med. Phys.* 21 (4): 581–618.

Nath, R., L. L. Anderson, G. Luxton, K. A. Weaver, J. F. Williamson, and A. S. Meigooni. 1995. Dosimetry of interstitial brachytherapy sources: Recommendations of the AAPM Radiation Therapy Committee Task Group No. 43. American Association of Physicists in Medicine. *Med. Phys.* 22 (2): 209–234.

Nath, R., L. L. Anderson, J. A. Meli, A. J. Olch, J. A. Stitt, and J. F. Williamson. 1997. Code of practice for brachytherapy physics: Report of the AAPM Radiation Therapy Committee Task Group No. 56. American Association of Physicists in Medicine. *Med. Phys.* 24 (10): 1557–1598.

Rivard, M. J., B. M. Coursey, L. A. DeWerd, W. F. Hanson, M. S. Huq, G. S. Ibbott, M. G. Mitch, R. Nath, and J. F. Williamson. 2004. Update of AAPM Task Group No. 43 Report: A revised AAPM protocol for brachytherapy dose calculations. *Med. Phys.* 31 (3): 633–674.

Rivard, M. J., J. L. Venselaar, and L. Beaulieu. 2009. The evolution of brachytherapy treatment planning. *Med. Phys.* 36 (6): 2136–2153.

Thomadsen, B. 2008. Critique of traditional quality assurance paradigm. *Int. J. Radiat. Oncol. Biol. Phys.* 71 (1 Suppl): S166–S169.

Thomadsen, B. R. 2000. *Achieving Quality in Brachytherapy.* Bristol (UK): Institute of Physics Publishing.

Thomadsen, B. R., J. F. Williamson, M. J. Rivard, and A. S. Meigooni. 2008. Anniversary paper: Past and current issues, and trends in brachytherapy physics. *Med. Phys.* 35 (10): 4708–4723.

Tölli, H., A. F. Bielajew, O. Mattsson, G. Sernbo, and K. A. Johansson. 1997. Fluence non-uniformity effects in air kerma determination around brachytherapy sources. *Phys. Med. Biol.* 42 (7): 1301–1318.

Venselaar, J. L. M., and J. Pérez-Calatayud, eds. 2004. *A Practical Guide to Quality Control of Brachytherapy Equipment.* 1st ed. Brussels: ESTRO.

Williamson, J. F. 2006. Brachytherapy technology and physics practice since 1950: A half-century of progress. *Phys. Med. Biol.* 51 (13): R303–R325.

Williamson, J. F. 2008. Current brachytherapy quality assurance guidance: Does it meet the challenges of emerging image-guided technologies? *Int. J. Radiat. Oncol. Biol. Phys.* 71 (1 Suppl.): S18–S22.

Williamson, J. F., G. A. Ezzell, A. Olch, and B. R. Thomadsen. 1994. Quality assurance for high dose rate brachytherapy. In *High Dose Rate Brachytherapy: A Textbook.* Library of Congress Cataloging-in-Publication Data, ed. S. Nag. Armonk, NY: Futura Publishing Company Inc. Chapter 7, pp. 147–212.

Yu, Y., L. L. Anderson, Z. Li, D. E. Mellenberg, R. Nath, M. C. Schell, F. M. Waterman, A. Wu, and J. C. Blasko. 1999. Permanent prostate seed implant brachytherapy: Report of the American Association of Physicists in Medicine Task Group No. 64. *Med. Phys.* 26 (10): 2054–2076.

Electronic Brachytherapy

Jessica Hiatt
Rhode Island Hospital

Introduction

Electronic brachytherapy (eBx) is a technological advancement for brachytherapy, during which a microminiature, water-cooled 50-kV x-ray source is used to deliver the radiation. Only recently introduced as a commercially available system, the Axxent Electronic Brachytherapy System (Xoft, Inc., Sunnyvale, CA) has been relatively unstudied thus far in the clinical setting and is unfamiliar to many medical physicists and radiation oncologists. Because this modality provides a high dose rate treatment using an x-ray source, eBx is unregulated by the Nuclear Regulatory Commission. Consequently, no modality-specific QA procedures have been formulated by either the Nuclear Regulatory Commission or the American Association of Physicists in Medicine (AAPM). The AAPM has formed the Task Group 182 to address this and other relevant eBx issues.

The eBx modality has several advantages. Because the device has relatively low energy, treatments can be delivered in an unshielded room in contrast to the significant shielding required for HDR, iridium-192 brachytherapy. The low exposure rate allows staff to remain near the treatment couch during dose delivery and an opportunity to provide comfort and encouragement while near the patient. It also facilitates installation in satellite clinics, mobile services, or rural locations that lack heavily shielded rooms. The high dose-rate, low-energy source approximates the therapeutic dose of iridium-192 at the prescription point. It generates, however, a more targeted dose with very rapid dose fall-off, resulting in significantly less dose to healthy tissue beyond the prescription point.

System components consist of the controller, which includes a touch-screen monitor, USB port, pull-back arm (adjustable arm with a high voltage port for the source connection), barcode scanner, x-ray source cooling pump, a Standard Imaging well chamber, a Standard Imaging Max-4000 electrometer, and 50 kV miniature x-ray sources (Figure 64.1). For exposure-to-air kerma strength conversions, the University of Wisconsin Accredited Dosimetry Calibration Laboratory has provided a source calibration coefficient.

The initial application of the system was for the treatment of breast cancer, adapting one of the most widely accepted techniques for accelerated partial breast irradiation. The Xoft Axxent balloon differs from other accelerated partial breast irradiation balloons in several ways. The balloon wall is impregnated with barium; thus, the need for contrast inside of the balloon, as is required with other breast accelerated partial breast irradiation intracavitary devices, is eliminated. The balloon applicator is also equipped with drainage ports for extraction of air and seroma to create more conformal treatment geometries and is shown in Figure 64.2.

Gynecologic applicators, in the form of vaginal cylinders, are also available (Figure 64.3). The applicators are compatible with both CT and fluoroscopy and range in diameter from 2.0 to 3.5 cm. The endometrial applicator kit includes a specialized base plate and clamp assembly for fixation of the device within the patient.

The company has recently developed other applicators for the treatment of superficial skin lesions. In addition, several patients have been treated thus far intraoperatively with the breast applicator.

Although eBx uses new technological approaches for the management of breast, gynecologic, and skin cancers, the equipment is relatively easy to implement clinically. By virtue of its many advantages, it is likely that in the near future, eBx will emerge as the brachytherapy modality of choice for an increasing number of institutions.

Commissioning an eBx System

At our institution, we developed a comprehensive quality assurance process for the commissioning of an eBx system (Hiatt et al. 2008). In general, eBx commissioning tests should include eight elements: (1) well-chamber constancy, (2) beam stability,

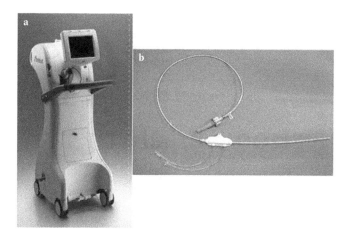

FIGURE 64.1   Controller and x-ray source.

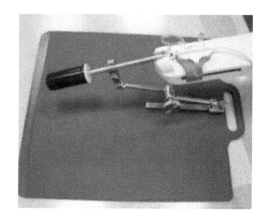

FIGURE 64.3   Endometrial applicator.

(3) source positional accuracy, (4) output stability, (5) timer linearity, (6) dummy marker/source position coincidence, (7) controller functionality and safety interlocks, and (8) treatment planning data verification following the AAPM Task Group 43 (TG43) data requirements and TG56 code of practice recommendations (Nath et al. 1995, 1997; Rivard et al. 2004). This methodology provides a comprehensive eBx system check for medical physicists commissioning such a device.

Well-Chamber Constancy

The first step in commissioning the eBx system is to verify the proper functioning of the manufacturer-provided well chamber/electrometer measurement system that is integrated into the Axxent treatment console. This can be accomplished by comparing the provided system with an independently calibrated system of the same make and model. At our institution, both systems consisted of a Standard Imaging HDR 1000 Plus well chamber and a Standard Imaging Max 4000 electrometer. The comparison test was performed using a 4-mg radium-equivalent

cesium-137 source, but can be any known source standard of relatively long half-life, and a fabricated source calibration jig that allowed for reproducible placement of the source within the well chamber. In the absence of a similar system, a source constancy test is essential in establishing a well-working measurement system.

Beam Stability

Data, including beam current and voltage and filament current, from a simulated treatment should be extracted and analyzed to determine the stability of the beam throughout a treatment fraction.

Source Positional Accuracy

A test of the coincidence between the planned and actual source positions should be performed. First, a "dwell file" is created in a Microsoft Excel spreadsheet using the Xoft-provided software add-in. The dwell file defines the source position and dwell time information for each plan. It is created in Excel and then transferred to the treatment console via a USB flash drive.

The dwell file contains data directing the source to stop, for example, for 10-s intervals at distances 24.5, 23.5, 22.5, 21.5, and 20.5 cm. The dwell file data can then be imported into the treatment console, the source calibrated, and the system connected to the QA test fixture provided by the manufacturer (Figure 64.4).

The QA test fixture consists of two acrylic blocks in a lead housing. The lower slab contains a channel fitted for the x-ray source with a second, perpendicular channel 1 cm below the source channel for the placement of either a Farmer or pinpoint chamber. The top portion of the lower slab includes thin horizontal lead graduations. The graduations are separated by 1 cm and correspond to the regularly spaced dwell positions. A sheet of GafChromic film should be placed 1 cm above the source and beneath the second acrylic slab. By positioning the film in this arrangement, the beam is attenuated at the programmed dwell positions by the lead graduations and a visual inspection of planned versus actual dwell position can be performed.

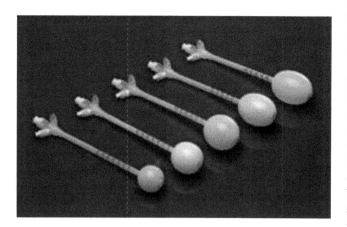

FIGURE 64.2   Balloon applicator.

FIGURE 64.4　QA test fixture.

A second test to assess the expected/actual source position coincidence is also performed. A dwell file can be created consisting of fifteen 10-s dwell positions separated by 0.5 cm. With the x-ray source located at the first dwell position, the distance from the source hub to the proximal end of the QA test fixture is recorded for each dwell position using digital calipers to ensure it retracts the correct amount.

Output Stability

The next test to perform is a beam output stability test. Relative current (nA) measurements can be obtained with a suitable instrument such as a PTW pinpoint chamber (Model 31006, SN0299) placed 1 cm from the source within the QA test fixture. The instantaneous current is measured when the source arrives at each of its programmed positions.

Timer Linearity

The linear response of the charge collected as a function of time is determined by recording the collected charge for set dwell times. For testing of timer linearity, we created four dwell files, with times varying from 30 to 120 s at a dwell position of 22 cm. This dwell position placed the source directly above the pinpoint chamber within the test fixture. Collected charge and dwell time can be graphed. The correlation coefficient should be with 1% of 1.000 (Table 64.1). The timer error can then be estimated using linear regression and solving for the intercept. The example shown in Table 64.1 was estimated to be 2.83 s. The error would

be considered significant for dwell times less than 10 times the timer error, that is, 28 s for this example.

Marker/Source Position Coincidence

To ensure the dummy marker and source positions are coincident within the applicator, to provide an accurate treatment plan when defining the first dwell position, some sort of film verification is necessary. To this end, we stapled XV film to a piece of GafChromic film and affixed a Xoft balloon applicator. A three-dwell position file was specified. With the dummy marker in place, a large field on the film was exposed using a conventional x-ray unit. The treatment was then administered to the balloon/film arrangement. Low-exposure XV (fast speed) film was required to visualize the dummy markers after conventional x-ray exposure; high-exposure GafChromic (slow speed) film was required to visualize the source positions after exposure using the Xoft Axxent 50-kV source. The processed films were then overlaid revealing coincidence between planned and actual dwell positions (Figure 64.5).

Controller Functionality and Safety Interlocks

A number of tests are performed to evaluate overall system performance and the safety of the device including radiation status indicator lights, treatment interrupt, treatment recovery, and emergency off.

Treatment Planning Data Verification of TG-43 Parameters

The manufacturer provides TG-43 physics data by Rivard et al. (2004) for input into the PLATO (Nucletron Corporation, Columbia, MD) and the BrachyVision (Varian Medical System, Palo Alto, CA) treatment planning systems. These data include S_k, air kerma strength; Λ, dose rate constant; $g_P(r)$, radial dose function; $G_P(r,\theta)$, line source geometry function; and $F(r,\theta)$, anisotropy function. During the treatment planning phase, a nominal source strength is used. Before treatment, the dwell times are adjusted to account for the source strength at that time; in this way, there are no source decay issues in the treatment planning systems. It is recommended that some variety of independent dose check be used to verify doses to specified points in the treatment area. At our institution, an in-house HDR second check program is used to verify the dose calculated by our PLATO treatment planning systems. The second check program applies the TG-43 parameters of the source—requiring the input of dwell times, dwell position coordinates, dose point coordinates, and the planned dose at those points—and outputs the independently determined dose for comparison using a 3D inverse square vector analysis.

TABLE 64.1　Timer Linearity

Timer Accuracy and Linearity						
Time (s)	120	90	60	30	Correlation	1.000
Average reading (nC)	27.62	20.92	14.18	7.36	Timer error (s)	2.83

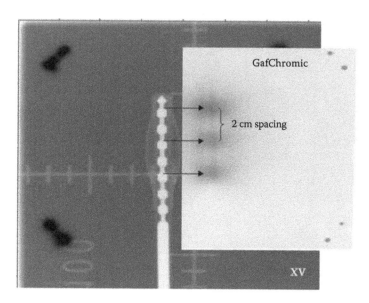

FIGURE 64.5   Screen shot of second check.

Commissioning eBx System Applicators

As eBx grows in popularity, evolves, and its equivalence to HDR Ir-192 is proven clinically, new applicators and niches for eBx treatment are sure to arise. It is the duty of the physicist to commission each applicator before clinical use. At a minimum, for each new applicator, the marker/source position coincidence should be verified and some form of dose distribution verification should be obtained (TLDs, diodes, etc.) for the source within the applicator.

Routine QA for an eBx System

Although no standardized QA procedures and/or recommendations exist at the time of publication of this text, it is our institution's policy to treat eBx as we would HDR Ir-192 requiring daily, monthly, and annual QA to be performed and a board-certified physicist to be present for all treatments.

Daily QA

Daily quality assurance of an eBx system is a cursory system check easily completed in a matter of minutes. One ensures the radiation survey meter is functional, the system passes self-checks, radiation indicator lights on the controller operate, source positioning and dwell times are accurate, and the log file is saved appropriately to the USB drive. This QA procedure needs to only be performed on the day of an eBx treatment. An example of a daily QA form is shown in Figure 64.6.

There is a large potential for error insomuch as the controller currently has no network connection to allow treatment data transfer from the treatment planning systems. The physicist is required to manually transfer the data and could inadvertently enter source positions or dwell times incorrectly. It

is recommended to have redundant second checks throughout the workflow process to avoid mistakes such as these. We created a treatment checklist for the second-checking physicist, the treating physicist, and the therapist to complete pretreatment (Figure 64.7).

Monthly QA

eBx QA performed on a monthly basis includes tests to ensure well chamber constancy, source stability, source positional accuracy, and timer accuracy/linearity. A spreadsheet example is shown in Figure 64.8. The daily QA is attached to the monthly report as proof of overall system functionality.

Annual QA

Annual eBx QA requires more comprehensive tests of the source positional accuracy and timer accuracy/linearity over the practical treatment range. To further assess the source positional accuracy, the marker catheters are checked for their overall condition and the reliability with which they indicate source positioning. It is recommended that the marker/source position procedure be applied as described above for several applicators. A radiation survey should be performed for the areas surrounding the controller during a simulated treatment.

QA for a Source Replacement

While each source undergoes rigorous testing by the Xoft laboratories before its release for human use, it is good practice for the physicist to check a new source's distribution before a patient treatment. A reasonable method for verifying the distribution is to use the QA test fixture with GafChromic film positioned

Rhode Island Hospital
Department of Radiation Oncology
Daily Electronic Brachytheraphy QA Instructions

The electronic brachytherapy controller Model # 100_ Serial # 041508-034_ is made by Xoft.
The electrometer and well chamber are manufactured by Standard Imaging, Serial #'s E080355 and A080397, respectively.

1. Survey meter functional and battery charged *Use dedicated check source; meter should read 0.2 mRfh*	Pass	Fail
2. Controller startup self-test passes *Turn on Xoft controller (electrometer should have been left on).* *After self-test, verify that the Patient Information screen appears.*	Pass	Fail
3. Verify date and time are correct *Upper right corner of Patient Information screen.*	Pass	Fail

4. Check source positioning, emergency interrupts and radiation monitors
Using the barcode scanner, enter the barcodes on the cover of the eBx QA binder into the Patient Information screen.
Use the Morning Warm-up USB drive to transfer the dwell file to the controller.
Ensure the following dwell positions and dwell times are loaded:

(cm)	25	24	23	22	21
(5)	40	4	4	4	20

On the Calibration Setup screen scan the barcode attached to the source.
Complete the source calibration
Enter the ambient temperature and pressure.
Record the calibration factor: _____ from the Calibration Completed screen.
Connect the pullback arm to the QA Test Fixture and place the source inside the QA Test Fixture connector.
Complete the tasks on the Applicator Setup screen
Confirm the new dwell times equal the original times multiplied by the calibration factor.
Press the "Load" button to advance the source to dwell position #1.

Looking through the Test Fixture ensure: Source advances to first dwell position *(most distal tick on graduated plate)*	Pass	Fail
Start treatment: Source remains at first position for 40 sec x cal. factor *Use stopwatch; start timing after 3 beeps indication beam-on*	Pass	Fail
Console radiation light functions	Pass	Fail
Allow source to move through the next 3 positions. *When the source reaches 21 cm. press the "Pause" button ensure:* X-ray production ceases	Pass	Fail
X-ray production restarts when "Resume" button is pressed	Pass	Fail

Complete the treatment.
Save treatment log to USB drive; print log file, attach to this form sign and file in eBx QA binder

_____   _____
Physics Personnel   Date

_____   _____
Authorized Medical Physicist's Signature   Date

FIGURE 64.6 Daily QA form.

Xoft Treatment Checklist:

Patient Name: ——————————————   Patient Number:
Fraction #: ————————————————

Temperature (C): ——————————   Pressure (mmHg):

☐ Xoft Dwell File times match the Plato generated dwell times in the patient chart.

    Checked by: ——————————————   Date: ———————————

☐ Xoft Dwell File positions in the patient chart match those on the treatment console screen.

    Checked by: ——————————————   Date: ———————————

☐ Xoft Dwell File times in the patient chart multiplied by the calibration factor are equal to those on the treatment console screen. (See calculation below)

    Calibration Factor: ————————

Source Position (cm)	Dwell Times (s)	Corrected Dwell Times (s)

    Checked by: ——————————————   Date: ———————————

☐ Treatment record is copied to USB key after treatment, printed and placed in Patient chart.

    Checked by: ——————————————   Date: ———————————

FIGURE 64.7   Treatment checklist.

above the source. A short treatment can then be delivered to the fixture and the film compared to an exposed film deemed as the standard. The overall testing time is generally less than 2 min—a source-on-time of more than 2 min registers as a source use. For each given source, only 10 uses are permitted before a source change is required. In the event of a source change during a patient treatment, verification of the source before clinical use is not feasible. It is reasonable to advance with the treatment and perform a source check immediately after treatment completion.

Department of Radiation Oncology, Rhode Island Hospital
Xoft Axxent Monthly Quality Assurance

Xoft Axxent Source Information:			Source No:	**101283**
			* used for Controller #1 data	

Equipment Information:			
Xoft Unit: **1**	Xoft Axxent Controller, Model X100 SN: 02206-014		
Electrometer:	SI MAX-4000, SN: E060753	Well Chamber:	SI HDR-1000, SN:060722

Rhode Island Hospital (RIH) Equipment Information

Electrometer:	SI MAX-4000, SN: E993471	Well Chamber:	SI HDR-1000, SN: 982851

Xoft Electrometer and Well Chamber Calibration factors

Electrometer (LOW Scale):	**1.000**	pA/Rdg		Calibration Date:	**04/27/06**
Electrometer (HIGH Scale):	**1.000**	nA/Rdg		Calibration Date:	**04/27/06**
Well Chamber: (6711 I-125)	**6.918**	$\times 10^7$ cGy.m^2/(hr.A)		Calibration Date:	**04/19/06**

RIH Electrometer and Well Chamber Calibration factors

Electrometer (LOW Scale):	**1.013**	pA/Rdg		Calibration Date:	**03/09/05**
Electrometer (HIGH Scale):	**1.001**	nA/Rdg		Calibration Date:	**03/09/05**
Well Chamber: (6711 I-125)	**2.357**	$\times 10^7$ cGy.m^2/(hr.A)		Calibration Date:	**04/05/07**

Temperature and Pressure Correction

Temperature:	**23.0**	Pressure:	**769.0**	Cpt:	**0.9916**

Well Chamber Constancy Check

Using Cs-137 source (3M 6D6 C 4 mg-Ra. Eq., SN: 0869). Insert the source tube into the chamber with the centering jig.
The center of the source is 6 cm from the bottom of the chamber well. Electrometer on LOW scale. Take 3 readings.

Half Life:	**30.00**	Year			Standard Reading on	**04/29/02**	**46.762**

RIH Unit:

Readings (pA):	**41.928**	**41.892**	**41.885**	% Diff:	1.65	Constancy Check:	Pass

Xoft Unit 1:

Readings (pA):	**42.510**	**42.494**	**42.530**	% Diff:	1.49	Constancy Check:	Pass

Source Stability Measurements

After source calibration, position 250 mm source in QA test fixture with pinpoint chamber 1 cm from catheter. 5 runs of a 5 dwell position plan (10 seconds at positions 2435, 23.5, 22.5, 21.5 and 20.5 cm) were completed. The electrometer was set to HIGH.

Rate (nA)	Trial Number							
Dwell Position (cm)	1	2	3	4	5	Average	Standard Deviation	Relative SD
24.5	0.007	0.007	0.007	0.007	0.007	0.007	0.000	0.00
23.5	0.024	0.024	0.024	0.025	0.024	0.024	0.000	0.02
22.5	0.114	0.114	0.111	0.112	0.114	0.113	0.001	0.01
21.5	0.205	0.206	0.198	0.199	0.202	0.202	0.004	0.02
20.5	0.081	0.081	0.080	0.080	0.080	0.080	0.001	0.01
Total Rate (nA)	0.431	0.432	0.420	0.423	0.427	0.427	0.005	0.01

Timer Accuracy and Linearity

5F, 1500 mm lumencath, 5F adapter. Program setting: step size: 2.5 mm; dwell position: Max. reading, index length: 1500mm.

Time (sec):	120	90	60	30		Correlation:	1.0000
Reading (nC):	27.40	20.65	14.00	7.26		Time Error (sec):	2.50

Daily QA Passed (Copy Attached):			**YES**	

Physicist:	**Gene A. Cardarelli, PhD, MPH, D**	Date:	03/03/06
	Jessica R. Hiatt, MS		

FIGURE 64.8 Monthly QA spreadsheet.

Summary

Although eBx is not under the jurisdiction of the Nuclear Regulatory Commission and currently does not have AAPM published QA recommendations, upon acquiring an eBx system, it is the responsibility of the physicist to ensure the system is and remains safe for the treatment of patients. An eBx system should be treated as an HDR remote afterloader with robust and rigorous daily, monthly, and annual QA procedures. This chapter can serve as a guide for baseline eBx QA until the report of the AAPM TG-182 is published.

References

Hiatt, J., G. Cardarelli, J. Hepel, D. Wazer, and E. Sternick. 2008. A commissioning procedure for breast intracavitary electronic brachytherapy systems. *J. Appl. Clin. Med. Phys.* 9 (3): 2775.

Nath, R., L. L. Anderson, G. Luxton, K. A. Weaver, J. F. Williamson, and A. S. Meigooni. 1995. Dosimetry of interstitial brachytherapy sources: recommendations of the AAPM Radiation Therapy Committee Task Group No. 43. American Association of Physicists in Medicine. *Med. Phys.* 22 (2): 209–234.

Nath, R., L. L. Anderson, J. A. Meli, A. J. Olch, J. A. Stitt, and J. F. Williamson. 1997. Code of practice for brachytherapy physics: report of the AAPM Radiation Therapy Committee Task Group No. 56. American Association of Physicists in Medicine. *Med. Phys.* 24 (10): 1557–1598.

Rivard, M. J., B. M. Coursey, L. A. DeWerd, W. F. Hanson, M. S. Huq, G. S. Ibbott, M. G. Mitch, R. Nath, and J. F. Williamson. 2004. Update of AAPM Task Group No. 43 Report: A revised AAPM protocol for brachytherapy dose calculations. *Med. Phys.* 31 (3): 633–674.

65

Technical Guidance Documents

Tommy Knöös
*Skåne University Hospital
and Lund University*

Introduction

Several organizations have published documents on the requirements of performance of equipment used in the radiotherapy process. The general idea of all of these documents is to ensure the delivery of an accurate treatment to the patient through accuracy in absorbed dose and spatial position. Having international steering documents allows the local users to rely on performance standards agreed on by international well-recognized expertise. It is also very convenient for the manufacturers to be able to base their production standards on requirements that are agreed upon by both users and vendors.

International Electrotechnical Commission

First of all, we have the recommendations from the International Electrotechnical Commission (IEC) (http://www.iec.ch). The IEC is a worldwide leading organization preparing and publishing international standards for all electrical, electronic, and related technologies known as electrotechnology. IEC standards cover electrotechnology in all areas, including the medical field, and, in particular, equipment used in medical physics for diagnostic and therapeutic purposes. The international standards published by the IEC form the basis for national standardization and for recommendations given by professional organizations and by others. It is also common to refer to these standards in tenders and contracts between users and manufacturers. More than 70 technical committees are active within the IEC, and they cover about 700 different projects (see, e.g., IEC 1987, 1993, 1994, 1997a, 1997b, 1997c, 1998, 1999, 2000, 2002a, 2002b, 2004a, 2004b, 2005a, 2005b, 2005c, 2005d, 2007, 2008a, 2008b, 2008c, 2009a, 2009b, 2009c, 2009d, 2009e; IEC/TR, 1997, 1998, 2001a, 2001b, 2002, 2005, 2006; IEC/TR 2002; IEC/TS 1993).

Of interest for the medical physics profession is the technical committee for "Electrical Equipment in Medical Practice" and, in particular, the subcommittees "SC62B—Diagnostic Imaging Equipment" and "SC62C—Equipment for Radiotherapy, Nuclear Medicine and Radiation Dosimetry." For example, the latter group has published recommendations and guidelines on simulators, linear accelerators, and record and verify systems. Guidance on DICOM RT objects and treatment planning systems are included in about 25 publications. It is also important to note that the committees of the IEC work in close collaboration with such organizations as the International Atomic Energy Agency (IAEA 2000, 2004), ICRP (2000), ICRU, and so on. The reports are reviewed regularly and amendments and/or new editions are released.

One example of IEC recommendations is the 61217 report on coordinates, movements, and scales for radiotherapy equipment. This recommendation has probably decreased the risk of delivering an erroneous treatment at many radiotherapy centers because of interpretation or transfer problems between equipment of different types and from different vendors.

Other important recommendations that can be found in the IEC documents are performance standards for accepting and commissioning of new equipment and also tolerances for periodic quality control.

International Atomic Energy Agency

The International Atomic Energy Agency provides the medical field with recommendations mostly from a radiation protection perspective, but they have also issued standards for reference dosimetry, treatment planning systems, and so on. Tolerances and performance recommendations are mainly based on the IEC standards but, when standards do not exist, working groups are put together by the agency for developing recommendations. An example of this is the comprehensive report on quality assurance

TABLE 65.1 Periodic Quality Control Tolerances for a Selection of Linear Accelerator Parameters, Which Are Recommended by Organizations and in Scientific Papers

	AAPM	IPEM	WHO	Brahme et al.[a]	IEC
Dose /monitor unit (X and e^-)	2%	±2%	±2%	1%	2%
Photon energy	2%	±2%	±2%	2 mm[b]	
Electron energy	±2 mm[c]	±2 mm[d]	±2 mm[e]	2 mm	
Flatness (X)	±2%	±3%	±3%	±2.5%	±3%
Symmetry (X)	±3%	±3%	±3%	±1.5%	±3%

[a] Brahme 1984; Brahme et al 1988.
[b] Constancy of depth dose.
[c] At therapeutic depth.
[d] D_z, where z is any value between 30 and 80.
[e] At therapeutic depth.

of treatment planning systems that was published a few years before the IEC published their standards in 2009.

Professional Organizations

Professional organizations, both national, such as the American Association of Physicists in Medicine (AAPM) (US) and the Institute of Physics and Engineering in Medicine (IPEM) (UK), and international, such as the European Society of Therapeutic Radiology and Oncology (ESTRO) (Europe), have also issued guidance documents and recommendations regarding the radiotherapy process and for specific procedures. Especially the AAPM has issued a large number of reports for both diagnostic and therapeutic procedures (AAPM Report 86 2004; AAPM 40 1994; AAPM Report 106 2008; AAPM Report 108 2006; AAPM Report 119 2009; AAPM Report 128 2008; AAPM Report 13 1984; AAPM Report 14 1985; AAPM Report 142 2009; AAPM Report 15 1985; AAPM Report 16 1986; AAPM Report 31 1990; AAPM Report 34 1992; AAPM Report 35 1992; AAPM Report 39 1993; AAPM Report 46 1994; AAPM Report 47 1994; AAPM Report 48 1994; AAPM Report 62 1998; AAPM Report 74 2002; AAPM Report 76 2001; AAPM Report 82 2003; AAPM Report 83 2003; AAPM Report 99 2009). The organization has a well-developed system for assigning task groups to new topics that have to be looked into and, if needed, guidelines are issued. Their reports cover both acceptance and commissioning procedures as well as periodic quality control. ESTRO has published reports covering both quality management in radiotherapy and more specific guidelines for quality assurance of different steps in the process. This includes treatment planning, in vivo dosimetry, independent monitor unit calculation, and the verification of intensity-modulated radiotherapy (IMRT). Practical guidelines for the implementation of a quality system in radiotherapy have been published by ESTRO (Leer et al. 1998). The IPEM has also published technical guidance documents (IPEM 2005a, 2005b, 2006, 2007, 2008, 2009).

Summary

Typical tolerances for established equipment such as conventional linear accelerators are shown in Table 65.1. However, a common problem for all organizations issuing guidelines and recommendations is the lead time for the process from the initial release of a new product or functionality in radiotherapy until the documents are published. This includes gathering a sufficient number of experienced users for meetings, the weighing of all opinions, and the publishing of relevant recommendations.

References

AAPM 40. 1994. Comprehensive QA for Radiation Oncology. Report of the AAPM Task Group 40. *Med. Phys.* 21: 581–618.

AAPM Report 13. 1984. *Physical Aspects of Quality Assurance in Radiation Therapy.* New York, NY: American Institute of Physics, Inc.

AAPM Report 14. 1985. *Performance Specifications and Acceptance Testing for X-ray Generators and Automatic Exposure Control Devices.* New York, NY: American Institute of Physics, Inc.

AAPM Report 15. 1985. *Performance Evaluation and Quality Assurance in Digital Subtraction Angiography.* New York, NY: American Institute of Physics, Inc.

AAPM Report 16. 1986. *Protocol for Heavy Charged-Particle Therapy Beam Dosimetry.* New York, NY: American Institute of Physics, Inc.

AAPM Report 31. 1990. *Standardized Methods for Measuring Diagnostic X-ray Exposures.* New York, NY: American Institute of Physics, Inc.

AAPM Report 34. 1992. *Acceptance Testing of Magnetic Resonance Imaging Systems.* New York: American Institute of Physics, Inc.

AAPM Report 35. 1992. *Recommendations on Performance Characteristics of Diagnostic Exposure Meters (Reprinted from Medical Physics, Vol. 19, Issue 1).* New York, NY: American Institute of Physics, Inc.

AAPM Report 39. 1993. *Specification and Acceptance Testing of Computed Tomography Scanners.* New York, NY: American Institute of Physics, Inc.

AAPM Report 46. 1994. *Comprehensive QA for Radiation Oncology (Reprinted from Medical Physics, Vol. 21, Issue 4).* New York, NY: American Institute of Physics, Inc.

AAPM Report 47. 1994. *AAPM Code of Practice for Radiotherapy Accelerators (Reprinted from Medical Physics, Vol. 21, Issue 7).* New York, NY: American Institute of Physics, Inc.

AAPM Report 48. 1994. *The Calibration and Use of Plane-Parallel Ionization Chambers for Dosimetry of Electron Beams (Reprinted from Medical Physics, Vol. 21, Issue 8).* New York, NY: American Institute of Physics, Inc.

AAPM Report 62. 1998. *Quality Assurance for Clinical Radiotherapy Treatment Planning (Reprinted from Medical Physics, Vol. 25, Issue 10).* New York, NY: American Institute of Physics, Inc.

AAPM Report 74. 2002. *Quality Control in Diagnostic Radiology.* New York, NY: American Institute of Physics, Inc.

AAPM Report 76. 2001. AAPM protocol for 40–300 kV x-ray beam dosimetry in radiotherapy and radiobiology. *Med. Phys.* 28 (6): 868–893.

AAPM Report 82. 2003. Guidance document on delivery, treatment planning, and clinical implementation of IMRT: Report of the IMRT subcommittee of the AAPM radiation therapy committee. *Med. Phys.* 30 (8): 2089–2115.

AAPM Report 83. 2003. Quality assurance for computed-tomography simulators and the computed-tomography-simulation process. *Med. Phys.* 30 (10): 2762–2792.

AAPM Report 86. 2004. *Quality Assurance for Clinical Trials: A Primer for Physicists.* New York, NY: American Institute of Physics, Inc.

AAPM Report 99. 2009. Recommendations for clinical electron beam dosimetry: Supplement to the recommendations of Task Group 25. *Med. Phys.* 36 (7): 3239–3279.

AAPM Report 106. 2008. Accelerator beam data commissioning equipment and procedures: Report of the TG-106 of the Therapy Physics Committee of the AAPM. *Med. Phys.* 35 (9): 4186–4215.

AAPM Report 108. 2006. AAPM Task Group 108: PET and PET/CT shielding requirements. *Med. Phys.* 33 (1): 4–15.

AAPM Report 119. 2009. IMRT commissioning: Multiple institution planning and dosimetry comparisons, a report from AAPM Task Group 119. *Med. Phys.* 36 (11): 5359–5373.

AAPM Report 128. 2008. AAPM Task Group 128: quality assurance tests for prostate brachytherapy ultrasound systems. *Med. Phys.* 35 (12): 5471–5489.

AAPM Report 142. 2009. Task Group 142 report: Quality assurance of medical accelerators. *Med. Phys.* 36 (9): 4197–4212.

Brahme, A. 1984. Dosimetric precision requirements in radiation therapy. *Acta Radiol. Oncol.* 23: 379–391.

Brahme, A., J. Chavaudra, T. Landberg, E. C. McCuliough, F. Nüsslin, A. Rawlinson, G. Svensson, and H. Svensson. 1988. Accuracy requirements and quality assurance of external beam therapy with photons and electrons. *Acta Oncol. Suppl. 1.*

IAEA. 2000. Absorbed dose determination in external beam radiotherapy: An international code of practice for dosimetry based on standards of absorbed dose to water. In *Technical Reports Series No. 398.* Vienna: International Atomic Energy Agency (IAEA).

IAEA. 2004. *Commissioning and Quality Assurance of Computerized Planning Systems for Radiation Treatment of Cancer, Technical Report Series.* Vienna: International Atomic Energy Agency.

ICRP. 2000. *Prevention of Accidental Exposure to Patients Undergoing Radiation Therapy.* Oxford: International Commission on Radiological Protection.

IEC. 1987. Medical Electrical Equipment—Part 2: Particular Requirements for the Safety of Therapeutic X-ray Equipment Operating in the Range 10 kV to 1 MV (60601-2-8). International Electrotechnical Commission.

IEC. 1993. Radiotherapy Simulators—Functional Performance Characteristics (61168). International Electrotechnical Commission.

IEC. 1994. Medical Electrical Equipment—Radionuclide Calibrators—Particular Methods for Describing Performance (61303). International Electrotechnical Commission.

IEC. 1997a. Medical Electrical Equipment—Dosimeters with Ionization Chambers and/or Semi-Conductor Detectors as Used in X-ray Diagnostic Imaging (61674). International Electrotechnical Commission.

IEC. 1997b. Medical Electrical Equipment—Dosimeters with Ionization Chambers as used in Radiotherapy (60731). International Electrotechnical Commission.

IEC. 1997c. Medical Electrical Equipment—Part 2: Particular Requirements for the Safety of Gamma Beam Therapy Equipment (60601-2-11). International Electrotechnical Commission.

IEC. 1998. Radionuclide Imaging Devices—Characteristics and Test Conditions—Part 3: Gamma Camera Based Wholebody Imaging Systems (61675-3). International Electrotechnical Commission.

IEC. 1999. Medical Electrical Equipment—Part 2–8: Particular Requirements for the Safety of Therapeutic X-ray Equipment Operating in the Range 10 kV to 1 MV (60601-2-8). International Electrotechnical Commission.

IEC. 2000. Medical Electrical Equipment—Dose Area Product Meters (60580). International Electrotechnical Commission.

IEC. 2002a. Amendment 1 (60731-am1). International Electrotechnical Commission.

IEC. 2002b. Amendment 1 (61674-am1). International Electrotechnical Commission.

IEC. 2004a. Amendment 1—Medical Electrical Equipment—Part 2–11: Particular Requirements for the Safety of Gamma Beam Therapy Equipment (60601-2-11-am1). International Electrotechnical Commission.

IEC. 2004b. Medical Electrical Equipment—Part 2–17: Particular Requirements for the Safety of Automatically-Controlled Brachytherapy Afterloading Equipment (60601-2-17). International Electrotechnical Commission.

IEC. 2005a. Medical Diagnostic X-ray Equipment—Radiation Conditions for Use in the Determination of Characteristics (61267). International Electrotechnical Commission.

IEC. 2005b. Medical Electrical Equipment—Characteristics and Test Conditions of Radionuclide Imaging Devices—Anger Type Gamma Cameras (60789). International Electrotechnical Commission.

IEC. 2005c. Medical Electrical Equipment—Safety of Radiotherapy Record and Verify Systems (62274). International Electrotechnical Commission.

IEC. 2005d. Radionuclide Imaging Devices—Characteristics and Test Conditions—Part 2: Single Photon Emission Computed Tomographs (61675-2). International Electrotechnical Commission.

IEC. 2007. Medical Electrical Equipment—Medical Electron Accelerators—Functional Performance Characteristics (60976). International Electrotechnical Commission.

IEC. 2008a. Medical Electrical Equipment—Part 2-29: Particular Requirements for the Basic Safety and Essential Performance of Radiotherapy Simulators (60601-2-29). International Electrotechnical Commission.

IEC. 2008b. Radionuclide Imaging Devices—Characteristics and Test Conditions—Part 1: Positron Emission Tomographs (61675-1). International Electrotechnical Commission.

IEC. 2008c. Radiotherapy Equipment—Coordinates, Movements and Scales (61217). International Electrotechnical Commission.

IEC. 2009a. Corrigendum 1—Medical Electrical Equipment—Characteristics and Test Conditions of Radionuclide Imaging Devices—Anger Type Gamma Cameras (60789 Corr. 1). International Electrotechnical Commission.

IEC. 2009b. Medical Electrical Equipment—Dosimetric Instruments as Used in Brachytherapy—Part 1: Instruments Based on Well-Type Ionization Chambers (62467-1). International Electrotechnical Commission.

IEC. 2009c. Medical Electrical Equipment—Dosimetric Instruments Used for Non-Invasive Measurement of X-ray Tube Voltage in Diagnostic Radiology (61676). International Electrotechnical Commission.

IEC. 2009d. Medical Electrical Equipment—Part 2-1: Particular Requirements for the Basic Safety and Essential Performance of Electron Accelerators in the Range 1 MeV to 50 MeV (60601-2-1). International Electrotechnical Commission.

IEC. 2009e. Medical Electrical Equipment—Requirements for the Safety of Radiotherapy Treatment Planning Systems (62083). International Electrotechnical Commission.

IEC/TR. 1997. Guidelines for Radiotherapy Treatment Rooms Design (61859). International Electrotechnical Commission.

IEC/TR. 1998. Medical Electrical Equipment—Digital Imaging and Communications in Medicine (DICOM)—Radiotherapy Objects (61852). International Electrotechnical Commission.

IEC/TR. 2001a. Nuclear Medicine Instrumentation—Routine Tests—Part 1: Radiation Counting Systems (61948-1). International Electrotechnical Commission.

IEC/TR. 2001b. Nuclear Medicine Instrumentation—Routine Tests—Part 2: Scintillation Cameras and Single Photon Emission Computed Tomography Imaging (61948-2). International Electrotechnical Commission.

IEC/TR. 2005. Nuclear Medicine Instrumentation—Routine Tests—Part 3: Positron Emission Tomographs (61948-3). International Electrotechnical Commission.

IEC/TR. 2006. Nuclear Medicine Instrumentation—Routine Tests—Part 4: Radionuclide Calibrators (61948-4). International Electrotechnical Commission.

IEC/TR. 2008. Medical Electrical Equipment—Medical Electron Accelerators—Guidelines for Functional Performance Characteristics (60977). International Electrotechnical Commission.

IEC/TR, (2002–03) Ed. 1.0 2002. Medical Electrical Equipment—Guidelines for Implementation of DICOM in Radiotherapy (62266). International Electrotechnical Commission.

IEC/TS. 1993. Radiotherapy Simulators—Guidelines for Functional Performance Characteristics (61170). International Electrotechnical Commission.

IPEM. 2005a. Report 89: Commissioning and Routine Testing of Mammographic X-ray Systems. Institute of Physics and Engineering in Medicine.

IPEM. 2005b. Report 91: Recommended Standards for the Routine Performance Testing of Diagnostic X-ray Imaging Systems. Institute of Physics and Engineering in Medicine.

IPEM. 2006. Report 93: Guidance for Commissioning and QA of a Networked Radiotherapy Department. Institute of Physics and Engineering in Medicine.

IPEM. 2007. Report 94: Acceptance Testing and Commissioning of Linear Accelerators Institute of Physics and Engineering in Medicine.

IPEM. 2008. Report 96: Guidance for the Clinical Implementation of Intensity Modulated Radiation Therapy. Institute of Physics and Engineering in Medicine.

IPEM. 2009. Report 97: Electrical Safety Testing: A Workbench Guide. Institute of Physics and Engineering in Medicine.

Leer, J. W. H., A. L. McKenzie, P. Scalliet, and D. I. Thwaites. 1998. Practical guidelines for the implementation of a quality system in radiotherapy ESTRO physics for clinical radiotherapy. In *ESTRO Quality Assurance Committee sponsored by "Europe Against Cancer," Physics for Clinical Radiotherapy, Booklet No. 4.*

VI

Quality Control: Equipment

66

Dosimetry Equipment and Phantoms

Andrea Molineu
Texas M. D. Anderson Cancer Center

Introduction

Many types of dosimetry equipment and phantoms exist for use in quality assurance activities in radiation therapy physics. Information about which equipment should be chosen for which measurements can be found in the report of the American Association of Physicists in Medicine (AAPM) Task Group 106 (Das et al., 2008). The equipment that is used for quality assurance measurements also needs to be tested to ensure it is functioning properly. Inaccurate results could cause much undue frustration if they indicate a problem exists in treatment delivery when the problem is really with the dosimeter. Conversely, malfunctioning equipment could incorrectly assure the user that treatments are adequate when they are actually outside the action criteria.

Dosimetry Equipment

Ion Chambers

Ion chambers consist of an air cavity with an internal electrode. The charge in the cavity created by ionizing radiation is collected, read with an electrometer, and can be converted to dose through the application of energy- and chamber-dependent factors. Protocols describing the conversion to dose have been published by the AAPM and the International Atomic Energy Agency (Almond et al. 1999; Andreo et al. 2000). Relative charge readings are used to describe beam characteristics such as percent depth dose, field size dependence, and wedge factors.

Different sizes of ionization chambers are used for different purposes. Small field measurements, for example, require smaller volume chambers than do output measurements in a 10×10 cm field (Das et al. 2008).

Ion chambers are used for absolute calibration of linear accelerators. They are also commonly used to take many commissioning and quality assurance measurements used for treatment planning systems. Because so many basic machine parameters are measured using ionization chambers, it is important to know that the chamber being used is working properly. Typically, farmer-type chambers are used for absolute calibration measurements for radiation beams to which TG-51 applies. All chambers used for absolute calibration should have a chamber factor that is traceable to a national standard, for example, National Institute of Standards and Technology in the United States (Kutcher et al. 1994). In the United States, this is typically achieved by calibration by an ADCL at least every 2 years. The calibration factor itself should not have significant changes over the life of the chamber.

Chambers sometimes suffer damage during shipping. To verify that the chamber was not damaged during shipping, it can be irradiated with a stable beam such as ^{60}Co before and after shipping. After decay and differences in temperature and pressure are accounted for, the responses of the chamber can be compared to ensure that the chamber's behavior after shipment has not changed. If ^{60}Co is not available, the response of the chamber in a beam from a linear accelerator can be compared to the response of another chamber in that same beam. This can also be done before and after shipment. When this is the case,

411

the second chamber should not be shipped for calibration at the same time as the first.

Chambers may also suffer damage in the course of regular use. If an institution has multiple chambers, periodic comparisons can be made to ensure that the response of the chambers used for calibration has not changed over time. The leakage can be measured every time the chamber is used. The leakage detected should be orders of magnitude smaller than charges expected for the measurement session. Chambers can be mailed to the manufacturer for repair when a problem is detected.

Thermoluminescent Dosimeters

After an exposure to radiation, thermoluminescent dosimeters (TLDs) have the ability to release photons when they are heated. The amount of light released is proportional to the amount of radiation received. A picture of a TLD reader can be seen in Figure 66.1. TLDs are used in a variety of dose measurements including personal monitors, patient skin dose, verification of machine output, and brachytherapy source characterization. They are commonly seen in chip, rod, and powder form. In all cases, the response per weight is proportional to the amount of radiation received (Cameron, Suntharalingam, and Kenney 1968).

Rods and chips can fairly easily be annealed and reused in future measurements. Because each separate rod or chip may be slightly different, a chip factor can be applied to each one by giving a known dose of irradiation to all TLDs and then reading them and finding the difference in readings. Powder, which is not as easily used multiple times, can be weighed at the time of reading so that the response per weight is used to calculate dose.

Other things must also be taken into account in a TLD program such as response of the reader over time, the sensitivity of the reader to light output from the TLD, the linearity of the dose response, and the correct annealing procedures. If results are questionable, measurements could be taken with more precise dosimetry equipment such as ion chambers. TLD programs in radiation departments must take all of the above into consideration (Cameron, Suntharalingam, and Kenney 1968).

There are some TLD mailing programs available to institutions. The Radiological Physics Center (RPC) sends TLD to all institutions that participate in National Cancer Institute–sponsored clinical trials on an annual basis. These TLDs are used to measure the output of the photon and electron beams in a standard setup. Radiation Dosimetry Services offers TLD for output measurements as a for-fee service at intervals chosen by the institution. Personnel monitoring is also commonly done through an external company.

Film

Radiographic and radiochromic films are often used for quality assurance measurements in radiation therapy. Both types of film consist of a base with an emulsion on one or both sides. The emulsion darkens when it is irradiated. The optical density of the film is dependent on the dose received. In many cases, film is used for relative measurements such as percent depth dose, output factors, and complex dose distributions for intensity-modulated radiotherapy (IMRT) measurements. Film is sometimes used for absolute dose measurements. If using it for absolute dose, much care must be taken to ensure adequate dose conversion (Niroomand-Rad et al. 1998; Pai et al. 2007).

FIGURE 66.1 A Harshaw TLD reader.

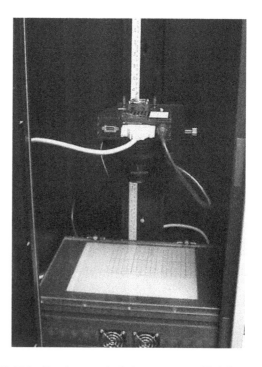

FIGURE 66.2 Densitometer including camera and lightbox.

The optical density of the film is read by a film scanner or densitometer (see Figure 66.2). A characteristic curve that shows the optical density response with dose should be made for each film batch/reader system. Even when film is used for relative measurements, tests should be done to determine the response per dose for the film and film reader combination being used. If there is no linear response, a nonlinearity correction should be applied before analyzing relative dose. When reading a film, care should be taken in the way the film is oriented in the reader. The use of masking is sometimes appropriate (Niroomand-Rad et al. 1998; Pai et al. 2007).

The reader system should be commissioned and checked on a periodic basis to ensure that it is functioning properly. One method of doing a periodic check is to read a standard film and to compare the reading with previous results for that same piece of film. If the accuracy of measurements made with film is questionable, the user should check the film setup during irradiation and reading, factors used in conversion from optical density to dose, film batch, and perhaps rescan the film. The user should also ensure that film is the appropriate dosimetry tool for the job.

Output Check Devices

A fourth common quality assurance measurement tool is the output check device. The devices can be as simple as one small ion chamber or diode embedded in a table that can support buildup material or an array of such devices. The devices are often used to check the daily output of the machine and for patient-specific IMRT measurements. Information on beam flatness and symmetry can also be gleaned from these devices. The performance of these devices should be thoroughly investigated before implementing their use clinically. A publication on such testing of a newly introduced device can be combined with spot testing of the features being used in your clinic. When these devices are being used for absolute measurements, they need to be calibrated and calibration should be checked on a periodic basis. The readings from these devices can be checked at the time of the monthly output check of the machine and baselines can be adjusted if necessary. They should also be checked on an annual basis when a more robust system is used to calibrate the beams. A comparison to the monthly or annual system can be done at any time that the accuracy of the output check device is questionable.

Phantoms

All of the previously described dosimetry equipment can be used in conjunction with radiological phantoms. Two categories of phantoms are commonly used in radiation therapy department. Geometrical phantoms are often cubic or cylindrical and are typically made of water or plastic. Anthropomorphic phantoms can be made to emulate any part of the body and are made of a combination of substances chosen to be radiologically similar to the different body parts represented.

Geometric Phantoms

Geometric phantoms are used in almost all aspects of quality assurance measurement. They are used for measurements as basic as calibration or PDD measurements and also for verification of complex IMRT treatment plans. Water phantoms used for calibration are big enough to provide adequate scatter in all directions for a measurement done at a depth of 10 cm in a field that is 10 × 10 cm. They are generally used in conjunction with ion chambers. Much larger water phantoms are often used to take other kinds of machine characterizing measurements required for treatment planning system commissioning. These can include percent depth dose as well as beam profile measurements for different energies and field sizes. Water phantoms have also been used in conjunction with TLD to make measurements of brachytherapy (Tailor, Tolani, and Ibbott 2008).

Other geometric phantoms are made of water-equivalent plastics or other plastics. These phantoms can be milled to accommodate specific types of ion chambers and TLD. They are often formed by a series of slabs between which film can be placed. The slabs themselves can be used as buildup for output check devices. Typically, the radiological and physical depths are the same for water equivalent plastics, but this relationship does not hold for other plastics. When using plastics that are not water equivalent, correct scaling factors should be applied if the depth of measurement is important (Task Group 21 1983). Current calibration protocols require that calibration measurements for megavoltage radiation beams be taken in water phantoms (Almond et al. 1999; Andreo et al. 2000). However, using plastic phantoms for monthly output checks is more convenient than setting up a bulky water phantom. When plastic phantoms are used for monthly constancy checks, a conversion factor should be assigned to the plastic phantom at the time of the annual calibration in water. The same phantom configuration that was used when the factor was assigned should be used monthly because differences between batches of plastic can affect measurement results (Tello, Tailor, and Hanson 1995).

Anthropomorphic Phantoms

Anthropomorphic phantoms are designed to closely resemble humans radiologically and physically. They can house an array of the dosimetry equipment discussed above. These phantoms are used to verify complex treatment deliveries rather than specific machine parameters. Often they are used to verify new technology used in treatments or new dose calculation algorithms. They also can be used in doing quality assurance measurements of the entire treatment process from initial scanning to planning to treatment delivery.

Anthropomorphic phantoms can be designed and built by an institution. There are also companies that sell anthropomorphic phantoms. Purchasing or building one allows an institution to use the phantom multiple times and on their own time table. Many institutions however do not have either the capabilities to build a phantom or the funds to purchase one for permanent

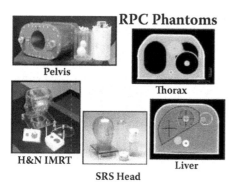

RPC Phantoms

Pelvis

Thorax

H&N IMRT

SRS Head

Liver

FIGURE 66.3 The RPC's family of anthropomorphic phantoms.

use. Or they may not have the ability to sustain adequate dosimetry equipment programs. An alternative to purchasing an anthropomorphic phantom is to use a service, for example, RDS, that provides the phantom and analysis for a fee. The phantoms are mailed to the institution loaded with dosimeters and are returned to RDS for analysis.

The RPC has a phantom mailing program that is used to credential institutions that participate in National Cancer Institute–sponsored clinical trials for advanced technology protocols. The RPC's phantom family consists of stereotactic head phantoms, IMRT head and neck phantoms, IMRT and SBRT thorax phantoms with lung or spine targets, SBRT liver phantoms, and IMRT prostate phantoms (Balter, Stovall, and Hanson 1999; Followill et al. 2007; Molineu, Hernandez et al. 2005) (see Figure 66.3). All of the phantoms house target volumes and organs at risk, and they use TLD and radiochromic film as dosimeters. An example of the target and organ at risk for the head and neck phantom is shown in Figure 66.4. They are shipped with dosimeters to the institution interested in participating in clinical trials.

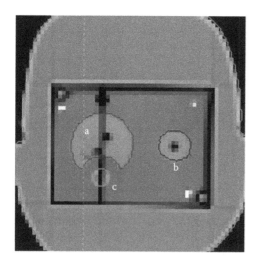

FIGURE 66.4 A CT image of the RPC's head and neck phantom with the targets labeled: (a) the primary PTV; (b) the secondary PTV; (c) the organ at risk. TLDs are located inside both targets and the OAR. Radiochromic film is placed in the axial plane shown and in the sagittal plane that goes through the center of the primary PTV.

TABLE 66.1 Pass/Fail Results through 2008 for a Subset of the RPC's Anthropomorphic Phantoms

	Head and Neck	Prostate	Thorax
Number of irradiations	637	127	81
Number of passes	491	104	56
Number of failures	146	23	25

The institution personnel fill the phantom shell with water and image the phantom, produce a treatment plan for the phantom, set up the phantom in the treatment location, and deliver the radiation according to the plan. They then ship the phantom back to the RPC where the film and TLD analysis is done.

These phantoms have been shipped numerous times. The pass/fail results through 2008 for many of the phantoms can be seen in Table 66.1. The results for the head and neck phantom show that institutions do not deliver what they plan to within 7% and 4 mm approximately 25% of the time. Approximately 15% of the failures are attributed to setup errors at the time of irradiation. These could be caused by the lack of knowledge of the standard setup and marking procedures of the person (perhaps the physicist) setting up the phantom (Molineu, Followill et al. 2005; Ibbott, Molineu, and Followill 2006; Ibbott et al. 2008). Presumably these setup errors would not occur as often on real patients as they do with the phantoms. The remaining errors (85% of the total errors) are treatment delivery errors that could very well occur on patients. They include errors in TPS modeling (Cadman et al. 2002), inaccurate data (PDD, small field FSD, jaw transmission, etc.) entered into the TPS, and machine calibration errors.

Results from the lung phantom show that the convolution superposition and analytic anisotropic algorithms more accurately predict the dose to the lung and target volumes within the lung than do the Clarkson or pencil beam algorithms. Most of the lung phantom failures are attributed to the inaccuracies in the algorithm used. It is interesting to note, however, that not all of the less accurate algorithms failed the phantom criteria irradiations (Alvarez et al. 2006, 2008; Davidson et al. 2007, 2008). Results from the prostate phantom appear to be dominated by setup errors. A possible explanation for this is that the prostate phantom does not stress the IMRT system to the extent that the head and neck phantom does. IMRT dose deliveries with larger segments are easier to model and deliver accurately than the smaller segments more often seen in the head and neck plans.

Summary

The frequent presence of errors demonstrates the importance of adequate quality assurance processes for all of the treatment methods. Dose delivery errors of greater than 5% may affect clinical outcomes. Appropriate knowledge of the working of dosimetry equipment is necessary to be able to interpret the results of measurements accurately so that dose delivery errors can be found and corrected. Furthermore, it is important to choose the right measurement tool for the job and to ensure that the tool is working properly.

References

Almond, P. R., P. J. Biggs, B. M. Coursey, W. F. Hanson, M. S. Huq, R. Nath, and D. W. Rogers. 1999. AAPM's TG-51 protocol for clinical reference dosimetry of high-energy photon and electron beams. *Med. Phys.* 26 (9): 1847–1870.

Alvarez, P., A. Molineu, N. Hernandez, F. Hall, D. Followill, and G. Ibbott. 2008. TU-C-AUD B-03: A comparison of heterogeneity correction algorithms. *Med. Phys.* 35: 2888.

Alvarez, P., A. Molineu, N. Hernandez, D. Followill, and G. Ibbott. 2006. TU-E-224A-01: Evaluation of heterogeneity corrections algorithms through the irradiation of a lung phantom. *Med. Phys.* 33: 2214.

Andreo, P., D. T. Burns, K. Hohlfeld, M. S. Huq, T. Kanai, F. Laitano, V. G. Smythe, and S. Vynckier. 2000. Absorbed dose determination in external beam radiotherapy. In *IAEA Technical Report Series No. 398.* Vienna: International Atomic Energy Agency.

Balter, P., M. Stovall, and W. F. Hanson. 1999. PO-T-196: An anthropomorphic head phantom for remote monitoring of stereotactic radiosurgery at multiple institutions. *Med. Phys.* 26: 1164.

Cadman, P., R. Bassalow, N. P. Sidhu, G. Ibbott, and A. Nelson. 2002. Dosimetric considerations for validation of a sequential IMRT process with a commercial treatment planning system. *Phys. Med. Biol.* 47 (16): 3001–3010.

Cameron, J. R., N. Suntharalingam, and G. N. Kenney. 1968. *Thermoluminescent Dosimetry.* Madison, WI: University of Wisconsin Press.

Das, I. J., C. W. Cheng, R. J. Watts, A. Ahnesjo, J. Gibbons, X. A. Li, J. Lowenstein, R. K. Mitra, W. E. Simon, and T. C. Zhu. 2008. Accelerator beam data commissioning equipment and procedures: Report of the TG-106 of the Therapy Physics Committee of the AAPM. *Med. Phys.* 35 (9): 4186–4215.

Davidson, S. E., G. S. Ibbott, K. L. Prado, L. Dong, Z. Liao, and D. S. Followill. 2007. Accuracy of two heterogeneity dose calculation algorithms for IMRT in treatment plans designed using an anthropomorphic thorax phantom. *Med. Phys.* 34 (5): 1850–1857.

Davidson, S. E., R. A. Popple, G. S. Ibbott, and D. S. Followill. 2008. Technical note: Heterogeneity dose calculation accuracy in IMRT: Study of five commercial treatment planning systems using an anthropomorphic thorax phantom. *Med. Phys.* 35 (12): 5434–5439.

Followill, D. S., D. R. Evans, C. Cherry, A. Molineu, G. Fisher, W. F. Hanson, and G. S. Ibbott. 2007. Design, development, and implementation of the radiological physics center's pelvis and thorax anthropomorphic quality assurance phantoms. *Med. Phys.* 34 (6): 2070–2076.

Ibbott, G. S., D. S. Followill, A. Molineu, J. R. Lowenstein, P. E. Alvarez, and J. E. Roll. 2008. Challenges in credentialing institutions and participants in advanced technology multi-institutional clinical trials. *Int. J. Radiat. Oncol. Biol. Phys.* 71 (1): S71–S75.

Ibbott, G. S., A. Molineu, and D. S. Followill. 2006. Independent evaluations of IMRT through the use of an anthropomorphic phantom. *Technol. Cancer Res. Treat.* 5 (5): 481–487.

Kutcher, G. J., L. Coia, M. Gillin, W. F. Hanson, S. Leibel, R. J. Morton, J. R. Palta et al. 1994. Comprehensive QA for radiation oncology: Report of AAPM Radiation Therapy Committee Task Group 40. *Med. Phys.* 21 (4): 581–618.

Molineu, A., N. Hernandez, P. Alvarez, D. Followill, and G. Ibbott. 2005. SU-FF-T-148: IMRT head and neck phantom irradiations: Correlation of results with institution size. *Med. Phys.* 32: 1983.

Molineu, A., D. S. Followill, P. A. Balter, W. F. Hanson, M. T. Gillin, M. S. Huq, A. Eisbruch, and G. S. Ibbott. 2005. Design and implementation of an anthropomorphic quality assurance phantom for intensity-modulated radiation therapy for the Radiation Therapy Oncology Group. *Int. J. Radiat. Oncol. Biol. Phys.* 63 (2): 577–583.

Niroomand-Rad, A., C. R. Blackwell, B. M. Coursey, K. P. Gall, J. M. Galvin, W. L. McLaughlin, A. S. Meigooni, R. Nath, J. E. Rodgers, and C. G. Soares. 1998. Radiochromic film dosimetry: Recommendations of AAPM Radiation Therapy Committee Task Group 55. American Association of Physicists in Medicine. *Med. Phys.* 25 (11): 2093–2115.

Pai, S., I. J. Das, J. F. Dempsey, K. L. Lam, T. J. Losasso, A. J. Olch, J. R. Palta, L. E. Reinstein, D. Ritt, and E. E. Wilcox. 2007. TG-69: Radiographic film for megavoltage beam dosimetry. *Med. Phys.* 34 (6): 2228–2258.

Tailor, R., N. Tolani, and G. S. Ibbott. 2008. Thermoluminescence dosimetry measurements of brachytherapy sources in liquid water. *Med. Phys.* 35 (9): 4063–4069.

Task Group 21. 1983. A protocol for the determination of absorbed dose from high-energy photon and electron beams. *Med. Phys.* 10 (6): 741–771.

Tello, V. M., R. C. Tailor, and W. F. Hanson. 1995. How water equivalent are water-equivalent solid materials for output calibration of photon and electron beams? *Med. Phys.* 22 (7): 1177–1189.

67

Conventional Simulators

Cheng B. Saw
Penn State Hershey
Cancer Institute

Introduction

Conventional simulators, more commonly called treatment simulators, have been developed to facilitate the radiographic visualization of treatment fields (Greene, Nelson, and Gibb 1964). Radiographs taken using the therapeutic source of treatment machines are of poor quality because megavoltage photon beams result in mostly Compton scattering. A conventional simulator will have identical machine parameters to those of a megavoltage treatment unit, including geometrical, mechanical, and optical properties, except it is fitted with a diagnostic x-ray tube as shown in Figure 67.1. The exposure from the diagnostic x-ray tube provides high image contrast so that anatomical structures can be clearly defined for the assessment of the appropriateness of field placement and field shaping to minimize the irradiation of the surrounding normal tissues. Besides having superior radiographic quality, conventional simulators are used for time-consuming patient position verification procedures and for correcting unforeseen setup problems, hence freeing the treatment machines for other clinical use (e.g., treating patients), which results in higher patient throughput.

The use of conventional simulators has evolved as the practice of radiation oncology has changed over the years. As developed, it is intended for the verification of treatment fields. Its use has been expanded into treatment planning because it emulates the treatment machine characteristics and it generates high-quality images at a lower cost than on a modern treatment unit. Patient simulation is typically performed in a conventional simulator room involving patient immobilization, patient setup, localizing the treatment volume, setting up treatment fields, recording patient positioning, and designing shielding blocks. This implementation has also increased the capabilities of conventional simulators to include fluoroscopy. The simulator room also has added features including laser lights, contour taker, and measuring tools to perform patient simulation. The availability of fluoroscopy, superior radiographic quality, and preservation of geometrical properties has allowed the conventional simulator

to be used for brachytherapy procedures including applicator placement, dosimetry, and applicator position verification. Often conventional simulators are found in brachytherapy suites for these purposes. The introduction of three-dimensional treatment planning systems has transformed the practice of radiation oncology into an image-based paradigm. Instead of patient simulation, patient data acquisition is performed in a CT scanning room to generate an image data set for treatment planning. The CT scanner used for the purpose of patient setup and patient image data set acquisition for treatment planning is called a CT simulator. This shift in paradigm has led to a decline in the use of conventional simulators. However, the limited bore size of CT scanners has caused difficulty in acquiring patient image data sets especially for breast cancer patients on inclined breast boards and large patients. This has lead manufacturers to include CT capabilities into modern simulators now called simulator CT.

Description of a Conventional Simulator

A conventional simulator will have (1) treatment machine mechanical, geometrical, and optical characteristics; (2) diagnostic radiograph capability; and (3) fluoroscopic capability. The conventional simulator will have a defined beam axis and the ability to rotate the radiation source about an isocenter just like a treatment machine. Other critical simulator characteristics are: (1) mechanical alignment tools such as laser lights and optical distance indicators; (2) beam defining tools such as field defining wires, and gantry and collimator rotations; and (3) patient positioning tools such as a digital treatment couch with rotational capabilities. The radiographic capability includes a kilovoltage x-ray source coupled with a good image receptor for taking radiographs. The fluoroscopic capability involves the use of the kilovoltage x-ray source with an imaging device that may be an image intensifier or a solid-state detector. The positioning of patients at extended distances may not be feasible for conventional simulators because of the limited source to detector distance.

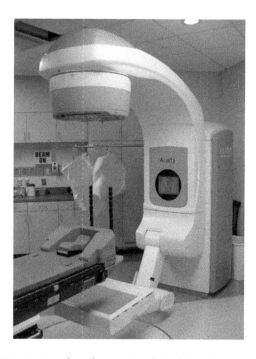

FIGURE 67.1 An adapted conventional simulator (Varian Acuity).

Quality Control

Quality control for conventional simulators can be divided into three sections: radiation safety checks, mechanical integrity checks, and exposure or output checks. Because conventional simulators are radiation-producing machines, they are regulated and hence must comply with state and federal

TABLE 67.1 Conventional Simulator Quality Control—Daily Tests

Test	Tolerance (mm)
Localizing lasers	2
Optical distance indicator	2

TABLE 67.2 Conventional Simulator Quality Control—Monthly Tests

Test	Tolerance
Field size indicator	2 mm
Gantry and collimator angle indicators	1°
Cross-hair centering	2 mm diameter
Focal spot axis indicator	2 mm
Fluoroscopic image quality	Baseline
Emergency/collision avoidance	Functional
Light/radiation field coincidence	2 mm or 1%
Film processor sensitometry	Baseline

TABLE 67.3 Conventional Simulator Quality Control—Annual Tests

A. Mechanical Checks	
Collimator rotation isocenter	2 mm diameter
Gantry rotation isocenter	2 mm diameter
Couch rotation isocenter	2 mm diameter
Coincidence of collimator, gantry, and couch axes with isocenter	2 mm diameter
Table top sag	2 mm
Vertical travel of couch	2 mm
B. Radiographic Checks	
Exposure rate	Baseline
Table top exposure with fluoroscopy	Baseline
kVp and mAs calibration	Baseline
High and low contrast resolution	Baseline

safety requirements. For example, the x-ray use indicator at the door and deadman switches must be checked on a daily basis. Functionality of emergency switches must be checked at a frequency in compliance with the regulatory, usually on a quarterly basis. Because conventional simulators are designed to reproduce the geometric conditions of the treatment machines, mechanical integrity checks are subject to the same performance frequency as imposed on treatment machines as listed in Tables 67.1 to 67.3 (Kutcher et al. 1994). The mechanical checks intended to ensure the integrity of the equipment such as the rotational capabilities should be within 1 degree in both the mechanical and digital readouts. Other mechanical checks are intended to ensure the alignment of the various components to the rotational axes and isocenter of the equipment. In addition, the simulators should be checked for image quality according to established guidelines for diagnostic radiographic units. The frequency of performance of quality control checks for diagnostic radiological equipment can be found in the AAPM Report No. 74 (AAPM Report 74 2002).

References

AAPM Report 74. 2002. *Quality Control in Diagnostic Radiology*. New York, NY: American Institute of Physics, Inc.

Greene, D., K. A. Nelson, and R. Gibb. 1964. The use of a linear accelerator "simulator" in radiotherapy. *Br. J. Radiol.* 37: 394–397.

Kutcher, G. J., L. Coia, M. Gillin, W. F. Hanson, S. Leibel, R. J. Morton, J. R. Palta et al. 1994. Comprehensive QA for radiation oncology: Report of AAPM Radiation Therapy Committee Task Group 40. *Med. Phys.* 21 (4): 581–618.

68

Computed Tomography, Positron Emission Tomography, and Magnetic Resonance Imaging

Sasa Mutic
*Washington University
School of Medicine*

Osama R. Mawlawi
MD Anderson Cancer Center

Jian-Ming Zhu
University of North Carolina

Introduction

Multimodality imaging is one of the fundamental ingredients of modern radiotherapy (RT). Although kilovoltage x-ray computed tomography (CT) imaging continues to be the primary imaging modality used in RT, the use of magnetic resonance imaging (MRI) and nuclear medicine imaging with single photon emission computed tomography (SPECT) and especially positron emission tomography (PET) is common in treatment planning for many patients. Optimal design of RT treatments often requires anatomical and biological information about target, volumes, and no single imaging modality provides all of the information that can be gathered and used. Each of the above-listed imaging modalities offers unique information, and complementary use of two or more modalities is needed. The use of multimodality imaging in RT has been extensively described in the literature (Kessler et al. 1991; Ling et al. 2000; Mutic 2006). Multimodality imaging is used for improved detection, diagnosis, staging, therapy selection, target definition, prognosis, response evaluation, and follow-up.

Computed tomography imaging has high spatial resolution and fidelity and is typically used to define a patient's general anatomy. When properly calibrated and free of image artifacts, CT images can also provide electron-density information for heterogeneity-based dose calculations. CT imaging has limited soft tissue contrast. MRI provides excellent soft tissue contrast, which allows better differentiation between normal tissues and many tumors, and MRI can also provide biological information. In addition, MRI is not limited to axial plane image acquisition. Disadvantages of MRI include susceptibility to spatial distortion and image intensity values that are not easily translated to electron density values. Spatial distortion may alter positions of target volumes with respect to other anatomical landmarks, patient skin, or external fiducial markers, rendering an MRI study unsuitable for treatment planning.

PET and SPECT also provide unique information about patient physiology rather than about anatomy. These modalities have been used for improved staging, tumor delineation, and response evaluation. Limitations of PET and SPECT imaging include poor spatial resolution, which render these imaging modalities inadequate for delineation of external patient contours and other normal structures.

To provide maximum benefit in the RT process, imaging studies have to be of the highest quality, artifact-free, and spatially and quantitatively correct. Quality assurance (QA) requirements for imaging equipment used in RT planning are similar to those found in diagnostic radiology. The main difference is the emphasis on QA of scanner geometry, spatial accuracy, and quantitative image performance. This chapter provides an overview of the major components of QA programs for CT, MRI, and PET scanners, which are used in the RT process. This chapter is not intended to be a comprehensive reference addressing the details of individual QA procedures, but rather it provides a summary of QA standards for these imaging modalities in RT in one single location. All three sections provide references for more comprehensive descriptions of individual tests.

Quality Assurance of CT in RT

Quality assurance of CT simulators includes testing of safety components, mechanical/geometric parameters, image quality, data transfer, virtual simulation software, and motion-correlated CT acquisition. Important references for QA of CT scanners in RT are the American Association of Physicists in Medicine (AAPM 1993), AAPM Task Group 53 (Fraass et al. 1998), and Task Group 66 (Mutic et al. 2003) reports. CT technology is continually evolving with numerous improvements in detector configurations and design, x-ray tube technology, motion-correlated imaging, scanner geometry, and reconstruction algorithms. Although the above-mentioned reports have been published before the development of many features of modern CT scanners, the concepts presented in those reports should be adequate for QA of most modern scanner features.

A typical CT simulator QA program consists of:

1. Evaluation of Radiation and patient safety
2. Evaluation of CT dosimetry
3. Evaluation of electromechanical components
4. Evaluation of image quality

Radiation and Patient Safety

Radiation safety concerns for CT simulators are addressed during commissioning through site surveys (Mutic et al. 2003). Safety interlocks and emergency off buttons are usually evaluated annually. The AAPM TG-66 report recommended that a CT simulator not be interlocked with the room door. If the opening of the door interrupts a CT scan, the patient procedure may be compromised and may have to be rescheduled while the exposure to the person who accidentally opened the door would likely be minimal. If the door has to be interlocked due to regulations, then the access to the door should be controlled during scanning.

CT Dosimetry

Computed tomography dose index is used to evaluate doses received by patients from CT scanning. Verification of the CT dose index involves the estimation of body and head doses using custom-designed phantoms and ionization chambers (Mutic et al. 2003). Radiation doses from a CT scan depend on the scanner model and can vary significantly. CT dosimetry is verified initially by comparing measured CT dose index values with manufacturer specifications. There are several CT dose index definitions (Mutic et al. 2003), and the user manual should be consulted to identify which definition was used by the manufacturer. Periodic QA verifications can include comparisons with manufacturer specifications or with baseline data acquired during commissioning.

Electromechanical Components

Performance evaluation of electromechanical components includes x-ray generator tests and verification of the geometry of laser systems and of the CT gantry and table. AAPM reports (AAPM 1993; Mutic et al. 2003) provide tolerance recommendations for electromechanical tests and detailed procedures for their performance evaluation. X-ray generator tests include evaluation of peak potential (kVp), half-value layer, mAs linearity, mAs reproducibility, and timer accuracy. Generator tests are generally performed during commissioning and after replacement of a major component of the x-ray generator system. Several manufacturers offer testing equipment that allows noninvasive evaluation of x-ray generator performance. This equipment is relatively inexpensive and generally affordable by radiation oncology facilities that own CT scanners.

Alignment

CT scanner gantry, couch, and the laser marking system are interdependent with respect to their geometric alignment. Alignment verification of these components requires specialized test tools. The AAPM TG-66 (Mutic et al. 2003) report described an alignment tool that can be used for alignment verification of a CT simulator (Figure 68.1). A similar tool is now commercially available. The verification process for alignment of these components begins with the gantry. The alignment and orientation of couch and laser systems is referenced to the imaging plane and not to the world, meaning couch and laser installation and motion has to be parallel/orthogonal to the imaging plane and not necessarily to the world. For example, the tabletop does not have to be leveled with a level as long as it is orthogonal to the CT imaging plane. Therefore, most alignment verification tests must be performed radiographically.

The CT simulator patient support assembly consists of two parts, the couch base and the flat tabletop. The second part, the flat tabletop, is typically added as an afterthought to a couch base that is used for diagnostic CT scanners. As such, flat tabletops often do not fit accurately and reproducibly on the couch base and are a source of most common misalignments. The CT simulator QA program should include rigorous verification of tabletop alignment. The following are performance requirements for the CT scanner couch and tabletop specified by the TG-66 (Mutic et al. 2003) report:

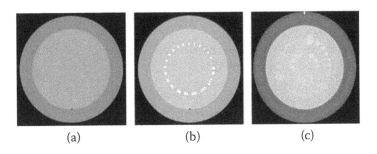

FIGURE 68.1 Image performance evaluation phantom: (a) noise–image uniformity section; (b) line-pair section; (c) low-contrast resolution section.

1. Flat tabletop should be level and orthogonal with respect to the imaging plane.
2. Table vertical and longitudinal motion according to digital indicators should be accurate and reproducible.
3. Table indexing and position under scanner control should be accurate.
4. Flat tabletop should not contain any objectionable artifact-producing objects (screws, etc.).
5. The scan volume and scan location, as prescribed from the scout (pilot, topograph) image, should be accurate within 1 mm.

The CT scanner patient marking/positioning lasers consist of three separate components: gantry lasers, wall-mounted lasers (which may be mobile in the vertical direction), and an overhead mobile sagittal laser (mobile in the lateral direction or lateral and longitudinal directions). As already noted, lasers are aligned and referenced to the scanner imaging plane. All lasers should be either parallel or orthogonal to the imaging plane. The accuracy of the lasers directly affects the ability to localize treatment volumes relative to patient skin marks and the reproducibly of patient positioning from the CT scanner to the treatment machine. The TG-66 (Mutic et al. 2003) report outlines the following performance requirements for CT scanner lasers:

1. Gantry lasers should accurately identify the scan plane within the gantry opening.
2. Gantry lasers should be parallel and orthogonal to the scan plane and should intersect in the center of the scan plane.
3. Vertical side-wall lasers should be accurately spaced from the imaging plane.
4. Wall lasers should be parallel and orthogonal to the scan plane and should intersect at a point that is coincident with the center of the scan plane.
5. The overhead (sagittal) laser should be orthogonal to the imaging plane.
6. The overhead (sagittal) laser movement should be accurate, linear, and reproducible.

The TG-66 (Mutic et al. 2003) report provides detailed instructions for alignment QA of the gantry, couch, and laser system.

Evaluation of Image Quality

CT image quality directly affects delineation accuracy of the target and normal structures for RT planning. Because of this, CT images used in RT should be of the highest achievable quality. Optimal image performance typically means that the scanner meets or exceeds manufacturer image performance recommendations at the time of commissioning and that this performance can be maintained throughout the life of the scanner. Common minimum standards for image performance indicators for all scanners do not exist and manufacturers' specifications are used as performance standards for individual scanners. The AAPM Report No. 39 (AAPM 1993) addresses in detail image performance tests required for commissioning of CT scanners. Although this document was published almost 20 years ago, its recommendations are still applicable. Outlined below are image quality tests that were discussed in the AAPM (1993) and TG-66 (Mutic et al. 2003) reports.

Random Uncertainty in Pixel Value—Noise

One of the fundamental image quality and the overall scanner performance indicators is image noise. A CT scan of a homogenous phantom results in an image that does not have uniform pixel values (CT numbers) throughout the uniform part of the phantom. This variation in pixel intensities has random and systematic components. The random component is noise. The standard deviation of pixel values in a region of interest (ROI) within a uniform phantom is an indication of image noise. An example of a uniform phantom is shown in Figure 68.1a. Image noise can be assessed as outlined in AAPM Report No. 39 (AAPM 1993) or using the manufacturer's image performance phantom and software. Because of its ability to evaluate the stability of the overall scanner performance, noise measurements should be part of the CT scanner periodic QA program.

Systematic Uncertainty—Field Uniformity

Report No. 39 (III.B.2) provides a detailed discussion of various causes of field nonuniformity and measurement procedures for evaluation of field uniformity. CT images should be free of systematic artifacts, and an image of a uniform phantom should have uniform appearance without streaking and artifacts. Uniformity phantoms are liquid-filled or solid plastic cylinders (Figure 68.1a), which are typically centered in the scan field for image uniformity testing. The difference in the mean HU values for ROIs sampled throughout a uniform phantom determines image uniformity. Typical sampling points for a cylindrical phantom are in the center, and at the 12, 3, 6, and 9 o'clock positions. Numerous uniformity phantoms are commercially available. Most CT scanner manufacturers provide a uniformity phantom or a test section in a multipurpose phantom.

Uniformity testing for CT scanners that have large scan field of view (FOV) or extrapolated FOV can be performed by scanning the uniformity phantom at different locations in the scan FOV. This allows evaluation of field uniformity at locations other than in the center of the scan FOV and can actually mimic several situations encountered in RT scanning (breast, extremity, etc.).

Quantitative CT

Several commercial electron density phantoms are available to evaluate quantitative CT accuracy and consistency. The AAPM TG-66 report (Mutic et al. 2003) provides a detailed procedure for assessment of quantitative CT accuracy. This test is especially important for facilities that rely on heterogeneity-corrected dose calculations. This evaluation parameter depends on scanner performance and calibration and should be tested relatively frequently, as recommended by the TG-66 (Mutic et al. 2003).

Spatial Integrity

One of the advantages of CT imaging in RT treatment planning is the high spatial integrity. Image distortions can potentially cause dosimetric errors by causing delivery of inappropriate radiation doses or treatment of the wrong area. Modern CT scanners should have spatial integrity within ±1 mm across the entire scan field. This test is performed by scanning an object of known dimension and measuring its size in the acquired images.

Spatial Resolution

An imaging system's ability to distinguish between two very small objects placed closely together is quantified with spatial (high-contrast) resolution. High-contrast resolution is measured by scanning a resolution pattern (Mutic et al. 2003; AAPM 1993) (line-pair phantom with a range of spatial frequencies, Figure 68.1b) or with a modulation transfer function (Mutic et al. 2003) method, which is a bit more complicated. Spatial resolution measurements are performed with objects that have a high contrast (contrast difference of 12% or greater) from a uniform background. CT scanner manufacturers commonly supply phantoms that can be used to assess spatial resolution and a number of third-party commercial phantoms are available for this purpose as well.

Contrast Resolution

Contrast resolution can be defined as the CT scanner's ability to distinguish relatively large objects that differ only slightly in density from background (AAPM 1993; Mutic et al. 2003). Contrast resolution is often referred to as low-contrast resolution. Low-contrast resolution is typically evaluated with a phantom that contains low-contrast objects of varying sizes and contrast (Figure 68.1c). Just as with high-contrast testing, low-contrast phantoms can come from the manufacturer or from a third party. The AAPM Report No. 39 (AAPM 1993) provides a detailed measurement procedure.

QA of Positron Emission Tomography in RT

Quality control (QC) and assurance in PET/CT imaging, like other diagnostic modalities, involves a set of ongoing measurements and analyses, which are designed to ensure that the performance of the scanner is within a defined acceptable range. Usually, QA is associated with the process of generating the final product, whereas QC is associated with the final product itself. In this regard, PET/CT manufacturers have developed several QA tests to assess the different steps involved in generating a PET/CT image as well as QC tests to evaluate the final image itself. To understand the basis of these tests, a brief description of the different steps involved in the detection and generation of a PET signal/image is first necessary.

PET Signal Detection Process

In PET imaging, a radioactive compound that decays by positron emission is first administered intravenously or inhaled by the patient. Upon decay, a positron (β+) (in addition to a neutrino) is ejected from the nucleus of the radioactive nucleus along with a transmutation of a proton to a neutron. The positron, after multiple interactions, then annihilates with an electron transforming the combined mass of the positron and the electron to energy according to $E = mc^2$. This energy appears as two simultaneous gamma rays that are separated by approximately 180°. The energy of each gamma ray is 511 keV. These gamma rays, upon exiting the patient's body, are then detected by scintillation detectors organized in a circular design that surrounds the patient. The process of gamma ray detection then occurs according to the following steps:

1. The gamma ray energy is deposited in a scintillation detector.
2. The scintillation detector generates light that is proportional to the deposited energy.
3. The generated light is channeled toward a photomultiplier tube (PMT) that is coupled to the scintillator.
4. The PMT transforms the light to a proportionally amplified electrical signal.
5. The electrical signal is further amplified and shaped for further processing.

In this regard, the resultant integrated electrical signal is proportional to the generated light and hence the energy deposited in the detector.

Upon detection, an electrical circuit (programmable threshold comparator) is used to record the beginning occurrence of an event as well as to assess whether the energy of the detected event is equivalent to 511 keV. Statistical variations in the number of light photons generated in the scintillator as well as the electrical signal and its associated amplification among other factors result in spectral blurring of the detected energy. In other words, a source of 511-keV photons does not produce an electrical signal of consistent amplitude but rather one that fluctuates around a mean energy of 511 keV. In this regard, lower and upper energy thresholds (which depend on the characteristics of the scintillator) are used to assess if the energy of the detected event is equivalent to 511 keV as well as to discriminate against background, scattered, and pileup events. A detected event that falls within this energy window is then known as a "singles" event.

Current PET scanners are designed based on a block detector. In this case, a rectangular or square block of scintillator detector is divided into multiple smaller detector elements, all of which are then coupled to a PMT. With such a design, we no longer have one detector element coupled to a single PMT but rather a large number of detector elements coupled to a smaller number of PMTs. Such a design allows for lower cost among other advantages but requires that some form of decoding the overall light output from the detector block to a specific detector element be in place. To identify which detector element within a block detected an event, the proportion of the signal coming from each PMT within that block is evaluated. However, the accuracy of this positioning information is not precise enough to exactly identify the location of the detector element due to physical variations within the PMT itself as well as transitions

between adjacent detector elements. In this regard, a detector position map within each block detector is created to convert the measured detector element position to a unique location within that block.

As mentioned before, in PET imaging, two gamma rays, which are approximately 180° apart, are emitted from each decay event and hence can potentially be detected as "singles" events. In PET imaging, if both of these single events are detected within a short time span of one another, then the line along which they are detected is known as a line of response (LOR) and the two detected events are known as a coincidence event. The timing window to test for coincidence depends on the scintillator characteristics and is usually on the order of a few nanoseconds. Each detected coincidence event (LOR) is then stored in a unique location within a two-dimensional memory matrix known as a sinogram. LORs that are parallel to one another at a given angle are stored along a single row of the sinogram. The width of the sinogram is then equal to the number of parallel LORs per view angle. Parallel LORs at other angles are stored along other rows of the sinograms. The range of allowable angles is 0–180° only because coincidence events are composed of two back-to-back gamma rays. Angles greater than 180° will encompass the same detectors in the original 0–180° range. Based on this arrangement, if one considers all coincidences that a single detector forms with a set of detectors on the opposite side of a detector ring, these will follow a diagonal line in the sinogram (Figure 68.2).

Detector QA

QA tests for the PET signal detection process then involve assessing the gain of the PMTs, the energy range discrimination of the various detector elements, position maps, and coincidence

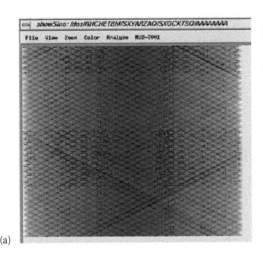
(a)

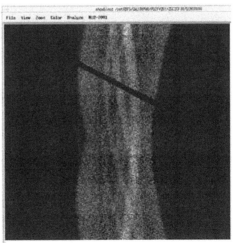
(b)

FIGURE 68.2 (a) A sinogram of a single slice from a daily QA scan showing the response of different detector pairs. Each pixel in the image corresponds to a detector pair (LOR). (b) Sinogram of a single slice from a phantom scan showing a block of detectors (black strip) that are not active (dead).

timing variations. PMT gain is mainly done to ensure that the conversion of light to an electrical signal has minimal drift and fluctuations. In this regard, variations in signal output will no longer be dependent on the PMTs but rather on the incident light falling on the surface of the PMT. Energy range discrimination is performed to normalize against differences in light detection efficiency across the surface of the PMT. With block detectors, detector elements that are at the edge of a PMT might have lower light collection efficiency. Energy range discrimination adjusts for this variation based on the precise detector element location. The role of the position maps is then to precisely identify the position of these detector elements within the block and hence apply the proper energy range adjustment. Coincidence timing, on the other hand, ensures that variations in the detection time of a coincidence event (two associated single events) are not due to intra- or inter-PMT variability nor due to different delays caused by different cable lengths or electronic circuit delays along the path of each single event.

The operational setting of all of these parameters is usually set at system installation. The recalibration of these settings is usually performed whenever a detector module or electronic signal processing board is replaced. The assessment of the performance of these settings is usually done on a daily basis (daily QA) by measuring the singles of each detector element and the coincidences between pairs of detector elements. This process is performed by exposing all the detectors in the scanner to a source of 511-keV gamma rays (usually one or several Ge-68 rod source(s)

are extended inside the scanner and rotated in front of the detectors) for a fixed period. Detected singles and coincidence events are then displayed in a manner that can be easily used to decipher if a detector element, block, or module is malfunctioning. One such display for coincidences is the sinogram that was previously discussed. In addition, this same process can be used to assess the timing differences along each detector pair as well as the energy drift of each detector element. Figure 68.3 shows a plot of a daily QA from a GE PET/CT scanner showing the performance of all of these parameters as well as the allowable upper and lower limits of each parameter. If the daily measured value falls out of these bounds, then that parameter is flagged (one manufacturer uses color coding to indicate the state of the parameter; red for outside the limits, yellow for at the boundary of the limits, and green for within limits) suggesting that a recalibration of that parameter might be necessary. In addition, statistical evaluation of these parameters across all detectors as well as across different dates can also be provided. Alternatively, Figure 68.2a shows the sinogram of the daily QA (usually known as the blank scan) from a series of slices. In either case, if a detector or a block is underperforming (via visual inspection of a red flag or dark streak on the sinogram—Figure 68.2b), then gain, position maps, and energy discrimination recalibration are performed to adjust its performance. This process is iterative in nature and is repeated until the detector/block performance is brought back to an acceptable range set by the manufacturer. In addition, coincidence timing should also be updated to ensure

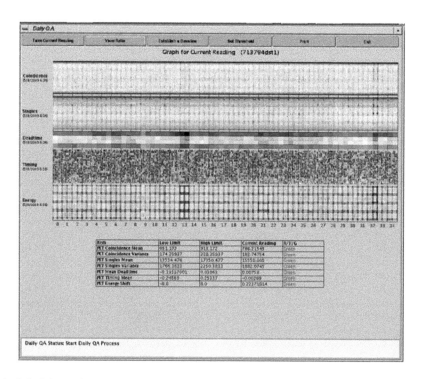

FIGURE 68.3 Output of a daily QA scan from a GE DISCOVERY PET/CT scanner showing the performance of the scanner for different parameters as well as a table with the allowable upper and lower limits and the current reading of each parameter. Each section in this graph corresponds to all the detector blocks in the scanner.

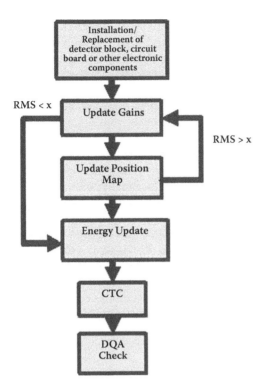

FIGURE 68.4 Flow chart of the process for updating the detector calibration. RMS is the root mean square error set for this variable by the manufacturer.

that delays along the paths of each single event in a coincidence pair falls within acceptable limits. A flow chart of this process is shown in Figure 68.4. On all commercially available PET scanners, all of these processes are usually automated using software tools that are provided on the scanner console.

System QA

In addition to daily QA, several other tests must be performed to ensure optimal performance of the scanner. These additional tests pertain to normalization and well counter calibration. Current PET scanners are composed of thousands of detectors. The performance of these detectors cannot be assumed to be exactly similar to one another. In this regard, some means that normalize against the differences in detector performance should be implemented and updated on a frequent schedule. This process is known as detector normalization. Usually, it is performed by acquiring a large number of events in each detector pair (using either rotating Ge-68 rod sources or a prefilled phantom that contains a positron emitter with a long half-life such as Ge-68) and hence requires a long scan duration. The performance of each detector pair is then normalized against the mean of all detector pairs in the scanner and a correction factor for that detector pair is calculated. This process is repeated for all detector pairs in the scanner to generate a normalization sonogram, which is subsequently used during image reconstruction to suppress variations in measured activity concentration.

Figure 68.5 shows a sinogram of a phantom with a uniform activity distribution before and after normalization correction. Detector normalization is a vital process to accurately quantify PET images because variations in detector performance might mask true variations in radioactivity distribution. In this regard, it is recommended to update the normalization map on a quarterly or at least semiannual basis. More frequent updates might be needed in situations where there is a large variability in ambient temperature and humidity because that affects the performance of the scintillation detectors.

Well counter calibration is the process of transforming the detected count rate to radioactivity concentration. This is also a very important calibration step because PET images are evaluated in activity concentration rather than in counts per second. To perform this calibration, a known source of uniform radioactivity concentration is positioned in the center of the FOV of the scanner and imaged. The measured count rate is then directly related to the activity concentration in the phantom. The ratio of these values is the required calibration factor. This process assumes that corrections for attenuation, scatter, sensitivity, normalization, and deadtime have been accurately applied. The well counter calibration can be checked by imaging another phantom with a known uniform activity concentration. If the measured activity concentration in a region of interest drawn on several of the slices through the phantom is equal to the known activity concentration in the phantom, then the scanner is considered to be properly calibrated. This calibration should be updated whenever a detector module or an electronic circuit board in the scanner has been replaced. Furthermore, this calibration should be updated after every detector normalization update. In this regard, it is recommended that this process be performed on a quarterly basis or at least on a semiannual basis.

Scanner Alignment

Most current PET scanners are hybrid systems composed of a PET and a CT scanner that are attached to one another. One of the uses of the CT in PET imaging is for attenuation correction. This application then necessitates that the CT number calibration (see section "Quantitative CT") is accurate and

FIGURE 68.5 Sinogram of one slice of a phantom before (left) and after (right) normalization correction. The detectors with lower/higher response have been adjusted after normalization.

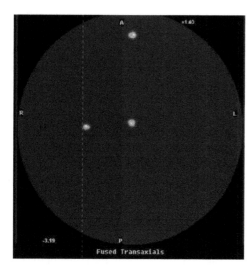

FIGURE 68.6 Fused PET and CT images of three capillary tubes at different transverse locations showing good alignment between the PET (color image) and CT (BW image) scanners.

routinely performed to ensure optimal attenuation correction. Furthermore, the design of these hybrid systems necessitates that the PET and the CT are accurately aligned. This process ensures that the PET and CT images are properly registered to one another but more importantly that the CT scan can accurately attenuate correct the PET scan with no bias. In this regard, a QC test is usually performed to assess the alignment accuracy of the PET and CT subsystems. This test should be performed whenever the PET and CT subsystems have been decoupled for service purposes as well as on an annual basis. Different manufacturers have different techniques on how to perform a PET and CT scanner alignment. One method to assess the alignment accuracy is to image a radioactive source that is also radiopaque in CT imaging. Capillary tubes filled with a solution of F-18 mixed with contrast media can be used as such sources. If placed in different locations in the FOV of the scanner, then the alignment accuracy can be easily evaluated. Figure 68.6 is the fused PET and CT images of three such sources showing good alignment between the PET and CT images. Manufacturers usually have preset limits on the allowable misalignment between the PET and CT subsystems. During installation, this misalignment (three translations and three rotations) is measured to assess if it is within the acceptable limits. The misalignment values are then stored in a file that is subsequently applied to every acquired image set to align the PET and CT image sets to one another.

Scanner Performance Testing

After scanner installation, a series of tests should be performed to ensure that the system is operating within its manufacturer's specifications. Acceptance tests for PET scanners have been identified by the National Electrical Manufacturers Association

(NEMA) in the NU2-2001 (NEMA 2001) and NU2-2007 (NEMA 2007) documents. These tests cover scanner resolution, sensitivity, scatter and count rate performance, image quality, and accuracy of corrections for count losses. A description of these tests can be found in the NEMA (2001, 2007) documents. It is recommended that all, or a subset, of these tests be performed on an annual basis to ensure a consistent scanner performance. The American College of radiology (ACR) has also devised a test to assess scanner performance that is the basis for ACR PET scanner accreditation. The test evaluates the scanner uniformity, contrast, and resolution (MacFarlane 2006; Mawlawi 2008). In addition, the test also evaluates the quantitative accuracy of PET images and its variability across objects of decreasing cross sections. These parameters encompass the overall PET image quality; hence, the test can be considered as a QC test for scanner performance. In this regard, it is recommended that an ACR test be performed on at least a semiannual basis. Because the ACR phantom is composed of multiple compartments, mixing the radioactivity with a CT contrast medium allows the evaluation of the scanner alignment of the PET and CT subsystems as well.

PET QA/QC Assessment and Performance Schedule

There are no specific criteria for performing QA and QC tests on PET scanners; however, the following schedule will ensure the optimal operation of the system with minimal down time or quantitative bias.

On a daily basis, a PET QA that assesses the performance of all detectors across the whole PET system should be performed. This process takes about 20 min to complete and evaluates the singles, coincidences, and their associated timing, as well as energy drifts. The results should be inspected visually by looking at the sinograms (or equivalent display) while checking for any detector or block malfunction that appears as a dark or bright streak on one or more sinograms. In such a case, gain, position maps, and energy discrimination should be updated followed by coincidence timing evaluation.

On a quarterly basis, detector normalization and a well counter correction should be performed. These tests should also be performed whenever a detector module or electronic board is replaced.

On a semiannual basis, an ACR phantom should be imaged as a QC assessment of system performance.

On an annual basis, a full NEMA acceptance testing should be performed to evaluate scanner drift from the manufacturer's specifications. A scanner alignment test should also be performed annually as well as whenever the PET and CT subsystems are decoupled from one another.

QA/QC of Other Associated Equipment

The use of PET scanners relies on other devices such as a dose calibrator that measures the total amount of activity in a source.

In this regard, QA tests for the dose calibrator should also be performed regularly to ensure its optimal operation. These tests include constancy, accuracy, linearity, and geometry. Constancy evaluates the reproducibility of measuring the same source over a period with decay correction. This test is usually done using Cs-137 because of its long half-life. Accuracy is the determination of the error between the measured value and the true value of a NIST-traceable radioactive source. Linearity is the evaluation of the proportionality of the measurement result over a range of activity values. This process is usually evaluated over a period during which the source has decayed substantially from its initial value. Finally, geometry is the evaluation of the change in measurement value with the volume of the source. Constancy and accuracy are usually done daily, whereas linearity is done on an annual basis. Geometry on the other hand is performed only at installation or when the dose calibrator has been moved to a new location.

QA of Magnetic Resonance Imaging in RT

The QA program for MRI scanners developed along with the clinical implementation of MRI systems for diagnostic radiology started in the 1980s (Covell et al. 1986). Early QA programs for MRI scanners focused primarily on the MRI image quality issues related to the measurement of the signal-to-noise ratio (SNR) (Murphy et al. 1993), mainly for evaluating and monitoring the stability of MRI systems (Duina, Mascaro, and Moretti 1992; Miller and Joseph 1993). For this purpose, specially designed MRI phantoms were introduced for system calibration, preventative maintenance, and routine QA tests, mainly by different scanner manufacturers. In 1990, the AAPM Task Group No. 1, Nuclear Magnetic Resonance committee, in collaboration with the ACR subcommittee on Magnetic Resonance, published the first report on QA methods and phantoms for MRI (Price et al. 1990). In this report, assessments of image quality tests such as geometric accuracy, spatial resolution, signal uniformity, SNR, contrast-to-noise ratio, and slice thickness were included. New QA protocols and materials for making testing phantoms were also suggested (Och et al. 1992). This led to the development of the ACR phantom (ACR 1998), which is currently being widely used for ACR MRI accreditation tests (www.acr.org). In addition to the ACR phantom, many MRI facilities also use QA phantoms provided by different vendors, and some even develop and maintain site-specific QA procedures. Most of these QA procedures have been primarily developed for diagnostic radiology applications.

MRI has recently gained much increased use in RT simulation for treatment planning and guidance (Dawson and Jaffray 2007; Khoo and Joon 2006), taking advantage of excellent soft tissue contrasts provided by MRI. Most of MRI system hardware, image acquisition software, and pulse sequences, when applied for RT simulation, are in many aspects similar to conventional diagnostic-use MRI. The primary differences are the requirements and use of patient positioning and immobilization systems, and treatment planning–specific imaging protocols, which often require increased scan time to cover large regions of interest, the use of contrast agents, and the placement of external localization marks on the patient's skin. In brachytherapy treatment planning, MRI simulation of patients is typically carried out after the implantation of brachytherapy radiation applicators (Nag, Cardenes, and Chang 2004). These differences make it important to have an enhanced QA program for applications of MRI in RT treatment planning and delivery guidance. Currently, there is no standard QA program for RT application, and most radiation oncology centers either use the diagnostic radiology QA protocols or have developed their own program based on existing diagnostic MRI QA programs.

The goal of an MRI simulation QA program is to ensure the safe and accurate operation of MRI systems for RT simulation, and 3D MR image data acquired for RT simulation are an accurate representation of the patient's in vivo anatomical structures. The QA program should include regular daily and weekly tests performed by technologists and reviewed by medical physicists, with additional quarterly and annual performance tests conducted by medical physicists. Any substantial MRI system upgrade or preventative maintenance services should also be followed by QA performance tests to ensure no system performance deviated from the established baselines. MRI QA programs should be supervised and regularly reviewed by qualified medical physicists or MRI scientists, according to ACR (2006) recommendations. A large part of the QA program is to include test procedures that will ensure accurate image acquisition, determination of the accuracy and precision for target, and critical structure localization when the images are used for treatment planning.

Typical QA tests and evaluation for MRI should include, but not be limited to, the following tests (ACR 2001):

1. Physical and mechanical inspection
2. Phase stability
3. Magnetic field homogeneity
4. Magnetic field gradient calibration
5. Radiofrequency (RF) calibration for all coils
6. Image SNR ratio for all coils
7. Intensity uniformity for all volume coils
8. Slice thickness and location accuracy
9. Spatial resolution and low contrast object detectability
10. Artifact evaluation
11. Film processor QC
12. Hardcopy fidelity
13. Softcopy fidelity
14. Evaluation of MRI safety—implementation of MRI safety rules, environment, and posting

If the QA procedure includes MR spectroscopy applications, separate scanning, testing, and measurement need to be performed. Proton MRS for brain imaging applications has been developed with a detailed AAPM report, TG No. 9, released in 2002 (AAPM 2002). In the following sections, we will discuss some pertinent MRI QA features and performance tests that are important for radiation therapy simulation.

MRI Safety

Although MRI procedures are considered noninvasive with no ionizing radiation involvement, the strong magnet field used in MRI for generating nuclear magnetization is considered a huge safety hazard (Kanal et al. 2002, 2004). Regular MRI safety training, review, and procedure guidelines should become part of the MRI QA program. Strong magnetic fields produced by the magnet are always on and invisible; its enormous force will cause any metal objects containing ferrous materials to fly into the magnet with huge impact. Ferrous materials or objects should not be brought into the magnet room. Safety signs and warnings at the magnet room door are a requirement for all MRI sites. It is important for each MRI magnet to have an MRI safety program implemented, including training procedures. The MRI QA program should include routine magnet safety evaluation, which is to be performed on a weekly basis. In addition, metal detector devices used for metal screening are to be tested for full functionality on a daily basis by the imaging technologists as part of the daily QA. Patients' MRI questionnaire, including safety screening forms, need to be reviewed and approved by qualified medical physicists or a clinical safety officer to ensure accuracy and full coverage before presenting to patients or scanning subjects.

Because RT simulation typically uses patient positioning and immobilization devices, it is extremely important to test and assess MRI compatibility of these devices before their use in the magnet room. Noncompatible devices and tools should never be brought into the MRI scanner room, and they should be labeled as noncompatible. All equipment such as patient monitoring devices should be screened and cleared for safety before their use in the MRI scanner, and they should be segregated from standard clinical supplies and labeled with "MRI-compatible" sticks. In many cases, equipment that is MRI-safe could still produce large artifacts in the images and could even obscure the entire imaging volume, and, if this is the case, the equipment should stay outside the magnet room during scanning.

In addition to the MRI magnet safety check, RF power monitoring should be part of QA checks as well. The US Food and Drug Administration has established specific requirements for levels of RF power deposition by monitoring the patient-specific absorption rate. The NEMA publication is an excellent resource for information on specific absorption rate (NEMA 1993).

Performance Tests for Electromechanical Components

The MRI simulation scanners used for RT require a flat top moving table, similar to radiation therapy treatment machines. The positioning and movement of the tabletop must be precisely controlled under constant load and can be checked for positioning accuracy with cross-calibrated laser systems. The laser system is

also used to ensure reproducible patient positioning if a patient is scanned repeatedly for multiple treatment fractions with adaptive treatment planning setup. Because the performance requirements for an MRI scanner laser system and patient table are the same as CT systems, one can refer the specific tests for CT simulators (Mutic et al. 2003) for more information on QA tests for patient table and external laser systems.

Scanner Performance Tests

Scanner performance tests are typically used for general evaluation of MRI systems, including magnet stability, RF system SNR, and magnet shimming performance. These are evaluated with tests on SNR, center frequency, and magnetic field homogeneity. System performance tests can typically be accomplished by using spherical phantoms doped with copper sulfate solutions, with spherical volume diameters ranging from 5 to 20 cm. These phantoms are usually available from vendors or can be prepared on site. The frequency of scanner performance tests can be determined by the medical physicists based on initial system performance evaluation, the age of the magnet and RF systems if the system is a used one, and the frequency of vendor-provided engineering services. These tests used to be very frequent on older systems, but for new generations of scanners, these can be performed on weekly or monthly schedules.

SNR Measurement

The SNR of both T1- and T2-weighted image series can be measured on T1/T2 images of the center slices from a spherical phantom by using the ROI measurement tool on the scanner console, with mean signal intensity calculated from the central 80% area of the phantom. Consequently, noise can be determined as the standard deviation of another ROI that is placed outside the phantom within an adequate area. To avoid ghosting artifacts outside the phantom, the location of the noise measurement needs to be kept away from the phase-encoding direction. The SNR can then be obtained by dividing the standard deviation of the outside ROI from the mean signal intensity of the inside ROI.

Magnetic Field Homogeneity Measurement

This measurement can be performed by using the spectral peak method with a uniform, spherical phantom. The phantom is placed at the center of the magnetic field and a spectrum of the phantom (Figure 68.7) is obtained by a single RF pulse sequence. The spectrum is then displayed, and the full width at half-maximum (FWHM) of the spectrum peak can be measured and recorded. The inhomogeneity of the magnetic field can then be calculated using the Larmor equation, FWHM (ppm) = FWHM (Hz)/42.567 B0 (T), where B0 is the magnet field strength measured in Tesla. In the recording, the spherical diameter size needs to be recorded in centimeters, with FWHM (parts per million).

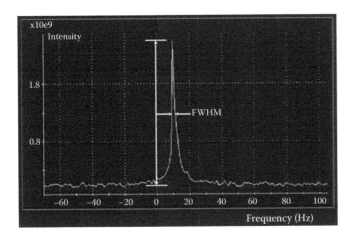

FIGURE 68.7 MR spectrum display for measuring magnetic field homogeneity and central frequency.

Central Frequency Measurement

The stability of the static magnetic field can be determined by the central frequency provided by the spectral peak (Figure 68.7) from the MRI scanner console.

MRI Image Quality Test Procedures

Image quality directly affects the ability to identify and delineate target volumes and surrounding critical structures. For accurate radiation treatment planning, the MRI images acquired must have high geometrical accuracy and spatial integrity. These tests typically include geometric accuracy, high-contrast spatial resolution, slice thickness accuracy, slice position accuracy, image intensity uniformity, signal ghosting, and the detectability of low-contrast objects.

Quantitative analysis of image quality can be performed on the scanner console or a computer workstation capable of basic image manipulation and display such as window and level

FIGURE 68.8 ACR MRI phantom.

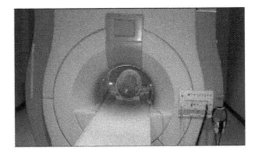

FIGURE 68.9 ACR setup in a head coil for QA measurement.

adjustments, magnification (zoom), mean and standard deviation measurements in a region of interest, and length measurements. For specific treatment planning application, these QA measurements of MRI image quality can also be performed on the RTP system. This step is necessary and required to integrate with treatment planning QA procedures where MRI will be used as the primary image sets for treatment planning such as in brachytherapy.

Many of the image quality tests can be accomplished by using the ACR MRI phantom. The ACR MRI phantom (Figure 68.8) is a short, hollow cylinder of acrylic plastic closed at both ends. The inside length is 148 mm; the inside diameter is 190 mm. It is filled with a solution of nickel chloride and sodium chloride (10 mM $NiCl_2$ and 75 mM NaCl). The outside of the phantom has the words "NOSE" and "CHIN" etched into it to help orient the phantom when scanning inside a head RF coil. The phantom is designed with several inside structures to facilitate some of the image quality tests.

Figure 68.9 shows the scanning setup for image quality tests using ACR phantom within a head coil. After delicate positioning of the phantom at the center of the head coil, localizer images are acquired using a 20-mm-thick single-slice spin-echo acquisition through the center of the phantom. This is followed by acquiring a series of image sets that are going to be evaluated for image qualities. Specific requirements and acquisition parameters can be found in the ACR Phantom Test Guidance (ACR 2001).

The image quality tests require at least two sets of ACR MRI phantom scans; for example, the conventional spin echo T1- and T2-weighted images for qualitative and quantitative analysis. The resulting measurement will, in fact, only reflect the performance of these two pulse sequences. These two pulse sequences, however, represent the most popular pulse sequences used in clinical MRI studies. For the purpose of completeness of the QA program, other disease site–specific pulse sequences or treatment-specific MRI protocols can also be added for image quality evaluation. For simplified discussion, we only chose to analyze the T1 and T2 imaging for QA purposes.

Geometric Accuracy

The geometric accuracy test determines the accuracy with which the image represents distance in the imaged subject. This

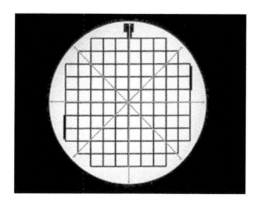

FIGURE 68.10 Geometric accuracy measurement.

test is done to ensure that there are no geometrical distortions presented in images in any direction. The geometry accuracy measurement consists of making length measurements on the images, between readily identified locations in the phantom, and comparing the results with the known values for those lengths. A failure means that dimensions in the images differ from the true dimensions substantially more than a specification.

Seven measurements of known distance within the ACR phantom can be made using the scanner console's distance measurement tool. The display window and level settings can affect the length measurements, so it is important to set them properly. Measure the diameter of the phantom in four directions: top to bottom, left to right, and both diagonals, as shown in Figure 68.10.

High-Contrast Spatial Resolution

As described earlier, the high-contrast spatial resolution test assesses the scanner's ability to resolve small objects when the contrast-to-noise ratio is sufficiently high that it is not a limiting factor. A failure of this test means that for a given FOV and acquisition matrix size, the scanner is not resolving small details. One can visually assess if individual small bright spots in arrays of closely spaced small bright spots can be distinguished. The test approach is similar to the one for other imaging modalities.

Slice Thickness Accuracy

The slice thickness accuracy test assesses the accuracy with which a slice of specified thickness is achieved. The prescribed slice thickness is compared with the measured slice thickness from the images. A failure of this test indicates that the scanner is producing slices of substantially different thickness from that being prescribed. A cause of failure is typically due to RF amplifier nonlinearity. Malfunctions in the high-power RF amplifiers can also cause distorted RF pulse shapes. Any possible causes for failure require RF calibration and tests by the service engineer.

Slice Position Accuracy Test

The slice position accuracy test assesses the accuracy with which slices can be prescribed at specific locations using the localizer image for positional reference. This test will also verify the accuracy of imaging slice locations in reference to the initialized magnet isocenter. A failure of this test will show that the actual locations of acquired slices differ from their prescribed locations by substantially more than the specified values. In the test, the differences between the prescribed and the actual slice locations from the acquired images are measured.

It is important to prescribe the slices carefully in this test because errors introduced here can add to other sources of error and lead to an acceptable level of performance. Many scanners shift the patient table position in the inferior–superior direction to place the center of a prescribed stack of images at gradient isocenter. This table shift occurs after the localizer images are acquired, and thus errors in the table positioning mechanism could lead to slice position errors. Bad gradient calibration or poor B0 homogeneity can also cause failure of this test.

Image Intensity Uniformity Test

The image intensity uniformity test measures the uniformity of the image intensity over a large region of the uniform phantom lying near the middle of the imaged volume; this is typically used for measuring the image uniformity for the volume head coil. Head coils for clinical use, such as for gamma knife radiosurgery imaging, require highly uniform spatial sensitivity in the central part of the coil when loaded with a human head. Lack of image intensity uniformity indicates a deficiency in the scanner, often a defective head coil or problems in the RT transmitting and receiving subsystems.

Degraded image intensity uniformity can result from failure of hardware components in the head coil and from failure of the mechanisms for inductive decoupling of the body coil from the head coil. In these cases, images usually become noticeably lower in SNR ratio, and they usually appear grainier. A service engineer is required to diagnose and correct these problems.

Percent-Signal Ghosting Test

The percent-signal ghosting test will evaluate the level of signal ghosting in acquired images. Ghosting is an artifact in which a faint display (ghost) of the imaged object appears superimposed on the image, displaced from its true location. There are sometimes many low-level ghosts that may not be recognizable from the scanned objects because they appear in the phase encode direction from the brighter regions of the image. Ghosting is a result of signal instability between RF pulses phase cycling repetitions. In this test, the ghost signal level is measured and reported as a percentage of the signal level in the primary image. Ghosts are most noticeable in the background areas of an image where there should be no signal, but usually they could overlap with the main portions of the image. A failure of this test means that there is significant signal ghosting.

FIGURE 68.11 Low-contrast object detectability test.

Low-Contrast Object Detectability Test

As discussed earlier, the low-contrast object detectability test (Figure 68.11) determines the extent to which objects of low contrast are discernible in the images. For this purpose, the phantom has a set of low-contrast objects of varying size and contrast. The ability to detect low-contrast objects is primarily determined by the contrast-to-noise ratio achieved in the image and may be degraded by the presence of artifacts such as ghosting. Scanners at different field strengths differ widely in their contrast-to-noise ratio performance, and scan protocols are normally adjusted to take these differences into account. Therefore, this low-contrast object detectability test needs to be applied in both ACR images and other site-specific images. A failure of this test means the images produced by the specific pulse sequences showed significantly lower contrasts.

Summary

Quality assurance for CT simulators in RT is well established and has been addressed by several reports. The maturity of CT simulator QA is largely because of the widespread presence of CT scanners in RT where most radiation oncology facilities own a CT scanner. The QA program for CT simulators is built on procedures and tests developed for diagnostic radiology scanning with significant expansion of scanner geometry and quantitative CT accuracy tests. Although CT technology continually expands and improves, the reports and recommendations on QA of CT scanners continue to be applicable and should be a resource for anyone who is developing a CT simulator QA program.

The use of PET and MR imaging in RT is continually growing. Most of these studies are acquired on scanners housed within diagnostic radiology facilities and only a small fraction of radiation oncology facilities own a PET or an MR scanner. As such, the specific issues for QA of PET and MR scanners in RT remain to be systematically addressed. As described in this chapter, the QA of PET and MR in RT is also very similar to the processes used in diagnostic radiology. The concerns with the geometric accuracy and integrity of these scanners in RT are the same as for CT scanners and should be formally addressed in the QA program. The similarity of the PET scanning process for diagnostic radiology and RT make the two QA programs practically identical. The need for special coils and the significant concern with MR spatial distortions in RT make the MR QA protocols for diagnostic and RT purposes somewhat different. At a point where significant numbers of PET or MR scanners are housed within RT facilities, formal recommendations from professional societies and other organizations will be developed to formally address the QA requirements of PET and MR in RT.

References

AAPM. 2002. Report of AAPM Task Group No. 9: Proton magnetic resonance spectroscopy in the brain. *Med. Phys.* 29: 2177–2197.

AAPM. 1993. Report No. 39. *Specification and Acceptance Testing of Computed Tomography Scanners.* New York, NY: American Institute of Physics.

ACR. 2001. *American College of Radiology MRI Quality Control Manual.* Reston, VA: American College of Radiology.

ACR. 2006. *American College of Radiology Technical Standard for Diagnostic Medical Physics Performance Monitoring of Magnetic Resonance Imaging (MRI) Equipment.* Reston, VA: American College of Radiology.

ACR. 1998. *Phantom Test Guidance for the American College of Radiology MRI Accreditation Program.* Reston, VA: American College of Radiology.

NEMA. 1993. *Characterization of SAR for MRI Systems.* Washington, DC: National Electrical Manufacturers Association.

NEMA. 2001. *Standards Publication NU2-2001: Performance Measurements of Positron Emission Tomographs.* Washington, DC: National Electrical Manufacturers Association.

NEMA. 2007. *Standards Publication NU2-2007: Performance Measurements of Positron Emission Tomographs.* Washington, DC: National Electrical Manufacturers Association.

Covell, M. M., D. O. Hearshen, P. L. Carson, T. P. Chenevert, P. Shreve, A. M. Aisen, F. L. Bookstein et al. 1986. Automated analysis of multiple performance characteristics in magnetic resonance imaging systems. *Med. Phys.* 13: 815–823.

Dawson, L. A., and D. A. Jaffray. 2007. Advances in image-guided radiation therapy. *J. Clin. Oncol.* 25: 938–946.

Duina, A., L. Mascaro, and R. Moretti. 1992. Results of the quality control of magnetic resonance images. *Radiol. Med.* 83: 276–281.

Fraass, B., K. Doppke, M. Hunt, G. Kutcher, G. Starkschall, R. Stern, and J. Van Dyke. 1998. American Association of Physicists in Medicine Radiation Therapy Committee Task Group 53: Quality assurance for clinical radiotherapy treatment planning. *Med. Phys.* 25 (10): 1773–1829.

Kanal, E., J. P. Borgstede, A. J. Barkovich, W. G. Bradley, J. P. Felmlee, J. W. Froelich, E. M. Kaminski et al. 2002. American College of Radiology white paper on MR safety. *AJR* 178: 1335–1347.

Kanal, E., J. P. Borgstede, A. J. Barkovich, C. Bell, W. G. Bradley, S. Etheridge, J. P. Felmlee et al. 2004. American College of Radiology white paper on MR safety: 2004 update and revisions. *AJR* 182: 1111–1113.

Kessler, M. L., S. Pitluck, P. Petti, and J. R. Castro. 1991. Integration of multimodality imaging data for radiotherapy treatment planning. *Int. J. Radiat. Oncol. Biol. Phys.* 21 (6): 1653–1667.

Khoo, V. S., and D. L. Joon. 2006. New developments in MRI for target volume delineation in radiotherapy. *BJR* 79: S2–S15.

Ling, C. C., J. Humm, S. Larson, H. Amols, Z. Fuks, S. Leibel, and J. A. Koutcher. 2000. Towards multidimensional radiotherapy (MD-CRT): Biological imaging and biological conformality. *Int. J. Radiat. Oncol. Biol. Phys.* 47: 551–560.

MacFarlane, C. 2006. ACR Accreditation of nuclear medicine and PET imaging departments. *J. Nucl. Med. Tech.* 34 (1): 18–24.

Mawlawi, O. 2008. PET ACR Accreditation and acceptance testing. In *The Physics and Applications of PET/CT Imaging.* Proceedings of the American Association of Physicists in Medicine 2008 Summer School, June 25–27, 2008, Houston, TX. CD-ROM Edition, 2008. ISBN: 978-1-888340-76-1.

Miller, A. J., and P. M. Joseph. 1993. The use of power images to perform quantitative analysis on SNR of MR images. *Magn. Reson. Imaging* 11: 1051–1056.

Murphy, B. W., P. L. Carson, J. H. Ellis, Y. T. Zhang, R. Hyde, and T. L. Chenevert. 1993. Signal-to-noise measures for magnetic resonance imagers. *Magn. Reson. Imaging* 11: 425–428.

Mutic, S. 2006. Use of imaging systems for patient modeling: PET, SPECT. In *Integrating New Technologies into the Clinic: Monte Carlo and Image Guided Radiation Therapy*, ed. B. H. Curran, I. J. Chetty, and J. M. Balter. Madison, WI: Medical Physics Publishing Corporation.

Mutic, S., J. R. Palta, E. Butker, I. J. Das, M. S. Huq, L. D. Loo, B. J. Salter, C. H. McCollough, and J. Van Dyk. 2003. Quality assurance for CT simulators and the CT simulation process: Report of the AAPM radiation therapy committee task group No. 66. *Med. Phys.* 30: 2762–2792.

Nag, S., H. Cardenes, and S. Chang. 2004. Proposed guidelines for image-based intracavitary brachytherapy for cervical carcinoma: Report from image-guided brachytherapy working group. *Int. J. Radiat. Oncol. Biol. Phys.* 60: 1160–1172.

Och, J. G., G. D. Clarke, W. T. Sobol, C. W. Rosen, and S. K. Mun. 1992. Acceptance testing of MRI systems: Report of AAPM NMR TG No. 6. *Med. Phys.* 19: 217–229.

Price, R. R., L. Axel, T. Morgan, R. Newman, W. Perman, N. Schneiders, M. Selikson et al. 1990. Quality assurance methods and phantoms for MRI: Report of AAPM NMR TG No. 1. *Med. Phys.* 17: 287–295.

69

Stand-Alone External Beam Treatment Planning Systems

Jean-Claude Rosenwald
Institut Curie, Paris, France

Introduction

Stand-alone treatment planning systems (TPSs) are used to prepare treatment plans to be used on general-purpose accelerators. They are generally networked to other pieces of equipment used in the radiotherapy department. The main data exchanges are with imaging equipment as input and record-and-verify systems (RVSs) as output. Their main functionalities are the following:

- Calculation of the dose distribution and monitor units for a given beam setup
- Optimization of the beam parameters (particularly the intensity modulation) to match predefined criteria
- Preparation of the beam parameters and reference images to be exported to the treatment machine (generally through an RVS)

The TPS is a critical component of the radiotherapy process because any incorrect beam parameter—including monitor units—that would be obtained from a computed plan could be used as such for treatment. Actually, several severe accidents have been reported as a consequence of the improper use of a TPS.

TPSs are becoming very complex. They must be compatible with the large variety of therapeutic approaches found in existing departments and should support the new treatment techniques as they become available. In spite of the trend to develop dedicated TPSs for very specialized techniques (e.g., stereotactic treatments, tomotherapy, Cyberknife), a reference multipurpose stand-alone TPS used for "standard" treatments is still needed. Stand-alone TPSs help guarantee (as much as possible) homogeneity of practice, facilitate training, ensure data exchanges, and ensure implementation of quality assurance (QA) procedures.

The aims of a QA program for a TPS are twofold:

1. To prevent errors (and accidents)
2. To ensure dosimetric and geometrical accuracy within acceptable tolerances

The accuracy that can be obtained is strongly dependent on the quality and stability of the data used to model the treatment beams, on the care taken to adjust the model parameters (i.e., beam parameterization), and on the performance and limitations of the algorithms. The assessment of the TPS accuracy for several situations encountered in clinical practice is part of the commissioning process (see section "Commissioning").

The risk of dosimetric or geometrical error leading to an incorrect treatment is mostly related to the improper design, parameterization, or use of the TPS. This risk is minimized if the following steps are taken:

- Just after TPS delivery, a formal and comprehensive acceptance process, using generic beam data, conducted jointly by the vendor and the user
- After acceptance and before clinical use, a comprehensive commissioning process including beam parameterization, staff training, and hands-on practice for theoretical and clinical cases
- After clinical implementation, an ongoing quality control (QC) approach including permanent vigilance in the daily use of the TPS

Several situations may be encountered in practice when setting up a new QA program. Each of them requires a different approach as summarized in Table 69.1.

TABLE 69.1   Suggested Approaches to the Implementation of a Generic TPS QA Program

Situation	Recommended Steps
1. Setting up a QA program for a TPS already in clinical use	• Open a dedicated logbook • Collect and organize all basic data (geometry and dose) previously used to parameterize the TPS • If some data are missing or are not reliable, repeat critical measurements and perform partial recommissioning according to Table 69.3 • Collect and organize documents supplied by the vendor • Check the adequacy of user's training; if necessary, prepare a new training plan • Check the existence of user's documentation; where required, complement with written procedures describing how to use the TPS in all clinically encountered situations • Set up an ongoing QC program, according to Table 69.4 • Reassess regularly the value of this program
2. Purchasing a new TPS (installation of a new center or a new clinical activity)	Follow the different steps of the process: acceptance, commissioning, periodic QC, and QC of individual plans, as described in the following sections
3. Replacing an existing TPS (or introducing a new dose calculation algorithm)	Perform acceptance and commissioning as for a new TPS but, in the commissioning process, include a comparison with the replaced TPS based on the following approach: • Determine deviations of MU calculations and dose distributions in a flat water phantom for various SSD, field sizes, and accessories; if the parameterization of the old and new systems are correct, the deviations should be negligible • Determine deviations of MU calculations and dose distributions for several representative clinical cases (e.g., pelvis, breast, lung) and assess the order of magnitude of the modification for each group of patients • If significant deviations (i.e., larger than 2% or 3%) are observed for some situations or groups of patients (e.g., due to an improved dose calculation algorithm in the lung), inform clinicians about it and take jointly a decision about changing or not the prescribed dose; document carefully any modification of clinical practice

Acceptance Testing

Acceptance testing of a TPS is performed jointly by the vendor and by the user. It consists mainly of two types of verification:

- Check that the hardware and software delivery is consistent with the order.
- Check that the system behaves according to the vendor specifications.

Acceptance is made much easier if the order has been based on predefined specifications agreed upon between the vendor and the user as part of the purchase process. These specifications could include compliance with the IEC 62083 standard (IEC 2000), presented in a tabular form in the TECDOC 1540 document from the International Atomic Energy Agency (IAEA 2007).

The main steps of the acceptance testing process are summarized in Table 69.2.

TABLE 69.2   Main Steps of the Acceptance Testing Process

Aim	Items to Check/Tests to be Done

Consistency with the order, based on the delivery documents and on visual inspection	• Hardware (CPU and peripherals)
	• Software (check license agreements and keep track of release numbers)
	• Documentation (paper or electronic files)
	    • Installation manual
	    • Physics manual (measurement and parameterization procedures)
	    • User manual
	    • List of tests done at the factory ("type tests" - see below)
	    • Generic beam data files
	    • Typical examples of treatment plans
	• Integration within hospital network and check of data exchanges
	• Training schedule

Proof by the vendor that the system behaves according to specifications	• "Type tests": the system should be compliant with the IEC 62083 standard, and this must have been thoughtfully investigated and tested at the factory; the vendor should provide a list of these tests and explain them to the user; by cosigning the acceptance document, the vendor guarantees that the tests have been done and the user acknowledges that he has been informed • "Site tests": a subset of the "type tests" used to actually demonstrate to the user the conformity with specification; by signing the acceptance document, the user confirms that these tests have been done on site and passed successfully

TABLE 69.3 Main Steps of the Commissioning Process

Step	Description

Beam parameterization ("modeling")	• Define machine and accessories names, consistent with the rest of the network (i.e. RVS)
	• Define machine geometrical and mechanical data, including interlocks at limits; check by performing network export to actual treatment machines followed with visual inspection of each scale and accessory placement
	• Adjust dose related parameters:
	  • Acquire (or gather) a consistent set of measurements according to the requirements of the dose calculation algorithms
	  • Export these data to the TPS
	  • Tune parameters using the tools supplied with the TPS and the procedures described in the physics manual

Assessment of the anatomical and geometrical functionalities of the TPS	• Calibrate and test the imaging devices used for planning:
	  • Use a calibrated multi density phantom for acquisition of the CT calibration curve expressed in relative electronic density; input these data into the TPS
	  • Check the absence of geometrical distortion based on the same phantom or another simple geometrical one (e.g. with fiducial markers)
	• Export the phantom images to the TPS and measure distances and density; check distances on print-out
	• Repeat the procedure for all patient orientations allowed by the imaging device and recognize head, feet, right, left, ant, post orientations at the TPS console; check if they have any influence on the machine parameters (i.e., gantry angle for a patient treated "feet first")
	• Repeat the procedure with specific phantoms for other imaging modalities (i.e., MRI, PET); use and test tools for image registration
	• Use and test, for these phantoms, the various options of the contouring tools (include complex shapes with several contours of the same structure in the same slice)
	• Use and test tools for bolus definition
	• Use and test the various options of the tools for structure expansion and Boolean operations
	• Use and test various options for beam shaping (including automatic beam shaping and asymmetry) and beam orientation (including table rotation); check consistency with RVS and machine setup
	• Use and test addition of accessories (e.g., blocks, wedges); check orientation and interlocks
	• Generate typical beam's eye views and DRRs from phantoms and patient data; check orientation information and scales; export DRRs for comparison with portal images

Assessment of dose and monitor unit (MU) calculations (nonmodulated beams)	• Compare measured and calculated dose distributions (depth dose, profiles) for a simple water phantom and various beam parameters for situations where measurements are available; this may include beam data used for parameterization but should also include other field sizes and shapes (rectangular, asymmetric) and distances; some examples may be found in the ESTRO report (Mijnheer et al. 2004) or in Appendix E of the IAEA TECDOC 1583 document (IAEA 2008)
	• Use a semi anatomical phantom, with inhomogeneity embedded, as described, for instance, in the IAEA TECDOC 1583 document to simulate the full process from imaging to treatment
	  • Acquire CT images of the phantom and export to TPS
	  • Prepare a limited number of plans using several beam setups (e.g. as in TECDOC 1583), prescribe the dose at a reference point (typically 2 Gy), calculate MUs and doses at some points where ionization chamber measurements may be carried out
	  • Irradiate the phantom with the calculated number of MU and compare measurements with calculations
	  • The deviation between calculated and measured dose at the various points is expressed in % of the measured dose at the reference point; the tolerance is typically between 2 and 4%, depending on the location of the point and complexity of the phantom/beam setup
	• If a tool is used for "independent" calculation of MU, the same plans may be tested to allow MU comparison for various clinical situations
	• Additional checks may be performed in complex situations (e.g., lack of scatter, perturbations from inhomogeneities), without measurements, using the quality index methodology (Caneva, Rosenwald, and Zefkili 2000; Tsiakalos et al. 2004; Caneva et al. 2006)
	• The consistency between dose volume histogram calculations and isodose representations may be carried out through specific tests in which several voxel sizes should be investigated (Panitsa, Rosenwald, and Kappas 1998; Mijnheer et al. 2004)

Test of clinical cases after the full planning process	• Select several image sets of real patients with typical representative anatomical sites and carry out the full process of planning, from image import to export to the RVS • Analyze the various steps of the planning process and recognize the critical issues (e.g., interpretation of the beam weighting, wedge filters handling, plan approval method, etc.) • Scrutinize the output documents to identify the information that is available, make sure that it is understood, and decide how it will be used to visually check the patient plans • Prepare written protocols, according to the various clinical sites • Complete the staff training

The list of the "type tests" and "site tests" is not given here, but it can be found in the IAEA TECDOC 1540 document (IAEA 2007).

The accuracy of the dose calculation algorithm is investigated using generic beam data and typical beam setups. It is recommended to use the data and test cases accompanying the IAEA TECDOC 1540 document (CD-ROM or files downloading from the IAEA Web site at http://www-naweb.iaea.org/nahu/dmrp/zip/CDROM TECDOC 1540.zip). Tolerances (between 2% and 4%) have been set for each of the tests and Excel tables are supplied, which allow the user to check easily if the calculated dose is within tolerance.

At the end of the acceptance process, the acceptance document supplied by the vendor is signed jointly, and this means that the warranty period starts. In case of disagreement on some points, it should be stated on the document and special arrangements could be negotiated. By no means is acceptance of the TPS sufficient to put it into clinical use.

Commissioning

The commissioning process is the user's responsibility. It may last several weeks before the TPS is allowed to be used clinically. The main tasks to be carried out during commissioning are summarized in Table 69.3.

More practical details on these procedures may be found in the literature (AAPM Report 62 1998; Van Dyk et al. 2004; Mijnheer et al. 2004; Rosenwald 2007). It should be recognized that a fundamental aspect of the commissioning process is the acquisition of a deep understanding of the TPS behavior through systematic completion of the various tests. As a matter of fact, most of the severe TPS-related accidents that occurred in the past years were the result of a misinterpretation in how to use the system (ICRP 2000, 2009).

Most of the tools used for conventional treatment planning are also available and used for IMRT. Commissioning a TPS with an IMRT option should therefore follow the same approach as above. One difference is the presence of an inverse planning module. The quality of the optimization might be software (and user)-dependent but presently, to our knowledge, there are no specific recommendations to check this part of the software. Although the risk of using a suboptimal plan exists, it is most important to check that the planned dose distribution is actually delivered to the patient. This is carried out on a phantom as part of the commissioning process. It is often repeated for each patient as part of the QC of individual plans.

Periodic QC

To avoid any drift of the quality throughout the clinical use of the TPS, it is mandatory to perform a periodic QC of the performances. It could be based on two main principles:

- Constancy check of the TPS output at regular intervals
- Partial recommissioning after modification

To simplify the procedures, it is best to use the results of the clinical commissioning tests as a reference for ongoing QC. Therefore, the results of some significant tests based on phantom or patient plans should be carefully documented and saved, and the corresponding computer files should be cautiously archived in the proper format.

The aim of the constancy check is to ensure integrity of hardware, software, and data transfer. The suggested frequency for these tests is given in Table 69.4.

The method for carrying out these tests is straightforward: The CPU test consists simply in rebooting the system and looking carefully at the displayed message or applying any procedure recommended by the manufacturer. The other tests are a repetition of commissioning tests. The expected result is a strict coincidence of the results if the TPS has not been altered.

However, relatively frequent modifications are likely to occur during the clinical use of a TPS. The main causes are software updating or upgrading, hardware upgrading, changes in machine/beam parameters due to new equipment/accessories or new techniques, and changes in the equipment with which the TPS communicates (imaging devices, RVS). It must be guaranteed that if any change occurs, the information is given to the physicist responsible for the TPS. It is then his responsibility to prepare, based on the description of the modifications (e.g., release notes), a validation procedure that must be followed before clinical use is resumed. This procedure is difficult to design because one can hardly anticipate the consequences of a modification that could appear very limited. It could be based on the reference plans used for regular QC, with special focus on the results of monitor unit calculations. In addition, the first patient plans calculated after such modifications should be carefully reviewed by an experienced physicist.

TABLE 69.4 Example of QC Checks and Corresponding Frequency

Test	Frequency
Hardware: CPU and digitizer	Monthly (weekly for sonic digitizer)
Hardware: plotter and backup recovery	Quarterly
Integrity of anatomical data handling: data transfer, geometry and density, patient representation	Quarterly
Integrity of dose and MU calculation: MU, details on printout, electronic plan transfer	Quarterly (once a year, closer look to full dose distribution)

Source: IAEA, *Commissioning of Radiotherapy Treatment Planning Systems: Testing for Typical External Beam Treatment Techniques*, TECDOC 1583, International Atomic Energy Agency, Vienna, 2008.

Large modifications such as the implementation of a new algorithm imply partial recommissioning with strict adherence to the steps described in Table 68.1 (situation 3).

QC of Individual Plans

Performing an initial and regular verification of the TPS does not prevent errors occurring for individual patients. These errors have several causes:

- Incorrect data input (including the prescription)
- Misinterpretation in the use of the TPS for some special situations
- Bug in the TPS software, undetected previously and occurring only in peculiar circumstances

Such events can affect a single patient or a group of patients.

It is then mandatory that all plans are reviewed by well-trained specialists (in principle physicists) and receive formal approval before being used clinically. The aspects of QC of individual plans are discussed in detail in Chapter 84.

References

AAPM Report 62. 1998. *Quality Assurance for Clinical Radiotherapy Treatment Planning (Reprinted from Medical Physics, Vol. 25, Issue 10)*. New York, NY: American Institute of Physics, Inc.

Caneva, S., J. C. Rosenwald, and S. Zefkili. 2000. A method to check the accuracy of dose computation using quality index: Application to scatter contribution in high energy photon beams. *Med. Phys.* 27 (5): 1018–1024.

Caneva, S., M. F. Tsiakalos, S. Stathakis, S. Zefkili, A. Mazal, and J. C. Rosenwald. 2006. Application of the quality index methodology for dosimetric verification of build-up effect beyond air–tissue interface in treatment planning system algorithms. *Radiother. Oncol.* 79 (2): 208–210.

IAEA. 2004. *Commissioning and Quality Assurance of Computerized Planning Systems for Radiation Treatment of Cancer* – Technical Reports Series n° 430 International Atomic Energy Agency, Vienna. http://www-pub.iaea.org/MTCD/publications/PDF/TRS430_web.pdf.

IAEA. 2007. *Specification and Acceptance Testing of Radiotherapy Treatment Planning Systems* - TECDOC n° 1540 International Atomic Energy Agency, Vienna. http://www-pub.iaea.org/MTCD/publications/PDF/te_1540_web.pdf.

IAEA. 2008. *Commissioning of Radiotherapy Treatment Planning Systems: Testing for Typical External Beam Treatment Techniques* – TECDOC n° 1583 International Atomic Energy Agency, Vienna. http://www-pub.iaea.org/MTCD/publications/PDF/te_1583_web.pdf.

IEC. 2000. *Requirements for the Safety of Radiotherapy Treatment Planning Systems. Report 62C/62083*. Geneva, Switzerland: International Electrotechnical Commission.

Mijnheer, B., A. Olszewska, C. Fiorino, G. Hartmann, T Knoos, J. C. Rosenwald, and H. Welleweerd. 2004. *Quality Assurance of Treatment Planning Systems: Practical Examples of Non-IMRT Photon Beams*. Brussels: European Society of Therapeutic Radiation Oncology (ESTRO). ESTRO 2004 Quality assurance of treatment planning systems; practical examples for non-IMRT photon beams - Booklet N°7 http://www.estro.org.

Panitsa, E., J. C. Rosenwald, and C. Kappas. 1998. Quality control of dose volume histogram computation characteristics of 3D treatment planning systems. *Phys. Med. Biol.* 43 (10): 2807–2816.

Rosenwald, J. C. 2007. Quality assurance of the treatment planning process, Chapter 39. In *Handbook of Radiotherapy Physics*, Mayles, Nahum and Rosenwald ed., Taylor and Francis, CRC Press, 841–866.

Tsiakalos, M. F., K. Theodorou, C. Kappas, S. Zefkili, and J. C. Rosenwald. 2004. Analysis of the penumbra enlargement in lung versus the quality index of photon beams: A methodology to check the dose calculation algorithm. *Med. Phys.* 31 (4): 943–949.

70

Conventional Linear Accelerators

Jiajin Fan
Fox Chase Cancer Center

Robert A. Price Jr.
Fox Chase Cancer Center

Introduction

The medical linear accelerator (or linac) has become the dominant radiation therapy treatment unit and accounts for more than 80% of all operational megavoltage treatment units. Some linear accelerators provide x-rays only, whereas most provide both x-rays and electrons at various megavoltage energies. Recommendations for general quality assurance (QA) testing for medical linear accelerators were first provided in the AAPM Task Group Report 40 in 1994 (Kutcher et al. 1994) and are updated in Task Group Report 142 in 2009 (Klein et al. 2009). The latter specifies tests for new technologies such as dynamic/virtual wedges and new procedures such as stereotactic radiosurgery and stereotactic body radiation therapy.

Goal of a Linear Accelerator QA Program

The goal of a linac QA program is to ensure that the machine characteristics do not deviate significantly from their baseline values acquired at the time of acceptance and commissioning. Many of these baseline values are used by the treatment planning systems or clinical dose calculation data books, and therefore deviation from the baseline values could result in suboptimal patient treatment on that machine. Machine parameters can deviate from their baseline values for many reasons. Gradual changes may occur as a result of aging of the machine components, or unexpected changes in machine performance may occur due to component failure, mechanical breakdown,

TABLE 70.1 Medical Accelerator QA Daily Test

Procedure	Description	Tolerance Non-SRS/SRT	SRS/SRT
Mechanical			
Laser/crosswire	Alignment of crosswire and appropriate lasers	2 mm	1 mm
Optical distance indicator	Gantry angle 0° and at 100 cm Source-to-Surface Distance (SSD)	2 mm	2 mm
Field size indicator	Gantry angle 0°, collimator angle 0°, 100 cm SSD, field sizes of 5 × 5 cm², 10 × 10 cm², and 20 × 20 cm²	2 mm	1 mm
Dynamic/universal/virtual wedges	Run for at least one angle	Functional	
Dosimetry			
Output constancy—photons	All energies	3%	
Output constancy—electrons	All energies	3%	
Safety			
Door interlock		Functional	
Motion interlock		Functional	
Audiovisual monitors		Functional	
Room radiation monitors (if used)		Functional	
Beam status indicator		Functional	
Beam interrupt		Functional	

Note: SRS, stereotactic radiosurgery; SRT, stereotactic radiotherapy.

439

TABLE 70.2 Medical Accelerator QA Monthly Test

Procedure	Description	Tolerance	
		Non-SRS/SRT	SRS/SRT
	Mechanical		
Optical distance indicators	Gantry angle 0°, 90 cm, 100 cm, and 110 cm SSD	2 mm	2 mm
Treatment couch position indicators	Mechanical and digital couch position readouts at 90 and 110 cm SSD to the surface of couch	2 mm	1 mm
	Mechanical and digital couch position readouts at couch angle 90° and 270°	1°	0.5°
Field size indicators	Gantry angle 0°, collimator angle 0°, 100 cm SSD, field sizes of 5 × 5 cm², 10 × 10 cm², and 20 × 20 cm². Different field sizes may be examined at different gantry angles if appropriate and efficient	2 mm for the summation of total, 1 mm on a side	
Light/radiation field coincidence	Geometric alignment of the radiation and optical field edges	2 mm for the summation of total, 1 mm on a side	
Laser/crosswire	Distance check device for lasers compare with front pointer. Alignment of crosswire and appropriate lasers	1 mm	
Collimator angle indicators	Digital and mechanical collimator angle readouts must be verified using a spirit lever or other appropriate leveling devices at 0°, 90°, and 270°	1.0°	
Gantry angle indicators	Digital and mechanical gantry angle readouts must be verified using a spirit lever or other appropriate leveling device at 0°, 90°, 180°, and 270°	1.0°	
Couch angle indicators	The couch angle must be verified over an appropriate clinical range, usually between 90° and 270°	1.0°	0.5°
	Dosimetry		
Output constancy—photons	Using a calibrated dosimetry system to check the output of all available photon beams against yearly reference dosimetry	2%	
Output constancy—electrons	Using a calibrated dosimetry system to check the output of all available electron beams against yearly reference dosimetry	2%	
Backup monitor chamber constancy	Readouts compare to the primary monitor chamber	2%	
Dose rate constancy	Dose monitoring as a function of dose rate	2% of normal treatment dose rate	2% of stereo dose rate
Output linearity	Dose monitoring as a function of monitor units	2%	
Energy constancy—photons	Measurements at two depths in an appropriate phantom to confirm that the depth dose has not changed since commissioning of this unit	2%	
Energy constancy—electrons	Measurements at two depths in an appropriate phantom to confirm that the depth dose has not changed since commissioning of this unit	2 mm	
Beam flatness and symmetry constancy—photons	Flatness and symmetry change from baseline	1%	
Beam flatness and symmetry constancy—electrons	Flatness and symmetry change from baseline	1%	
Dynamic/universal/virtual wedges	Spot-check wedge factors for at least one wedge/energy	2%	
	Safety		
Wedge, tray, cone interlocks		Functional	
Accessories and centering		Functional	
Laser guard-interlock test		Functional	

Note: SRS, stereotactic radiosurgery; SRT, stereotactic radiotherapy.

TABLE 70.3 Medical Accelerator QA Annual Test

Procedure	Description	Tolerance	
		Non-SRS/SRT	SRS/SRT
Mechanical			
Collimator/gantry/couch rotation isocenter	Using film, star, or spoke patterns are produced. Tolerance levels refer to the maximum diameters of the measured intersections for each spoke.	1 mm from baseline	
Coincidence of radiation with optical and mechanical isocenters	The radiation, optical, and mechanical isocenters are determined with reference to the laser system	2 mm from baseline	1 mm from baseline
Couch top sag	Couch top sag is measured with 70 kg at the end with the couch extended to the isocenter	2 mm from baseline	
Electron applicator interlocks		Functional	
Stereotactic accessories, lockouts, etc.		Functional	
Dosimetry			
Photons flatness and symmetry reproducibility vs. gantry angle	Film or other appropriate devices are used to measure the beam flatness and symmetry at gantry angles 0°, 90°, 180°, and 270°	1% from baseline	
Electrons flatness and symmetry reproducibility vs. gantry angle	Film or other appropriate devices are used to measure the beam flatness and symmetry at gantry angles 0°, 90°, 180°, and 270°	1% from baseline	
Spot-check of field size dependent output factors—photons	Check two or more field sizes for each energy	2% from baseline for field size $<4 \times 4$ cm^2; 1% from baseline for field size $\geq 4 \times 4$ cm^2	
Spot-check of applicator size-dependent output factors—electrons	Check at least one applicator/energy	2% from baseline	
Beam energy check—photons	Check PDD$_{10}$ or TMR$_{20/10}$	1% from baseline	
Beam energy check—electrons	Check R$_{50}$	1 mm from baseline	
Output linearity—photons	Dose monitoring as a function of monitor unit	5% (2–4 MU); 2% $\geq$5 MU	
Output linearity—electrons	Dose monitoring as a function of monitor unit	2% $\geq$5 MU	
Output constancy vs. dose rate—photons	Dose monitoring as a function of dose rate	2%	
Output constancy vs. gantry angle—photons	Gantry 0°, 90°, 180°, and 270°; an ion chamber with buildup cap may be positioned at the isocenter for these measurements	1% from baseline	
Output constancy vs. gantry angle—electrons	Gantry 0°, 90°, 180°, and 270°; an ion chamber with buildup cap may be positioned at the isocenter for these measurements	1% from baseline	
Wedge transmission factor constancy		2% from baseline	
Dynamic/universal/virtual wedges	Check wedge angle for 60°, full field and spot-check for intermediate angle, field size	2% from baseline	
Accessory transmission factor constancy		2% from baseline	
SRS arc rotation mode: monitor units set vs. delivered		1.0 MU or 2% (whichever is greater)	
SRS arc rotation mode: gantry arc set vs. delivered		1.0° or 2% (whichever is greater)	
Photon/electron output calibration	Perform a full TG-51 calibration		
Safety			
Emergency off		Functional	
All other manufacturer's test procedures		Functional	

Note: SRS, stereotactic radiosurgery; SRT, stereotactic radiotherapy.

or physical accidents. These patterns of failure need to be considered when establishing periodic QA programs. Although the roles of performing the various tasks are typically divided among physicists, dosimetrists, therapists, and engineers, the overall responsibility for linac QA always lies with the qualified medical physicists.

Summary

Recommendations and frequencies for linac QA are summarized in Table 70.1 (daily), Table 70.2 (monthly), and Table 70.3 (annually). Each table has specific recommendations based on TG-40 and TG-142, which are differentiated into non–stereotactic radiosurgery/stereotactic radiotherapy and stereotactic radiosurgery/stereotactic radiotherapy machines. There are also simple descriptions of the experimental conditions for performing these QA tests. Finally, the quality control record contains the results of the tests, the date(s) on which they were performed, and the identification of the tester and the supervising physicist.

References

Klein, E. E., J. Hanley, J. Bayouth, F. F. Yin, W. Simon, S. Dresser, C. Serago et al. 2009. Task Group 142 report: Quality assurance of medical accelerators. *Med. Phys.* 36 (9): 4197–4212.

Kutcher, G. J., L. Coia, M. Gillin, W. F. Hanson, S. Leibel, R. J. Morton, J. R. Palta et al. 1994. Comprehensive QA for radiation oncology: Report of AAPM Radiation Therapy Committee Task Group 40. *Med. Phys.* 21 (4): 581–618.

71

Linear Accelerator–Based MV and kV Imaging Systems

William Y. Song
University of California San Diego

Jean-Pierre Bissonnette
Princess Margaret Hospital

Introduction

Image-guided radiation therapy has emerged in the past decade as the next paradigm in state-of-the-art care in radiation oncology (Herman et al. 2001; Jaffray et al. 2002; McBain et al. 2006). Arguably, the core imaging technologies that triggered the widespread practice are electronic portal imaging devices followed by cone-beam computed tomography (CBCT) systems that are attached to linear accelerators such that MV and/or kV in-room imaging of patients in the treatment position can be taken (see Figure 71.1). As such, the necessity to ensure optimal functioning of these imaging systems is essential in quality patient care, especially in the era of extreme hypofractionation and stereotactic body radiotherapy (Song et al. 2006; Timmerman et al. 2007). As of this writing, however, there exist no consensus guidelines for comprehensive quality assurance (QA) of CT-based image guidance systems, whereas abundant information on the topic exists in the literature (Bissonnette 2007; Bissonnette et al. 2008; Yoo et al. 2006; Menon and Sloboda 2004; Gayou and Miften 2007; Balter and Antonuk 2008), potentially creating confusion among practitioners and, worse, creating disparity in standards of care. The American Association of Physicists in Medicine recognizes this and a task group (TG-179) has recently been formed to resolve this issue. The general themes consist of safety, geometric accuracy, image quality, and dose. Consistent with these themes, in this chapter, we evaluate and suggest a set of generic QA tests and frequencies to be carried out to maintain a proper level of performance for routine clinical use. The discussions and recommendations will be limited to MV and kV active matrix flat panel imagers that are attached to conventional linear accelerators only and, hence, systems such as helical tomotherapy, Cyberknife, and CT-on-rails will not be reviewed but are covered elsewhere in this book.

MV In-Room Imager

The American Association of Physicists in Medicine TG-58 report (Herman et al. 2001) recommends a set of daily and monthly QA tasks for electronic portal imaging devices that is helpful but is now mostly outdated because the ion chambers and CCD camera–based systems are largely replaced by the high-performance active matrix flat panel imagers (Antonuk et al. 1996). For example, the Las Vegas phantom is very limited in its scope to evaluate the various image quality aspects that have now become important with the current imagers, which offer much improved portal image quality. With this in mind, the following suggestions are made and are listed in Table 71.1. These suggestions conform to the general themes of safety, geometric accuracy, and image quality. The dose measurements are not listed though because the current linear accelerators are not only quite stable in output but also the monthly QA of the machine will catch any drifts. This is not true for kV x-rays, as will be discussed in the next section.

Daily QA includes safety checks only and can be performed only on days that the imager is used. In weekly QA, however, the end-to-end testing of imaging, registration, and couch shift should be performed for both 2D and 3D systems. If one has an MV-CBCT–capable system, then perhaps such a procedure can be implemented where first a 2D image is taken, the image registered, the couch shifted, and then the MV-CBCT can be acquired to verify the correct couch shift. A simple, cubic phantom with a metal ball bearing at the center can be used for this test. For monthly QA, much more comprehensive testing should include (1) detector position accuracy/reproducibility, (2) imager calibration and response constancy, and (3) image quality. The detector position accuracy/reproducibility can be performed either with in-room lasers and/or physical rulers such as meter sticks.

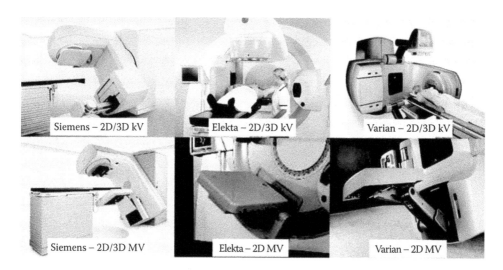

FIGURE 71.1 MV and kV in-room imaging systems that are currently under development and/or in the market for sale. Siemens electronic portal imaging device system (bottom left) is the only MV imager that can perform both 2D and 3D imaging. Siemens Artiste® system (top left) is the latest kV/MV imager but is currently unavailable in kV mode, whereas Elekta Synergy® (top middle) and Varian Trilogy® (top right) systems are in clinical practice.

Imager calibration includes taking dark and flood fields for each energy and dose rate to correct for stationary nonuniformities in detector response and dark currents. If the electronic portal imaging device is used for dosimetry, including IMRT QA, the central axis pixel response to a standard MU and field size should be measured and corrected for as well. For MV-CBCT systems, an additional geometry calibration outlined by Pouliot et al. (2005) should be performed. For image quality tests, for 2D imaging, phantoms that are developed around the concept of contrast-to-noise ratio such as the QC-3 phantom (Standard

TABLE 71.1 MV In-Room Imager QA Recommendations

	Imaging Type and Tolerance	
	2D Radiograph	3D CBCT
Daily		
Door interlock (T)	Functional	
Collision interlocks (T)	Functional	
Warning lights and sounds (T)	Functional	
Weekly		
2D match and couch shift (T)	2 mm	
3D match and couch shift (T)		2 mm
Monthly		
Detector position accuracy/reproducibility (P)		2 mm
Significant artifacts (P)	None	
Imager calibration—dark/flood fields (P or E)	Performed	
Central axis pixel response for dosimetry (P or E)	1%	
Spatial resolution (P)	Baseline	Baseline
Low contrast resolution (P)	Baseline	Baseline
Noise (P)	Baseline	Baseline
Uniformity (P)		Baseline
Slice thickness (P)		Baseline
CT number accuracy (P)		Baseline
Yearly		
Review of daily/weekly/monthly results and trends (P)	Satisfactory	

Note: P, physicist; E, engineers; T, therapist.

Imaging, Middleton, WI) should be used. This type of phantom will allow calculations of spatial resolution, contrast resolution, and noise with a single image. For MV-CBCT imaging, in addition to the standard image quality indices, the CT number accuracy and reproducibility should be measured as well because the importance of dose calculation on the CBCT images is ever increasing with the implementation of sophisticated adaptive radiotherapy protocols (Gayou and Miften 2007; Pouliot et al. 2005). The CatPhan phantoms (The Phantom Laboratory, Salem, NY) can be used, but because of the high energy of the imaging beams, specialized phantoms should really be devised (Gayou and Miften 2007). It is advised that reference images (2D and 3D) at the time of acceptance/commissioning be taken to establish baselines. The yearly QA can then simply be the review of all QAs done for the year.

kV In-Room Imager

As stated earlier, there is currently no consensus guideline for kV in-room imaging systems, either in 2D or 3D mode. However, a number of investigators have thoroughly studied the two commercial systems (i.e., Varian OBI® and Elekta XVI®) and their recommendations also revolve around safety, geometric accuracy, image quality, and dose (Bissonnette 2007; Bissonnette et al. 2008; Yoo et al. 2006; Song et al. 2008; Islam et al. 2006). In accordance with the literature, the following suggestions are listed in Table 71.2.

Daily QA includes safety checks, x-ray tube warm-up, and 2D radiograph alignment with the isocenter. The 2D image alignment, which is absent in MV imager QA, is necessary because the kV isocenter is not shared directly with the treatment beam. For this test, a cubic phantom with a metal ball at the center should suffice (usually supplied by the vendor). The phantom is first aligned to the in-room lasers, followed by AP and LAT radiographs; subsequently, the digital graticules can be used to evaluate the amount of metal ball offset. For Elekta XVIs, because digital graticules are not available, a full 3D-CBCT can be taken instead. For weekly QA, full end-to-end testing of imaging, registration, and couch shift should be performed for both 2D and 3D systems. This test is necessary to simultaneously check the imager geometry, image registration accuracy, and remote-controlled couch movements. For monthly QA, a comprehensive

TABLE 71.2 kV In-Room Imager QA Recommendations

	Imaging Type and Tolerance	
	2D Radiograph	3D CBCT
Daily		
AP and LAT images with digital graticule (T)	2 mm	
X-ray warm-up (T)	Performed	
Door interlock (T)	Functional	
Collision interlocks (T)	Functional	
Warning lights and sounds (T)	Functional	
Weekly		
2D match and couch shift (T)	2 mm	
3D match and couch shift (T)		2 mm
Monthly		
MV/kV isocenter coincidence (T)	2 mm	
Detector/source position accuracy/reproducibility (P)	2 mm	
Significant artifacts (P)	None	
Spatial resolution (P)	Baseline	Baseline
Low contrast resolution (P)	Baseline	Baseline
Noise (P)	Baseline	Baseline
Uniformity (P)		Baseline
Slice thickness (P)		Baseline
CT number accuracy (P)		Baseline
Yearly		
Measure half-value layer for each kV (P or E)	Baseline	
Measure reference dose (P)	2%	
Flexmap measure and update (P or E)	Performed	
Review of daily/weekly/monthly results and trends (P)	Satisfactory	

Note: P, physicist; E, engineers; T, therapist.

test that includes (1) detector position accuracy/reproducibility, (2) MV/kV isocenter coincidence, and (3) image quality is to be performed. The detector position accuracy/reproducibility test can be performed similarly to that for the MV imager. For the MV/kV isocenter coincidence test, a common method consists in placing a metal ball bearing near the isocenter and using portal imaging to compare the centroid of the image of the ball bearing to the field edges. The ball bearing can thus iteratively be moved to the true radiation isocenter. Once this is accomplished, the subsequent kV images can be taken to measure the offset. For image quality tests, for 2D imaging, phantoms that are designed for kV imagers such as the QCkV-1 phantom (Standard Imaging) should be used. For kV-CBCT imaging, the CatPhan phantoms or the American Association of Physicists in Medicine CT performance phantoms (CIRS, Norfolk, VA) can be supplied by the vendor for use. If quantitative CT is to be used with kV-CBCT images, the CT number accuracy and reproducibility may also be measured, as in the MV-CBCT images. For annual QA, the half-value layers for each energy and dose for each clinical imaging protocol should be measured and compared to the baseline. The half-value layer measurements are necessary to ensure that the x-ray spectrum does not change, which may indicate wearing out of the x-ray tubes. For dose measurements, an ion chamber reading at the center of a stack of solid water can be used. In addition to this, for Elekta XVIs, the flexmap should be measured and refreshed at least six-monthly or after service interventions because image quality depends critically on this where the projections are digitally shifted according to this map (Jaffray et al. 2002). For the Varian OBI, the imager is physically moved instead to compensate for the flexes and thus the image quality should be monitored carefully to detect any drifts.

References

Antonuk, L. E., J. Yorkston, W. Huang, H. Sandler, J. H. Siewerdsen, and Y. el-Mohri. 1996. Megavoltage imaging with a large-area, flat-panel, amorphous silicon imager. *Int. J. Radiat. Oncol. Biol. Phys.* 36 (3): 661–672.

Balter, J. M., and L. E. Antonuk. 2008. Quality assurance for kilo- and megavoltage in-room imaging and localization for off- and online setup error correction. *Int. J. Radiat. Oncol. Biol. Phys.* 71 (1 Suppl): S48–S52.

Bissonnette, J. P. 2007. Quality assurance of image-guidance technologies. *Semin. Radiat. Oncol.* 17 (4): 278–286.

Bissonnette, J. P., D. Moseley, E. White, M. Sharpe, T. Purdie, and D. A. Jaffray. 2008. Quality assurance for the geometric accuracy of cone-beam CT guidance in radiation therapy. *Int. J. Radiat. Oncol. Biol. Phys.* 71 (1 Suppl.): S57–S61.

Gayou, O., and M. Miften. 2007. Commissioning and clinical implementation of a mega-voltage cone beam CT system for treatment localization. *Med. Phys.* 34 (8): 3183–3192.

Herman, M. G., J. M. Balter, D. A. Jaffray, K. P. McGee, P. Munro, S. Shalev et al. 2001. Clinical use of electronic portal imaging: Report of AAPM Radiation Therapy Committee Task Group 58. *Med. Phys.* 28: 712.

Islam, M. K., T. G. Purdie, B. D. Norrlinger, H. Alasti, D. J. Moseley, M. B. Sharpe, J. H. Siewerdsen, and D. A. Jaffray. 2006. Patient dose from kilovoltage cone beam computed tomography imaging in radiation therapy. *Med. Phys.* 33 (6): 1573–1582.

Jaffray, D. A., J. H. Siewerdsen, J. W. Wong, and A. A. Martinez. 2002. Flat-panel cone-beam computed tomography for image-guided radiation therapy. *Int. J. Radiat. Oncol. Biol. Phys.* 53 (5): 1337–1349.

McBain, C. A., A. M. Henry, J. Sykes, A. Amer, T. Marchant, C. M. Moore, J. Davies, J. Stratford, C. McCarthy, B. Porritt, P. Williams, V. S. Khoo, and P. Price. 2006. X-ray volumetric imaging in image-guided radiotherapy: The new standard in on-treatment imaging. *Int. J. Radiat. Oncol. Biol. Phys.* 64 (2): 625–634.

Menon, G. V., and R. S. Sloboda. 2004. Quality assurance measurements of a-Si EPID performance. *Med. Dosim.* 29 (1): 11–17.

Pouliot, J., A. Bani-Hashemi, J. Chen, M. Svatos, F. Ghelmansarai, M. Mitschke, M. Aubin, P. Xia, O. Morin, K. Bucci, M. Roach 3rd, P. Hernandez, Z. Zheng, D. Hristov, and L. Verhey. 2005. Low-dose megavoltage cone-beam CT for radiation therapy. *Int. J. Radiat. Oncol. Biol. Phys.* 61 (2): 552–560.

Song, W. Y., S. Kamath, S. Ozawa, S. A. Ani, A. Chvetsov, N. Bhandare, J. R. Palta, C. Liu, and J. G. Li. 2008. A dose comparison study between XVI and OBI CBCT systems. *Med. Phys.* 35 (2): 480–486.

Song, W. Y., B. Schaly, G. Bauman, J. J. Battista, and J. Van Dyk. 2006. Evaluation of image-guided radiation therapy (IGRT) technologies and their impact on the outcomes of hypofractionated prostate cancer treatments: A radiobiologic analysis. *Int. J. Radiat. Oncol. Biol. Phys.* 64 (1): 289–300.

Timmerman, R. D., B. D. Kavanagh, L. C. Cho, L. Papiez, and L. Xing. 2007. Stereotactic body radiation therapy in multiple organ sites. *J. Clin. Oncol.* 25 (8): 947–952.

Yoo, S., G. Y. Kim, R. Hammoud, E. Elder, T. Pawlicki, H. Guan, T. Fox, G. Luxton, F. F. Yin, and P. Munro. 2006. A quality assurance program for the on-board imagers. *Med. Phys.* 33 (11): 4431–4447.

Elekta Multileaf Collimator

Chihray Liu
University of Florida

Darren Kahler
University of Florida

Thomas Simon
Sun Nuclear Corporation

Introduction

The concept of the multileaf collimator (MLC) system was introduced by Proimos (1960), Trump et al. (1961), and Takahashi (1965) in the early sixties and systems became available in the United States in the late eighties (Moeller 1989; Galvin, Smith, and Lally 1993; Jordan 1991; Jordan and Williams 1991). Initially, MLC systems were used only as a replacement for Cerrobend blocks in the delivery of conventional radiation therapy. Dosimetry issues were not of great concern until relatively recently, when MLCs became used in the delivery of intensity-modulated radiation therapy (IMRT). IMRT uses multiple treatment-field segments to form a desired radiation intensity pattern. Radiation dose to surrounding critical structures is minimized and a conformal dose distribution is delivered to the target structures. Because the IMRT field segments contain multiple field edges, the requirements for MLC mechanical precision and for TPS modeling accuracy of the MLC penumbra are more stringent than the requirements for simple block replacement. In this chapter, we describe the characteristics of the Elekta Synergy MLC (Elekta Ltd, Crawley, West Sussex, UK). A number of issues need to be addressed to model this MLC's characteristics accurately in the TPS. Because the MLC leaf ends are rounded, the radiation does not follow the light field projection at the MLC's edge in the same way that it does for a collimator jaw. This rounded leaf end construction causes the radiation field-edge position to deviate slightly from its nominal value, with the degree of deviation dependent on the off-axis position of the leaf. In addition, the leaves are of tongue-and-groove design, which introduces several effects that should be accounted for. If opposite sides of a radiation field are defined

by the sides of two leaves, the tongue-and-groove construction of the leaves will cause the field to be slightly smaller than, and slightly asymmetric compared to, the same field with the sides defined by the edge of the collimator. The radiation transmission characteristics of the MLC are also different from those of either the collimator jaws or of Cerrobend blocks. Another issue for the Elekta device is that a given MLC leaf cannot interdigitate with the adjacent leaf of the opposing leaf bank, so it is sometimes necessary to place the inherent 6-mm junctions of opposing leaf pairs beneath one of the backup jaws to minimize the radiation exposure outside of the treatment field. The backup jaws are only 3 cm thick and have, per our measurements, a radiation transmission of approximately 10%. Dose calculation accuracy in the area under these jaws is therefore important, especially because the area may overlie critical structures. All of these effects can adversely impact the dosimetric accuracy of IMRT delivery if they are not correctly modeled by the TPS. Below, we discuss the various MLC characteristics and parameters that must be modeled by the TPS and we describe an effective MLC QA and calibration method for maintaining the positioning accuracy of the MLC leaves using a diode array QA device. An example of the QA method is given and the findings are discussed.

Elekta MLC Characteristics

The Synergy's gantry head has three major components that are used for field shaping: a 40-leaf-pair MLC, an upper "backup" jaw housed immediately above the MLC that tracks the MLC's movement to decrease the interleaf radiation transmission, and a lower jaw that is housed below and moves orthogonal to the MLC (Precise Treatment System Clinical Mode Users Manual R50 2003).

An optical video camera system is used to calculate the MLC leaf positions in real time. The advantage of the camera system is that the leaf positioning accuracy is not affected by the gantry rotation because the absolute position of each leaf is monitored by the camera system, which is rigidly attached to the gantry head.

Field Size Definition

The radiation field width in the direction of leaf movement is defined by both the MLC leaves and the backup jaw. During the Linac acceptance testing, the backup jaw is typically programmed to always remain 1–2 mm behind the MLC leaf ends that define the outermost edge of the treatment field. The radiation field width will change if the backup jaw is moved away from this position. This will occur if, for example, a field is defined by the MLC only, with the backup jaws retracted.

Rounded Leaf-End Offset

The Synergy's MLC leaf ends are rounded, so the radiation field edge does not follow the divergence of the light field projection in the same way that it does for a collimator jaw. The difference between the light field projection and the actual radiation field projection has been discussed (Jordan and Williams 1994). It is therefore very important to scan radiation fields defined only by the MLC leaves at different off-axis positions when collecting commissioning data for the TPS. This is done to characterize the difference between the nominal MLC leaf positions and the measured radiological positions for the TPS model. It is also important to scan the backup-jaw defined fields alone so that the TPS can model the primary radiation transmission through the backup jaws correctly.

Tongue and Groove

The purpose of the tongue-and-groove construction of the MLC is to minimize interleaf radiation transmission. However, this construction causes the width of each leaf to project to 1.1 cm in the isocenter plane rather than the nominal 1.0-cm value. If this is not taken into account in the TPS model, a significant error in the calculated dose will result. The tongue-and-groove effect is demonstrated by comparing field edges defined by the sides of the MLC leaves to collimator-defined field edges. A nominal 20.0-cm-wide radiation field whose field-width edges are defined by the sides of two MLC leaves has a full width at half maximum value of 19.8 cm. The 2-mm deficit occurs on the side of the field defined by a leaf tongue, whereas the grooved side of the leaf on the opposite side of the field projects the same field edge position as would the collimator jaw. Although the TPS model will account for this 2-mm deficit, it will be modeled symmetrically, that is, the profile generated by the treatment planning system will have a difference, with respect to the actual treatment field, of 1 mm on each side. In addition, the characteristics of the radiation-defined field shape depend on whether the field edges are defined by the leaf ends or by the leaf sides (Butson, Yu, and Cheung 2003).

Leaf Abutment and Backup Jaw Interaction

The Elekta MLC system always maintains, at minimum, a 6-mm gap between opposing leaf ends. Interdigitation with the adjacent leaf of the opposing leaf bank is not allowed. The purpose of the backup jaw is to minimize the interleaf radiation transmission. The backup jaws are only 3 cm thick and their transmission, which is approximately 10%, should be explicitly taken into account in the dose calculation algorithm of the TPS.

Radiation Field Edge Effects—MLC and Backup Jaw

The backup jaws are designed to track the MLC leaves. Incorrect modeling of the interaction between the MLC leaves with the backup jaws near the field edge can result in dose calculation inaccuracies. As mentioned above, the backup jaws are most often programmed to remain 1 to 2 mm behind the MLC leaf ends that define the outermost edges of the field. The TPS dose calculation engine should take this into account. In addition, during IMRT delivery, the leaf pair that lies just outside of the treatment field and is closest to the edge of the lower jaw will be automatically retracted, with the result that the field edge will be defined by the lower jaw alone for the actual radiation delivery. The TPS needs to model this; otherwise, the tongue-and-groove effect will be overestimated.

MLC Intraleaf and Interleaf Transmission

MLC leaf radiation transmission measured using film is approximately 1.5% (Huq et al. 2002). We have confirmed this with measurements taken with a small-volume chamber placed directly under the MLC. The interleaf leakage adds another 0.7% to the leaf transmission.

Output Factors

The MLC-defined field is closely correlated to that of the area encompassed by the backup jaws because the backup jaws track the MLC leaf bank. The output factor is accurately predicted by the equivalent square formalism (Palta, Yeung, and Frouhar 1996). For cases in which the backup jaw is far from the field edges defined by the MLC leaves, the output factor is better approximated by calculating the equivalent square field size using the detector's eye view method proposed by Kim et al. (1997).

Sources of Error

Based on the above discussion, we can summarize the potential sources of error in MLC system TPS modeling as follows:

1. Leaf-end field edge modeling at different off-axis positions
2. Leaf-side modeling
3. Backup jaw transmission modeling
4. Backup jaw and MLC leaf field-edge interaction

5. Lower jaw and MLC interaction modeling in the direction orthogonal to the MLC leaf bank movement
6. MLC intra- and interleaf transmission

Electronic Portal Imaging Device

The Elekta electronic portal imaging device system is a Perkin Elmer Amorphous Silicon Detector. It has a resolution of 1024 × 1024 16-bit pixels, with a 41 × 41 cm² detector panel size (approximately 26 × 26 cm²) at isocenter. The system is fixed at an approximate 160-cm source to detector distance.

Radiation-Defined Reference Line Method

We termed our MLC calibration technique the *radiation-defined reference line* (RDRL) method because the radiation-defined edge of the backup jaw is used as the calibration reference line. Application of the method is based on three assumptions. First, that the leading edge of an MLC leaf bank is parallel to the leading edge of its backup jaw. Second, that the backup jaw can provide a reproducible and uniform radiation field edge, the RDRL. Third, that the measured radiation field edge created by each leaf end is representative of that leaf's position.

Inherent in the third assumption is the requirement that the detectors have a spatial measurement resolution that does not suffer from signal averaging in the high spatial frequency of the penumbra. Dempsey et al. (2005) have shown that measurements with a detector size of 2 mm or smaller are sufficient for IMRT fields shaped with MLCs. The detector size of the electronic portal imaging device satisfies this requirement; however, not all of the MLC leaves can be measured with the electronic portal imaging device at one time, and going in and out of the room to adjust the position of the electronic portal imaging device is cumbersome and time-consuming. An easier, more efficient way to perform the MLC calibration and/or QA using the RDRL method is with the Profiler2 device (Sun Nuclear Inc., Melbourne, FL). The Profiler2 is a two-dimensional diode array device that has a diode spacing of 4 mm, with 57 diodes lying along the *x* axis and 83 diodes lying along the *y* axis of the device. This gives the device the capability of measuring a 20 × 30 cm radiation field at 100 cm source-to-surface distance. The inherent buildup of the device is 1 g/cm³. The diode detector size is 0.64 mm², which satisfies the spatial measurement resolution requirement; however, the detector location must be known with a precision that is better than the desired leaf position accuracy.

MLC Leaf Calibration

Elekta MLCs have traditionally been calibrated using standard measurement tools, for example, film and scanning water tanks, to determine what are termed *major and minor leaf offsets*, as illustrated in Figure 72.1. A reference leaf pair, leaf pair no. 20, is used in the control of the MLC. The major leaf offset is a calibration value that defines the field size created by the reference

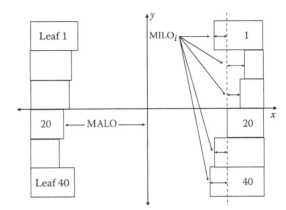

FIGURE 72.1 Minor leaf offsets are defined as the spatial offset of each leaf in a leaf bank relative to the reference leaf, leaf no. 20. Major leaf offsets are defined as the distance of the reference leaf to the radiation center of the beam.

leaf pair. Minor leaf offsets are the position alignment errors of the other leaves in relation to the reference leaf and are the main focus of the RDRL method.

The method uses the backup jaw edge to precisely locate all of the *y*-axis detectors' offsets relative to the reference detector, which is positioned in the reference leaf's direction of movement as illustrated in Figure 72.2. These relative detector positions are referred to as RDO_j, where *j* is the *y*-axis diode number $1 \le j \le 83$; they effectively create a uniform RDRL. Once these detector positions are known, the detector array is used to measure the position of each leaf's projected edge at detector *j*.

In addition, there are major and minor gain calibrations for MLCs. The gain calibration is used to ensure the linear tracking of the radiation field width when it changes. The major gain calibration is again relative to leaf pair no. 20, and it is usually adjusted during the machine commissioning using a scanning water tank. The minor gain calibration is used to track the linear

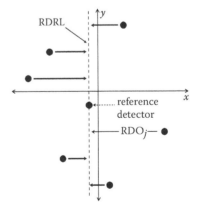

FIGURE 72.2 The RDRL method requires precisely known detector positions; small positional errors in the array's detectors can disrupt this method by introducing false offsets into the leaf positions. The detector offsets are corrected for, which effectively creates uniform detector positions.

movement of all leaves with respect to leaf no. 20, and it is also adjusted during the machine commissioning. Typical major and minor gain calibrations do not vary much over a long period, and so they only need to be checked annually.

The beam edge defined by leaf no. 20 is characterized in the TPS with a beam modeling parameter. It is important to ensure that appropriate position offset corrections are used to characterize the field edge at the off-axis positions. The TPS assumes that all 40 leaf pairs behave the same as the leading leaf. Therefore, the MLC QA should ensure that the major and the minor leaf calibrations are accurate to within 1 mm, which is the manufacturer's specification. Finally, gravity can cause the position of the MLC leaves to change with both gantry and collimator rotation. The reported reproducibility of the MLC leaf positions is within 0.3 mm, which is the inherent uncertainty in the leaf position determined with the optical camera system used in the Elekta Linac (Sastre-Padro, van der Heide, and Welleweerd 2004).

Delivery: MLC-Specific QA/Calibration

The Synergy MLC system QA is divided into two parts: (1) the reference leaf position at different off-axis locations is compared with a baseline value; (2) the relative positions of all of the other leaves of the same bank are checked to verify that they fall within a predefined specification with respect to the reference leaf position. As discussed previously, the RDRL method uses the backup jaw to define a radiation field edge to be used as a reference line, and each leaf position is determined with respect to this line. For the first part of the QA, determining the reference leaf position at different off-axis locations, the same principle is applied. The offsets of the reference leaf with respect to the RDRL at different off-axis locations are determined and compared to baseline values that are determined during the linear accelerator commissioning. We believe that this method is better than the commonly used "picket fence" test, which fails to adequately discern positional inaccuracies of leaf pairs for the two banks. The remainder of this section details the implementation of the RDRL method using the Profiler2 device, with the description of the minor leaf offset calibration as the main focus.

Step 1. The Profiler2 is placed on the treatment table at a source-to-surface distance of 80 cm and its x–y axis is aligned with the collimator crosshair so that the positive y direction of the device is pointing toward the gantry. The collimator head is then rotated to 180° so that the ascending order of the leaf index, 1 through 40, matches the ascending order of the y-axis diode index, 1 through 83. With this setup, every other diode will lie at the center of an MLC leaf. If QA is being performed on the Synergy-S model, the device should be placed at a source-to-surface distance of 100 cm, rather than 80 cm, because the Synergy-S leaf width is 4 mm rather than 1 cm. The Profiler2 is then set to record in a continuous mode at a 100-ms data recording rate.

Step 2. A backup-jaw-only field (MLC leaves retracted) is delivered using 10 control points, with 30 MU delivered per control point. The first control point is for a 20 × 40 cm open field,

with the backup jaws set at ±10 cm with respect to the machine's central axis and with the lower jaws defining the 40-cm-field length. The remaining nine control points are delivered in a step-and-shoot fashion, with one backup jaw moving in 1-mm increments from a position of −4 mm to +4 mm with respect to the machine's central axis and the other jaw settings remaining the same. The ±4 mm range is used to ensure that the radiation-defined field edge can be determined in the event of an imperfect Profiler2 and/or collimator crosshair alignment.

Step 3. Each y-axis diode reading obtained for each of the nine control points (−4 to +4 mm) is then normalized with respect to its 20 × 40 cm open field reading. These readings are used to determine the position of 50% radiation intensity, with respect to the open field value, for each y-axis diode. These positions define a line of 50% radiation intensity, which is the RDRL for one leaf bank.

Step 4. Steps 2 and 3 are repeated for the other backup jaw to obtain the RDRL for the other leaf bank. The same steps are then performed using MLC-only fields (backup jaws retracted) for both leaf banks to determine the 50% radiation intensity position for each of the individual MLC leaves of both banks. All of these steps combined result in four Profiler2 data files, one for each backup jaw–only field and one for each MLC–only field.

Step 5. To obtain the corresponding position of the 50% radiation intensity value of the jaws/MLCs near each diode position, a linear interpolation between the reading lying above, and the reading lying below, the 50% open field value for the same diode is performed. These readings correspond to two of the nine −4 to +4 mm control points. This interpolation is done for all of the y-axis diodes for all four of the field setups. The results obtained for each of the backup jaw–only fields give the RDRL for each leaf bank, and the results obtained for the MLC-defined fields give a 50% radiation intensity position for each of the individual MLC leaves. The measurements for the Synergy MLC indicate that the change in slope in the 50% radiation intensity region is 15%/mm and 13.5%/mm for the backup jaws and the MLC leaves, respectively. The 50% position of each individual leaf of a leaf bank is then subtracted from the RDRL position corresponding to that leaf to obtain an offset value. The minor leaf offset values, that is, the differences between the offset values for each leaf with respect to the offset value of the reference leaf, leaf no. 20, are then calculated. Figure 72.3a shows the detector offsets, with respect to the central axis detector, based on the backup jaw position determined using three profiles. The scatter in the data represents both the misalignment of the diode array and the variable diode response. In our experiments, the data show that the reproducibility of the measured RDRL and MLC leaf offsets is better than 0.1 and 0.3 mm, respectively. Figure 72.3b shows the MLC alignment result using the technique for one leaf bank.

An in-house program (available upon request) was written to read the four Profiler2 data files (from step 4) and generate a file that contains the correction units, 12 units per millimeter, of minor leaf offsets. This text file, Minor_Offsets.txt, can simply be imported to the Elekta Linac control software, RT-Desktop (version 6.0 or above), to perform all of the leaf corrections at once.

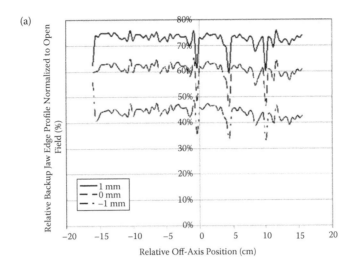

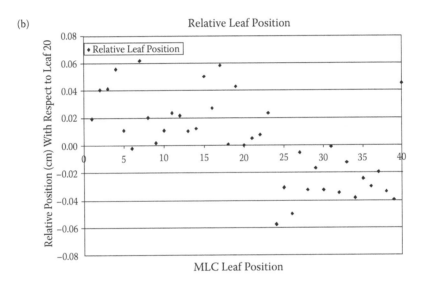

FIGURE 72.3 (a) Example of relative backup jaw edge profile normalized to an open field. (b) Example of relative MLC leaf position with respect to leaf no. 20 for the same bank.

The minor gain offset file, Minor_Gains.txt, can be imported to RT_Desktop as well.

For the absolute MLC off-axis positions, the reference leaf position at different off-axis positions can be established as the baseline during the machine commissioning. The RDRL method can then be used for the MLC QA to obtain the MLC reference leaf position to compare to the baseline. Minor gain calibrations can be acquired using a 15 × 40 cm field with the Profiler2 set up as described in step 1. The RDRL calibration procedure is then followed with the table moved ±60 mm in the direction of the *x* axis of the Profiler2. The average minor gain offsets at these two positions are obtained and imported into the RT-Desktop for the leaf alignment. A 3D water tank data acquisition system should be used to check and/or to adjust the major leaf offset and major gain.

MLC leaf positions on the Elekta machine are very stable. We recommend performing leaf position QA quarterly as long as patient-specific QA is being performed. If patient-specific QA is not being done, then MLC QA should be performed on a weekly basis.

Profiler2 Data Analysis

When taking the readings for the MLC QA, the cumulated data counts for all of the diodes are collected continuously every 100 ms. A typical total count for a diode that sees the full radiation field is shown in Figure 72.4a and the differential plot is shown in Figure 72.4b. The background count shown in Figure 72.4b is used to distinguish between readings obtained for different control points. The first control point is for the open field and all diode readings for all other control points are normalized to this open field. The field edges (50% of the open field radiation intensity) for both backup jaws and for the MLC leaves of both leaf

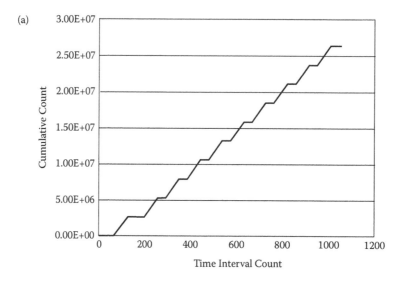

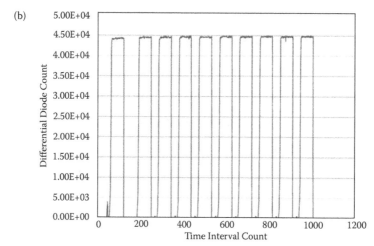

FIGURE 72.4 (a) Cumulative count for a diode of the Profiler2 that sees full radiation during the multiple control points delivery using continuous mode. (b) Differential count of a diode of the Profiler2 that sees full radiation during the multiple control points delivery using continuous mode.

banks are determined. Typically, the 50% location is determined from the interpolation between the two control points whose readings lie just above and just below the 50% value.

Discussion

To test the accuracy of the RDRL, three sets of measurements were performed using a different Profiler2 diode each time to obtain the data for the reference leaf. The maximum difference between the three resulting sets of minor leaf offset values was 0.34 mm, which is in line with expected MLC reproducibility. All major leaf offset and gains should be performed using a 3D water tank data acquisition system because these sets of data are used for the TPS modeling. For routine MLC QA, only a test of minor leaf offsets is needed; minor gains only need to be checked annually because they are not expected to change.

The MLC camera system is an optical system that uses reflectors on each leaf to determine the leaf's position. If maintenance is performed on the components in the gantry head, QA of MLC should be performed to determine both the major and minor leaf offsets.

The RDRL method can be used for all of the tests because positioning of the backup jaw does not rely on the optical tracking system and will therefore not be affected by the gantry head maintenance.

Summary

MLC-based IMRT requires a rigorous clinical commissioning of the MLC system in the TPS. The parameters that must be explicitly characterized in the TPS include the following: rounded leaf ends, tongue-and-groove leaf sides, leaf transmission, backup jaws, and small field output. The most important QA parameter for the MLC system is leaf positioning accuracy and reproducibility, which can be measured efficiently and accurately using the methods described in this chapter.

References

Butson, M. J., P. K. Yu, and T. Cheung. 2003. Rounded end multileaf penumbral measurements with radiochromic film. *Phys. Med. Biol.* 48 (17): N247–N252.

Dempsey, J. F., H. E. Romeijn, J. G. Li, D. A. Low, and J. R. Palta. 2005. A fourier analysis of the dose grid resolution required for accurate IMRT fluence map optimization. *Med. Phys.* 32 (2): 380–388.

Galvin, J. M., A. R. Smith, and B. Lally. 1993. Characterization of a multi-leaf collimator system. *Int. J. Radiat. Oncol. Biol. Phys.* 25 (2): 181–192.

Huq, M. S., I. J. Das, T. Steinberg, and J. M. Galvin. 2002. A dosimetric comparison of various multileaf collimators. *Phys. Med. Biol.* 47 (12): N159–N170.

Jordan, T. J. 1991. Commissioning of a multisegment collimator. Abstract of proceedings: dosimetry protocols, quality assurance and asymmetric collimators. *Phys. Med. Biol.* (36): 669–670.

Jordan, T. J., and P. C. Williams. 1991. Commissioning and use of a multileaf collimator system. Paper read at 1st ESTRO Meeting on Physics in Clinical Radiotherapy, Budapest.

Jordan, T. J., and P. C. Williams. 1994. The design and performance characteristics of a multileaf collimator. *Phys. Med. Biol.* 39 (2): 231–251.

Kim, S., T. C. Zhu, and J. R. Palta. 1997. An equivalent square field formula for determining head scatter factors of rectangular fields. *Med. Phys.* 24 (11): 1770–1774.

Moeller, R. D. 1989. *New Concept for Multileaf Collimation. Proc. 6th Varian European Users Meeting.* Palo Alto, CA: Varian.

Palta, J. R., D. K. Yeung, and V. Frouhar. 1996. Dosimetric considerations for a multileaf collimator system. *Med. Phys.* 23 (7): 1219–1224.

Precise Treatment System Clinical Mode Users Manual R50. 2003. Crawley, UK: Elekta Oncology Systems Ltd.

Proimos, B. S. 1960. Synchronous field shaping in rotational megavolt therapy. *Radiology* 74: 753–757.

Sastre-Padro, M., U. A. van der Heide, and H. Welleweerd. 2004. An accurate calibration method of the multileaf collimator valid for conformal and intensity modulated radiation treatments. *Phys. Med. Biol.* 49 (12): 2631–2643.

Takahashi, S. 1965. Conformation radiotherapy. Rotation techniques as applied to radiography and radiotherapy of cancer. *Acta Radiol. Diagn. (Stockh.)* Suppl 242: 1+.

Trump, J. G., K. A. Wright, M. I. Smedal, and F. A. Salzman. 1961. Synchronous field shaping and protection in 2-million-volt rotational therapy. *Radiology* 76: 275.

Varian Multileaf Collimator

Thomas J. LoSasso
*Memorial Sloan Kettering
Cancer Center*

Introduction

Quality assurance (QA) methodology for IMRT with the Varian multileaf collimator (MLC) is derived mainly from treatment centers that have extensive clinical IMRT experience. A variety of tests has been proposed to perform MLC-specific QA for the Varian MLC (LoSasso, Chui, and Ling 2001; Low et al. 2001; Essers et al. 2001; Van Esch et al. 2002; LoSasso 2003a, 2003b). Many tests are redundant, but these are useful for independently confirming routine QA results; other tests are not appropriate for routine QA because of their time and effort requirements. As such, it is practice at the Memorial Sloan Kettering Cancer Center (MSK) to use these alternative methods, but at less frequent intervals. This chapter will present the QA procedures that are routinely used at MSK for IMRT with the Varian MLC.

QA programs should be efficient without compromising safety. The number of IMRT patients on treatment is a factor in formulating the QA protocol at MSK, where approximately 70% of XRT patients receive IMRT. Under these conditions, the QA approach emphasizes verification of individual components of the process. From the start, the value of properly designing and commissioning the treatment planning system for IMRT has been a priority, minimizing calculation inaccuracies that can be confused with delivery errors. QA tools have been developed to target known MLC delivery problems, specifically wear and tear of the MLC drive components. It is important to acknowledge the inherent features within the Varian software that are designed to detect leaf positioning errors during treatment. Finally, comprehensive end-to-end testing is performed periodically using a reference set of plans to maintain consistency among planning and delivery systems over time.

The standard Varian MLC hardware and software has undergone changes since their introduction in the early 1990s and its first use for IMRT in 1995 (Ling et al. 1996). The shapes of the leaves and the tongues and grooves have been tailored to address the need for better resolution and lower transmission while maintaining the sensitive round edge geometry. As a result, only modest changes in commissioning and QA were incurred for the Millennium MLC relative to the earlier Mark MLC versions. Patterns of failure of leaf motors have changed over time, most notably in 2001, because of software revisions imposing a 0.5 mm minimum gap between opposing leaves in motion and improved pulse width modulation (LoSasso 2003a, 2003b).

Dose Calculation—Commissioning TPS

For IMRT, uncertainty of calculated dose in an IMRT field is an inverse function of the average gap width in an IMRT field (LoSasso, Chui, and Ling 1998). Transmissions through the multifaceted structure of the leaves may contribute up to 20% of the delivered dose for highly modulated fields, but they may not be modeled in treatment planning systems correctly, if at all. Although assigning values for the transmission parameters is the responsibility of the individual user, it should be some comfort that these transmissions should not vary among Varian Millennium MLCs, and their dependence on energy and position within the field is insignificant.

The most sensitive parameter in dose accuracy is referred to as the dosimetric leaf gap (DLG), which corrects for the transmission through the round edges of the leaves. Several methods using ion chambers and/or film have been proposed to quantify the DLG. Some involve direct derivation from measurements as shown in Figure 73.1; others rely on modulation of the DLG during calculation until agreement with measurement is achieved (LoSasso, Chui, and Ling 1998; Arnfield et al. 2000; Low et al. 2001; Van Esch et al. 2002; Tangboonduangjit et al. 2004). Notably, if the abutting field technique is used, the appropriate goal is a uniform integral dose over the abutment region of two uniform fields comparable to the dose at the center of each field. As shown in Figure 73.2, if other criteria are used, an erroneous

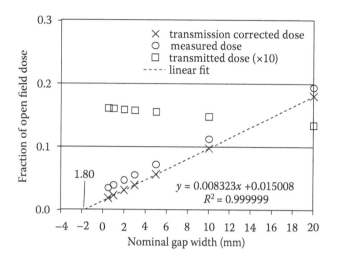

FIGURE 73.1 Direct derivation of the DLG from measurements.

value for the DLG may be obtained. At MSK the DLG has been derived by a number of the above methods to be ~1.8 mm for 6-MV x-rays; it increases slightly to ~1.9 mm for 15 MV. Thus, for typical clinical IMRT fields at MSK with an average gap of 2 cm, the transmission through the round edge will contribute ~9% of the dose, and a 0.1-mm deviation of the DLG corresponds to a 0.5% change in the calculated dose.

Narrow beam attenuation for the Millennium MLC is ~1.2% for 6- and 15-MV x-rays increasing to ~1.5% for a 10 × 10 jaw setting due to scatter from the leaves. Depending on the duty cycle for the IMRT field, the actual MLC transmission can be two to three times these values, and the variation with field size can be more than 1%.

Interleaf effects, that is, tongue-and-groove transmissions, add to the delivered dose when adjacent leaves are traveling side by side and reduce the dose when adjacent leaf gaps are offset during delivery. Combined, interleaf effects have altered the average target dose by as much as 4% in highly modulated IMRT fields at MSK, hastening their modeling in our treatment planning system. Leaf sequencing strategies have been developed (van

Santvoort and Heijmen 1996; Kamath et al. 2004); they minimize, but do not eliminate, tongue-and-groove effects because the residual tongue or groove will still produce an underdose.

Accurate pencil beam kernels and source distributions are crucial to the penumbra for the small subfields in IMRT treatments. Discrepancies between measurements and calculations have been traced to scatter kernels designed for, and adequate for, static fields (Van Esch et al. 2002; Chui, LoSasso, and Palm 2003). However, for IMRT, these parameters require higher resolution affordable either with Monte Carlo calculations or with deconvolution of measurements obtained with small-diameter chambers or solid-state detectors.

Delivery—Machine-Specific QA

The principal sources of dose uncertainty during IMRT delivery are derived from leaf position inaccuracy. A shift in leaf position affects the width of the gap between opposed leaves, unquestionably the most sensitive parameter for IMRT dose delivery, and the position of the gap; these are issues for segmented or step and shoot (SMLC) and dynamic or sliding window (DMLC) deliveries. Fractional errors in the gap width in a DMLC or SMLC treatment lead to proportional errors in the integral dose delivered comparable in magnitude to those described above for the DLG. For the same plan, the average dose error in SMLC fields is similar to that in DMLC fields, although it is concentrated at the edges of the subfields. The errors are greater when average gap widths decrease usually as the degree of modulation increases. The offset of the leaf banks with respect to centerline is defined at 0° gantry angle and is affected by backlash due to mechanical tolerances in the leaf and carriage assemblies as the gantry is rotated (LoSasso 2008).

One should understand the sources of leaf positioning error for the specific MLC design and implement QA tests and frequencies to detect these mechanical problems before dose errors become significant. The QA for an MLC used in static mode, conventional 3DCRT, is relatively simple, considering that these parameters only define the dose near the borders of the

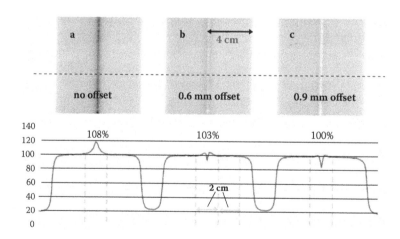

FIGURE 73.2 Data showing that an erroneous value for the DLG may be obtained if inappropriate criteria are used.

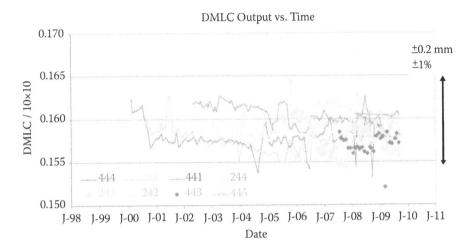

FIGURE 73.3 Dosimetric measurements for a 5-mm sliding window normalized to open field measurements with the gantry stationary.

field. Tests used for static QA of the MLC, for example, periodic checks of the projected leaf positions with graph paper at isocenter are sufficient to verify 1–2 mm tolerance, but these tests and tolerances are inadequate for IMRT. For IMRT, the leaves modulate the dose delivered throughout the target volume and they require more stringent tolerances. Regardless of MLC designs, or whether the delivery mode is segmented (SMLC) or dynamic (DMLC), these tests should stress the precise execution of the gap width defined by opposing leaf positions.

Leaf positions can vary from intended positions due to calibration error or wear and tear in the leaf and carriage drive assemblies. Hardware tolerances also introduce backlash as the gantry is rotated. Routine QA should identify these calibration errors and gradual shifts due to deterioration in a timely manner. The picket fence test, which allows for rapid visual inspection of individual leaves relative to other leaves, detects aberrations less than 0.5 mm (Chui, Spirou, and LoSasso 1996). Dosimetric measurements for a 5-mm sliding window normalized to open field measurements with the gantry stationary (Figure 73.3) and as the

gantry is rotated (Figure 73.4) show deviations of carriages from their average positions over time (LoSasso 2003a). These data are useful for intercomparing MLCs and for detecting carriage-related problems. During beam delivery, MLC software compares independent readouts of leaf position with each other and with the intended leaf position for each leaf and will interrupt the delivery if a deviation exceeds defined tolerances, thereby mitigating clinically significant dose errors between QA sessions.

For DMLC, the ability of the leaves to maintain their specified leaf speed is also important. Slow-moving leaves, noticeable as excessive beam holdoffs, are mostly a concern for DMLC delivery as they may affect the gap width during delivery. This symptom may be indicative of either dirt or grease buildup requiring cleaning of the sides of the affected leaves or deterioration of the brushes within the motors due to electrical arcing, requiring replacement of the motor. At MSK where 70% of treatments are IMRT, leaf motors are replaced at a rate of about one motor per month per MLC. Leaf acceleration is not a dosimetric problem (Chui, Spirou, and LoSasso 1996; Essers et al. 2001).

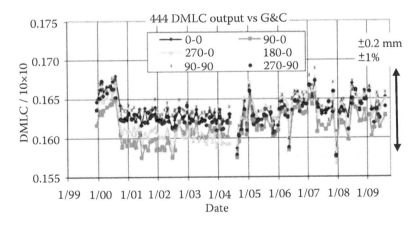

FIGURE 73.4 Dosimetric measurements for a 5-mm sliding window normalized to open field measurements with the gantry stationary as the gantry is rotated showing deviations of the carriages from their average positions over time.

Latency effects due to communication delays between MLC and linac control systems can also introduce leaf position errors. Small MU per segment (Ezzell and Chungbin 2001; Xia, Chuang, and Verhey 2002) at high dose rates for SMLC and improper leaf sequencing resulting in excessive beam holdoffs for DMLC will affect the intended leaf position relative to the cumulative MU index specified in the leaf sequence file. However, these effects produced less than 2% dose error in a variety of clinical plans (Xia, Chuang, and Verhey 2002; Zygmanski et al. 2003). Nevertheless, small MU/segment coupled with high dose rate in step and shoot mode and dose rate modulation in dynamic mode should be avoided.

Calibration of the leaf positions of the Varian MLC using the Varian Field Alignment Tool, a feeler gauge, or a dosimetric measurement properly places the emphasis on the accuracy of the gap width between opposed leaves.

Clinical Test Fields

Accurate IMRT treatment requires careful and ongoing assessment of the dose delivery as compared with the dose calculations. Dose calculation models and IMRT-specific planning parameters may be thoroughly tested; nevertheless, they are based on assumptions and approximations, and irregular-shaped or highly modulated fields can present unexpected problems. End-to-end testing should be a standard component of any QA program. Early in the MSK IMRT experience, we identified the potential sources of uncertainty and error in both the planning and delivery aspects of the treatment. Although specific tests designed to address known mechanical problems must be the mainstay of safe and efficient QA for IMRT, redundancy in the form of a comprehensive test that evaluates planning software, transfer pathways to the treatment workstation, and delivery mechanics provides another layer of assurance. Such testing is appropriate for the commissioning process and also when a new treatment site or an unusual modulation pattern is encountered, orchestrating the sum of all the pieces. For ongoing QA, end-to-end testing verifies the stability of the system over time and allows an intercomparison of delivery systems.

Test cases and clinical dose distributions selected from actual patients' IMRT fields for a variety of simple to complex targets can be planned, delivered, and measured to evaluate the overall accuracy of the system (Xing et al. 1999; Essers et al. 2001; Van Esch et al. 2002; LoSasso 2003a, 2003b). Alternatively, a suite of standardized test cases with a detailed description of the planning, measurement, and analysis process can be downloaded from the AAPM Web site and used to evaluate planning and delivery systems (Ezzell et al. 2009). Verification of these fields using film, EPID, or one of the low-resolution matrix detectors evaluates the overall performance of the planning system and the MLC at the level of dose and dose variation actually received by the patient. Dose overlays and differences are more familiar than standard QA data to many physicians, therapists, and physicists alike. Repeated use of the same fields demonstrates the stability of the delivery system over time, and it can also be the basis for

an IMRT intercomparison of MLC. Discrepancies can be indicative of irregularities in the delivery system, the dose calculation algorithm, and/or the leaf sequencer.

Summary

IMRT performance with the Varian MLC depends on the accuracy with which MLC parameters are set in the treatment planning system at the time of commissioning, and other parameters are maintained at the MLC over its lifetime as per periodic QA. These two aspects of IMRT, commissioning and QA, are codependent as it is often difficult to distinguish errors in delivery from errors in calculation, and dose delivery is gauged by dose calculations. When physicists understand the nature of these potential problems, QA programs can be designed efficiently, and recommended tests and testing frequencies can be implemented and analyzed to categorize deviations from normalcy and to minimize dose uncertainty.

References

Arnfield, M. R., J. V. Siebers, J. O. Kim, Q. Wu, P. J. Keall, and R. Mohan. 2000. A method for determining multileaf collimator transmission and scatter for dynamic intensity modulated radiotherapy. *Med. Phys.* 27 (10): 2231–2241.

Chui, C. S., S. Spirou, and T. LoSasso. 1996. Testing of dynamic multileaf collimation. *Med. Phys.* 23 (5): 635–641.

Chui, C. S., T. LoSasso, and A. Palm. 2003. Computational algorithms for independent verification of IMRT. In *A Practical Guide to Intensity-Modulated Radiation Therapy.* Madison, WI: Medical Physics Publishing. pp. 81–101.

Essers, M., M. de Langen, M. L. Dirkx, and B. J. Heijmen. 2001. Commissioning of a commercially available system for intensity-modulated radiotherapy dose delivery with dynamic multileaf collimation. *Radiother. Oncol.* 60 (2): 215–224.

Ezzell, G. A., J. W. Burmeister, N. Dogan, T. J. LoSasso, J. G. Mechalakos, D. Mihailidis, A. Molineu et al. 2009. IMRT commissioning: Multiple institution planning and dosimetry comparisons, a report from AAPM Task Group 119. *Med. Phys.* 36 (11): 5359–5373.

Ezzell, G. A., and S. Chungbin. 2001. The overshoot phenomenon in step-and-shoot IMRT delivery. *J. Appl. Clin. Med. Phys.* 2 (3): 138–148.

Kamath, S., S. Sahni, J. Palta, S. Ranka, and J. Li. 2004. Optimal leaf sequencing with elimination of tongue-and-groove underdosage. *Phys. Med. Biol.* 49 (3): N7–N19.

Ling, C. C., C. Burman, C. S. Chui, G. J. Kutcher, S. A. Leibel, T. LoSasso, R. Mohan et al. 1996. Conformal radiation treatment of prostate cancer using inversely-planned intensity-modulated photon beams produced with dynamic multileaf collimation. *Int. J. Radiat. Oncol. Biol. Phys.* 35 (4): 721–730.

LoSasso, T. 2003a. Acceptance testing and commissioning of IMRT. In *A Practical Guide to Intensity-Modulated Radiation Therapy.* Madison, WI: Medical Physics Publishing. pp. 147–167.

LoSasso, T. 2003b. IMRT delivery system QA. In *Intensity Modulated Radiation Therapy: The State of the Art. Vol Medial Physics Monography no. 29*, ed. J. Palta. Colorado Springs, CO: Medical Physics Publishing. pp. 561–591.

LoSasso, T. 2008. IMRT delivery performance with a varian multileaf collimator. *Int. J. Radiat. Oncol. Biol. Phys.* 71 (1 Suppl.): S85–S88.

LoSasso, T., C. S. Chui, and C. C. Ling. 1998. Physical and dosimetric aspects of a multileaf collimation system used in the dynamic mode for implementing intensity modulated radiotherapy. *Med. Phys.* 25 (10): 1919–1927.

LoSasso, T., C. S. Chui, and C. C. Ling. 2001. Comprehensive quality assurance for the delivery of intensity modulated radiotherapy with a multileaf collimator used in the dynamic mode. *Med. Phys.* 28 (11): 2209–2219.

Low, D. A., J. W. Sohn, E. E. Klein, J. Markman, S. Mutic, and J. F. Dempsey. 2001. Characterization of a commercial multileaf collimator used for intensity modulated radiation therapy. *Med. Phys.* 28 (5): 752–756.

Tangboonduangjit, P., P. Metcalfe, M. Butson, K. Y. Quach, and A. Rosenfeld. 2004. Matchline dosimetry in step and shoot IMRT fields: A film study. *Phys. Med. Biol.* 49 (17): N287–N292.

Van Esch, A., J. Bohsung, P. Sorvari, M. Tenhunen, M. Paiusco, M. Iori, P. Engstrom, H. Nystrom, and D. P. Huyskens. 2002. Acceptance tests and quality control (QC) procedures for the clinical implementation of intensity modulated radiotherapy (IMRT) using inverse planning and the sliding window technique: Experience from five radiotherapy departments. *Radiother. Oncol.* 65 (1): 53–70.

van Santvoort, J. P., and B. J. Heijmen. 1996. Dynamic multileaf collimation without 'tongue-and-groove' underdosage effects. *Phys. Med. Biol.* 41 (10): 2091–2105.

Xia, P., C. F. Chuang, and L. J. Verhey. 2002. Communication and sampling rate limitations in IMRT delivery with a dynamic multileaf collimator system. *Med. Phys.* 29 (3): 412–423.

Xing, L., B. Curran, R. Hill, T. Holmes, L. Ma, K. M. Forster, and A. L. Boyer. 1999. Dosimetric verification of a commercial inverse treatment planning system. *Phys. Med. Biol.* 44 (2): 463–478.

Zygmanski, P., J. H. Kung, S. B. Jiang, and L. Chin. 2003. Dependence of fluence errors in dynamic IMRT on leaf-positional errors varying with time and leaf number. *Med. Phys.* 30 (10): 2736–2749.

74

Siemens Multileaf Collimator

Andrew Hwang
University of California, San Francisco

Lijun Ma
University of California, San Francisco

Introduction

In Siemens linear accelerators, the multileaf collimator (MLC) replaces the lower (x) jaws (Figure 74.1). A variety of MLC configurations has been available. The earliest models (3D-MLC) had 29 leaf pairs and these were followed by models with 41 leaf pairs. The first-generation 3D-MLC and the second-generation Optifocus™ designs both use a double-focused design as opposed to the more commonly found single-focused design. In other words, the edges of the Siemens MLC leaves are parallel to the beam divergence both parallel and perpendicular to the leaf motion direction. To accomplish this, the ends of the leaves effectively travel in an arc. This design reduces leaf end transmission penumbra and thus improves the light-to-radiation field coincidence. Each leaf has an independent motor and drive assembly, and leaf position is detected using both a potentiometer and an encoder. The Optifocus design has increased backlash between the motor drive assembly and the pot/encoder than the 3D-MLC, potentially resulting in larger variability in leaf positions.

The 80-leaf-pair MLC (160 MLC™) uses a single-focused design with rounded leaf ends. The leaves are divided into two types, upper and lower, and are arranged in an alternating pattern (Figure 74.2). In addition, the tongue-and-groove design is not used. Instead, interleaf transmission is minimized by tilting the MLC leaves slightly with respect to the ray from the source to the leaf center (Figure 74.3). Transmission through a closed leaf pair is also minimized via the sinusoidal-shaped leaf side interweaved in a tongue-and-groove design. The characteristics of the Siemens MLCs are summarized in Table 74.1.

Currently, Siemens linacs only deliver IMRT in step and shoot mode. The design of the Siemens MLC has been well described in the medical physics literature (Das et al. 1998; Tacke et al. 2006; Xia and Verhey 2001).

Calibration Procedure

Because the MLC leaves move in an arc, the relationship between the encoder/potentiometer readings and the field size is not linear. However, to simplify calculations for fast performance, the relationship between the encoder/potentiometer and field size is described using a piecewise linear approximation with three segments. Therefore, position calibration is performed at four points: +20 (fully open), +10, 0, and −10 cm. All MLC leaves for a given bank are individually moved to the correct position and the pot and encoder readings are captured. Because the MLC leaves have a flat end due to the double-focus design, the leaves are usually positioned using the light field. In this case, good light-radiation field coincidence is critical to accurate MLC calibration.

An automated motor speed calibration can also be performed. This calibration adjusts the motors driving an individual leaf such that the leaves move at the same speed. This calibration procedure is performed with the gantry and collimator head rotated such that the jaw being calibrated opens in the downward direction.

MLC Performance

The mechanical performance of various Siemens MLC designs has been characterized by several authors. Bayouth (2008) measured the 3D-MLC to have a leaf position precision of 0.3–0.5 mm. A study by Sastre-Padro et al. (2009) found the long-term leaf positioning reproducibility to be 0.2–0.4 mm for the Optifocus MLC. For the 160 MLC, Tacke et al. (2008) found the leaf position accuracy to be within 0.6 mm and the short-term position reproducibility to be less than 0.1 mm. Leakage and other dosimetric characteristics for the different Siemens MLCs have also been characterized.

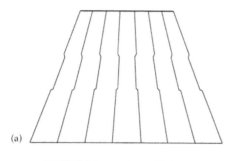

FIGURE 74.1   Schematic end (a) and top (b) views of the Siemens MLC, which illustrate how the leaf edges are focused on the source. Note that the leaf shapes are designed to minimize interleaf transmission and that the center leaf is unique.

Dose Modeling

Because the MLC replaces the lower jaw on Siemens linacs, the collimator scatter factor must be determined carefully (Bayouth 2008).

QA Procedure

At UCSF, a mixture of Optifocus and 3D-MLC models is currently in use. A 160 MLC is currently being installed. The MLC QA procedure is performed monthly and includes the following tests based on the AAPM TG142 recommendations:

1. Light-to-radiation coincidence films
2. Jaw concentricity checks
3. Picket fence
4. MLC leakage

TABLE 74.1   Siemens MLC Configurations

	3D-MLC	Optifocus	160 MLC
Number of leaves	29 × 2	41 × 2	80 × 2
Width at isocenter	1.0 cm[a]	1.0 cm	0.5 cm
Max overtravel (cm)	10	10	20
Leaf height (cm)	7.6	7.6	9.5
Interdigitation	No	No	Yes
Max leaf speed (cm/s)	1.5	2	4
Focus	Double	Double	Single

Note: Models: 58/82/160-leaf.
[a] Two end leaf pairs project to 6.5 cm.

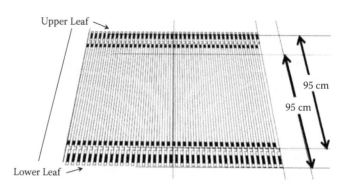

FIGURE 74.2   Profile of the latest 160 MLC showing the alternating upper and lower leaves.

The light-to-radiation coincidence is critical because the light field is commonly used for MLC calibration. Therefore, this is checked using film for several field sizes. Jaw concentricity is checked for each jaw. A picket fence test is performed as is MLC leakage. Because the y-jaws are closed behind the MLC, small openings are required at the field edges to force the y-jaws to stay open to allow proper evaluation of the MLC leakage (Figure 74.4). In addition, MLC position accuracy should be checked if there is a loss of power while the MLC leaves are in motion as the MLC calibration may be lost.

Summary

Like competing MLC systems from other vendors, the Siemens MLC has evolved significantly over time. The latest shift from the traditional double-focused design to a single-focused design was largely driven by clinical application as well as enhanced positional control as any small deviation of the MLC leaf end

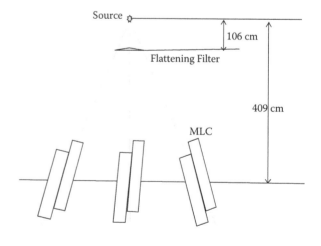

FIGURE 74.3   Schematic showing interleaf transmission being reduced by tilting the leaves relative to the ray running from the leaf center to the source.

FIGURE 74.4 Film taken to evaluate MLC leakage. Note that an opening is left at the end of the field to force the y jaws to remain fully open. The bright line is a scanner artifact that can be used as vertical reference for checking uniformity in MLC leaf motions.

from the beam divergence can now be offset easily. More importantly, as major applications of MLC started departing from shaping a few fields of radiation as in 3D conformal delivery toward overlapping a high number of radiation fields such as in the rotational intensity-modulated delivery, matching beam divergence has become less of a priority compared to the need to rapidly move the MLC leaf to the desired position. As a result, future quality assurance of the Siemens MLC will likely focus more on the dosimetric aspects of its operation (such as consistency in leakage profiles) versus mechanical precision (such as light-radiation field edge matching).

References

Bayouth, J. E. 2008. Siemens multileaf collimator characterization and quality assurance approaches for intensity-modulated radiotherapy. *Int. J. Radiat. Oncol. Biol. Phys.* 71 (1 Suppl.): S93–S97.

Das, I. J., G. E. Desobry, S. W. McNeeley, E. C. Cheng, and T. E. Schultheiss. 1998. Beam characteristics of a retrofitted double-focused multileaf collimator. *Med. Phys.* 25 (9): 1676–1684.

Sastre-Padro, M., C. Lervåg, K. Eilertsen, and E. Malinen. 2009. The performance of multileaf collimators evaluated by the stripe test. *Med. Dosim.* 34 (3): 202–206.

Tacke, M. B., S. Nill, P. Haring, and U. Oelfke. 2008. 6 MV dosimetric characterization of the 160 MLC, the new Siemens multileaf collimator. *Med. Phys.* 35 (5): 1634–1642.

Tacke, M. B., H. Szymanowski, U. Oelfke, C. Schulze, S. Nuss, E. Wehrwein, and S. Leidenberger. 2006. Assessment of a new multileaf collimator concept using GEANT4 Monte Carlo simulations. *Med. Phys.* 33 (4): 1125–1132.

Xia, P., and L. J. Verhey. 2001. Delivery systems of intensity-modulated radiotherapy using conventional multileaf collimators. *Med. Dosim.* 26 (2): 169–177.

75

Niko Papanikolaou
*University of Texas Health
Sciences Center San Antonio*

Alonso N. Gutierrez
*University of Texas Health
Sciences Center San Antonio*

Sotirios Stathakis
*University of Texas Health
Sciences Center San Antonio*

Carlos Esquivel
*University of Texas Health
Sciences Center San Antonio*

Chengyu Shi
*University of Texas Health
Sciences Center San Antonio*

Integrated Systems: Tomotherapy

Introduction

Quality assurance (QA) protocols for conventional linear accelerators, encompassing both mechanical and dosimetric tests, have been widely studied and reported in the literature (Almond et al. 1999; Kutcher et al. 1994). The introduction of helical tomotherapy (HT) as a novel approach for the delivery of image-guided, intensity-modulated radiation therapy has instigated the development of new QA protocols, which test not only the unique radiation delivery geometry and properties of HT but also the integrated, volumetric megavoltage CT onboard imaging.

Some of the existing protocols for conventional linear accelerators can be modified so that they are applicable to the HT; however, these protocols do lack procedures to evaluate the intrinsic dynamic nature of HT. A comprehensive QA protocol for HT should include tests for the dynamic functionality and synchrony among the gantry rotation, couch translation, and binary multileaf collimator because the synchronization of those motions is the core operational principle of HT. Recommendations for HT QA procedures have been published by several groups (Mahan, Chase, and Ramsey 2004; Fenwick et al. 2004; Balog, Holmes, and Vaden 2006) following the testing structure originally proposed in the TG 40 report. In 2008, the AAPM commissioned a task group (TG-148) to develop a comprehensive QA methodology for HT (Langen et al. 2010). Elements of that report, are reproduced in this chapter.

In addition to regular QA testing, replacement of major radiation-producing or shaping components on the HT unit necessitates supplemental QA and acceptance testing before resuming clinical use of the unit. Some of the major components that require replacement include the magnetron, solid-state modulator (SSM), linear accelerator, and binary multileaf collimator bank. The target, depending on the machine use, needs to be replaced on average twice a year. Exchange of any of these components mandates testing after replacement because each component significantly impacts the radiation output, quality, and intensity distribution. A significant deviation from the traditional protocols is the output calibration of the HT unit. Although the methodology of calibration is not discussed here, it is important to note that it deserves special attention by the user.

With consideration of the challenges regarding QA of an HT unit, the following tables provide recommended QA protocols for daily, monthly, annual, and post-component replacement testing (Tables 75.1 through 75.4). These protocols are not intended to be all inclusive but rather serve to establish the minimum required actions.

Summary

A comprehensive QA program for an HT unit is a key element of the overall success of a radiation therapy department outfitted with such unit. The recommended tomotherapy-specific QA protocols listed in the tables are derived from many years of collective experience of the authors as well as the AAPM TG-148 report. Tolerance thresholds for the proposed tests stem from the necessity to minimize deviations between calculated and delivered doses to the patient. Ultimately, careful and timely QA testing of the HT unit ensures that the system is functioning according to specifications, and treatment is delivered accurately and precisely.

TABLE 75.1 Recommended Daily HT QA Tests

Test		Evaluation Criteria	Tolerance
Safety tests	Door interlock	Functional	Pass/fail
	Beam status indicators (audio/visual)	Functional	Pass/fail
	Patient audiovisual monitors	Functional	Pass/fail
MVCT artifacts		Constancy	Pass/fail
MVCT registration		Constancy	1 mm
Red laser initialization		Congruency with green lasers	1 mm
Red laser movement		Constancy	2 mm
Radiation output—static		Constancy	3%
Radiation output—rotational		Constancy	3%
Radiation energy—static		Constancy	2%

Note: MVCT, megavoltage CT.

TABLE 75.2 Recommended Monthly HT QA Tests

Test	Evaluation Criteria	Tolerance
Safety tests	See Table 75.1	Pass/fail
MVCT alignment	Correct position	2–1 mm (non-SRS/SBRT–SRS/SBRT)
MVCT uniformity	Constancy	<25 HU
MVCT noise for water	Constancy	30 HU within baseline
MVCT spatial resolution	Constancy	>1.6 mm
MVCT dose	Constancy	±30%
Couch drive	Absolute	1 mm
Couch speed	Uniform	2%
Couch/gantry synchrony	Absolute	1 mm
Interlock/Interrupt button	Functional	Pass/fail <(3% from no interrupt)
Red laser movement	Absolute	1 mm
Radiation output—static	Constancy	2%
Radiation output—rotational	Constancy	2%
Radiation energy—static	Constancy	2%
Transverse profile (5.0 cm field width)	Constancy	2%/1 mm
Longitudinal profiles (all field widths)	Constancy	2%/DTA: 1% of field width

TABLE 75.3 Recommended Annual HT QA Tests

Test	Evaluation Criteria	Tolerance
All monthly tests	See Table 75.2	Pass/fail
PDD measurements (each field width)	Agreement with model	1%/1 mm
Transverse profiles (each field width)	Agreement with model	1%/1 mm
Longitudinal profiles (each field width)	Agreement with model	1%/DTA: 1% of field width
Y-jaw centering[a]	Alignment with source	0.3 mm at isocenter
Y-jaw divergence/beam centering[a]	Alignment with axis of rotation	0.5 mm at isocenter
Jaw/gantry rotation plane alignment[a]	Alignment with axis of rotation	0.5°
Treatment beam field centering[a]	Alignment with center	0.5 mm at isocenter
MLC lateral offset[a]	Alignment with axis of rotation	0.75 mm at isocenter
MLC twist[a]	Alignment with beam plane	0.5°
Axial green laser	Distance to radiation isocenter	1 mm/0.3°
Sagittal/coronal green laser	Alignment with radiation isocenter	1 mm
Factory on-axis/off-axis phantom tests[a]	Consistency with plans	3%/3 mm

Note: MLC, multileaf collimator.

[a] Tests in which required XML and sinogram files are supplied by TomoTherapy, Inc., as part of the machine quality validation.

TABLE 75.4 Software System Maintenance and Acceptance

Test	Period/ Evaluation Criteria
Patient backup	As needed or once a month
Protocol backup	Once a month
Machine backup	Once a month
Software system reboot	Twice a month
QA procedure cleanup	Once a month
Software Upgrades	
DICOM import/export	Functional
Geometric tests	Agreement within 1 voxel
Optimization	Functional
Plan printout	Functional
QA of an existing patient	Pass/fail
Data server communication	Functional

TABLE 75.5 Recommended Quality Control Procedures Post-Component Replacement

Component	Test	Evaluation Criteria	Tolerance
Magnetron/SSM	Radiation output—static	Constancy	2%
	Radiation output—rotational	Constancy	2%
	Radiation energy—static	Constancy	2%
	Transverse profile (largest field width)	Constancy	2%/1 mm
	Longitudinal profiles (all field width)	Constancy	2%/DTA: 1% of field width
	DQA/phantom plan dose verification[a]	Agreement with calculated dose	3%
Linac/Target	Y-jaw centering[a]	Alignment with source	0.3 mm at source
	Y-jaw divergence[a]	Alignment with axis of rotation	0.5 mm at isocenter
	MLC lateral offset[a]	Alignment with axis of rotation	0.5 mm at isocenter
	Radiation output—static	Constancy	2%
	Radiation output—rotational	Constancy	2%
	Radiation energy—static	Constancy	2%
	Transverse profile (largest field width)	Constancy	2%/1 mm
	Longitudinal profiles (all field width)	Constancy	2%/DTA: 1% of field width
	DQA/phantom plan dose[a]	Calculated dose	3%
MLC bank	MLC lateral offset[a]	Alignment with axis of rotation	1.5 mm at isocenter
	MLC twist[a]	Alignment with beam plane	0.5°
	DQA/phantom plan dose (multiple plans)[a]	Calculated dose	3%

Note: MLC, multileaf collimator.

[a] Tests in which required XML and sinogram files are supplied by TomoTherapy, Inc., as part of the machine quality validation.

References

Almond, P. R., P. J. Biggs, B. M. Coursey, W. F. Hanson, M. S. Huq, R. Nath, and D. W. Rogers. 1999. AAPM's TG-51 protocol for clinical reference dosimetry of high-energy photon and electron beams. *Med. Phys.* 26 (9): 1847–1870.

Balog, J., T. Holmes, and R. Vaden. 2006. A helical tomotherapy dynamic quality assurance. *Med. Phys.* 33 (10): 3939–3950.

Fenwick, J. D., W. A. Tome, H. A. Jaradat, S. K. Hui, J. A. James, J. P. Balog, C. N. DeSouza et al. 2004. Quality assurance of a helical tomotherapy machine. *Phys. Med. Biol.* 49 (13): 2933–2953.

Kutcher, G. J., L. Coia, M. Gillin, W. F. Hanson, S. Leibel, R. J. Morton, J. R. Palta et al. 1994. Comprehensive QA for radiation oncology: Report of AAPM Radiation Therapy Committee Task Group 40. *Med. Phys.* 21 (4): 581–618.

Langen, K., N. Papanikolaou, J. Balog, R. Crilly, D. Followill, S. Goddu, W. Grant, W. Grant III, G. Olivera, and C. Shi. AAPM Task Group 148: QA for helical tomotherapy. *Med. Phys.* 37 (9): 4817–4853.

Mahan, S. L., D. J. Chase, and C. R. Ramsey. 2004. Technical note: Output and energy fluctuations of the tomotherapy Hi-Art helical tomotherapy system. *Med. Phys.* 31 (7): 2119–2120.

Integrated Systems: Gamma Knife

Paula L. Petti
*Washington Hospital
Healthcare System*

Introduction

Lars Leksell, a neurosurgeon, conceptualized stereotactic radiosurgery based on his experience with stereotactic neurosurgical techniques (Leksell 1951). In 1967, the first patient was treated by Leksell and his group on a gamma stereotactic device in Stockholm, Sweden, and in 1987, the first Leksell Gamma Knife® (LGK) Model U (Elekta Instruments AB, Stockholm, Sweden) was installed at the University of Pittsburgh Medical Center in Pittsburgh, PA. Since then, the LGK has evolved over several generations, the most recent design being the LGK Perfexion™. Other companies, for example, Cancer Care International, LTD (CCI), have developed their own line of gamma stereotactic radiosurgery (GSR) devices; however, most GSR systems in use today are LGK units manufactured by Elekta. Although GSR technology has evolved over the years with marked improvements in automation and treatment planning, certain design aspects of LGK radiosurgery have remained constant. For example, all LGK models use approximately 200 ^{60}Co sources collimated toward the isocenter. The focal point of the radiation is fixed, and the patient is moved to place the targeted area precisely at the radiation focus. In older LGK units (models U and B), the patient's head (affixed to a stereotactic frame) is positioned manually via devices called "trunnions." In mid-generation models (LGK models C and 4C), the automatic patient positioning system (APS) was introduced to automate patient treatment in many, but not all, situations. The CCI Rotating Gamma System Vertex 360 uses a similar patient positioning system. In the latest LGK model, Perfexion, patient positioning is fully automated via precise couch motions. In all GSR devices, the patient is positioned with submillimeter precision, which is one of the defining features of stereotactic radiosurgery.

Early LGK models have 201 stationary ^{60}Co sources, a fixed internal primary collimator, and four different-sized interchangeable external collimators (4-, 8-, 14- and 18-mm diameter) that are mounted manually. The Perfexion, on the other hand, has one internal fixed collimator that has three different-sized collimating channels: 4-, 8-, and 16-mm diameter. For the Perfexion, the ^{60}Co sources are arranged on eight moveable sectors with 24 sources per sector, amounting to a total of 192 sources. The sectors are positioned over the desired collimating channels automatically as prescribed by the treatment plan. The CCI Rotating Gamma System Vertex 360 uses only 30 sources that reside in an arc that rotates around the patient, and the patient couch moves along the longitudinal axis of the patient's body. Another GSR unit, the CCI GammaArt-6000TB, which, to date, is licensed only in China, can perform stereotactic body radiation as well as radiosurgery of the brain. Six collimators are available in this unit, ranging in size from 6×6 to 14×60 mm^2.

Radiation Safety

In the United States, the Nuclear Regulatory Commission (NRC) regulates the use of all GSR devices either directly or through an agreement with the state in which the device is located. Specific regulations for older model GSR units are set forth in 10 CFR Part 35.600, Subpart H, "Photon Emitting Remote Afterloader Units, Teletherapy Units and Gamma Stereotactic Radiosurgery Units," which can be found on the NRC's Web site (http://www.nrc.gov/reading-rm/doc-collections/cfr/part035). Shortly after the first LGK Perfexion was installed in the United States, the NRC issued a licensing guidance document specifically for the Perfexion (10 CFR 35.1000, found at http://www.nrc.gov/materials/miau/med-use-toolkit/perfexion-guidance.pdf). To date, the NRC has not issued specific recommendations for the Rotating Gamma Unit Vertex 360 or for the GammaArt-6000TB, but prudence

TABLE 76.1 Daily Tests: To Be Performed on Treatment Days Before Treatments

		Performance	
Description of Test	GSR Device		Action
System emergency alarm	All		Functional
Radiation monitors	All		Functional
Door interlock	All		Functional
Video monitors	All		Functional
Audio communication	All		Functional
Emergency procedures	All		Posted
Emergency release tools	H		Available
Interlock for left and right guard rails on couch	All		Functional
Helmet cover interlock	H		Functional
Mattress lock/unlock	All		Functional
Frame adapter docks correctly for all gamma angles	PFX		Functional
Frame adapter can be attached correctly to coordinate frame	PFX		Functional
Radiation survey meter	All		Available and functional
Treatment couch retraction mechanism (e.g., treatment pause sequence)	All		Functional: treatment couch retracts from unit, shielding doors close; for PFX, sources move to shielded "off" position; treatment can be resumed after pause sequence is cleared
Emergency stop (procedure to stop couch and shielding door motion)	All		Functional: confirm that all motion stops when the emergency stop sequence is initiated. For PFX, sources move to shielded "home" position; motion resumes when stop sequence is cleared
Proper termination of treatment timer	All		Test run terminates at the correct time
System date and time	All		Agrees with actual date and time
Dose rate for largest collimator	All		Value in TPS on the day in question agrees with the decayed dose rate from the last full calibration
Patient imaging QA	All		On patient images, confirm known distances between fiducials

dictates that one follows, as closely as is possible, the NRC's general GSR requirements until specific guidelines for these newer GSR devices become available.

QA Guidelines

Comprehensive QA guidelines for radiation therapy are provided in the American Association of Physicists in Medicine Task Group Report 40 (Kutcher et al. 1994). QA specific to GSR has been addressed in American Association of Physicists in Medicine Task Group Report 54 (Schell et al. 1995) and by many other authors (e.g., Goetsch 2008 and references cited therein).

Recently, the American Association of Physicists in Medicine organized a new task group, TG-178, to review and update the recommendations of American Association of Physicists in Medicine Task Group Report 54.

In Tables 76.1 through 76.4, the designation "H" refers to GSR devices that use external "helmets" as collimators. This includes the LGK models U, B, C, and 4C as well as any similar units sold by other manufacturers. The designation "PFX" refers to the LGK Perfexion, and the designation "All" refers to all GSR units, regardless of manufacturer. One anticipates that QA items will be added to these tables once the NRC issues specific recommendations for the newer GSR devices. Other parameters in the tables are defined as:

TABLE 76.2 Weekly Tests: To Be Performed on Average Once Every 7 Days and at Intervals Between 5 and 9 Days

		Performance	
Description of Test	GSR Device	Tolerance	Action
Perform all daily QA tests	All	All tests pass	
Diode focus position test	PFX	$\Delta r < 0.3$ mm	$\Delta r < 0.5$ mm
Helmet cap sensor	H	Presence or absence of helmet cap accurately recognized	
Helmet ID sensor	H	Helmet size (4-, 8-, 14- or 18-mm) accurately identified	
Helmet microswitches	H	Adjust to manufacturer-specified tolerance	
Helmet changer	H	Functional	
Couch release handle	H	Functional	
Helmet trunnion centricity (measured using trunnion test tool supplied by manufacturer)	H	$\Delta X < 0.1$ mm	$\Delta X < 0.2$ mm
APS QA test run	H	Test run finishes without error	

TABLE 76.3   Monthly Tests: To Be Performed on Average Once Every 4 Weeks and at Intervals Between 3 and 5 Weeks

Description of Test	GSR Device	Performance	
		Tolerance	Action
Perform all daily QA tests	All	All tests pass	
Confirm that weekly QA has been performed	All	All tests pass	
Dose rate for largest available collimator	All	Within ±1% of the TPS value	Within ±3% of the TPS value
Timer accuracy over range of use	All	±0.2% (the stated accuracy in LGK manuals)	±0.3%
Timer linearity over range of use	All	±1%	±2%
Timer constancy	All	±0.2%	±0.3%
Timer error	All	<0.01 min	<0.03 min
Emergency stop buttons inside GSR treatment suite	All	Functional	
Control system failure test	All	Treatment stops following simulated system failure; can resume treatment once error clears	
Clearance tool test procedure	PFX	Manufacturer-supplied test executes successfully	
Extreme position test procedure	PFX	Manufacturer-supplied test executes successfully	
Transit dose for largest collimator	PFX	≤0.03 Gy	≤0.05 Gy
UPS battery tests	All	Functional	

1. X, Y, and Z refer to the LGK coordinate system. X is along the patient right–left axis, Y is in the posterior–anterior direction, and Z defines the superior–inferior direction with respect to a supine patient.
2. ΔX, ΔY, and ΔZ are the deviations between measured and expected values for the central position of the LGK coordinate system.
3. $\Delta r = \sqrt{\Delta X^2 + \Delta Y^2 + \Delta Z^2}$.
4. FWHM denotes full width at half maximum.

Unless otherwise noted, the action levels in these tables were taken from Elekta's LGK user manuals, and the tolerances levels reflect the anticipated range of variation if appropriate measurement techniques are used. If any of the tests listed in the tables fail to meet the action criteria, then patients should not be treated until a service engineer trained by the manufacturer of the GSR unit identifies and corrects the underlying problem.

Note that the NRC requires that records of certain QA test results be retained for at least 3 years.

Notes Pertaining to Weekly Tests

1. The Perfexion system requires that the diode-focus-position test is successfully executed at least once every month; otherwise, the system will not allow patient treatment. However, because this is a quick and simple check of the coincidence between the patient positioning system and the radiation focal point, it seems reasonable to perform this test weekly.
2. The APS QA test run is specific to LGK models C and 4C and, besides being performed weekly, should also be performed whenever the system configuration changes from trunnion to APS mode.

TABLE 76.4   Annual Tests: To Be Performed on Average Every 12 Months and at Intervals of Between 10 and 14 Months

Description of Test	GSR Device	Performance	
		Tolerance	Action
Perform daily QA	All	All tests pass	
Confirm that weekly QA has been performed	All	All tests pass	
Confirm that monthly QA has been performed	All	All tests pass	
Irradiate and analyze films to confirm coincidence of PPS and RFP for all collimators	All	$\Delta r < 0.4$ mm	$\Delta r < 0.5$ mm
Irradiate and analyze films to confirm that dose profiles for all collimators agree with those calculated by the TPS	All	FWHM within ±0.5 mm of value in the TPS	FWHM within ±1.0 mm of value the TPS
Relative output factors: for all collimators relative to the largest collimator	All	Within ±3% of the value in the TPS	Within ±5% of the value in the TPS
Wipe test (performed on external collimating helmets or on the outer surface of the collimator cap for PFX)	All	2 kBq	2 kBq
TPS	All	Produces consistent results	

Note: PPS, patient positioning system; RFP, radiation focal point; ROF, relative output factor; TPS, treatment planning system; GSR, gamma stereotactic radiosurgery.

Notes Pertaining to Monthly Tests

1. If the difference between the measured and expected dose rate for the largest collimator exceeds the action level (3%), or if the average of three consecutive measurements differs from the expected dose rate by between 1% and 3%, then the recommended action is to change the dose rate stored in the treatment planning system (TPS).
2. The measured dose rate should be compared to (a) the decayed dose rate from the last full calibration, (b) the stated dose rate on a test plan calculated on the day of measurement, and (c) the dose rate calculated by dividing the prescribed dose to the center of the dosimetry phantom by the time calculated by the TPS to deliver this dose.
3. The tolerance and action levels for the timer error and transit dose reflect the author's experience with LGK units.
4. In addition to these monthly tests, the NRC Licensing Guidance for Perfexion requires that approximately every 6 months the vendor confirms that, within tolerance limits, each sector moves correctly to each sector position. If, in the future, the manufacturer provides the user with a means of checking sector positioning, this test should be added to the monthly QA.

Notes Pertaining to Annual Tests

1. Relative output factors are notoriously difficult to measure for small field sizes. Elekta's relative output factor values stored in the TPS are derived from detailed Monte Carlo calculations. If a measured relative output factor value differs from the value stored in the TPS by an amount greater than the action level, the user can, at his/her discretion, change the values stored in the TPS. However, before doing so, he/she should confirm that appropriate measurement techniques were used.
2. Besides being checked at least annually, the TPS should be tested after all software upgrades.
3. Besides being checked annually, the tests listed in Table 76.4 should be performed after a ^{60}Co source change or after a major repair on the GSR unit.

References

Goetsch, S. J. 2008. Linear accelerator and gamma knife–based stereotactic cranial radiosurgery: Challenges and successes of existing quality assurance guidelines and paradigms. *Int. J. Radiat. Oncol. Biol. Phys.* 71 (1 Suppl): S118–S121.

Kutcher, G. J., L. Coia, M. Gillin, W. F. Hanson, S. Leibel, R. J. Morton, J. R. Palta et al. 1994. Comprehensive QA for radiation oncology: Report of AAPM Radiation Therapy Committee Task Group 40. *Med. Phys.* 21 (4): 581–618.

Leksell, L. 1951. The stereotaxic method and radiosurgery of the brain. *Acta Chir. Scand.* 102 (4): 316–319.

Schell, M. C., F. J. Bova, D. A. Larson, D. D. Leavitt, W. R. Lutz, E. B. Podgorsak et al. 1995. *Stereotactic Radiosurgery: Report of Task Group 42 Radiation Therapy Committee.* Woodbury, NY: American Institute of Physics.

77

Integrated Systems: CyberKnife

Christos Antypas
Aretaieion Hospital, University of Athens and CyberKnife Center Iatropolis - Magnitiki Tomografia

Evaggelos Pantelis
Medical School, University of Athens and CyberKnife Center Iatropolis - Magnitiki Tomografia

Introduction

The CyberKnife (Accuray Inc., Sunnyvale, CA) is a frameless stereotactic radiosurgery/radiotherapy device (Figure 77.1) that combines the principles of stereotactic techniques, image guidance, and robotic technology aimed at the delivery of highly conformal dose distributions to intracranial and extracranial lesions with submillimeter accuracy (Antypas and Pantelis 2008; Murphy and Cox 1996; Adler et al. 1999).

Principles of Operation and System Components

The major components of the CyberKnife system are:

- A robotic manipulator with 6 degrees of freedom initially designed for industrial applications. In each CyberKnife installation and during commissioning, the robot's available workspace is charted into the manipulator's driving software, and every available treatment path is calibrated. A treatment path consists of series of nodes through which the robot travels during a treatment course. Each node is a potential point in space from which the linac can be directed toward the target and is represented by a set of Cartesian coordinates in the space around the target (Murphy and Cox 1996). Depending on the clinical application, head or body, the robot travels through coordinates at different distances from the target: 650, 800, or 1000 mm.
- A 6-MV linear accelerator mounted on the robotic arm. The small dimensions of the clinical radiosurgery beams eliminate the need for a flattening filter in the linac head. These result in reduced-size components of the linac and

consequently less weight to mount on the robotic arm. The linac beam is collimated using either 12 fixed circular collimators of 5, 7.5, 10, 12.5, 15, 20, 25, 30, 35, 40, 50, and 60 mm nominal diameter defined at SAD = 800 mm or a variable-aperture IRIS™ collimator simulating 12 beam sizes with characteristics similar to the fixed ones.
- A computer-controlled couch able to remotely position the patient to the desired location with fine maneuvers. The treatment couch, in its standard version, has 5 degrees of freedom with three translational movements (lateral, vertical, longitudinal) and two rotational movements (left/right lateral roll, head up/down). The robotic version of the treatment couch has 6 degrees of freedom incorporating an extra yaw rotational movement.
- The kV-imaging subsystem for patient alignment and target tracking consists of two pairs of diagnostic x-ray tubes and amorphous silicon digital flat panel detectors (cameras A and B). The x-ray tubes are installed on the ceiling of the treatment room at 45° angles with beams orthogonal to each other. In the G3 system, the image detectors are installed on the ground, whereas in the latest version, the G4 system, the image detectors are sitting flat on the floor.

Treatment procedure, planning and delivery, is based on an axial CT of the patient in the treatment position. Digitally reconstructed radiographs for the respective x-ray camera's geometric configuration are created. Initially and during treatment, image pairs taken from both x-ray cameras are used by the image guidance software to compare the reference digitally reconstructed radiographs and live x-ray images. Depending on the tracking method, the target is located and 6D differences (translations and rotations) between the real and the planned

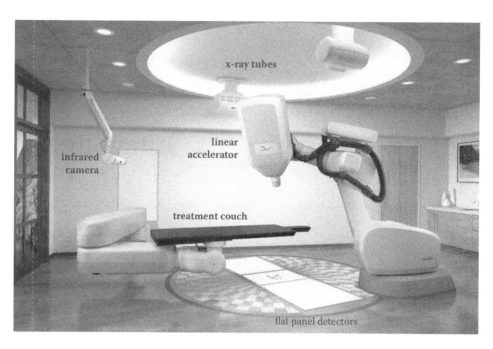

FIGURE 77.1 CyberKnife system and components.

target position are calculated by the image registration software (Fu and Kuduvalli 2008). The calculated deviations are fed back into the robot manipulator to realign the linac before the delivery of each beam.

Mainly, three tracking methods are used in CyberKnife clinical practice. For extracranial applications, gold fiducials are implanted in the target region and the fiducial tracking method is used (Murphy 2002). A skull tracking method, based on rigid body registration of bony landmarks of the head, is used for intracranial applications (Fu and Kuduvalli 2008). The Xsight spine tracking method is based on nonrigid body registration of spinal bony landmarks and is used for spine applications (Ho et al. 2007).

Tracking of moving targets is achieved with the Synchrony® system and implanted fiducials. LED markers are attached to the patient's chest and their motion, due to the patient's respiration, is continuously tracked by an infrared camera. Before treatment, a correlation model is created between the LED markers and target motion as tracked by the fiducials with x-ray images. During treatment, the robot motion is synchronized to the target motion according to the continuously updated correlation model (Schweikard, Shiomi, and Adler 2004).

Quality Assurance of the CyberKnife System

The concept of CyberKnife is unique compared to other external beam image-guided delivery systems. Robotic technology, image guidance, small fields, inverse planning, and treatment delivery with stereotactic accuracy are incorporated into the CyberKnife. QA procedures for the CyberKnife are based not only on existing recommended procedures (AAPM TG-40, TG-45), adapted for CyberKnife clinical implementation, but also on some of the

manufacturer's procedures designed to implement the system's technological aspects. Following the general QA guidelines, individual quality control checks for each of the CyberKnife subsystems and components are required.

Quality Assurance of the Robot

In CyberKnife's nomenclature, the isocenter is physically represented by a small crystal situated on top of a floor-mounted post, called "isopost" (Murphy and Cox 1996). The central axes of the kV beams are aligned to intersect at the isocenter and all treatment paths of the robot are calibrated against it.

Mechanical accuracy of the robot is evaluated by instructing the robot to pass through all nodes of a treatment path, determining at each node, the linac beam direction, as represented by a laser beam, over a fine mesh of points surrounding the isopost. Laser light striking the isopost is transmitted by an optical fiber to a light-sensitive diode. The signal created in the diode is then amplified and monitored. The closer the laser beam strikes the center of the crystal, the proportionately larger signal is created inside the diode. The manipulator coordinates that correspond to the maximum signal created by the crystal are stored and compared with the corresponding values calculated during calibration. Robot mechanical accuracy is defined as the radial displacement error, dr_i, of the node coordinates obtained during the evaluation and calibration procedures. The maximum individual radial displacement, dr_{max}, and the mean radial displacement, $\overline{dr}$, of the nodes are recorded for each of the available treatment paths, and these have been reported to be less than 1 mm (Murphy and Cox 1996; Antypas and Pantelis 2008).

Mechanical accuracy of the robot must be quantitatively evaluated initially for all available treatment paths and checked annually thereafter. A qualitative check of the robot mechanical accuracy can be done by delivering an isocentric plan in simulation mode without radiation and the laser beam on and pointing toward the isopost. This test may be performed on a monthly basis to visually inspect consistency of the robot mechanical accuracy.

Quality Assurance of Treatment Couch

Quality control testing of the standard treatment couch demonstrates whether the positioning mechanical accuracy of the couch and the safety interlocks are within the manufacturer's specifications and working properly. Using a mechanical ruler and a digital level, the mechanical accuracy of digital and measured values of individual table movements is evaluated. The test should be performed on a monthly basis.

Quality Assurance of the Image Guidance Subsystem

The image guidance subsystem is responsible for measuring and reporting patient position. Accuracy of the image guidance subsystem depends on the quality of the radiographs recorded by the cameras (Fu and Kuduvalli 2008) and the accurate output of the tracking algorithm. QA of the imaging guidance subsystem must include tests for the mechanical alignment of the cameras, the performance of the x-ray generators, evaluation of the imaging characteristics of the flat panel detectors, and the performance of the tracking algorithm.

Mechanical Alignment of the Imaging Cameras

Mechanical accuracy of the alignment of each camera is evaluated by mounting the isopost on the treatment room floor and imaging its position. The coordinates of the projected crystal should be in the central pixel of both images A and B within a maximum tolerance of ±1 mm in both imaging directions (Physics Essentials Guide in CyberKnife System Manual 2006).

Quality Assurance of the X-ray Generators

The accuracy and precision of kV and time exposure settings should be checked. A noninvasive multifunction meter must be used to measure kVp values and exposure times. The multifunction meter must be positioned at the isocenter facing the x-ray tube being tested and a set of measurements can be taken for each generator using nominal kV values in the range from 70 to 140 kV and times from 20 to 500 ms. Kilovoltage beam exposure output, precision, and linearity should be checked. AAPM report 74 (Shepard et al. 2002) recommends the precision of the kV beam output in mR/mAs to be less than 10%. Exposure linearity between all mA or mAs settings must be within 10% (Rossi et al. 1985). A high-precision diagnostic electrometer connected to a detector with a flat energy response and calibrated in kV diagnostic beams must be used. The detector must be positioned at the isocenter facing the tested x-ray tube. At selected kV settings (e.g., 80–140 kV) and constant exposure time, for example,

100 ms, a set of exposures (mR) must be performed for mAs values ranging from 5 to 30 mAs by independently changing the mA setting. Although accuracy and precision of kV parameters should be less than 10% and 5% in diagnostic installations (Rossi et al. 1985; Shepard et al. 2002), the user may reduce them to 5% for the accuracy and 2% for the precision, respectively, for the kV parameters, mostly used in CK clinical practice (Antypas and Pantelis 2008). Quality assurance of the kV-imaging subsystem parameters should be checked during acceptance of the system and checked quarterly thereafter.

Quality Assurance of the Flat Panel Detectors

Quality assurance of the imaging characteristics of the digital detectors may be performed in terms of high-contrast resolution, low-contrast resolution, and geometrical distortion. An appropriate multipurpose diagnostic QA tool should be used to quantitatively evaluate the above characteristics (Antypas and Pantelis 2008; Antypas et al. 2009). Each camera must be tested individually with the QA tool appropriately positioned to face each camera. Imaging characteristics must be measured on the raw data of the x-ray images, which means that those images must be extracted from the system software by the user. Image processing of the raw images may be done using freeware or commercially available software (Antypas et al. 2009). High-contrast resolution in terms of spatial resolution f50, that is, the frequency at 50% of the relative modulation transfer function and low-contrast resolution in terms of contrast-to-noise ratio may be calculated. Geometric distortion may be evaluated either qualitatively or quantitatively by measuring the ratio of known phantom dimensions, for example, length/width, on the x-ray images (Figure 77.2a and b). Baseline values of the imaging characteristics should be determined during installation of the system and checked quarterly thereafter.

Quality Assurance of the Tracking Algorithm

The tracking algorithm is checked in correspondence with the table movements and should be performed for all tracking methods. A plan on a phantom is created for each tracking method. In simulation mode, the phantom on the couch is aligned using the appropriate tracking algorithm to get couch offsets close to zero. Using the couch digital readouts, the phantom is moved away to several positions in both translational and rotational axes. At each position, x-ray images are acquired and the tracking algorithm results are recorded and compared with the actual couch offsets. After subtracting the "zero-position" offsets from the data, the RMS of all the errors should be less than 1 mm. The test must be performed during installation and repeated quarterly thereafter. Every tracking method should be checked at least once per year.

Quality Assurance of the CyberKnife Linear Accelerator

General recommendations on linac QA and beam data commissioning can be followed for the CyberKnife linear accelerator

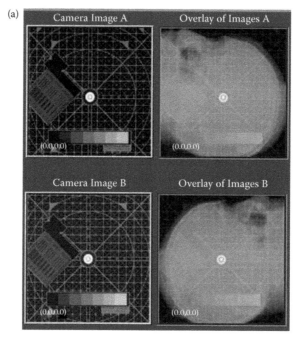

FIGURE 77.2 (a) Qualitative analysis of geometrical distortion using an imaging QA tool. The grid pattern with orthogonal and diagonal marks reveals no geometrical distortion on the images of either detector. (b) Quantitative analysis of the flat panel detectors' imaging characteristics using an imaging QA phantom and commercially available image analysis software.

QA (AAPM TG-40, TG-45, TG-106). In the CyberKnife, there is no light field representing the radiation beam. Linac beam axis and pointing direction is represented by a laser beam. Axes coincidence must be checked using film, either radiographic or radiochromic, irradiated for at least two system collimators and at three different SADs, for example, 650, 800, and 1000 mm. The laser denoting the linear accelerator beam axis is appropriately marked on both types of films. Coincidence is quantitatively evaluated by film processing (Physics Essentials Guide in CyberKnife System Manual 2006; Antypas and Pantelis 2008) and should be less than 1 mm.

According to absorbed dose determination protocols (Almond et al. 1999; Andreo et al. 2000), beam quality specification is directly measured on the traditional 10×10 cm² reference field. Because the CyberKnife maximum collimator size is 60 mm at SAD = 800 mm, a modified way according to existing protocols or new literature recommendations must be followed to specify beam quality with measurement with the 60-mm collimator (Physics Essentials Guide in CyberKnife System Manual 2006; Antypas and Pantelis 2008; Kawachi et al. 2008; Alfonso et al. 2008). Basic dosimetric measurements of the CyberKnife

linear accelerator beam, to be used also for TPS beam data input, should include beam quality specification, beam profiles, PDD/TPR measurements, output factors, output calibration, reproducibility, linearity, constancy versus linac orientation, collimator transmission, leakage radiation, and end effect. Furthermore, small field dosimetry issues such as lack of electronic equilibrium and finite dimensions of the detector compared to the radiation field must be taken into account for detector choice and reliable measurements especially for collimator sizes less than 12.5 mm (Francescon et al. 1998; Pantelis et al. 2008).

Quality Assurance of Targeting Accuracy and Precision

Targeting accuracy refers to the ability of the system to deliver radiation dose to a target defined on a set of CT images. The so-called end-to-end (E2E) test integrates all components of the therapeutic procedure including CT scan, treatment planning (CT data import, contouring, dose calculation), software generating digitally reconstructed radiographs and treatment delivery using the robot, the registration algorithm, the linear

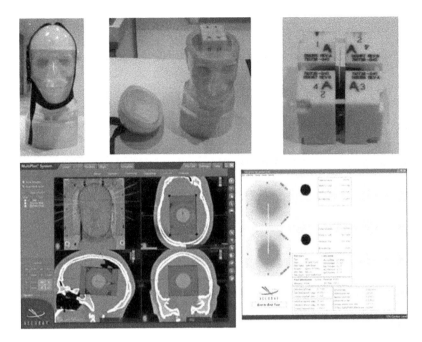

FIGURE 77.3 E2E test steps for targeting accuracy QA.

accelerator, and the patient safety components. A head and neck phantom with intracranial and spine ball cubes loaded with precut MD-55 radiochromic films in an orthogonal configuration is used for E2E test. In the TPS, the 70% isodose line is planned to cover isotropically the ball inside the cube. Analysis of the exposed films is performed using a transparency scanner and the manufacturer's film analysis software for the E2E test (Figure 77.3). The difference between the centroid of the actual delivered dose distribution on the exposed films and the center of the ball inside the cube, that is, the center of the planned dose distribution is a measure of the total error in targeting accuracy. Targeting accuracy E2E test is performed during commissioning to establish accuracy of the system for each treatment delivery modality and has been reported to be less than 1 mm (Chang et al. 2003; Ho et al. 2007; Antypas and Pantelis 2008). The E2E test should be repeated on a monthly basis to document constancy of the system accuracy by delivering a predesigned treatment plan for each treatment modality. Each treatment mode must be tested and documented at least three times annually. On an annual basis, the complete E2E test should be conducted for each modality separately including new CT scans, planning, and delivery.

Reproducibility of the system's targeting accuracy is performed on a daily basis with a simplified test using the Automated Quality Assurance (AQA) tool, which is based on the gantry-linac stereotactic QA isocentric technique (Schell et al. 1995). The test is modified for CyberKnife nonisocentric techniques by loading a radiopaque ball inside a plastic phantom containing fiducials and two radiochromic EBT films in an orthogonal arrangement. The image guidance of the system is used to align the phantom and deliver two orthogonal treatment beams to the phantom. The shadow of the radiopaque ball is exposed on the films. Images are analyzed for concentricity of the beam and shadow with the Automated Quality Assurance film analysis tool (Physics Essentials Guide in CyberKnife System Manual 2006).

Targeting precision is associated with the combined performance of the robot, the imaging system, the couch positioning accuracy, and the image guidance system. It represents the ability of the system to track any offset in the patient's position and then to realign the linac beam to that offset position. Treatment delivery precision is evaluated by delivering repeatedly the same E2E plan on the phantom (e.g., by deliberately setting an offset by 8 mm along a coordinate axis). The test should be performed on an annual basis and at least once per year for each tracking method.

Quality Assurance of Motion Tracking Method—Synchrony System

Ideally, a QA method for the Synchrony system must incorporate such equipment that simulates as accurately as possible the patient's respiratory pattern and ensures accurate recording and tracking of the position of a moving target. For routine QA purposes, the Synchrony QA motion table provided by the manufacturer is used. The Synchrony motion table is able to linearly move a target in the SI direction with variable phase shifts between the moving target and the moving chest in the AP direction. The test has been designed to evaluate the performance of the Synchrony system by comparing the targeting accuracy and dose profiles of static and moving E2E tests. The standard ballcube, loaded with two orthogonal MD55 Gafchromic films, is fitted inside a spherical phantom assembly and both are placed together on the Synchrony motion table. Two LED markers are used to track the chest motion. Two Synchrony/fiducial E2E tests are generated and delivered for a static target and a full moving target with 10° phase shift. The 10° phase shift is considered to represent approximately 85% of patient respiratory motion. The exposed films are scanned with a flatbed transparency table and analyzed with the E2E analysis software for targeting accuracy evaluation. DICOM RT dose file, exported from the TPS, and film images are further analyzed with freeware image analysis software or another software application able to analyze film and DICOM RT files. Calculated dose profiles along the center of the target are compared with film dose profiles along the same axes. When the Synchrony system is used properly, targeting accuracy for moving targets may be in the order of 1.5 mm (Seppenwoolde et al. 2007). Synchrony targeting accuracy should be established during commissioning of the system and routinely checked thereafter following the system's targeting accuracy QA recommendations.

References

Adler, Jr., J. R., M. J. Murphy, S. D. Chang, and S. L. Hancock. 1999. Image-guided robotic radiosurgery. *Neurosurgery* 44 (6): 1299–1306; discussion 1306–1307.

Alfonso, R., P. Andreo, R. Capote, M. S. Huq, W. Kilby, P. Kjall, T. R. Mackie et al. 2008. A new formalism for reference dosimetry of small and nonstandard fields. *Med. Phys.* 35 (11): 5179–5186.

Almond, P. R., P. J. Biggs, B. M. Coursey, W. F. Hanson, M. S. Huq, R. Nath, and D. W. Rogers. 1999. AAPM's TG-51 protocol for clinical reference dosimetry of high-energy photon and electron beams. *Med. Phys.* 26 (9): 1847–1870.

Andreo, P., D. T. Burns, K. Hohlfeld, M. S. Huq, T. Kanai, F. Laitano, V. G. Smythe, and S. Vynckier. 2000. Absorbed dose determination in external beam radiotherapy. In *IAEA Technical Report Series No. 398.* Vienna: International Atomic Energy Agency.

Antypas, C., and E. Pantelis. 2008. Performance evaluation of a CyberKnife G4 image-guided robotic stereotactic radiosurgery system. *Phys. Med. Biol.* 53 (17): 4697–4718.

Antypas, C., E. Pantelis, L. Sideri, K. Verigos, A. Tzouras, and N. Salvaras. 2009. Quality assurance of the imaging subsystem G4 CyberKnife. Evaluation of the flat panel imaging characteristics. Poster presented in CyberKnife® Users Meeting, Hollywood, FL.

Chang, S. D., W. Main, D. P. Martin, I. C. Gibbs, and M. P. Helibrun. 2003. An analysis of the accuracy of the CyberKnife: A robotic frameless stereotactic system. *Neurosurgery* 52: 140–146.

Francescon, P., S. Cora, C. Cavedon, P. Scalchi, S. Reccanello, and F. Colombo. 1998. Use of a new type of radiochromic film, a new parallel-plate micro-chamber, MOSFETs, and TLD 800 microcubes in the dosimetry of small beams. *Med. Phys.* 25 (4): 503–511.

Fu, D., and G. Kuduvalli. 2008. A fast, accurate, and automatic 2D–3D image registration for image-guided cranial radiosurgery. *Med. Phys.* 35 (5): 2180–2194.

Ho, A. K., D. Fu, C. Cotrutz, S. L. Hancock, S. D. Chang, I. C. Gibbs, C. R. Maurer Jr., and J. R. Adler Jr. 2007. A study of the accuracy of cyberknife spinal radiosurgery using skeletal structure tracking. *Neurosurgery* 60 (2 Suppl 1): ONS147–ONS156; discussion ONS156.

Kawachi, T., H. Saitoh, M. Inoue, T. Katayose, A. Myojoyama, and K. Hatano. 2008. Reference dosimetry condition and beam quality correction factor for CyberKnife beam. *Med. Phys.* 35 (10): 4591–4598.

Murphy, M. J. 2002. Fiducial-based targeting accuracy for external-beam radiotherapy. *Med. Phys.* 29 (3): 334–344.

Murphy, M. J., and R. S. Cox. 1996. The accuracy of dose localization for an image-guided frameless radiosurgery system. *Med. Phys.* 23 (12): 2043–2049.

Pantelis, E., C. Antypas, L. Petrokokkinos, P. Karaiskos, P. Papagiannis, M. Kozicki, E. Georgiou, L. Sakelliou, and I. Seimenis. 2008. Dosimetric characterization of CyberKnife radiosurgical photon beams using polymer gels. *Med. Phys.* 35 (6): 2312–2320.

Physics Essentials Guide in CyberKnife System Manual 2006. Sunnyvale, CA: Accuray™ Inc.

Rossi, R. P., P. J. P. Lin, P. L. Rauch, and K. J. Strauss. 1985. Performance Specifications and Acceptance Testing for X-ray Generators and Automatic Exposure Control Devices, AAPM Report 14. New York: American Association of Physicists in Medicine. American Institute of Physics.

Schell, M. C., F. J. Bova, D. A. Larson, D. D. Leavitt, W. R. Lutz, E. B. Podgorsak et al. 1995. *Stereotactic Radiosurgery: Report of Task Group 42 Radiation Therapy Committee.* Woodbury, NY: American Institute of Physics.

Schweikard, A., H. Shiomi, and J. Adler. 2004. Respiration tracking in radiosurgery. *Med. Phys.* 31 (10): 2738–2741.

Seppenwoolde, Y., R. I. Berbeco, S. Nishioka, H. Shirato, and B. Heijmen. 2007. Accuracy of tumor motion compensation algorithm from a robotic respiratory tracking system: A simulation study. *Med. Phys.* 34 (7): 2774–2784.

Shepard, S., P.-J. P. Lin, J. Boone, D. Cody, J. Fisher, G. Frey, H. Glasser et al. 2002. Quality control in diagnostic radiology. In *AAPM Report No. 74.*

78

Stand-Alone Localization Systems: Ultrasound

Laura Drever
Kingston General Hospital

Michelle Hilts
British Columbia Cancer Agency

Introduction

Ultrasound (US) localization systems can be used in radiation therapy as a part of an image-guided radiation therapy (IGRT) program. The goal of IGRT is to pinpoint the target location at time of treatment and hence eliminate, or reduce, the risk of a geometric miss. As radiotherapy treatments become more conformal and treatment margins are reduced to allow for dose escalation, the risk of a geometric miss of the treatment target increases. This risk is especially high in sites where the target organs can move relative to the bony anatomy and traditional electronic portal image techniques, which use bony anatomy to correct for patient setup variation, do not accurately represent target location. The process of IGRT locates the soft tissue target each day of treatment and allows for adjustment of patient position to correctly place the target within the radiation treatment fields.

US localization is one method of performing IGRT. Its advantages include the use of nonionizing radiation, good soft-tissue imaging, and a relatively low cost. US images cannot be acquired through bone or through air; thus, these localization systems have been used primarily for prostate localization (Cury et al. 2006; Fuller et al. 2006; Johnston et al. 2008; Langen et al. 2003; Scarbrough et al. 2006; Serago et al. 2006); however, work with other sites is beginning (Berrang et al. 2009).

How US Localization Systems Work

There are several commercially available US IGRT systems with varying features and potential advantages and disadvantages; however, all operate on the same basic principles and consist of the same basic components. A US localization system consists of (1) a diagnostic US imaging system, (2) a tracking system that positions the US images accurately with respect to the treatment

room, and (3) a system, either physical or software based, for accurately repositioning the patient. The US imaging component consists of an imaging probe, a viewing console, and the necessary software to view the US images as well as to realign the patient relative to the baseline images. Obtained US images are positioned within the coordinate system of the treatment room by tracking the US probe position. There are currently three methods used to track the US probe. Early localization systems, such as the original BAT system (NOMOS, Chatsworth, CA), tracked the US probe by mounting it on a mechanical arm. Two more recent methods track the probe using infrared light–emitting diodes and a camera system or by fitting the probe with a camera that tracks a known pattern at a known location. Examples of the former method are the SonArray (Varian, Palo Alto, CA), ExacTrac (BrainLAB, Heimstetten, Germany), Restitu (Resonant, Montreal, Canada), and BATCAM (NOMOS) systems, whereas an example of the latter method is the I-Beam (CMS/Elekta, Maryland Heights, MO). Regardless of the tracking system used, absolute location of the target is determined by relating the position of the probe to the room coordinates. Then, relating the location of the US images, and structures within the images, to the probe coordinates. This absolute target localization is achieved through a calibration procedure where images are acquired of a calibration phantom placed at a precisely known location in the treatment room. Each manufacturer provides a system-specific calibration procedure (not discussed here). A thorough explanation of how calibration techniques work is outlined in the paper by Bouchet et al. (2001).

Quality Assurance Tests

There are three classes of quality assurance (QA) tests that are required to verify that a US localization system is functioning properly: (1) image quality and display tests, (2) localization and data transfer tests, and (3) physical QA tests. The first series of tests are related to image quality and display tests of the diagnostic US system. The second series of tests relate to the localization of the US images, and patient, with respect to the room coordinates. This series also tests the transfer of data between the simulation system, either both CT and treatment planning systems or simulation US, to the US system in the treatment room. The third series of QA tests check physical and mechanical aspects of the system for wear and tear.

Image Quality and Display Tests

Image quality and display tests ensure that images produced by the US system are of sufficient quality to accurately visualize and segment the target (typically the prostate). These tests are not unique to US localization systems and several published guides to diagnostic US image QA are applicable (e.g., ACR [Bushberg 2002] and AAPM [Goodsitt et al. 1998; Pfeiffer et al. 2008]). However, the frequency of the tests may need to be increased to take into account that these US systems are used to target the location for treatment, not just for diagnostic purposes. Presented here is a summary of the image quality test equipment (i.e., phantom) and procedures required for US localization systems.

Image quality tests are conducted using a phantom constructed of a material that echogenically mimics human tissue, that is, speed of sound in the material of 1540 ± 10 m/s at 22°C and an attenuation coefficient of 0.5–0.7 dB/cm/MHz. The phantom should include at least one pair of horizontally and vertically spaced filaments, one anechoic structure (sphere or ellipse), and other structures, such as rods at varying depths. It may be possible to use a single well-designed phantom for all imaging and localization tests, or a series of phantoms may be required. See the AAPM report (Goodsitt et al. 1998) for other possible phantom designs. Furthermore, it may be desirable to have at least two phantoms, one for calibration, monthly and weekly QA and a second less complex phantom for daily QA (Drever and Hilts 2007).

Image quality tests required to ensure system functionality are (1) grayscale visibility, (2) depth visualization, (3) image

uniformity, (4) distance accuracy, and (5) anechoic object imaging. To test grayscale visibility, a grayscale test pattern (step-wedge) image is displayed and the number of visible steps is recorded. In addition, text on the display monitor is checked to ensure it is not blurry. Note that the remaining tests (2) through (5) require the use of a US phantom. Depth of visualization tests record how deep into the phantom structures or filaments can be visualized. Image uniformity is checked to ensure that there are no significant streaks in an image of a uniform phantom. Vertical and horizontal distance accuracy tests are performed by measuring the distance between pairs of filaments embedded in the phantom and comparing these measures to the known separation. Finally, anechoic object imaging is tested to ensure that reconstruction of objects of known size is done correctly. For a 2D US system, the area of the anechoic structure can be measured and compared to baseline, or for 3D US systems, a volume can be recorded. For more information on these tests, consult the two AAPM reports (Goodsitt et al. 1998; Pfeiffer et al. 2008).

Localization and Data Transfer Tests

US systems for target localization in radiation therapy treatment require thorough QA testing of the accuracy of the localization of US images within the treatment room (Bouchet et al. 2001; Tome et al. 2002). As discussed above, the mapping of the US images to the treatment room coordinate system is achieved by a calibration process. It is the purpose of QA testing to ensure that this calibration remains correct.

To align the patient daily, the US images acquired in the treatment room must be compared to a baseline image set. Once the two data sets have been correctly registered, the localization system will report the distance and direction of the shifts required to align the patient. The predicted shift and direction of shift must be tested regularly. This is achieved by performing three tests: (1) alignment of isocenters, (2) data transfer with shift, and (3) realignment tests. The alignment of isocenters test is similar to the anechoic object imaging test but confirms, given a well-defined phantom setup, that the anechoic object is at the expected location within the treatment room. The phantom is

TABLE 78.1 Monthly US QA Tests

Description of Test	Action Level
Physical and mechanical check of probes, cables, etc.	Damaged
Grayscale visibility (using step wedge, a change from baseline)	Two steps or 10%
Depth of visualization (change from baseline)	10 mm
Image uniformity	Significant change from baseline
Vertical distance accuracy	2 mm or 2%
Horizontal distance accuracy	3 mm or 3%
Anechoic object imaging (either area or volume)	5%
Phantom QA (mechanical and CT)	Damaged
Alignment of isocenters (for each axis)	±2 mm
Data transfer with shift test	±2 mm
Realignment test	±1 mm

TABLE 78.2   Weekly US QA Tests

Description of Test	Action Level
Vertical distance accuracy	2 mm or 2%
Horizontal distance accuracy	3 mm or 3%
Anechoic object imaging	5%
Alignment of isocenters	±2 mm
Data transfer with shift test	±2 mm
Realignment test	±1 mm

TABLE 78.3   Daily US QA Test

Description of Test	Action Level
Alignment of isocenters	±2 mm

aligned to room lasers, scanned to produce a quality image set (particularly of the anechoic object) that is analyzed to verify that it matches a reference image set obtained under identical conditions. The data transfer with shift test is a variation of the alignment of isocenters test in which a known shift off isocenter is introduced at the time of simulation. The phantom is set up in the treatment room using the lasers and the location of the anechoic volume is recorded. If the system is performing correctly, this location should reflect the shift required to realign the phantom to the simulation position. Both direction and magnitude of shift should be verified. Finally, the realignment test uses the device, or software, that is provided with the system to make the required shift. A follow-up scan of the phantom is then used to ensure that the anechoic volume is now at the correct location.

Physical QA Tests

Physical QA tests are pass/fail tests to verify that the US localization system and US phantoms have not been physically damaged. Each of the US probes should be checked for any sign of damage, video monitors checked for scratches and dirt, electrical cables for cracks or discolorations, wheels for smooth rotation and wheel locks for functionality, and mechanical arms for sinking or drifting from placed position. In addition, all surfaces of the scanner housing should be checked for dents or scrapes.

Finally, testing of the phantoms used for US QA is required because the tissue mimicking material that the phantoms are made of is water based and the phantoms are susceptible to desiccation (Drever and Hilts 2007). In addition, the layers of gel in the phantom can separate or pull away from the phantom wall even with gentle handling. Therefore, it is important to check that any phantoms used for QA remain undamaged over time (Goodsitt et al. 1998; Pfeiffer et al. 2008). This can be achieved by CT scanning the phantoms and comparing resultant images against baseline images acquired of each phantom when it is new.

Summary of Recommended Testing and Test Frequency

Testing is divided into recommended monthly, weekly, and daily QA tests. Monthly QA is a comprehensive set of tests that should be performed by the medical physics staff using a calibration phantom (typically supplied by the manufacturer). It includes image quality, localization, and physical QA tests. Weekly testing is designed to be a rapid yet precise test of the spatial accuracy and localization of targets in the US images. A highly accurate phantom (typically the calibration phantom) must be used for these tests. Because there is a large amount of overlap between monthly and weekly QA, it may not be required to do weekly QA the same week as monthly QA. Daily testing consists of a single rapid test of target localization. This test is designed to be conducted by radiation therapists daily before patient treatments using the US localization system. A simplified phantom is recommended for these tests (Drever and Hilts 2007). Monthly, weekly, and daily QA tests are summarized in Tables 78.1 through 78.3, respectively.

References

Berrang, T. S., P. T. Truong, C. Popescu, L. Drever, H. A. Kader, M. L. Hilts, T. Mitchell et al. 2009. 3D ultrasound can contribute to planning CT to define the target for partial breast radiotherapy. *Int. J. Radiat. Oncol. Biol. Phys.* 73 (2): 375–383.

Bouchet, L. G., S. L. Meeks, G. Goodchild, F. J. Bova, J. M. Buatti, and W. A. Friedman. 2001. Calibration of three-dimensional ultrasound images for image-guided radiation therapy. *Phys. Med. Biol.* 46 (2): 559–577.

Bushberg, J. T. 2002. *The Essential Physics of Medical Imaging.* 2nd ed. Philadelphia, PA: Lippincott Williams & Wilkins.

Cury, F. L., G. Shenouda, L. Souhami, M. Duclos, S. L. Faria, M. David, F. Verhaegen, R. Corns, and T. Falco. 2006. Ultrasound-based image guided radiotherapy for prostate cancer: Comparison of cross-modality and intramodality methods for daily localization during external beam radiotherapy. *Int. J. Radiat. Oncol. Biol. Phys.* 66 (5): 1562–1567.

Drever, L. A., and M. Hilts. 2007. Daily quality assurance phantom for ultrasound image guided radiation therapy. *J. Appl. Clin. Med. Phys.* 8 (3): 2467.

Fuller, C. D., C. R. Thomas, S. Schwartz, N. Golden, J. Ting, A. Wong, D. Erdogmus, and T. J. Scarbrough. 2006. Method comparison of ultrasound and kilovoltage x-ray fiducial marker imaging for prostate radiotherapy targeting. *Phys. Med. Biol.* 51 (19): 4981–4993.

Goodsitt, M. M., P. L. Carson, S. Witt, D. L. Hykes, and J. M. Kofler Jr. 1998. Real-time B-mode ultrasound quality control tests procedures. Report of AAPM Ultrasound Task Group No. 1. *Med. Phys.* 25 (8): 1385–1406.

Johnston, H., M. Hilts, W. Beckham, and E. Berthelet. 2008. 3D ultrasound for prostate localization in radiation therapy: A comparison with implanted fiducial markers. *Med. Phys.* 35 (6): 2403–2413.

Langen, K. M., J. Pouliot, C. Anezinos, M. Aubin, A. R. Gottschalk, I. C. Hsu, D. Lowther et al. 2003. Evaluation of ultrasound-based prostate localization for image-guided radiotherapy. *Int. J. Radiat. Oncol. Biol. Phys.* 57 (3): 635–644.

Pfeiffer, D., S. Sutlief, W. Feng, H. M. Pierce, and J. Kofler. 2008. AAPM Task Group 128: Quality assurance tests for prostate brachytherapy ultrasound systems. *Med. Phys.* 35 (12): 5471–5489.

Scarbrough, T. J., N. M. Golden, J. Y. Ting, C. D. Fuller, A. Wong, P. A. Kupelian, and C. R. Thomas Jr. 2006. Comparison of ultrasound and implanted seed marker prostate localization

methods: Implications for image-guided radiotherapy. *Int. J. Radiat. Oncol. Biol. Phys.* 65 (2): 378–387.

Serago, C. F., S. J. Buskirk, T. C. Igel, A. A. Gale, N. E. Serago, and J. D. Earle. 2006. Comparison of daily megavoltage electronic portal imaging or kilovoltage imaging with marker seeds to ultrasound imaging or skin marks for prostate localization and treatment positioning in patients with prostate cancer. *Int. J. Radiat. Oncol. Biol. Phys.* 65 (5): 1585–1592.

Tome, W. A., S. L. Meeks, N. P. Orton, L. G. Bouchet, and F. J. Bova. 2002. Commissioning and quality assurance of an optically guided three-dimensional ultrasound target localization system for radiotherapy. *Med. Phys.* 29 (8): 1781–1788.

79

Stand-Alone Localization Systems: Transponder Systems

Twyla Willoughby
M. D. Anderson Cancer Center Orlando

Amish Shah
M. D. Anderson Cancer Center Orlando

Lakshmi Santanam
Washington University School of Medicine

Introduction

The Calypso® Medical 4D localization system uses radiofrequency for tracking during radiation therapy. Small (2 × 8 mm) Beacon™ transponders are implanted in or near the target. Each electronic transponder consists of an alternating current electromagnetic resonance circuit encapsulated in glass. At the time of the writing of this chapter, the device is FDA cleared for use in prostate, but there are plans underway for applications on other body sites.

Localization of the transponders is achieved using an electromagnetic array consisting of radiofrequency signaling and receiving coils. Each of the three transponders that are implanted has a unique resonant frequency. The transponders absorb some of the radiofrequency energy and reemit that energy in the form of a decaying signal that is detected by the electromagnetic array. The transponder position is then detected relative to the array, which is calibrated to the room reference frame by three rigidly mounted infrared cameras. A misalignment of the target can be detected by a proprietary algorithm that identifies shifts from its prescribed location anytime throughout the treatment. The accuracy of the detection of a target in phantom is less than 1 mm. The Calypso system displays real-time graphs instantaneously highlighting shifts in position that exceed a user-specified threshold. A radiation monitoring device inside the room makes it possible to generate reports synchronized to the radiation delivery. Figure 79.1 shows the beacon transponders and the Calypso system along with a sample report.

There have been many publications of the clinical use of the Calypso system. These references more thoroughly described the system and have given guidance to implantation technique, beacon stability, and use of the beacons in other applications (Murphy et al. 2008; Noel et al. 2009; Mayse et al. 2008; Santanam et al. 2008; Kupelian et al. 2007; Willoughby et al. 2006). The purpose of this chapter is to discuss the quality assurance (QA) and safety aspects of this device.

System Integration

Define/Measure Isocenter

Before the installation of an RF localization system, it is recommended that a full quality control of the mechanical aspects of the linear accelerator be performed with documentation of the machine isocenter including a Winston and Lutz type of film (Lutz, Winston, and Maleki 1988). It is also appropriate to ensure that the lasers match the radiation isocenter while performing this QA because the laser system is used to calibrate the Calypso system.

Table Attachment and Collision Issues

Because the Calypso device is a large piece of equipment that is placed in the treatment room, the user should spend some time becoming familiar with the movement of the console within the treatment room as well as the positioning of the array in the area surrounding the gantry and couch. There are currently

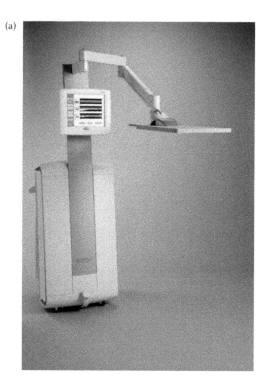

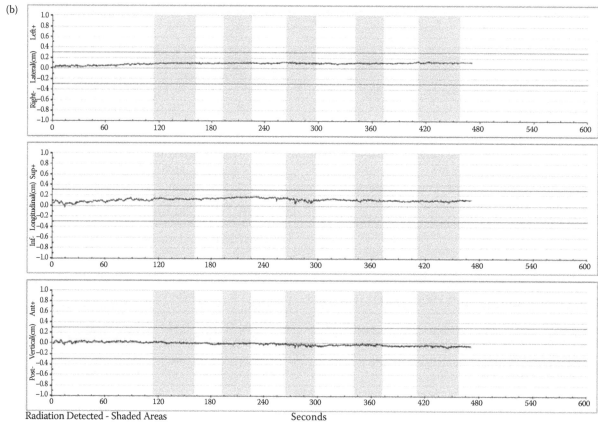

Radiation Detected - Shaded Areas   Seconds

FIGURE 79.1 (a) Calypso system, (b) sample report, and (c) schematic showing the configuration of the system within the treatment control area and treatment room.

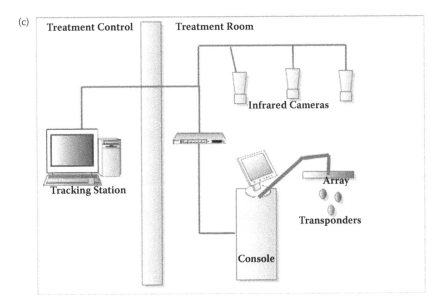

FIGURE 79.1   (Continued)

limitations to couch angles that can be used because of the close proximity of the console to the treatment couch. In some instances, the array may also be too close to the gantry or to the patient so certain gantry angles cannot be used. Special attention must be given when moving the gantry, the Calypso console, or the array so that injury does not occur and equipment is not damaged.

In some instances, an additional table overlay may be necessary to remove any carbon fiber in the vicinity of the array or the beacons. Carbon fiber, which is typically used in radiation therapy because of its ideal radiographic properties, is highly reflective of radiofrequency signals and can cause noise in the signal being collected by the array. A table overlay made out of Kevlar is often used to position the patient away from any possible carbon fiber in the treatment couch. When using the table overlay, the user must be conscious of gantry position relative to the table because there may be limitations due to the added table height.

Record and Verify, and Treatment Planning

The current release of the Calypso system has a limited ability to interface with treatment planning or record and verify systems. Initially, it is recommended to test several orientations (prone/supine) where Beacon transponder locations are identified and transferred from a treatment planning CT to the localization system. This transfer may be either electronic or manual. Errors that can occur in clinical situations include misidentification of beacon location or frequency, coordinate transformation errors, and mistyping of information. At least one of the test cases should simulate each error to make sure that a quality control program is created for clinical use to identify each type of error. In addition, the transfer of treatment alignment data and reports from the Calypso system to the record and verify system or offline storage should be tested for functionality.

Radiation Delivery Integration

Several different treatment delivery scenarios (complete treatment plans such as IMRT, arcs, and conformal beams) should be delivered in a phantom with and without the Calypso system in place at the time of commissioning. These dosimetric tests are done to ensure that there is no change to the radiation delivered due to the array being in place or due to the operation of the system.

The current release of the Calypso system includes a radiation monitor that will generate a report indicating when radiation beam-on occurred during beacon tracking. An example is seen on the sample report in the insert in Figure 79.1. The functionality of this can be tested by recording the beacon transponder location while timing the radiation delivery and beam-off times for several different beams. The report should properly record the radiation on time. This is not meant for dose calculations but as an aid in reporting approximately when the radiation was delivered. Specifically, this radiation monitor does not turn on and off for different segments of IMRT fields.

Future releases of the Calypso system may interface (gate) with the radiation delivery system. For those purposes, more testing would be necessary to test gating efficiency and system latency.

Calibration

Infrared Camera Calibration (i.e., Wanding)

The camera calibration establishes the field of view for all three rigidly mounted infrared cameras. The recommendation for camera calibration is monthly; however, this frequency may only be necessary if camera calibration is lost. A detailed procedure of the calibration of the system is described in the Calypso system's user manual. Two kinds of fixtures with optical reflectors (L-frame and T-frame) are used for this purpose. Whereas the L-frame fixture is placed on the couch and tracked by the cameras, the T-frame fixture is used to establish the optical field of view by waving it (like a wand) near the isocenter.

Overall System Calibration

The calibration of the Calypso system is performed using a large phantom (Pharos) that includes infrared reflectors on the surface as well as implanted beacon transponders in a known configuration. This phantom has external marks and can be placed at the machine isocenter using the room lasers. The Calypso system calibration is performed through software that uses the infrared cameras to detect the infrared reflectors on the Pharos phantom and then calibrates the array position and the beacon transponder positions in the phantom relative to the isocenter. This process must be completed anytime that the infrared cameras are calibrated [see section "Infrared Camera Calibration (i.e. Wanding)"] and the software currently forces these two steps together. The initial recommendations are to perform this monthly; however, this may be reduced to quarterly or annually.

For a final verification, this phantom contains a high-Z material target that is exactly in the center of location of the external indicators on the Pharos phantom. This is ideal for a "hidden target" for taking portal films to verify the calibration of the Calypso RF tracking as well as the infrared tracking as it compares to the radiation isocenter. This is done by taking AP and lateral portal images at the end of the calibration procedure. If the target is outside of tolerance (about 1 mm), the calibration should be repeated and the room lasers should be checked.

Quality Assurance

General Operation

The software has two different modes of operation. The first is localization mode used when initially aligning the patient.

After localization, the user can select the tracking mode. When a patient is first selected in localization mode, the algorithm checks the locations and distances between each of the frequency beacons compared to the expected locations and calculates the angles of the beacons relative to the isocenter. The fit of the beacons to the original location is called the geometric residual. The Calypso user manual gives some recommendations regarding the tolerance for the geometric residual as well as the angles of the beacons. Typically, large changes in geometric residual indicate that one or more of the transponders have moved. Large angles can also indicate that a transponder has moved or that the anatomy does not match the anatomy from the treatment planning CT scan. It is recommended that the user design a QA program to include imaging (CT or kV x-ray) to determine if the beacon geometry is appropriate for treatment localization anytime there is a question about the geometric residual or the rotations. In tracking mode, the system does not use as many checks for beacon locations and only checks the relative offset of the beacons from the initial tracking location.

Drift and Reproducibility

The optical cameras that are used to track the array as well as the RF coils within the array can be susceptible to spatial drift. Testing procedures specific to optical systems and expected tolerances can be found in the literature (Bova et al. 1997; Meeks et al. 2005; Tome et al. 2001; Phillips et al. 2000). Drift will most often be exhibited within the first 30 minutes of the cameras being turned on. General operation of the Calypso system recommends leaving the infrared cameras on at all times. If for some reason the cameras must be powered off, it is recommended to allow them to warm up before use.

Localization Accuracy (End-to-End Test)

The end-to-end test is performed as a QA of the entire treatment process. A phantom containing the transponders as well as a high-Z target such as the Pharos phantom is used (see section "Overall System Calibration"). A CT scan should be taken at the slice thickness that will be used for treatment. A treatment plan

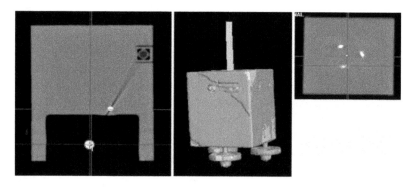

FIGURE 79.2 An example of a phantom used for end-to-end testing. This particular phantom is embedded with Calypso beacons with a tungsten ball for performing portal film verification.

TABLE 79.1 Results of Measurements Using the End-to-End Phantom Shown in Figure 79.2

	Anterior–Posterior	Lateral	Axial	3D Vector
Phantom measurements (*n* = 47)				
Average (mm)	−0.14	0.09	−0.18	0.78
Standard deviation (mm)	0.63	0.46	0.29	0.34

is generated by identifying the three beacon transponder coordinates and a proposed isocenter placed at the radiographic target. The Calypso system is then used to position the phantom on the treatment machine. At this point, orthogonal portal images are taken with a small radiation field (typically a stereotactic cone or a 2-cm square) to identify how accurately the target is centered within the phantom relative to the machine isocenter.

An example of the use of the end-to-end test for the Calypso system is described and shown in Figure 79.2 with the corresponding results from measurements shown in Table 79.1. Planning CT images (left) show the "hidden target" (radiopaque sphere) inside the Calypso QA phantom, with a simulated beam passing through the sphere (middle). AP and lateral verification films are obtained using a 10-mm circular collimator. The sphere and radiation field are automatically identified using a software application, and the offsets between the corresponding centers are calculated. The scan-plan-localize process should be repeated multiple times and the RMS error for these values calculated to establish the system accuracy. In addition, this test should be repeated monthly and compared to the RMS error to determine if the system is performing within the acceptable tolerance for the user's institution.

Accuracy of Localizing (Shifting Patient)

In general, the Calypso system has a limited area under the array where the Beacon transponders can be excited and their signal collected. This is roughly 14 × 14 × 23 cm³ (w × l × d) in the current configuration. Patients positioned with the beacons outside of this volume will not be identified by the system. The user should roughly test this operating volume by using a small phantom with Calypso beacons that can be placed close to and far from the array to become familiar with the different messages presented by the system.

The accuracy of the range of motion should be tested in both localization and tracking modes of the software. The daily QA device can be used for these tests by positioning it at known distances from the isocenter to determine the shift calculated by the Calypso system. To test the tracking function, the same phantom can be localized to the isocenter and then, during tracking, displaced by known amounts to test that the Calypso system is accurately calculating the expected offsets. A precision table will give a more accurate testing of this process and should be done at commissioning. On a monthly basis, it is adequate to use the couch readout or a ruler. This test should encompass the range of shifts expected in patients and should be accurate to better than 2 mm (depending on the method used for testing). Portal films

of the radiographic target can be used to verify that the shift was accurate.

Rotation corrections can also be tested. This is most accurately tested using a precision phantom that can be rotated to a set location. If the user plans to use the rotation information from the Calypso system with a 6D robotic couch, then this

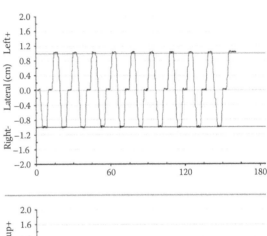

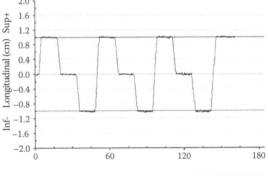

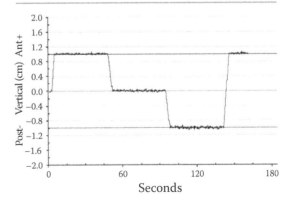

FIGURE 79.3 Comparison of a programmed track and a measured Calypso track.

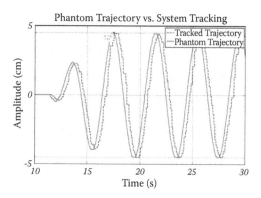

FIGURE 79.4 Measurement of latency between a 4D phantom and the collected trajectory from the Calypso system.

rotation calculation and correction should be tested. Rotations can be simulated by moving one of the three beacon transponders in a specific geometry. It is unclear if rotational corrections are applicable when using fixed points, so caution is necessary in applying this information in the clinical situation.

Tracking Accuracy

Tracking accuracy can be tested by shifting the fixed target to a known location in all three coordinates to verify that the real-time tracking system can properly measure the introduced shifts. Ideally, this is tested using a motion phantom. Some motion phantoms can be programmed to deliver a precise pattern while simultaneously tracking the phantom to verify tracking accuracy. For more information on 4D phantoms specific to respiratory motion measurements, one should look at the AAPM report of Task Group 76 on respiratory motion management (Serban et al. 2008; van Sornsen de Koste et al. 2007; Keall et al. 2006). An example of a comparison of a programmed track and a measured Calypso track are shown in Figure 79.3 (unpublished observation). This example shows a maximum reported excursion (as programmed in the 4D phantom) within a few millimeters and that there are no drifts in the trajectory.

The system latency and time (frequency) of tracking should also be tested. The Calypso system specifies a reporting frequency of 10 Hz. The easiest way to test the latency is again by using the programmable motion phantom. The phantom is programmed to move at a specific time from an initial coordinate. A tracking of the target is recorded for the initial set point up through the time point that the phantom was programmed to move. Comparing the internal computer clocks of the programmable phantom and the report from the Calypso system can indicate the latency in the time for motion and the detected time of the motion. Another method to test the latency is with a video camera that could simultaneously image the phantom motion and Calypso tracking display. A frame-by-frame analysis of the video file can be performed to compare the spatial phantom position with the concurrent Calypso readout displayed on the tracking monitor. Figure 79.4 (unpublished observation) shows

the Calypso trajectory lagging behind the phantom trajectory and the average system latency was measured to be 300 ms.

Ongoing QA Schedules

On a monthly basis, an end-to-end test of accuracy should be performed as well as camera and system calibrations, as mentioned above. Tracking and localization shifts of the system should be checked using the couch readout or a ruler on a monthly basis. On a daily basis, the basic operation of the system should be tested including system startup and a quick check of the calibration to isocenter using a daily QA phantom that has embedded Calypso beacons. Patient-specific QA should be performed for each new patient to ensure that the data are transferred properly from the planning CT data into the Calypso system and that for each treatment the correct patient is selected in the localization system. This may require a "time-out" procedure where the treating therapist stops to identify the patient in the localization system compared to the patient on the treatment couch.

During preventative maintenance, the user should verify that the field service engineer tests the camera calibration, system calibration, and a total system QA (similar to a patient test plan). In addition, the beam-on detection system should be verified by the engineer and the end user through multiple static fields delivered 30 s apart as well as an IMRT delivery.

Summary

It is extremely important that the coordinates of the Calypso beacons be identified correctly in the Calypso system because this can lead to mistreatment of patients if they are incorrectly identified. A QA program must be in place that provides a means for checking these data before the start of the patient's treatment. In addition, it is recommended that some methods of verifying large changes in Beacon locations within the patient be carried out. This can be done through a repeat CT scan or some other radiographic check. The Calypso system will be able to indicate some large changes in beacon location or identify if a beacon is missing; however, it is possible for the beacons to move in such a way that the system may still calculate a location for treatment that is different than the planned location.

References

Bova, F. J., J. M. Buatti, W. A. Friedman, W. M. Mendenhall, C. C. Yang, and C. L. Liu. 1997. The Univeristy of Florida frameless high-precision stereotactic radiotherapy system. *Int. J. Radiat. Oncol. Biol. Phys.* 38: 875–882.

Keall, P. J., G. S. Mageras, J. M. Balter, R. S. Emery, K. M. Forster, S. B. Jiang, J. M. Kapatoes et al. 2006. The management of respiratory motion in radiation oncology report of AAPM Task Group 76. *Med. Phys.* 33 (10): 3874–3900.

Kupelian, P., T. Willoughby, A. Mahadevan, T. Djemil, G. Weinstein, S. Jani, C. Enke et al. 2007. Multi-institutional clinical experience with the Calypso System in localization and continuous, real-time monitoring of the prostate gland during external radiotherapy. *Int. J. Radiat. Oncol. Biol. Phys.* 67 (4): 1088–1098.

Lutz, W., K. R. Winston, and N. Maleki. 1988. A system for stereotactic radiosurgery with a linear accelerator. *Int. J. Radiat. Oncol. Biol. Phys.* 14 (2): 373–381.

Mayse, M. L., P. J. Parikh, K. M. Lechleiter, S. Dimmer, M. Park, A. Chaudhari, M. Talcott, D. A. Low, and J. D. Bradley. 2008. Bronchoscopic implantation of a novel wireless electromagnetic transponder in the canine lung: A feasibility study. *Int. J. Radiat. Oncol. Biol. Phys.* 72 (1): 93–98.

Meeks, S. L., W. A. Tome, T. R. Willoughby, P. A. Kupelian, T. H. Wagner, J. M. Buatti, and F. J. Bova. 2005. Optically guided patient positioning techniques. *Semin. Radiat. Oncol.* 15 (3): 192–201.

Murphy, M. J., R. Eidens, E. Vertatschitsch, and J. N. Wright. 2008. The effect of transponder motion on the accuracy of the Calypso Electromagnetic localization system. *Int. J. Radiat. Oncol. Biol. Phys.* 72 (1): 295–299.

Noel, C., P. J. Parikh, M. Roy, P. Kupelian, A. Mahadevan, G. Weinstein, C. Enke, N. Flores, D. Beyer, and L. Levine. 2009. Prediction of intrafraction prostate motion: Accuracy of pre- and post-treatment imaging and intermittent imaging. *Int. J. Radiat. Oncol. Biol. Phys.* 73 (3): 692–698.

Phillips, M. H., K. Singer, E. Miller, and K. Stelzer. 2000. Commissioning an image-guided localization system for radiotherapy. *Int. J. Radiat. Oncol. Biol. Phys.* 48 (1): 267–276.

Santanam, L., K. Malinowski, J. Hubenshmidt, S. Dimmer, M. L. Mayse, J. Bradley, A. Chaudhari et al. 2008. Fiducial-based translational localization accuracy of electromagnetic tracking system and on-board kilovoltage imaging system. *Int. J. Radiat. Oncol. Biol. Phys.* 70 (3): 892–899.

Serban, M., E. Heath, G. Stroian, D. L. Collins, and J. Seuntjens. 2008. A deformable phantom for 4D radiotherapy verification: Design and image registration evaluation. *Med. Phys.* 35 (3): 1094–1102.

Tome, W. A., S. L. Meeks, T. R. McNutt, J. M. Buatti, F. J. Bova, W. A. Friedman, and M. Mehta. 2001. Optically guided intensity modulated radiotherapy. *Radiother. Oncol.* 61 (1): 33–44.

van Sornsen de Koste, J. R., J. P. Cuijpers, F. G. de Geest, F. J. Lagerwaard, B. J. Slotman, and S. Senan. 2007. Verifying 4D gated radiotherapy using time-integrated electronic portal imaging: A phantom and clinical study. *Radiat. Oncol.* 2: 32.

Willoughby, T. R., P. A. Kupelian, J. Pouliot, K. Shinohara, M. Aubin, M. Roach 3rd, L. L. Skrumeda et al. 2006. Target localization and real-time tracking using the Calypso 4D localization system in patients with localized prostate cancer. *Int. J. Radiat. Oncol. Biol. Phys.* 65 (2): 528–534.

80

Stand-Alone Localization: Surface Imaging Systems

Laura Cerviño
University of California, San Diego

Todd Pawlicki
University of California, San Diego

Introduction

Optical imaging has established itself in this era of image-guided radiation therapy as an accurate imaging technique for patient setup and monitoring (Meeks et al. 2005). Some optical systems reconstruct the three-dimensional (3D) coordinates of markers fixed to the patient (Baroni et al. 2006; Liebler et al. 2009), whereas others reconstruct the complete 3D surface of the patient without the need for markers (Smith et al. 2003; Djajaputra and Li 2005). The main advantage of optical imaging is that it constitutes a nonionizing imaging technique and may therefore be applied daily, for example, before and after correcting the setup of the patient without concern for any additional imaging dose to the patient. Optical imaging, when available in real time, is also useful for continuously monitoring the patient for intrafractional movement. This provides useful information to interrupt the radiation beam when motion thresholds are exceeded. The fact that optical imaging does not require additional radiation may make it a very appealing option for breast and pediatric radiotherapy treatments. Among all optical imaging techniques, surface imaging is the only one that provides 3D surface information of the patient. Surface imaging is achieved by means of stereo cameras. A pair of cameras separated by some specified distance is triggered at the same time, acquiring two different images that mimic human vision. These two pictures can be combined to obtain 3D information of the object imaged, in this case, the patient. There have been several studies on different commercially available systems (Djajaputra and Li 2005; Bert et al. 2005). In this chapter, we will focus on the AlignRT system (VisionRT Ltd., London, UK).

System Description

A picture of an AlignRT camera pod is shown in Figure 80.1. Each pod consists of two data cameras used for stereo photography, one texture camera for obtaining texture images that can be overlapped to the 3D surface image, one speckle projector that projects a red patterned light on the patient's surface during static or dynamic image acquisition, and one speckle light flash for static image acquisition.

The system works as follows. A calibrated camera system fixed to the treatment room ceiling acquires an image of the patient with the speckle pattern projected onto the skin surface. The images from the two data cameras are then analyzed to calculate the surface coordinates in 3D space. The topology of the surface is reconstructed by identifying in the stereo images the pattern from the projector, which is known by the AlignRT software. The computer then extracts the motion or shift with respect to a reference image by comparing the acquired image with the reference image. The reference image can be obtained from the CT scan contours, from another AlignRT system installed in the CT room, or from an image acquired previously, for example, in a previous session. Using a proprietary registration algorithm that minimizes the point-to-surface distance between the reference image and the real-time-generated surface, AlignRT determines the 3D transformation (rotations and translations) that relates one surface to the other. This transformation is then decomposed into the three rotations and three translations along the longitudinal, lateral, and vertical axes, which give the real-time difference between the reference and acquired images of the patient's position (deltas). The software displays these real-time deltas.

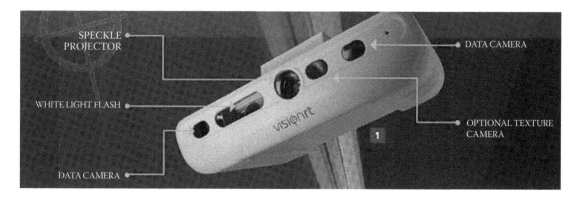

FIGURE 80.1   AlignRT camera system.

To obtain a more extensive coverage of the patient's surface, some systems are composed of two or three camera pods. These ceiling-mounted stereo camera pods work together to build a 3D surface image of the patient. The two pods in the two-pod system or two of the three pods in the three-pod camera system are located laterally to the treatment couch. The other camera in the three-pod camera system is located centrally at the foot of the treatment couch, as shown in Figure 80.2. For some gantry angles, when AlignRT is used for monitoring in real time, the gantry might occlude one of the lateral pods, resulting in coverage of a smaller surface and potentially leading to inaccurate monitoring if only the lateral pods are used. This effect can be more evident when there are couch rotations involved in the treatment. The use of the central pod together with the lateral pods eliminates this problem. For every gantry angle, there are at least two camera pods (central plus one lateral) available for imaging, allowing for accurate monitoring in all situations.

The accuracy of AlignRT in detecting and quantifying patient shifts has been previously studied by Bert et al. (2005) who obtained submillimeter accuracy (0.75 mm) for the three translational degrees of freedom, and less than 0.1° for each rotation.

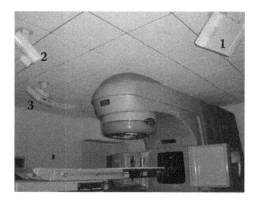

FIGURE 80.2   Surface imaging system consisting of three camera pods: two lateral cameras (labeled "1" and "3" in the figure) and one central camera (labeled "2" in the figure).

Installation, Acceptance, and Commissioning

Before installation can be performed, a visit to perform a site survey is necessary to determine where the AlignRT pods will be installed in the room ceiling. It may not always be possible to install in the predetermined optimal positions and in this case the ceiling nearby will be analyzed and alternative locations determined. A simulation of the coverage of the patient surface that can be obtained will then be performed and the positions validated. In addition, appropriate locations for the PC, in-room and control-room monitors and keyboards, electrical power requirements, and interfacing with third-party devices can be assessed and discussed with the end user.

The actual system installation can then be performed usually over 1 or 2 days depending on room access. Mechanically, this consists of installation of camera mounts, running cables, and connecting the PC followed by adjustment of pod orientation. Then, the optics contained within each pod are adjusted. Software configuration for DICOM server connection, interfacing with third-party devices, and remote access for support are also completed.

System Acceptance

The following steps are required for system acceptance for the basic AlignRT system. Additional modules for third-party interfaces may require further commissioning steps.

1. A monthly calibration (described in detail later) is performed followed by the daily QA, which should pass with a root mean square error of less than 1 mm.
2. A test object, such as a cube, is imaged and the coverage and quality of the resulting 3D surface should be acceptable.
3. A test to confirm that the correct machine coordinate system has been set within the AlignRT system is completed.
4. A system accuracy validation test should pass (see below for details).
5. User training is performed.

System Accuracy Validation

To compute the precision and accuracy of AlignRT for patient positioning involves moving a phantom into a number of known positions relative to a calibration reference and computing accuracy measures by comparing the surface positions generated by AlignRT against a fixed reference position. These known positions are created by shifting and rotating the treatment couch.

The couch positions and surfaces created by the AlignRT software are analyzed within a separate software module that outputs statistics related to the overall accuracy of the system for positioning. The pass/fail criteria are shown in Table 80.1. These errors are computed as follows:

1. Couch coordinate errors: it is possible to compare directly, between both systems, the parameters that define the motion of the subject as follows:

$$\theta_e = \theta_{VRT} - \theta_{RT} \atop t_e = t_{VRT} - t_{RT}, \qquad (80.1)$$

where θ is the angle of rotation about the vertical axis (the treatment couch can rotate about a single axis only); t are the translations along the longitudinal, lateral, and vertical axes of the couch; and the subscripts e, VTR, and RT are abbreviations for error, VisionRT, and radiotherapy couch, respectively.

2. Target registration error: target registration error (TRE) is a metric that was first conceived for image-guided surgery and determines the error in localizing the actual target (e.g., tumor inside patient), which may be larger than the localization errors of reference fiducials (Fitzpatrick, West, and Maurer 1998). For surface matching–based localization, the TRE is defined as follows: if the phantom moves from position A to B, let the corresponding motion of the target, recorded from the treatment couch readings, be called S_{RT} and that derived from the AlignRT system be called S_{VRT}. Thus, given a target at position r (the treatment isocenter in our case), the following TRE may be computed:

$$TRE^2(r) = \left\| S_{RT} - S_{VRT} \right\|^2 \qquad (80.2)$$

TABLE 80.1 Measures and Criteria Used to Determine a Pass and Fail Scenario

Measure	Criteria
Mean TRE at isocenter (mm)	<1.0 mm
StdDev TRE at isocenter (mm)	<0.5 mm
RMS Vrt_Error (mm)	<1.0 mm
RMS Lng_Error (mm)	<1.0 mm
RMS Lat_Error (mm)	<1.0 mm
RMS Rtn_Error (deg)	<0.5°

Note: TRE, target registration error; RMS, root mean square; Vrt, vertical; Lng, longitudinal; Lat, lateral; Rtn, rotation.

FIGURE 80.3 Calibration board showing imaging template with four numbered corners (large dots on the calibration board).

Quality Assurance—Daily QA

Daily quality assurance (QA) is performed by therapists with the help of an image template imprinted on a calibration board provided with the AlignRT system. This calibration plate is shown in Figure 80.3. The calibration plate should be positioned on the couch, lined up so that the laser beams in the room intersect the cross on it. The calibration plate has a "G" printed on one side, which is to be positioned toward the gantry. The couch height should be set so that the top of the plate is coincident with the isocenter, and the central cross on the plate should be aligned to the light crosshairs. An image from each camera is acquired, and the system automatically verifies the position of the four numbered points on the calibration board by comparing the new acquired image to the image acquired during monthly calibration. If the verification is successful (position of the isocenter as seen by the cameras has not changed), the user may proceed to use the system. The verification is successful when the root mean square error is less than 1 mm. If the verification fails, meaning that the isocenters of the camera and the linac are different, the system will not be able to be used until it is recalibrated. The calibration, which is also part of the monthly QA, is explained in detail in the next section.

Quality Assurance—Monthly QA

One of the major potential problems related to QA of video surface imaging systems is motion of the cameras from day to day (Johnson et al. 1999; Milliken et al. 1997). During installation, the cameras are mounted on the ceiling in stable positions to minimize camera motion. However, motion might occur, and thus it needs to be checked. In addition, it is good practice to check the accuracy of the system, in case some other unanticipated problems appear. Monthly QA should include at least two tests: (1) system calibration and (2) accuracy of shifts tests.

System Calibration

The AlignRT software has a mode that allows performing a camera calibration by following the on-screen instructions. The camera calibration involves acquiring image data of a calibration board from all cameras. It is a procedure similar to the daily QA, but more comprehensive. The same calibration board is used here. The plate shown in Figure 80.3 has four numbered corner points, which are to be used during the QA. The calibration plate is placed on the treatment couch horizontally with the "G" toward the gantry and the central cross aligned with the light crosshairs. For a better alignment, the collimator should be fully opened and at 0°. The top of the plate should be aligned with the isocenter: for better precision during the monthly QA, a calibrated rod can be used to ensure this requirement. Poor positioning of the plate will result in poor isocenter calibration and reduced positioning accuracy. In addition, note that the coincidence accuracy of the crosshairs with the mechanical and radiation isocenter should be separately verified as part of a routine linear accelerator QA program. It is recommended that the room lights be slightly dimmed for this procedure, as well as later for treatment, so that image quality will be improved and light reflections will be minimized.

For each data camera, an image of the calibration plate is acquired. AlignRT will ask the user to select a proper gray-scale value to threshold the images such that the circles in the plate are seen as black and the rest as white. Then, the user is asked to identify the center of the four numbered points on the plate in increasing order (this order is important for 3D coordinate orientation). If the order of the points is selected in the wrong order, the images might appear rotated. Once the points are selected, the software displays crosses that should coincide with the black circles on the calibration plate, and if everything looks fine, the user approves the calibration.

A verification test (daily QA test) is recommended after the calibration of the cameras to ensure that the calibration files have been properly saved and the calibration has been properly made.

Accuracy of Shifts

Shifts provided by AlignRT should be checked on a monthly basis. The test consists of applying known shifts to a phantom and checking that the shifts (or real-time deltas if monitoring in real time) provided by AlignRT are the same. Different equipment can be used for these tests. The ideal equipment to perform a test on the accuracy of the shifts would be a precise motion platform with 6 degrees of freedom (to check all the translations and rotations) and a phantom for surface imaging. However, this platform is not generally available in the clinic, and therefore, other techniques have to be used.

The easiest way to check shifts accuracy is to use the couch coordinates to apply the known shifts, as is done during acceptance of AlignRT. A phantom is positioned on top of the treatment couch and a reference image is acquired. Then, known shifts are applied based on the couch coordinates. The shifts calculated with AlignRT are then compared to the shifts applied, and both should agree within 1 mm. This QA method has, however, two pitfalls. First, it can only check translations and couch rotation, and therefore misses the check of the two other rotational

degrees of freedom. Second, most couches do not have a perfect rotation around the isocenter (normally within a circle of 1 mm radius) and have an inherent translation associated with the rotation, which should be taken into account when validating the AlignRT shifts.

Another 3D imaging method that can be used for comparison is cone beam CT. In this case, a phantom undergoes a CT scan, and the isocenter is defined. The CT scan is then exported from the treatment planning system and the surface contour is imported in AlignRT. The phantom is placed on the treatment couch not too far from the isocenter (up to a centimeter). A cone beam CT image is taken, and the shifts to take the phantom to the proper isocenter are given by the treatment console. The user should enable the 6 degrees of freedom registration in the cone beam CT image. At the same time, an image is acquired with AlignRT and compared to the reference image, which was extracted from the CT scan. AlignRT will also provide shifts to take the phantom to the isocenter. Both shifts, from cone beam CT and from AlignRT, should be compared and agree within 1 mm and 0.5 deg.

References

Baroni, G., C. Garibaldi, M. Riboldi, M. F. Spadea, G. Catalano, B. Tagaste, G. Tosi, R. Orecchia, and A. Pedotti. 2006. 3D optoelectronic analysis of interfractional patient setup variability in frameless extracranial stereotactic radiotherapy. *Int. J. Radiat. Oncol. Biol. Phys.* 64 (2): 635–642.

Bert, C., K. G. Metheany, K. Doppke, and G. T. Chen. 2005. A phantom evaluation of a stereo-vision surface imaging system for radiotherapy patient setup. *Med. Phys.* 32 (9): 2753–2762.

Djajaputra, D., and S. Li. 2005. Real-time 3D surface-image-guided beam setup in radiotherapy of breast cancer. *Med. Phys.* 32 (1): 65–75.

Fitzpatrick, J. M., J. B. West, and C. R. Maurer Jr. 1998. Predicting error in rigid-body point-based registration. *Med. Imag. IEEE Trans.* 17 (5): 694–702.

Johnson, L. S., B. D. Milliken, S. W. Hadley, C. A. Pelizzari, D. J. Haraf, and G. T. Chen. 1999. Initial clinical experience with a video-based patient positioning system. *Int. J. Radiat. Oncol. Biol. Phys.* 45 (1): 205–213.

Liebler, T., M. Hub, C. Sanner, and W. Schlegel. 2009. An application framework for computer-aided patient positioning in radiation therapy. *Med. Inf. Internet Med.* 28 (3): 161–182.

Meeks, S. L., W. A. Tome, T. R. Willoughby, P. A. Kupelian, T. H. Wagner, J. M. Buatti, and F. J. Bova. 2005. Optically guided patient positioning techniques. *Semin. Radiat. Oncol.* 15 (3): 192–201.

Milliken, B. D., S. J. Rubin, R. J. Hamilton, L. S. Johnson, and G. T. Chen. 1997. Performance of a video-image-subtraction-based patient positioning system. *Int. J. Radiat. Oncol. Biol. Phys.* 38 (4): 855–866.

Smith, N., I. Meir, G. Hale, R. Howe, L. Johnson, P. Edwards, D. Hawkes, M. Bidmead, and D. Landau. 2003. Real-time 3D surface imaging for patient positioning in radiotherapy. *Int. J. Radiat. Oncol. Biol. Phys.* 57 (2, Suppl. 1): S187.

81

Mobile Electron Beam Intraoperative Radiation Therapy

A. Sam Beddar
*The University of Texas MD
Anderson Cancer Center*

Introduction

Intraoperative radiation therapy (IORT) has been customarily performed either in a shielded operating suite located in the operating room (OR) or in a shielded treatment room located within the Department of Radiation Oncology. In both cases, this treatment modality uses stationary linear accelerators. With the development of new technology, mobile linear accelerators have recently become available for IORT and can be used in existing ORs with reduced shielding requirements, which make the cost and logistics of setting up an IORT program much easier. These mobile systems have provided a stimulus to the field and a renewed interest in this special procedure. There are three electron mobile linac models on the market at the present time: the Mobetron manufactured by Intraop Medical Incorporated (Santa Clara, CA) (Figure 81.1), the Novac7 manufactured by Hitesys (Aprilia, Italy) (Figure 81.2), and the LIAC distributed by Sordina (Italy) (http://www.sordina.it/download/Catalogo_IORT.pdf). This chapter discusses briefly the quality assurance (QA) for mobile systems and summarizes the recommendations made by the Radiation Therapy Committee Task Group No. 72 (Beddar et al. 2006b) and published in their full report, the AAPM report No. 92 (Beddar et al. 2006a).

Recommended QA

Individual regulations require certain QA practices for medical linacs; these requirements differ from state to state within the United States, may be different from one country to another, and some may not be well suited to these special-purpose devices. The physicist *must* ensure that the use of mobile IORT equipment complies with any relevant regulations within their state or country and/or apply for exemptions where justified.

Previous QA recommendations for medical linear accelerators must acknowledge the recommendations published in the TG-40 report (Kutcher et al. 1994) and in the TG-142 report regarding QA for medical linacs in general and the TG-48 report (Palta et al. 1995), which discussed specific QA issues for linacs used for IORT. The pertinent recommendations of previous reports have been discussed and summarized in the TG-72 report (Beddar et al. 2006a).

When implementing a QA program for mobile systems, the clinical physicist needs to deal with some conflicting considerations. These units are partially disassembled and transported each day of use. Their simplified design with a specific set of applicators (no adjustable collimator jaws) and with no bending magnets, to reduce weight and radiation leakage, make the electron beam energy more dependent on variations in RF power generation and coupling to the accelerator. Therefore, on the one hand, there are reasons to perform more frequent beam measurements than with conventional installations. On the other hand, the equipment is used in ORs with little or no added shielding, so radiation safety considerations argue for limiting the beam time for QA as much as possible. These competing concerns can be partially resolved by developing an efficient QA process, but they do present an ongoing challenge. It is imperative to ascertain a reasonable workload as to not exceed the limits for radiation exposure to personnel. This workload, calculated for ORs, should include the MUs for machine warm-up, daily QA, and monthly QA. It is strongly advised to consult the recommendations made by Task Group 72 (Beddar et al. 2006a) and to determine the workloads for each OR and its surroundings.

Interlocks Check

All interlocks and safety features should be tested following the manufacturer's acceptance testing procedures. One should

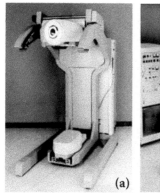

FIGURE 81.1 The Mobetron: (a) transport mode and (b) treatment mode configuration.

make sure that these interlocks and safety features, including the emergency off switches, are operational during the normal mode of operation of the unit (e.g., the clinical mode). The mechanical testing includes verifying the full range of gantry motion, including rotational and translational movements. The mechanical movements and controls of mobile IORT units differ significantly in design and function from those of a conventional accelerator. The mechanical movements of the gantry and treatment couch for a conventional accelerator are designed to rotate and translate with respect to a fixed isocenter. In contrast, a mobile accelerator will have no isocenter per se, and the geometric accuracy of treatment delivery using a mobile unit will depend solely on the accuracy of the docking. Once the electron applicator is placed inside the patient and aligned to the intended treatment area, the operator should be able to control the gantry movement in all directions available to achieve docking. The proper operation of the beam stopper, if any, should also be verified.

When judging how many MUs to apply when checking interlocks, the physicist needs to ensure that the beam runs long enough to enable all interlocks. (Some machines disable some dosimetry interlocks during an initial period.)

Output and Energy Check

Output and energy can be checked efficiently with the use of a dedicated solid phantom in which a dosimeter can be placed at two depths: near the depth of dose maximum and at a point on the depth dose curve in the 50–80% range. The output constancy is taken from the measurement near d_{max} and the energy constancy from the ratio of the two readings. The electron output constancy *must* be checked each day of use. The electron energy check *should* also be checked daily. If it proves to be sufficiently consistent, then the physicist may judge it reasonable to reduce the frequency to monthly after properly documenting the energy consistency. A typical arrangement for QA and beam calibration using a dedicated solid phantom for mobile IORT units is shown in Figure 81.3.

Docking Mechanism and Flatness and Symmetry

Both the accelerator characteristics and the docking mechanism affect the flatness and symmetry of the treatment fields. This is especially true for machines using soft-docking mechanisms. The TG-72 recommends that field flatness *should* be checked monthly. The docking apparatus *should* be checked for basic integrity each day of use, in accordance with TG-48. The TG-72 further recommends that the alignment of soft-docking systems *should* be checked at least monthly. For systems that use special attachments for routine QA at least annually, the flatness and symmetry of the beams *should* be checked in the soft-docked configuration used clinically.

For mobile systems, the practical question of when to set up the machine and do the QA checks takes on added significance. As with any multidisciplinary, single-dose procedure, the tolerance for machine downtime is very low, but the need to move and set up the machine adds complexity and the possibility for malfunctions. Consequently, there can be value to setting up and testing the basic operation of the machine on the day before its intended use. Full dosimetric QA can follow the next day, preferably early enough to permit some troubleshooting if needed. This is a labor-intensive process that can be simplified if experience with the machine demonstrates its reliability. The stability of any mobile system (Beddar 2005) should be determined to some extent to

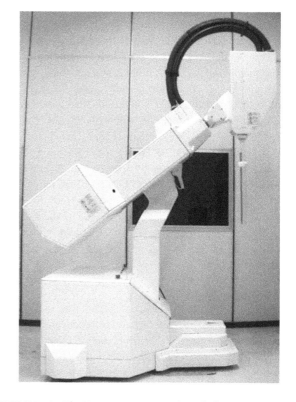

FIGURE 81.2 The Novac7 in its normal mode for treatment, with a treatment electron applicator attached to its gantry head.

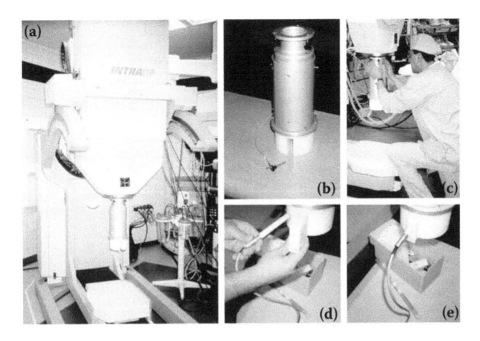

FIGURE 81.3 Typical arrangement used for QA and beam calibration for the Mobetron. (a) The mobile unit with the specialized QA electron applicator attached to it. (b) Close-up view of the specialized applicator. A dedicated plastic phantom, shown without inserts, is mounted at the bottom of the applicator. (c) Attaching the applicator to the gantry. (d) Placement of energy and depth-specific inserts into the dedicated phantom. (e) The applicator is ready for measurement. (The QA electron applicator, phantom, and inserts are supplied with the unit.)

TABLE 81.1 Summary of the QA Recommendations for Mobile Electron Accelerators Used for IORT

Parameter	Tolerance	Action level
	Day of use	
Output constancy	3%	Recommended
Energy constancy	Range of energy ratios corresponding to 2-mm shift in depth dose	Recommended
Door interlocks	Functional	Recommended
Mechanical motions	Functional	Recommended
Docking system	Functional	Recommended
	Monthly	
Output constancy	2%	Recommended
Energy constancy	Range of energy ratios corresponding to 2-mm shift in depth dose	Recommended
Flatness and symmetry constancy	3%	Recommended
Docking system	Functional	Recommended
Emergency off	Functional	Recommended
	Annually	
Output calibration for reference conditions	2%	Required
Percent depth dose for standard applicator	2 mm in depth over the range of clinical interest	Required
Percent depth dose for selected applicators	2 mm in depth over the range of clinical interest	Recommended
Flatness and symmetry for standard applicator	3%	Required
Flatness and symmetry for selected applicators	3%	Recommended
Applicator output factors	3%	Recommended
Monitor chamber linearity	1%	Recommended
Output, PDD, and profile constancy over the range of machine orientations	As above	Recommended
Inspection of all devices normally kept sterile	Functional	Recommended

TABLE 81.2   Sample of a Monthly Calibration Worksheet

DATE_____ O.R. #_____

MOBETRON IORT Monthly Calibration

I. Warm up:

SF6 pressure before _____ and after re-charging _____.

Start the warm-up with 12 MeV setting 500 MU at 1000 MU/min. Repeat with 9, 6, and 4 MeV.

II. Calibration equipment

   NEL Graphite Farmer chamber, model 2571, SN 140.

   PRM Electrometer, Model SH-1, SN 8883.

   Intraop Polyethylene QA phantom.

Readings for 200 MU, 1000 MU/min, insert ion chamber into QA block, take 3 readings,

Replace dmax block with d50 block for energy check, insert chamber into block, take 2 reading.

Adjust the dose rate to be within ±3 %. Then adjust Dose 2 (MU2 to be within 3%).

   Dose Rate = Average Reading at dmax x Overall Factor x C_{TP}.   T = _____ °C

   Energy check = Ratio of readings at d50 and dmax. P = _____ mm Hg

   $C_{TP} = [(T + 273) / 295] \times (760 / P)$ C_{TP} = _____

4 MeV calibration:

   Readings at dmax: _____ _____ _____ _____   Avg = _____

   Dose 2 (MU2):   _____ _____ _____ _____

   Dose rate = _____ $\times$ **0.02265** $\times$ _____ = _____ cGy/MU

   Rdg at d50 _____ Ratio D50 / Dmax _____ Expected 0.536 (____ %)

6 MeV calibration:

   Readings at dmax: _____ _____ _____ _____   Avg = _____

   Dose 2 (MU2):   _____ _____ _____ _____

   Dose rate = _____ $\times$ **0.02237** $\times$ _____ = _____ cGy/MU

   Rdg at d50 _____ Ratio D50 / Dmax _____ Expected 0.626 (____ %)

9 MeV calibration:

   Readings at dmax: _____ _____ _____ _____   Avg = _____

   Dose 2 (MU2):   _____ _____ _____ _____

   Dose rate = _____ $\times$ **0.02220** $\times$ _____ = _____ cGy/MU

   Rdg at d50 _____ Ratio D50 / Dmax _____ Expected 0.706 (____ %)

12 MeV calibration:

   Readings at dmax: _____ _____ _____ _____   Avg = _____

   Dose 2 (MU2):   _____ _____ _____ _____

   Dose rate = _____ $\times$ **0.02218** $\times$ _____ = _____ cGy/MU

   Rdg at d50 _____ Ratio D50 / Dmax _____ Expected 0.676 (____ %)

III. Safety Checks:

  1. Beam interrupt test:

  2. Emergency off test:

Comments:

List the energies that were adjusted with their corresponding dose rate before adjustment.

Medical Physicist _____

---

gain confidence on the stability and reproducibility of the output considering the hours of inactivity that are the norm for this procedure, especially for those who opt to perform the output check the day before the procedure. The author's preference would be to perform the QA the night before the procedure and then repeat just the output measurements on the day of the procedure, which can then be used for MU calculations. In case of any machine malfunction that cannot be resolved by 8:00 P.M., the radiation oncologist or the surgeon should be notified so that a decision can be made to cancel the case and notify the patient by 10:00 P.M.

Workload Limitations

For a machine having four energies, a typical protocol is to warm up the machine and dosimeter with about 400 MU and then check the output and the electron energy for each energy with single 200-MU readings. Given that the machine is prepared for use more often than it is actually used, more beam time (and ambient radiation) may be allocated to QA than to treatment. Hence, there is good reason to carefully assess which readings and how many MUs per reading are necessary for QA. Use of a dual-channel dosimeter to simultaneously acquire readings at two depths would be advantageous.

As for any radiation treatment equipment, annual QA checks *should* repeat critical elements of the initial acceptance testing and commissioning. The task is likely to be complicated by workload limitations, however, unless the unit can be moved into a shielded environment. For example, it may be necessary to use film instead of a scanning water phantom. If the initial commissioning and subsequent annual QA tests are to be done in different environments with different dosimeters, then appropriate baseline measurements should be done during the commissioning. Table 81.1 summarizes the QA recommendations made by TG-72 for mobile electron accelerators used for IORT. The term *constancy* refers to the agreement with original commissioning data. Table 81.2 provides a monthly calibration worksheet.

Other Important Considerations to QA

Training of personnel for in-house maintenance and QA of a mobile linac is an important consideration. In-house capabilities should include the ability to perform quick repairs in cases where the accelerator breaks down during an IORT treatment. Otherwise, the patient will not benefit from an IORT treatment because a second operation would be unlikely. Therefore, special training should be considered.

As part of a comprehensive QA program, the TG-72 recommended that the annual review should include an assessment of clinical procedures and radiation safety procedures, maintenance history, and spare parts inventory.

The management of cancer using IORT is limited to the delivery of one single large dose during surgery, which is an occasional modality that can hardly be postponed or repeated. Therefore, machine interlocks, which exclude patient treatment, should be restricted to those that are necessary to warrant patient safety and to avoid machine damage. Interlocks of lower priority (e.g., those that are triggered by slight applicator misalignments or machine instability) should be well documented and their effect on the dose distribution well quantified so that, if necessary, an override can be considered at the time of treatment.

A staff in-service session and review for IORT should be conducted yearly. This in-service session consists of training and reviewing the IORT treatment procedure and equipment used and should include the emergency procedures and safety features of the mobile linac. A team composed of a medical physicist, a radiation oncologist, and a certified OR nurse should conduct these review and training sessions (Beddar et al. 2001).

A sign-up sheet to document the in-service session and training of hospital employees involved in this special procedure should be provided, which could be used to fulfill hospital and state regulations required for special procedures.

Summary

Another important consideration would be getting an independent check on the institution's dosimetry. The TG-72 strongly recommends institutions using mobile accelerators to use the Radiological Physics Center services, before the initiation of treatments if possible. Other institutions outside of the United States should use a similar calibration service center to provide an independent check of the output calibration of the machine.

References

Beddar, A. S. 2005. Stability of a mobile electron linear accelerator system for intraoperative radiation therapy. *Med. Phys.* 32 (10): 3128–3131.

Beddar, A. S., M. L. Kubu, M. A. Domanovic, R. J. Ellis, T. J. Kinsella, and C. H. Sibata. 2001. A new approach to intraoperative radiation therapy. *AORN J.* 74 (4): 500–505.

Beddar, A. S., P. J. Biggs, S. Chang, G. A. Ezzell, B. A. Faddegon, F. W. Hensley, and M. D. Mills. 2006a. Intraoperative Radiation Therapy Using Mobile Electron Linear accelerators: Report of AAPM Radiation Therapy Committee Task Group No. 72. http://www.aapm.org/pubs/reports/RPT_92.pdf.

Beddar, A. S., P. J. Biggs, S. Chang, G. A. Ezzell, B. A. Faddegon, F. W. Hensley, and M. D. Mills. 2006b. Intraoperative radiation therapy using mobile electron linear accelerators: Report of AAPM Radiation Therapy Committee Task Group No. 72. *Med. Phys.* 33: 1476–1489.

http://www.sordina.it/download/Catalogo_IORT.pdf.

Kutcher, G. J., L. Coia, M. Gillin, W. F. Hanson, S. Leibel, R. J. Morton, J. R. Palta et al. 1994. Comprehensive QA for radiation oncology: Report of AAPM Radiation Therapy Committee Task Group 40. *Med. Phys.* 21 (4): 581–618.

Palta, J. R., P. J. Biggs, D. J. Hazle, M. S. Huq, R. A. Dahl, T. G. Ochran, J. Soen, R. R. Dobelbower, and E. C. McCullough. 1995. Intraoperative electron beam radiation therapy: Technique, dosimetry, and dose specification: Report of Task Group 48 of the Radiation Therapy Committee, American Association of Physicists in Medicine. *Int. J. Radiat. Oncol. Biol. Phys.* 33: 725–746.

82

Proton Radiotherapy

Wayne Newhauser
The University of Texas

Sandeep Hunjan
Texas Oncology

Introduction

Although proton therapy was proposed more than 65 years ago (Wilson 1946) and first performed clinically in 1954 (ICRU 1998), by 2009, a mere 28 proton therapy centers were operating worldwide, only seven of which were in the United States. Today, less than 1% of all radiotherapy patients receive proton therapy. Consequently, the literature contains comparatively few reports on quality assurance (QA) for proton therapy.

Many proton therapy QA procedures can be readily adapted from well-documented methods for photon and electron beam therapy (Ezzell et al. 2003; Fraass et al. 1998; Kutcher et al. 1994; Marbach et al. 1994; Nath et al. 1994); some of these were discussed in earlier chapters and will not be described in detail here. However, proton therapy differs from photon therapy in several ways that necessitate special consideration in a QA program (Maughan and Farr 2008).

This chapter mainly focuses on QA for passively scattered proton therapy because this technique is used for more than 99% of all proton treatments. At this time, facilities treating with magnetically scanned beams are few and technologically diverse; more experience is needed before generally applicable QA procedures for magnetically scanned beams can be developed. Nonetheless, some of the QA procedures for scanned-beam proton therapy will be conceptually similar to those for passively scattered proton therapy. For these reasons, the recommendations provided here are intentionally general; most will have to be adapted to accommodate the needs of a particular treatment machine and clinical practice.

The following discussion will focus mainly on daily, monthly, and annual machine QA for passively scattered systems. Patient-specific QA for proton therapy is discussed in more detail in Chapter 96, along with QA for image-guided patient positioning systems. The comprehensive QA program at each facility will have to address the unique needs of the clinical practice, the treatment delivery unit, and the treatment planning system. Therefore, this chapter can only provide illustrative examples and guidelines on QA; it does not attempt to convey an exhaustive list of tests that must be made at any particular facility. To that end, the chapter identifies additional resources of relevance to proton QA and highlights areas for which new or refined QA procedures are still needed.

Description of the Proton Therapy Machine

In our general description of proton therapy machines, we will use the unit at the University of Texas M. D. Anderson Cancer Center as an example. This unit is similar in design and function to most of those in clinical use. The proton therapy equipment consists of a synchrotron accelerator that can deliver protons at a range of energies (100–250 MeV) to one or more rotating gantries and one or more horizontal fixed beam lines. Passively scattered protons are typically delivered with a source-to-axis distance of 235–270 cm. The proton beam energy delivered to the treatment head can vary, either continuously or in increments of approximately 20 MeV. In the latter case, fine control over beam energy occurs inside the treatment head. Users can select the treatment field sizes for most units. Typically, systems have interchangeable treatment applicators (snouts), which hold the field-shaping range compensator and aperture. The treatment unit has three snout sizes providing maximum field sizes of 37, 25, and 14 cm in diameter (before final collimation). A longitudinally spread-out Bragg peak (SOBP) is produced by using a range modulator wheel (RMW) to spread the proton dose longitudinally. The SOBP width is adjusted by gating the beam on and off synchronously with the RMW rotation angle. The RMW also acts as a primary

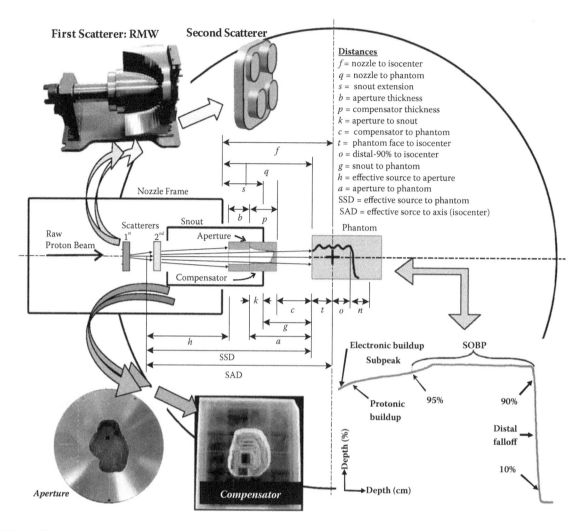

FIGURE 82.1 Major components in the treatment head (left half), distances between components (top right quadrant), and SOBP descriptors (lower right quadrant) for the passively scattered proton therapy units at the University of Texas M. D. Anderson Cancer Center.

scatterer to spread the beam laterally. A downstream secondary scattering foil performs additional beam spreading and flattening. Most treatment heads allow the RMW and second scatterer to be selected from a small library of devices, where the devices are installed either manually or automatically. Some proton treatment heads contain several retractable energy absorbers (i.e., range shifters) for making fine adjustments to the beam range. These hardware components must be properly selected, in operable condition, and correctly aligned for treatment at a particular energy and depth. The integrity of this alignment can be inferred by analyzing depth dose curves and lateral profiles in water. Figure 82.1 shows a schematic of the major components in the treatment head as well as SOBP descriptors.

Overall QA Needs

Data acquired during acceptance and commissioning tests provide benchmark values and intervention thresholds to be used during periodic QA. For proton therapy machines, some of

the major treatment-field characteristics that need to be tested include dose output constancy, SOBP width, beam range, collimated field size, field flatness and symmetry, radiation isocenter, and light field versus radiation field coincidence. Other aspects that require testing include all safety features (e.g., radiation, mechanical, electrical), gantry mechanical and radiation isocentricity, couch precision and accuracy (especially if pseudo-isocentric positioning is to be used), x-ray tube alignment and coincidence between x-ray and proton beams, and laser alignment. For radiation field testing, up to 24 combinations of the RMW and second scatterer may be required. For each of these combinations, many of the field properties must be measured for various SOBP widths, that is, where the SOBP width varies from a few centimeters to the maximum range of the beam. This large number of measurements requires efficient measuring instruments and methods.

Although all proton treatment heads contain basic beam-monitoring instruments, some treatment heads are equipped with sophisticated position-sensitive detector arrays, for example, multisegment ionization chambers. Using built-in and/or stand-alone multidetector instruments, the QA team may rapidly measure multiple beam characteristics, such as field flatness, field symmetry, field size, and output. If a built-in instrument reports a value that exceeds a predetermined intervention threshold, the treatment control system may pause or interrupt the irradiation. Because the importance of delivering beams with the appropriate characteristics is paramount, the QA program must include tests of both the field properties and measurement equipment. A description of the QA procedures and their tolerances (in parentheses) follows.

Daily QA

Parameters checked daily include mechanical integrity, safety interlocks, the patient positioning system, output consistency, range, and SOBP width. The digital gantry angle indicator is checked at one cardinal angle using a calibrated level (±1° tolerance). The integrity of the couch positioning system is checked at arbitrarily chosen intervals (e.g., 10-cm increments) using wall-mounted lasers incident on orthogonal rulers placed on the couch (±1 mm tolerance). Safety features, including interlocks, emergency buttons, radiation monitors, and audiovisual equipment, are checked for functionality. Output consistency for a reference condition is verified to confirm correct operation of the beam monitoring system and to monitor unit calibration (±2% tolerance). For example, one energy per day at one gantry angle at the center of the SOBP with a standard 10×10 cm^2 field is verified using an ionization chamber in a water tank. Figure 82.1 plots an example of a percent–depth–dose curve with descriptors of the features of an SOBP. Confirmation of the beam range is a sensitive test of the consistency of the beam delivery system (accelerator and energy degrading system) and may be checked by measuring depths of the 90% and 10% dose levels on the distal edge in water using the same tank (±1 mm tolerance). The SOBP width is defined as the distance between the proximal 95% dose level (±2% tolerance) and the distal 90% level (±10% tolerance). The SOBP is measured daily at water-equivalent depths in solid water phantoms (±5 mm tolerance).

Monthly QA

Parameters that are typically checked monthly include mechanical integrity, the patient positioning system, output constancy versus gantry angle, and field flatness and symmetry. Integrity of gantry and couch isocenter may be checked using the projection of a crosshair in the snout (Arjomandy et al. 2009). If no such device is supplied by the therapy system vendor, the user should construct one. The projection of this crosshair at different gantry angles can then be compared to a mechanical pointer placed at the isocenter to give an indication of the gantry's isocentricity (≤1 mm radius tolerance). The couch isocenter accuracy is measured by projecting the crosshair onto the couch with the gantry at 0° and then rotating through various couch angles and marking the position of the crosshair (≤1 mm radius tolerance). The couch translational accuracy is measured using a device with orthogonal markings at 10-cm intervals (±1 mm). Output consistency versus gantry angle may be measured using an air-filled ionization chamber attached to the snout, positioned such that the chamber's sensitive volume is on the beam axis (±2% tolerance). This typically includes measurements at several representative beam energies and at the four cardinal angles. In-plane and cross-plane flatness and symmetry are measured at each cardinal angle for all beam energies and snout sizes (±2% tolerance). Although film can be irradiated and scanned for this purpose (Nichiporov et al. 1995; Vatnitsky 1997), a less laborious approach is to use two-dimensional ion chamber or diode arrays, which provide an instantaneous readout with acceptable accuracy (Arjomandy et al. 2008). Calibrated arrays can also be used for output measurements, allowing output plus flatness and symmetry to be measured in one exposure. Once confidence in the machine's performance has been established, users may alter the scope and number of pretests in the monthly QA program.

Annual QA

All daily and monthly tests are typically performed as described above during an annual QA test. Annually, all components that could become physically damaged or worn out with frequent handling, such as removable RMWs and compensator holders, should be inspected visually. Additional parameters to be checked and validated yearly are described below.

Absolute dosimetry should be performed under reference conditions, which will vary with the particular beam line (Newhauser, Burns, and Smith 2002; Newhauser et al. 2002) and the accuracy requirements for the types of treatments being delivered. The reference conditions at our institution are a 10×10 cm^2 field with a medium snout, resulting in a beam with a range of 28.5 cm in water at a beam energy of 250 MeV. The calibration point is in the middle of a 10-cm-wide SOBP (therefore, the depth of calibration is 23.5 cm in water), resulting in a dose of 1 MU/cGy. Reference dosimetry is usually performed by measuring the ionization charge collected in an air-filled ionization chamber. The interpretation of the charge measurement may be accomplished with a dosimetry protocol, such as that from the International Commission on Radiation Units and Measurements (ICRU 1998). The overall experience with this method has been positive (Newhauser, Burns, and Smith 2002; Newhauser et al. 2002; Vatnitsky et al. 1996, 1999). Although some facilities have adopted the more recent International Atomic Energy Agency's TRS-398 protocol (IAEA 2000), it had not been validated in any large-scale proton dosimetry intercomparisons at the time of this writing.

Typically, treatment field settings differ from reference conditions, thereby changing the output. To take this into account, several dosimetry factors are measured under nonreference conditions at the time of commissioning and rechecked annually for possible changes. For passively scattered protons, these factors are related to changes in beam energy, lateral scattering, range modulation width, range shifter thickness, and scatter from the patient and range compensator (Sahoo et al. 2008; Fontenot et al. 2007). The corresponding measurable dosimetry factors are the relative output factor, SOBP factor, range shifter factor, SOBP off-center factor, off-center ratio, compensator scatter and patient scatter factor, field size factor, and inverse square factor. These multiplicative factors convert the dose/MU value for reference conditions to the dose/MU value under nonreference conditions. The measurement of these output-modifying factors allows for standard constancy checks and, importantly, an independent secondary MU calculation method that can be compared with ionization chamber measurements of treatment fields, as described in detail previously (Sahoo et al. 2008; Koch et al. 2008). Relative output factor accounts for the change in output relative to reference conditions caused by changing the proton range and field size (which are influenced by the selection of RMW and second scatterer). SOBP factor accounts for changes in the SOBP width relative to the reference SOBP width, which is created by changing the proton beam path length through range-modulating material (increased SOBP width results in decreased dose/MU). Range shifter factor accounts for the component of the decrease in output resulting from increased amount of range shifter in the path, that is, to make a fine adjustment to the maximum proton range. It is measured at the center of the SOBP of interest and is determined by taking the ratio of outputs with and without the range shifter in place, while maintaining the SOBP at isocenter. SOBP off-center factor accounts for the fact that there are small changes in output at displacements away from the center of the SOBP in the longitudinal beam direction. Off-center ratio accounts for the change in D/MU when the point of measurement is located lateral to the central axis; it can usually be determined from beam profile measurements. Field size factor accounts for changes in output that occurs when the collimated field size is different than the reference field size. Field-size effects are typically negligible for field sizes larger than 5 × 5 cm². Compensator scatter and patient scatter factor can be expressed as a product of two terms: the compensator scatter factor and the patient scatter factor. Compensator scatter factor takes into account the effect of the range compensator and is calculated as the ratio of the output with the compensator to the output without the compensator at the same water-equivalent depth. The patient scatter factor accounts for differences in dose due to differences in scattering that occur within a patient and a homogeneous water phantom. It is recommended to estimate compensator scatter and patient scatter factor in an integral way and not to estimate it from the product of separate determinations of compensator scatter factor and patient scatter factor. Compensator scatter and patient scatter factor can be determined automatically using a treatment planning system, the patient's treatment plan, and a verification plan in which the treatment fields are applied to a homogeneous water-box phantom. Our work in this area (Giebeler et al., Uncertainty in D/MU estimates dose passively scattered proton therapy: Part I. Prostate; Giebeler et al., Uncertainty in D/MU estimates for passively scattered proton therapy: Part II. Thorax) has been submitted for publication.

The calibration of computed tomography (CT) Hounsfield units (HU) to proton stopping powers must be confirmed annually because of the dependence of proton range on the correct calibration of electron density. Proton therapy requires greater accuracy of HU measurement than photon therapy to accurately predict the beam's end of range. Detailed procedures have been reported for calibrating CT scanners using kilovoltage photons (Schaffner and Pedroni 1998; Schneider, Pedroni, and Lomax 1996) and megavoltage photons (Newhauser et al. 2008) for use with proton treatment planning systems. A CT scanner (and treatment technique) is characterized by the following procedure. First, an image of a special calibration phantom is acquired containing a few tissue substitute materials of known elemental composition and mass density. Then, the parameters that characterize the CT scanner are deduced from a comparison of measured and theoretically estimated HU values. Next, these deduced parameters are used to generate HU values for a wide variety of tissues and tissue substitute materials. Finally, the latter set of HU values is fitted to a trilinear model. The trilinear model is used to generate the data points entered into the treatment planning system's CT calibration curve. The treatment planning system uses the calibration curve to convert HU values in individual CT voxels to relative linear proton stopping power values, which enables range and dose calculations. It is the constancy of the trilinear curve that must be checked annually using the calibration phantom. In addition, appropriate QA is required of all imaging modalities used, for example, positron emission tomography, magnetic resonance imaging, and ultrasound.

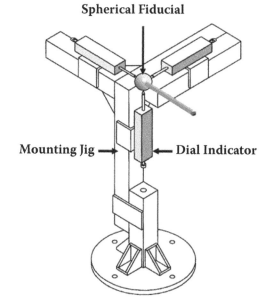

FIGURE 82.2 Measurement apparatus for testing gantry isocentricity, including three orthogonally mounted dial indicators, a mounting jig, and a spherical fiducial. (From Newhauser et al., *Mechanical Isocentricity Tests of the Hitachi Gantry Number 1 for the University of Texas M. D. Anderson Cancer Center, Houston*, University of Texas, Houston, TX, 2004. With permission.)

Gantry isocentricity also must be measured annually with submillimeter accuracy. This test is performed using a 5-cm-diameter sphere placed at the isocenter and three orthogonally mounted dial indicators (Figure 82.2) in contact with the sphere (Newhauser et al. 2004). This apparatus allows the rapid determination of the isocenter in three dimensions with high precision and has been used successfully at our institution for routine clinical QA (Arjomandy et al. 2009). Proton beam isocentricity (±1 mm radius tolerance) may also be confirmed, with less precision, using the familiar star-shot technique, in which film is placed perpendicular to the gantry rotation axis, and several exposures are taken using a narrow-slit-collimated beam of protons. The film is marked with the mechanical isocenter, which is compared to the proton beam isocenter (the intercept of the multiple exposed strips). This test should also be performed on the patient positioning system. Many other annual QA procedures can be readily adapted from well-documented methods for photon and electron beam therapy (Kutcher et al. 1994). Arjomandy et al. (2009) have published a detailed account of several QA procedures for proton therapy.

Patient-Specific QA

Patient-specific QA for proton therapy is described in more detail in Chapter 96. In brief, patient-specific QA checks include manual calculations of the MU, range checks, and, most importantly, independent verification of the dose/MU value using an ionization chamber. The objective of the D/MU verification is to ensure that the delivered dose (based on an electronic prescription that is transferred to the treatment unit) is equal to the dose prescribed by the radiation oncologist. The tolerance value will depend on the prescribed dose and a variety of other patient- and treatment-dependent factors. Dose/MU measurements should be made with an air-filled ionization chamber inside a tissue-mimicking phantom. Water phantoms are commonly used because water-equivalence corrections are small or negligible. In some situations, solid plastic phantoms may provide greater convenience and precision (Newhauser et al. 2002; Arjomandy et al. 2009). In many, but not all cases, the dose/MU measurements should be made without the range compensator (Fontenot et al. 2007; Giebeler et al. 2010). Consideration should be given to the dosimetric effects of protons that scatter off the edge of the collimator (Titt et al. 2008), which may be important in some cases. Patient-specific QA is typically performed separately for each treatment field and should be performed before the first therapeutic use of each field.

References

Arjomandy, B., N. Sahoo, X. Ding, and M. Gillin. 2008. Use of a two-dimensional ionization chamber array for proton therapy beam quality assurance. *Med. Phys.* 35 (9): 3889–3894.

Arjomandy, B., N. Sahoo, R. Zhu, J. R. Zullo, R. Y. Wu, M. Zhu, X. Ding, C. Martin, G. Ciangaru, and M. Gillin. 2009. An overview of the comprehensive proton therapy machine quality assurance procedures implemented at The University of Texas M. D. Anderson Cancer Center Proton Therapy Center–Houston. *Med. Phys.* 36 (6): 2269–2282.

Ezzell, G. A., J. M. Galvin, D. Low, J. R. Palta, I. Rosen, M. B. Sharpe, P. Xia, Y. Xiao, L. Xing, C. X. Yu, IMRT subcommittee, and AAPM Radiation Therapy committee. 2003. Guidance document on delivery, treatment planning, and clinical implementation of IMRT: Report of the IMRT Subcommittee of the AAPM Radiation Therapy Committee. *Med. Phys.* 30 (8): 2089–2115.

Fontenot, J. D., W. D. Newhauser, C. Bloch, R. A. White, U. Titt, and G. Starkschall. 2007. Determination of output factors for small proton therapy fields. *Med. Phys.* 34 (2): 489–498.

Fraass, B., K. Doppke, M. Hunt, G. Kutcher, G. Starkschall, R. Stern, and J. Van Dyke. 1998. American Association of Physicists in Medicine Radiation Therapy Committee Task Group 53: Quality assurance for clinical radiotherapy treatment planning. *Med. Phys.* 25 (10): 1773–1829.

Giebeler, A., Zhu, X. R., Titt, U., Lee. A., Tucker, S., Newhauser, W. D. 2010. Uncertainty in dose per monitor unit estimates for passively scattered proton therapy, Part I: The role of FCSPS in the prostate. *Phys. Med. Biol.* 18(23): (in press so page numbers are not yet available).

IAEA. 2000. Absorbed Dose Determination in External Beam Radiotherapy, Technical Reports Series No. 398, IAEA. In *Technical Reports Series*. Vienna: IAEA.

ICRU. 1998. Clinical Proton Dosimetry—Part I: Beam Production, Beam Delivery and Measurement of Absorbed Dose. ICRU Report No. 59 Bethesda, MD: ICRU. Oxford University Press, Oxford, UK.

Koch, N., W. D. Newhauser, U. Titt, D. Gombos, K. Coombes, and G. Starkschall. 2008. Monte Carlo calculations and measurements of absorbed dose per monitor unit for the treatment of uveal melanoma with proton therapy. *Phys. Med. Biol.* 53 (6): 1581–1594.

Kutcher, G. J., L. Coia, M. Gillin, W. F. Hanson, S. Leibel, R. J. Morton, J. R. Palta, J. A. Purdy, L. E. Reinstein, and G. K. Svensson. 1994. Comprehensive QA for radiation oncology: Report of AAPM Radiation Therapy Committee Task Group 40. *Med. Phys.* 21 (4): 581–618.

Marbach, J. R., M. R. Sontag, J. Van Dyk, and A. B. Wolbarst. 1994. Management of radiation oncology patients with implanted cardiac pacemakers: Report of AAPM Task Group No. 34. American Association of Physicists in Medicine. *Med. Phys.* 21 (1): 85–90.

Maughan, R., and J Farr. 2008. Quality assurance for proton therapy. In *Proton and Charged Particle Radiotherapy*, ed. T. F. Delaney and H. Kooy. Philadelphia, PA: Lippincott Williams & Wilkins. 823–869.

Nath, R., P. J. Biggs, F. J. Bova, C. C. Ling, J. A. Purdy, J. van de Geijn, and M. S. Weinhous. 1994. AAPM code of practice for radiotherapy accelerators: Report of AAPM Radiation Therapy Task Group No. 45 [see comment]. *Med. Phys.* 21 (7): 1093–1121.

Newhauser, W. D., J. Burns, and A. R. Smith. 2002. Dosimetry for ocular proton beam therapy at the Harvard Cyclotron

Laboratory based on the ICRU Report 59. *Med. Phys.* 29 (9): 1953–1961.

Newhauser, W. D., A. Giebeler, K. M. Langen, D. Mirkovic, and R. Mohan. 2008. Can megavoltage computed tomography reduce proton range uncertainties in treatment plans for patients with large metal implants? *Phys. Med. Biol.* 53 (9): 2327–2344.

Newhauser, W. D., K. D. Myers, S. J. Rosenthal, and A. R. Smith. 2002. Proton beam dosimetry for radiosurgery: Implementation of the ICRU Report 59 at the Harvard Cyclotron Laboratory. *Phys. Med. Biol.* 47 (8): 1369–1389.

Newhauser, W., M. Zhu, X. Ding, M. Bues, and A. Smith. 2004. *Mechanical Isocentricity Tests of the Hitachi Gantry Number 1 for the University of Texas M. D. Anderson Cancer Center, Houston.* Houston, TX: University of Texas.

Nichiporov, D., V. Kostjuchenko, J. M. Puhl, D. L. Bensen, M. F. Desrosiers, C. E. Dick, W. L. McLaughlin, T. Kojima, B. M. Coursey, and S. Zink. 1995. Investigation of applicability of alanine and radiochromic detectors to dosimetry of proton clinical beams. *Appl. Radiat. Isot.* 46 (12): 1355–1362.

Sahoo, N., R. Zhu, B. Arjomandy, G. Ciangaru, M. Lii, R. Amos, R. Wu, and M. T. Gillin. 2008. A procedure for calculation of monitor units for passively scattered proton radiotherapy beams. *Med. Phys.* 35 (11): 5088–5097.

Schaffner, B., and E. Pedroni. 1998. The precision of proton range calculations in proton radiotherapy treatment planning: Experimental verification of the relation between CT-HU and proton stopping power. *Phys. Med. Biol.* 43 (6): 1579–1592.

Schneider, U., E. Pedroni, and A. Lomax. 1996. The calibration of CT Hounsfield units for radiotherapy treatment planning [see comment]. *Phys. Med. Biol.* 41 (1): 111–124.

Titt, U., Y. Zheng, O. N. Vassiliev, and W. D. Newhauser. 2008. Monte Carlo investigation of collimator scatter of proton-therapy beams produced using the passive scattering method. *Phys. Med. Biol.* 53 (2): 487–504.

Vatnitsky, S. M. 1997. Radiochromic film dosimetry for clinical proton beams. *Appl. Radiat. Isot.* 48 (5): 643–651.

Vatnitsky, S., M. Moyers, D. Miller, G. Abell, J. M. Slater, E. Pedroni, A. Coray et al. 1999. Proton dosimetry intercomparison based on the ICRU report 59 protocol [erratum appears in *Radiother. Oncol.* 1999 Sept; 52(3): 281]. *Radiother. Oncol.* 51 (3): 273–279.

Vatnitsky, S., J. Siebers, D. Miller, M. Moyers, M. Schaefer, D. Jones, S. Vynckier et al. 1996. Proton dosimetry intercomparison. *Radiother. Oncol.* 41 (2): 169–177.

Wilson, R. R. 1946. Radiological use of fast protons. *Radiol.* 47: 487–491.

83

High-Dose-Rate Brachytherapy

Jeffrey Bews
CancerCare Manitoba

Introduction

The process by which a high-dose-rate (HDR) brachytherapy afterloader delivers a treatment is simple compared to most other modalities used presently in radiation therapy. As such, the development of a quality control (QC) program to assess the accuracy and safety of its operation is uncomplicated. The HDR brachytherapy treatment unit's utilization of a radioactive source does add to the scope but not to the complexity of the QC program. This simplicity of operation allows for the development of a QC program that can test most aspects of operation and safety on a daily or even a per-patient basis. Of course, the accurate and safe operation of an HDR brachytherapy unit is contingent on regular preventative maintenance, which is usually delivered by the treatment machine vendor and coordinated with source replacement.

Day-of-Treatment QC Measurements

The distribution of radiation dose delivered by an HDR treatment unit will depend on the strength of the radioactive source, where it is positioned within the patient (dwell locations), and the length of time it resides at those positions (dwell times). Verification of the accuracy of the source strength, dwell locations, and dwell times will therefore ensure an accurate treatment delivery. All three can be assessed easily as part of a treatment-day QC program. A QC program for the day of treatment that uses at least one test that runs the source into a catheter ultimately verifies that the unit actually can deliver a treatment before any patient preparation takes place. The measurements discussed in this section can be completed within about 20 min if procedures are designed to be efficient.

The strength of the radioactive source used by the HDR unit should be accurately measured at the time it is installed. HDR brachytherapy treatment units typically use ^{192}Ir because of its high specific activity and photon energy. With a half-life of approximately 74 days, the radioactive source should be replaced quarterly. The principles guiding source strength measurements are discussed in Chapter 63. At the time of calibration, it is useful to generate a decay table describing source strength as a function of time (often this can be created using the HDR treatment planning computer software). Once this table has been verified to be correct by manually checking a few of its entries, it can be used daily to assess the correctness of the source strength value used by the HDR treatment unit to calculate dwell times. The decay table should be created with a resolution of two entries per day. A practical approach is to display a 6:00 A.M. source strength and a 6:00 P.M. source strength. Assessing the correctness of the source strength value stored in the HDR treatment unit can be accomplished by simply verifying that it lies within the 6:00 A.M. and 6:00 P.M. decay table values on that particular treatment day (assuming that this test is conducted after 6:00 A.M. the morning of treatment).

Dwell location accuracy can be assessed by running the radioactive source into a catheter to a predetermined position. The difference between the actual and intended source position quantifies dwell location accuracy. Although there are many techniques that can be used to accomplish this test, the most direct will define the intended position using the same x-ray markers used to identify potential source positions during the treatment planning process and the actual source positions using x-ray film. For example, a typical approach would fix a clear catheter to a piece of GafChromic EBT film (International Specialty Products, Wayne, NJ), insert a strand of x-ray markers, and label with pen on the film the location of three markers. A treatment run will then be initiated in which the HDR unit has been instructed to run the source out to the previously defined locations. By comparing the centers of the resulting exposure spots on the film with the previously identified pen marks, an average dwell location error can be calculated (Figure 83.1).

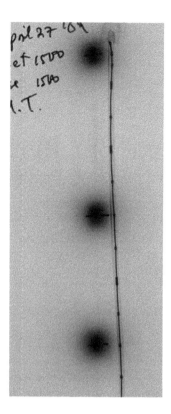

FIGURE 83.1 Dwell location accuracy film showing actual source positions (exposure spots) and intended source positions (vertical pen lines most visible on second exposure spot) as determined using x-ray markers. The catheter/x-ray marker combination has been shifted off the exposure spots for illustration purposes only.

An alternative strategy uses a ruler-like test tool with markings that designate source positions. The radioactive source can be run into a catheter embedded in the test tool to a predefined position and its actual location assessed by viewing the tool with a treatment room camera. This approach constitutes a less direct assessment and is a measurement of the position of the physical encapsulation of the source as opposed to its radiation emitting structure (Kutcher et al. 1994). In addition, the intended position is not defined by x-ray markers as is the case during an actual patient planning session. A third method makes use of a radiation detector system that is sensitive to source position (e.g., a well chamber with a shielded insert containing a small window).

The precision of the HDR unit's timer can be assessed daily through a measurement of source strength. At the time of source calibration, the source is run into a well chamber to a predetermined location (usually the center of the sensitive volume of the chamber) for a known period of time (20 s works well). The electrometer reading associated with this source run (including charge measured during transit) is recorded, corrected for the temperature and pressure dependence of the chamber and used as a baseline. Day-of-treatment timer precision can be assessed by repeating the source run, correcting for temperature and pressure, adjusting the electrometer response for source decay,

and comparing to the baseline. Because the response of the well chamber will be insensitive to small changes in source position and because source decay can be accurately calculated, this measurement constitutes a relative check of the HDR unit timer (Glasgow et al. 1993). An absolute check is discussed as part of the quarterly QC measurements.

In addition to the accuracy of dose delivery, safe design and operation of the radiation facility must also be verified. This includes testing not only some of the safety features inherent in the design of the treatment unit but also ancillary equipment that must be available to ensure staff and patient safety during routine operation and emergency situations.

Safety Checks

Clearly not all of the safety design features of an HDR unit can be tested as part of a QC program because of the destructive nature of the testing that would be required. However, most models run self-tests as part of their start-up procedures (including emergency battery backup power) and before patient treatments (catheter restriction). The unit will not deliver a treatment if any of these tests fail. Over and above these internal checks, it is recommended that the user run daily tests to check the emergency off circuit of the unit (by sending the source into a catheter and verifying that it retracts upon activating the emergency off button at the operator's console), the treatment room door interlock (by running the source into a catheter and ensuring that the treatment terminates when the door is opened and verifying that a treatment cannot be initiated when the door is open), the treatment channel interlock (by attempting to initiate a treatment through a channel to which a catheter has not been attached), and the unit's ability the identify a path of travel that is too "curved" for the radioactive source to maneuver before initiating treatment (by attempting to send the source through a catheter with a loop approximately 3 cm in diameter).

As with any radiation therapy delivery technique in which the patient must be left alone in the treatment room, properly functioning treatment room video and audio systems are essential for safe operation. The independent room radiation monitor (permanent mount) and the handheld radiation monitor, both essential ancillary equipment, also must be tested on a treatment-day basis. The availability of source retraction failure tools such as long forceps, cutters, and a shielded storage container must be established.

Quarterly QC Measurements

Typically, a new HDR source will be installed every 3–4 months and this will initiate preventative maintenance and calibration. It is prudent at this time to perform the daily tests discussed above to ensure that the unit is accurate and safe before it is brought back into clinical use. The installation of a new source is also a good time to trigger an assessment of the linearity of the timer, the absolute accuracy of the timer, and the transit dose reproducibility.

Timer linearity can be quantified by running the HDR source into a well chamber for different dwell times and measuring the charge associated with each (including transit dose). The goodness of fit of a straight line to the data will quantify timer linearity. The *y* intercept (time equal to zero) represents the transit dose of this particular geometry. By dividing the intercept by source strength and comparing the result to that obtained at the time of acceptance testing, transit dose reproducibility can be assessed.

The absolute accuracy of the HDR unit timer can be quantified by using two of the data points collected during the timer linearity measurements. By subtracting these two points (e.g., the charge associated with a 60-s and a 30-s dwell time), the charge associated with the difference in dwell times is obtained. This charge difference will not contain the effects of transit dose. Next, the source is run into the well chamber and charge is collected using the electrometer timer set for a collection time equal to the difference in the two dwell times associated with the charge difference calculation (e.g., 30 s). A comparison of this charge with the charge difference calculation is a measure of the absolute accuracy of the HDR unit timer.

The calibration of a new source is also a good time to check the accuracy of the treatment planning computer (assuming that all software upgrades are performed at the time of source replacement). Although much of the planning computer's performance can be assessed as part of a patient-specific QC process, a reproducibility test of dose calculation and computer peripherals (plotter, digitizer, and data transfer) on a quarterly basis is prudent. This can be accomplished by identifying one patient image data set as a test case and using it to generate a new treatment plan quarterly. The output from this session (including printed materials, plots, and transferred data) can be compared with results obtained previously.

All standard treatment plans (both those stored on the treatment unit and those stored in the treatment planning computer) need to be checked for consistency. This is accomplished by printing the plans and comparing the output with that generated when the standard plans were first created.

An emergency response drill should also be conducted quarterly. This drill should simulate the response of staff to a source that failed to retract into the shielded storage container of the HDR unit. The drill should be designed to be as realistic as possible including a staff member playing the role of a patient (within reason), patient monitoring equipment, and an applicator attached to the HDR unit. A stopwatch can be used to monitor the time of response.

Annual QC Measurements

It is prudent to perform a repeat acceptance test on equipment once a year. This testing will constitute a thorough examination of performance and safety to ensure that the equipment is

TABLE 83.1 HDR Afterloader Brachytherapy QC Tests

QC Test	Frequency (Reference)	Tolerance
HDR treatment unit source strength	Treatment day (Ezzell 1994; Barnett and Cygler 2006; Nath et al. 1997)	Between 6:00 A.M. and 6:00 P.M. decay table value
Dwell location accuracy	Treatment day (Barnett and Cygler 2006; Nath et al. 1997)	1 mm (Ezzell 1994; Barnett and Cygler 2006; Nath et al. 1997; Kutcher et al. 1994)
Timer precision	Treatment day (Barnett and Cygler 2006; Nath et al. 1997)	1% (Barnett and Cygler 2006)
Emergency off circuit	Treatment day (Ezzell 1994; Barnett and Cygler 2006)	Functional
Treatment room door interlock	Treatment day (Ezzell 1994; Barnett and Cygler 2006; Nath et al. 1997)	Functional
Treatment channel interlock	Treatment day	Functional
Path of travel excess curvature	Treatment day	Functional
Treatment room video/audio monitoring	Treatment day (Ezzell 1994; Barnett and Cygler 2006; Nath et al. 1997)	Functional
Independent room radiation monitor	Treatment day (Ezzell 1994; Nath et al. 1997)	Functional
Handheld radiation monitor	Treatment day (Ezzell 1994; Nath et al. 1997)	Functional
Source retraction failure tools	Treatment day (Ezzell 1994; Nath et al. 1997)	Available
Timer linearity	Quarterly (Barnett and Cygler 2006; Nath et al. 1997)	1% (Barnett and Cygler 2006)
Timer accuracy	Quarterly (Nath et al. 1997; Kutcher et al. 1994)	1% (Kutcher et al. 1994)
Transit dose reproducibility	Quarterly	1% (Barnett and Cygler 2006)
Treatment plan consistency	Quarterly	Identical[a]
Standard plan consistency	Quarterly	Identical
Source retraction failure response drill	Quarterly	N/A
X-ray marker length	Yearly (Barnett and Cygler 2006; Nath et al. 1997; Kutcher et al. 1994)	1 mm (Ezzell 1994; Barnett and Cygler 2006)
Radiation license	Yearly (Ezzell 1994; Nath et al. 1997)	N/A
QC review	Yearly (Barnett and Cygler 2006; Nath et al. 1997)	N/A

[a] Should software upgrades be expected to introduce changes, the differences between output need to be assessed and, if appropriate, accepted.

functioning at the same level it was at the time of installation. However, tests performed on a treatment day or quarterly basis need not be repeated at this time. This section will identify a few of the important tests that are unique to HDR afterloader annual QC.

Once a year, the imaging x-ray markers should be checked for length against the manufacturer's specifications. In addition, the distance of the first x-ray marker from the distal end of all non-disposable applicators should be determined using x-ray imaging and compared to that obtained at the time of acceptance testing. It should be noted that these measurements should be conducted with transfer tubes (source guides that serve to connect the applicator to the treatment unit) connected to the applicators so that the position of the x-ray marker is assessed under conditions that represent source travel in a clinical treatment.

Although catheters should be inspected for damage or wear before insertion into a patient and all connections assessed as they are performed, a more robust evaluation of hardware integrity should be conducted annually.

Summary

Although beyond the scope of this chapter, radiation safety license issues should be reviewed. Compliance to license conditions should be reported including a review of workload to ensure that radiation barrier design assumptions have not been violated. Finally, the entire QC program (policies, procedures, and recorded data) should be reviewed once a year to ensure that it is current, fulfilling its objectives, and being followed. A list of all QC tests is provided in Table 83.1.

References

Barnett, R., and J. Cygler. 2006. Quality Control standards: Brachytherapy Remote Afterloaders. www.medphys.ca/media.php?mid=125 (accessed April 29, 2009).

Ezzell, G. 1994. Commissioning of single stepping-source remote applicators. In *Brachytherapy Physics*, ed. J. Williamson, B. Thomadsen, and R. Nath. Madison, WI: Medical Physics Publishing pp 557–576.

Glasgow, G., J. Bourland, P. Grigsby, J. Meli, and K. Weaver. 1993. *Remote Afterloading Technology: Report of AAPM Task Group No. 41*. New York, NY: American Institute of Physics Inc.

Kutcher, G. J., L. Coia, M. Gillin, W. F. Hanson, S. Leibel, R. J. Morton, J. R. Palta, J. A. Purdy, L. E. Reinstein, G. K. Svensson, et al. 1994. Comprehensive QA for radiation oncology: Report of AAPM Radiation Therapy Committee Task Group 40. *Med. Phys.* 21 (4): 581–618.

Nath, R., L. L. Anderson, J. A. Meli, A. J. Olch, J. A. Stitt, and J. F. Williamson. 1997. Code of practice for brachytherapy physics: Report of the AAPM Radiation Therapy Committee Task Group No. 56. American Association of Physicists in Medicine. *Med. Phys.* 24 (10): 1557–1598.

VII

Quality Control: Patient-Specific

External Beam Plan and MU Checks

Robin L. Stern
UC Davis Medical Center

Introduction

Quality assurance of the complete treatment planning process is necessary for safe and accurate treatment of patients. Comprehensive written procedures should be in place which cover all aspects of planning and plan checking, and results of all testing should be fully documented. These aspects include topics discussed in detail in other chapters. Connectivity needs to be established between the different systems (e.g., simulator and planning system; planning system and record-and-verify system), and the fidelity of data exchange should be regularly tested. In addition to these system tests, checks should be performed on each individual patient plan, as described in the following two sections.

Treatment Plan Quality Control

Each and every patient plan should receive an initial review by a medical physicist. Guidelines are given in the AAPM Task Group 40 report (Kutcher et al. 1994) and in the ACR Practice Guidelines and Technical Standards (http://www.acr.org /SecondaryMainMenuCategories/quality_safety/guidelines .aspx). The purposes of this present review are threefold: (1) to ensure that the plan will deliver treatment as specified in the physician's prescription, (2) to check that the plan is clinically reasonable, and (3) to verify that the dosimetric calculations are correct. Thorough quality control of the treatment planning system does not negate the need to perform the initial plan review.

Table 84.1 gives a list of tasks involved in the initial plan review. The review should be performed by an independent physicist who was not involved in the planning process. The physicist should sign and date the plan to document the review. The review should be performed before the first treatment. If that is not possible, it should be performed before delivery of the third fraction or 10% of the prescribed dose, whichever occurs first (Kutcher et al. 1994).

An independent verification of the monitor unit (or timer) calculation is a fundamental part of the initial plan review. The MU calculation is verified by a separate calculation of either the MU or the dose to a point within the plan, preferably near the center of the target volume and away from field edges and tissue–air or tissue–metal interfaces. For MU generated by a treatment planning system or a calculation program, a different program or a hand calculation should be used for the verification. All parameters influencing the monitor unit calculation (see Table 84.2) should be reviewed and verified. Parameters may be downloaded directly from the primary planning system to the verification calculation system as long as the fidelity of the download is confirmed. If heterogeneity correction is used in the primary calculation, correction should also be used in the verification calculation, even if the methodologies are not the same. Action level criteria should be set on the agreement between the primary and verification calculations. These criteria can be dependent on the complexity of the calculation and the relative complexities of the calculation algorithms, but typically range from 3–5%. The American Association of Physicists in Medicine formed Task Group 114 to reassess the MU verification process and provide guidelines. Their report is expected to be published in 2010.

IMRT Patient-Specific Quality Control

Because of its increased complexity, IMRT requires a more thorough validation of the proper delivery of the planned dose distribution, in addition to the plan QC tasks described above. For each plan, absolute dose should be verified at one or more points, and relative or absolute dose should be verified either at a large number of individual points or along one or more planes. The AAPM, ASTRO, and ACR all recommend that this verification be done by direct measurement of delivered dose (Ezzell

TABLE 84.1 Treatment Plan Review Quality Control Tasks

Confirm treatment planner signs and dates plan
Confirm physician signs and dates plan
Confirm Rx is complete and signed by physician
Confirm plan agrees with Rx: anatomical site, modality, technique, energy, prescribed dose (fractional and total), dose constraints
Confirm plan is consistent with accepted clinical practices
Review dose distributions, including evaluation of hot and cold spots
Review DVHs
Verify MU
Confirm correct transcription/download of treatment parameters (e.g., gantry/table/collimator angle, MU) to patient chart and/or record-and-verify system
Confirm beam modifiers are included and are correct in the treatment directions
Confirm dose recording in the record-and-verify system is enabled and correct

et al. 2003; Galvin et al. 2004; Hartford et al. 2009). Verification could potentially be done by calculation; however, commercial independent calculation methods are under development and not yet in widespread use. Furthermore, measurement will catch errors in the dose delivery that a verification calculation would not catch, for example, due to MLC leaf positions out of calibration. Many different techniques and detectors are used for measurement, the most common being planar diode or ion chamber arrays and phantoms with one or two ion chambers plus film. If a calculational verification is performed, the leaf sequence files should be downloaded from the record-and-verify/linac control system if possible rather than from the treatment planning system to verify fidelity of those files.

For measurement tests, the treatment fields with their MLC segmentation and MU are applied within the planning system to a standard phantom, and the resultant dose distribution is calculated. The fields either retain their original gantry angles or are all to be set to the nominal AP direction. This latter method has evolved for planar arrays that do not accurately measure oblique fields and is purely pragmatic with no clinical or scientific basis. The patient's treatment is then delivered to the phantom on the linac. If possible, the phantom should be irradiated using the exact MLC leaf sequencing files in the record-and-verify/linac control system that will be used for patient treatment. Because some calculation errors, particularly those related to the patient-specific geometry, will not be detected with phantom plans, an additional MU verification check as previously described is recommended for each beam. For calculational verification, the standard phantom may be used or the patient geometry can be

TABLE 84.2 Parameters Affecting MU Calculation

Linac
Beam modality and energy
Absolute beam calibration
Patient surface contour
SSD
Physical depth
Radiological depth
Collimator setting
Beam aperture shape
Beam modifiers (wedges, compensators)
Blocking tray

transferred to the verification system for the calculation. For both measurement and calculational verification, comparisons with the planning system calculation can be performed for each beam individually or for the entire treatment as a whole. Planar dose distributions are usually evaluated using a software tool such as the gamma index (Low et al. 1998), which combines percentage dose difference and distance to agreement techniques. There are currently no published consensus acceptance criteria or recommendations for agreement between measurement and calculation for IMRT patient-specific QC. However, there is an indication that 3%/3 mm (used in many clinics) is not adequate to identify some MLC errors (Yan et al. 2009).

End-to-End Tests

The radiation therapy planning/treatment process has become increasingly complex and interconnected. End-to-end tests provide an effective method for testing the entire process as a whole rather than testing different parts separately. In an end-to-end test, a phantom is imaged, planned, checked, and treated as if it were a patient, with measurements taken to ensure treatment is delivered as planned, dosimetrically as well as geographically. End-to-end tests should be performed as part of the commissioning of the treatment planning system and of new treatment modalities. The same tests should be repeated annually and also when major changes/upgrades occur in the planning system or the planning/treatment process.

References

Ezzell, G. A., J. M. Galvin, D. Low, J. R. Palta, I. Rosen, M. B. Sharpe, P. Xia, Y. Xiao, L. Xing, and C. X. Yu. 2003. Guidance document on delivery, treatment planning, and clinical implementation of IMRT: Report of the IMRT Subcommittee of the AAPM Radiation Therapy Committee. *Med. Phys.* 30 (8): 2089–2115.

Galvin, J. M., G. Ezzell, A. Eisbrauch, C. Yu, B. Butler, Y. Xiao, I. Rosen et al. 2004. Implementing IMRT in clinical practice: A joint document of the American Society for Therapeutic Radiology and Oncology and the American Association of Physicists in Medicine. *Int. J. Radiat. Oncol. Biol. Phys.* 58 (5): 1616–1634.

Hartford, A. C., M. G. Palisca, T. J. Eichler, D. C. Beyer, V. R. Devineni, G. S. Ibbott, B. Kavanagh et al. 2009. American Society for Therapeutic Radiology and Oncology (ASTRO) and American College of Radiology (ACR) practice guidelines for intensity-modulated radiation therapy (IMRT). *Int. J. Radiat. Oncol. Biol. Phys.* 73 (1): 9–14.

Kutcher, G. J., L. Coia, M. Gillin, W. F. Hanson, S. Leibel, R. J. Morton, J. R. Palta et al. 1994. Comprehensive QA for radiation oncology: Report of AAPM Radiation Therapy Committee Task Group 40. *Med. Phys.* 21 (4): 581–618.

Low, D. A., W. B. Harms, S. Mutic, and J. A. Purdy. 1998. A technique for the quantitative evaluation of dose distributions. *Med. Phys.* 25 (5): 656–661.

Yan, G., C. Liu, T. A. Simon, L. C. Peng, C. Fox, and J. G. Li. 2009. On the sensitivity of patient-specific IMRT QA to MLC positioning errors. *J. Appl. Clin. Med. Phys.* 10 (1): 2915.

85

Monte Carlo

Indrin J. Chetty
Henry Ford Health System

Introduction

A novel class of Monte Carlo (MC)–based codes optimized for photon and electron beams in patient-specific geometries has invigorated interest in the use of MC-based dose calculations for radiotherapy treatment planning. These codes, for example, Macro MC (MMC), VMC++, XVMC, DPM, among others, have made it possible to perform MC-based photon beam dose calculations within minutes even on a single processor (Chetty et al. 2007; Fragoso et al. 2009; Gardner, Siebers, and Kawrakow 2007; Hasenbalg et al. 2008). The main advantage of these "second-generation codes" over "first-generation codes" such as EGS4, MCNP, GEANT4, Penelope, and so on, is that they are optimized for radiation transport in the therapeutic energy ranges and over a range of material atomic numbers and densities characteristic of human tissues. The transport mechanics and boundary crossing implementations with second-generation codes are more efficient, with the result that they converge faster, that is, fewer condensed history steps are required for the same precision compared to first-generation codes. The development of "second-generation" MC codes has also significantly reduced the need for massive parallel computer clusters for routine MC dose calculations in the clinic, as was typically required in the past. Commercially available, FDA-approved, MC-based treatment planning systems for photon and electron beams, including Nucletron (Oncentra), Varian (Eclipse), BrainLab (iPlan), Elekta (PrecisePlan), among others, use second-generation code systems implemented on multiple CPUs, typically four to eight, running in parallel on a single workstation.

With the impending widespread accessibility of MC-based dose computation systems in the clinical setting comes the need to understand the differences between MC and conventional calculation algorithms, ultimately to facilitate efficient and accurate clinical treatment planning and delivery procedures. The focus of this chapter will be to establish the feasibility of using MC-based algorithms in the quality control (QC) of various aspects of clinical treatment planning and delivery.

Review of the MC Method

With MC-based calculation, one simulates the trajectories of individual particles (primarily photons and electrons in the therapeutic energy range) as they traverse from the linear accelerator (linac) treatment head to the patient. A more formal definition of the MC method has been provided by Rogers and Bielajew (1990), as follows: "The Monte Carlo technique for the simulation of the transport of electrons and photons though bulk media consists of using knowledge of the probability distributions governing the individual interactions of electrons and photons in materials to simulate the random trajectories of individual particles. One keeps track of physical quantities of interest for a large number of histories to provide the required information about the average quantities." A typical transport scheme involves explicit or "analog" simulation of photon interactions. Because of the numerous collisions that electrons undergo as they traverse matter, analog transport of electrons is not practical; the condensed history technique (Berger 1963) is used. The condensed history technique was a pioneering development by Berger (1963) and without it MC simulations for radiotherapy dose calculations today would not be practical. The condensed history technique is based on the observation that most electron interactions lead to changes in the electron's energy and/or angle. These interactions can therefore be grouped into relatively few condensed history "steps" and their cumulative effect be taken into account by sampling the relevant distributions for energy loss and angular change at each "step."

In simulating physical processes (i.e., particle tracks produced by a linac and depositing energy in a patient), it is apparent that the MC method is conceptually much more straightforward than conventional algorithms. It does not involve major approximations nor rely on dose deposition models as is the case with conventional algorithms. It only requires knowledge of the physics of particle interactions, which has been well understood for many years.

Application of the MC Method in Radiotherapy Dose Calculations

As is the case with any dose algorithm, the first step in calculating dose to the patient is to develop a "model" of particle interactions within the linear accelerator treatment head. The AAPM Task Group No. 105 (TG-105) has classified a beam model as an "entity" that produces a *phase space description* of interactions in the treatment head (Chetty et al. 2007). The phase space is composed nominally of the position, energy, direction, and particle type of each particle crossing a plane perpendicular to the beam central axis, typically at a location below the "patient-independent" treatment head components, that is, at a plane just above the secondary collimators. According to TG-105 (Chetty et al. 2007), an MC-based beam model can be derived from one of the following three possible "routes": (1) direct use of phase space information obtained by simulation of the treatment head; (2) development of multiple-, virtual-source models reconstructed from the treatment head simulation with or without enhancement from measurements; or (3) other models derived exclusively from measurements (measurement-driven models). Advantages of the first two approaches include the fact that phase space information is derived from explicit simulation of the treatment head. This establishes a connection between a physical process (i.e., simulating interactions in the treatment head components) and the model used for dose calculations within the patient. In the past, direct simulation approaches have been considered inefficient and therefore impractical for routine clinical use. However, recent advances in MC code systems involving, in particular, the use of aggressive variance reduction techniques has lead to substantial improvements in calculation speed (Kawrakow 2006a, 2006b). MC simulations of the entire treatment head, collimating jaws, and patient-specific CT geometry are now possible within minutes even on a single processor (Fragoso et al. 2009; Gardner, Siebers, and Kawrakow 2007; Hasenbalg et al. 2008). Direct simulation or virtual approaches require detailed knowledge of the constituents of the treatment head components, which may ultimately limit their applicability in routine clinical treatment planning. The third approach, in which the incident energy spectra are derived from measurements, is an established method used with many conventional algorithms and does not require information on the treatment head design. The use of measurement-driven beam models has been adopted by several popular vendor systems. The literature is replete with publications involving the use of MC methods in modeling linac treatment heads. Comprehensive reviews of such studies are provided in the articles by Ma and Jiang (1999)

(electron beams) and Verhaegen and Seuntjens (2003) (photon beams).

The second step in an MC dose calculation involves transport within the patient-dependent treatment head components (such as the collimating jaws and the multileaf collimator [MLC]) and the patient-specific CT geometry. Patient-dependent components are simulated using either explicit transport methods or approximate transport methods before detailed transport in the patient. Detailed reviews of studies concerning MC-based patient-specific dose calculations are provided in the TG-105 report (Chetty et al. 2007) as well as in the article by Reynaert et al. (2007). As with conventional dose algorithms, proper commissioning (the requirements for which are described in other relevant documents; IAEA 2004; Fraass et al. 1998; Van Dyk et al. 1993) is required before an MC-based algorithm can be used in the clinical setting. Experimental verification in more complex fields and in heterogeneous geometries, especially under conditions of electron disequilibrium, will be useful to verify the expected improved accuracy of the MC method in these situations.

Application of the MC Method in QC

MC-Based Simulation in IMRT Planning and Delivery

The ability to perform detailed simulations of complex delivery techniques provides one of the most compelling reasons for consideration of MC as a tool in the IMRT QC process. A large number of studies is available in the literature related to detailed simulation of the IMRT optimization and planning process, including detailed simulations of complex MLC-delivered fields. Here, examples of such studies will be highlighted toward the goal of establishing the feasibility of MC as a tool in the treatment and delivery process.

The BEAM code system (Rogers et al. 1995) served as a major breakthrough in our ability to perform detailed simulations of the treatment head and beam modifying devices (e.g., the MLC). Using BEAM, studies were performed to understand, on a fundamental level, the role of various effects, such as inter- and intraleaf transmission in the MLC, the tongue-and-groove effect, transmission through the leaf tips, and leaf scattering, on IMRT planning and delivery. For instance, Heath and Seuntjens (2003) developed a component module to perform detailed simulations of the geometry of the Varian Millennium 120-leaf MLC (Varian Medical Systems, Palo Alto, CA). Calculations of the basic leaf characteristics as well as complex delivery patterns showed good agreement with measurements (Heath and Seuntjens 2003). The work of Ma and collaborators (Deng, Lee, and Ma 2002; Deng et al. 2001; Fan et al. 2006; Lee et al. 2001; Li, Boyer, and Ma 2001; Li et al. 2004; Luo et al. 2006; Ma 1998; Ma, Ding et al. 2003; Ma, Jiang et al. 2003; Ma, Li et al. 2000; Ma et al. 1999; Ma, Pawlicki, Jiang et al. 2000) on the use of MC in IMRT planning and QC has contributed significantly to our understanding of the value of a well-commissioned MC dose algorithm in IMRT QC. Using the

MCDOSE (Ma, Li et al. 2000) code system to model the details of the MLC, Deng et al. (2001) showed differences up to 10% in the maximum dose in one example IMRT field between calculations performed with and without inclusion of the tongue-and-groove effect. The study by Deng et al. (2001) provided a comprehensive example of the intricate details one is able to incorporate using MC-based simulation of the MLC. Toward the goal of verifying IMRT dose distributions using the MCDOSE code system, Ma, Pawlicki, Jiang et al. (2000) showed discrepancies of greater than 5% (relative to the prescribed target dose) in the target region and larger than 20% in the critical structures, for example, IMRT patient calculations computed using MC and the Corvus system (which uses a finite-size pencil beam algorithm). It should also be noted in this study (Ma, Pawlicki, Jiang et al. 2000) that the MC-calculated IMRT dose distributions agreed with the measurements (in homogeneous and heterogeneous phantoms) to within 2% of the maximum dose for all the beam energies and field sizes tested.

Another novel clinical application involving the use of MC dose calculations has been the development of modulated electron beam radiation therapy (Lee et al. 2001; Ma, Ding et al. 2003; Ma, Pawlicki, Lee et al. 2000). With modulated electron beam radiation therapy, electron beam energy and intensity are optimized to produce dose distributions that have been shown to be superior to IMRT and tangential photon beams in terms of dose conformity and normal tissue sparing for breast cancer treatments (Ma, Ding et al. 2003). A specially designed MLC was originally proposed to deliver the optimized fluence distributions, and an inverse-planning system based on MC dose calculations was developed to optimize electron beam energy and intensity to achieve dose conformity for target volumes near the surface (Ma, Pawlicki, Lee et al. 2000). Dosimetric characteristics of the MLC were studied by performing MC-based simulations of the detailed geometry. These characteristics included analysis of the required leaf widths to deliver complex field shapes, analysis of scatter distributions using rounded versus straight leaf ends, and electron intra- and interleaf leakage (Ma, Pawlicki, Lee et al. 2000).

In developing the MCV MC code system (Siebers et al. 2000), and applying it to various problems in IMRT, Siebers and collaborators have made significant contributions to our understanding of the importance of a well-benchmarked MC algorithm in inverse planning (Dogan et al. 2006; Jang, Liu et al. 2006; Jang, Vassiliev et al. 2006; Mihaylov et al. 2006, 2007; Mihaylov and Siebers 2008; Sakthi et al. 2006; Siebers and Mohan 2003; Siebers, Kawrakow, and Ramakrishnan 2007; Siebers et al. 2002). An accurate model of the MLC applicable to both dynamic and segmental IMRT beam delivery was incorporated within the MCV code system to perform patient IMRT dose calculations as well as pre- and post-IMRT treatment delivery verification (Siebers et al. 2002). By computing dose prediction errors for a group of head-and-neck cancer patients treated with dynamic MLC IMRT, it was demonstrated that superposition/convolution (SC) and MC, IMRT dose calculations, which use MC-derived intensity matrices for fluence prediction, are in relatively good agreement with

full MC dose computations (Mihaylov et al. 2007). It was pointed out that the use of pencil beam algorithms may result in clinically significant dose deviations (Mihaylov et al. 2007). The importance of accurate incident fluence prediction was also observed in a study comparing SC and MC dose calculations for head-and-neck IMRT patients treated with the simultaneous integrated boost technique (Sakthi et al. 2006). Here, it was found that the ability to account for details in the MLC delivery with MC resulted in better agreement of fluence prediction on flat phantom measurements in comparison with the SC algorithm. Differences of greater than 5% between SC and MC in the patient geometries (in several cases) were attributed to improved fluence modulation prediction with MC (Sakthi et al. 2006). In studying optimization convergence errors, in addition to dose prediction errors, using both SC- and MC-based optimization for a large group of head-and-neck IMRT patients, Siebers and colleagues (Dogan et al. 2006; Mihaylov and Siebers 2008) demonstrated that calculations based on full MC-computed optimization resulted in lower normal tissue doses and therefore were considered advantageous for head-and-neck IMRT.

Another example study demonstrating the ability of an MC-based calculation tool to account for radiation transport within the detailed MLC geometry, found to be in good agreement with measurements, is that of Tyagi et al. (2007). In this work, the DPM MC code (Chetty, Tyagi et al. 2003; Sempau, Wilderman, and Bielajew 2000; Tyagi et al. 2007) was used to develop a simulation model of the Varian 120-leaf MLC (Tyagi et al. 2007). Full photon transport through the detailed Millennium MLC (including air spaces between leaves) was performed, with electrons depositing energy locally (Tyagi et al. 2007). Figure 85.1 (from Tyagi et al. 2007) shows verification results for a complex IMRT head-and-neck split field case, where excellent agreement between MC calculations and film measurements is noted.

The above examples form part of a large body of work (see TG-105 [Chetty et al. 2007] and Reynaert et al. [2007] for more detailed reviews) substantiating the possible role of MC in QC of complex delivery techniques, such as IMRT. The ability to model intricate details within the patient-specific delivery system, as well as accurate radiation transport within heterogeneous patient geometries, suggests that a well-commissioned and properly verified MC dose algorithm can be used as an effective tool in IMRT QC. The presence of complex and, often, sharp gradients in IMRT dose distributions may also limit accurate dose measurements with detectors, such as ion chambers, because of volume averaging effects. Accurate MC-based dose calculations may help mitigate detector limitations in such situations, and others (Bouchard et al. 2009; Das, Ding, and Ahnesjo 2008).

MC-Based Simulation Incorporating Detectors and Dosimetry Systems

The ability to perform radiation transport simulation within detector geometries such as ion chambers (Kawrakow 2000; Ma and Nahum 1991, 1995; Wulff, Zink, and Kawrakow 2008) and

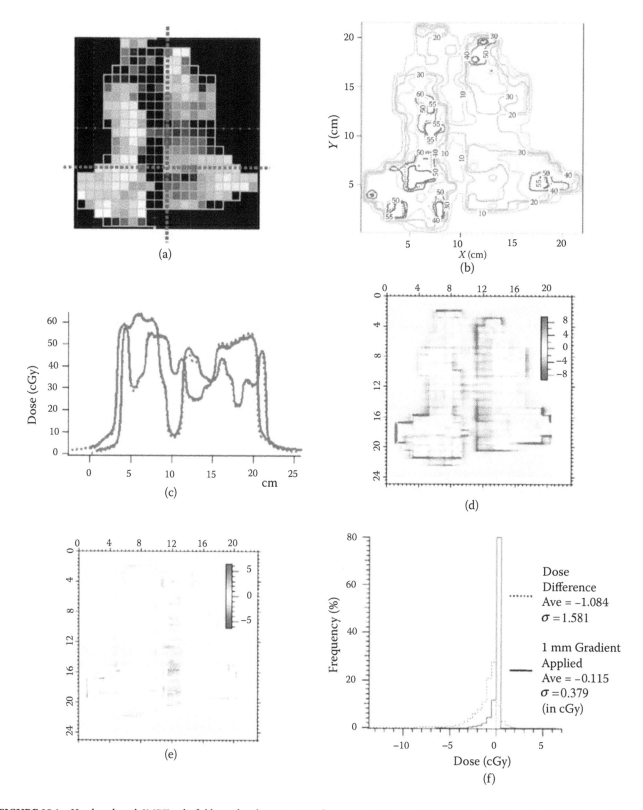

FIGURE 85.1 Head-and-neck IMRT, split field simulated using an MC dose computation module (DPM-based) for static-MLC delivery (Tyagi et al. 2007). (a) Beam intensity map. (b) Isodose display for film measurement and DPM calculation; film is shown in solid and DPM in dashed lines. (c) One-dimensional profile comparisons between film measurement and DPM calculation; film is shown in solid and DPM in dashed lines. (d) Dose difference map in cGy: DPM-film. (e) Dose difference map (in cGy) generated by applying 1 mm gradient compensation. (f) Dose difference histogram of the dose difference map (dotted line) and the gradient-compensated dose difference map (solid line). (Reproduced from Tyagi et al., *Med. Phys.* 34 (2), 651–663, 2007. With permission.)

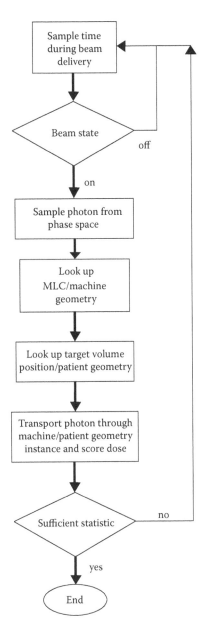

FIGURE 85.2 Flow chart for the synchronized dynamic dose reconstruction method implemented in the DPM Monte Carlo dose calculation algorithm. (Reproduced From Litzenberg et al., *Med. Phys.* 34 (1), 91–102, 2007. With permission.)

other devices (Ma and Nahum 1993; Mainegra-Hing, Reynaert, and Kawrakow 2008) using MC may have beneficial implications for the role of MC in IMRT QC. For example, recent work involving careful MC-based simulations of different ionization chamber geometries has led to an improved understanding of the effective point of measurement correction in ion chamber depth dose and profile measurements (Kawrakow 2006a, 2006b; McEwen, Kawrakow, and Ross 2008). Although direct

MC simulation of ion chamber geometries is a difficult problem, requiring the use of an accurate transport code (Kawrakow 2000), it is conceivable that responses and associated perturbation factors for various ion chambers may be computed (at calibration laboratories or research centers, for example, National Research Council of Canada) and applied broadly for use in IMRT QC (Bouchard et al. 2009).

Dosimetry systems, such as flat-panel imagers, may also be characterized using MC-based computation. Siebers and colleagues (Siebers et al. 2004; Wang et al. 2009) demonstrated the effectiveness of MC in IMRT electronic portal imaging device–based dosimetry by developing and testing a method to compute dosimetric images from an amorphous silicon flat panel detector. In the original study, an EGS4-based simulation of a Varian aS500 (Varian Medical Systems) flat panel imager was performed, incorporating the details of the imager geometry, to investigate various dosimetric characteristics of the imager (Siebers et al. 2004). Experimental verification of calibration flood fields as well as an IMRT patient test field showed good agreement between the MC-based virtual imager calculations and measurements (Siebers et al. 2004). This work demonstrated that MC may be considered a useful tool in IMRT electronic portal imaging device–based dosimetry.

MC-Based Dose Calculation for Nonequilibrium Conditions

Accurate measurements under conditions of nonequilibrium, including small fields, tissue interfaces, buildup dose regions, and so on, are fraught with challenges (Das, Ding, and Ahnesjo 2008; Seuntjens 2006). As pointed out by Das, Ding, and Ahnesjo (2008), the use of small fields in beamlet-based IMRT and clinical programs, such as stereotactic radiosurgery and stereotactic body radiotherapy, can result in significant uncertainty in the accuracy of clinical dosimetry associated with these technologies. Das, Ding, and Ahnesjo (2008) point out that when the field size is reduced to the level that the radiation source is eclipsed from the detector's point of view, then there is a potential for significant output reduction and, subsequently, a condition in which traditional methods for field size determination break down. Conditions of loss of charged-particle equilibrium can also be created when the field size is reduced such that the lateral ranges of the secondary electrons become comparable to (or greater than) the field size (Das, Ding, and Ahnesjo 2008). This situation is exacerbated in tissues of density much lower than that of water (e.g., lung or air-equivalent tissues) where the ranges of the lateral electrons are increased, as encountered in treatment planning of patients with lung cancer treated with stereotactic body radiotherapy (Ding et al. 2007). Moreover, in low-density, lung-equivalent tissues, the range of the secondary electrons along (parallel to) the beam axis contributes to the dose "build-down" effect at the edges of the tumor (at the lung–tumor interface if the tumor is located proximal to the lung), an effect which increases with beam energy. Under such nonequilibrium conditions, detectors introduce perturbations, which are difficult to quantify. Excellent reviews of the perturbation conditions and other issues associated with reliable measurements under nonstandard conditions are presented in the chapter by Seuntjens (2006) and the article by Das, Ding, and Ahnesjo (2008). In addition, conventional dose algorithms, which are typically based on analytical methods and do not account for transport of secondary electrons, are severely limited under nonequilibrium conditions. The article by Reynaert et al. (2007) and the TG-105 (Chetty et al. 2007) provide examples of numerous studies reported on the inaccuracies associated with conventional algorithms for dose calculations in patient-specific treatment sites, such as the lung.

Based on our present knowledge of nonequilibrium dosimetry, it is likely that MC-based methods will become more widely used in the dose computation and verification of clinical technologies involving small fields and treatment sites with low-density tissues, especially as some of these technologies (e.g., stereotactic body radiotherapy for early-stage lung cancers) become more popular. Das, Ding, and Ahnesjo (2008) provided the following summary of the role of MC: "It is also expected that the Monte Carlo techniques will increasingly be used in assessing the accuracy, verification, and calculation of dose, and will aid perturbation calculations of detectors used in small and highly conformal radiation beams." It is appealing to consider a properly commissioned MC dose algorithm in the QC of small-field IMRT treatment plans for lung cancer. Such a QC device would provide more accurate and realistic estimates of the doses delivered to the actual patient target(s) and normal tissues than current methods for IMRT verification, often involving measurements in a solid–water phantom.

MC-Based Dose Calculation Incorporating Motion and Time Dependence

In recognizing a reciprocal relationship between the radiation source and the patient, Beckham, Keall, and Siebers (2002) proposed a fluence-convolution method to incorporate random setup errors using MC-based dose calculation. With fluence convolution, the radiation fluence distribution is convolved with a Gaussian function representing random setup errors (Beckham, Keall, and Siebers 2002). A "fluence-translation" approach to incorporate both random and systematic setup errors in MC-based dose calculation was proposed by Chetty, Tyagi et al. (2003). In their implementation, each particle sampled from a matrix of the collimated fluence distribution was translated a distance determined by randomly sampling an arbitrary distribution function (Chetty, Tyagi et al. 2003). To account for random setup errors, the arbitrary distribution corresponded to a Gaussian function representative of the setup error. By translating, as opposed to convolving the fluence distribution, systematic setup errors were also accounted for, that is, a systematic setup error corresponded effectively to a translation of the isocenter of the collimated fluence matrix (Chetty, Tyagi et al. 2003). This approach was further developed to account for respiratory-induced motion of thoracic tumors in planning dose

calculations, where the fluence translation was determined by sampling distributions corresponding to the respiratory-induced motion (Chetty et al. 2004; Chetty, Rosu et al. 2003). It should be pointed out that the MC-based fluence translation computation incorporating setup errors and/or respiratory-induced motion is performed in a single dose calculation. Such an approach implemented using a conventional, analytical dose calculation algorithm would require n dose calculations, where n is dependent on the number of points sampled in the distribution. Therefore, in addition to enhanced accuracy in regions such as the thorax, MC-based dose calculation offers a significant advantage in terms of efficiency over analytical dose algorithms when fluence translation is used to account for motion. Verification experiments of the fluence convolution approach using a moving platform demonstrate it to be potentially useful in accounting for motion (Naqvi and D'Souza 2005).

MC has been proposed as a tool for four-dimensional (4D) treatment planning (Keall et al. 2004) and also has been used to evaluate the dosimetric effects of respiratory motion in 4D versus three-dimensional (3D) treatment planning (Rosu et al. 2007). In a study by Keall et al. (2004), a 4D CT scan consisting of multiple 3D CT data sets acquired at different respiratory phases was used in conjunction with deformable image registration for automatic contouring of the target and normal structures at each phase. MC dose calculation was then performed for each of the eight CT data sets, and the dose distributions were mapped back to the end-inhale phase for analysis (Keall et al. 2004). For MC computations on each of the n, 3D CT data sets, Keall et al. (2004) used $1/n$ fewer particles than that required for a full 3D plan and noted that using deformable image registration to merge all n dose distributions onto the reference CT data set yielded a 4D MC calculation with statistical uncertainty approximately equivalent to a 3D calculation (with a similar calculation time for the 3D and 4D methods). Consequently, they found that, if deformable image registration is used, the calculation time with MC is approximately independent of the number of 3D CT data sets constituting a 4D CT, which is a distinct advantage over analytical algorithms for which the calculation time scales linearly with the number of 3D CT data sets (Keall et al. 2004). The use of MC-based dose calculation in 4D treatment planning has also facilitated the development of novel techniques for dose accumulation (Heath and Seuntjens 2006; Siebers and Zhong 2008).

In developing and verifying an approach for synchronized dynamic dose reconstruction, Litzenberg et al. (2007) used the inherent temporal nature of MC-based dose calculation. By translating the incident fluence according to the patient motion (determined by tracking the tumor using the Calypso localization system; Calypso Medical Systems, Seattle, Washington) in a synchronized manner with the actual delivered MLC leaf sequences (acquired from the Varian Dynalog files), Litzenberg et al. (2007) demonstrated that the MC method can be used to account for interplay effects between target and MLC motion as well as MLC-based delivery errors (e.g., dropped segments, faulty motors) in a single dose computation. The method developed

TABLE 85.1 Role of Well Commissioning and Properly Verified MC-Based Calculation Tools in QC of Various Components of Radiotherapy Planning and Delivery

Component of RT Planning and Delivery	Role of MC Dose Calculation
Complex delivery techniques	Modeling of detailed geometry of MLC; accurate incident fluence prediction for complex IMRT fields. Can be used to verify incident fluence distributions for IMRT QC.
Detectors and dosimetry systems	Modeling of detailed geometry of detector responses and perturbations; modeling of detailed geometry of dosimetry measurement systems, such as flat panel detector of an EPID. Can be used to incorporate detector perturbations in dose calculations as a tool for verification and QC of complex delivery fields, where detectors may introduce considerable perturbation.
Nonequilibrium conditions	Accurate radiation transport under conditions of nonequilibrium, e.g., small fields, low-density tissues, buildup dose regions, high-gradient regions in IMRT. Can be used in QC of small-field IMRT treatment plans in lung, and IMRT monitor unit calculation verification.
Motion compensation	Incorporation of motion into dose distributions, efficient 4D dose computation, dose accumulation. Can be used as a tool in QC of the overall 4D planning and delivery process.
Synchronization of delivery and motion	Temporal dose synchronization of monitored tumor trajectories with delivered MLC leaf sequences to account for interplay effects and possible implementation errors. Can be used as a QC tool of patient-specific IMRT treatment plans (by accounting for delivered sequences and real-time target motion in synchrony).

for synchronized dynamic dose reconstruction using MC in a single dose calculation is shown in Figure 85.2 (Litzenberg et al. 2007). As with other fluence-translation approaches, this method assumes rigid motion of the target and does not account for shape changes in the target and/or normal structures during respiration. Excellent agreement was found between MC calculations and film measurements, which included doses delivered to a platform moving with a sinusoidal function (Litzenberg et al. 2007). Results were suggestive that MC can be used as a tool to verify all aspects of motion and delivery for patient-specific IMRT. The method of Litzenberg et al. (2007) can also be applied to technologies involving dynamic MLC tracking of target motion (Poulsen et al. 2010).

Summary

A summary of the possible role of MC in the QC of various components of patient-specific treatment planning and delivery, as discussed in this chapter, is presented in Table 85.1. At this juncture, the MC method is viewed as an emerging technology in the field of radiation therapy. With the increased availability of FDA-approved, commercial MC-based systems, it is likely that MC-based dose algorithms will become part of the routine clinical treatment planning procedure over the next 5 years. It is important that clinical implementation of MC-based systems be performed thoughtfully. Successful implementation of clinical MC algorithms will require strong support from the entire clinical team and an understanding of the paradigm shift with MC algorithms, as noted, for instance, in the viewing of statistically uncertain dose distributions. A well-commissioned and properly verified MC dose algorithm, in addition to improving dose calculation accuracy in 3D-CRT and IMRT, can be a useful tool in patient-specific QC.

Acknowledgment

The author acknowledges support grant no. R01 CA106770 from the NIH (NCI).

References

Beckham, W. A., P. J. Keall, and J. V. Siebers. 2002. A fluence-convolution method to calculate radiation therapy dose distributions that incorporate random set-up error. *Phys. Med. Biol.* 47 (19): 3465–3473.

Berger, M. J. 1963. Monte Carlo calculations of the penetration and diffusion of fast charged particles. In *Methods in Computational Physics*, vol. 1, ed. S. F. B. Alder, M. Rothenberg. New York, NY: Academic Press.

Bouchard, H., J. Seuntjens, J. F. Carrier, and I. Kawrakow. 2009. Ionization chamber gradient effects in nonstandard beam configurations. *Med. Phys.* 36 (10): 4654–4663.

Chetty, I. J., B. Curran, J. E. Cygler, J. J. DeMarco, G. Ezzell, B. A. Faddegon, I. Kawrakow et al. 2007. Report of the AAPM Task Group No. 105: Issues associated with clinical implementation of Monte Carlo–based photon and electron external beam treatment planning. *Med. Phys.* 34 (12): 4818–4853.

Chetty, I. J., N. Tyagi, M. Rosu, P. M. Charland, D. L. McShan, R. K. Ten Haken, B. A. Fraass, and A. F. Bielajew. 2003. Clinical implementation, validation and use of the DPM Monte Carlo code for radiotherapy treatment planning. In *Nuclear Mathematical and Computational Sciences: A Century in Review, A Century Anew, Gatlinburg, TN*. LaGrange Park, IL: American Nuclear Society.

Chetty, I. J., M. Rosu, D. L. McShan, B. A. Fraass, J. M. Balter, and R. K. Ten Haken. 2004. Accounting for center-of-mass target motion using convolution methods in Monte Carlo–based dose calculations of the lung. *Med. Phys.* 31 (4): 925–932.

Chetty, I. J., M. Rosu, N. Tyagi, L. H. Marsh, D. L. McShan, J. M. Balter, B. A. Fraass, and R. K. Ten Haken. 2003. A fluence convolution method to account for respiratory motion in three-dimensional dose calculations of the liver: A Monte Carlo study. *Med. Phys.* 30 (7): 1776–1780.

Das, I. J., G. X. Ding, and A. Ahnesjo. 2008. Small fields: Nonequilibrium radiation dosimetry. *Med. Phys.* 35 (1): 206–215.

Deng, J., M. C. Lee, and C. M. Ma. 2002. A Monte Carlo investigation of fluence profiles collimated by an electron specific MLC during beam delivery for modulated electron radiation therapy. *Med. Phys.* 29 (11): 2472–2483.

Deng, J., T. Pawlicki, Y. Chen, J. Li, S. B. Jiang, and C. M. Ma. 2001. The MLC tongue-and-groove effect on IMRT dose distributions. *Phys. Med. Biol.* 46 (4): 1039–1060.

Ding, G. X., D. M. Duggan, B. Lu, D. E. Hallahan, A. Cmelak, A. Malcolm, J. Newton, M. Deeley, and C. W. Coffey. 2007. Impact of inhomogeneity corrections on dose coverage in the treatment of lung cancer using stereotactic body radiation therapy. *Med. Phys.* 34 (7): 2985–2994.

Dogan, N., J. V. Siebers, P. J. Keall, F. Lerma, Y. Wu, M. Fatyga, J. F. Williamson, and R. K. Schmidt-Ullrich. 2006. Improving IMRT dose accuracy via deliverable Monte Carlo optimization for the treatment of head and neck cancer patients. *Med. Phys.* 33 (11): 4033–4043.

Fan, J., J. Li, L. Chen, S. Stathakis, W. Luo, F. Du Plessis, W. Xiong, J. Yang, and C. M. Ma. 2006. A practical Monte Carlo MU verification tool for IMRT quality assurance. *Phys. Med. Biol.* 51 (10): 2503–2515.

Fraass, B., K. Doppke, M. Hunt, G. Kutcher, G. Starkschall, R. Stern, and J. Van Dyke. 1998. American Association of Physicists in Medicine Radiation Therapy Committee Task Group 53: Quality assurance for clinical radiotherapy treatment planning. *Med. Phys.* 25 (10): 1773–1829.

Fragoso, M., I. Kawrakow, B. A. Faddegon, T. D. Solberg, and I. J. Chetty. 2009. Fast, accurate photon beam accelerator modeling using BEAMnrc: A systematic investigation of efficiency enhancing methods and cross-section data. *Med. Phys.* 36 (12): 5451–5466.

Gardner, J., J. Siebers, and I. Kawrakow. 2007. Dose calculation validation of VMC++ for photon beams. *Med. Phys.* 34 (5): 1809–1818.

Hasenbalg, F., M. K. Fix, E. J. Born, R. Mini, and I. Kawrakow. 2008. VMC++ versus BEAMnrc: A comparison of simulated linear accelerator heads for photon beams. *Med. Phys.* 35 (4): 1521–1531.

Heath, E., and J. Seuntjens. 2003. Development and validation of a BEAMnrc component module for accurate Monte Carlo modelling of the Varian dynamic Millennium multileaf collimator. *Phys. Med. Biol.* 48 (24): 4045–4063.

Heath, E., and J. Seuntjens. 2006. A direct voxel tracking method for four-dimensional Monte Carlo dose calculations in deforming anatomy. *Med. Phys.* 33 (2): 434–445.

IAEA. 2004. *Technical Report Series No. 430: Commissioning and Quality Assurance of Computerized Planning Systems for Radiation Treatment of Cancer.* Vienna: International Atomic Energy Agency.

Jang, S. Y., H. H. Liu, X. Wang, O. N. Vassiliev, J. V. Siebers, L. Dong, and R. Mohan. 2006. Dosimetric verification for intensity-modulated radiotherapy of thoracic cancers using experimental and Monte Carlo approaches. *Int. J. Radiat. Oncol. Biol. Phys.* 66 (3): 939–948.

Jang, S. Y., O. N. Vassiliev, H. H. Liu, R. Mohan, and J. V. Siebers. 2006. Development and commissioning of a multileaf collimator model in monte carlo dose calculations for intensity-modulated radiation therapy. *Med. Phys.* 33 (3): 770–781.

Kawrakow, I. 2000. Accurate condensed history Monte Carlo simulation of electron transport: II. Application to ion chamber response simulations. *Med. Phys.* 27 (3): 499–513.

Kawrakow, I. 2006a. Efficient photon beam treatment head simulations. *Radiother. Oncol. Suppl.* 81: 82.

Kawrakow, I. 2006b. On the effective point of measurement in megavoltage photon beams. *Med. Phys.* 33 (6): 1829–1839.

Keall, P. J., J. V. Siebers, S. Joshi, and R. Mohan. 2004. Monte Carlo as a four-dimensional radiotherapy treatment-planning tool to account for respiratory motion. *Phys. Med. Biol.* 49 (16): 3639–3648.

Lee, M. C., J. Deng, J. Li, S. B. Jiang, and C. M. Ma. 2001. Monte Carlo based treatment planning for modulated electron beam radiation therapy. *Phys. Med. Biol.* 46 (8): 2177–2199.

Li, J. S., A. L. Boyer, and C. M. Ma. 2001. Verification of IMRT dose distributions using a water beam imaging system. *Med. Phys.* 28 (12): 2466–2474.

Li, J. S., G. M. Freedman, R. Price, L. Wang, P. Anderson, L. Chen, W. Xiong, J. Yang, A. Pollack, and C. M. Ma. 2004. Clinical implementation of intensity-modulated tangential beam irradiation for breast cancer. *Med. Phys.* 31 (5): 1023–1031.

Litzenberg, D. W., S. W. Hadley, N. Tyagi, J. M. Balter, R. K. Ten Haken, and I. J. Chetty. 2007. Synchronized dynamic dose reconstruction. *Med. Phys.* 34 (1): 91–102.

Luo, W., J. Li, R. A. Price Jr., L. Chen, J. Yang, J. Fan, Z. Chen, S. McNeeley, X. Xu, and C. M. Ma. 2006. Monte Carlo based IMRT dose verification using MLC log files and R/V outputs. *Med. Phys.* 33 (7): 2557–2564.

Ma, C.-M. 1998. Characterization of computer simulated radiotherapy beams for Monte-Carlo treatment planning. *Radiat. Phys. Chem.* 53 (3): 329–344.

Ma, C.-M., and S. B. Jiang. 1999. Monte Carlo modelling of electron beams from medical accelerators. *Phys. Med. Biol.* 44 (12): R157–R189.

Ma, C.-M., J. S. Li, T. Pawlicki, S. B. Jiang, and J. Deng. 2000. MCDOSE—A Monte Carlo dose calculation tool for radiation therapy treatment planning. In *Proceedings of the 13th ICCR,* ed. T. B. W. Schlegel. Heidelberg: Springer-Verlag, 123–125.

Ma, C.-M., E. Mok, A. Kapur, T. Pawlicki, D. Findley, S. Brain, K. Forster, and A. L. Boyer. 1999. Clinical implementation of a Monte Carlo treatment planning system. *Med. Phys.* 26 (10): 2133–2143.

Ma, C.-M., T. Pawlicki, S. B. Jiang, J. S. Li, J. Deng, E. Mok, A. Kapur, L. Xing, L. Ma, and A. L. Boyer. 2000. Monte Carlo verification of IMRT dose distributions from a commercial treatment planning optimization system. *Phys. Med. Biol.* 45 (9): 2483–2495.

Ma, C. M., M. Ding, J. S. Li, M. C. Lee, T. Pawlicki, and J. Deng. 2003. A comparative dosimetric study on tangential photon beams, intensity-modulated radiation therapy (IMRT) and modulated electron radiotherapy (MERT) for breast cancer treatment. *Phys. Med. Biol.* 48 (7): 909–924.

Ma, C. M., S. B. Jiang, T. Pawlicki, Y. Chen, J. S. Li, J. Deng, and A. L. Boyer. 2003. A quality assurance phantom for IMRT dose verification. *Phys. Med. Biol.* 48 (5): 561–572.

Ma, C. M., and A. E. Nahum. 1991. Bragg-Gray theory and ion chamber dosimetry for photon beams. *Phys. Med. Biol.* 36 (4): 413–428.

Ma, C. M., and A. E. Nahum. 1993. Dose conversion and wall correction factors for Fricke dosimetry in high-energy photon beams: Analytical model and Monte Carlo calculations. *Phys. Med. Biol.* 38 (1): 93–114.

Ma, C. M., and A. E. Nahum. 1995. Monte Carlo calculated stem effect correction for NE2561 and NE2571 chambers in medium-energy x-ray beams. *Phys. Med. Biol.* 40 (1): 63–72.

Ma, C. M., T. Pawlicki, M. C. Lee, S. B. Jiang, J. S. Li, J. Deng, B. Yi, E. Mok, and A. L. Boyer. 2000. Energy- and intensity-modulated electron beams for radiotherapy. *Phys. Med. Biol.* 45 (8): 2293–2311.

Mainegra-Hing, E., N. Reynaert, and I. Kawrakow. 2008. Novel approach for the Monte Carlo calculation of free-air chamber correction factors. *Med. Phys.* 35 (8): 3650–3660.

McEwen, M. R., I. Kawrakow, and C. K. Ross. 2008. The effective point of measurement of ionization chambers and the build-up anomaly in MV x-ray beams. *Med. Phys.* 35 (3): 950–958.

Mihaylov, I. B., F. A. Lerma, M. Fatyga, and J. V. Siebers. 2007. Quantification of the impact of MLC modeling and tissue heterogeneities on dynamic IMRT dose calculations. *Med. Phys.* 34 (4): 1244–1252.

Mihaylov, I. B., F. A. Lerma, Y. Wu, and J. V. Siebers. 2006. Analytic IMRT dose calculations utilizing Monte Carlo to predict MLC fluence modulation. *Med. Phys.* 33 (4): 828–839.

Mihaylov, I. B., and J. V. Siebers. 2008. Evaluation of dose prediction errors and optimization convergence errors of deliverable-based head-and-neck IMRT plans computed with a superposition/convolution dose algorithm. *Med. Phys.* 35 (8): 3722–3727.

Naqvi, S. A., and W. D. D'Souza. 2005. A stochastic convolution/superposition method with isocenter sampling to evaluate intrafraction motion effects in IMRT. *Med. Phys.* 32 (4): 1156–1163.

Poulsen, P. R., B. Cho, A. Sawant, and P. J. Keall. 2010. Implementation of a New Method for Dynamic Multileaf Collimator Tracking of Prostate Motion in Arc Radiotherapy Using a Single kV Imager. *Int. J. Radiat. Oncol. Biol. Phys.* 76 (3): 914–923.

Reynaert, N., S. C. van der Marck, D. R. Schaart, W. Van der Zee, C. Van Vliet-Vroegindeweij, M. Tomsej, J. Jansen, B. Heijmen, M. Coghe, and C. De Wagter. 2007. Monte Carlo treatment planning for photon and electron beams. *Radiat. Phys. Chem.* 76: 643–686.

Rogers, D. W. O., and A. F. Bielajew. 1990. Monte Carlo techniques of electron and photon transport for radiation dosimetry. In *The Dosimetry of Ionizing Radiation*, ed. B. B. K. Kase and F. Attix. New York, NY: Academic Press.

Rogers, D. W. O., B. A. Faddegon, G. X. Ding, C. M. Ma, J. We, and T. R. Mackie. 1995. BEAM: A Monte Carlo code to simulate radiotherapy treatment units. *Med. Phys.* 22 (5): 503–524.

Rosu, M., J. M. Balter, I. J. Chetty, M. L. Kessler, D. L. McShan, P. Balter, and R. K. Ten Haken. 2007. How extensive of a 4D dataset is needed to estimate cumulative dose distribution plan evaluation metrics in conformal lung therapy? *Med. Phys.* 34 (1): 233–245.

Sakthi, N., P. Keall, I. Mihaylov, Q. Wu, Y. Wu, J. F. Williamson, R. Schmidt-Ullrich, and J. V. Siebers. 2006. Monte Carlo–based dosimetry of head-and-neck patients treated with SIB-IMRT. *Int. J. Radiat. Oncol. Biol. Phys.* 64 (3): 968–977.

Sempau, J., S. J. Wilderman, and A. F. Bielajew. 2000. DPM, a fast, accurate Monte Carlo code optimized for photon and electron radiotherapy treatment planning dose calculations. *Phys. Med. Biol.* 45 (8): 2263–2291.

Seuntjens, J. 2006. Measurement Issues in Commissioning and Benchmarking of Monte Carlo Treatment Planning Systems. In *Integrating New Technologies into the Clinic: Monte Carlo and Image-Guided Radiation Therapy*, ed. B. Curran, Balter, J. M., Chetty, I. J. Madison, WI: Medical Physics Publishing.

Siebers, J., and R. Mohan. 2003. Monte Carlo and IMRT. In *Intensity Modulated Radiation Therapy, The State of the Art, Proceedings of the 2003 AAPM Summer School*, ed. T. R. Mackie and J. R. Palta. Madison, WI: Advanced Medical Publishing.

Siebers, J. V., I. Kawrakow, and V. Ramakrishnan. 2007. Performance of a hybrid MC dose algorithm for IMRT optimization dose evaluation. *Med. Phys.* 34 (7): 2853–2863.

Siebers, J. V., P. J. Keall, J. Kim, and R. Mohan. 2000. Performance benchmarks of the MCV Monte Carlo System. In *Proceedings of the 13th ICCR*, ed. W. Schlegel and T. Bortfeld, 129–131. Heidelberg: Springer Verlag.

Siebers, J. V., P. J. Keall, J. O. Kim, and R. Mohan. 2002. A method for photon beam Monte Carlo multileaf collimator particle transport. *Phys. Med. Biol.* 47 (17): 3225–3249.

Siebers, J. V., J. O. Kim, L. Ko, P. J. Keall, and R. Mohan. 2004. Monte Carlo computation of dosimetric amorphous silicon electronic portal images. *Med. Phys.* 31 (7): 2135–2146.

Siebers, J. V., and H. Zhong. 2008. An energy transfer method for 4D Monte Carlo dose calculation. *Med. Phys.* 35 (9): 4096–4105.

Tyagi, N., J. M. Moran, D. W. Litzenberg, A. F. Bielajew, B. A. Fraass, and I. J. Chetty. 2007. Experimental verification of a Monte Carlo–based MLC simulation model for IMRT dose calculation. *Med. Phys.* 34 (2): 651–663.

Van Dyk, J., R. B. Barnett, J. E. Cygler, and P. C. Shragge. 1993. Commissioning and quality assurance of treatment planning computers. *Int. J. Radiat. Oncol. Biol. Phys.* 26 (2): 261–273.

Verhaegen, F., and J. Seuntjens. 2003. Monte Carlo modelling of external radiotherapy photon beams. *Phys. Med. Biol.* 48 (21): R107–R164.

Wang, S., J. K. Gardner, J. J. Gordon, W. Li, L. Clews, P. B. Greer, and J. V. Siebers. 2009. Monte Carlo–based adaptive EPID dose kernel accounting for different field size responses of imagers. *Med. Phys.* 36 (8): 3582–3595.

Wulff, J., K. Zink, and I. Kawrakow. 2008. Efficiency improvements for ion chamber calculations in high energy photon beams. *Med. Phys.* 35 (4): 1328–1336.

Patient Setup

Uulke A. van der Heide
University Medical Center Utrecht

Introduction

With modern radiotherapy delivery techniques, it is possible to realize dose distributions that conform precisely to the shape of the target volume while avoiding healthy tissue. However, even the most advanced delivery techniques are ineffective if they deliver the dose to the wrong position in the patient. As geometrical misses may cause large deviations from the prescribed dose to the target volume, margins are applied to ensure adequate coverage. Unfortunately, such margins result in the involvement of healthy tissue in the planning target volume. Thus, to take full advantage of the delivery techniques that are now at our disposal, high-quality position verification is essential.

In this chapter, the principles of patient setup quality assurance are discussed. Offline and online correction strategies are described. The results in our clinic show that, with a relatively minor effort, a large gain in treatment accuracy can be achieved.

What to Verify?

The purpose of patient setup verification is to ensure that the target for irradiation is in fact irradiated during the course of a treatment. To this end, images are made in the treatment room, using an electronic portal imaging device or a cone-beam CT scanner, mounted on the linear accelerator gantry (Figure 86.1). The contrast obtained from portal images using MV photons is too limited to visualize the tumor directly for most tumors. For this reason, position verification is often based on the bony anatomy. A drawback of this approach is that any displacement of the actual target volume relative to the bony anatomy is missed and must be accommodated using PTV margins. Although this

is not a large problem for tumors in the brain, large internal motions are observed for tumors in the pelvis, such as prostate and cervical tumors (Figure 86.2) (Nederveen et al. 2003; Chan et al. 2008; van de Bunt et al. 2008). Even for patients with head–neck tumors, displacements occur between tumors and bony anatomy (van Kranen et al. 2009).

For some treatment sites, such as the prostate, it is possible to implant radiopaque fiducial markers inside the target volume (Balter et al. 1995; Dehnad et al. 2003; van der Heide et al. 2007). In this way, a surrogate of the actual target volume can be visualized that reflects its translations and rotations. Deformations of the target volume are more difficult to detect and must again be dealt with using PTV margins. Full anatomical imaging using a cone-beam CT scanner may be helpful in reducing these errors as well. Although the soft-tissue contrast of such scans is limited, it is possible to recognize tumors and organs in the lung and head and neck (Borst et al. 2007; van Kranen et al. 2009). When implantation of fiducial markers is not feasible, the use of cone-beam CT may be advantageous in pelvic tumors as well (Lotz et al. 2005; Nijkamp et al. 2009).

Position correction strategies based on in-room imaging of the patient are highly successful in reducing the setup uncertainty. However, it is important to realize that other sources of uncertainty remain. We discussed the internal motion and deformations of the target volume relative to the bony anatomy. The detection accuracy of the reference structure (bony anatomy, fiducial marker, or soft tissue) on portal images or cone-beam CT is limited and should be taken into account (Smitsmans et al. 2005; van der Heide et al. 2007). Finally, variations in the field shaping of the linear accelerator as well as uncertainties in delineation of the target volume by the physician must be considered when choosing appropriate margins (Dehnad et al. 2003).

FIGURE 86.1 Linear accelerator equipped with an amorphous silicon flat panel portal imaging device and a kV cone-beam CT scanner.

Correction Protocols

During a fractionated treatment, daily variations in patient setup will occur. When a deviation from the correct patient setup is identified, a correction protocol is used to determine the optimal course of action. In many cases, it is sufficient to minimize the systematic setup error, defined as the average of the deviations over all treatment fractions. Systematic errors lead to a shift of the entire dose distribution creating the risk of severe underdosage of the target or overdosage of organs at risk. The random errors, characterized by the variance of the deviations over all treatment fractions, cause a blurring of the dose distribution, which creates a much smaller risk of dose errors. This is reflected in the so-called margin recipes in which the systematic setup errors contribute to a much larger extent to the margin size than the random setup errors (Stroom et al. 1999; van Herk et al. 2000). To characterize the setup accuracy for a particular group of patients, the parameters Σ and σ are widely used. Here, the systematic error, Σ, is defined as the standard deviation of systematic errors of all patients. The random error, σ, is defined as the square root of the average of the variances of errors of all patients.

Offline Protocols

Various correction protocols have been proposed to minimize the systematic error for a patient at a minimal effort (Bel et al. 1993; De Boer and Heijmen 2001). The simplest approach is proposed in the "no action level" (NAL) protocol. Here, the setup deviation is determined in the first n fractions, where n typically has a value of 3 or 4. After these fractions, a correction is applied that is equal to the average deviation observed so far. It is important that n is not chosen lower than 3, as the estimate of the systematic setup error gets highly inaccurate. On the other hand, a value of n that is too large results in a very accurate estimate of the systematic error; however, a correction is applied too late in the course of the treatment to be effective. An illustration

of the impact of various choices is given by De Boer and Heijmen (2001).

Several studies have observed that deviations in patient setup may exhibit time trends. Then, a systematic error determined in the first few fractions of the treatment is not representative for the later fractions. To deal with this problem, an extended NAL protocol (eNAL) was proposed (De Boer and Heijmen 2007). After n fractions, the setup correction is determined. From then on, the deviation is determined once a week. A linear fit is made through all points, with the applied corrections removed. This provides the necessary correction for the next week.

The main advantage of offline protocols such as the NAL and eNAL is that an effective reduction in systematic setup error can be achieved at a minimal work load. In addition, the radiation exposure from the fields used for portal imaging or cone-beam CT scan remains limited. On the other hand, a higher accuracy of patient positioning can be obtained when daily position verification is applied. In an adaptation of the shrinking action level protocol (Bel et al. 1993; van der Heide et al. 2007), at each fraction a decision is made if a setup correction is warranted for the next day. Large deviations, above an action level α, are considered significant and are corrected immediately; smaller deviations are corrected only if they persist over some fractions. The action level is set to $\alpha /\sqrt{n}$, with n the number of fractions considered. After a correction, n is reset to 1. The maximum number of fractions over which the position deviation is averaged is n_{max}. After n_{max} is reached, a running average over the past n_{max} fractions is determined each day and compared to the action level. Such a protocol typically can be applied when the actual treatment fields can be used for portal imaging, avoiding additional dose to the patient. Although the work load is higher than for the NAL and eNAL protocols, the offline character of the protocol still allows processing of the portal image data at a convenient time.

Online Corrections

The offline correction protocols described above minimize the systematic setup error but do not reduce the random daily setup variations. Only online corrections, where the patient position is imaged and corrected before starting the irradiation in

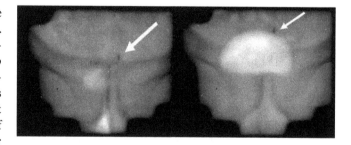

FIGURE 86.2 Portal image of a PA prostate field during IMRT. Left: fraction 8; right: fraction 23. Note the shift of the fiducial marker relative to the bony anatomy due to rectal gas.

each fraction, minimize both systematic and random interfraction setup errors. Online corrections are particularly useful in hypofractionated treatments, where the blurring of the dose over many fractions does not take place. They also have a benefit when steep dose gradients are applied in combination with dose levels that far exceed the tolerance of nearby healthy structures. Nevertheless, even with this correction strategy, it is important to be aware of remaining sources of positioning uncertainty. In particular, intrafraction motion of the target volume is hard to avoid and must be considered in choosing appropriate treatment margins. For target volumes in the pelvic area, intrafraction motions can be considered (Kotte et al. 2007; Chan et al. 2008). Thus, it is worthwhile to study the benefit of online over offline correction procedures in detail, for example, by investigating the impact of all uncertainties on the effectively delivered dose distribution (Astreinidou et al. 2005; van Haaren et al. 2009).

Offline Position Verification Based on Bony Anatomy

In this section we describe the clinical results for four categories of patients who are treated with fractionated external-beam radiotherapy and offline position verification based on imaging of the bony anatomy. For irradiation of the brain and head–neck, patients are stabilized in a posicast thermoplastic mask, using a posifix supine head rest (Civco, Kalona, IA). For irradiation of the lung, patients lie on a flat table top with their arms above their head in a PET arm support (Civco). Patients irradiated in the pelvic area are positioned in a supine position on a flat table top with a knee roll to prevent rotation of the hip.

Patients are imaged, as indicated by the eNAL protocol, in the first three fractions followed by a correction. The remainder of the treatment images is made once per week. The images are made with an amorphous silicon portal image device (Elekta, iView, Crawley, UK) before the delivery of the fraction using two orthogonal fields of five monitor units each. Registration of the portal images to the reference digitally reconstructed radiograph is done manually in the iView software package with the help of delineated reference structures. The results are shown in Table 86.1.

To avoid the influence of the first three fractions on the results of the weekly imaged fractions after correction, only the latter are taken into account. Although for all patients the eNAL protocol was applied, the uncorrected results are listed as well for comparison. In the absence of position correction, the systematic error as indicated by Σ is larger for lung and pelvis than for brain and head–neck, suggesting an improved setup by the application of a mask. Upon application of the eNAL protocol, Σ is reduced to less than 1 mm for all categories and all directions. In contrast, the random error (as indicated by σ) increases, albeit by less than 1 mm. Because systematic errors require much larger margins than random errors, the application of a simple offline protocol results in a substantial reduction in setup uncertainty that may permit a reduction in overall PTV margin.

Offline Position Verification of the Prostate Using Fiducial Markers

Patients with prostate cancer receive an IMRT treatment combined with fiducial marker–based position verification. Three gold fiducial markers (1 mm diameter, 5 mm length) are implanted in the prostate by a radiation oncologist. The 18G needles are placed transperineally through a template in a procedure similar to ^{125}I-seed implantation. All patients receive prophylactic antibiotics. Patients under antiplatelet therapy who are unable to stop this for a period of 1 week are excluded and will receive position verification using cone-beam CT. To allow for some migration of the markers in the first days after implantation, a planning CT scan is made after 1 week (van der Heide et al. 2007).

During treatment, patients are positioned in a supine position on a flat table top, with a knee roll to prevent rotation of the hip.

TABLE 86.1 Results of the eNAL Protocol for Four Groups of Patients

Category	Number of Patients	Σ Anterior–Posterior (mm)	Σ Left–Right (mm)	Σ Superior–Inferior (mm)	σ Anterior–Posterior (mm)	σ Left–Right (mm)	σ Superior–Inferior (mm)
Brain	46						
	Uncorrected	2.1	1.8	1.7	1.1	0.9	1.0
	eNAL	0.5	0.3	0.5	1.4	1.2	1.3
Head–neck	108						
	Uncorrected	2.0	2.2	1.9	1.1	1.3	1.0
	eNAL	0.7	0.6	0.5	1.5	1.8	1.4
Lung	60						
	Uncorrected	3.2	2.4	4.2	1.7	1.9	2.3
	eNAL	0.7	0.8	0.9	2.3	2.6	3.1
Pelvis	130						
	Uncorrected	3.3	2.9	2.3	2.1	2.4	1.5
	eNAL	0.8	0.9	0.5	3.0	3.2	2.2

TABLE 86.2 Results of the Adapted Shrinking Action Level Protocol for 453 Patients with Prostate Cancer with Implanted Fiducial Markers

Translations	Number of Patients	Σ Anterior–Posterior (mm)	Σ Left–Right (mm)	Σ Superior–Inferior (mm)	σ Anterior–Posterior (mm)	σ Left–Right (mm)	σ Superior–Inferior (mm)
	Uncorrected	4.8	2.2	2.9	3.5	2.0	2.3
	eNAL	0.7	0.8	0.8	4.0	2.3	2.5
Rotations	Number of Patients	Σ Anterior–Posterior (deg)	Σ Left–Right (deg)	Σ Superior–Inferior (deg)	σ Anterior–Posterior (deg)	σ Left–Right (deg)	σ Superior–Inferior (deg)
	Uncorrected	2.8	6.8	2.8	1.7	3.1	2.0

Portal images are taken in all 35 treatment fractions, using the largest segment of each of the five IMRT beams. In this way, no additional dose is given for position verification.

The adapted shrinking action level protocol is used as previously described above. In Table 86.2, the results of the position verification and correction are shown. In particular in the anterior–posterior direction, the systematic error (characterized by Σ) is reduced significantly by the correction protocol from 4.8 to 0.7 mm. Importantly, rotations of the prostate occur frequently as well, with a Σ of 6.8° around the lateral axis. Although a correction of rotations falls outside the scope of this chapter, it is clear that such uncertainties must be taken into account when defining PTV margins.

Discussion

Position verification using electronic portal imaging or cone-beam CT scanning is quite effective in reducing the setup errors that occur in fractionated radiotherapy. Even relatively simple offline correction protocols, such as the NAL and the eNAL protocols, are highly successful in reducing systematic setup errors while requiring a minimal effort. For patients irradiated on brain and head–neck, the use of masks results in relatively small setup errors, but the offline protocols reduce these errors further, to a Σ of less than 1 mm.

Most position verification techniques are based on bony anatomy rather than the target volume itself. Therefore, internal target motion remains a source of uncertainty. In particular, for treatments of tumors in the pelvis, this can be large. For the cervix, internal motions of up to 2 cm relative to the bony anatomy have been reported. Measures to minimize this variation such as controlling the bladder or rectum filling tend not to be effective (van de Bunt et al. 2008). Thus, a direct measurement of the position of the target volume by either fiducial markers or 3D soft-tissue imaging is required. The cone-beam CT scanner is an important step in this direction, but probably the combination of an MRI scanner with a radiation device will be most effective given the superior soft-tissue contrast of MRI (Lagendijk et al. 2008).

The success of position verification can result in the reduction of PTV margins required to ensure adequate coverage of the tumor. In particular when comparing to a treatment without position verification, the benefits can be large. However, when the remaining setup uncertainties are of the same order of magnitude as other treatment uncertainties, a further margin reduction becomes risky. As a rule of thumb, the various sources of uncertainty can be added quadratically if they are independent. In this way, the contribution of various systematic errors can be pooled to one overall systematic error for which a margin recipe can be applied. Similarly, sources of various random errors can also be added quadratically. Following this approach, it is clear that if systematic errors are reduced to less than 1 mm, a further margin reduction is difficult to achieve by further quality assurance of the patient setup. This should be the result of a thorough inspection of the remaining sources of errors in the entire treatment.

Summary

Advanced treatment techniques in radiotherapy lose much of their benefit if no adequate position verification is applied. Fortunately, even with relatively simple procedures such as the offline NAL protocol, substantial improvements in positioning accuracy can be achieved. Good quality assurance of patient setup must be an integral part of modern radiation therapy.

References

Astreinidou, E., A. Bel, C. P. Raaijmakers, C. H. Terhaard, and J. J. Lagendijk. 2005. Adequate margins for random setup uncertainties in head-and-neck IMRT. *Int. J. Radiat. Oncol. Biol. Phys.* 61 (3): 938–944.

Balter, J. M., K. L. Lam, H. M. Sandler, J. F. Littles, R. L. Bree, and R. K. Ten Haken. 1995. Automated localization of the prostate at the time of treatment using implanted radiopaque markers: Technical feasibility. *Int. J. Radiat. Oncol. Biol. Phys.* 33 (5): 1281–1286.

Bel, A., M. van Herk, H. Bartelink, and J. V. Lebesque. 1993. A verification procedure to improve patient set-up accuracy using portal images. *Radiother. Oncol.* 29 (2): 253–260.

Borst, G. R., J. J. Sonke, A. Betgen, P. Remeijer, M. van Herk, and J. V. Lebesque. 2007. Kilo-voltage cone-beam computed tomography setup measurements for lung cancer patients; first clinical results and comparison with electronic portal-imaging device. *Int. J. Radiat. Oncol. Biol. Phys.* 68 (2): 555–561.

Chan, P., R. Dinniwell, M. A. Haider, Y. B. Cho, D. Jaffray, G. Lockwood, W. Levin, L. Manchul, A. Fyles, and M. Milosevic. 2008. Inter- and intrafractional tumor and organ movement in patients with cervical cancer undergoing radiotherapy: A cinematic-MRI point-of-interest study. *Int. J. Radiat. Oncol. Biol. Phys.* 70 (5): 1507–1515.

De Boer, H. C. J., and B. J. Heijmen. 2007. eNAL: An extension of the NAL setup correction protocol for effective use of weekly follow-up measurements. *Int. J. Radiat. Oncol. Biol. Phys.* 67: 1586–1595.

De Boer, H. C. J., and B. J. Heijmen. 2001. A protocol for the reduction of systematic patient setup errors with minimal portal imaging of implanted markers. *Int. J. Radiat. Oncol. Biol. Phys.* 50: 1350–1365.

Dehnad, H., A. J. Nederveen, U. A. van der Heide, R. J. van Moorselaar, P. Hofman, and J. J. Lagendijk. 2003. Clinical feasibility study for the use of implanted gold seeds in the prostate as reliable positioning markers during megavoltage irradiation. *Radiother. Oncol.* 67 (3): 295–302.

Kotte, A. N., P. Hofman, J. J. Lagendijk, M. van Vulpen, and U. A. van der Heide. 2007. Intrafraction motion of the prostate during external-beam radiation therapy: Analysis of 427 patients with implanted fiducial markers. *Int. J. Radiat. Oncol. Biol. Phys.* 69 (2): 419–425.

Lagendijk, J. J., B. W. Raaymakers, A. J. Raaijmakers, J. Overweg, K. J. Brown, E. M. Kerkhof, R. W. van der Put, B. Hardemark, M. van Vulpen, and U. A. van der Heide. 2008. MRI/linac integration. *Radiother. Oncol.* 86 (1): 25–29.

Lotz, H. T., M. van Herk, A. Betgen, F. Pos, J. V. Lebesque, and P. Remeijer. 2005. Reproducibility of the bladder shape and bladder shape changes during filling. *Med. Phys.* 32 (8): 2590–2597.

Nederveen, A. J., H. Dehnad, U. A. van der Heide, R. J. van Moorselaar, P. Hofman, and J. J. Lagendijk. 2003. Comparison of megavoltage position verification for prostate irradiation based on bony anatomy and implanted fiducials. *Radiother. Oncol.* 68 (1): 81–88.

Nijkamp, J., R. de Jong, J. J. Sonke, P. Remeijer, C. van Vliet, and C. Marijnen. 2009. Target volume shape variation during hypo-fractionated preoperative irradiation of rectal cancer patients. *Radiother. Oncol.* 92 (2): 202–209.

Smitsmans, M. H., J. de Bois, J. J. Sonke, A. Betgen, L. J. Zijp, D. A. Jaffray, J. V. Lebesque, and M. van Herk. 2005. Automatic prostate localization on cone-beam CT scans for high precision image-guided radiotherapy. *Int. J. Radiat. Oncol. Biol. Phys.* 63 (4): 975–984.

Stroom, J. C., H. C. de Boer, H. Huizenga, and A. G. Visser. 1999. Inclusion of geometrical uncertainties in radiotherapy treatment planning by means of coverage probability. *Int. J. Radiat. Oncol. Biol. Phys.* 43 (4): 905–919.

van de Bunt, L., I. M. Jurgenliemk-Schulz, G. A. de Kort, J. M. Roesink, R. J. Tersteeg, and U. A. van der Heide. 2008. Motion and deformation of the target volumes during IMRT for cervical cancer: What margins do we need? *Radiother. Oncol.* 88 (2): 233–240.

van der Heide, U. A., A. N. Kotte, H. Dehnad, P. Hofman, J. J. Lagenijk, and M. van Vulpen. 2007. Analysis of fiducial marker-based position verification in the external beam radiotherapy of patients with prostate cancer. *Radiother. Oncol.* 82 (1): 38–45.

van Haaren, P. M., A. Bel, P. Hofman, M. van Vulpen, A. N. Kotte, and U. A. van der Heide. 2009. Influence of daily setup measurements and corrections on the estimated delivered dose during IMRT treatment of prostate cancer patients. *Radiother. Oncol.* 90 (3): 291–298.

van Herk, M., P. Remeijer, C. Rasch, and J. V. Lebesque. 2000. The probability of correct target dosage: Dose-population histograms for deriving treatment margins in radiotherapy. *Int. J. Radiat. Oncol. Biol. Phys.* 47 (4): 1121–1135.

van Kranen, S., S. van Beek, C. Rasch, M. van Herk, and J. J. Sonke. 2009. Setup uncertainties of anatomical sub-regions in head-and-neck cancer patients after offline CBCT guidance. *Int. J. Radiat. Oncol. Biol. Phys.* 73 (5): 1566–1573.

Motion Management

Geoffrey D. Hugo
Virginia Commonwealth
University

Introduction

"Motion" of the intended radiotherapy target can be broadly separated into two categories: interfraction positional error of the bony anatomy, termed *setup error*; and shifting of the target or other anatomy in relation to the bony anatomy, termed *internal motion*. The focus of this chapter is the latter, specifically, the intrafraction motion of soft tissue anatomy relative to bony anatomy. Intrafraction motion management for respiratory-influenced sites in the thorax and upper abdomen is most developed and will be the focus of this chapter.

AAPM Task Group 76, "The Management of Respiratory Motion in Radiation Oncology" is a useful reference providing an overview of the many available motion-management techniques (Keall et al. 2006). Each method requires specific equipment varying in technological sophistication and relies on particular assumptions, and can be implemented with various combinations of technologies. As such, it is impossible to broadly prescribe a quality control process for respiration management that is applicable to all such systems and clinics. Instead, this chapter is intended to serve as an example of quality control process development and implementation using a clinical respiration gating process as an example strategy.

Quality Control for Respiration Gating

Motion Management Technique Selection

A solitary motion management strategy does not exist that is suitable for every site, patient, and treatment goal. For this reason, it is useful to have several motion management strategies available. Even for broadly applicable strategies, a clinic should develop (at least) one secondary strategy as a backup. For example, a patient could develop comorbidities that preclude further use of the primary strategy, or required equipment could fail. A set of fixed criteria should be used to stratify patients between strategies. Example criteria include site; treatment intent, fractionation scheme, and

modality; range of tumor motion; patient compliance; tumor volume and limitations in prescription and normal tissue dose.

Respiration Gating Clinical Process

In respiratory gating, the tumor position is measured during the respiration cycle. Assuming the tumor position reliably returns to a fixed position at a certain respiration state, the radiation beam can be "gated" to this respiration state such that the beam is on only when the tumor is at the fixed reference position. Because of the difficulty in directly measuring the tumor position in real time, surrogate signals such as spirometry, optical tracking of the patient surface, or external markers are used. A significant portion of the quality control process is designed to verify and update the strength of the relationship between the surrogate and the tumor position.

Simulation

Because of the large variability in tumor motion between patients, the range of tumor motion should be measured for each patient individually with some form of dynamic imaging. A widely available method is 4DCT; if unavailable, dynamic MRI, "slow" CT, breath-hold CT, or fluoroscopy may be used. Conventional reconstruction of 4DCT images generally assumes a reproducible, stable breathing pattern. Irregularity in breathing (from cycle to cycle) has been demonstrated to have a detrimental effect on 4DCT image quality (Keall 2004) through the introduction of artifacts in the reconstructed images. Examples of 4DCT images for regular and irregular breathing are shown in Figure 87.1. For extremely irregular breathing, 4DCT currently cannot accurately represent the dynamic anatomy, and procedures (see example below) should be in place to modify the motion management process appropriately.

A protocol such as the following can be used for quality control of the 4D simulation process:

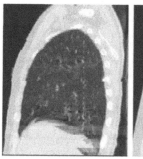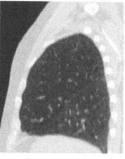

FIGURE 87.1 End-of-exhalation images from 4DCT scans of two patients. (Left) A patient with irregular breathing. (Right) A patient with regular breathing. The irregular breathing introduces artifacts in the image reconstruction (shown in the lower part of the left figure at the patient's diaphragm) due to inconsistencies in the breathing pattern between subsequent table positions.

1. Assess stability of the patient breathing signal (surrogate signal) and set appropriate 4DCT acquisition parameters.
   a. Record average duration of patient breathing period, average amplitude of breathing surrogate, and standard deviation of these quantities.
   b. Based on these values, set 4DCT scan parameters. Guidelines suggested by the manufacturer should be followed. 4DCT acquisition parameters are also discussed by Pan et al. (2004) among others.
2. Acquire 4D imaging study.
   a. Evaluate breathing signal for irregularity.
   b. Verify that all requested phase images have been reconstructed and appropriately labeled (e.g., end inhalation is labeled as the 0% phase).
   c. Visually inspect reconstructed 4DCT for irregularity-induced artifacts (Figure 87.1).
3. Evaluate acceptability of 4DCT images.
   a. If acceptable—measure the range of tumor motion.
   b. If unacceptable—follow guidelines below for irregular breathing.
4. Select motion management strategy based on fixed criteria (listed above).

To account for irregular breathing, a protocol such as the following can be used:

1. Determine if the scan was acquired with appropriate parameters to limit artifacts. Rescan with higher sampling to try to reduce artifacts.
2. If (1) fails, attempt to induce regular breathing (techniques include audiovisual biofeedback [George et al. 2006] or some form of abdominal compression). Reverify the stability of the respiration surrogate signal.
3. If (2) fails, evaluate patient compliance for alternative active-compensation strategies (e.g., breath hold). If compliant, acquire planning images consistent with the requirements of the alternative strategy.

4. If (3) fails, acquire a slow CT scan at low pitch or fluoroscopy over several breathing cycles. The patient's treatment will be planned and the patient will be treated with a motion envelope to encompass the range of motion (internal target volume).

Planning

Quality Control of Structure Delineation for Respiration Gating

A high-quality image representing the anatomy at near the average position within the gating window is selected from the 4DCT for use as the primary planning image on which to delineate organs at risk, define reference points, and calculate dose. A secondary image is used to incorporate residual target motion (which could be substantial depending on the size of the gating window) into a residual-motion internal target volume. The process for planning structure generation is listed as follows:

1. Select gating window. Generally, the window is set to 30% of the range of motion, centered on end of exhalation. Adjustments are made to narrow this window if residual tumor motion within the window is too large.
2. Based on the gating window, select a representative phase image for structure delineation and isocenter placement. For the gating window listed above, the end-of-exhalation phase is selected.
3. Create a maximum intensity projection image only for phases within the gating window. Delineate the residual-motion internal target volume on the maximum intensity projection image.

Patient-Specific Quality Assurance of Respiration-Gated Treatment Plans

In addition to standard patient-specific quality control of treatment planning, the following applies for motion management planning using this procedure:

1. Quality assurance tests should be in place to evaluate transfer and use of *derived* images such as 4DCT, maximum intensity projection, and average images.
2. A standard nomenclature for 4DCT, maximum intensity projection, and other derived images should be documented and implemented to avoid confusion (e.g., the end-of-inhalation image from a 4DCT should be defined as the 0% phase).
3. Procedures should be in place to select and verify appropriate image sets for generation of image-guidance data such as digitally reconstructed radiographs and reference CTs for CT guidance.
4. Multileaf collimator sequences should be reviewed to determine expected accuracy of delivery in the presence of residual motion. A number of studies (Hugo, Agazaryan, and Solberg 2002; Court et al. 2008; Seco et al. 2007) detail

specific recommendations regarding leaf speed, monitor units per segment, and number of segments, among other factors.

Treatment

Ample evidence exists that lung volume variation from fraction to fraction can produce large shifts between the average tumor position and the bony anatomy (Sonke, Lebesque, and van Herk 2008). Such variations can cause interfraction target localization errors, even if the surrogate signal remains in phase with the target position between fractions. Changes in the amplitude, pattern, or direction of motion may also affect the accuracy of gated delivery. Thus, motion management should be implemented along with an appropriate image guidance strategy, which can range from portal or radiographic imaging to verify the bony anatomy and a tumor surrogate such as the diaphragm to daily online cone-beam CT guidance.

For respiration gating, the important factors to evaluate are the correspondence between the surrogate and internal anatomical motion (if a surrogate is used) and the location and residual motion of the target while the surrogate is within the gating window. The simplest method to evaluate these factors—and to do so simultaneously—is to acquire localization images of the target within the gating window. The current target position can be compared to the planned target position, and residual motion of the tumor within the gating window can be compared to that measured during simulation and planning.

Examples of image-guided motion management include:

- Portal or radiographic imaging to capture the tumor position or a surrogate such as the diaphragm. Images should be captured within the gating window to allow for localization and assessment of residual motion within the window.
- Online or offline dynamic imaging. 4DCT, MRI, fluoroscopy, or 4D cone-beam CT can be used to assess changes in respiration. A method should be in place to enable quantification of the target position within the gating window. For offline assessment (such as with MRI), a method for online target localization is still required.

If it is not possible to acquire gated images for localization, care must be taken to eliminate error in the guidance process. For example, most commercial cone-beam CT systems do not allow for imaging within a gating window, but rather all imaging is performed during free breathing. If a "gated" image, such as the end-of-exhalation phase image, is used as the reference image for registration with a free breathing image, a systematic error could be introduced into the localization process. Instead, for this type of localization, the estimated end-of-exhalation position of the target from the free-breathing cone-beam CT should be manually aligned with the end-of-exhalation position from the reference CT.

Routine Process Evaluation

In addition to routine quality assurance for 4D equipment, an end-to-end process evaluation should be performed for each motion management method at regular intervals. The goal of this test is to evaluate the motion management system as a whole, including all procedures and equipment. This test should be performed on a phantom with a known target object capable of holding a dosimeter and capable of simulating breathing motion. An excellent example is the Radiological Physics Center's anthropomorphic liver phantom, which has predefined targets and risk structures, and is compatible with localization imaging. An example end-to-end test for respiration gating is provided below:

- Acquire a 4DCT scan of the dynamic phantom using the clinical protocol. Many commercially available phantoms can be used with a surrogate tracking system such as an external optical marker system. If compatible, the dynamic phantom should be interfaced with the surrogate tracking system and the surrogate used to sort the 4DCT using the clinical methods.
- Evaluate the reconstructed images for accuracy of target position and shape.
- Transfer 4DCT images to the treatment planning system following the clinical protocol and delineate the target object and any risk structures. Create a gated treatment plan. Transfer the plan and any required localization images to the treatment system.
- Set up the dynamic phantom using the clinical localization procedure, including localization imaging, while the phantom is in motion. Place dosimeters in the phantom and deliver the treatment plan with gated delivery while the phantom is in motion. Evaluate the clinical localization accuracy and dosimetric accuracy.

References

Court, L. E., M. Wagar, D. Ionascu, R. Berbeco, and L. Chin. 2008. Management of the interplay effect when using dynamic MLC sequences to treat moving targets. *Med. Phys.* 35 (5): 1926–1931.

George, R., T. D. Chung, S. S. Vedam, V. Ramakrishnan, R. Mohan, E. Weiss, and P. J. Keall. 2006. Audio-visual biofeedback for respiratory-gated radiotherapy: Impact of audio instruction and audio-visual biofeedback on respiratory-gated radiotherapy. *Int. J. Radiat. Oncol. Biol. Phys.* 65 (3): 924–933.

Hugo, G. D., N. Agazaryan, and T. D. Solberg. 2002. An evaluation of gating window size, delivery method, and composite field dosimetry of respiratory-gated IMRT. *Med. Phys.* 29 (11): 2517–2525.

Keall, P. 2004. 4-Dimensional computed tomography imaging and treatment planning. *Semin. Radiat. Oncol.* 14 (1): 81–90.

Keall, P. J., G. S. Mageras, J. M. Balter, R. S. Emery, K. M. Forster, S. B. Jiang, J. M. Kapatoes et al. 2006. The management of respiratory motion in radiation oncology report of AAPM Task Group 76. *Med. Phys.* 33 (10): 3874–3900.

Pan, T., T. Y. Lee, E. Rietzel, and G. T. Chen. 2004. 4D-CT imaging of a volume influenced by respiratory motion on multi-slice CT. *Med. Phys.* 31 (2): 333–340.

Seco, J., G. C. Sharp, J. Turcotte, D. Gierga, T. Bortfeld, and H. Paganetti. 2007. Effects of organ motion on IMRT treatments with segments of few monitor units. *Med. Phys.* 34 (3): 923–934.

Sonke, J. J., J. Lebesque, and M. van Herk. 2008. Variability of four-dimensional computed tomography patient models. *Int. J. Radiat. Oncol. Biol. Phys.* 70 (2): 590–598.

88

Total Skin Electron Therapy

Antonella Bufacchi
*S. Giovanni Calibita
Fatebenefratelli Hospital*

Introduction

Development of any total skin electron therapy (TSET) program is heavily dependent on the specific technique chosen and on the particular equipment on which it is carried out. The techniques are often complex and time-consuming to develop and to execute on a routine basis. A rigorous quality assurance program should be an integral part of a TSET program, particularly because high electron dose rates at isocenter are used to minimize the treatment time in a plane some meters from the isocenter. This requires that the accelerator operates at beam currents comparable to those used in x-ray therapy, which are greater than those required for small-field and 100-cm SSD electron treatments. Consequently, special attention to safety measures such as interlocks, beam monitoring, and so on, are necessary.

It is recommended that a local, written procedure be provided for changing from conventional modalities to TSET and vice versa and it should be available at the console. Whether a medical physicist should be present for TSET treatments depends on the complexity of the procedure and the relevant staff's experience in using it. However, it is the responsibility of the medical physicist to work on commissioning, beam data measurements,

calibration, and routine quality assurance (AAPM Task Group 1988).

Routine quality control objectives must be consistent with the measures that have been performed during the commissioning of such an irradiation technique. In vivo dosimetry measurements are important for TSET to determine the distribution of the dose to the patient's skin and to verify that the prescribed dose is correctly administered. Several types of dosimeters may be considered for use in these measurements: diodes, film, and thermoluminescent dosimeters (TLDs).

The quality assurance program developed in our center regarding the in vivo dosimetry measurements is provided in the following section. Figure 88.1 shows the TSET geometry used in our center and Table 88.1 describes the tests performed during TSET commissioning and as part of our quality assurance program.

In Vivo Dose Measurements

In vivo dosimetry is recommended for each patient at the beginning of the treatment to verify the absorbed dose calibration and calculate accurately the monitor units required to deliver the

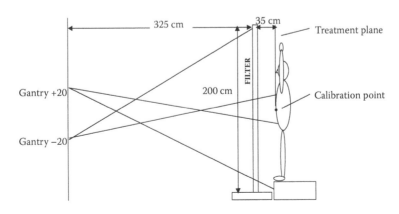

FIGURE 88.1 TSET geometry.

TABLE 88.1 Description of TSET QA Tests, Their Frequency, and Tolerance Levels

Description of Test	Frequency	Performance Tolerance	Action
Output constancy—electrons			
Electron beam used for TSET with the configuration of the accelerator console on TSET modality. A practical dosimetry device is used to provide the reading at SSD = 100 cm for gantry angle of 0° and MUs equal to 100 and 200.	Daily—at the beginning of the treatments and separated by at least 12 h[a].	±2%	±3%
Output reproducibility in TSET geometry			
A parallel-plate ionization chamber is positioned at the *calibration point* with the reference point chosen at the maximum depth dose. The absorbed dose is performed in accordance with IAEA TRS n. 398 (IAEA 2000).	Weekly—on average once every 7 days and at intervals of between 5 and 9 days.	±2%	±3%
Depth dose reproducibility in TSET geometry			
Depth dose is acquired with a parallel-plate ionization chamber overlaid with varying thickness of an appropriate water equivalent phantom. The practical range R_p and the depth at which the dose reaches its 50% maximum value, that is, R_{50}, are estimated and compared with those measured at commissioning.	Six-monthly—on average once every 6 months and at intervals of between 4 and 8 months.	±2 mm	±3 mm
Beam flatness in TSET geometry			
Films are used and positioned on a plane corresponding to the *treatment plane*. The values measured along vertical and horizontal profiles are compared with those measured at commissioning.	Annual—on average once every 12 months and at intervals of between 10 and 14 months.	±3%	±4%

[a] Output constancy check: a daily instrument reading (corrected for temperature and pressure) taken under reproducible geometrical conditions designed to check that the radiation output (e.g., cGy/UM) value in TSE modality use is not grossly in error.

prescribed dose. At least two other phases of measurements are performed during the entire treatment to identify areas requiring local boost fields.

Dosimeters

Historically, the measurements are performed with TLDs. Recently, however, EBT Gafchromic film has also been used (Bufacchi et al. 2007). To perform a comparison, at the beginning of the treatment, the simultaneous use of TLD and EBT films is recommended.

Calibration

The calibration of the TLD and EBT films is performed against a parallel-plate ion chamber with the electron energy used for the treatment and in TSET geometry.

Mapping of the Skin Dose

Figure 88.2 shows the 20 different points on the patient's body where the dose is measured for our TSET protocol. At the beginning of the treatment for each patient, the average of the six

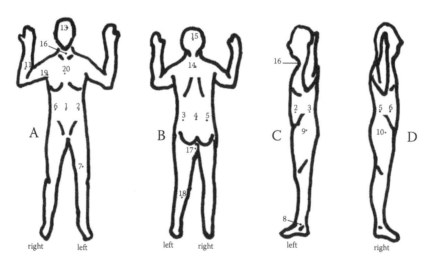

FIGURE 88.2 Twenty different points on the patient's body where the dose is measured for TSET.

TABLE 88.2 Expected Precision of In Vivo TLD Readings for TSET

	Expected Standard Deviation
Standard deviation calculated on the six readings on the belt	4%
Standard deviation calculated on the remaining 14 readings	10%

readings on the belt is obtained to verify the absorbed dose calibration and accurately calculate the monitor units required to deliver the prescribed dose. The belt is shown in Figure 88.2a as points 1, 2, and 6; in Figure 88.2b as points 3, 4, and 5; in Figure 88.2c as points 2 and 3; and in Figure 88.2d as points 5 and 6. If one uses TLDs as the in vivo dosimeter, the expected precision of the TLD readings is shown in Table 88.2.

The results of the mapping of the skin dose must be written in a dated report that is part of the patient's documentation.

References

AAPM Task Group. 1988. Total skin electron therapy: Technique and dosimetry. In *Report 23*. New York, NY: American Institute of Physics.

Bufacchi, A., A. Carosi, N. Adorante, S. Delle Canne, T. Malatesta, R. Capparella, R. Fragomeni, A. Bonanni, M. Leone, L. Marmiroli, and L. Begnozzi. 2007. In vivo EBT radiochromic film dosimetry of electron beam for total skin electron therapy (TSET). *Phys. Med.* 23 (2): 67–72.

IAEA. 2000. Absorbed dose determination in external beam radiotherapy: An international code of practice for dosimetry based on standards of absorbed dose to water. In *Technical Reports Series No. 398*. Vienna: International Atomic Energy Agency (IAEA).

89

Total Body Irradiation and Intensity-Modulated Total Marrow Irradiation

Bulent Aydogan
The University of Chicago

Introduction

Total body irradiation (TBI) has been used in conditioning regimens prior to bone marrow transplantation to provide immunosuppression and additional malignant cell kill for the chemotherapy inaccessible sanctuary sites. Bone marrow transplant protocols, in general, use supralethal doses of both chemotherapy and radiation, inducing major normal tissue toxicity. Many clinical investigations have been conducted with the goal of devising less toxic therapy techniques while achieving prolonged remission. However, lung toxicity still continues to be the major dose-limiting complication, particularly with therapy regimens that include TBI. This prevents the delivery of higher prescription doses that are suggested to reduce the relapse rate after bone marrow transplant (Clift et al. 1990, 1991). There is a renewed interest in adjuvant radiation therapy in bone marrow transplant protocols in the form of total marrow irradiation. Early studies (Wilkie et al. 2008; Hui et al. 2005) demonstrated that an intensity-modulated radiation therapy (IMRT) irradiation technique may allow higher doses to the bone marrow while reducing the dose to the surrounding organs at risk (OARs). Large-field irradiations such as TBI and total marrow irradiation are very challenging and are subject to many technical limitations and requirements. A brief review of these limitations and requirements will be presented first.

Analysis of dose response data and evaluation of errors in radiotherapy dose delivery concluded that an overall accuracy of ±5% in dose delivery is required in radiotherapy (ICRU 1976). This raises the question of what accuracy can be achieved with

TBI treatments. Associated uncertainties in absolute dosimetry and lack of reliable in vivo dosimetry limit developing a reasonable dosimetric quality assurance program for TBI applications. Nevertheless, the 5% suggested accuracy per "as precise as readily achievable" principle (Van Dyk et al. 1986).

In addition, TBI should deliver a uniform dose within ±10% of the prescribed dose (Van Dyk et al. 1986). Delivery of homogenous dose distribution through the whole body is very challenging with TBI. Dose inhomogeneity in TBI is inevitable because of the following factors: (1) irregular body thickness over the irradiated volume causes the midline dose to differ; (2) variation in the tissue density along the beam axis changes the photon flux at a given point increasing the dose in the less dense parts of the body such as lungs; and (3) summation of entrance and exit beam doses causes the radiation dose near the skin to be much larger than the midline dose (Van Dyk et al. 1986). A thorough understanding of these limitations helps in devising improved methods to reduce dose inhomogeneity. These methods include but are not limited to (1) treating at extended distance with higher-energy photons to reduce the shallow depth-to-midline dose ratio. This may have the disadvantage of reducing the dose to skin and can be offset by using beam spoilers. (2) Compensators can be used to offset the missing tissue or tissue density changes in the irradiated volume. It should be noted that each method has advantages and disadvantages and can be tailored to fit existing limitations of the institution. There exists a wide range of TBI techniques used in bone marrow transplant protocols making it very challenging to provide detailed quality assurance guidelines. Therefore, in this work we will only discuss the

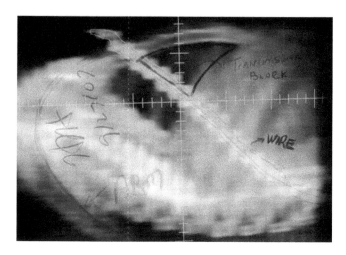

FIGURE 89.1 Port film showing the lung, the wire, the arm and the partial transmission block.

opposed lateral TBI and linac-based intensity-modulated total marrow irradiation (IMTMI) techniques that have been used in our center.

Total Body Irradiation

In our institution, an adapted bilateral irradiation technique using a custom-made TBI table at a source-axis distance of 410 cm is used. To improve the dose homogeneity, 18-MV x-ray is preferred in order to improve dose homogeneity in TBI treatment.

Simulation

The patient is simulated in the supine position with the arms partially blocking the lungs and hands under the buttocks for a better reproducibility. The anterior contours of both arms are defined with wires. A CT scan of the patient is then obtained using a 1-cm slice thickness. CT image data are then transferred and a 3D image is reconstructed and DRRs for patient setup are created using an Eclipse treatment planning system (Varian Medical Systems, Palo Alto, CA). The lungs, arms, and wires are contoured and the shape of the lung block is drawn on the DRR shown in Figure 89.1. The contours of arms, wires, and lungs simplify, considerably, the process of defining the shape and the size of the blocks to protect the unshielded part of the lungs. Body separations at the level of shoulder, chest, pelvis, and umbilicus are measured. MU calculation is performed with the average separation and the corresponding 18-MV tissue/phantom ratio value obtained during the commissioning at the treatment distance (source-axis distance) of 410 cm.

Lung Block

Our TBI protocol allows a dose up to 1000 cGy to the lungs. A partial transmission block to be used in every fraction is designed to reduce the lung dose to less than the permissible lung dose. The use of partial transmission blocks in every fraction may be expected to reduce overall error due to misplacement or motion during the treatment due to averaging effect. The lung block is designed to provide approximately 83% transmission and its thickness depends on the amount of tissue and lung present in the beam's path. For instance, for an arm-to-arm width of 44.5 cm and lung thickness of 24 cm, a lung block with 2-mm lead and 3.5-cm wax will provide the desired 83% transmission for 18-MV x-rays. Wax is used to prevent skin reaction from the forward scattered electrons created by the lead.

Treatment Setup

An custom made TBI treatment table shown in Figure 89.2 has been used in our center. The dimensions of the table are 52 and

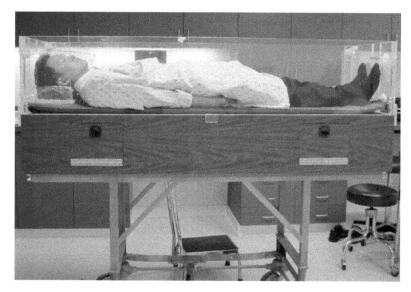

FIGURE 89.2 Example patient setup using the custom-made TBI table, lung blocks to allow partial transmission, and rice bags to even out irregular body contour.

180 cm in width and length, respectively and it is raised to a fixed height of 110 cm from the floor. The TBI table consists of 1.5-cm lucite windows and extra lucite head jigs as spoilers and a 2-cm-thick table pad to provide additional comfort for the patient during the long treatment process. Storage space under the table is to conveniently keep the rice bags when they are not in use. The H&N jig is used both to add extra bolus on the outside and to hold rice bags that used to even out the irregular body contour around the head and neck (H&N) region. H&N jigs are adjusted to the level of the shoulder as the arms are considered to be natural shielding for the lungs and not part of the body. Figure 89.2 shows the treatment setup with the rice bags compactly piled around the H&N down to the shoulder and legs up to the mid femur providing approximately the same depth for the whole treatment volume. The table is placed at 410 cm source-axis distance from the umbilicus level identified by an extra sagittal laser attached to the wall for TBI purposes. This laser is checked by measuring the distance from the wall with a custom-made measuring rod and verified to be within 1 mm of the commissioning value before each treatment. The next step is the verification of the lung block position. A port film is taken after placing the lung block for the side that will be treated. The lung block and the arm position are then adjusted using the port film to provide desired shielding. Small adjustments on the block shape and size are occasionally needed for the correct positioning to provide optimal blocking of the lungs.

In Vivo Dosimetry

Thermoluminescent dosimeters (TLDs) placed on the patient's skin at different levels are used to verify the dose homogeneity. Three TLDs are placed on the umbilicus, both sides (i.e., right and left) of the neck, under the block, inner thighs about the mid-thighs, and inner ankles. The TLDs placed on the umbilicus are covered with 2 cm of bolus to improve the accuracy of the readings. The average values of two TLD readings per each location and the calibration reading obtained under a known setup condition are used to calculate the entrance and the exit skin dose. These two readings are expected to be within ±5% of the prescription dose unless the packing is not uniform. If this is the case, the TLD measurement should be repeated at the next fraction paying more attention to packing.

Intensity-Modulated Total Marrow Irradiation

One of the most important components of all radiation treatments is a reliable and reproducible immobilization and verification method. For IMTMI, an ideal immobilization device would be a fully indexed whole body frame similar to those used for stereotactic body radiosurgery. Such a device is in development at our institution. In addition, because the IMTMI plan is composed of three separate IMRT plans, extra care is needed for

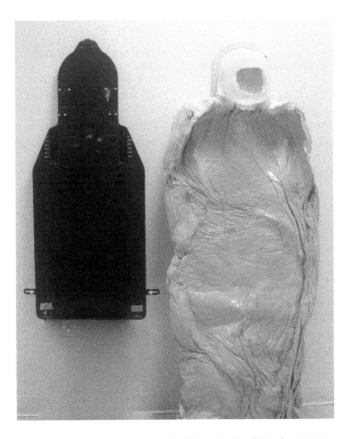

FIGURE 89.3 Components of IMTMI immobilization system: CDR H&N board with table index, H&N aquamask, and full-body alpha cradle.

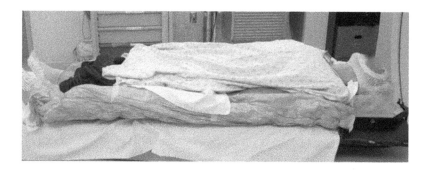

FIGURE 89.4 Patient setup with the IMTMI immobilization system.

correct setup and delivery. A small patient rotation can have a large effect on the doses delivered to the target and critical structures. For instance, a 1° rotation at the level of the neck may easily result in a 1-cm translation in the pelvis (Aydogan, Mundt, and Roeske 2006). This small rotation may overdose critical structures while decreasing the desired coverage of the target. Such rotations may not be evident with portal imaging, but CBCT may be used to properly address it. CBCT not only may improve the accuracy of the setup but also may reduce the time it takes to set up and verify the treatment fields. We currently immobilize IMTMI patients using a full body alpha cradle indexed to the treatment table with a head and neck board (CDR Systems, Calagary Alberta Canada) shown in Figure 89.3. Treatment verification is done by comparing the DRR created from the planning CT and the port films obtained using an electronic portal imaging device.

Simulation

A custom made indexed immobilization jig and bars are used in combination with a full body alpha cradle and CDR H&N board to immobilize and to index the patient to the table (Figure 89.4). An aquamask is used to immobilize the H&N as a part of the CDR system. A scout CT scan is first taken to straighten the patient, isocenters are then placed based on the length of the treatment volume to be covered. A planning CT scan with a maximum of 5 mm slice thickness covering the treatment volume from the crown to the mid femur is acquired for treatment planning.

Planning

IMTMI consists of three IMRT plans, one to treat the bones of the H&N region, one for the thoracic region, and another one for the pelvis down to mid femur. These three plans are electronically feathered using the base-plan function in Eclipse (Varian Medical Systems). Because of complexity of the target and OAR shapes and locations, IMTMI plans are, in general, highly modulated and the dose gradient is observed to be very steep to provide good target coverage while protecting OARs. As such, it is very common to have small areas of hot spots in the order of

30–40%. This can be acceptable as long as they are neither within nor near an OAR. Each axial slice should be reviewed very carefully to identify hot and cold spots. As a last resort minimal fluence editing can be used to reduce hot or cold spots. IMTMI being essentially an IMRT technique, all the relevant IMRT quality assurance discussions presented in Chapters 92 and 93 apply to the individual IMTMI plans as well. For instance, each IMRT field should be delivered and verified using an appropriate dosimetric method. Currently, we use MATRIXX (IBA Dosimetry, Bartlett, TN), a 2D ion chamber array for IMTMI plan verification at our institution. Our IMTMI verification QA standard is 95% gamma analysis pass rate with the 3% dose and 3mm distance-to-agreement criteria. An example MATRIXX dosimetry report for an IMTMI treatment field is shown in Figure 89.5.

Isocenter Verification and Treatment

A few days before the treatment, isocenter placement and treatment verification is performed. Three isocenters placed during the CT simulation serve as the surrogates to verify the relative position of external marks with respect to the treatment isocenters defined during the planning process. Generally small

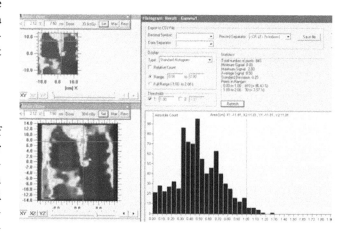

FIGURE 89.5 Example MATRIXX dosimetry report for an IMTMI treatment field.

adjustments are required and permanent external marks are placed on the patient skin. One common reason for adjustment in the order of 3 mm in the vertical direction is the table sag when the table is fully extended to treat the pelvic field. Before the treatment, junction plans may be verified with respect to each other to ensure the setup, specifically the rotational (yaw) accuracy of the whole treatment volume in the absence of CBCT capability. For instance, verifying the thorax isocenter right after verifying the H&N and returning back and reverifying the H&N before the treatment should provide a good degree of accuracy in the setup. Similarly the pelvis may be verified with respect to the thorax before the treatment. Our current IMTMI protocol (http://clinicaltrials.gov/ct2/show/NCT00988013.) does not require the treatment of the legs. Nevertheless, legs can easily be treated with AP/PA open fields clinically matched with the pelvic IMRT fields.

References

Aydogan, B., A. J. Mundt, and J. C. Roeske. 2006. Linac-based intensity modulated total marrow irradiation (IM-TMI). *Technol. Cancer Res. Treat.* 5 (5): 513–519.

Clift, R. A., C. D. Buckner, F. R. Appelbaum, S. I. Bearman, F. B. Petersen, L. D. Fisher, C. Anasetti et al. 1990. Allogeneic marrow transplantation in patients with acute myeloid leukemia in first remission: A randomized trial of two irradiation regimens. *Blood* 76 (9): 1867–1871.

Clift, R. A., C. D. Buckner, F. R. Appelbaum, E. Bryant, S. I. Bearman, F. B. Petersen, L. D. Fisher et al. 1991. Allogeneic marrow transplantation in patients with chronic myeloid leukemia in the chronic phase: A randomized trial of two irradiation regimens. *Blood* 77 (8): 1660–1665.

Hui, S. K., J. Kapatoes, J. Fowler, D. Henderson, G. Olivera, R. R. Manon, B. Gerbi, T. R. Mackie, and J. S. Welsh. 2005. Feasibility study of helical tomotherapy for total body or total marrow irradiation. *Med. Phys.* 32 (10): 3214–3224.

ICRU. 1976. *Determination of Absorbed Dose in a Patient Irradiated by Beams of X or Gamma Rays in Radiotherapy, Report No. 24*. International Commision on Radiation Units and Measurements. Washington, DC.

Van Dyk, J., M. Galvin, G. Glasgow, and E. B. Podgoršak. 1986. *The Physical Aspects of Total and a Half Body Photon Irradiation, AAPM Report 17*. New York, NY: American Institute of Physics.

Wilkie, J. R., H. Tiryaki, B. D. Smith, J. C. Roeske, J. A. Radosevich, and B. Aydogan. 2008. Feasibility study for linac-based intensity modulated total marrow irradiation. *Med. Phys.* 35 (12): 5609–5618.

90

Tomotherapy

Alexander Usynin
Thompson Cancer Survival Center

Chester Ramsey
Thompson Cancer Survival Center

Introduction

The current standard of practice for helical tomotherapy (HT) is to perform dosimetric verification for each patient by performing patient-specific quality assurance (QA) for treatment delivery. Each institution sets up its own implementation of the QA program based on available equipment and established preferences. This chapter will describe the phantom-based approach that has been adopted in our institution.

Being a highly conformal intensity-modulated radiation therapy technique, HT imposes multiple requirements for various aspects of a particular implementation of the intensity-modulated radiation therapy QA program. Table 90.1 summarizes the major components of the HT QA program established in our institution.

This chapter leaves the aspects of the delivery system QA out of scope and focuses on the patient-specific QA component. Functionally, any tomotherapy patient-specific QA procedure consists of three sequentially performed steps: (1) calculate the reference dose map using the patient treatment plan that has been approved for delivery and the CT scan of the phantom; (2) deliver the radiation fluence to the phantom; (3) assess the agreement between the calculated dose map and the measured one. In the following sections, these steps will be discussed in a thorough manner.

Planning the HT Patient-Specific QA Procedure

Given an HT patient treatment plan, for which the QA procedure should be performed, the physicist picks a QA phantom that is most appropriate for the given anatomical site. The criterion of appropriateness may depend on how well the phantom geometry mimics the patient geometry at the anatomical site. The choice of the phantom also may be dictated by the spatial characteristics of the radiation fluence maps. In practice, the most universal approach applied by HT users is the cylindrical "cheese" phantom, whose shape and dimensions are well suited for most anatomical sites.

The only deficiency associated with the usage of the cylindrical QA phantom is that the relative dosimetry check is limited to the use of film only. The phantom is not designed to be used with an ionization chamber array, although the phantom has multiple channels for miniature thimble chambers.

A CT scan of the phantom should be taken and imported into the tomotherapy treatment planning system (TPS). The spatial orientation of the QA phantom should remain the same as if the phantom was placed onto the HI-ART unit treatment table for radiation delivery. The imported CT scan of the phantom will be used for calculation of the phantom dose maps in conducting the institutional HT patient-specific QA program.

Having decided what phantom to use, the physicist picks the CT density curve that sets the relationship between the Hounsfield units found in the phantom CT scan and the phantom's electron density. Special attention should be paid to the calibration of the CT density curve because any CT density curve miscalibration would adversely affect every single treatment delivery plan created in the tomotherapy TPS.

The physicist determines what portion of the target volume will be of particular interest in terms of QA criteria. In addition, the physicist identifies all organs at risk whose dose should be checked as well in the course of the QA routine. For example, in performing patient-specific QA for a cranial case, it is important to check the dose agreement for the following normal structures: the optic chiasm, optic nerves, pituitary gland, and brain stem.

The tomotherapy TPS allows the user to freely move the virtual phantom (the CT scan of the phantom) across the field of view. The physicist should place and orient the virtual phantom at an appropriate location relative to the organs at risk and the target volume. Usually one needs to only translate the virtual phantom rather than rotate the phantom to preserve the phantom orientation relative to the treatment table.

TABLE 90.1 Major Components of the HT IMRT QA Program

HT IMRT QA			
Delivery System QA		Patient-Specific QA	
Treatment delivery	Treatment imaging	Absolute dosimetry	Relative dosimetry
–Mechanical QA	–Spatial/geometry QA	Measurement at single or	Measurement of the dose distributions
–Beam parameter QA	–Image quality tests	multiple points in a QA	delivered to a plane in a QA phantom
–TG-51 calibration	–MVCT dosimetry QA	phantom	
–Synchrony tests			

Note: IMRT, intensity-modulated radiation therapy; MVCT, megaelectron voltage computed tomography.

The phantom should be placed such that the most important organs at risk are enclosed in the virtual phantom volume. The dose planes that will be used for the relative dosimetry checks are selected such that they include a representative cross section of the target volume and, if possible, the adjacent normal structures. However, oftentimes the physicist is limited to only one dose plane available for measurement at a time, for example, in the case of using a chamber array. In this situation the physicist has to repeat the QA delivery procedure for different locations of the chamber array as many times as it needs to be done to ensure a good dose agreement for the important normal structures.

The physicist creates a reference point that will aid in placing the QA phantom onto the HI-ART unit treatment table. The reference point is defined using the movable red lasers that are set according to certain fiducial marks on the phantom's surface. For example, in the case of the tomotherapy cylindrical phantom, the red lasers are set such that they equally dissect the phantom in the coronal, sagittal, and axial planes (see Figure 90.1).

The physicist picks an appropriate voxel size for the dose calculation grid, which varies from coarse to fine. The dose calculation is performed without changing dosimetrically important parameters found in the original patient plan approved for the treatment delivery. The beamlets that have been optimized for the given treatment plan are cast onto the virtual phantom.

It is recommended to use the fine calculation grid for treatment plans whose dose maps have a great deal of high dose gradient regions. In other cases the normal grid will suffice. In addition, it is important to match the dose voxel size and the resolution of the radiosensitive imaging device that will be used in the QA procedure delivery (film, ionization chamber array). From the experience in our institution, it can be concluded that an ionization chamber array such as the I'mRT MatriXX (IBA Dosimetry, Germany) provides the spatial resolution that is inferior to the voxel size of the fine calculation grid used in the tomotherapy TPS. As such, the use of the fine calculation grid is considered to be computationally inefficient when accompanied with the use of the I'mRT MatriXX.

Upon completion of the dose calculation, the physicist creates a new QA delivery procedure that will be loaded and delivered at the tomotherapy operator console. The physicist reads off the calculated dose for the absolute dosimetry check at the point(s) of interest where the ionization chamber(s) will reside during the QA procedure delivery. Because of possible setup errors, it is recommended to account for a few voxels adjacent to the point(s) of interest to obtain the average and standard deviation of the reference dose at the point(s) of interest.

Delivering Radiation to the Phantom

In this section the aspects of the radiation fluence delivery to the QA phantom are described, which include the physical phantom setup onto the HI-ART unit treatment table, the usage of fiducial marks, and image guidance for better accuracy.

The QA phantom is physically placed at the reference location on the HI-ART unit treatment table. The phantom fiducial

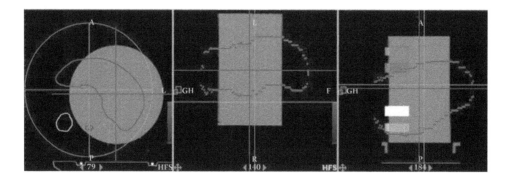

FIGURE 90.1 Axial, coronal, and sagittal projections of the cylindrical QA phantom virtually placed in the field of view of the HI-ART tomotherapy unit. The darker crosshair defines the reference point according to which the cylindrical QA phantom will be placed and oriented relative to the isocenter (the lighter crosshair). The contour is the target volume.

marks are aligned along with the red lasers according to the planned phantom setup.

The physicist performs a megaelectron-voltage computed tomography scan of the QA phantom to further improve the positional accuracy of the QA phantom setup through registering the newly obtained megaelectron-voltage computed tomography scan with the CT scan used for the QA dose calculation.

Usually only a few slices of the phantom volume are to be scanned to get an acceptably good registration. However, for more consistent registration results, one may need to scan the entire volume of the phantom. Although the use of image guidance for precise positioning of the phantom is optional, it is recommended to use it to account for possible setup errors such as those due to imperfect positioning of the phantom, misalignment of the red movable lasers, and gravitational sag of the treatment table. If, for example, the red lasers are misaligned relative to the machine true isocenter, the image-guided phantom setup is the only means to account for this type of error.

It should be noted that the radiosensitive material, such as a piece of film used for relative dosimetry, is to be placed into the phantom only after performing the megaelectron voltage computed tomography scan to avoid unnecessary exposure of the film.

The physicist applies the registration corrections to the phantom setup. Longitudinal, vertical, and roll corrections are applied automatically, whereas the lateral correction is applied manually using the in-room control panel.

Having applied the registration shifts, the physicist places the radiosensitive material, such as a piece of film, at its planned location. It is important not to disturb the phantom orientation and position when placing the measurement tools; otherwise, the megaelectron voltage computed tomography–based spatial correction may become invalid. If an ionization chamber array is used, this step is skipped.

Having the phantom and measurement tools set up at the planned position and orientation, the physicist delivers radiation to the phantom as if it was done during the regular treatment delivery. If the machine interrupts during the QA delivery, a completion procedure should be created and then delivered using the same phantom setup. The ionization chamber reading should be corrected accordingly.

Upon completion of the QA procedure delivery, the physicist reads the point dose and relative dose measurements. If an ionization chamber array is used, the accompanying software should be used to obtain the measured dose map in the format acceptable for import to the tomotherapy TPS. If a piece of film is used for the relative dosimetry check, one needs to create fiducial marks in the piece of film to register the measured dose map within the tomotherapy TPS coordinate system.

The most commonly adopted fiducial marks for film-based dosimetry are pin-pricked holes in the film envelope. The punctures are usually made along the green lasers, for example, coronal and lateral, so that the projection of the true isocenter onto the film plane can be easily identified. If an ionization chamber array is used, one needs to make sure that the dose image produced by the chamber array has fiducial marks whose spatial location can be easily determined in the ionization chamber array CT scan.

If the ionization chamber array has no distinct landmarks that correspond to the fiducial marks found in the dose map image, the spatial correspondence between the fiducial marks and the array volume should be determined empirically. For example, the I'mRT MatriXX software produces a dose map for import to the tomotherapy TPS that has two fiducial marks in opposite corners of the image (see Figure 90.2b). Each fiducial mark is a cross-like glyph depicted in either white or black in contrast to the background color. However, there is no evident correspondence between these fiducial marks and particular voxels

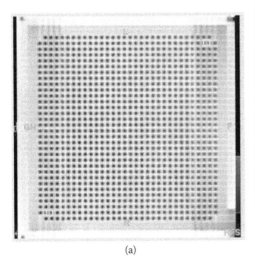

(a)   (b)

FIGURE 90.2 (a) Coronal view of the I'mRT MatriXX at the level of the ionization chambers. The yellow squares are the voxels that correspond to the fiducial marks depicted as cross-like glyphs at the right-upper and left-lower corners in the dose image (b) produced by the I'mRT MatriXX software.

inside the volume of the I'mRT MatriXX. The correspondence is determined in a "trial-and-error" manner.

It is recommended to use a mock treatment plan with an artificially defined target volume whose size and shape will allow for quick and easy registration without using fiducial marks. Having achieved a good agreement between the measured and calculated dose maps of the mock treatment plan, the physicist is able to identify the spatial correspondence between the I'mRT MatriXX volume and the cross-like fiducial marks.

Figure 90.2a shows the landmark points adopted in our institution as the counterparts of the cross-like fiducial marks found in a dose image file produced by the I'mRT MatriXX.

Agreement Assessment of Radiation Fluence Maps and Absolute Dosimetry

The physicist imports the relative dose image file and the corresponding calibration file into the tomotherapy TPS. The dose image is then converted to a dose map according to the calibration file. At present, the tomotherapy TPS allows the physicist to use nine different methods of registration, which are essentially all the possible combinations of the following parameters (see Table 90.2).

If an ionization chamber array is used for the relative dosimetry check, the easiest registration method is one that makes use of the coronal view and no laser marks (General Coronal Registration). If the ionization chamber array was set up either sagittally or transversely, the corresponding method of registration should be used.

Once the dose maps are co-registered, the physicist calculates the gamma indices and visually assesses the 2D isodose lines. The 2D isodose lines should appear in a distinct and clear manner so that one would be able to easily assess the agreement between the measured and calculated dose maps (see Figure 90.3). In most cases, three isodose lines at 95%, 75%, and 50% of the maximum dose will suffice for reliable assessment of the dose map agreement.

These parameters are usually combined into a single metric called the gamma index coefficient (Low et al. 1998). The pixels, where the gamma index exceeds the value of 1, fail to meet the distance to agreement and dose tolerance criteria. For a quantitative analysis of the dose agreement, it is recommended to use the following gamma analysis parameters: distance to agreement of 3 mm and dose tolerance of 3%. However, for QA procedures that are performed using an ionization chamber array, it is reasonable to loosen the distance to agreement and dose tolerance parameters up to 4 mm and 4%, respectively, because of the

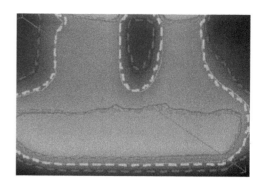

FIGURE 90.3 Example of the isodose lines that allow for an easy assessment of the dose agreement between the calculated and measured dose maps.

ionization chamber array's decreased spatial resolution relative to film.

Using a "normal" dose grid and film, the gamma analysis should yield a pass rate of 90%. Figure 90.4 shows two examples of good and poor dose agreement shown in terms of the gamma index.

As can be seen in Figure 90.4a, the gamma index map has a few spots depicted in yellow, in which the gamma index exceeds the value of 1. However, these spots are mostly outside of the target volume and are likely to be caused by random factors. Figure 90.4b shows a large number of pixels that do not meet the gamma criterion. The red spots indicate the pixels where the gamma index even exceeds the value of 2, which is considered to be fully unacceptable for tomotherapy QA.

For the absolute dosimetry check, the physicist verifies that the point dose measurement is no more than 3% different from the reference dose computed by the tomotherapy TPS. To achieve a good agreement, one needs to pick the reference dose in a low dose gradient region; otherwise, any inevitable random errors in the phantom setup may cause a substantial dose discrepancy that would invalidate the QA procedure. It is also important that the ionization chamber used for the tomotherapy QA absolute dosimetry check has little angular dependency because tomotherapy beamlets are usually coming to the target volume from the entire 360° range of gantry angles.

It is highly recommended to keep track of the ionization chamber data to establish a baseline of typical deviations observed in the HT patient-specific QA program. Usually, the ionization chamber data are expected to have a Gaussian distribution because the absolute dosimetry errors are an outcome of multiple random factors such as phantom setup errors, the linac output fluctuation, and the uncertainty attributed to the measurement tools including the ionization chamber, electrometer, thermometer, and barometer.

Figure 90.5 shows the temporal trend of the absolute dosimetry measurements observed in our institution conducting the HT patient-specific QA program. As can be seen, the mean error value does not significantly deviate from 0 ($P = 0.009$, $df = 115$), which implies that the observed absolute dosimetry errors are

TABLE 90.2 Parameters for the Film Registration Methods Found in the Tomotherapy TPS

Orientation of the Film Plane	Number of Lasers
Sagittal	None
Coronal	Three
Transverse	Four

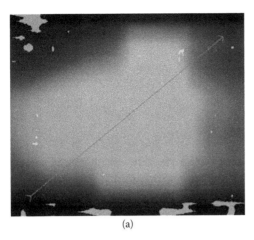 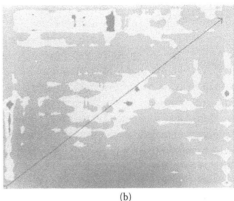

(a)   (b)

FIGURE 90.4 Example of (a) good and (b) poor agreement between the calculated and measured dose maps in terms of the gamma index.

mostly due to random factors rather than to some systematic factor such as the linac miscalibration or an error in the beam calculation model.

Another important aspect of the HT patient-specific QA program is to computationally verify the amount of radiation to be delivered to the patient. However, this quantity in the tomotherapy TPS is expressed in terms of time rather than conventional MUs. In our institution, there has been adopted a basic, yet efficient, method for computational verification of the HT beam-on time. The verification is performed according to the following equation:

$$t = \frac{G \times L}{p \times W} \tag{90.1}$$

where t is the estimated beam-on time (seconds), G is the HT gantry rotation period (seconds), L is the distance that the treatment table is to travel longitudinally during the treatment delivery (centimeters), p is the value of pitch, and W is the beam longitudinal width (centimeters). Given these parameters that

are found in the tomotherapy treatment plan report, the physicist calculates the beam-on time according to Equation (90.1) and then compares the calculated value against the treatment time produced in the tomotherapy TPS. The discrepancy is normally expected not to exceed 1%.

Summary

To conclude the chapter, the following points should be emphasized:

- The complexity of the HT treatment delivery necessitates a thorough and accurate validation of the HT treatment plans. Conducting the HT patient-specific QA program, the physicist should ensure that the delivered radiation maps are in good agreement with those computed by the tomotherapy TPS. The agreement criteria are set in terms of the absolute dose discrepancy and the relative dose agreement expressed in the gamma index values.
- The usage of image guidance in the HT patient-specific QA procedures may greatly facilitate precise positioning

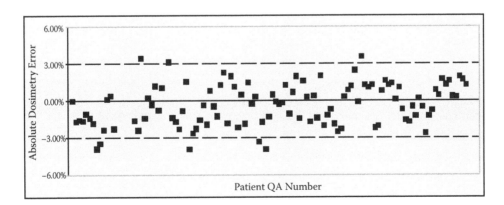

FIGURE 90.5 Temporal trend of the absolute dosimetry deviations observed in the HT patient-specific QA program. The mean value of −0.42% is not significantly different from zero ($P = 0.009$, $df = 115$).

of the QA phantom, thus providing a reliable comparison of the measured and planned dose maps.

- The QA software tool built into the tomotherapy TPS allows the physicist to seamlessly perform the QA analysis, although it limits the user to just a few methods of film registration. This limitation may lead to situations where the QA dose map registration appears to be an art rather than a well-designed routine.
- Tracking the numerical outcomes of the HT patient-specific QA procedures may help the physicist to identify in time potential faults in the tomotherapy delivery system as well as in the treatment planning computers.

- Directly performed QA measurements should be accompanied with computational methods of dosimetry verification. The usage of independent computational methods for HT patient-specific QA purposes is yet to be established in the clinical setting.

Reference

Low, D. A., W. B. Harms, S. Mutic, and J. A. Purdy. 1998. A technique for the quantitative evaluation of dose distributions. *Med. Phys.* 25 (5): 656–661.

91

CyberKnife

Cheng Yu
University of Southern California

Introduction

This report intends to provide guidelines on setting up a comprehensive quality assurance/quality control (QA/QC) program for the CyberKnife System (Accuray Inc., Sunnyvale, California). The ultimate responsibility for the QA/QC program lies with the physicist in charge of the CyberKnife System in the facility or institution. In CyberKnife radiosurgery (Adler et al. 1997), most of the treatment plans are developed using non-isocentric beams. For each patient, the directions and placements of the non-isocentric beams are unique. At present, there is no easy way for medical physicists to verify that the non-isocentric beam direction targeted by the robot is the direction designed by the planning system. Because of the unique features of the CyberKnife System and the lack of a robust and rapid method for patient-specific QA currently available for the system, performance of a patient-specific QA with experimentally measured dose distributions, like that performed with IMRT, would be very time consuming for every patient. It is recommended that such rigorous patient-specific QA procedures should be performed for the first few patients for every tracking modality and should be checked periodically afterward (Dieterich et al.).

Pretreatment Setup

Patient Immobilization

It is essential that patient alignment should replicate as closely as possible the patient position during the primary CT scan. The same head and/or body immobilization system that the patient wears during the CT scan should be used for treatment. It is important to verify that the treatment couch is free of unnecessary objects.

Scanning or Imaging Guidelines

The CyberKnife System uses two types of imaging studies: primary CT images and secondary images. The secondary images include MR, PET, 3D angiography, and additional CT scans. For intracranial lesions, all devices for patient immobilization that will be used during patient treatment must be used during CT scanning. Those devices include the headrest, CT table top, head mount, index bar, and any pads. The field of view should include the entire circumference of the skull and part of the headrest. It should also extend 1 cm anterior past the tip of the nose or forehead and 1 cm superior to the patient skull. For extracranial lesions, the patient must be centered left/right and anterior/posterior in the field of view. The lesions should be centered in the inferior/superior direction with minimum 10–15 cm scanned in either direction.

Treatment Planning

Comprehensive guidelines and recommendations for commissioning the treatment planning system can be found in the AAPM Task Group 53 (Fraass et al. 1998). This present report will focus on the unique software QA/QC issues associated with the MultiPlan System of the CyberKnife System.

Imaging Fusion

The image fusion process combines the information from two or more imaging studies into one volume. The fusion process can be done either manually or automatically. The algorithm of the automatic fusion is based on the relative intensities of the common information of interest. During the automatic intensity-based fusion, the MultiPlan System computes a transformation based on user-determined seed points on the two imaging studies. If the fusion or refinement process is successful, the final accuracy of the fusion is not affected by the initial choice of seed points. However, if the selection of the seed points differs too greatly between the two imaging studies, the refinement process will fail. In that case, the selection of the seed points needs to be adjusted and the fusion process repeated.

CT Center

The CyberKnife System uses two orthogonal kV x-ray beams to track the lesion during treatment delivery. The relationship between the machine center and the CT images must be defined correctly during treatment planning to ensure that the diagnostic x-ray system and tracking can function accurately during treatment delivery. For skull tracking, the CT image should be aligned so that the silhouette of the patient has a 10- to 15-mm gap at the superior and anterior sides of the patient anatomy in each view. For fiducial tracking, all fiducials should be within the field of view of the cameras or x-ray imagers. For Xsight spinal tracking, the CT center should be properly selected to make sure that the CT center is within 5 cm of the target and is in an area that has rich skeletal structures for image registration. The anatomical regions to be treated (i.e., cervical spine, thoracic spine, or lumbar spine) should be selected.

Density Models

The MultiPlan System provides several preset CT density models including water/air, lung standard, and body standard for treatment planning. These preset CT density models cannot be modified. A custom CT model based on electron or mass density is available for some treatment planning systems to allow the use of a custom CT density model for computation of corrections for heterogeneity. However, the user should be aware of the density type the system uses and fluctuations in Hounsfield numbers for the respective tissue types.

Inhomogeneity Corrections

The accurate delivery of the desired radiation dose to a region associated with tissue inhomogeneity is a challenging task in stereotactic radiosurgery (Wilcox and Daskalov 2008). The dose accuracy in a region such as the lung could be less than 5%. The inhomogeneity correction options available in the MultiPlan should be evaluated for their respective accuracy on a phantom.

An appropriate inhomogeneity model for an anatomical location should be chosen.

Phantom Overlay Plan

Patient-specific QA, or verification between a phantom overlay and treatment delivery, should be performed for a patient-like plan when the CyberKnife System is newly installed, or as needed. The phantom overlay mode with the MultiPlan System creates a new phantom overlay plan and enables the CyberKnife System to deliver the overlay plan to a phantom to perform dose distribution analysis. The orientation and number of monitor units for each beam are the same for the phantom overlay plan as for the original patient plan. When a phantom overlay plan is created, the beam data and treatment parameters from the original patient plan are copied to the new phantom overlay plan. However, the volumes of interest and fusion information are not copied to the new plan.

Fine Tuning

The Finetune feature with the MultiPlan System can be used to review existing treatment beams and to make detailed changes to certain properties of the beams, such as removing beams with small monitor units or beams intersecting critical structures. After making changes, the MultiPlan System checks certain conditions and displays appropriate error or warning messages. The dose distribution, based on the current beam data, needs to be recalculated.

Independent MU Checks

Independent MU checks should be performed for all beams developed by the MultiPlan System. A typical CyberKnife plan

TABLE 91.1   CyberKnife Sample MU Calculation Check

Name	Value
Total number of nonzero beams	144
Beam n	79
u (collimator size)	7.5
SAD (mm)	814.8
d (CAX depth)	85.8
d' (radiological depth)	94.3
r (off-axis radius at CalibSAD)	0.54
InvSqr $\cdot$ TPR $\cdot d_m$	0.533
Dose (cGy)	107.35
MU	201.42
InvSqr = (CalibSAD^2/SAD^2)	0.964
TPR(u,d')	0.660
OF(u)	0.836
InvSqr $\cdot$ TPR $\cdot$ OF	0.532
OCR (u,r,d)	0.997
MU = dose/(TPR $\cdot$ OF $\cdot$ OAR $\cdot$ InvSqr)	202.44
Ratio (MU/MU from plan)	1.005

consists of a couple hundred beams. An in-house developed MU check program or commercially available program capable of extracting beam data from the treatment plan file directly and verifying MUs for all beams at once would be ideal. The difference for any beam should be less than 5%. Table 91.1 shows an example of an MU check.

Patient Safety

Patient Safety Zone

The proximity detection program (PDP) monitors movements of the treatment couch and the treatment manipulator or the robot to prevent collisions during treatment. The PDP model is based on specific room geometry and CyberKnife System components. There is no temporary object allowed inside the treatment room. The PDP program sets a patient safety zone to determine the boundaries within which the patient must be positioned for safe treatment. During the process of treatment planning, beams are disabled if they would cause a PDP error. If the robot is operated under manual model, the PDP is not functional and cannot prevent a violation of the patient exclusion zone and a subsequent collision.

Treatment Simulation or Animation

A fully integrated robot simulation program provides a treatment animation for previewing the exact motion of the robot to check for potential collision hazards. The program accurately models the kinematics of the robot as well as the geometry of the treatment room, including all the equipment. Simulation of a treatment plan is optional. However, it is strongly recommended that a simulation should be performed if a treatment area is near the lower torso of a patient. If a treatment plan is not simulated, warning messages will be displayed when it is prescribed and treated.

Emergency and Safety Procedures

Safety systems incorporated into the facility design must be checked and verified regularly as part of daily or weekly QA. These systems include emergency interruption for robot movement, emergency power off, door interlocks, collimator assembly

collision detector, robot perch position, and audio and visual monitors. The list of safety checks is shown in Table 91.2. Once activated, interlocks should occur immediately and remain on until the condition is resolved. A safety feature built into the CyberKnife System automatically checks that the collimator inserted in the system matches the one in the plan before start of treatment. However, it is recommended that an incorrect collimator be inserted periodically to verify that a fault occurs and that the fault cannot be cleared until a correct one is in place.

Daily Warm-Up and Calibration

Power-Up Checklist

It is recommended that the system status check be performed daily to ensure that the CyberKnife System functions within specifications and to monitor any drift in operation parameters. Table 91.3 shows a list of checking procedures needed to be performed daily before patient treatment.

X-Ray Tubes Warm-Up

The CyberKnife System relies on the diagnostic x-ray to track the patient under treatment. It is very important to condition and warm up the x-ray tubes every morning. The intent is to minimize target (anode) cracking due to thermal shock. If the x-ray unit has been inactive for a period of 8 h or more, it is necessary to start the warm-up at a kVp considerably lower than

TABLE 91.2 CyberKnife Safety Checks

Safety Checks	Frequency	Status
Control panel OFF button	Daily	Pass/fail
Control panel E-stop	Weekly	Pass/fail
E-stops inside room	Weekly	Pass/fail
Collimator check	Weekly	Pass/fail
Door interlock	Daily	Pass/fail
Laser at perch position	Daily	<±1 mm
Tech pendent set to external	Daily	Pass/fail
Audio and video	Daily	Functional

TABLE 91.3 Daily System Status Check Procedures

	Units	Range
Status: System Off		
SF6 gas pressure	psi	28–32
Gun filament	V	3.5–5
Ion pump	μA	<5
Humidity	%	Normal
Status: System On		
Gun filament	V	5–6.5
Mag filament	V	8–11
Steering 1	A	No drift
Steering 2	A	No drift
Mag tuner	A	0.2–0.8
Water temperature	°C	18–20
Water flow rate	L/min	>3
Status: Beam On		
Deliver 6000 MU	MU	6000
HV current	mA	No drift
High voltage	kV	No drift
Mag filament	V	No drift
Ion pump	μA	<5
Mag tuner	A	0.2–0.8
Dose rate	MU/min	350–600

the highest kVp normally used on a given tube to realign the internal distribution of charges.

Linac Warm-Up

The ion chambers inside the CyberKnife System are currently open to the atmosphere. The linear accelerator needs to be warmed up sufficiently before any treatment or calibration. A recommended warm-up of 6000 MU will bring the chambers to a temperature close to the average chamber temperature during a typical treatment. After the recommended warm-up, the radiation output should be within 2% of the intended delivery value.

Linac Output Constancy Check

Changes in atmospheric pressure and temperature affect the radiation output because of the ion-chamber design of the CyberKnife System. The output of the linear accelerator needs to be checked or verified daily. Additional checks throughout the day may be required if dramatic environmental changes occur. It is recommended that no adjustment to the calibration factor be made if the variation is within a tolerance level of 2% (Dieterich et al.). However, if the variation exceeds this limit, a qualified medical physicist must change the calibration factor by following the steps in calibration adjustment provided by the CyberKnife System. Afterward, additional verification measurements are required.

Automated QA and End-to-End Test

The automated QA and the end-to-end tests are routinely performed to verify the overall clinical delivery accuracy of the CyberKnife System. The automated QA test checks the robot by delivering a two-path treatment to films placed inside the Shadow Phantom provided by Accuray. The nodes are located anterior to and to the right of the phantom. The end-to-end test consists of approximately 50 beams and is designed to verify the geometric targeting accuracy of the CyberKnife throughout all phases of treatment planning, tracking, and delivery. The targeting errors for both tests should be less than 1 mm. The automated QA test is less time consuming and could be performed daily or weekly. The end-to-end test needs to be performed at least monthly for one intracranial and one extracranial mode.

Treatment Delivery

Alignment and Tracking

In the patient alignment process, features such as bony landmarks and fiducials in digitally reconstructed radiographs (DRRs) of the planning CT images align and track with those from the live x-ray images (Fu and Kuduvalli 2008; Fu, Kuduvalli, and Maurer 2006). By minimizing the offsets between the DRR and the live x-ray images, the treatment robot is more likely to compensate accurately for patient movement during treatment. In other words, the smaller the alignment error, the fewer errors are likely to occur during treatment delivery (Murphy et al. 2003; Yu et al. 2004). It is recommended that the patient be aligned (i.e., positioned as in the treatment planning CT) until the total detected alignment error is 1 mm or less for translation error, and as close to 0° as possible for rotation error. Couch motion is not monitored by the collision detection system during the setup or patient alignment. Visual observation of all movement of the treatment couch is required to prevent any potential collisions. It should be cautioned that while the automatic couch alignment is in progress, no one except the patient under treatment is allowed to enter the treatment room. Otherwise, exposure to x-rays will occur.

During treatment delivery, the robot repositions the linear accelerator to compensate patient offset or displacement between the DRR and live x-ray images, that is, the robot corrects for both translational and rotational displacements. The maximum corrections that the robot can make along each translation and rotation axis are shown in Table 91.4.

Imaging Frequency

In general, imaging frequency can vary with treatment modality, area of treatment, immobilization, tumor margin, and proximity of organs at risk. Taking x-ray images for every beam or every other beam might be warranted to achieve higher delivery accuracy. The secondary radiation dose from the diagnostic x-ray is up to 0.1 cGy per exposure inside the radiation field (Yu 2006). Because of extensive use of the diagnostic x-ray beam for treatment tracking, there is concern for potential secondary radiation risks for patients with benign lesions. It is advisable to first image more frequently (every beam) for the initial treatment to

TABLE 91.4 Maximum Corrections by Treatment Robot

Translation	X Direction (mm)	Y Direction (mm)	Z Direction (mm)
Standard couch without synchrony	±10	±10	±10
RoboCouch without synchrony	±10	±10	±10
Either couch with synchrony	±25	±25	±25
Rotation	Pitch (Head Up/Down) (°)	Roll (Left/Right) (°)	Yaw (CW/CCW) (°)
Standard couch	±1	±1	±1
RoboCouch	±1.5	±1.5	±1.5

determine whether the patient is stable and then to image less frequently (every three to five nodes) if the patient is stable. In other words, the frequency of x-ray acquisition should be carefully evaluated establishing a balance between clinical necessity and potential radiation risks.

References

Adler, Jr., J. R., S. D. Chang, M. J. Murphy, J. Doty, P. Geis, and S. L. Hancock. 1997. The Cyberknife: A frameless robotic system for radiosurgery. *Stereotact. Funct. Neurosurg.* 69 (1–4 Pt 2): 124–128.

Dieterich, S., C. Cavedon, C. F. Chuang, A. B. Cohen, J. A. Garrett, C. L. Lee, J. R. Lowenstein et al. Report of AAPM TG 135: Quality assurance for robotic sadiosurgery. *Med. Phys.* (in review).

Fraass, B., K. Doppke, M. Hunt, G. Kutcher, G. Starkschall, R. Stern, and J. Van Dyke. 1998. American Association of Physicists in Medicine Radiation Therapy Committee Task Group 53: Quality assurance for clinical radiotherapy treatment planning. *Med. Phys.* 25 (10): 1773–1829.

Fu, D., G. Kuduvalli, C. J. Maurer, J. W. Allision, and J. R. Adler. 2006. 3D target localization using 2D local displacements of skeletal structures in orthogonal x-ray images for image-guided spinal radiosurgery. *Int. J. CARS* 1: 198–200.

Fu, D., and G. Kuduvalli. 2008. A fast, accurate, and automatic 2D-3D image registration for image-guided cranial radiosurgery. *Med. Phys.* 35 (5): 2180–2194.

Murphy, M. J., S. D. Chang, I. C. Gibbs, Q. T. Le, J. Hai, D. Kim, D. P. Martin, and J. R. Adler Jr. 2003. Patterns of patient movement during frameless image-guided radiosurgery. *Int. J. Radiat. Oncol. Biol. Phys.* 55 (5): 1400–1408.

Wilcox, E. E., and G. M. Daskalov. 2008. Accuracy of dose measurements and calculations within and beyond heterogeneous tissues for 6 MV photon fields smaller than 4 cm produced by Cyberknife. *Med. Phys.* 35 (6): 2259–2266.

Yu, C. 2006. Secondary radiation doses from CyberKnife SRS/RT. *Med. Phys* 33 (6): 2135.

Yu, C., W. Main, D. Taylor, G. Kuduvalli, M. L. Apuzzo, and J. R. Adler Jr. 2004. An anthropomorphic phantom study of the accuracy of Cyberknife spinal radiosurgery. *Neurosurgery* 55 (5): 1138–1149.

92

Multileaf Collimator-Based Intensity-Modulated Radiotherapy

Parminder S. Basran
*BC Cancer Agency-Vancouver
Island Center*

Introduction

Multileaf collimator (MLC)-based intensity-modulated radiotherapy (IMRT) provides the most accessible and cost-effective means of sculpting radiation dose to the tumor while minimizing dose to normal tissues. With increasing complexity of IMRT planning and delivery methods comes the need for more sophisticated quality assurance (QA) processes that can capture errors in the radiation therapy delivery chain. These QA processes must strike a careful balance between accurate estimation of delivered dose and efficiency. Before discussing patient-specific IMRT QA methodologies or processes, acceptance testing, and QA of CT simulation, treatment planning and delivery systems must be well defined, controlled, and their associated errors in the treatment delivery chain understood. Although potential sources of systematic error have been addressed in other sections of this chapter, their importance in patient-specific QA is vital in MLC-based IMRT. Details of linac and MLC (Kutcher et al. 1994; Klein, Hanley, and Bayouth 2009), IMRT acceptance testing (Ezzell et al. 2003, 2009), and inverse treatment planning (Fraass et al. 1998) will not be discussed in this chapter. The purpose of this chapter is to describe the important aspects of QA of patient-specific MLC-based IMRT.

The process of IMRT QA is not simply limited to a series of phantom measurements and verifying that the results are within some tolerance: the process should capture all vital sources of potential errors from CT simulation to the final fractionated treatment. In this chapter, considerations for pre-treatment patient QA are provided, followed by a discussion of IMRT treatment validation by way of measurements and calculation, mid-treatment QA issues, and some final comments about retrospective analysis of patient QA data.

Pretreatment Patient QA

Most literature on IMRT QA focuses on measurements or calculations that validate the intended treatment planning dose with the "true" patient dose. Although this step is an important one, it is equally—if not more—important to ensure that the intended plan is logical and consistently deliverable over many fractions. An independent check of the IMRT plan by a physicist not involved in planning the patient is strongly advised. Table 92.1 provides an example of a checklist for an IMRT treatment plan before any dosimetric measurement or validation. Key considerations when checking IMRT plans are very similar to conventional plan checks.

Patient-Specific IMRT Dosimetric Validation

Measurement-Based IMRT Dosimetric Validation

Because IMRT is a complex procedure with many potential sources of error, comprehensive system tests are performed on a case-by-case basis before treatment to validate the intended treatment dose. These system tests generally require recalculating the clinical IMRT fields on a water equivalent phantom and using some type of detector to validate the calculated dose.

TABLE 92.1 Example of Patient-Specific IMRT QA Checklist

Task	Description
General	Patient demographics/ID correct for all electronic/hardcopy documentation
	Current treatment planning software version is valid on documentation
	Clinical prescription corroborates with treatment planning prescription
	Special patient considerations, such as pacemakers, defibrillators, neural stimulators, etc.
	Appropriate signatures/electronic approvals on all documentation
	Treatment plan is "locked" or locked pending physicists plan approval
CT and contours	CT simulation/electron density table selected correctly
	Secondary images properly QA'd before contouring (e.g., correct series, registration)
	Contours for CTV/GTV adequately defined
	PTV volume is reasonable • expansion done correctly and within limit of external patient contour (e.g., within 5 mm of patient surface) or bolus applied • Assess whether density heterogeneities in PTV are reasonable (e.g., air cavities in rectum) • Bolus is defined accurately and applied for all relevant beams
	External contour includes patient and relevant patient accessories
	Treatment couch properly accounted for in treatment plan • Modeled explicitly in the TPS or removed • If modeled, done so with the correct machine type
	CT image quality OK and no streak artifacts • Streaks are minor (e.g., dental work), apply appropriate corrective measures • Streaks are major (e.g., prosthetic), apply appropriate corrective measures
	Top and bottom CT slices do not invoke a dosimetric artifact
	Normal tissue contours and avoidance structures correctly defined, overlapping structures identified
Planning and objectives	Appropriate target volume or point defined in the prescription
	Correct objectives template for tumor and normal tissue volumes used with no conflicting objectives
	Dose and normalization points used correctly
	Isocenter placement • Reference marks on patient corroborate with reference coordinates on CT • Position of isocenter corroborates with necessary shifts • Laterality of target confirmed with applied shifts • Isocenter location is not too low (couch collision), too high (patient collision, or too lateral (patient/couch collision)
	Collimator angles optimized to minimize tongue–groove effect/improve plan
	Non-coplanar beams avoid couch or patient collision
	No opposing beams optimized
	Any non-IMRT beams used for treatment do not have fluence patterns
	All beam angles • Do not transmit needlessly through normal tissues (e.g., shoulders) • Do not "skim" the couch if couch is not modeled
	Correct segmentation, leaf-motion calculation, or aperture optimization used
	Highest dose target structure clinical objectives achieved; for example: • 95% of volume receives 99% of prescription dose • Mean dose is within 2% of prescription dose • Maximum and minimum dose to the target is within 95–107% of prescription dose
	Secondary target structure(s) clinical objectives achieved
	Normal tissue tolerances *or* maximum permissible dose not exceeded
	Dose volume histograms of target and normal tissue structures are reflective of clinical and optimization objectives
	Evaluate dose to noncontoured structures (e.g., skin dose)
Dose and monitor units	Appropriate field size of segments, monitor unit per segments, and number of segments (segmented MLC delivery) for this treatment site
	Dose grid resolution is correct and dose grid extent covers all relevant organs
	Final dose calculation algorithm correct
	Monitor units per beam are reasonable
	Permissible MLC leaf and jaw positions used
Treatment QA	Treatment plan is locked/unalterable
	Electronic treatment planning data is recorded and transferred to the record-and-verify system correctly

(continued)

TABLE 92.1 Example of Patient-Specific IMRT QA Checklist (Continued)

Task	Description
	Measurement and or calculation QA of the patient plan
	Appropriate hand-off to radiation therapy and/or RT administrator
	Sufficient time given to RT for their QA process before treatment

Ion chamber measurements provide the simplest method of dosimetric validations but can be sensitive to partial volume effects (Yang and Xing 2003) and positioning errors (Woo and Nico 2005; Yan et al. 2009). Smaller detectors are useful for micro-MLC-based IMRT (Basran and Yeboah 2008). Newer emulsions (Chetty and Charland 2002) for film dosimetry can be used over typical treatment doses to provide unparalleled spatial resolution and relatively robust absolute dosimetry for IMRT QA (Tangboonduangjit et al. 2003). Diode (Letourneau et al. 2004), ion chamber (Spezi et al. 2005) detector arrays, and electronic portal imaging dosimetry (Steciw et al. 2005; Ansbacher 2006) provide a practical and efficient balance between robust absolute and relative dosimetry. Many commercial detectors systems are accompanied with powerful software that can quantitatively compare measured and calculated doses from treatment planning systems.

Calculation-Based IMRT Dosimetric Validation

Independent calculations can also be performed to validate the patient dosimetry before or after the patient's treatment. The intended gantry angles, MLC leaf sequences (segmented or dynamic), monitor units, and the expected dose per field can be

extracted from treatment files (e.g., DICOM RT or RTP files) and the dose to one or several points can be calculated independently with commercial software (Yang et al. 2003; Georg, Nyholm et al. 2007), another treatment planning system (Georg, Stock et al. 2007), or a Monte Carlo dose engine (Fan et al. 2006). These systems can be useful to identify systematic discrepancies between measured and primary treatment planning dose calculations, while also providing a means of validating MLC segment shapes and expected fluence distributions. Whether such independent calculations are sufficient to validate patient-specific IMRT QA is a subject of debate.

In vivo 3D IMRT dose validations are currently being investigated with EPID (Wendling et al. 2009; van Elmpt et al. 2009) and ancillary detector arrays (Islam et al. 2009) during and after each fraction. These methods, in conjunction with daily image guidance, can be helpful in adaptive IMRT strategies. Real-time in vivo point dose measurement can serve as a useful means of tracking changes in patient dosimetry and detecting gross errors in delivery (Higgins et al. 2003). In addition, post treatment delivery details, such as the delivered gantry positions, monitor units, and MLC leaf positions, can be extracted from record-and-verify log files or machine-specific log files of the MLC controller. These files can be imported into Monte Carlo dose engines (Pawlicki and Ma 2001) or secondary treatment planning systems (Stell et al. 2004) to compute the delivered dose to the patient or to compare true and actual MLC positions during the course of treatment. Pre- and post-treatment MLC checks from log files or portal imaging can also be used to examine possible changes in treatment delivery or data corruption (Zeidan et al. 2004).

Establishing Tolerances for IMRT Processes

All dosimetric validation techniques attempt to examine the fidelity of the intended and actual dose delivered to the patient. Useful dose evaluation metrics, such as the absolute and relative percent discrepancy, gamma-statistic (Low et al. 1998), and

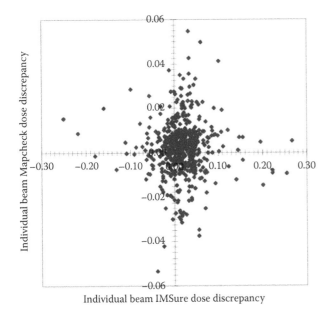

FIGURE 92.1 Plot of 555 individual IMRT beam discrepancies from diode array (Mapcheck) measurements and a secondary monitor unit (IMSURE) checking program. Retrospective analysis of large collections of data is useful in determining expected tolerances from IMRT QA procedures.

TABLE 92.2 Example Expected Tolerance Values as Derived from a Systematic Audit of IMRT QA Checking Program for Head and Neck and Non-Head and Neck IMRT Plans

Diode Array Measurements	Non-HN (%)	HN (%)
Total dose discrepancy	2	2
Individual beam discrepancy	3	3
Individual beam gamma statistic (3 mm/3% with 10% threshold)	95	88

Note: HN, head and neck.

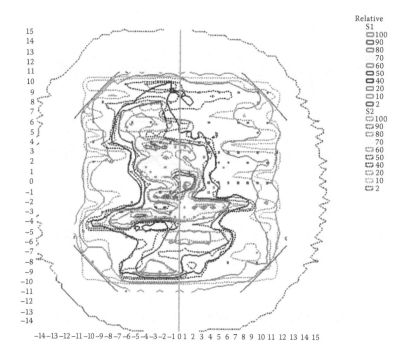

FIGURE 92.2 Comparison of measured (solid) and calculated (hatched) isodose lines from a random sequence of 12 MLC segments. Blue and red dots indicate point dose discrepancies of –3%, and +3%, respectively. Although not clinically useful, the gamma statistic (74% passing a 2%, 2 mm tolerance) places clinical gamma values into perspective.

chi-statistic (Bakai, Alber, and Nusslin 2003), can distill large amounts of data and assist in this assessment, and are well summarized by Jiang et al. (2006). Establishing tolerances levels of a metric can be achieved with a retrospective analysis of plans deemed safe to deliver (Basran and Woo 2008) (see Figure 92.1 and Table 92.2) and by surveying the literature (Both et al. 2007). A useful exercise in understanding practical tolerances is to examine the metrics obtained from QA test plans typically measured during IMRT commissioning (Ezzell et al. 2003), such as test fluence patterns as recommended by the AAPM report on IMRT commissioning (see Figure 92.2). Finally, the use of control charts as derived from process control methods also serves as an efficient and robust means of establishing tolerances in IMRT QA processes (Pawlicki et al. 2008; Breen et al. 2008).

It is important to note that passing a tolerance does not sufficiently prove a plan is safe to deliver. It is not unreasonable, for example, to reject an IMRT plan that passes an IMRT QA test (e.g., gamma is within tolerance) but systematically fails in, for example, in spatial regions where the spinal cord might reside (e.g., due to MLC miscalibration). Results from an IMRT QA process should be judged independently, according to the patient's clinical indication.

Patient-Specific QA during Treatment

CT simulation assumes that a single CT provides a surrogate of the patient anatomy throughout the course of treatment. However, the planning CT is only a single representation of the patient. Therefore, continuous monitoring of the patient by way

of frequent source to skin distance checks, image guidance procedures (e.g., portal images or cone-beam CT), and continuous patient monitoring for adverse effects is strongly advised for IMRT treatments. Many rapidly responding tumors are treated with IMRT and the changes in patient geometry can be dramatic during the treatment (Ding et al. 2007; Lee, Le, and Xing 2008). In such instances, if radiation treatment is continued without intervention, geographical miss of the tumor or prohibitive doses to normal tissues can jeopardize the treatment. Although adaptive treatment planning processes can be time consuming, efficiencies can be applied to help streamline the process. Having clear and simple tolerances in place for specific sites can prove to be efficient and beneficial for the patient (see Figure 92.3).

Summary

Although not commonly done, retrospective auditing of patient IMRT QA results can identify systematic discrepancies in the delivery process and can be helpful in establishing or refining tolerances, identifying outliers in the QA processes, and provide guidance on IMRT treatment planning objectives. Retrospective auditing can facilitate continuous quality improvement in IMRT delivery.

Any IMRT QA methodology imposed in a clinic should not be considered unique or ubiquitous and each clinic should develop their own QA processes and tolerances based on their equipment and needs. Passing a tolerance or completing a checklist does not imply that an IMRT plan is safe to deliver: the experience and knowledge of qualified medical physicists, radiation

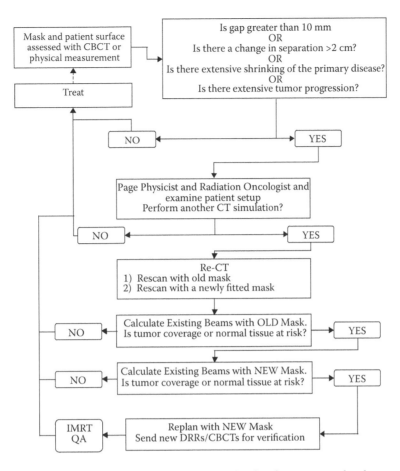

FIGURE 92.3 Example of midtreatment (adaptive) IMRT QA process for head and neck cancer treated with an aquaplast mask. In this process, gaps between the bolus or mask and the patient skin surface are assessed daily by radiation therapy staff. If a specific tolerance is exceeded, another CT simulation is obtained only if deemed clinically necessary. Note also that two CT simulations are obtained to minimize the need for creating another IMRT plan if the dosimetric repercussions of the gap are minimal.

oncologists, and radiation therapists is crucial in determining whether an IMRT plan is safe for treatment.

References

Ansbacher, W. 2006. Three-dimensional portal image-based dose reconstruction in a virtual phantom for rapid evaluation of IMRT plans. *Med. Phys.* 33 (9): 3369–3382.

Bakai, A., M. Alber, and F. Nusslin. 2003. A revision of the gamma-evaluation concept for the comparison of dose distributions. *Phys. Med. Biol.* 48 (21): 3543–3553.

Basran, P. S., and M. K. Woo. 2008. An analysis of tolerance levels in IMRT quality assurance procedures. *Med. Phys.* 35 (6): 2300–2307.

Basran, P., and C. Yeboah. 2008. Dosimetric verification of micro-MLC based intensity modulated radiation therapy. *J. Appl. Clin. Med. Phys.* 9 (3): 2832.

Both, S., I. M. Alecu, A. R. Stan, M. Alecu, A. Ciura, J. M. Hansen, and R. Alecu. 2007. A study to establish reasonable action limits for patient-specific quality assurance in intensity-modulated radiation therapy. *J. Appl. Clin. Med. Phys.* 8 (2): 1–8.

Breen, S. L., D. J. Moseley, B. Zhang, and M. B. Sharpe. 2008. Statistical process control for IMRT dosimetric verification. *Med. Phys.* 35 (10): 4417–4425.

Chetty, I. J., and P. M. Charland. 2002. Investigation of Kodak extended dose range (EDR) film for megavoltage photon beam dosimetry. *Phys. Med. Biol.* 47 (20): 3629–3641.

Ding, G. X., D. M. Duggan, C. W. Coffey, M. Deeley, D. E. Hallahan, A. Cmelak, and A. Malcolm. 2007. A study on adaptive IMRT treatment planning using kV cone-beam CT. *Radiother. Oncol.* 85 (1): 116–125.

Ezzell, G. A., J. W. Burmeister, N. Dogan, T. J. LoSasso, J. G. Mechalakos, D. Mihailidis, A. Molineu et al. 2009. IMRT commissioning: Multiple institution planning and dosimetry comparisons, a report from AAPM Task Group 119. *Med. Phys.* 36 (11): 5359–5373.

Ezzell, G. A., J. M. Galvin, D. Low, J. R. Palta, I. Rosen, M. B. Sharpe, P. Xia, Y. Xiao, L. Xing, and C. X. Yu. 2003. Guidance document on delivery, treatment planning, and clinical implementation of IMRT: Report of the IMRT Subcommittee of the AAPM Radiation Therapy Committee. *Med. Phys.* 30 (8): 2089–2115.

Fan, J., J. Li, L. Chen, S. Stathakis, W. Luo, F. Du Plessis, W. Xiong, J. Yang, and C. M. Ma. 2006. A practical Monte Carlo MU verification tool for IMRT quality assurance. *Phys. Med. Biol.* 51 (10): 2503–2515.

Fraass, B., K. Doppke, M. Hunt, G. Kutcher, G. Starkschall, R. Stern, and J. Van Dyke. 1998. American Association of Physicists in Medicine Radiation Therapy Committee Task Group 53: Quality assurance for clinical radiotherapy treatment planning. *Med. Phys.* 25 (10): 1773–1829.

Georg, D., T. Nyholm, J. Olofsson, F. Kjaer-Kristoffersen, B. Schnekenburger, P. Winkler, H. Nystrom, A. Ahnesjo, and M. Karlsson. 2007. Clinical evaluation of monitor unit software and the application of action levels. *Radiother. Oncol.* 85 (2): 306–315.

Georg, D., M. Stock, B. Kroupa, J. Olofsson, T. Nyholm, A. Ahnesjo, and M. Karlsson. 2007. Patient-specific IMRT verification using independent fluence-based dose calculation software: Experimental benchmarking and initial clinical experience. *Phys. Med. Biol.* 52 (16): 4981–4992.

Higgins, P. D., P. Alaei, B. J. Gerbi, and K. E. Dusenbery. 2003. In vivo diode dosimetry for routine quality assurance in IMRT. *Med. Phys.* 30 (12): 3118–3123.

Islam, M. K., B. D. Norrlinger, J. R. Smale, R. K. Heaton, D. Galbraith, C. Fan, and D. A. Jaffray. 2009. An integral quality monitoring system for real-time verification of intensity modulated radiation therapy. *Med. Phys.* 36 (12): 5420–5428.

Jiang, S. B., G. C. Sharp, T. Neicu, R. I. Berbeco, S. Flampouri, and T. Bortfeld. 2006. On dose distribution comparison. *Phys. Med. Biol.* 51 (4): 759–776.

Klein, E. E., J. Hanley, J. Bayouth, F. F. Yin, W. Simon, S. Dresser, C. Serago et al. 2009. Task Group 142 report: Quality assurance of medical accelerators. *Med. Phys.* 36 (9): 4197–4212.

Kutcher, G. J., L. Coia, M. Gillin, W. F. Hanson, S. Leibel, R. J. Morton, J. R. Palta et al. 1994. Comprehensive QA for radiation oncology: Report of AAPM Radiation Therapy Committee Task Group 40. *Med. Phys.* 21 (4): 581–618.

Lee, L., Q. T. Le, and L. Xing. 2008. Retrospective IMRT dose reconstruction based on cone-beam CT and MLC log-file. *Int. J. Radiat. Oncol. Biol. Phys.* 70 (2): 634–644.

Letourneau, D., M. Gulam, D. Yan, M. Oldham, and J. W. Wong. 2004. Evaluation of a 2D diode array for IMRT quality assurance. *Radiother. Oncol.* 70 (2): 199–206.

Low, D. A., W. B. Harms, S. Mutic, and J. A. Purdy. 1998. A technique for the quantitative evaluation of dose distributions. *Med. Phys.* 25 (5): 656–661.

Pawlicki, T., and C. M. Ma. 2001. Monte Carlo simulation for MLC-based intensity-modulated radiotherapy. *Med. Dosim.* 26 (2): 157–168.

Pawlicki, T., S. Yoo, L. E. Court, S. K. McMillan, R. K. Rice, J. D. Russell, J. M. Pacyniak et al. 2008. Moving from IMRT QA measurements toward independent computer calculations using control charts. *Radiother. Oncol.* 89 (3): 330–337.

Spezi, E., A. L. Angelini, F. Romani, and A. Ferri. 2005. Characterization of a 2D ion chamber array for the verification of radiotherapy treatments. *Phys. Med. Biol.* 50 (14): 3361–3373.

Steciw, S., B. Warkentin, S. Rathee, and B. G. Fallone. 2005. Three-dimensional IMRT verification with a flat-panel EPID. *Med. Phys.* 32 (2): 600–612.

Stell, A. M., J. G. Li, O. A. Zeidan, and J. F. Dempsey. 2004. An extensive log-file analysis of step-and-shoot intensity modulated radiation therapy segment delivery errors. *Med. Phys.* 31 (6): 1593–1602.

Tangboonduangjit, P., I. Wu, M. Butson, A. Rosenfeld, and P. Metcalfe. 2003. Intensity modulated radiation therapy: Film verification of planar dose maps. *Australas. Phys. Eng. Sci. Med.* 26 (4): 194–199.

van Elmpt, W., S. Nijsten, S. Petit, B. Mijnheer, P. Lambin, and A. Dekker. 2009. 3D in vivo dosimetry using megavoltage cone-beam CT and EPID dosimetry. *Int. J. Radiat. Oncol. Biol. Phys.* 73 (5): 1580–1587.

Wendling, M., L. N. McDermott, A. Mans, J. J. Sonke, M. van Herk, and B. J. Mijnheer. 2009. A simple backprojection algorithm for 3D in vivo EPID dosimetry of IMRT treatments. *Med. Phys.* 36 (7): 3310–3321.

Woo, M. K., and A. Nico. 2005. Impact of multileaf collimator leaf positioning accuracy on intensity modulation radiation therapy quality assurance ion chamber measurements. *Med. Phys.* 32 (5): 1440–1445.

Yan, G., C. Liu, T. A. Simon, L. C. Peng, C. Fox, and J. G. Li. 2009. On the sensitivity of patient-specific IMRT QA to MLC positioning errors. *J. Appl. Clin. Med. Phys.* 10 (1): 2915.

Yang, Y., and L. Xing. 2003. Using the volumetric effect of a finite-sized detector for routine quality assurance of multileaf collimator leaf positioning. *Med. Phys.* 30 (3): 433–441.

Yang, Y., L. Xing, J. G. Li, J. Palta, Y. Chen, G. Luxton, and A. Boyer. 2003. Independent dosimetric calculation with inclusion of head scatter and MLC transmission for IMRT. *Med. Phys.* 30 (11): 2937–2947.

Zeidan, O. A., J. G. Li, M. Ranade, A. M. Stell, and J. F. Dempsey. 2004. Verification of step-and-shoot IMRT delivery using a fast video-based electronic portal imaging device. *Med. Phys.* 31 (3): 463–476.

93

Compensator-Based Intensity-Modulated Radiotherapy

Sha Chang
University of North Carolina Medical School

Introduction

Compensator-based intensity modulation is a well-accepted approach for intensity-modulated radiotherapy (IMRT) treatment delivery on any linac or Co-60 unit. First implemented in the early 1990s at a few academic institutions such as the University of North Carolina (Chang et al. 2004), compensator-IMRT is currently used by more than 250 clinics in the United States. The widespread use of this approach is promoted by commercial companies such as .decimal Inc. (Sanford, FL) that offer convenient mail-order services for IMRT compensators. There are a number of different compensator-based IMRT techniques described in the literature (Chang et al. 2004; Chang 2006; Sasaki and Obata 2007; Nakagawa et al. 2007; Nakagawa, Fukuhara, and Kawakami 2005; Xu, Al-Ghazi, and Molloi 2004; O'Daniel et al. 2004; Yoda and Aoki 2003). Despite their differences, all share the same basic dissimilarity with multileaf collimator (MLC)-based IMRT techniques in intensity modulation generation (Williams 2003). Compensator-IMRT converts the open-field uniform photon fluence map to the intended intensity-modulated map via a custom-made compensator as shown in Figure 93.1 that is often manually placed in the wedge or block holder of the treatment machine. MLC-IMRT techniques rely on the accelerator's built-in MLC to sequentially deliver different field segments, which form the intended intensity-modulated map during treatment delivery (Figure 93.2). The fundamental difference in intensity modulation formation between compensator-based and MLC-based IMRT delivery approaches dictates the major differences in their commissioning and patient-specific QA procedures.

For MLC-IMRT treatment, a recent study by the Radiological Physics Center has reported that approximately 30% of 250 accelerators has failed to deliver the IMRT dose distribution planned

by their own treatment planning systems to a head and neck phantom to within 7% or 4 mm (Ibbott et al. 2008). The phantom irradiations, as part of credentialing, have identified the following errors found in the study: (1) incorrect output factors and percentage depth dose data entered by the institution; (2) inadequate modeling of the penumbra at MLC leaf ends (Jiang and Ayyangar 1998); (3) incorrect application of QA calculations or measurements; (4) inadequate QA of MLC; (5) incorrect patient positioning, including couch indexing errors with a serial tomography system; and (6) errors in treatment-planning software. At least two of the six types of errors are MLC related.

Compensator-IMRT has very different quality assurance concerns that are less studied and understood compared to MLC-IMRT approaches. Academic institutions that developed their in-house compensator-IMRT programs often have the resources and expertise to develop a comprehensive QA program that is specific to their own system. Independent of the specific compensator-IMRT technique used, there is a basic dissimilarity between MLC-based and compensator-based IMRT in intensity

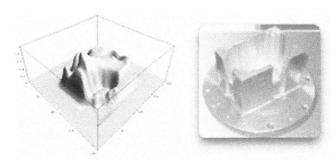

FIGURE 93.1 (Left) Smooth intensity-modulation map to be produced by UNC's compensator-IMRT technique. (Right) Mail-order compensator produced by .decimal Inc.

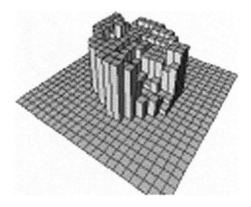

FIGURE 93.2 Discrete intensity-modulation map to be produced by the segmental MLC-IMRT technique.

map generation and thus in IMRT QA. In this chapter, some of the basic compensator-specific IMRT QA issues are reviewed and discussed.

Compensator-IMRT Commissioning

There are generally three components in the compensator-IMRT commissioning process: (1) commissioning of the treatment planning system for compensator-IMRT, (2) commissioning of the compensator fabrication, and (3) the commissioning of compensator clinical usage procedure including routine QA procedure. Although the first and the last components affect all compensator-IMRT users, the second component is less relevant to most users who use commercial compensator services, which should have completed the component already. Regardless, users should have a good understanding of all three components of compensator-IMRT commissioning before clinical use. It is very important that the commissioning process is performed under realistic and real clinical situations and thus the outcome is representative of the real patient treatment. For this reason, simple geometry phantoms alone are not adequate for compensator-IMRT commissioning, and clinical or realistic patient IMRT compensators of different shapes and sizes should be used in the commissioning process.

Treatment Planning Commissioning

Today, practically all commercial treatment planning systems offer the compensator-IMRT option. However, the arbitrary shape and size of customized compensators pose unique challenges for dose computation in many treatment planning systems because the existing dose code structure was not originally designed for such applications (Chang et al. 2004; du Plessis and Willemse 2006; Bakai, Laub, and Nüsslin 2001). Compensator beams are not commissioned during the conventional accelerator and treatment planning system commissioning. Thus, one of the first steps in establishing a compensator-IMRT program is to commission the treatment planning system. Specifically, the intensity modulation map resulting from the inverse planning

dose optimization needs to be converted into a custom compensator file for fabrication. The accuracy of such a conversion is crucial to the quality of the compensator-IMRT patient-specific QA and, thus, patient treatment. The accuracy of the intensity map to compensator conversion depends largely on the ability of the dose algorithm to model the photon beam as it traverses an arbitrarily shaped compensator. As the photon beam passes through the compensator, it suffers from attenuation and scattering, and its energy spectrum changes. Dimitriadis and Fallone (2002) studied the beam hardening effect of a cerrobend compensator design and reported significant changes in measured linear attenuation coefficient as a function of thickness for a 6-MV beam using in air measurement with a 1.5-cm building-up cap. The measurement showed a significant decline in linear attenuation coefficient value decreasing by 15–17% for cerrobend thickness increasing from 0.3 to 6 cm. The data show the considerable field size dependence of the linear attenuation coefficient for all thicknesses of cerrobend. The group also studied cerrobend scatter and demonstrated an increase with increasing field size and compensator thickness. The Dimitriadis and Fallone study illustrates the significant perturbation the photon beam experiences when traversing the cerrobend compensator, and those changes must be taken into consideration in the dose algorithm for accurate compensator-IMRT dosimetry calculation. However, the perturbation occurs in a complex manner that it is difficult to model in most commercial treatment planning systems.

The common approach, in existing treatment planning systems, to modeling the compensator perturbation is by introducing a compensator linear attenuation coefficient. Chang et al. used an empirical approach that works well for compensators of tin and tungsten granule and solid cerrobend materials of maximum thickness 5 cm. In their in-house treatment planning system (called PLUNC), the conversion of the intensity map $I(x,y)$ to compensator thickness $t(x,y)$ is computed using Equation (93.1). The linear attenuation coefficient through the compensator

$$I^{\mathrm{IMRT}}(x,y) = I^{\mathrm{open}}(x,y) \cdot \exp[-\mu \cdot t^{\mathrm{comp}}(x,y)] \qquad (93.1)$$

where

$$\mu = \mu_0 + c_1 \cdot t^{\mathrm{comp}}(x,y) + c_2 \cdot r + c_3 \cdot S \qquad (93.2)$$

is an empirical function (Equation 93.2) of the photon beam energy and the effective density (μ_0) of the compensator material, compensator thickness traversed by the pencil beam [$t^{\mathrm{comp}}(x,y)$], beam energy change with off-axis distance (r), and field size (S). The dependence on each of the above parameters within the range of compensator clinical application is represented by the coefficients (c_1, c_2, and c_3) in Equation (93.2), and they are iteratively determined by fitting the calculation to the measurement for a range of shapes and sizes of the anticipated clinical or clinically relevant IMRT compensators. Figures 93.3 illustrates the fitting using this empirical approach and compares the measured and PLUNC-calculated dose profile for compensator-IMRT. Many commercial treatment planning systems,

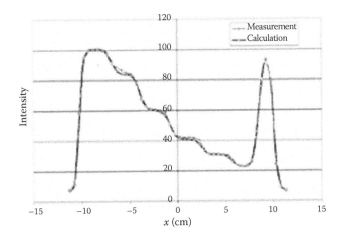

FIGURE 93.3 Dose profiles of PLUNC calculation and compensator-IMRT commissioning measurement using Profiler (Sun Nuclear). A step-like test compensator was used. Note that the radiation field used is larger than the compensator. There is excellent agreement between calculation and measurement in and outside the field.

however, have less flexibility in the dose algorithm structure to modify the linear attenuation coefficient. Oguchi and Obata (2009) reported a study using the Xio treatment planning system (CMS-Eleckta, Stockholm, Sweden) that used a single effective linear attenuation coefficient value for .decimal's mail-order brass compensators. They found that the effective coefficient did not change significantly with depth and source-to-surface but it did change with field size. As a remedy, they used two sets of effective linear attenuation coefficient per photon energy. In a study by Bakai, Laub, and Nüsslin (2001), a poor agreement was reported, especially in the low-dose region between the measured compensator-IMRT dose profile with the calculation from the KonRad treatment planning system (Siemens Medical Solutions, Concord, California) that also uses a single value linear attenuation coefficient for a cerrobend compensator. The measured dose profile is in better agreement with their EGS4 Monte Carlo simulation. In a more detailed Monte Carlo simulation study using OMEGA-BEAM code, Jiang and Ayyangar (1998) reported that a simple formula is adequate to convert a 6-MV photon intensity modulation map to a cerrobend compensator file. The study also indicated that the perturbation by the compensator of the photon beam is insignificant in terms of beam hardening, scatter dose, and surface dose in stark contrast to the results of Dimitriadis and Fallone who showed that a cerrobend compensator has a beam hardening effect and scatter effect that can cause a 6% error. The apparent disagreement between these two Monte Carlo studies may be attributed to differences in measurement. Dimitriadis and Fallone used in-air measurement (with a building-up cap) that is more sensitive to the low-energy component of the energy spectrum; Jiang and Ayyangar studied dose at depths that is less sensitive to low-energy region spectrum change. The latter study pointed out that although the percentage depth dose is not significantly altered by the compensator, the photon beam energy spectrum is.

A more recent Monte Carlo study, using the DOSXYZnrc code by du Plessis and Willemse (2006), shows that to achieve good agreement between measured and computed dose profiles at different depths and field sizes of compensators, similar to those found in actual compensator-IMRT clinical application, requires the consideration of both beam hardening through and scattering from the compensator. The study was conducted using 6-, 8-, and 15-MV polyenergetic pencil beams and the compensators studied included wax, brass, copper, and lead material and step-like shaped compensators. du Plessis and Willemse found that the single-value linear attenuation coefficient approach can only replicate the DOSXYZnrc data at limited depth range (10 cm in their case) and it overestimates dose at shallower depths and underestimates dose at greater depths. They conclude that exclusion of compensator-generated scatter can lead to dose errors as high as 7–8% for a lead compensator at 6 MV.

The quality of compensator-IMRT dosimetry heavily depends on the fidelity of the intensity map to compensator file conversion afforded by the treatment planning system. Compensators of high specific density material, large thickness, and large field size can generate more beam perturbations. Treatment planning systems that are less well equipped to handle beam perturbations may need ad hoc remedies such as the use of field size–dependent linear attenuation coefficients to improve the quality of compensator-IMRT dose calculation. Today, practically none of the treatment planning systems' dose algorithms, including those that are Monte Carlo based, are solely based on first principles and approximations are always made. Thus, obtaining theoretical parameters such as narrow beam linear attenuation coefficients may not lead to better compensator-IMRT dosimetry and approximations should be made under clinical application conditions. Using clinically realistic and actual clinical compensators for commissioning measurements and achieving the best fit of calculation to measurement for a variety of compensator shapes and sizes and tissue depths is the most reliable approach for compensator-IMRT commissioning. Independent of the specific compensator-IMRT technique, this approach optimizes the performance of the treatment planning system in compensator-IMRT clinical use and provides a reliable baseline for patient-specific compensator-IMRT QA.

IMRT Compensator Fabrication Commissioning

Each compensator-IMRT technique has its specific QA concerns in commissioning and in clinical use. Most of the techniques today use either a positive compensator or a negative mold of the compensator that is fabricated using a milling machine. The maximum thickness, maximum field size, and maximum weight of the compensator design must be measured/verified, recorded, and set up in the treatment planning system if possible in the commissioning process. The milling machine drill bit diameter and other milling machine setup parameters define the spatial resolution of the compensator and thus that of the intensity map it produces. Any intensity map feature that is smaller than drill bit diameter will be lost in the milling process. Smoothing out high spatial frequency features in the intensity map in the dose computation and optimization process that cannot be reproduced by the milling system will improve the agreement between compensator-IMRT dose calculation and measurement.

For compensators that are made of granular materials, there are additional quality assurance steps in the commissioning process. These compensators are fabricated by filling the negative compensator mold, fabricated first based on the compensator file, with granular compensator material. A rigorous filling procedure needs to be established and proven to give a consistent effective density of the packed compensator granule material before compensator-IMRT commissioning data collection. Chang et al. (2004) described their granule compensator commissioning technique and reported that the impact of dose delivery errors for the extreme cases of a packing density is 2–3%. Menon and Sloboda (2004) described an electronic portal imaging device–based method to verify their steel granule compensator fabrication quality and positioning accuracy. Different than commercial compensators that are milled directly from solid metal block via high precision computer-controlled milling machine, the quality of granule-based compensators is highly dependent on the fabrication process of the individual compensator. For this reason, it is crucial to establish a detailed fabrication procedure and limit the fabrication process to several designated and well-trained individuals.

For compensators that are fabricated via solidification of liquid cerrobend in a negative mold, the specific concerns during commissioning include internal voids and surface irregularity of the compensator. An uncertainty of 1 mm cerrobend thickness can lead to a 4% uncertainty in photon dose. The solidification problems can be resolved or lessened by modifying the heat sinks (metal plates) and cerrobend cooling rate. There are two heat sinks, or cooling sources, for liquid cerrobend that is inside a Styrofoam mold and placed on a metal plate—the metal plate underneath and the air above. Voids occur when the liquid cerrobend starts to solidify and shrink in volume simultaneously from bottom and top. Increasing the cooling rate of the metal plate can effectively reduce the impact of the air cooling and thus solidify liquid cerrobend in single direction from the bottom up. This can significantly reduce the chance of forming voids inside compensator and surface irregularity.

Compensator fabrication commissioning is a crucial step to produce stable and high-quality compensators for commissioning and for clinical use. This step needs to be carried out before compensator-IMRT treatment planning system commissioning.

Compensator-IMRT Clinical Use Procedure Commissioning

There are special considerations in using compensator-IMRT for patient treatment. Although there are automated compensator-IMRT techniques (Xu, Al-Ghazi, and Molloi 2004; Yoda and Aoki 2003), most compensator-IMRT patient treatments today rely on manual exchange of compensators between fields. The recent multi-institutional study by Chang et al. (2009) has shown that the manual compensator-IMRT used comparable, if not less, time compared to segmental MLC-IMRT treatments on Siemens, Elekta and Varian accelerators. Because it lacks the automation and the accelerator integration of the MLC-IMRT, a safe clinical use of compensator-IMRT requires a compensator-IMRT clinical use procedure before its first clinical use. Clinical physicists and therapists should work together to establish such procedures and conduct a dry run before the first patient treatment. One potential error is that the wrong compensator is used during treatment. One solution is to code each compensator tray (that is attached to the wedge/block holder of linac) with specific codes that can be recognized by the accelerator control and specify treatment record-and-verify system. When such patient-specific compensator verification system is not available, a usage procedure and training are needed to ensure the correct compensator is used for each patient and each treatment field. For granule material–based compensators, precautions are needed in the design and use of the compensator to prevent potential leakage of granule material during daily patient treatment.

Compensator-IMRT Patient-Specific QA

Patient-specific compensator-IMRT QA should follow the same guidelines as MLC-IMRT QA (Nelms and Simon 2007). A commonly used method of IMRT QA is 2D intensity map measurement using array detectors such as the diode-based MapCheck (Sun Nuclear, Melbourne, FL) (Jursinic and Nelms 2003) or ion chamber–based PTW 2D array (PTW, Freiburg, Germany) system. A radiographic film and ion chamber combination together with a film scanner and image analysis system such an RIT (RIT, Colorado) is an alternative to the 2D IMRT QA approach (Schneider et al. 2009; Srivastava and De Wagter 2007). Because the intensity modulation is static, compensator-IMRT QA measurement has the advantage of using a smaller number of MUs than actually used for the patient treatment and this speeds up the IMRT QA time compared with MLC-IMRT.

Although MLC-IMRT QA has specific concerns related to the MLC, compensator-IMRT QA concerns are related to compensator beam hardening and scatter, especially if the measurement technique itself has energy dependence. For instance, EDR2 film is more sensitive to lower-energy photons than higher-energy photons and thus film dosimetry may be affected by the scattered photons from compensators. Srivastava and De Wagter (2007) have studied the impact of MCP-96 cerrobend compensators on 6-MV beam EDR2 film dosimetry. They found that EDR2 film has a maximum deviation of 2% compared to a diamond detector that has practically no energy dependence. For a 25-MV beam, the maximum deviation is larger (5%) and there is a beam hardening effect below the depth of 20 cm and a beam softening effect at depths greater than 30 cm. Therefore, it is recommended to use measurement devices with minimum energy independence for compensator-IMRT QA.

A benchmark survey study on head and neck IMRT QA that involved 45 clinics using commercial compensators from .decimal Inc. and segmental MLC-IMRT is underway. Using the CRIS head and neck phantom, a seven-field IMRT plan was generated in each of the radiation oncology clinics using the site's linac and treatment planning system. The Mapcheck diode array was used for the IMRT QA measurement on all sites. The IMRT QA passing rate is a measure of the agreement between delivered and calculated doses, which is more an indication of the IMRT treatment planning quality than of the quality of the accelerator or MLC system. Preliminary results for this 50-clinic benchmark IMRT QA survey study indicate that commercial compensators are likely to have patient-specific IMRT QA that is comparable, if not better than, MLC-IMRT treatment.

Summary

Compensator-IMRT is a well-accepted alternative IMRT delivery approach that has provided high-quality patient IMRT treatment in more than 60 cancer treatment clinics in the United States. The quality of a compensator-IMRT program in clinical use is determined by the quality of treatment planning system commissioning, compensator fabrication commissioning, compensator-IMRT application procedure commissioning, and patient-specific IMRT QA. For most users that rely on commercial mail-order compensator services, the most important QA steps are careful treatment planning system and compensator-IMRT clinical use procedure commissioning.

References

Bakai, A, W. U. Laub, and F. Nüsslin. 2001. Compensators for IMRT—An investigation in quality assurance. *Z. Med. Phys.* 11: 15–22.

Chang, S. X. 2006. Compensator-intensity-modulated radiotherapy: A traditional tool for modern application. *US Oncol. Dis.* (II): 80–83.

Chang, S. X., T. J. Cullip, K. M. Deschesne, E. P. Miller, and J. G. Rosenman. 2004. Compensators: An alternative IMRT delivery technique. *J. Appl. Clin. Med. Phys.* 5 (3): 15–36.

Chang, S. X., K. M. Deschesne, H. Chen, K. Weeks, C. Sibata, E. Carey, P. Hill, T. Mackie, and L. Marks. 2009. IMRT treatment delivery efficiency—A multi-institutional retrospective study. *Med. Phys.* 36: 2551.

Dimitriadis, D. M., and B. G. Fallone. 2002. Compensators for intensity-modulated beams. *Med. Dosim.* 27 (3): 215–220.

du Plessis, F. C., and C. A. Willemse. 2006. Inclusion of compensator-induced scatter and beam filtration in pencil beam dose calculations. *Med. Phys.* 33 (8): 2896–2904.

Ibbott, G. S., D. S. Followill, H. A. Molineu, J. R. Lowenstein, P. E. Alvarez, and J. E. Roll. 2008. Challenges in credentialing institutions and participants in advanced technology multi-institutional clinical trials. *Int. J. Radiat. Oncol. Biol. Phys.* 71 (1 Suppl): S71–S75.

Jiang, S. B., and K. M. Ayyangar. 1998. On compensator design for photon beam intensity-modulated conformal therapy. *Med. Phys.* 25 (5): 668–675.

Jursinic, P. A., and B. E. Nelms. 2003. A 2-D diode array and analysis software for verification of intensity modulated radiation therapy delivery. *Med. Phys.* 30 (5): 870–879.

Menon, G. V., and R. S. Sloboda. 2004. Compensator thickness verification using an a-Si EPID. *Med. Phys.* 31 (8): 2300–2312.

Nakagawa, K., N. Fukuhara, and H. Kawakami. 2005. A packed building-block compensator (TETRIS-RT) and feasibility for IMRT delivery. *Med. Phys.* 32 (7): 2231–2235.

Nakagawa, K., K. Yoda, Y. Masutani, K. Sasaki, and K. Ohtomo. 2007. A rod matrix compensator for small-field intensity modulated radiation therapy: A preliminary phantom study. *IEEE Trans. Biomed. Eng.* 54 (5): 943–946.

Nelms, B. E., and J. A. Simon. 2007. A survey on planar IMRT QA analysis. *J. Appl. Clin. Med. Phys.* 8 (3): 2448.

O'Daniel, J. C., L. Dong, D. A. Kuban, H. Liu, N. Schechter, S. L. Tucker, and I. Rosen. 2004. The delivery of IMRT with a single physical modulator for multiple fields: A feasibility study for paranasal sinus cancer. *Int. J. Radiat. Oncol. Biol. Phys.* 58 (3): 876–887.

Oguchi, H., and Y. Obata. 2009. Commissioning of modulator-based IMRT with XiO treatment planning system. *Med. Phys.* 36 (1): 261–269.

Sasaki, K., and Y. Obata. 2007. Dosimetric characteristics of a cubic-block-piled compensator for intensity-modulated radiation therapy in the Pinnacle radiotherapy treatment planning system. *J. Appl. Clin. Med. Phys.* 8 (1): 85–100.

Schneider, F., M. Polednik, D. Wolff, V. Steil, A. Delana, F. Wenz, and L. Menegotti. 2009. Optimization of the gafchromic EBT protocol for IMRT QA. *Z. Med. Phys.* 19 (1): 29–37.

Srivastava, R. P., and C. De Wagter. 2007. The value of EDR2 film dosimetry in compensator-based intensity modulated radiation therapy. *Phys. Med. Biol.* 52 (19): N449–N457.

Williams, P. C. 2003. IMRT: Delivery techniques and quality assurance. *Br. J. Radiol.* 76 (911): 766–776.

Xu, T., M. S. Al-Ghazi, and S. Molloi. 2004. Treatment planning considerations of reshapeable automatic intensity modulator for intensity modulated radiation therapy. *Med. Phys.* 31 (8): 2344–2355.

Yoda, K., and Y. Aoki. 2003. A multiportal compensator system for IMRT delivery. *Med. Phys.* 30 (5): 880–886.

94

Cranial Stereotactic Radiosurgery and Stereotactic Radiotherapy

Christopher F. Serago
Mayo Clinic

Siyong Kim
Mayo Clinic

Ashley A. Gale
Mayo Clinic

Laura A. Vallow
Mayo Clinic

Introduction

Patient-specific quality control measures are essential for both stereotactic cranial radiosurgery and radiotherapy because of the risks associated with the procedures. For the purpose of this discussion, stereotactic radiosurgery (SRS) will mean a single-fraction, high-dose, highly focused radiotherapy treatment and stereotactic radiotherapy (SRT) will mean hyperfractionated, high-dose, highly focused radiotherapy treatment. Both cranial SRS and SRT are treatments that entail risk of severe consequences to the patient should a deviation from the intended treatment occur because of the proximity of eloquent areas of the brain or critical intracranial structures such as the optic chiasm, optic nerves, or brain stem to the therapeutic high-dose region. Quality control measures must be in place for patient immobilization, imaging, treatment planning, and treatment. General quality assurance references may be found in the published literature: quality assurance for linac-based stereotactic radiosurgery (Drzymala 1991), AAPM Task Group Report 54, stereotactic radiosurgery (Schell and Bova 1995), and quality assurance programs on stereotactic radiosurgery (Hartmann and Lutz 1995). An earlier report of quality assurance methods for a linac-based SRS program with some patient-specific tests is given by Tsai et al. (1991). This chapter describes our institution's quality control measures for *specific* patients with intracranial lesions. Extracranial targets require some different quality control measures that are not described within this chapter. A conventional linear accelerator for SRS and SRT is used at our institution; thus, we will not be describing quality control measures appropriate for other devices such as GammaKnife, CyberKnife, or Tomotherapy, although some of the concepts discussed are relevant across all platforms.

Immobilization

Quality control measures for SRS and SRT differ somewhat and are dependent on the immobilization and treatment technique used for the specific patient. At our institution, patient immobilization for SRS procedures may be performed either with a stereotactic head frame (Integra Radionics BRW) rigidly fixed with screws to the patient's skull or with a custom-fit thermoplastic mask. All of our SRT procedures use a thermoplastic mask for immobilization.

If the stereotactic head ring is used for immobilization and circular collimators are used for treatment, proper placement of the ring is required for successful immobilization, imaging, and treatment. The tilt of the head ring on the patient must position the patient's head/neck in a comfortable orientation for treatment. An improper tilt might force the patient's chin down to such an extent that breathing may become difficult. It is desirable but not always possible to choose the position of the screws attached to the skull so that computed tomography (CT) artifacts from the high-density screw tips do not interfere with visualization of the target lesion. The superior/inferior position of the ring must be sufficiently inferior to avoid potential collisions with the circular collimator housing during treatment. By not violating the clearance dimensions given in the checklist below, the treatment couch and gantry angles that may be used for treatment will be limited. The head ring must be oriented correctly so the ring can be correctly inserted into the couch mount device during imaging and treatment and so that localization hardware may be correctly attached to the ring. A checklist for the head ring placement used at our institution is as follows:

- Patient ID confirmed
- Ring tilt okay
- Screws not in the same CT plane as target volume
- Ring >10 mm below the inferior border of target volume
- Ring >30 mm below the center of the target volume
- Anterior of ring aligned with patient anterior
- Localizer mounting holes facing superior

Immobilization with a thermoplastic mask is less invasive for the patient but slightly less rigid and reproducible. For those reasons, we always use image guidance before treatment to ensure the patient is in the same position as they were for CT simulation. The thermoplastic mask does not have the high-density material associated with the stereotactic head ring, and the mask attached to the base frame (Type-S, Civco, Kalona, IA) is more compact than the head ring so the risk of collision is less. We use a mouthpiece with an impression of the upper dentures to assist patient immobilization in a reproducible position.

Thermoplastic mask checklist:

- Patient head position neutral, straight alignment
- Mask conforms to patient head, nose bridge, chin
- Mouthpiece comfortable and stabilizes patient cephalad/caudad head tilt

Imaging

The overall accuracy of the SRS or SRT treatment is dependent on the parameters of the CT and magnetic resonance imaging. Therefore, small slice thickness, small field of view, and high matrix value (i.e., 512 × 512) parameters are desirable. When the stereotactic head ring is used, it is explicitly assumed that the ring does not move on the patient's head from the time of CT simulation through the completion of treatment. To validate this assumption, a depth helmet is used to measure the position of the patient's head at the time of the CT simulation and just before the initiation of treatment. If mask immobilization is chosen, then the patient head position is verified before treatment with x-ray imaging. A physical CT-visible marker is attached to the right side of the patient's head and later used to verify if the software-generated "patient right" is correct. CT and magnetic resonance images are very commonly used for SRS and SRT treatment planning. Thin-slice magnetic resonance images improve the quality of subsequent image fusion of these two modalities. The checklist for imaging is as follows:

- Patient ID confirmed
- Depth helmet measurements complete (head ring immobilization)
- Stereotactic head ring correctly attached to couch (head ring immobilization)
- Mask correctly attached to base frame (mask immobilization)

- Radiographic "patient right" marker present
- CT slice thickness ≤1.25 mm
- Magnetic resonance imaging slice thickness 1 mm
- Scout view inferior/superior extent reviewed by radiation oncologist
- Field of view minimized
- CT localizer rods visible on CT image (head ring immobilization)

Treatment Planning

Treatment planning quality control measures do not differ significantly for SRS and SRT from conventional external beam radiotherapy planning, so are not discussed in detail here. Two items are worth mentioning, however. First, the couch and gantry orientations used for SRS and SRT may use the entire useful range of motion of both the couch and gantry. It is possible to inadvertently choose a combination of gantry and couch position that will result in a collision with either the patient or treatment couch. Therefore, it is recommended to review these parameters during the treatment planning process. Second, at our institution, we calculate the expected position of the treatment couch for each isocenter during the planning process and use this during the patient treatment setup procedure as a verification step. All treatment planning checks are completed before treatment.

Treatment planning checklist:

- Prescription verification
- Monitor unit check
- Reference DRR created (mask immobilization)
- Gantry not in a collision position
- Couch position calculated for treatment verified
- Plan check complete

Treatment

As described above, we may choose one of two alternative immobilization techniques. If the stereotactic head ring is used, then the depth helmet measurements are repeated before treatment to verify the ring has not moved. A couch mount with positional verniers is used with the stereotactic head ring, and the coordinates set on the couch mount are double checked. We expect depth helmet measurements to be consistent within 1 mm. If a thermoplastic mask is used, kilovoltage orthogonal x-rays, cone beam, and/or megavoltage x-rays are used for patient position verification. We consider image position verification of less than 2 mm to be acceptable for mask immobilization. Typically, it is less than 1 mm.

SRS and SRT treatments at our institution may be given via two alternative beam-shaping techniques. The first option is to use circular collimators with additional beam shaping with the secondary X and Y collimator jaws. The second option is to use the accelerator's multileaf collimator. The clearance between the end of the circular collimator and the linac isocenter is less than the clearance when the multileaf collimator is used, which increases the concern for a possible collision. In either instance, all gantry and couch orientations are checked before treatment. When circular collimators are used, a special collimator housing is attached to the gantry head, and we verify that the room lasers, collimator, collimator housing, and linac central axis beamline are properly aligned using a film technique first described by Lutz, Winston, and Maleki (1988). If the linac multileaf collimator is used for beam collimation, the coincidence of the room lasers with the accelerator isocenter is checked. The coincidence of the intersection of the room lasers with the accelerator isocenter is confirmed to be within 0.5 mm. Our accelerator does not have an interlock for the circular collimator size; thus, a second check of the correct collimator is imperative. For either treatment option, we calculate the expected position of the treatment couch and confirm the actual position is within a tolerance of 5 mm (head ring) or 6 mm (mask). This is not the accuracy of the treatment, but it rather allows for variation of the couch position dependent on the patient weight, the position of the couch mount adaptor, or the mask base frame. A visual confirmation of the isocenter position within the patient's head is completed before treatment. This consists of comparing the location of the lasers as they project on the patient's head to the known location of the target. This is intended to check gross positioning errors (e.g., right vs. left). After the patient is positioned for treatment, the lateral, longitudinal, and vertical couch motors are disabled so unintended movements in those directions cannot occur during treatment. These couch positions are monitored during treatment for consistency. After each couch rotation, the coincidence of the linac isocenter with the treatment position is verified and the couch position is adjusted when necessary.

Treatment: head ring checklist:

- Patient ID confirmed
- Confirm laser alignment to isocenter <0.5 mm
- Second staff member verifies collimator size
- Film test isocenter confirmation
- Set couch mount position verniers to center of travel
- Stereotactic coordinates double checked
- Verify the patient couch position is within tolerance of the expected couch position (<5 mm)
- Perform gantry rotation through all treatment arcs and confirm that no collisions will occur
- Repeat depth helmet measurements
- Visual confirmation of isocenter location
- Disable appropriate Varian couch movement motors

- Verify couch position with ceiling laser at each couch rotation angle
- Record Varian couch position. Verify that couch coordinates do not change during treatment

Treatment: frameless checklist:

- Patient ID confirmed
- Check overhead laser alignment to isocenter <0.5 mm
- Perform gantry rotation through all treatment arcs or beam orientations and confirm that no collisions will occur
- Verify the patient couch position is within tolerance of the expected couch position (<6 mm)
- Record couch coordinates. Verify couch coordinates do not change during treatment
- Visual confirmation of isocenter location
- Disable appropriate Varian couch movement motors
- Verify patient is securely immobilized
- Cone beam imaging completed if appropriate
- Kilovoltage lateral and AP (or PA) lateral setup verification x-rays taken
- Megavoltage port films for verification taken
- Setup verification x-rays match DRR with less than 2 mm difference
- Check couch position with ceiling laser at each couch rotation

References

Drzymala, R. E. 1991. *Quality Assurance in Radiotherapy Physics*, ed. G. Starkschall and J. Horton. Madison, WI: Medical Physics Publishing.

Hartmann, G. H., and W. Lutz. 1995. *Quality Assurance Program on Stereotactic Radiosurgery: Report from a Quality Assurance Task Group*. Berlin: Springer.

Lutz, W., K. R. Winston, and N. Maleki. 1988. A system for stereotactic radiosurgery with a linear accelerator. *Int. J. Radiat. Oncol. Biol. Phys.* 14 (2): 373–381.

Schell, M. C., F. J. Bova, D. A. Larson, D. D. Leavitt, E. B. Podgorsak, and A. Wu. 1995. *Stereotactic Radiosurgery: Report of Task Group 42 Radiation Therapy Committee*. Woodbury, NY: American Institute of Physics.

Tsai, J. S., B. A. Buck, G. K. Svensson, E. Alexander 3rd, C. W. Cheng, E. G. Mannarino, and J. S. Loeffler. 1991. Quality assurance in stereotactic radiosurgery using a standard linear accelerator. *Int. J. Radiat. Oncol. Biol. Phys.* 21 (3): 737–748.

95

Stanley H. Benedict
University of Virginia Health System

Jing Cai
Duke University Medical Center

Bruce Libby
University of Virginia Health System

Michael Lovelock
Memorial Sloan-Kettering Cancer Center

David Schlesinger
University of Virginia Health System

Ke Sheng
University of Virginia Health System

Wensha Yang
University of Virginia Health System

Stereotactic Body Radiation Therapy

Introduction

Patient-specific quality assurance is a critical component in the overall quality control of radiation treatment because it provides assurance that the delivered dose will accurately match the planned dose distribution. In particular, it is important to verify that the treatment device is physically capable of delivering the planned dose distribution and that patient setup in the treatment room matches as closely as possible the patient setup at time of simulation.

Patient-specific QA becomes even more critical for stereotactic body radiation therapy (SBRT) than for traditional treatments. SBRT differs from traditional radiation therapy in that it features highly conformal, hypofractionated dose delivery. Current SBRT protocols generally involve three to five treatments with a dose of 6–22 Gy per fraction to sites such as the spine, liver, and lung (Fukumoto et al. 2002; Gerszten et al. 2004, 2005; Hara et al. 2002; Herfarth et al. 2001; Timmerman et al. 2003; Whyte et al. 2003). The desired biological effect is achieved both by fractionation and perhaps more importantly by the differential dose delivered to targeted and normal tissue; the goal is to minimize the volume of normal tissue exposed to a high dose of radiation. Therefore, in SBRT, the traditional gross tumor volume and clinical tumor volume as described in ICRU 62 (ICRU1999) are often treated interchangeably (Timmerman 2008; Wulf et al. 2001, 2003). SBRT is also highly dependent on image guidance for localization and repositioning. The requirements of large doses and highly accurate targeting in SBRT mean that special attention needs to be paid to all aspects of the treatment for each patient, including immobilization, localization, pretreatment dose verification, and review of in-room imaging by the physician.

Designing a Comprehensive QA Program for SBRT

Patient-specific QA procedures for SBRT should be developed as an integrated part of a comprehensive ongoing QA program in the clinic. Therefore, before implementing an SBRT program, the clinic first needs to determine which system(s) will be used and to develop QA procedures to match. SBRT-enabled systems often have specialized equipment such as immobilization systems, localization systems, and on-board imaging systems that are not always found in the clinic. In other cases, the entire system is specialized for SBRT (e.g., the Accuray Cyberknife). Table 95.1 summarizes the stereotactic localization and image guidance strategies used by commercially available systems. These specialized components require detailed and specialized QA procedures, over and beyond the general guidelines for external beam radiotherapy as specified in the AAPM Reports of TG 40 and 45 (Kutcher et al. 1994; Nath et al. 1994). Table 95.4 summarizes recommendations for annual, monthly, and daily QA activities for SBRT clinics that enable verification of overall device accuracy.

As an example specifically relevant to SBRT, Table 95.2 lists published repositioning accuracies of various SBRT immobilization schemes. The reported accuracies vary significantly depending on the particulars of the approach. In general, errors in target localization and patient repositioning can be sorted into two categories: (1) setup errors, which can be greatly reduced with proper localization procedures; and (2) organ motion, especially motion due to the respiratory cycle, which can be highly dependent on the specific geometry and respiratory characteristics of the patient. QA procedures must therefore be developed that can address repositioning errors at the device level as well as at the patient level.

TABLE 95.1 Commercial Systems for SBRT and Their Associated Stereotactic Coordinate Systems

Stereotactic Coordinate System Definition	Commercial System	Image Guidance and Treatment Delivery
Frame/fiducial-based systems	Elekta body frame	Any accelerator
	Leibinger body frame	Any accelerator
	Medical Intelligence BodyFIX	Any accelerator
Tracked IR systems	Accuray Cyberknife	Robotic-assisted linear accelerator
	BrainLAB ExacTrac	Any gantry-based accelerator
In-room imaging-based systems	Elekta Synergy	Elekta linear accelerator
	Varian Trilogy	Varian linear accelerator
	Tomotherapy	Tomotherapy accelerator
	Accuray Cyberknife	Robotic-assisted linear accelerator
	BrainLAB ExacTrac	Any gantry-based accelerator
External fiducial/imaging combination systems	Medical Intelligence BodyFIX/ Radionics Localization device with GE or Siemens CT on rails	Varian Exact Target LINAC/CT-on-rails Siemens LINAC/CT-on-rails

Table 95.3 summarizes these two types of errors and lists strategies for verifying target localization, for both inter- and intrafraction variations. Image guidance can often be used to correct for setup errors. However, organ motion is not as easily accommodated, and further strategies that include compensation for this motion must be used.

Simulation, Treatment Planning, and Pretreatment Dose Verification

As discussed earlier, one of the characteristics of SBRT is that treatments are simulated and planned with the goal of providing the smallest possible margins to minimize the dose to normal

TABLE 95.2 Reported Accuracy of SBRT Immobilization and Repositioning Systems

Author (Year)	Site	Immobilization/Repositioning	Reported Accuracy
Lax (1994)	Abdomen	Wood frame/stereotactic coordinates on box to skin marks	3.7 mm lat 5.7 mm long
Sato (1998)	Abdomen	Frameless/combination CT, x-ray, and linac	Not reported
Hamilton et al. (1995)	Spine	Screw fixation of spinous processes to box	2 mm
Murphy (1997)	Spine	Frameless/implanted fiducial markers with real time imaging and tracking	1.6 mm radial
Lohr et al. (1999)	Spine	Body cast with stereotactic coordinates	≤3.6 mm mean vector
Yenice et al. (2003)	Spine	Custom stereotactic frame and in-room CT guidance	1.5 mm system accuracy, 2–3 mm positioning accuracy
Chang et al. (2004)	Spine	MI™ BodyFix with Stereotactic frame/ linac/CT on rails with 6D robotic couch	1 mm system accuracy
Tokuuye (1997)	Liver	Prone position Jaw and arm straps	5 mm
Nakagawa et al. (2000)	Thoracic	MVCT on linac	Not reported
Wulf et al. (2000)	Lung, liver	Elekta™ body frame	3.3 mm lat 4.4 mm long.
Fuss and Thomas (2004)	Lung, liver	MI™ BodyFix	Bony anatomy translation 0.4, 0.1, 1.6 mm (mean X, Y, Z); tumor translation before image guidance 2.9, 2.5, 3.2 mm (mean X, Y, Z)
Herfarth et al. (2001)	Liver	Leibinger body frame	1.8–4.4 mm
Nagata et al. (2002)	Lung	Elekta™ body frame	2 mm
Fukumoto et al. (2002)	Lung	Elekta™ body frame	Not reported
Hara et al. (2002)	Lung	Custom bed transferred to treatment unit after confirmatory scan	2 mm
Hof et al. (2003)	Lung	Leibinger body frame	1.8–4 mm
Timmerman et al. (2003)	Lung	Elekta™ body frame	Approximately 5 mm
Wang et al. (2006)	Lung	Medical Intelligence body frame stereotactic coordinates/CT on rails	0.3 ± 1.8 mm AP −1.8 ± 3.2 mm Lat 1.5 ± 3.7 mm SI

TABLE 95.3 Error-Analysis Strategy for Setup and Organ Motion

		Strategy		
		Immobilization/Setup Aids	Offline	Online
Setup errors	Interfraction	Alignment/constraint Standard procedures Lasers/light field on Tattoos Thermoplast masks Tape	Conventional weekly port film practice Statistical approaches: (1) population-based thresholds (2) individual-based thresholds	MV radiographs (conventional pre-ports) EPID MV radiographs Online kV radiographs (with/without markers) Optical video monitoring
	Intrafraction	Bite blocks Vacu-Form molds/casts Thermoplast body casts Stereotactic head frame Stereotactic body frame	[a]	MV fluoroscopic kV fluoroscopic Optical video monitoring Optical reflectors/markers
Organ motion	Interfraction	Breath-hold Consistent time-of-day Active breathing control Specifications (bladder/rectum, full/empty) Patient position (prone/supine)	Offline strategies based on repeat CT scans	Online computed tomography (CT-on-a-rail, cone-beam CT, tomotherapy) Ultrasound Other imaging modalities (MRI, ultrasound)
	Intrafraction	Breath-hold Compression plate Active breathing control	[b]	Respiratory gating Cardiac gating MV/kV fluoroscopy of surrogates for organ motion

[a] Devices and procedures often serve not only to provide accurate inter-fraction alignment (setup aids) but also to constrain against intrafraction motion (immobilization devices); hence, the distinction between inter- and intrafraction strategies is blurred in this case.

[b] Although offline correction strategies do not address intrafraction variability directly, such strategies may provide margins that better accommodate such variability provided the inter- and intrafraction motions are from the same distribution.

TABLE 95.4 Summary of QA Recommendations for SBRT Systems

Source	Purpose	Proposed Test	Proposed Tolerance	Proposed Frequency
Ryu et al. (2001)	End-to-end localization accuracy	Stereo x-ray/DRR fusion	1.0 to 1.2 mm root mean square	Initial commissioning and annually thereafter
Ryu et al. (2001)	Intrafraction targeting variability	Stereo x-ray/DRR fusion	0.2 mm average, 1.5 mm maximum	Daily (during treatment)
Verellen et al. (2003)	End-to-end localization accuracy	Hidden target (using stereo x-ray/DRR fusion)	0.41 ± 0.92 mm	Initial commissioning and annually thereafter
Verellen et al. (2003)	End-to-end localization accuracy	Hidden target (using implanted fiducials)	0.28 ± 0.36 mm	Initial commissioning and annually thereafter
Potters et al. (2004)	Comprehensive QA program for SBRT	N/A	N/A	Various
Yu et al. (2004)	End-to-end localization accuracy	Dosimetric assessment of hidden target (using implanted fiducials)	0.68 ± 0.29 mm	Initial commissioning and annually thereafter
Sharpe et al. (2006)	CBCT mechanical stability	TBD	TBD	TBD
Galvin and Bednarz (2008)	Overall positioning accuracy, including image registration (frame-based systems)	Winston–Lutz test	≤ 2 mm for multiple couch angles	Initial commissioning and monthly thereafter
Galvin and Bednarz (2008)	Overall positioning accuracy, including image registration (frame-based systems)	Winston–Lutz test modified to make use of the in-room imaging systems	≤ 2 mm for multiple couch angles	Initial commissioning and monthly thereafter
Palta, Liu, and Li (2008)	MLC accuracy	Light field, radiographic film, or EPID	<0.5 mm (especially for IMRT delivery)	Annually
Solberg et al. (2008)	End-to-end localization accuracy	Hidden target in anthropomorphic phantom	1.10 ± 0.42 mm	Initial commissioning and annually thereafter
Jiang, Wolfgang, and Mageras (2008)	Respiratory motion tracking and gating in 4D CT	Phantoms with cyclical motion	N/A	N/A
Bissonnette et al. (2008)	CBCT geometric accuracy	Portal image vs. CBCT image isocenter coincidence	± 2 mm	Daily

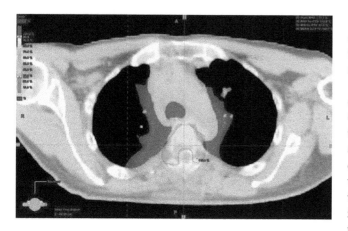

FIGURE 95.1 Axial CT slice for a spinal SBRT treatment showing the PTV (red contour) and isodose distribution.

tissue, account for motion of the tumor, and provide a repeatable setup. A variety of techniques for accounting for tumor motion within the PTV has been described and are summarized in the report of AAPM Task Group 76 (Keall et al. 2006). The efficacy of many of these techniques (for instance, breath-hold techniques) can vary on a patient-specific basis and therefore require patient-specific QA procedures to verify their appropriateness in any given situation.

Treatment planning may require patient-specific quality assurance procedures to ensure the dose distribution is within the appropriate dose-volume constraints for both the target and all relevant organs at risk.

The clinic should have established procedures for a second check and patient-specific QA for SBRT treatments. An axial CT slice is shown for a spinal SBRT plan in Figure 95.1, with the PTV contoured in red. Figure 95.2 provides an example of strategies that could be applied to patient-specific pretreatment dose verification. Figure 95.2a illustrates patient-specific QA that was performed for the spinal SBRT treatment in Figure 95.1 on the TomoTherapy® Hi-Art system® (TomoTherapy Inc.). Film is used to determine the measured isodose curves for a single plane, which is then compared to the planar dose determined from the planning system, along with a point dose at the plane of the film and the gamma distribution. In addition, the dose to a point (usually isocenter) is determined using an ionization chamber, which is then compared to the calculated dose at that point. An SBRT plan for the same patient was developed for delivery on a Varian Trilogy™ linear accelerator (Varian Medical Systems Inc.), with the QA performed using a MatriXX array (IBA Dosimetry GmbH). The isodose curves for a single plane for this treatment are shown in Figure 95.2b, along with the gamma distribution. The QA program used for SBRT needs to be well understood by the planning team, including changes in the plan (such as scaling of the monitor units so that film is not saturated) that need to be performed for the specific QA technique.

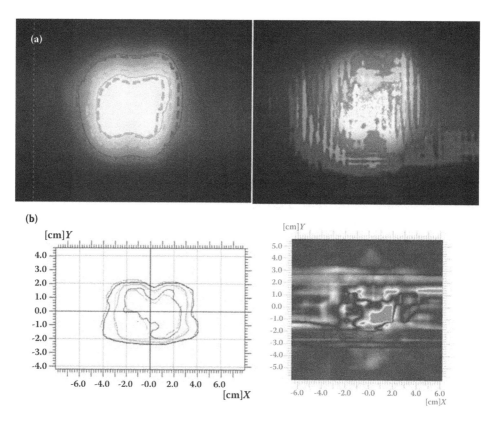

FIGURE 95.2 QA performed for the plan in Figure 95.1 on the (a) TomoTherapy Hi-Art and (b) Varian Trilogy linear accelerators.

Patient Repositioning and Treatment Delivery

Current SBRT systems rely on image guidance for patient repositioning at the beginning of each fraction of treatment. Typically, simulation images are transferred to the treatment console and are co-registered with kV and/or MV images acquired with the in-room imaging system from the treatment device. Offsets in the resultant co-registration are detected as setup shifts required to bring the patient into optimal setup correspondence with the simulation position. The clinic should have specific quality assurance procedures to evaluate the quality of the in-room imaging as well as the quality of the final patient setup. This may include a procedure that requires the on-treatment images be reviewed and approved by the physician for each fraction.

Summary

Four tables have been presented in this chapter that summarize the generalized patient-specific QA issues associated with SBRT. In addition, a dosimetric analysis for a spinal SBRT is presented in two figures that demonstrate the issues that must be addressed in verification of dose planning and delivery. For further and more comprehensive discussion on patient-specific QA for SBRT, the readers are advised to refer to the AAPM Task Group 101, which is currently in review by the AAPM with an anticipated publication date of 2010.

References

Bissonette, J. P., D. Moseley, E. White, M. Sharpe, T. Purdie, and D. A. Jaffray. 2008. Quality assurance for the geometric accuracy of cone-beam CT guidance in radiation therapy. *Int. J. Radiat. Oncol. Biol. Phys.* 71 (1 Suppl.): S57–S61.

Chang, E. L., A. S. Shiu, M. F. Lii, L. D. Rhines, E. Mendel, A. Mahajan, J. S. Weinberg et al. 2004. Phase I clinical evaluation of near-simultaneous computed tomographic image-guided stereotactic body radiotherapy for spinal metastases. *Int. J. Radiat. Oncol. Biol. Phys.* 59 (5): 1288–1294.

Fukumoto, S., H. Shirato, S. Shimzu, S. Ogura, R. Onimaru, K. Kitamura, K. Yamazaki, K. Miyasaka, M. Nishimura, and H. Dosaka-Akita. 2002. Small-volume image-guided radiotherapy using hypofractionated, coplanar, and noncoplanar multiple fields for patients with inoperable Stage I nonsmall cell lung carcinomas. *Cancer* 95 (7): 1546–1553.

Fuss, M., and C. R. Thomas Jr. 2004. Stereotactic body radiation therapy: An ablative treatment option for primary and secondary liver tumors. *Ann. Surg. Oncol.* 11 (2): 130–138.

Galvin, J. M., and G. Bednarz. 2008. Quality assurance procedures for stereotactic body radiation therapy. *Int. J. Radiat. Oncol. Biol. Phys.* 71 (1): S122–S125.

Gerszten, P. C., S. A. Burton, W. C. Welch, A. M. Brufsky, B. C. Lembersky, C. Ozhasoglu, and W. J. Vogel. 2005. Single-fraction radiosurgery for the treatment of spinal breast metastases. *Cancer* 104 (10): 2244–2254.

Gerszten, P. C., C. Ozhasoglu, S. A. Burton, W. J. Vogel, B. A. Atkins, S. Kalnicki, and W. C. Welch. 2004. CyberKnife frameless stereotactic radiosurgery for spinal lesions: Clinical experience in 125 cases. *Neurosurgery* 55 (1): 89–98; discussion 98–99.

Hamilton, A. J., B. A. Lulu, H. Fosmire, B. Stea, and J. R. Cassady. 1995. Preliminary clinical experience with linear accelerator-based spinal stereotactic radiosurgery. *Neurosurgery* 36 (2): 311–319.

Hara, R., J. Itami, T. Kondo, T. Aruga, Y. Abe, M. Ito, M. Fuse, D. Shinohara, T. Nagaoka, and T. Kobiki. 2002. Stereotactic single high dose irradiation of lung tumors under respiratory gating. *Radiother. Oncol.* 63 (2): 159–163.

Herfarth, K. K., J. Debus, F. Lohr, M. L. Bahner, B. Rhein, P. Fritz, A. Hoss, W. Schlegel, and M. F. Wannenmacher. 2001. Stereotactic single-dose radiation therapy of liver tumors: Results of a phase I/II trial. *J. Clin. Oncol.* 19 (1): 164–170.

Hof, H., K. K. Herfarth, M. Munter, A. Hoess, J. Motsch, M. Wannenmacher, and J. J. Debus. 2003. Stereotactic single-dose radiotherapy of stage I non-small-cell lung cancer (NSCLC). *Int. J. Radiat. Oncol. Biol. Phys.* 56 (2): 335–341.

ICRU. 1999. *ICRU Report 62. Prescribing, Recording and Reporting Photon Beam Therapy (Supplement to ICRU Report 50).* Bethesda, MD: International Commission on Radiation Units and Measurements.

Jiang, S. B., J. Wolfgang, and G. S. Mageras. 2008. Quality assurance challenges for motion-adaptive radiation therapy: Gating, breath holding, and four-dimensional computed tomography. *Int. J. Radiat. Oncol. Biol. Phys.* 71 (1 Suppl): S103–S107.

Keall, P. J., G. S. Mageras, J. M. Balter, R. S. Emery, K. M. Forster, S. B. Jiang, J. M. Kapatoes et al. 2006. The management of respiratory motion in radiation oncology report of AAPM Task Group 76. *Med. Phys.* 33 (10): 3874–3900.

Kutcher, G. J., L. Coia, M. Gillin, W. F. Hanson, S. Leibel, R. J. Morton, J. R. Palta, J. A. Purdy et al. 1994. Comprehensive QA for radiation oncology: Report of AAPM Radiation Therapy Committee Task Group 40. *Med. Phys.* 21 (4): 581–618.

Lohr, F., J. Debus, C. Frank, K. Herfarth, O. Pastyr, B. Rhein, M. L. Bahner, W. Schlegel, and M. Wannenmacher. 1999. Noninvasive patient fixation for extracranial stereotactic radiotherapy. *Int. J. Radiat. Oncol. Biol. Phys.* 45 (2): 521–527.

Murphy, M. J. 1997. An automatic six-degree-of-freedom image registration algorithm for image-guided frameless stereotaxic radiosurgery. *Med. Phys.* 24 (6): 857–866.

Nagata, Y., Y. Negoro, T. Aoki, T. Mizowaki, K. Takayama, M. Kokubo, N. Araki et al. 2002. Clinical outcomes of 3D conformal hypofractionated single high-dose radiotherapy for one or two lung tumors using a stereotactic body frame. *Int. J. Radiat. Oncol. Biol. Phys.* 52 (4): 1041–1046.

Nakagawa, K., Y. Aoki, M. Tago, A. Terahara, and K. Ohtomo. 2000. Megavoltage CT-assisted stereotactic radiosurgery for thoracic tumors: Original research in the treatment of thoracic neoplasms. *Int. J. Radiat. Oncol. Biol. Phys.* 48 (2): 449–457.

Nath, R., P. J. Biggs, F. J. Bova, C. C. Ling, J. A. Purdy, J. van de Geijn, and M. S. Weinhous. 1994. AAPM code of practice for radiotherapy accelerators: Report of AAPM Radiation Therapy Task Group No. 45. *Med. Phys.* 21 (7): 1093–1121.

Palta, J. R., C. Liu, and J. G. Li. 2008. Quality assurance of intensity-modulated radiation therapy. *Int. J. Radiat. Oncol. Biol. Phys.* 71 (1 Suppl.): S108–S112.

Sharpe, M. B., D. J. Moseley, T. G. Purdie, M. Islam, J. H. Siewerdsen, and D. A. Jaffray. 2006. The stability of mechanical calibration for a kV cone beam computed tomography system integrated with linear accelerator. *Med. Phys.* 33 (1): 136–144.

Solberg, T. D., P. M. Medin, J. Mullins, and S. Li. 2008. Quality assurance of immobilization and target localization systems for frameless stereotactic cranial and extracranial hypofractionated radiotherapy. *Int. J. Radiat. Oncol. Biol. Phys.* 71 (1 Suppl.): S131–S135.

Timmerman, R. D. 2008. An overview of hypofractionation and introduction to this issue of seminars in radiation oncology. *Semin. Radiat. Oncol.* 18 (4): 215–222.

Timmerman, R., L. Papiez, R. McGarry, L. Likes, C. DesRosiers, S. Frost, and M. Williams. 2003. Extracranial stereotactic radioablation: Results of a phase I study in medically inoperable stage I non-small cell lung cancer. *Chest* 124 (5): 1946–1955.

Wang, L., S. Feigenberg, L. Chen, K. Pasklev, and C. C. Ma. 2006. Benefit of three-dimensional image-guided stereotactic localization in the hypofractionated treatment of lung cancer. *Int. J. Radiat. Oncol. Biol. Phys.* 66 (3): 738–747.

Whyte, R. I., R. Crownover, M. J. Murphy, D. P. Martin, T. W. Rice, M. M. DeCamp Jr., R. Rodebaugh, M. S. Weinhous, and Q. T. Le. 2003. Stereotactic radiosurgery for lung tumors: Preliminary report of a phase I trial. *Ann. Thorac. Surg.* 75 (4): 1097–1101.

Wulf, J., U. Hadinger, U. Oppitz, B. Olshausen, and M. Flentje. 2000. Stereotactic radiotherapy of extracranial targets: CT-simulation and accuracy of treatment in the stereotactic body frame. *Radiother. Oncol.* 57 (2): 225–236.

Wulf, J., U. Hadinger, U. Oppitz, W. Thiele, and M. Flentje. 2003. Impact of target reproducibility on tumor dose in stereotactic radiotherapy of targets in the lung and liver. *Radiother. Oncol.* 66 (2): 141–150.

Wulf, J., U. Hadinger, U. Oppitz, W. Thiele, R. Ness-Dourdoumas, and M. Flentje. 2001. Stereotactic radiotherapy of targets in the lung and liver. *Strahlenther. Onkol.* 177 (12): 645–655.

Yenice, K. M., D. M. Lovelock, M. A. Hunt, W. R. Lutz, N. Fournier-Bidoz, C. H. Hua, J. Yamada et al. 2003. CT image-guided intensity-modulated therapy for paraspinal tumors using stereotactic immobilization. *Int. J. Radiat. Oncol. Biol. Phys.* 55 (3): 583–593.

96

Proton Radiotherapy

Zuofeng Li
*University of Florida Proton
Therapy Institute*

Roelf Slopsema
*University of Florida Proton
Therapy Institute*

Stella Flampouri
*University of Florida Proton
Therapy Institute*

Daniel Yeung
*University of Florida Proton
Therapy Institute*

Introduction

Proton therapy treatments may be delivered using a scattering technique, in which a proton pencil beam is scattered and optionally flattened laterally using physical scatterers, or a scanning beam technique, in which a proton pencil beam is magnetically scanned over the treatment field. Along the depth direction, pristine Bragg peaks of stepwise-reduced energies are stacked together to form the spread-out-Bragg-peak (SOBP). The intensity of a scanning pencil beam may be constant, delivering a cylindrical dose distribution just like in the scattering beam technique, resulting in the so-called wobbling or uniform scanning technique. The beam intensity may also be modulated during scanning to obtain intensity-modulated proton therapy. This chapter discusses patient-specific quality control issues of image-guided proton treatments using scattering and uniform scanning beams.

The selection of patient-specific QA tests for proton therapy, as in conventional x-ray photon therapy, is dependent on the completeness of system commissioning. In general, dosimetric parameters such as beam range and modulation width, machine output factors for a limited number of beam range and modulation width combinations, accuracy of treatment accessory calculation and fabrication, and the process of data transfer from the treatment planning system (TPS), via the record-and-verify system, to the proton therapy delivery system, have been extensively tested and confirmed during the system commissioning process. On the other hand, proton therapy systems are typically capable of delivering a nearly unlimited number of range and modulation combinations. It is therefore practically impossible to measure the dosimetric characteristics of all such combinations during system commissioning. Patient-specific QA tests therefore aim to identify the reproducibility of system commissioning results as applied to specific patient treatment fields and to complement these results where the specifics of the treatment fields fall beyond system commissioning. Table 96.1 lists the typical patient-specific QA tests in our clinical practice.

QA of Apertures and Compensators

The lack of ability to modulate beam intensity in scattering and uniform scanning techniques requires the use of compensators (Urie, Goitein, and Wagner 1984) such that the dose distribution conforms to the target distally, with the field shape defined by apertures of brass or similar high-Z material. The accuracy of these devices directly impacts the accuracy of treatment delivery, thus necessitating quality assurance before treatment.

Calculation of compensators, including the stopping power of the compensator material, should have been validated during the system commissioning process. Routine QA of a patient-specific compensator is therefore limited to the fidelity of its geometric shape and physical thicknesses. Compensator thicknesses at strategically selected points are measured, using an electronic height gauge, and compared to TPS calculated values, as shown in Figure 96.1. The geometric shape is verified by comparing the iso-thickness contours of the compensator to the beam's eye view plot from the TPS. QA tests of patient-specific apertures are performed by comparing their outlines to beam's eye view plots for geometric accuracy. In addition, for proton systems that accept multiple slabs of apertures, the number of slabs (total thickness) for each treatment field is verified to adequately stop all protons of the given energy of the beam. Apertures and compensators are checked for their labeling for patient and field names and their orientations. Bar code labels (when applicable) must be verified before their use for patient treatment.

Monitor Unit Determination and SOBP Verification

Unlike in conventional MV x-ray therapy, no consensus monitor unit (MU) calculation formalisms exist for proton therapy. Reported formalisms are usually specific to the mechanical designs of the beam scattering, flattening, and modulating elements of the proton treatment nozzle. Kooy et al. (2003, 2005)

TABLE 96.1 Image-Guided Proton Therapy Patient-Specific QA Tests

Description of Test	Tolerance
Aperture QA	
• Verify physical apertures against TPS plots of contours	1 mm
• Orientation and labeling	As planned
• Total thickness, when fabricated in multiple slabs	As planned
Compensator QA	
• Verify compensator thickness at multiple points using a height-gauge	0.5 mm for base plate
• Verify compensator iso-thickness contours against TPS iso-thickness plots	1.0 mm for all other points
• Orientation and labeling	1.0 mm
	As planned
MU measurement	As appropriate
SOBP measurement	As appropriate
Range and modulation measurement	As appropriate

reported a semiempirical approach that included a limited number of fitted parameters for each of the energy options of scattering proton beams. Moyers (1999) and Sahoo et al. (2008), on the other hand, adopted an approach similar to that used for conventional photon beam MU calculations. The method relies on measured tables of field size–dependent factors, modulation-dependent factors, and distance corrections of inverse square nature. The choice of MU calculation method for a given proton therapy system will be dependent on the design of the system, especially on how the energy options are structured. For example, output factors in the overlapping region of two energy options for the IBA System may differ by more than 20%, depending on which option is selected to deliver the beam. An output factor

table without consideration of energy options would therefore be inadequate.

Determination of MUs for a given treatment field can depend significantly on the homogeneity of the dose distribution around the calculation point, similar to the case of IMRT. Proton dose distributions can be significantly perturbed by the presence of sharp tissue interfaces of large inhomogeneities. The dose calculation point should be selected in a region of relatively homogeneous dose. A verification plan for QA purposes may be generated using the TPS. The plan is calculated without the compensator, using the same proton energy fluence of the field in the original patient plan, in a homogeneous water phantom. The geometry of the QA plan, as well as the location of the dose calculation point, must be reproduced for MU measurement.

Output factors of proton fields may be significantly dependent on field size and the beam's maximum range as well, as shown, for example, by Akagi et al. (2006). Such factors, if not measured and tabulated during system commissioning, will need to be measured during patient-specific QA tests.

Similarly, the flatness of a proton SOBP beam can be perturbed by the field size relative to the maximum range as well. Confirmation that the TPS-calculated SOBP of a proton beam of small field size agrees with the measured SOBP is necessary in such cases.

For uniform scanning beams, no MU calculation formalisms have been reported. Output factor and SOBP measurements for such beams are therefore necessary for each patient treatment field using this technique. In addition, the statistical reproducibility of scanning beams may need to be verified in repeated measurements.

Range and Modulation Width Measurement

The proton beam range and modulation width combination for a specific patient treatment field is frequently not encountered during system acceptance testing and commissioning, due to the potentially infinite number of such combinations of a proton

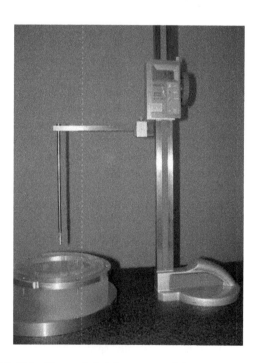

FIGURE 96.1 Electronic height gauge used for compensator thickness measurement.

therapy system. Delivery accuracy of a given combination, therefore, may be highly uncertain. Patient-specific QA tests of range and modulation width accuracy have to be considered in relation to initial system acceptance and commissioning, as well as to periodic system QA tests. Ideally, the commissioning and periodic QA tests provide adequate confidence in the accuracy of initial beam range and modulation calibration, and the consistency of system performance over time, such that patient-specific range and modulation width measurements are minimized. In addition, proton therapy systems are typically equipped with real-time beam range measurement devices that, although with coarser resolution than achievable with typical measurements, provide adequate monitoring that the system is functioning correctly. Although confidence in a proton therapy system may be less in the initial period of clinical operations using the said system, necessitating frequent patient-specific range and modulation width measurements, the frequency of such measurements may be reduced as experiences and confidence grows. In an established clinical operation, patient-specific range and modulation width measurements should be rare, limited to outlier combinations, or when range accuracy expectations exceed local system commissioning or periodic QA criteria.

References

Akagi, T., N. Kanematsu, Y. Takatani, H. Sakamoto, Y. Hishikawa, and M. Abe. 2006. Scatter factors in proton therapy with a broad beam. *Phys. Med. Biol.* 51 (7): 1919–1928.

Kooy, H. M., S. J. Rosenthal, M. Engelsman, A. Mazal, R. L. Slopsema, H. Paganetti, and J. B. Flanz. 2005. The prediction of output factors for spread-out proton Bragg peak fields in clinical practice. *Phys. Med. Biol.* 50 (24): 5847–5856.

Kooy, H. M., M. Schaefer, S. Rosenthal, and T. Bortfeld. 2003. Monitor unit calculations for range-modulated spread-out Bragg peak fields. *Phys. Med. Biol.* 48 (17): 2797–2808.

Moyers, M. F. 1999. Proton therapy. In *Modern Technology of Radiation Oncology*, ed. J. Van Dyk. Madison, WI.

Sahoo, N., X. R. Zhu, B. Arjomandy, G. Ciangaru, M. Lii, R. Amos, R. Wu, and M. T. Gillin. 2008. A procedure for calculation of monitor units for passively scattered proton radiotherapy beams. *Med. Phys.* 35 (11): 5088–5097.

Urie, M., M. Goitein, and M. Wagner. 1984. Compensating for heterogeneities in proton radiation therapy. *Phys. Med. Biol.* 29 (5): 553–566.

97

High-Dose-Rate Brachytherapy

Michael D. Thomas
Wake Forest University Baptist Medical Center

Megan M. Bright
Wake Forest University Baptist Medical Center

Introduction

High-dose-rate (HDR) brachytherapy enjoys several unique advantages as compared to external beam radiation therapy including an enhanced therapeutic ratio, smaller setup errors, decreased intrafractional organ motion, avoidance of anatomical changes over the treatment duration, and patient convenience because of fewer fractions. Thomadsen (2000) noted many of the disadvantages of HDR such as the utilization of a complicated treatment system, the compressed time frame for the procedure, and the delivery of a small number of large fractions, which makes dosimetric corrections unlikely. The disadvantages of HDR therapy can be ameliorated to a large degree by the generation of comprehensive protocols and the utilization of quality assurance (QA) checklists to ensure conformance to protocols. The HDR protocols and QA checklists define the patient-specific HDR QA activities and are a cornerstone of a comprehensive HDR QA program. This chapter summarizes the process and results of the development of a patient-specific QA program at a large community hospital, which uses written protocols and checklists.

Generation of Treatment Site–Specific Protocols

The generation of written protocols or procedures requires the gathering of input from the various team members in the HDR process including physicians, physicists, dosimetrists, nurses, computed tomography (CT) technicians, and others, and processing the input into a coherent and thoughtful document. The intent is not to generate step-by-step instructions but rather to form a prospective definition of critical parameters or procedural steps. In many ways, patient-specific QA is then a process of ensuring conformance to protocols and/or procedures.

Table 97.1 identifies a number of topics that may be addressed by the written treatment site–specific protocol. Although each topic is important, from a pure radiation therapy perspective, perhaps the treatment planning guidance warrants the most attention and forethought. It must be emphasized that the information within the protocol does not supplant the written directive but rather supplements the written directive by providing more detailed information. For instance, the protocol may stipulate the definition of gynecological treatment points per a specific industry standard or provide details of a physician's expectations. As an example, it is common for a physician to write the dose point for a vaginal cylinder as 0.5 cm from the surface of the applicator and provide the desired treatment length. The protocol may provide more details by stating that the dose point is considered to be the midpoint of the treatment length and that the planner is to ensure the dose received for the treatment length is within ±5% of the prescribed dose.

The planning portion of the protocol should also address the physician's delineation of the gross tumor volume as well as the margins used to define the clinical treatment volume and planning treatment volume. The identification of the dose volume histogram constraints to organs at risk for planning purposes must also be included. General guidance to the treatment planner for resolving potentially conflicting treatment goals may also be appropriate to increase the probability of physician approval of the treatment plan. For instance, for prostate HDR therapy, the physician may prefer a lower dose to the posterior aspect of the prostate to spare more rectal tissue.

Generation of QA Checklists

The HDR process is well suited for the use of patient-specific QA checklists. The checklists serve as a stimulus to ensure the proper completion of a critical task. The difficulty in the creation of QA

TABLE 97.1 Treatment Site–Specific Protocol Topics

Topic	Description
Patient procedure preparations	The documentation of necessary preoperative patient preparations may prevent undesirable conditions that may lead to an inferior procedure. For instance, for prostate HDR procedures, proper bowel preparation may prevent the formation feces/gas that may degrade the optimum geometry. For bronchial HDR, the use of an anesthetic to numb the airway may help prevent coughing and potential movement of the catheter.
Approved applicators, inspection, and handling instructions	The protocol should provide a description of the approved applicators including catheters and connectors; this description should define the critical parameters of the applicators to include the nominal catheter length to be used in the treatment planning system. The protocol should provide instructions for inspecting and ensuring the integrity of the applicator before its use. In addition, handling instructions, including sterilization requirements, are critical for the prevention of infection. It is recommended that the infection control department review these instructions.
Implantation of radiographic markers	The need for the implantation of radiographic markers and the intended region of implantation should be presented.
Image guidance requirements	It is recommended that the requirements for each image guidance modality be presented, including the expected image quality.
Applicator insertion guidelines	The instructions for the proper insertion of the applicator should be presented, including proper assembly, the use of build-up caps, and the optimum geometric relationships between components of the applicator. The placement of packing to limit radiation dose to normal tissues should be discussed.
Treatment position and immobilization	Depending on the treatment site, patient positioning may be important in the consistent application of the treatment plan. The use of immobilization may be beneficial to ensure consistent patient positioning.
CT parameters	The documentation of the CT parameters is necessary to ensure the CT planning set is adequate for treatment planning. Such parameters as slice width, inferior and superior boundaries, contrast requirements, use of catheter marker wires, and so on, should be addressed. In addition, the requirement that the physician approve the CT image set before patient removal from the CT bore should be stipulated.
Dose regimen	The anticipated fractional doses and total dose should be stated.
Planning guidance	Utilization of industry standard protocols should be identified and typical prescriptions on written directives fully explained. The protocol should provide contouring guidelines including margins used to derive PTV. It is recommended that dose goals to the PTV or prescription points are identified and dose constraints to critical targets stated.
Afterloader connection process	State the personnel responsible for the connections and the procedure to be followed. It should be stipulated whether physician final approval of setup is required before treatment.
Wound care	Patient education and physician orders for wound care should be described in the treatment site–specific protocol.

Note: PTV, planning treatment volume.

checklists is (1) the identification of the critical steps in the process that warrant a check and (2) the generation of a document that is useful but not burdensome or for which the completion is perfunctory. The following sections summarize four approaches that may prove useful to the designer of a patient-specific HDR QA program.

Dry Runs

Before the initial performance of the HDR procedure for a particular treatment site and after the generation of the treatment site–specific protocol, a dry run of the procedure should be performed. The dry run should be as realistic as possible with a "patient" and include activities performed in the operating room and radiology as applicable. The designer of the patient-specific HDR QA program should be an observer and, to the extent possible, not be an active participant in the procedure. This will give the designer an opportunity to map out the process as a flowchart and note personnel errors or confusion about the process, thus indicating the potential need for a task to be included on a checklist. The dry run may also identify shortcomings in the treatment site protocol that require correction.

Failure Mode and Effect Analysis

Failure mode and effect analysis is an analytical tool used by the designer of a QA program to help identify likely errors and to postulate the consequences of such errors. Using the process flowcharts from the dry run, the designer of the HDR QA program can identify critical steps in the process and potential failure modes within the steps that can be prevented by a checklist task. Depending on the experience of the designer, the failure mode and effect analysis may be more or less rigorous. As discussed by Huq et al. (2008), the American Association of Physicists in Medicine Task Group 100 has adopted failure mode and effect analysis and similar techniques for evaluating QA requirements for radiation therapy.

Professional Guidance

The American Association of Physicists in Medicine Task Guide 59 (Kubo et al. 1998) is the definitive publication from the medical physics community describing the many elements of a comprehensive HDR QA program. The document notes the benefit of QA checklists by stating, "Formalizing treatment execution by means of QA check-off forms is often a worthwhile enhancement of the basic 'integrated' QA program. Developing such forms forces one to systematically conceptualize the treatment planning process." Some recommendations include verification of the patient's name and date of treatment, source strength, source position reconstruction, prescription, and so on, and these tasks have been incorporated into the recommendations of this document.

Periodic Reviews

Periodic reviews of the protocols and checklists are recommended because the HDR treatment process may evolve and previously unidentified weakness may need to be addressed. In addition, some QA checklist tasks may be found to be unnecessary or provide little value in mitigating risk.

Example of Tasks for a Patient-Specific HDR QA Program

The following sections present example tasks for a patient-specific HDR QA program implemented at a large community hospital. The patient-specific HDR QA program consists of four checklists each having a distinct purpose as described below:

1. Pretreatment checklist: documents QA tasks to be performed that are generic in nature and not tied to a specific treatment site

2. Independent assessment: documents the QA activities performed by an independent reviewer who did not participate in the treatment planning process
3. Treatment site–specific checklist: documents the QA tasks that are unique to a particular treatment site
4. Posttreatment checklist: documents the surveys performed subsequent to the treatment procedure

Certainly, the checklists can be integrated into a single document, but the work flow in the department may necessitate multiple sections of the document. For instance, many activities may occur simultaneously at different locations and it would be counterproductive for personnel to have to locate the integrated checklist. The design of the patient-specific QA program must consider the department's workflow for a particular HDR procedure.

Pretreatment QA Checklist

The pretreatment QA checklist is designed to be a generic checklist to guide personnel (usually the physicist/treatment planner) through the HDR process irrespective of the treatment site. The QA tasks shown in Table 97.2 are those tasks that are primarily designed to ensure (1) the daily functional safety checklist for the afterloader has been successfully completed, (2) an independent assessment of the treatment plan has been performed, (3) the site-specific checklist is complete, and (4) the plan and the afterloader printout are consistent. The pretreatment checklist

TABLE 97.2 HDR Pretreatment Checklist

QA Task	Description
Daily functional and safety checklist complete	A check conducted to ensure successful performance of the daily functional and safety tests for the HDR afterloader.
Independent reviewer assessment complete	A check to ensure an independent assessment has been conducted. The independent reviewer assessment is intended to provide a mechanism for a detached, objective review to ensure the plan to be delivered satisfies the physician's prescription. See Table 97.3.
Site-specific checklist complete	A check to ensure successful completion of treatment site–specific QA tasks. The treatment site checklist specifies those QA tasks to be performed that are unique for the treatment of a particular anatomical site. See Table 97.4 for examples.
Prescription, plan, and isodose distribution signed by physician	A final verification to ensure the physician has documented approval for the plan to be delivered.
Plan and afterloader printout are for the patient being treated	A verification to ensure that the correct plan has been transferred for the patient to be treated. In addition, the identity of the patient must be verified before treatment (Code of Federal Regulations 2007).
Source activity matches between plan and afterloader printout	A verification that the source activity on the approved plan agrees with the source activity stated by the HDR afterloader computer.
Dwell locations and times match between plan and afterloader printout	The successful transfer of the plan from the treatment planning system to the HDR computer is verified. For plans with a large number of catheters/dwell positions, this is done by two persons in a reader–checker fashion.
Two-person verification that catheters connected per plan	Verification of the proper connection of the treatment catheters is critical to delivering the planned treatment. A two-person team connects the catheters; one person connects the catheters under the direct observation by the second individual.
Catheters verified to have no kinks or bends and routed to prevent unnecessary dose	To avoid unsuccessful treatment based on the detection of an obstruction, the absence of kinks and sharp bends must be verified. In addition, unnecessary dose to the patient is achieved by the use of towels to separate the catheters from the patient.
Catheter length on plan, afterloader printout, and site-specific checklist agree within ±1 mm	A verification is performed to ensure acceptability between the catheter length documented on the plan, afterloader printout, and treatment site–specific checklist.

also documents the verification of the proper connection of the treatment catheters to the afterloader and a final check that the catheter length on the treatment plan and afterloader are within ±1 mm of the recorded measurements.

HDR Independent Assessment

Subsequent to the completion of the treatment plan by the treatment planner (dosimetrist or physicist), the daily functional and safety checklist, prescription, and treatment plan are presented to a reviewer for an independent assessment. It is important that the reviewer has not been involved in the treatment planning process so that preconceived ideas are avoided. Table 97.3 provides a listing and description of the tasks performed by the independent reviewer. The primary function of the independent reviewer is to ensure that the treatment plan meets the requirements of the signed prescription and the guidelines in the department's previously approved planning protocols. The independent reviewer also assesses the treatment planner's reconstruction of the treatment catheters to ensure that the reconstruction is reasonable. This can be accomplished by reviewing the plan in the treatment planning system or by reviewing digitally reconstructed radiographs. Note the intent is not to have the reviewer re-perform the planning process but to review the result to ensure reasonableness.

The reviewer is also tasked with the responsibility of performing an independent assessment of the dose computational algorithm of the treatment planning system. This can be accomplished by various methods including the calculation of dose using spreadsheets, the utilization of mathematical models and dose indices, the exporting of dwell coordinates and times to a secondary computational platform, and the evaluation of the source strength and dwell times used to create an isodose volume. The use of spreadsheets to perform a computational check of point and line sources is very straightforward. The point source approximation and formalism of the American Association of Physicists in Medicine Task Guide 43 (Rivard et al. 2004) can be incorporated into a spreadsheet and the resultant dose calculation for the prescription point is usually within a few percent of

the treatment planning system. Before the widespread use of CT and 3D planning, the principal method of ensuring the reasonableness of an HDR treatment plan was comparison to mathematical models and the use of dose indices as discussed in the report of the American Association of Physicist in Medicine Task Group 59 (Kubo et al. 1998). The exporting of dwell coordinates and times from the treatment planning computer to a secondary computational platform can be accomplished by commercially available or institutionally developed software.

The use of CT planning facilitates a quick and straightforward method to ensure the computational reasonableness of a treatment plan. The relationship between the volume enclosed by the 100% isodose contour and the source strength and total dwell time can be used to perform a verification of the treatment planning system computation. The relationship can be evaluated as an arbitrary empirical fit (Rush and Thomas 2005; Wilkinson 2006) or can be based on a fit to a point source approximation (Das et al. 2006). Rush and Thomas (2005) and Wilkinson (2006) note that the resultant empirical fit has demonstrated the ability to predict the relationship for various HDR treatment plans with an average difference of less than ±3%, and Das et al. (2006) notes that the largest deviation was less than ±6% for various HDR plan geometries. Of note, the authors chose to use the volume of the 100% isodose as the parameter of choice rather than the volume of the planning treatment volume. If the planning treatment volume is used as the analyzed volume, two implant quality parameters are being evaluated simultaneously. These two implant quality parameters are (1) the computational correctness of the treatment planning system and (2) the dosimetric coverage consistency of the planning treatment volume as compared to the plan used to derive the empirical fit. By the use of the 100% isodose volume, only the computational accuracy is being evaluated and the dosimetric adequacy is assessed by the physician using dose volume histograms.

There is an explicit reliance on the volume calculation routine of the treatment planning system when one determines the volume of the 100% isodose volume. In addition, physicians judge the acceptability of treatment plans using dose volume histograms that also rely on the treatment planning system's volume

TABLE 97.3   HDR Independent Assessment

QA Task	Description
Physician prescription signed and complete	A verification that the physician has appropriately documented and signed the prescription.
Plan prepared per the prescription and treatment site protocol	An independent assessment is performed to ensure the plan meets the requirements of the prescription and criteria specified in protocols including margins. Deviations with approved protocols are discussed with the planner and physician to ensure acceptability.
Localization reconstruction evaluated and deemed reasonable	A qualitative assessment of the catheter reconstruction is performed to provide assurance that each catheter is correct. This can be performed by review of the CT in the treatment planning system or by use of digitally reconstructed radiographs from the CT and treatment planning system.
Plan source activity is correct	A verification to ensure the source activity of the plan matches the activity documented on the daily functional and safety checklist.
Independent calculation performed to verify reasonableness of plan	This is a quantitative assessment of the plan to ensure the computational algorithm of the treatment planning system is performing correctly. This check may use spreadsheets, empirical fits to isodose data, mathematical models or indices, or the exporting of dwell coordinates/times to a secondary computational platform.

Note: DRRs, digitally reconstructed radiographs.

TABLE 97.4 Examples of Checklist QA Tasks for Specific Treatment Sites

QA Task	Description
Prostate	
Needle length documented in operating room (OR)	The needle length used is documented in the OR and later used by the treatment planner to ensure the catheter length in the treatment planning system is correct.
Needle template form complete and needles scribed; template locked (operating room)	A form documenting the needle placement using a facsimile of the template is completed in the OR and the needles are numbered per the department protocol. A scribe mark is placed on each needle to indicate the insertion depth based on placement by ultrasound, fluoroscopy, and cystoscopy. Needle hubs can be marked with colored pens to indicate row to facilitate later connection. The template is locked to help prevent needle migration.
Needle depth verified by physician during CT imaging, subsequent fractions checked by fluoroscopy and needle scribe marks	For prostate HDR procedures, the depth of needle insertion is the geometric parameter most likely to change. The OR scribe marks are used to by the physician to confirm the proper needle depth during the CT imaging, but the needle depths and thus the scribe marks may be modified based on CT imaging. For subsequent fractions, fluoroscopy is used by the physician to ensure the needles are inserted to the correct depth based on the visualized anatomy. The scribe marks on the needles provide a means to check depth immediately before treatment.
Treatment plan prepared per written prescription and protocol	This is a verification performed by the treatment planner to ensure that the plan has been prepared according to the site protocol with attention to the prescribed dose and margins. In addition, a final check is performed to ensure the catheter length in the plan is correct based on the OR needle length and that the catheters are numbered per the needle template form.
Catheter measurement performed	This is a spot check measurement in the treatment room of one or two of the patient's catheters to ensure agreement (±1 mm) with the nominal lengths and the treatment plan.
Breast	
Documentation of balloon size, fill volume, and use of ellipsoidal balloon	The balloon characteristics are documented and used to assess the geometric adequacy of the implant.
CT imaging complete per protocol; initial measurement of catheter length	This is a verification that the CT imaging conducted per the department protocol includes optimal patient positioning and the use of a marker wire. In addition, an initial measurement of the catheter length is documented and provided as input for treatment planning.
Assessment of balloon conformance, balloon-to-skin distance, balloon diameter, balloon symmetry and deformation	This evaluation ensures conformance to the manufacturer's specifications, and these parameters are assessed to ensure acceptability according to set tolerances.
Treatment plan prepared per written prescription and protocol	This is a verification performed by the treatment planner to ensure that the plan has been prepared consistently with the treatment site protocol and dose volume histogram objectives. In addition, a check is performed to ensure the catheter length in the plan is correct based on the catheter length measurement performed during CT imaging.
Documentation/verification of balloon diameters using orthogonal films	With the patient in the treatment position (may use immobilization devices) on the first treatment day, the fluoroscopic simulator is used to image the treatment site and balloon. Orthogonal films are obtained and the diameters of the balloon recorded. Marks are placed on the patient so that on subsequent treatment days, the constancy of the balloon diameters can be easily verified to ensure consistent geometry.
Verification of catheter length	The catheter length is verified to be within ±1 mm of the plan value before the delivery of each fraction.
Vaginal Cylinder	
Diameter and length of vaginal cylinder documented, verification that the applicator is assembled properly	Because of simplistic nature of the applicator and procedure, vaginal cylinder treatments are typically delivered using a generic treatment plan without the aid of patient CT imaging. The documentation of the cylinder diameter and length is necessary to ensure the treatment is delivered per the physician's prescription, which is usually referenced to the surface of the applicator. The applicator is visually inspected to ensure proper assembly.
Documentation of fixation device measurements	Subsequent to the position approval by the physician under fluoroscopic imaging, measurements of the fixation device are recorded. On subsequent treatment fractions, the orientation of the vaginal cylinder is recreated based on these measurements. In addition, the measurements are used to re-verify the vaginal cylinder geometry in the treatment room because the patient is moved from the simulator to the treatment room (a transport board helps to minimize geometric changes).
Radiographic images reviewed	A visual verification by the treatment planner to ensure the dimensions of the applicator are correct and the applicator is properly assembled as compared to the radiographic baseline.
Treatment plan prepared per written prescription and protocol	This is a verification performed by the treatment planner to ensure that the plan has been prepared consistently with the treatment site protocol. In addition, a check is performed to ensure the catheter length in the plan is correct based on the nominal catheter length.
Verification of catheter length	The catheter length is verified to be within ±1 mm of the plan value before the delivery of each fraction.

(continued)

TABLE 97.4 Examples of Checklist QA Tasks for Specific Treatment Sites (Continued)

QA Task	Description
	Ring and Tandem
Documentation of applicator inserted including build-up cap used	This parameter is documented because the definition of the dose points may include the offset provided by the buildup caps.
Documentation of fixation device measurements	Subsequent to the position approval by the physician under CT imaging, measurements of the fixation device are recorded. On subsequent treatment fractions, the orientation of the ring and tandem is recreated based on the measurements and verification by fluoroscopy. In addition, the measurements are used to re-verify the geometry in the treatment room since the patient is moved from the simulator to the treatment room (a transport board helps to minimize geometric changes).
Treatment plan prepared per written prescription and protocol	This is a verification performed by the treatment planner to ensure that the plan has been prepared consistently with the treatment site protocol. In addition, a check is performed to ensure the catheter length in the plan is correct based on the nominal catheter length.
Verification of catheter length	The catheter length is verified to be within ±1 mm of the plan value before the delivery of each fraction.
	Bronchus
Proper catheter depth marked indicated with tape at the nares	A tape marker is placed in the OR and subsequently used as verification that the catheter insertion depth has not changed.
CT imaging complete per protocol; verification that the catheter placement is consistent with imaging performed during bronchoscopy	This is a verification that the CT imaging conducted per the department protocol includes optimal patient positioning and the use of a marker wire; this also verifies that the physician approves catheter placement consistent with imaging performed during bronchoscopy.
Initial measurement of catheter length	The catheter length for each treatment is unique because the catheter is cut to accommodate the measurement wire. This is the measurement value documented and later used as input in the treatment planning system. The end of the open catheter is colored with a marker to provide a method to detect cutting by hospital personnel between fractions.
Treatment plan prepared per written prescription and protocol	This is a verification performed by the treatment planner to ensure that the plan has been prepared consistently with the treatment site protocol. In addition, a check is performed to ensure the catheter length in the plan is correct based on the measured catheter length.
Verification of catheter length	The catheter length is verified to be within ±1 mm of the plan value before the delivery of each fraction.

calculation routine. For these reasons, it is considered prudent to test this routine on a periodic basis using a CT set with an object of a known volume. Constancy checks can also be performed for patient-specific QA by contouring an applicator object and comparing the result to previous patients. Rush and Thomas (2005) note that a constancy check using template holes for three consecutive slices yielded volumes within ±4% of the mean.

Treatment Site–Specific Checklist

The use of a treatment site–specific checklist affords the QA program designer a mechanism to generate a list of those QA tasks that are unique to a particular medical procedure. As an example, the scribing of the needles at the template for a prostate HDR procedure is a useful action that has no corollary with

TABLE 97.5 HDR Posttreatment Checklist

QA Task	Description
Documentation of the manufacturer, model number, and serial number used for the patient survey instrument	The manufacturer, model number, and serial number of the survey instrument are documented for regulatory purposes.
Documentation of the date of the last annual calibration for the survey instrument	The date of the last annual calibration of the survey instrument is documented for regulatory purposes.
Background exposure rate in the treatment room	The background exposure rate in the treatment room (6 ft or more from the patient) is documented. This measurement is used as a comparison to the patient's measurement to ensure the HDR source is no longer present.
Exposure rate in contact with the patient at the treatment site	The exposure rate in contact with the patient at the treatment site is conducted to ensure the HDR source has been removed.
Exposure rate in contact with the HDR at a predesignated point	The exposure rate in contact with the HDR unit is measured and documented as an additional verification that the source has been properly parked. The exposure rate reading is compared with the initial reading (indicated by a label at the predesignated point) recorded at the time of source replacement. Because of radioactive decay, the posttreatment measurement should always be lower than the initial reading.
Exposure rate of all catheters, connectors, and applicators used during treatment	This is a verification that the exposure rates for the catheters, connectors, and applicators are background.

a breast balloon implant. There are two similar QA tasks that appear on all treatment site–specific checklists, namely, verification that the treatment plan is prepared per the written directive/protocol and verification that the treatment room catheter measurement is acceptable. The reason for this is that the treatment site–specific QA checklist may document which protocol dose volume histogram goals were not met and the unique expected nominal catheter length is provided. In addition, the catheter length measurement may be complicated by whether a connector is to be attached during the measurement. Therefore, the expected nominal catheter length must specify whether or not the connector is to be attached for the measurement.

Table 97.4 shows examples of checklist QA tasks for specific treatment sites. As a matter of convenience, the treatment site-specific QA checklist can be integrated into a larger department administrative checklist; however, the QA tasks should be explicitly identified. The administrative checklist may contain such tasks as ensuring the signing of an informed consent, the sterilization of required equipment, the adequacy of supplies, the education of the patient with regard to wound care, and so on.

Posttreatment QA Checklist

The posttreatment QA checklist (Table 97.5) is used primarily to document the patient's radiation survey subsequent to the completion of the HDR procedure. An independent verification of the proper parking of the afterloader source can also be performed by measuring the exposure rate at a predesignated point on the afterloader housing. The posttreatment exposure rate reading is compared to an initial reading (indicated by a label at the predesignated point) recorded at the time of source replacement. Because of radioactive decay, the posttreatment reading should always be equal to or lower than the initial reading.

References

Code of Federal Regulations. 2007. 10CFR35. Medical Use of Byproduct Material, 72 Federal Register 45151 (Aug. 13).

Das, R. K., K. A. Bradley, I. A. Nelson, R. Patel, and B. R. Thomadsen. 2006. Quality assurance of treatment plans for interstitial and intracavitary high-dose-rate brachytherapy. *Brachytherapy* 5 (1): 56–60.

Huq, M. S., B. A. Fraass, P. B. Dunscombe, J. P. Gibbons Jr., G. S. Ibbott, P. M. Medin, A. Mundt et al. 2008. A method for evaluating quality assurance needs in radiation therapy. *Int. J. Radiat. Oncol. Biol. Phys.* 71 (1 Suppl.): S170–S173.

Kubo, H. D., G. P. Glasgow, T. D. Pethel, B. R. Thomadsen, and J. F. Williamson. 1998. High dose-rate brachytherapy treatment delivery: Report of the AAPM Radiation Therapy Committee Task Group No. 59. *Med. Phys.* 25 (4): 375–403.

Rivard, M. J., B. M. Coursey, L. A. DeWerd, W. F. Hanson, M. S. Huq, G. S. Ibbott, M. G. Mitch, R. Nath, and J. F. Williamson. 2004. Update of AAPM Task Group No. 43 Report: A revised AAPM protocol for brachytherapy dose calculations. *Med. Phys.* 31 (3): 633–674.

Rush, J. B., and M. D. Thomas. 2005. Quality assurance of HDR prostate plans: Program implementation at a community hospital. *Med. Dosim.* 30 (4): 243–248.

Thomadsen, B. R. 2000. *Achieving Quality in Brachytherapy*. Bristol (UK): Institute of Physics Publishing.

Wilkinson, D. A. 2006. High dose rate (HDR) brachytherapy quality assurance: A practical guide. *Biomed. Imaging Interv. J.* 2 (2): e34.

98

Low-Dose-Rate Brachytherapy

Yan Yu
Thomas Jefferson University

Tarun Podder
Thomas Jefferson University

Introduction

Brachytherapy (from the Greek *brachy*, meaning "short") is the form of radiotherapy where a radioactive source is placed inside or near the target volume requiring treatment. Example organ sites that are treated with low-dose-rate (LDR) brachytherapy are prostate (Nag et al. 1999, 2001), breast (Pignol et al. 2009), lung (Trombetta et al. 2008; Mutyala et al. 2010), eye (Jensen et al. 2005; Rivard and Melhus 2009), brain (Fernandez et al. 1995; Chen et al. 2007), and liver (Armstrong, Anderson, and Harrison 1994; Zhang et al. 2009). Some cases of esophageal cancer (Okawa et al. 1999) and gynecological cancer (McGuire, Frank, and Eifel 2008) treatment have also been reported. Of note, prostate is the most common site for LDR brachytherapy where sealed radioactive sources ("seeds") are implanted permanently. Two commonly used isotopes are ^{125}I and ^{103}Pd.

With the advancement of radiation therapy, modern prostate brachytherapy has become a multidisciplinary effort that requires involvement of radiation oncology, urology, and medical imaging. Rigorous quality assurance is a necessity to achieve and maintain a high-quality seed implant program. Recently, issues regarding image guidance (Cormack 2008) as well as quality assurance for LDR brachytherapy procedure have been reported in the literature (Williamson 2008; Williamson et al. 2008). Successful implementation and continued improvement of a prostate brachytherapy program rely on effective teamwork and ongoing quality assurance review of the entire program.

In an attempt to facilitate discussion on quality management issues, we have divided the entire LDR process into subtasks (or submodules) as (1) system calibration and imaging, (2) treatment planning system, (3) seed assaying, (4) dosimetric planning, (5) implanting procedure, (6) patient release, (7) postimplant analysis, (8) cleaning and decontamination.

System Calibration and Imaging

Although attempts were made to use CT-guided prostate seed implant (Fichtinger et al. 2002), the main image-guided modality is still transrectal ultrasonography (TRUS). Although biplanar (sagittal and transverse) TRUS probes are available for acquiring images, only transverse images are used for anatomical contouring and dosimetric planning. Image quality degrades significantly when transverse images are reconstructed from acquired sagittal images. During imaging, it is very important to have good contact between the TRUS probe and the anterior rectal wall. A water-filled balloon or gel-filled rubber condom may be used to improve contact. Moreover, care should be taken not to disturb (displace or deform) the prostate during the image acquisition. It is important to use two to three stabilization needles to minimize unwanted motion of the prostate (Feygelman et al. 1996; Dattoli and Waller 1997; Taschereau et al. 2000; Podder et al. 2008). Depending on needle geometry and placement configuration, a 25–60% reduction in movement of the prostate may be achievable (Podder et al. 2008).

Verification should be made to ensure that the graphical grid pattern on the ultrasound image corresponds to the physical locations of the implant template. If using a phantom (e.g., water) for system calibration, care should be taken to maintain the proper temperature so that the speed of sound in the medium (e.g., water) corresponds closely to that in living human tissue (about 1540 m/s). To achieve 1540 m/s speed, the water temperature should be about 48°C, which may be too high for the TRUS probe. However, water temperature in the range of 38–42°C can produce reasonably accurate results. A second approach is increasing the salinity of the water (approximately 45 parts per thousand), which would yield a speed of sound close to 1540 m/s. Details about ultrasound calibration and verification can be found in AAPM Task Group No. 128 (Pfeiffer et al. 2008). Any fluoroscopy unit used in the operating room should display minimal distortion in a screen area that encompasses the implant region.

If the preoperative plan is used for seed implantation, specific parameters that should be checked during image acquisition include (1) the angle of elevation of patients legs in the stirrups, (2) the alignment of the ultrasound probe with respect to the prostate in all of the ultrasound images, and (3) the superposition of the template hole pattern on the contours of the prostate. In particular, the most posterior row of template holes must be close enough to the rectum–prostate interface so that the most posterior aspects of the prostate can be adequately covered by the prescribed dose. The same patient setup is expected to be recreated in the operating room for the implant.

Treatment Planning System

Various factors that can affect the dosimetry of low-energy photon-emitting brachytherapy sources such as ^{125}I and ^{103}Pd are the source geometry, encapsulation, and internal structure due to self-absorption effects. Therefore, it is not appropriate to use the dose rate constant, radial dose function, anisotropy function, anisotropy factor, or constant of one source design for another source design. The manufacturer should provide a calibration of source strength that is traceable to a standard, and the medical physicist should ensure that the clinical dosimetry parameters have been validated by independent investigators other than the manufacturer.

The treatment planning system must comply with the dosimetric formalism recommended by the AAPM Task Group No. 43 (TG43) (Nath et al. 1995). The planning system should be verified for the dose at total decay from a ^{125}I seed and/or a ^{103}Pd seed model in the point source approximation consistently with the TG43 data and/or TG43 update (TG43U1) and/or TG43U1 supplement (TG43U1S1) (Rivard et al. 2004, 2007). In addition, the treatment planning system should be verified to perform the correct dose summation at one or more locations in a simple configuration of multiple seeds. As recommended by AAPM TG40 (Kutcher et al. 1994) regarding quality assurance of treatment planning systems, these tests shall be performed before the computer treatment planning system is put into clinical use, at each subsequent software release, and at any seed model change.

In conventional treatment planning software, the seed position (x, y, z coordinates) is defined based on the x–y coordinates of a particular grid location in the physical template together with the depth or z-coordinate along the straight path of the needle orthogonal to the template. Therefore, needle deflection from ideal straight line, seed deposition inaccuracy, organ movement, and deformation causing deviation from the ideal

scenario can potentially contribute to dosimetric deviation from the implanted dose distribution.

An important consideration in the planning process should be to assess the degree of pubic arch interference. The pubic arch may "shadow" the anterior and lateral portions of the prostate, making it difficult or impossible to implant seeds in these locations. If this restriction exists, a plan must be devised that can overcome this limitation, typically by angling the template and ultrasound probe or using angulated needles if using robotic assistance; otherwise, the implant will probably be suboptimal. Therefore, it is important to know whether pubic arch interference exists before initiating the seed placement procedure.

Seed and Applicator Preparation

Radioactive LDR seeds may be obtainable as loose seeds, ready-loaded cartridges, or absorbable suture. In whatever form the seeds are received, the manufacturer's assay must be independently confirmed using a well ionization chamber. It is recommended in AAPM TG56 (Nath et al. 1997) that a random sample of at least 10% of the seeds in the shipment be checked. According to the AAPM Brachytherapy Subcommittee of the Therapy Physics Committee, the use of third-party calibration services that provide independent source strength verification would appear to provide nominal adherence to TG-64, but not necessarily to TG-56, and, in addition, the use of such calibrations to replace TG-56 compliant end-user measurements raises questions and concerns (Butler et al. 2006).

The seeds may be supplied in sterilized packages, that is, sterilized loose seeds in a seed cartridge or sterilized stranded seeds. In either case, a random sample of seeds (10% of the seeds required for the plan) should be obtained from the same production batch for independent verification. Discrepancies of 3% or more between the mean of the assay and the manufacturer's calibration should be investigated. If the discrepancy is 5% or more, it should be reported to the manufacturer.

The well chamber needs to be calibrated for the assay of each type and model of seed used in the prostate brachytherapy program. The user's well chamber can be calibrated at an ADCL with direct traceability to NIST for the seed type and model. Alternatively, individually calibrated seeds may be obtained with direct traceability to NIST to establish a calibration factor for the particular geometry being used.

Three different approaches used for implanting the seeds are (1) the Mick applicator with which the seeds are implanted along the needle track one at a time, (2) preloaded needles in which a column of seeds and spacers in the needle are implanted at the same time, and (3) stranded in which a column of seeds are encased in biocompatible material that maintains 1 cm spacing between the seeds center-to-center or in a plastic polymer tube with required variable spacing.

The loose seeds are best handled with reverse action tweezers that reduce the occurrences of seeds being "ejected" by pressure when they are picked up by regular tweezers or uncontrolled force damaging the encapsulation. The seeds must be sterilized before use. This is most conveniently done by steam sterilization.

Dosimetric Planning

For each patient, contouring is followed by a dosimetric plan before seed placement. Here, treatment planning refers equivalently to intraoperative planning or traditional preoperative planning. The treatment plan consists of isodose distributions superposed on the contours of the prostate in selected planes and dose–volume histograms for the prostate and organs at risk. The main objectives of the plan are to (1) provide coverage of the entire target volume by the prescribed dose while keeping the rectal and urethral doses within acceptable tolerances, (2) minimize dose inhomogeneity, and (3) keep the implant as technically simple as possible.

Before implantation, the dosimetric plan should be checked using an independent procedure or by a second member of the physics staff, and reviewed by the radiation oncologist. Three-dimensional visualization of the dose distribution can be useful for the review process. As an example, a reasonably good plan for 145-Gy prescription dose should have dose coverage for urethra D_{10} (dose received by 10% of the volume) <200 Gy, rectum D_5 < 145 Gy (or V_{100} <1 cm³), and prostate V_{150} (fraction of volume receiving 150% of the prescribed dose) in the range of 40–60%, in addition to V_{100} > 90% and D_{90} in the range of 145–180 Gy. A 2- to 3-mm margin of the prescribed isodose line to the prostate periphery is desirable.

Implantation Procedure

There are four generally distinct techniques of implantation in current use: (1) Mick applicator and loose seeds, (2) preloaded needles (stranded or loose seeds and spacers), (3) stranded seeds (manually needle-by-needle), and (4) robotic implantation. For all these techniques, either a preoperative plan or an intraoperative plan can be used. Recently, several robotic systems have been developed for TRUS-guided prostate brachytherapy (Wei et al. 2004; Fichtinger et al. 2006; Yu et al. 2007; Meltsner, Ferrier, and Thomadsen 2007; Podder, Ng, and Yu 2007; Salcudean et al. 2008). Most of them are capable of positioning a needle guide against the patient's perineum with great flexibility and maneuverability by replacing the conventional physical template. Needle insertion and seed delivery are commonly performed manually, with the exception of the single-channel Euclidian robotic system (Yu et al. 2007) and a multichannel robotic system (Podder, Ng, and Yu 2007) developed at the Thomas Jefferson University. The Euclidian is a fully automated needle insertion and seed delivery system with provision for needle force sensing, manual takeover, and physician's override from automatic mode. Implantation under this new modality requires careful supervision by the clinician.

A medical physicist, familiar with the treatment plan and the dosimetric consequences of any deviation from the plan, should be present in the operating room during the seed implantation.

If the implantation is performed using a preoperative plan based on prior volume studies, the position of the prostatic gland relative to the template coordinates should be verified in more than one imaging plane. The physicist should evaluate whether modification to the setup and/or treatment plan is required and recommend corrective action, if any deviation from the planned position is detected. It is useful to monitor the needle path in the sagittal plane during insertion, if a planned graphics of the needle track is available with the software. Deployment of a marker cable identifying the base plane of the prostate is quite helpful for needle insertion as well as subsequent seed delivery.

Local movement and long distance migration, that is, embolization of implanted seeds, are known phenomena (Nag, Vivekanandam, and Martinez-Monge 1997; Tapen et al. 1998; Merrick et al. 2000; Davis et al. 2000; Miura et al. 2008). Although loose seeds are more prone to embolization, migration of stranded seeds is not uncommon (Saibishkumar et al. 2009). Deposition of seeds into blood vessels or puncturing of the bladder may create an escape route for seeds. The length of the procedure and the number of needle insertions are thought to be related to the extent of prostatic edema. This can potentially result in undesirable dose delivery (Butler et al. 2000; Waterman and Dicker 2000; Taussky et al. 2005).

It is important to have an account of the needles and seeds implanted as the procedure progresses. At the end of implantation and after cystoscopy, the physicist should confirm the total number of seeds implanted in the patient and the number of seeds remaining, which must add up to the total number brought into the operating room. At the completion of the procedure, a complete radiation survey must be conducted, which includes the vicinity of the implant area, the floor, waste, linen, and all used applicators. The exposure rate at the surface and at 1 m from the patient should be measured by a properly calibrated ion chamber survey meter and documented in accordance with pertinent federal and state regulations. A scintillation detector or Geiger–Müller counter should be available in the operating room to locate any stray seeds if necessary.

Patient Release

The physicist (medical or health) should routinely review the patient survey results postimplantation to confirm that the prostate seed implant program continually satisfies all pertinent federal and state regulations regarding the release of patients with radioactive sources. The exposure rate from the patient at 1 m should be less than 2 mR/h before release. The institution's accountability of radioactive sources for a permanent prostate seed implant ends at the time of patient release (Yu et al. 1999). However, basic instructions to the patient on identifying the seeds and on radiation protection principles should be provided (Michalski et al. 2003). In addition, it is not uncommon for the treating physician to provide the patient with a letter stating that the brachytherapy procedure would result in a detectable level of radiation (e.g., to the extent of triggering an alarm). It is not necessary to require the patient to strain urine and return dislodged seeds.

Postimplant Analysis

In prostate brachytherapy, it is challenging to deliver the seeds accurately according to the planned coordinates, and the seeds may not stay precisely at the deposited locations. Prostate edema, needle deflection, prostate displacement, and deformation can contribute to geometric deviation of the seed implantation. Therefore, a postimplantation quantitative dose analysis should be carried out for each patient. The postimplant analysis is very important for the purposes of multi-institutional comparison, improving techniques, evaluating outcome, and identifying patients who might benefit from supplemental therapy or be at risk for long-term morbidity.

The postimplant analysis should include two-dimensional dose distributions on which the target volume for dose evaluation is outlined. A 3D volumetric visualization of the isosurface could be useful. It is recommended to construct the dose-volume histograms for this target volume and to document the dose levels that cover 100% and 90% of the target volume for postimplant evaluation, that is, D_{100} and D_{90}, and the fractional volume receiving 200%, 150% 100%, and 90% of the prescribed dose, that is, V_{200}, V_{150}, V_{100}, and V_{90}. The timing of postimplant dosimetric evaluation has some direct impact on the dosimetry results and therefore should be recorded. A set of CT images acquired 30 days after implantation is currently considered most useful for dosimetric evaluation.

The V_{100} and D_{90} can be used as indicators of implant quality in dosimetric coverage of the prostate. A ^{125}I implant with good coverage is characterized by D_{90} equal to or greater than the prescribed and/or V_{100} greater than 90%. Dose to the rectum is also important for prediction of toxicity. D_5 less than the prescribed dose and/or V_{100} less than 1 cm^3 can be considered a good implant while the criteria for target coverage are satisfied. Of course, such dosimetric analysis is sensitiviu dependent on the definition of the target volume for postimplant evaluation. Therefore, a consistent imaging interpretation of the target volume should be used to facilitate intercomparison and analysis of the dosimetric outcome.

Cleaning and Decontamination

All the devices used for LDR brachytherapy must be cleaned and decontaminated after and/or before the procedures as applicable. Components such as the template, Mick applicator, tweezers, marking cable, and so on that come in contact with needle, seeds, or patient must be sterilized before use in the procedure. Needless to say, needles, implanted seeds, and any instrument in contact with the seeds must be sterilized before the implantation.

References

Armstrong, J. G., L. L. Anderson, and L. B. Harrison. 1994. Treatment of liver metastases from colorectal cancer with radioactive implants. *Cancer* 73 (7): 1800–1804.

Butler, W. M., M. S. Huq, Z. Li, B. R. Thomadsen, L. A. DeWerd, G. S. Ibbott, M. G. Mitch et al. 2006. Third party brachytherapy seed calibrations and physicist responsibilities. *Med. Phys.* 33 (1): 247–248.

Butler, W. M., G. S. Merrick, A. T. Dorsey, and J. H. Lief. 2000. Isotope choice and the effect of edema on prostate brachytherapy dosimetry. *Med. Phys.* 27 (5): 1067–1075.

Chen, A. M., S. Chang, J. Pouliot, P. K. Sneed, M. D. Prados, K. R. Lamborn, M. K. Malec, M. W. McDermott, M. S. Berger, and D. A. Larson. 2007. Phase I trial of gross total resection, permanent iodine-125 brachytherapy, and hyperfractionated radiotherapy for newly diagnosed glioblastoma multiforme. *Int. J. Radiat. Oncol. Biol. Phys.* 69 (3): 825–830.

Cormack, R. A. 2008. Quality assurance issues for computed tomography-, ultrasound-, and magnetic resonance imaging-guided brachytherapy. *Int. J. Radiat. Oncol. Biol. Phys.* 71 (1 Suppl): S136–S141.

Dattoli, M., and K. Waller. 1997. A simple method to stabilize the prostate during transperineal prostate brachytherapy. *Int. J. Radiat. Oncol. Biol. Phys.* 38 (2): 341–342.

Davis, B. J., E. A. Pfeifer, T. M. Wilson, B. F. King, J. S. Eshleman, and T. M. Pisansky. 2000. Prostate brachytherapy seed migration to the right ventricle found at autopsy following acute cardiac dysrhythmia. *J. Urol.* 164 (5): 1661.

Fernandez, P. M., L. Zamorano, D. Yakar, L. Gaspar, and C. Warmelink. 1995. Permanent iodine-125 implants in the up-front treatment of malignant gliomas. *Neurosurgery* 36 (3): 467–473.

Feygelman, V., J. L. Friedland, R. M. Sanders, B. K. Noriega, and J. M. Pow-Sang. 1996. Improvement in dosimetry of ultrasound-guided prostate implants with the use of multiple stabilization needles. *Med. Dosim.* 21 (2): 109–112.

Fichtinger, G., E. C. Burdette, A. Tanacs, A. Patriciu, D. Mazilu, L. L. Whitcomb, and D. Stoianovici. 2006. Robotically assisted prostate brachytherapy with transrectal ultrasound guidance—Phantom experiments. *Brachytherapy* 5 (1): 14–26.

Fichtinger, G., T. DeWeese, A. Patriciu, A. Tanacs, D. Mazilu, J. Anderson, K. Masamune, R. Taylor, and D. Stoianovici. 2002. Robotically assisted prostate biopsy and therapy with intra-operative CT guidance. *J. Acad. Radiol.* 9: 60–74.

Jensen, A. W., I. A. Petersen, R. W. Kline, S. L. Stafford, P. J. Schomberg, and D. M. Robertson. 2005. Radiation complications and tumor control after 125I plaque brachytherapy for ocular melanoma. *Int. J. Radiat. Oncol. Biol. Phys.* 63 (1): 101–108.

Kutcher, G. J., L. Coia, M. Gillin, W. F. Hanson, S. Leibel, R. J. Morton, J. R. Palta et al. 1994. Comprehensive QA for radiation oncology: Report of AAPM Radiation Therapy Committee Task Group 40. *Med. Phys.* 21 (4): 581–618.

McGuire, S. E., S. J. Frank, and P. J. Eifel. 2008. Treatment of recurrent vaginal melanoma with external beam radiation therapy and palladium-103 brachytherapy. *Brachytherapy* 7: 359–363.

Meltsner, M. A., N. J. Ferrier, and B. R. Thomadsen. 2007. Observations on rotating needle insertions using a brachytherapy robot. *Phys. Med. Biol.* 52: 6027–6037.

Merrick, G., Butler W., A. Dorsey, J. Lief, and M. Benson. 2000. Seed fixity in the prostate/periprostatic region following brachytherapy. *Int. J. Radiat. Oncol. Biol. Phys.* 46: 215–220.

Michalski, J. M., K. Winter, J. A. Purdy, R. B. Wilder, C. A. Perez, M. Roach, M. B. Parliament et al. 2003. Preliminary evaluation of low-grade toxicity with conformal radiation therapy for prostate cancer on RTOG 9406 dose levels I and II. *Int. J. Radiat. Oncol. Biol. Phys.* 56 (1): 192–198.

Miura, N., Y. Kusuhara, K. Numata, A. Shirato, K. Hashine, Y. Sumiyoshi, M. Kataoka, and S. Takechi. 2008. Radiation pneumonitis caused by a migrated brachytherapy seed lodged in the lung. *Jpn. J. Clin. Oncol.* 38: 623–625.

Mutyala, S., A. Stewart, A. J. Khan, R. A. Cormack, D. O'Farrell, D. Sugarbaker, and P. M. Devlin. 2010. Permanent iodine-125 interstitial planar seed brachytherapy for close or positive margins for thoracic malignancies. *Int. J. Radiat. Oncol. Biol. Phys.* 76 (4): 1114–1120.

Nag, S., D. Beyer, J. Friedland, P. Grimm, and R. Nath. 1999. American Brachytherapy Society (ABS) recommendations for transperineal permanent brachytherapy of prostate cancer. *Int. J. Radiat. Oncol. Biol. Phys.* 44 (4): 789–799.

Nag, S., R. J. Ellis, G. S. Merrick, R. Bahnson, K. Wallner, and R. Stock. 2001. American Brachytherapy Society recommendations for reporting morbidity after prostate. *Int. J. Radiat. Oncol. Biol. Phys.* 51: 1422–1430.

Nag, S., S. Vivekanandam, and R. Martinez-Monge. 1997. Pulmonary embolization of permanently implanted radioactive palladium-103 seeds for carcinoma of the prostate. *Int. J. Radiat. Oncol. Biol. Phys.* 39 (3): 667–670.

Nath, R., L. L. Anderson, G. Luxton, K. A. Weaver, J. F. Williamson, and A. S. Meigooni. 1995. Dosimetry of interstitial brachytherapy sources: Recommendations of the AAPM Radiation Therapy Committee Task Group No. 43. American Association of Physicists in Medicine. *Med. Phys.* 22 (2): 209–234.

Nath, R., L. L. Anderson, J. A. Meli, A. J. Olch, J. A. Stitt, and J. F. Williamson. 1997. Code of practice for brachytherapy physics: Report of the AAPM Radiation Therapy Committee Task Group No. 56. American Association of Physicists in Medicine. *Med. Phys.* 24 (10): 1557–1598.

Okawa, T., T. Dokiya, M. Nishio, Y. Hishikawa, and K. Morita. 1999. Multi-institutional randomized trial of external radiotherapy with and without intraluminal brachytherapy for esophageal cancer in Japan. Japanese Society of Therapeutic Radiology and Oncology (JASTRO) Study Group. *Int. J. Radiat. Oncol. Biol. Phys.* 45 (3): 623–628.

Pfeiffer, D., S. Sutlief, W. Feng, H. M. Pierce, and J. Kofler. 2008. AAPM Task Group 128: Quality assurance tests for prostate brachytherapy ultrasound systems. *Med. Phys.* 35 (12): 5471–5489.

Pignol, J. P., E. Rakovitch, B. M. Keller, R. Sankreacha, and C. Chartier. 2009. Tolerance and acceptance results of a palladium-103 permanent breast seed implant Phase I/II study. *Int. J. Radiat. Oncol. Biol. Phys.* 73 (5): 1482–1488.

Podder, T. K., W. S. Ng, and Y. Yu. 2007. Multi-channel robotic system for prostate brachytherapy. *Conf. Proc. IEEE Eng. Med. Biol. Soc.* 2007: 1233–1236.

Podder, T., J. Sherman, D. Rubens, E. Messing, J. Strang, W. S. Ng, and Y. Yu. 2008. Methods for prostate stabilization during transperineal LDR brachytherapy. *Phys. Med. Biol.* 53 (6): 1563–1579.

Rivard, M. J., and C. S. Melhus. 2009. An approach to using conventional brachytherapy software for clinical treatment planning of complex, Monte Carlo–based brachytherapy dose distributions. *Med. Phys.* 36: 1968–1975.

Rivard, M. J., W. M. Butler, L. DeWerd, M. Huq, G. Ibbott, A. Meigooni, C. Melhus, M. Mitch, R. Nath, and J. F. Williamson. 2007. Supplement to the 2004 update of the AAPM Task Group No. 43 Report. *Med. Phys.* 34: 2187–2205.

Rivard, M. J., B. M. Coursey, L. A. DeWerd, W. F. Hanson, M. S. Huq, G. S. Ibbott, M. G. Mitch, R. Nath, and J. F. Williamson. 2004. Update of AAPM Task Group No. 43 Report: A revised AAPM protocol for brachytherapy dose calculations. *Med. Phys.* 31 (3): 633–674.

Saibishkumar, E. P., J. Borg, I. Yeung, C. Cummins-Holder, A. Landon, and J. Crook. 2009. Sequential comparison of seed loss and prostate dosimetry of stranded seeds with loose seeds in 125I permanent implant for low-risk prostate cancer. *Int. J. Radiat. Oncol. Biol. Phys.* 73 (1): 61–68.

Salcudean, S. E., T. D. Prananta, W. J. Morris, and I. Spadinger. 2008. A Robotic Needle Guide for Prostate Brachytherapy. Paper read at IEEE Int. Conf. Robotics and Automation (ICRA), at Pasadena, CA.

Tapen, E. M., J. C. Blasko, P. D. Grimm, H. Ragde, R. Luse, S. Clifford, J. Sylvester, and T. W. Griffin. 1998. Reduction of radioactive seed embolization to the lung following prostate brachytherapy. *Int. J. Radiat. Oncol. Biol. Phys.* 42 (5): 1063–1067.

Taschereau, R., J. Pouliot, J. Roy, and D. Tremblay. 2000. Seed misplacement and stabilizing needles in transperineal permanent prostate implants. *Radiother. Oncol.* 55 (1): 59–63.

Taussky, D., L. Austen, A. Toi, I. Yeung, T. Williams, S. Pearson, M. McLean, G. Pond, and J. Crook. 2005. Sequential evaluation of prostate edema after permanent seed prostate brachytherapy using CT-MRI fusion. *Int. J. Radiat. Oncol. Biol. Phys.* 62 (4): 974–980.

Trombetta, M. G., A. Colonias, D. Makishi, R. Keenan, E. D. Werts, R. Landreneau, and D. S. Parda. 2008. Tolerance of the proximal aorta using intraoperative iodine-125 interstitial brachytherapy in cancer of the lung. *Brachytherapy* 7: 50–54.

Waterman, F. M., and A. P. Dicker. 2000. The impact of postimplant edema on the urethral dose in prostate brachytherapy. *Int. J. Radiat. Oncol. Biol. Phys.* 47 (3): 661–664.

Wei, Z., G. Wan, L. Gardi, G. Mills, D. Downey, and A. Fenster. 2004. Robot-assisted 3D-TRUS guided prostate brachytherapy: System integration and validation. *Med. Phys.* 31 (3): 539–548.

Williamson, J. F. 2008. Current brachytherapy quality assurance guidance: Does it meet the challenges of emerging image-guided technologies? *Int. J. Radiat. Oncol. Biol. Phys.* 71 (1 Suppl.): S18–S22.

Williamson, J. F., P. B. Dunscombe, M. B. Sharpe, B. R. Thomadsen, J. A. Purdy, and J. A. Deye. 2008. Quality assurance needs for modern image-based radiotherapy: Recommendations from 2007 interorganizational symposium on "quality assurance of radiation therapy: Challenges of advanced technology." *Int. J. Radiat. Oncol. Biol. Phys.* 71 (1 Suppl.): S2–S12.

Yu, Y., L. L. Anderson, Z. Li, D. E. Mellenberg, R. Nath, M. C. Schell, F. M. Waterman, A. Wu, and J. C. Blasko. 1999. Permanent prostate seed implant brachytherapy: Report of the American Association of Physicists in Medicine Task Group No. 64. *Med. Phys.* 26 (10): 2054–2076.

Yu, Y., T. K. Podder, Y. D. Zhang, W. S. Ng, V. Misic, J. Sherman, D. Fuller et al. 2007. Robotic system for prostate brachytherapy. *Comput. Aided Surg.* 12 (6): 366–370.

Zhang, F. J., C. X. Li, L. Zhang, P. H. Wu, D. C. Jiao, and G. F. Duan. 2009. Short- to mid-term evaluation of CT-guided 125I brachytherapy on intra-hepatic recurrent tumors and/or extra-hepatic metastases after liver transplantation for hepatocellular carcinoma. *Cancer Biol. Ther.* 8 (7): 585–590.

99

Brachytherapy with Unsealed Sources

Daniel J. Macey
Certified Medical Physics Services

Introduction

Brachytherapy with unsealed sources depends on the administration of soluble radioactive solutions to patients by intravenous injection or by ingestion. In 1946, the IAEA made radioisotopes available for medical use in the United States; and today, more than 3 million patients in the United States receive unsealed sources per year for diagnosis or therapy. A small proportion of these patients receive unsealed sources of selected radionuclides for radionuclide therapy (Macey et al. 2001), and quality assurance of these new methods is evolving. This chapter is a summary of the quality assurance and radiation safety procedures and requirements for the growing list of cancer treatments such as liver tumor ablation (Dezarn 2008) with Y-90-labeled microspheres.

The treatment of thyroid cancer with I-131, which started in 1942, remains the most efficacious treatment model for radiotherapy in terms of availability, minimal trauma, low toxicity, modest cost, and convenience especially because patients can now be discharged with more than 30 mCi I-131 (New Guidance on Release of Thyroid Patients 1997). Ratios of target to nontarget are expressed as tumor to whole body or tumor to organs at risk. Intratumoral injections of various radiopharmaceuticals have shown promise in animal models and remain to be tested in patient studies. Unlike I-131 therapy for thyroid cancer, many of the new radiolabeled molecules and antibodies cannot currently deliver an adequate therapeutic dose to a target tumor volume without exceeding toxic radiation dose limits to the marrow and any critical organs.

Quality Assurance and Management Program for Unsealed-Source Brachytherapy

The ACR (ACR 2005) and IAEA (IAEA 2006) have provided guidance on setting up clearly defined policies and procedures for the range of health professionals involved in using unsealed sources in radiation oncology. The medical physicist in radiation oncology is increasingly called upon to assume the responsibility of using unsealed sources in the clinic. This responsibility involves an important role in helping to set up a patient-specific treatment plan and providing advice on the safe use of the radiopharmaceuticals selected by the radiation oncologist for each patient. The companies supplying these radiopharmaceuticals provide in-house training and guidance on using their products. All procedures using unsealed sources follow a common series of steps that are designed to ensure that the patient receives the prescribed amount of radionuclide safely while minimizing the dose to staff and relatives. It is also important that the patient understands the procedures that are designed to reduce contamination.

Each radionuclide therapy procedure requires an inventory documenting the receipt, use, and disposal of all radioactive sources used for patient treatment as required by the local state and federal regulations. Ordering and maintaining an inventory for unsealed sources is a requirement so that the maximum amount of radionuclides approved on the state license is not exceeded. Because unsealed sources are taken off the inventory as they are administered to designated patients, the inventory must be maintained as accurately as possible. A fraction of all unsealed sources administered to therapy patients is excreted via urine and feces to the clinic drains, where dilution and decay can be estimated to ensure the concentration is reduced to less than limits set by the local city water authority. This fraction of the administered radionuclide activity can be estimated from the reciprocal of whole body retention measurements of the patient acquired at selected intervals after administration.

Most of the survey and patient dose monitoring and counting equipments are usually kept in radiation oncology. Counting wipes of areas for contamination can be accomplished with a pancake Geiger Mueller or scintillation probe of appropriate sensitivity.

TABLE 99.1   Daily Dose Calibrator QA/QC and Procedures

Check background level reading for each check source sensitivity button (usually Cs-137 and Co-57).

Check receipt of correct isotope and amount with order and wipe test results.

Record shipment information in inventory.

Store shipments in a fume hood to release any volatile isotope.

Check dose calibrator voltage reading is correct.

Measure activity of Cs-137 check source with correct sensitivity button selected and confirm agreement with predicted value to <1%. Repeat with Co-57 check source as appropriate.

Measure activity of therapy isotope with sensitivity matching the vendor's data for the dose calibrator. For beta-only sources (e.g., Y-90, P-32, etc.), check sensitivity setting and factors for correct assay.

Compare activity measured in-house with vendor's values on shipment after correcting for decay and predict source activity at administration time.

Ask patient to empty their bladder and bowel to minimize the risk associated with a full bladder, etc.

Administer the unsealed source carefully via an intravenous line without pressure buildup that could spatter and contaminate staff and the clinic.

Assay any residue of the isotope in vials, syringes, tubing, and any lines used to administer the therapy solution dose. Subtract this residual activity from the previously measured activity.

Record the corrected administered source activity in the patient's chart and inventory. Corrections for source position and volumes in postadministration containers are usually ignored if the residue is estimated to be less than 10% of the prescription.

For oral administration, follow the capsule source with about 100 mL water to reduce the I-131 capsule sticking to the stomach wall.

Measure and record the dose rate at 1 m from the patient's umbilicus and provide a card to the patient that includes clinic contact information, radioisotope, amount administered, etc.

Check all areas in lab and administration room for contamination with a suitable survey meter or wipe test.

Patient education and compliance with radiation safety requirements are an important part of quality assurance and public safety for unsealed-source brachytherapy. Most unsealed-source therapy procedures rely on the tracer principle that assumes the therapy dose can be predicted from a prior tracer study. The tracer study is performed with the identical mass and formulation as the planned therapy, and the amount of radionuclide administered for the tracer study conforms within the defined limits for diagnostic imaging studies. Today, most radionuclide therapy patients can be discharged with more than 30 mCi of the radioisotope provided they qualify for the radiation safety restrictions at home. This change can reduce radiation dose to hospital staff and reduce the cost of a clinically unnecessary prolonged hospital stay.

Quality Assurance in Assay of Radionuclide

The amount of radioisotope prescribed for each therapeutic procedure is independently measured using a dose calibrator in the clinic, and this is compared with the vendor's record included with the shipment. These readings should agree to within ±3%. The dose calibrator is usually a well-type ion chamber in the

nuclear medicine department, and the daily quality control, quarterly, and annual requirements are recorded for inspection by the state and federal officials. The quantity of unsealed isotope for each therapy patient is usually calculated as a fraction of total body weight or body mass index. It is also very important to confirm that the patient has good bladder and excretion control to minimize the risks of contaminating the clinic and their home environment. The prescribing MD usually can confirm these parameters before ordering the isotope. The daily dose calibrator QA/QC and procedures are described in Table 99.1.

Quarterly test for an unsealed-source program includes checking the linearity of the dose calibrator with a Tc-99m source of activity starting at approximately 800 mCi. A decay or attenuator should be used to confirm linearity to about 1 mCi.

Annual QA consists of checking absolute calibration of the dose calibrator using two to three radionuclides. Accuracy should be confirmed to within ±5%. The change in dose calibrator sensitivity with volume should be confirmed. Tc-99m sources in 1- to 50-mL volumes can be used for this purpose.

Table 99.2 describes the parameters that need to be verified when using unsealed sources in radiotherapy. The most common unsealed sources are listed in Table 99.3.

TABLE 99.2   Quality Assurance Parameters for Use of Unsealed Sources in Radiotherapy

Parameter	Description
Check accuracy of clinical information	Age, previous therapy, pregnancy, blood counts, marrow reserve
Analyze tracer study	Review uptake sites, find effective half-life, estimate of target tumor and critical organ doses, calculate therapy dose
Review accuracy of radionuclide assay	Agreement with vendor, QA and QC of dose calibrator
Review patient discharge criteria	Radiation safety education, discharge of excretions from home, appropriate home environment
Check availability of survey equipment	Calibration records, dose monitoring via body badges, or pocket dose meters

TABLE 99.3 Current List of Common Unsealed Sources Used in Radiation Oncology

Radiopharmaceutical	Cancer Treatment	Half-life (days)	Emissions	Product
I-131 NaI	Thyroid	8.0	beta/gamma	Amersham
I-131 antibody	Lymphoma	8.0	beta, gamma	Tostamoab; Lym-1
Y-90 antibody	Lymphoma	2.7	beta	Zevalin
Y-90 microspheres	Liver	2.7	beta	SirSpheres, TheraSpheres
Sm-153 EDTMP	Bone pain	1.9	beta, gamma	Cytogen

Review of Quality Assurance and Quality Management Program

Regular audits and review by state and federal radiation control inspectors are mandatory, and regular internal reviews of incidents, errors, and accidents encountered with unsealed-source therapy can help to mitigate repeating these issues.

Two typical examples of incidents encountered with unsealed-source therapy are as follows:

1. A patient receives a 37-mCi dose of Sm-153 EDTMP for bone pain palliation and is escorted to her car about 5 min after her intravenous injection to drive home. Another 10 min later, the patient reports that she has emptied her bladder while driving home.
   Q1: What should the physicist do?
   Q2: What should the clinic do to minimize repeating this accident?
2. A female patient with thyroid cancer is escorted to the entrance of the medical center approximately 30 min after ingesting a 100-mCi capsule of I-131 as NaI solution. While the patient is waiting for her ride, the patient empties her stomach on the pavement, and the clinic pages the physicist.
   Q1: What should be done to resolve this event?
   Q2: What is the recommendation to reduce repeating such an event?

An appropriate quality management program includes the review of errors, incidents/events recorded, and actions taken. Suggested changes, additional training, and future directions should be discussed and the procedures modified accordingly.

References

ACR. 2005. *American College of Radiology Report, Guidelines and Standards Section, Quality Assurance for Unsealed Sources. Available from http://www.acr.org/SecondaryMain MenuCategories/quality_safety/guidelines/ro/unsealed_ radiopharmaceuticals.aspx.*

Dezarn, W. A. 2008. Quality assurance issues for therapeutic application of radioactive microspheres. *Int. J. Radiat. Oncol. Biol. Phys.* 71 (1 Suppl.): S147–S151.

IAEA. 2006. Quality assurance for radioactivity measurement for nuclear medicine. In *IAEA Technical Report Series 454.* Vienna: IAEA. Available from http://www-pub.iaea.org/ MTCD/publications/PDF/TRS454_web.pdf.

Macey, D. J., L. J. Williams, H. B. Breitz, A. Liu, T. K. Johnson, and P. B. Zanzonico. 2001. *A Primer for Radioimmunotherapy and Radionuclide Therapy. AAPM Report 71.*

New Guidance on Release of Thyroid Patients. 1997. In *NRC Document 10 CFR 35.75.* Available from http://www.nrc .gov/reading-rm/doc-collections/cfr/part035/part035-0075 .html.

100

Intravascular Brachytherapy

Christian Kirisits
Medical University of Vienna

Introduction

Intravascular (or endovascular) brachytherapy is a treatment modality used to reduce the risk of restenosis after percutaneous revascularization in coronary or peripheral blood vessels. Beta- and gamma-emitting sources are used, nowadays always as a temporary irradiation with a catheter-based approach. For peripheral treatments, gamma-emitting sources are mainly transferred by modern conventional afterloading devices. Dedicated devices have been introduced for coronary applications. Therefore, specific QA procedures and dosimetric concepts, not directly related to conventional brachytherapy and the sources used there, have been recommended. Radiation safety has to be taken into account as manual afterloading equipment is used quite often. Clinical QA, especially not considered in the beginning phase of intravascular brachytherapy, can nowadays be based on recommendations and margin concepts based on clinical results.

Equipment

Dose Distribution around Sources

Several recommendations and guidelines to define, verify, and calculate the dose distribution around intravascular brachytherapy sources are available. At the international level, details related to calibration of photon and beta sources used in endovascular brachytherapy can be found in the IAEA-TECDOC-1274 (IAEA 2002). The ICRU Report 72 on "Dosimetry of Beta Rays and Low-Energy Photons for Brachytherapy with Sealed Sources" focuses explicitly on beta sources (ICRU 2004). At the national level, the AAPM TG60 and TG149 reports contain the most detailed information on formalisms and consensus dosimetric data for dose distribution calculation (Nath et al. 1999; Chiu-Tsao et al. 2007). A comprehensive overview, with very practical descriptions, is given

in the Netherlands NCS Report 14 "Quality Control of Sealed Beta Sources in Brachytherapy" (NCS 2004) and the German DGMP Report No. 16 on "Guidelines for Medical Physical Aspects of Intravascular Brachytherapy" (DGMP 2001). A recent comprehensive document on all dosimetric aspects of intravascular brachytherapy is ISO 21439:2009 on "Clinical Dosimetry—Beta Radiation Sources for Brachytherapy" (ISO 2009). Although there is general agreement on many issues, the reports differ in details of the recommended dosimetric concepts and equipment to perform acceptance and constancy checks. Here, only a rough overview is provided. For detailed information, the reader is referred to the original published literature cited above.

Source Strength

Almost all reports propose the reference air kerma rate as the quantity to specify the source strength for gamma devices. Dedicated standards for acceptance tests, similar to those for conventional brachytherapy sources, can be used (IAEA 2002). In daily clinical practice, the use of well-type ionization chambers is widely accepted. As for coronary irradiation systems with source trains consisting of several individual seeds, the well-type ionization chamber must have an appropriate size to include the whole source train. Measurements are also possible with ionization chambers and in air phantoms or dedicated solid phantoms using appropriate geometrical corrections.

For beta sources the source strength is specified as the reference absorbed dose rate to water in water at a reference distance of 2 mm from the source center perpendicular to the source axis (IAEA 2002; ICRU 2004; Nath et al. 1999; Pötter et al. 2001). The verification of this reference dose rate during acceptance testing is therefore different and more complex compared to checks of conventional gamma sources. Suitable detectors for such measurements are plastic scintillators, radiochromic films, diodes, and TLDs (ISO 2009). All detectors have to be calibrated in an adequate radiation field representative of the source to be used. Appropriate phantoms and experimental setups are needed to position the detector. Scaling algorithms are needed to define the effective point of measurement. The detector volume should not exceed 1 mm in any dimension.

Another procedure is to use a well-type ionization chamber with an appropriate insert and a calibration certificate for the type of source to be measured. Often, long line sources or seed trains are used; the appropriate position of the sources and an appropriate sweet length, then, have to be taken into account. Although this type of measurement is appropriate for the source strength, one additional detector is needed to verify the relative dose distribution (as described in the next section). The action level for the source strength measurement has been defined differently, especially for beta sources as indicated in Table 100.1. However, as the expanded uncertainty of the primary standard is already 7.5–11% ($k = 2$), and the measurement at the clinic has an uncertainty of at least 3% ($k = 1$), the threshold levels are generally high, but never more than ±20% (ISO 2009).

Relative Dose Distribution

To verify the whole 3D dose distribution as closely as possible, additional measurements are needed. Because line sources are used, it is mainly recommended to use a 2D approach consisting of the relative radial depth dose and the longitudinal dose profile also often referred to as source uniformity or homogeneity. These measurements can be performed with one of the above-mentioned detectors except the well-type ionization chamber. The radial dose profile is needed to calculate the dose at different

distances to the source axis depending on the vessel diameter and the depth for prescription or reporting.

The aim of the nonuniformity or homogeneity test is to (1) verify the active length of the source and the region that can be used to deliver an appropriate dose to the target and (2) check that the nonuniformity along this length is within certain limits. Although this is similar for all recommendations and guidelines for intravascular brachytherapy dosimetry, the detailed concept and terminology is slightly different between the various publications. Suitable definitions are related to the concept used for treatment planning with dose prescription and reporting. One simple straightforward concept is to follow the GEC ESTRO guidelines (Pötter et al. 2001). There the reference isodose length (RIL), which is the length at reference depth along the source axis where at least 90% of the prescribed dose is delivered. In daily clinical practice, the RIL is not very different to the RIL at a distance of 2 mm from the source axis. Therefore, the longitudinal dose profile at this distance is usually sufficient for acceptance testing. In case of homogenous activity distributions within the line source, the RIL is always smaller than the active source length (ASL) of the source. The margin at both source edges is essential for appropriate treatment planning and has been determined for various commercial systems (Kirisits et al. 2002; Kirisits, Pokrajac et al. 2004). The homogeneity of the dose within the RIL is a quality parameter.

Delivery Device

For intravascular afterloading devices, similar quality issues have to be considered as for conventional devices. These are

TABLE 100.1 Compilation of Recommendations on Quality Control

Parameter	Maximum Deviation	Test Frequency
Source strength for beta sources	±5%[a], ±10%[b], ±15%[c]	At source exchange[a,b,c,d] for long half-lives, additional measurements on a regular basis,[a] each few years,[c] or every 3 years[b]
ASL	±1 mm[a,b]	At source exchange[b]
Source nonuniformity	±10%[b,e], ±20%[a]	At source exchange[b]
Source positioning	±1 mm[a,b]	At source exchange[b]
Internal timer accuracy and linearity		At source exchange or annually, whichever is shorter[b]
Emergency source retraction system		At source exchange or quarterly, whichever is shorter[b]
Missing catheter		At source exchange or quarterly, whichever is shorter[b]
Catheter obstruction		For each patient[b]
Leakage radiation		At source exchange or annually, whichever is shorter[b]
Source integrity		For each patient[b]
Catheter integrity		For each patient[b]
Emergency equipment functionality		Quarterly[b]
Training of emergency procedures		Annually[b]

[a] EVA GEC ESTRO (Pötter et al. 2001).
[b] NCS report No. 14 (NCS 2004).
[c] ICRU 72 (ICRU 2004).
[d] DGMP Report No. 16 (DGMP 2001).
[e] AAPM TG-60 (Nath et al. 1999).
[f] ISO 21439 (ISO 2009).

mainly to check for a missing catheter interlock to avoid source loss in the cathlab (i.e., treatment room) and the accuracy of source positioning via a source cable or other transfer techniques. The accuracy of source positioning is essential. However, in contrast to conventional afterloading techniques, the actual source position is checked in vivo in the cathlab via fluoroscopy. If automatic afterloader devices are used, the irradiation timer (i.e., the interrupt button), the independent power supply, and the manual retraction facility have to be tested.

Radiation Safety

Radiation protection for intravascular brachytherapy procedures needs some special attention. Although remote afterloading brachytherapy with high-activity gamma sources is performed with remote-controlled afterloaders located in shielded rooms, many coronary devices are used manually and often with hand-held devices. Although mainly beta radiation is used, the dose in the vicinity should not be underestimated. High dose rates are present when the source is travelling outside of the delivery device (without any shielding) and outside the patient. In case of malfunction, high dose to the whole body and extremities has to be taken into account for the operating staff. Therefore, appropriate emergency plans and equipment have to be prepared. For beta sources, emergency containers, which consist of inner plastic shielding, are useful. The plastic shielding should be enclosed by lead to account for the remaining Bremsstrahlung x-rays. A dedicated report from the IAEA (2006) describes appropriate methods in detail.

Clinical QA

Quality for patient treatment is closely related to reproducible and clear procedures. The EVA GEC ESTRO working group proposed a concept for prescribing, recording, and reporting of intravascular brachytherapy treatments (Pötter et al. 2001). It can be used as a basis for an approach to treatment planning with appropriate clinical QA procedures. Treatment planning is linked to a concept of different lengths and radial distances describing the target volume, and retrospective analysis of endovascular brachytherapy is linked to the actual irradiated tissue volumes.

Determination of the Clinical Target Volume

The lesion length is defined as the stenotic or occluded length of the vessel segment describing the part of the vessel that is either stenotic or totally occluded. Sometimes the lesion length corresponds to the length of a restenotic stent. In the literature, the term is often replaced by the stent segment. The lesion length is not part of physical aspects of treatment planning but is needed when comparing the type of patient recruitment and for outcome analysis.

The interventional length (IL) is defined as the length from the most distal to the most proximal part of the vessel that is treated using PTCA/PTA or other interventional procedures (laser, rotablator, directional atherectomy). It can be measured using angiography. Care has to be taken that the measurement includes all inflated balloon positions or any other interventional devices. In many publications, the IL is called the injured length or injury length. The IL is the basis for the entire target volume concept. Therefore, it should be emphasized that this length is an important parameter, which has to be determined carefully. Documentation using cine-angiography of each interventional procedure is mandatory. Every cine has to be analyzed before each brachytherapy intervention. The use of automatic quantitative coronary angiograpy software should be performed only with caution as the lengths can be significantly underestimated due to the foreshortening of the balloon lengths in the angiographic documentation. The actual IL can be determined using the nominal balloon lengths and delivery catheter marker distances as a reference scale including the overlap of the balloons and their positions related to landmarks (e.g., stent edges, side branches) (Schmid et al. 2004).

Interventional procedures increasing the vessel lumen diameter or removing plaque may also injure the vessel wall proximal and distal to the visible IL. The clinical target length is a radiotherapy-related term taking into account possible subclinical invasion to make the treatment efficient. Currently, there is no evidence guiding the exact amount of this safety margin.

The lumen radius should be measured after angioplasty, by angiography or IVUS, in at least one representative plane within the clinical target length. This diameter serves as the reference lumen diameter. In case of little variation in lumen diameter, one measurement may be sufficient, but in case of large variations, at least three measurements (at a central, distal, and proximal plane) should be taken and a mean value should be recorded and reported, which is the mean reference lumen diameter. In such cases, the variation of the reference lumen diameter should also be indicated. The measurement technique should be clearly reported.

Determination of the Appropriate Treatment Parameters—Treatment Planning

Geometrical uncertainties such as inaccuracies in the positioning of the source, heart movements, or patient movements need to be compensated by additional safety margins proximal and distal to the clinical target length. Following the terminology, this length is the planning target length. The use of such margins has to be carefully determined on the amount of uncertainties present in the respective treatment modality. First experience suggests a total safety margin of at least 10 mm added to the IL to avoid a geographic miss due to uncertainties in catheter positioning (Schmid et al. 2004; Syeda et al. 2002).

According to ICRU concepts, the dose should be specified at a reference point related, and relevant to, the target volume and independent of the treatment technique or type of source. In endovascular brachytherapy, a certain depth in the arterial wall, at which the dose is specified, should be used both for beta and photon sources. Therefore, the concept of a general reference depth is strongly recommended, which should be specific for a given anatomical site. According to current knowledge, a depth of 1 mm in coronary and 2 mm in peripheral arteries is relevant for target dose assessment (lumen radius +1 mm or lumen radius +2 mm, respectively). Therefore, these reference depths are recommended to report the reference depth dose for a specific treatment (Pötter et al. 2001; ISO 2009).

The RIL is related to the reference depth dose. The reference isodose length is the vessel length at the reference depth (1 or 2 mm, respectively) enclosed by the isodose representing 90% of the reference depth dose. The geometry is illustrated in Figure 100.1. The reference isodose length is dependent on the lumen diameter, the source nuclide, and the source arrangement. A device with a suitable RIL is chosen to include the whole planning target length, or in other words, to treat the IL with sufficient margins. If this is not possible with one source position, then the treatment has to be performed for subsequent segments.

The treated length should never be confused with the ASL, which is one of the technical characteristics of the source(s). Therefore, it is recommended to avoid terms such as *radiated segment* or *treatment length* to describe a vascular brachytherapy device. The ASL is defined as the length of the radioactive source(s) or source train or the length of the active dwell positions of a stepping source. The ASL, which is necessary to treat a given target at a defined depth, is sometimes significantly different for the available source types and isotopes—depending on the dose fall-off around the sources and on the vessel diameter (difference between ASL and RIL).

The reference dose in the central plane is precisely defined as a flat dose profile along the source axis. The reference depth dose is an average value due to source arrangements as seed sources with spacing and stepping sources. The region to determine this mean value can be, for example, over the central two sources for an even number of seeds or over the central three sources for an odd number of seeds. The ISO and NCS recommend an average dose over the ASL minus a margin of 2.5 mm for ^{32}P and 3.0 mm

for ^{90}Sr sources on each side, which is slightly different from the GEC ESTRO recommendations as this region is close to the RIL representing at least 90% of the reference dose.

The prescribed dose is selected by the radiation oncologist based on their decision on the adequate dose to reach the aim of treatment. The dose prescription is done at the prescription point. This point can, in principle, be chosen without restrictions by the radiation oncologist. It is recommended, for reasons of clarity, to use the same point for prescription as for recording and reporting. Dose specification for the prescription point is then performed according to the same criteria as for the reference depth dose point in the central plane.

For reporting issues, additional parameters are recommended to estimate the maximum dose to the vessel wall and other variations of dose levels within the treated length. These variations can be high in the case of noncentered sources.

For a full three-dimensional dose calculation, allowing the determination of dose volume histograms for different vessel wall structures, a full dose calculation concept has to be used as introduced by AAPM TG60 and AAPM TG149 (Nath et al. 1999; Chiu-Tsao et al. 2007).

Steps in Clinical Practice

The following steps are important in clinical practice:

- Determination of the IL and reference lumen diameter
- Choice of a device with appropriate RIL to treat the IL including margins (or calculation of necessary source steps)
- Calculation of the treatment time to deliver the prescribed dose to the reference depth (taking into account relative depth dose curve and actual source strength)
- Angiography to check the appropriate source positions
- Radioprotection measurements after source retraction

A much more detailed quality management program in intravascular brachytherapy has been published by Chakri and Thomadsen (2002). Experiences with several commercial available devices have been reported (Kirisits, Stemberger et al. 2004; Kollaard et al. 2006).

References

Chakri, A., and B. Thomadsen. 2002. A quality management program in intravascular brachytherapy. *Med. Phys.* 29 (12): 2850–2860.

Chiu-Tsao, S. T., D. R. Schaart, C. G. Soares, and R. Nath. 2007. Dose calculation formalisms and consensus dosimetry parameters for intravascular brachytherapy dosimetry: recommendations of the AAPM Therapy Physics Committee Task Group No. 149. *Med. Phys.* 34 (11): 4126–4157.

DGMP. 2001. Guideline for medical physical aspects of intravascular brachytherapy. In *DGMP-Report No. 16. ISBN 3-925218-71-8.* Deutsche Gesellschaft für Medizinische Physik.

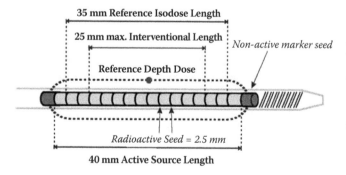

FIGURE 100.1 Schematic example of the most important parameters: the reference isodose length specifies the length where at least 90% of the reference depth dose is delivered. Taking into account safety margins for various uncertainties, each device can treat up to maximum intervention length with one single source step.

IAEA. 2002. *Calibration of Photon and Beta Ray Sources Used in Brachytherapy. Guidelines on Standardized Procedures at Secondary Standards Dosimetry Laboratories (SSDLs) and Hospitals.* Vienna: International Atomic Energy Agency.

IAEA. 2006. *Radiological Protection Issues in Endovascular Use of Radiation Sources. IAEA-TECDOC-1488.* Vienna: International Atomic Energy Agency.

ICRU. 2004. *Dosimetry of Beta Rays and Low-Energy Photons for Brachytherapy with Sealed Sources.* International Commission on Radiation Units and Measurements.

ISO. 2009. *Clinical Dosimetry—Beta Radiation Sources for Brachytherapy. In ISO 21439:2009.* International Organization for Standardization.

Kirisits, C., D. Georg, P. Wexberg, B. Pokrajac, D. Glogar, and R. Pötter. 2002. Determination and application of the reference isodose length (RIL) for commercial endovascular brachytherapy devices. *Radiother. Oncol.* 64: 309–315.

Kirisits, C., B. Pokrajac, D. Berger, E. Minar, R. Potter, and D. Georg. 2004. Treatment parameters for beta and gamma devices in peripheral endovascular brachytherapy. *Int. J. Radiat. Oncol. Biol. Phys.* 60 (5):1652–1659.

Kirisits, C., A. Stemberger, B. Pokrajac, D. Glogar, R. Potter, and D. Georg. 2004. Clinical quality assurance for endovascular brachytherapy devices. *Radiother. Oncol.* 71 (1): 91–98.

Kollaard, R. P., W. J. Dries, H. J. van Kleffens, T. H. Aalbers, H. van der Marel, H. P. Marijnissen, M. Piessens, D. R. Schaart, and H. de Vroome. 2006. Recommendations on detectors and quality control procedures for brachytherapy beta sources. *Radiother. Oncol.* 78 (2): 223–229.

Nath, R., H. Amols, C. Coffey, D. Duggan, S. Jani, Z. Li, M. Schell et al. 1999. Intravascular brachytherapy physics: report of the AAPM Radiation Therapy Committee Task Group no. 60. American Association of Physicists in Medicine. *Med. Phys.* 26 (2): 119–152.

NCS. 2004. *Quality Control of Sealed Beta Sources in Brachytherapy. Report 14 of the Netherlands Commission on Radiation Dosimetry.* Netherlands Commission on Radiation Dosimetry.

Pötter, R., E. Van Limbergen, W. J. Dries, Y. Popowski, V. Coen, C. Fellner, D. Georg et al. 2001. Recommendations of the EndoVAscular GEC ESTRO Working Group (EVA GEC ESTRO): Prescribing, recording, and reporting in endovascular brachytherapy. Quality assurance, equipment, personal and education. *Radiother. Oncol.* 59: 339–360.

Schmid, R., C. Kirisits, B. Syeda, P. Wexberg, P. Siostrzonek, B. Pokrajac, D. Georg, D. Glogar, and R. Poetter. 2004. Quality assurance in intracoronary brachytherapy. Recommendations for determining the planning target length to avoid geographic miss. *Radiother. Oncol.* 71 (3): 311–318.

Syeda, B., P. Siostrzonek, R. Schmid, P. Wexberg, C. Kirisits, S. Denk, G. Beran et al. 2002. Geographical miss during intracoronary irradiation: Impact on restenosis and determination of required safety margin length. *J. Am. Coll. Cardiol.* 40 (7): 1225–1231.

Index